Petroleum Geochemistry in Exploration of the Norwegian Shelf

Petroleum Geochemistry in Exploration of the Norwegian Shelf

Proceedings of a Norwegian Petroleum Society (NPF) conference *Organic Geochemistry in Exploration of the Norwegian Shelf* held in Stavanger, 22–24 October 1984

Edited by
B. M. Thomas et al.

Published by
Graham & Trotman Ltd
for the Norwegian Petroleum Society

First published in 1985 by
Graham & Trotman Limited
Sterling House
66 Wilton Road
London SW1V 1DE
UK

Graham & Trotman Inc
13 Park Avenue
Gaithersburg
MD 20877
USA

British Library Cataloguing in Publication Data
Petroleum geochemistry in exploration of the Norwegian Shelf: proceedings of a Norwegian Petroleum Society (NPF) conference held in Stavanger, 22-24 October, 1984.
1. Petroleum — Geology — North Sea 2. Geochemical prospecting — North Sea 3. Petroleum industry and trade — Norway
I. Thomas, B.M. *et al.* II. Norsk Petroleumsforening
622'.18282'0916336 TN874.N78

ISBN 0-86010-706-X

Typeset in Great Britain by Spire Print Services, Salisbury, UK
Printed and bound in Great Britain by The Alden Press, Oxford, UK

FOREWORD

This volume is the record of a three day symposium entitled "Organic Geochemistry in Exploration of the Norwegian Shelf", which was sponsored by the Norwegian Petroleum Society (Norsk Petroleumsforening) and held at the Rogalands Regional College, Stavanger on 22–24 October 1984. Twenty-nine papers were presented, and all but one are published in full herein.

The aim of the conference was to focus on the application of geochemical methods to the current and highly active exploration of the Norwegian offshore. Emphasis was on practical interpretation and case histories rather than laboratory methods and techniques, and a strong attendance was sought among geologists and seismic interpreters active in exploration in Norway and Northwest Europe generally. On all counts the symposium was a great success with a total of 213 participants registered.

In his opening address Mr Egil Bergsager, director of the Norwegian Petroleum Directorate, observed that during the 1970s petroleum geochemistry emerged from being a somewhat academic pursuit into a practical aid in exploration for hydrocarbons. This first stage, when many of the basic methods were developed, has now led in the 1980s to an expansion into applications in regional geological studies, including mathematical modelling of thermal history, hydrocarbon migration and basin development.

The papers which follow clearly reflect these trends. They also indicate the high level of geochemical research in Norwegian exploration, and a remarkable convergence of ideas and methods among the various companies and workers. The use of Rock-Eval pyrolysis data in routine source rock analysis is now almost universal. Mathematical modelling of thermal history and hydrocarbon generation is seen in many forms, most of which owe much to the early work of Karweil (1955) and Lopatin (1971). Regional mapping of thermal maturity on key source horizons is now routine, and some papers have followed with a detailed consideration of possible hydrocarbon migration paths.

The complex nature of the principal Upper Jurassic source interval in the North Sea was emphasized by several speakers, and the concept of organic facies variations within it is now firmly established. The use of GC–MS techniques in detailed oil–oil, and oil–source correlation studies seems to be progressing from an earlier experimental stage to one of practical applications in the exploration sphere. Several papers on marine surface geochemical and so-called "sniffing" techniques indicate increased sophistication in methods and interpretation, but, as before, still provoke healthy debate.

The symposium was strongly supported by companies active in exploration in Norway, both in terms of papers and delegates. The organizers are particularly grateful for the high level of technical disclosure represented in many of the contributions, which clearly indicate the strong emphasis given to geochemical methods in modern prospect and basin evaluation.

Many thanks also to the anonymous helpers involved in refereeing of papers for this volume, and to the Rogalands Regional College for making available to the NPF such a superb venue for the symposium.

B. M. Thomas
For the Conference and Editorial Committees

REFERENCES

Karweil, J. 1955. Die Metamorphose der Kohlen vom Standpunkt der physikalischen Chemie. *Dt. Geol. Ges. Z.* **107,** 132–139.

Lopatin, N.V. 1971. Temperature and geologic time as factors in coalification (in Russian). *Akad. Nauk SSSR, Izv. Ser. Geol.,* **3,** 95–106.

CONTENTS

METHODS

Part I
Regional Studies

1 Organic facies and hydrocarbon distributions in the Norwegian North Sea

B. M. Thomas, P. Møller-Pedersen, M. F. Whitaker and N. D. Shaw
A/S Norske Shell, Stavanger, Norway

Significant hydrocarbon accumulations have been found in North Sea reservoirs ranging in age from Devonian (Buchan Field) to Eocene (Frigg area). In Norwegian waters, the oldest reservoirs are at present Rhaeto–Liassic Statfjord Formation sands. Source rock development in the northern North Sea is restricted almost entirely to the Jurassic. Rich oil-prone source rocks occur mainly in the Humber Group, and particularly within the Kimmeridge Clay Equivalent (Upper Jurassic 'Hot Shale'). In basinal settings within the Viking and Central Grabens, Kimmeridge source rock sections may be over 1000 m thick. Net source rock thicknesses exceeding 400 m are documented in several wells.

North Sea oils appear to be sourced from these Humber Group anoxic shales, although the Liassic Dunlin Group may also contribute in parts of the northern Viking Graben. Biodegraded oils are found in fields where reservoir temperatures are less than 75 °C, and some possible low maturity crudes occur in the Egersund Sub-basin. Mature Jurassic 'kitchen' areas coincide closely with the Viking and Central Grabens and deeper parts of the East Shetland Basin, although a small oil kitchen also occurs in the Egersund Sub-basin.

Rock Eval pyrolysis and solvent extract studies of core material from the Kimmeridge Clay Equivalent suggest that the principal zone of oil generation for 'Type II' source rocks lies between VR/E = 0.7 and 1.0*, with the peak possibly at VR/E = 0.8.

Organic facies variations occur within the Kimmeridge Clay Equivalent. In thick basinal sections source rocks are rich and oil-prone, although the inertinite content may be locally high in turbidite sequences. The Kimmeridge Clay Equivalent on the basin margins and over some structural highs was deposited under predominantly dysaerobic conditions, often with some terrestrial influx. Source rocks from these settings are leaner and of mixed organic type. However, such facies variations are thought to be, at best, only a secondary influence on the distribution of apparent oil- and gas-prone areas in the North Sea. Other factors, such as thermal maturity of kitchen areas, timing relationships, migration paths and charge competition are considered to be more significant.

Hydrocarbon distributions and migration paths in the North Sea are complex. In some areas, e.g. Central Graben, hydrocarbon migration patterns seem to be dominantly vertical, whilst in other areas significant lateral migration is apparent, e.g. in Troll Field. On many occasions migration has occurred to a stratigraphically lower level. Geochemical data from deep, mature kitchen areas suggest that hydrocarbon charge vastly exceeds the available volume in adjacent traps. Many structures appear to be filled to spill-point, and in some areas long chains of fields are related to each other and a common 'kitchen' as a continuous fill-and-overspill sequence.

INTRODUCTION

Since the mid-1960s, the development of the North Sea as one of the world's major hydrocarbon-producing provinces has been paralleled by the emergence of applied organic geochemistry (and allied basin evaluation techniques) as new and effective tools in modern petroleum exploration. The concept of the 'generative basin' has rapidly come into widespread use in the exploration industry (Demaison, 1984). Hydrocarbons are known to originate from distinctive source beds where they are thermally mature within mappable 'kitchen' areas. Their subsequent migration paths are largely determined by the stratigraphic framework and basin geometry, which defines the 'drainage area' of the trapping structure.

The North Sea, with its large data base, rich and distinctive source horizons, and numerous hydrocarbon accumulations, provides an ideal calibration area for such methods, which may then be applied to the

---

**Abbreviations*:

VR = vitrinite reflectance, measured directly on teleocollinite,
VR/E = vitrinite reflectance equivalent, either measured with another maturity parameter (e.g. spore discolouration) or estimated by thermal modelling techniques (e.g. Lopatin method).

Petroleum Geochemistry in Exploration of the Norwegian Shelf © Norwegian Petroleum Society (Graham & Trotman, 1985) pp. 3–26.

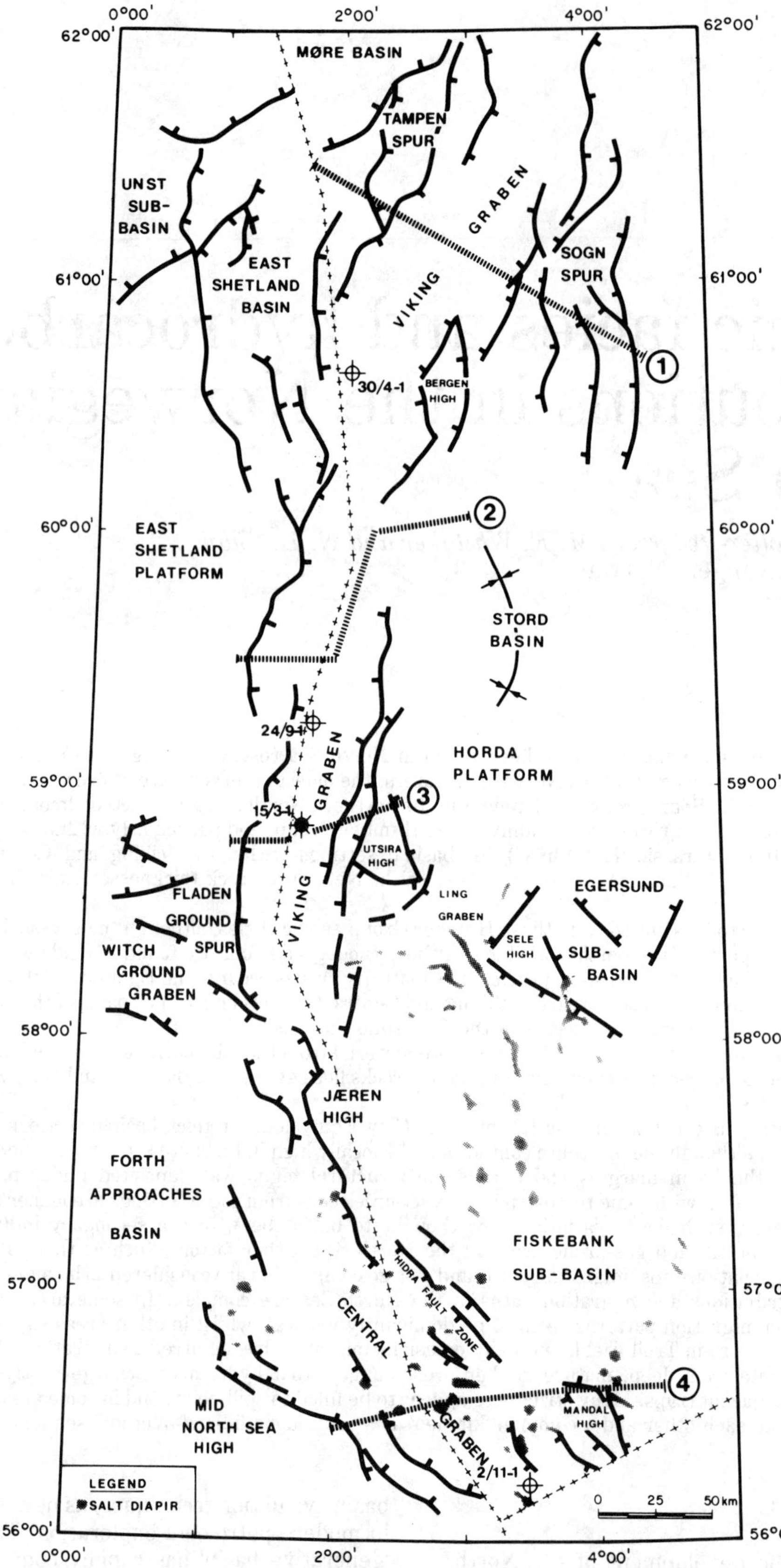

Fig. 1. Tectonic elements and well locations. For regional cross-sections, see Fig. 15.

appraisal of remaining prospects or be used as an analogue for the evaluation of other basins.

The following is a summary of some recent investigations of the hydrocarbon habitat of the Norwegian sector of the North Sea, with particular emphasis on the northern part of the Central Graben and the Viking Graben.

TECTONIC FRAMEWORK

The tectonic framework of the northern North Sea (Fig. 1) is dominated by the north-trending Viking Graben, which intersects in the south with the southeasterly-oriented Central Graben and the westerly Witch Ground Graben. The thick, mainly-

Mesozoic fill of the North Sea Graben system is overlain by gently downwarped Tertiary sediments, comprising a total sedimentary section of up to 9 km.

Development of the North Sea rift began in the Late Permian (Ziegler, 1981, 1982). The primitive graben system was accentuated by intermittent 'Kimmerian' tectonic events which occurred between the late Triassic and the early Cretaceous. These movements resulted in deep axial graben systems bounded by rotated fault block complexes. Such characteristics are most evident in the Viking Graben, as they are partially obscured by salt flowage and diapirism in the Central Graben and Southern North Sea.

Triassic sediments within the rift system are almost entirely non-marine clastics, largely 'red-beds', but by late Liassic time fully marine conditions were established in the North Sea area and have persisted, with minor interruptions, to the present. Rift movements gradually diminished during the late Cretaceous, and ceased altogether after the early Tertiary Laramide tectonic phase, to be replaced by broad regional downwarping.

HYDROCARBON OCCURRENCES

Principal reservoir intervals and the occurrence of hydrocarbons in the northern North Sea are summarized in Figs. 2 and 3. Significant hydrocarbon accumulations have been found in reservoirs ranging in age from Devonian (Buchan Field) to Eocene (Frigg area). In Norwegian waters, the oldest hydrocarbon-bearing reservoirs are Rhaeto-Liassic Statfjord sands of the Statfjord and Gullfaks Fields, and in a recent discovery in 34/4–4. The geographical distribution of North Sea fields is shown in Fig. 4.

In the Central Graben, the major oil fields occur in unique fractured Danian–Maastrichtian chalk reservoirs, the largest being Ekofisk (1.0 billion barrels recoverable) which was the first commercial oil field in Norway. Gas-condensate occurs in Paleocene sandstones in the Cod Field, but this reservoir sequence is most productive in UK waters where it hosts such oil fields as Forties (1.8 billion barrels recoverable), Maureen and Montrose. To date, Jurassic reservoirs of the Central Graben have yielded only a few small discoveries in Norwegian waters, the most significant being the Ula Field (150 million barrels recoverable), where oil occurs in marine sandstones of Oxfordian–Kimmeridgian age.

In the Sleipner area of the southern Viking Graben, gas-condensate occurs widely in Lower–Middle Jurassic sandstones of the Vestland Group, and in the 'Gamma' structure, also in overlying Paleocene sands. In contrast, in adjacent UK waters, oil is found in a string of fields (Brae, Tiffany, Toni, Thelma) where Upper Jurassic fault-scarp sandstones are productive. The Gudrun Field, in Norwegian waters, has oil in Upper Jurassic turbidites (presumably distal equivalents of the Brae sands) and gas in Middle Jurassic sands of the Vestland Group.

In the central Viking Graben, known accumulations are largely in the Tertiary, where Eocene sandstones are gas-productive in the Frigg area; the Paleocene contains gas at Heimdal and oil at Balder. Both Paleocene and Eocene reservoirs are believed to be deep-water sand fans shed from westerly sources on the East Shetland Platform. The Lower–Middle

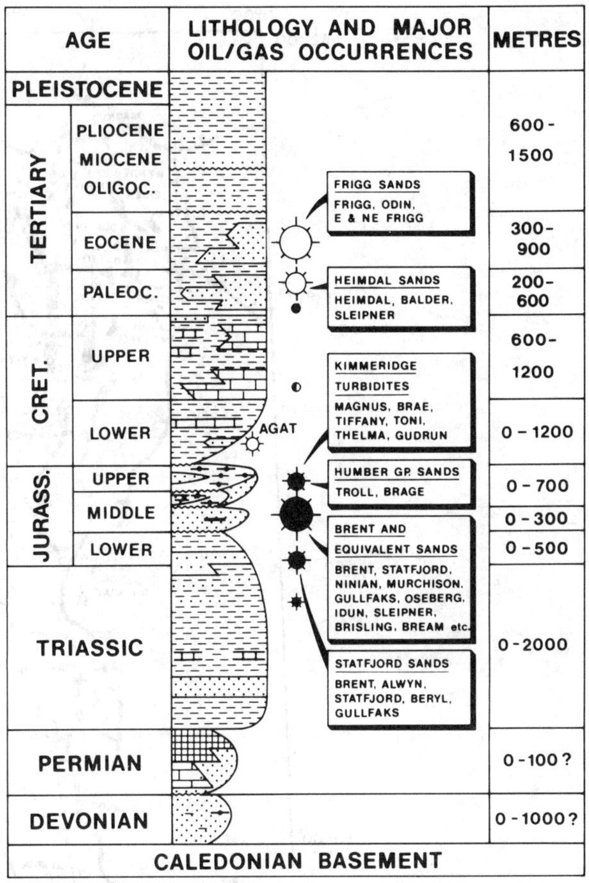

Fig. 2. Stratigraphic occurrence of hydrocarbons, Viking Graben (modified after Ziegler, 1980).

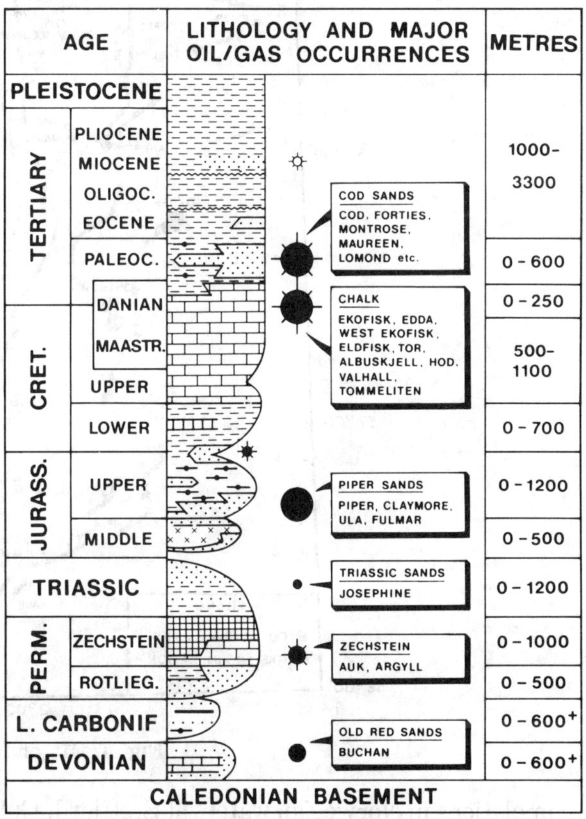

Fig. 3. Stratigraphic occurrence of hydrocarbons, Central Graben (modified after Ziegler, 1980).

Jurassic contains oil at Heimdal and Idun, and in UK fields Bruce and Beryl.

In the northern Viking Graben and on its adjacent flanks, Middle Jurassic paralic sands of the Brent Group are widely productive and host major

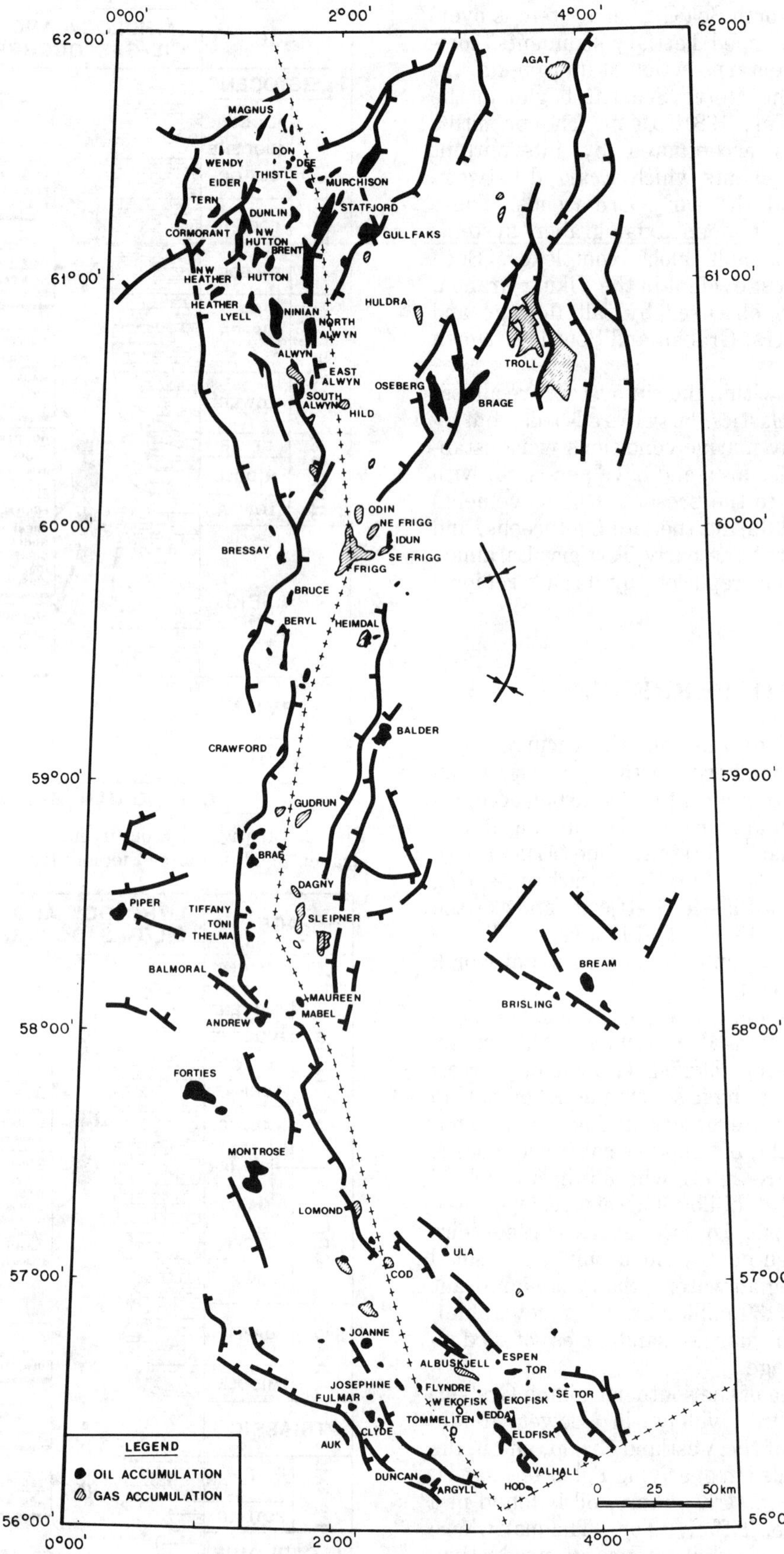

Fig. 4. Major fields, northern North Sea.

accumulations in Norwegian waters at Statfjord, Gullfaks and Oseberg. Oxfordian–Kimmeridgian 'Humber Group sands' are gas- and oil-bearing in the Troll Field, and contain oil at Brage. Rhaeto–Liassic Statfjord sands are also oil-productive at Statfjord and Gullfaks, as well as in the Brent Field in UK waters. A small gas discovery has been made in Lower Cretaceous submarine fan sands at Agat. Three small oil accumulations have also been found in Middle Jurassic sandstones of the Vestland Group in the Egersund Sub-basin.

SOURCE ROCKS

Geochemical profiles of North Sea wells indicate that the occurrence of significant source rock intervals is restricted almost entirely to the Jurassic section (see

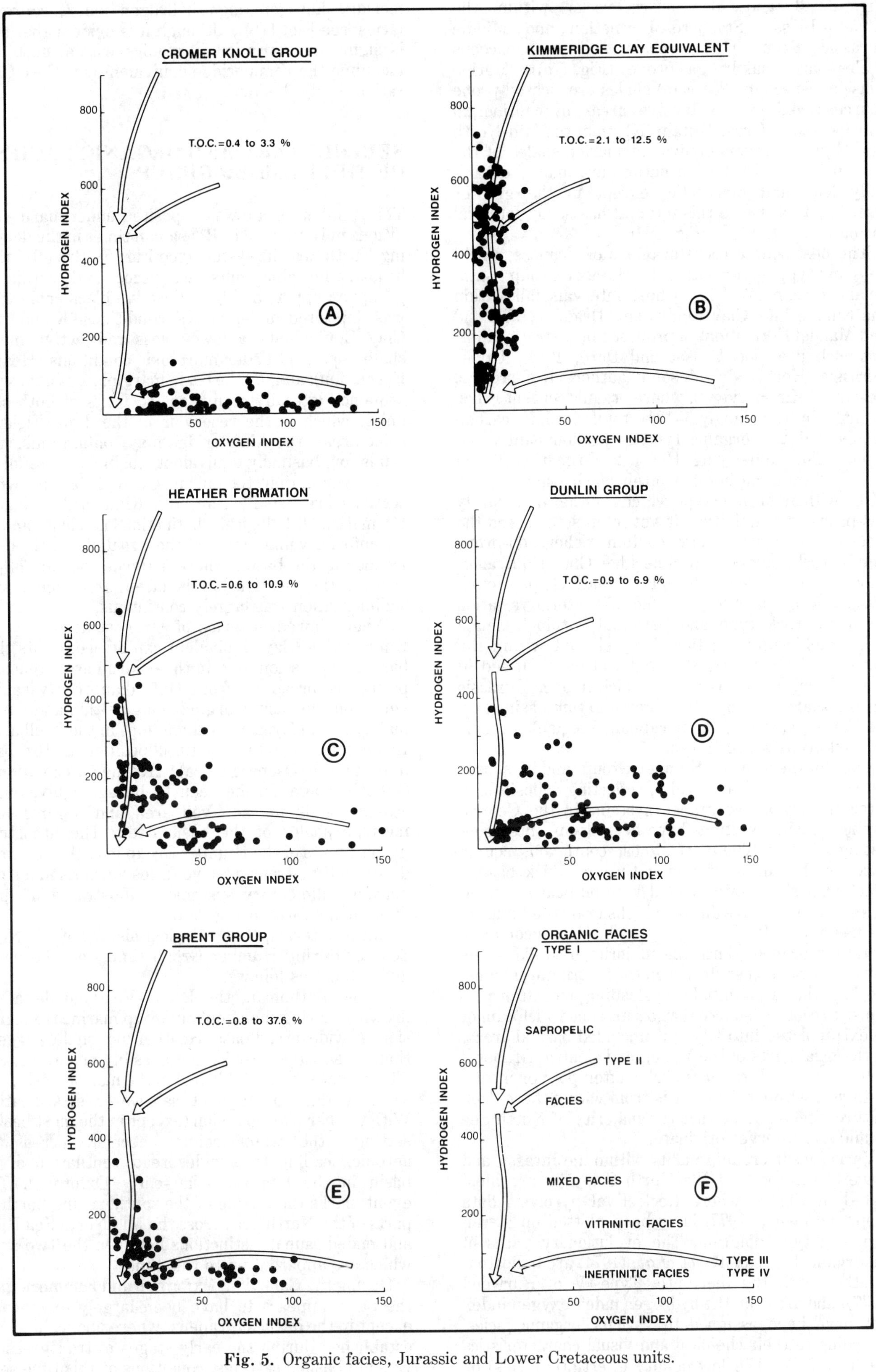

Fig. 5. Organic facies, Jurassic and Lower Cretaceous units.

Figs. 2, 3 and 5). Source rocks within the Rhaeto–Liassic Statfjord Formation and Middle Jurassic Brent Group—coals and carbonaceous shales—are mainly gas-prone (Fig. 5(e)). Marine Liassic (especially Toarcian) shales are rich oil-prone source rocks over extensive areas of continental Europe and in Great Britain. North of the Mid-North Sea High, however, equivalent marine shales of the Dunlin Group (Drake Formation) are mainly organically lean and inertinitic, except in the greater Oseberg area where this interval has some potential for oil generation, see Fig. 5(d).

The best source rocks in terms of richness, thickness and type, occur within the Humber Group (Bathonian–Ryazanian). The richest intervals fall within the Kimmeridge Clay Equivalent (Draupne, Farsund and Mandal Formations of proposed new stratigraphic nomenclature (see Vollset and Doré, 1984); Upper Jurassic 'Hot Shale' of some authors (referred to herein as 'Kimmeridge'), where organic content is frequently in the range 5–12% total organic carbon (TOC) and the organic type is predominantly oil-prone. The underlying Heather Formation ('Cold Shale' of some authors) is often much leaner (2–4% TOC) with an organic type which is either dominantly gas-prone or inertinitic. However, in some areas the Heather Formation may contain richer, oil-prone source rocks and the Kimmeridge Clay Equivalent may also include lean intervals of relatively poor quality. The concept of 'organic facies'—lateral variation in source rock type and potential within a single stratigraphic unit—is therefore relevant within the Jurassic of the North Sea, and will be discussed in more detail later. Within a thick Upper Jurassic marine shale section in the Egersund Sub-basin, only the Tau Formation (Kimmeridgian–Volgian) contains rich, oil-prone source rocks.

The thickness of the Humber Group (and its source intervals) varies dramatically, reflecting deposition in a tectonically active graben system. In the Central Graben, basinal sections have been documented consisting of over 1000 m of almost continuous source rock. In the southern Viking Graben, thick basinal 'black shales' are interbedded with oil-bearing turbidite sands. On active Jurassic highs or on the flanks of the basin, the Humber Group may be thin, incomplete or totally eroded. Since the majority of wells in the North Sea have been drilled on these structural highs, they provide a potentially misleading and often pessimistic record of source rock occurrence which cannot be extrapolated into adjacent undrilled basinal areas. In the axial parts of the Viking and Central Grabens, well control on Jurassic 'sinks' is often poor or incomplete yet, without doubt, it is from within these deep mature kitchen areas that the majority of North Sea oil and gas reserves originate.

Variations in organic quality within the Jurassic and Lower Cretaceous of the North Sea area are illustrated in Fig. 5, where Rock Eval pyrolysis data (Espitalie *et al.*, 1977) have been plotted on a 'van Krevelen'-type diagram. The evolutionary paths of the organic types of Tissot *et al.* (1974) are indicated, together with the inertinitic Type IV of Harwood (1977), and areas on the hydrogen index/oxygen index plot have been assigned to various 'organic facies' according to their chemical and visual characteristics (see Fig. 5(f)). The low quality (inertintic facies) of Cromer Knoll kerogen is immediately apparent (see Fig. 5(a)) and contrasts strongly with oil-prone Kimmeridge kerogen (high hydrogen index, sapropelic facies; see Fig. 5(b)), although it is evident that there is significant variation in quality within most units including the Kimmeridge equivalent (see Core Calibration Study, below).

SEDIMENTARY AND ORGANIC FACIES OF THE HUMBER GROUP

The Humber Group was deposited under marine conditions in Bathonian to Ryasanian times in the deepening North Sea rift system (see Fig. 6). Broadly, three lithostratigraphic units are recognized within the group: an upper, highly radioactive black shale which was deposited under anoxic conditions (Kimmeridge Clay Equivalent); a lower, less radioactive marine shale deposited under more oxic conditions (Heather Formation); and, on the basin margins, a coastal sandy sequence laterally equivalent to parts of both shale units, which is the reservoir in the Troll Field. In some areas, the Kimmeridge may contain fault scarp sands or basinally-equivalent turbidites, reflecting 'Late Kimmerian' tectonism. On high blocks, where sections are usually thin, the Kimmeridge/Heather Formation lithological distinction is clear and an unconformity may separate the two units (Kirk, 1980). Deeper in the basin, where a thicker section is preserved, the relationship is often gradational since sedimentation was largely continuous.

When viewed in terms of chronostratigraphy, the Kimmeridge Clay Equivalent can be seen in its thick basinal expression (e.g. in the Gudrun area) to span a period as broad as from Oxfordian to Ryazanian, whilst on the basin margin it may be thin and limited mainly to the Ryazanian stage (e.g. in the Troll area). In an intermediate structural position in the basin (e.g. in the Oseberg area), Heather Formation of Oxfordian age can be seen to be age-equivalent to coastal sands in the Troll area and organic Kimmeridge shales of the basin deep. The lithostratigraphic units of the Humber Group may thus be considered as time-transgressive facies variants in a single marine sedimentary system, as illustrated in simplified model form in Fig. 6(a).

In these terms, the geological history of the North Sea rift during Humber Group times may be better understood, as follows:

In the Bathonian, the Brent–Vestland delta was drowned during the Humber Group marine transgression. A widespread basal Heather 'hot shale' suggests that at least dysaerobic, and possibly anaerobic, conditions were soon established in the marine basin, presumably due to stagnant salt-water stratification. With further transgression (except in the most basinal settings) the water column became increasingly aerobic, leading to a widespread ventilation of the basin in the late Callovian–early Oxfordian. This event marks the linking of the southern and northern parts of the North Sea across the Mid North Sea High, and ended faunal distinctions between the two areas which are apparent up to this time.

During the Callovian, Oxfordian and Kimmeridgian, the basin appears to have been largely oxygenated except in the deepest grabens, where anoxia had been established during the early stages of transgression. Even in basinal settings, conditions at this time seem to have varied along the axis of the Viking Graben. Fully anaerobic conditions are apparent in the Gudrun

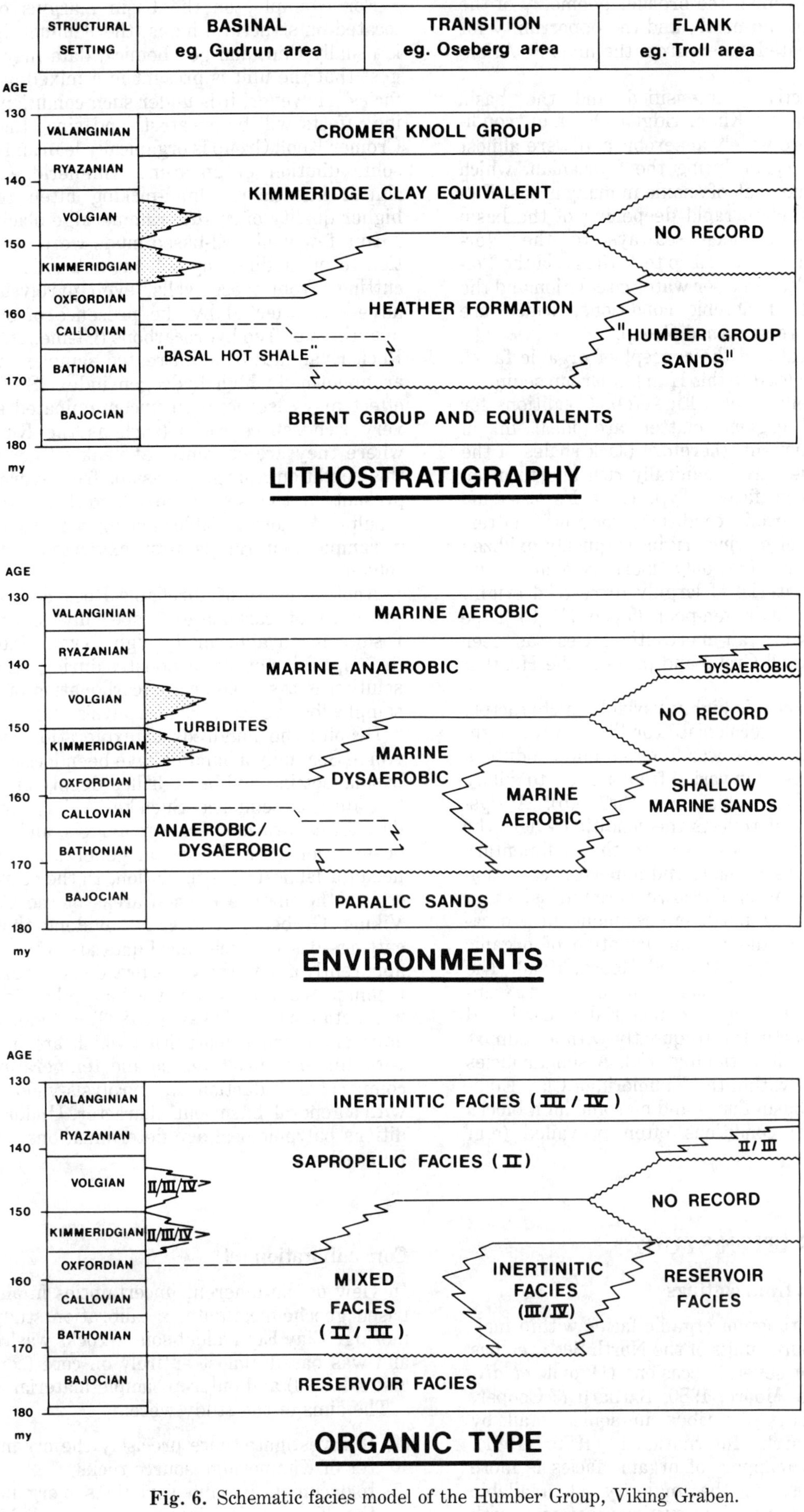

Fig. 6. Schematic facies model of the Humber Group, Viking Graben.

area as early as in the Oxfordian, whilst in the north the basin was largely aerobic until Volgian–Ryazanian times. This may reflect the broader geometry of the Viking Graben in the north, and the opportunity for access to oxygenated waters from the juvenile Atlantic–Arctic seas.

As rifting activity intensified and the basin deepened during the Kimmeridgian–Volgian, conditions became more widely anaerobic and were almost universal in the basin during the Ryazanian, which corresponds to the peak of anoxia in many areas. Following the Ryazanian, rapid deepening of the basin and opening of broader seaways to the Norwegian–Greenland sea and also to Tethys (via the Polish Trough) resulted in freer water circulation and the re-establishment of aerobic conditions, which are marked by the Cromer Knoll Group.

Source potential and the concept of organic facies may be simply related to this Humber Group sedimentary model as follows (see Fig. 6(c)). Conditions for preservation of organic matter are ideal in an anaerobic environment, therefore black shales of the Kimmeridge facies are organically rich and form an oil-prone sapropelic facies (Type II of Tissot *et al.* (1974)). Under aerobic conditions, organic matter, particularly that of marine origin, is quickly oxidized and destroyed, leaving only inert remnants and more-resistant material of largely terrestrial origin. This kerogen is hydrogen-poor (Type III to Type IV) and represents an inertinitic facies, as seen in the Cromer Knoll Group and parts of the Heather Formation.

The Heather facies is highly variable in character, and represents an intermediate condition in the basin where terrestrial input was limited but conditions within the water column varied from aerobic to mildly anaerobic (i.e. dysaerobic), see Fig. 6(b). A dysaerobic environment reflects the condition where the oxic/anoxic interface lies close to the sedimentary surface, providing a variable and imperfect condition for organic preservation. Burrowing organisms, which are absent in a truly anoxic environment, may occasionally be active, and partial oxidation of organic matter will occur (Demaison and Moore, 1980). The result is a 'mixed' organic facies where Type II (sapropelic) and degraded marine material is combined with Type IV (inertinite), frequently with an admixture of Type III (vitrinitic) material. A similar facies is also recognized within the Kimmeridge Clay Equivalent along the basin flanks and on some high blocks where dysaerobic conditions often prevailed (e.g. Troll area).

effects could at times be partly the result of such 'noise' in the data.

For example, on the basin margins or in wells located on structural highs, the Kimmeridge interval is usually thin, and geochemical data may often suggest that the unit is present in a mixed or inertinitic facies. However, it is under such conditions that caving effects will be greatest, and since the overlying Cromer Knoll Group is organically lean and inertinitic, some dilution of apparent Kimmeridge quality is almost inevitable. Hand-picking often reveals the higher quality of *in situ* Kimmeridge black shales.

In a few wells, oil-based muds were used. In addition to precluding any solvent extraction analysis of cuttings from these wells, even pyrolysis data are adversely affected by the presence of such excess quantities of free hydrocarbons (Clementz, 1979), and Rock Eval data from untreated samples may exhibit an abnormally high hydrogen index (S_2/TOC). This effect may also occur in uncontaminated samples of very rich source rocks (such as the Kimmeridge), where they are currently at peak oil generation. In this condition a large excess of free hydrocarbons is present in the source rock and leads to spurious results. A more reliable hydrogen index is obtained if sample material is first extracted with organic solvent.

Another cause of unreliable Rock Eval data is the presence of carbonates, especially siderite, which results in an abnormally high oxygen index due to decomposition of the carbonates during pyrolysis. The solution is a simple acid decarbonation of calcareous samples before Rock Eval analysis.

Despite the hundreds of exploration wells in the North Sea, only a handful have been located in a truly basinal setting and been drilled deeply into a generative Jurassic sequence. Such key wells provide critical data for calibration of charge models, and allow direct observation of hydrocarbon generation without the need for laboratory stimulation. In the search for control in the mature kitchen areas of the Central and Viking Grabens, it became evident that existing cuttings-based geochemical data are often unreliable, apparently due to the widespread use of turbo-drilling techniques in these deep synclinal wells. The effects of 'bit metamorphism' (Taylor, 1983) include elevation of apparent thermal maturities which are incompatible with current burial depths and temperatures, and a comparable reduction in pyrolysis-yield consistent with a general 'burnt-out' character. Under these conditions palynological age determinations may also be impossible.

CORE CALIBRATION STUDY

Geochemical data from cuttings

The concept of a range of organic facies within individual Jurassic source units of the North Sea area has been proposed on several occasions (Dypvik *et al.*, 1979; Demaison & Moore, 1980; Barnard & Cooper, 1981) and has been described in some detail by Robertson Research International (1979, 1981). Detailed regional mapping of organic facies is more difficult, however, as the majority of available geochemical data is based on cuttings material, with attendant problems of poor sample quality, caving and contamination. Subtle variations attributed to facies

Core calibration set

In view of the inherent uncertainties in all cuttings-based geochemical data, a calibration study of Kimmeridge Clay Equivalent source rocks was conducted, and was based almost entirely on core (conventional and sidewall) and outcrop sample material.

The aims of the study were:

(1) to investigate more precisely the organic character of Kimmeridge source rocks;
(2) to document possible variations in organic facies of the Kimmeridge interval using reliable sample material; and
(3) by choosing material from a wide range of thermal

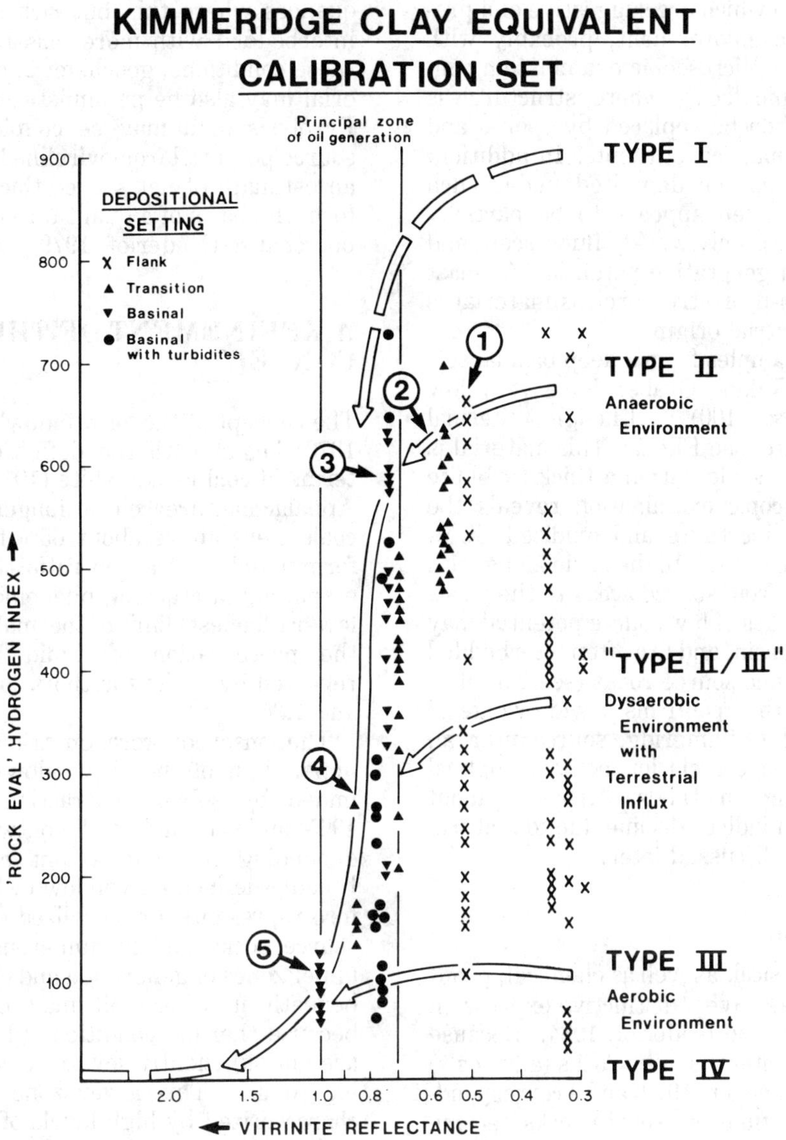

Fig. 7. Organic facies and maturation pathways, Kimmeridge Clay Equivalent. For extract data, points 1–5, see Fig. 8.

maturities, to study the generation of hydrocarbons within the Kimmeridge source rocks.

The calibration set covers a range of thermal maturities from VR/E = 0.3 to 1.0, and comes from a variety of structural positions in the basin. All samples are dark grey-black, organically-rich shales, often finely laminated and sparsely fossiliferous, consistent with deposition in a broadly anoxic environment.

Source potential and organic type were routinely determined by 'Rock Eval' pyrolysis. Since spot checks indicated the presence of some minor carbonate, many samples were rerun after decarbonation. Samples with a thermal maturity above VR/E = 0.7 were also re-analysed after solvent extraction. Duplicate samples were prepared using standard palynological methods, and were examined microscopically under both white and ultraviolet light.

When plotted on a standard van Krevelen-type plot (hydrogen index versus oxygen index), see Fig. 5(b), the calibration set exhibits a wide range in apparent organic types, but offers little distinction between immature source rocks of rather poor primary quality and post-mature source rocks whose past potential was probably high, since both have low hydrogen indices. A cross-plot of hydrogen index versus vitrinite reflectance (using a logarithmic scale) provides the required discrimination (see Fig. 7) and also overcomes the inherent unreliability of oxygen index and atomic O/C ratio in van Krevelen-type diagrams. Samples in the calibration set are categorized according to their depositional setting in the basin—'flank', 'transition', 'basinal', and 'basinal with turbidites' (refer also to Fig. 6).

For immature samples, the hydrogen index of Kimmeridge kerogen varies from less than 100 to more than 700 (see Fig. 7). There is little evidence of segregation into discrete organic types as proposed by Tissot *et al.* (1974); rather, a continuum exists between 'Type II' and 'Type IV'. Kimmeridge source rocks from deep basinal wells are organically rich (5–12% TOC) and mostly have very high hydrogen indices (500–700) which are characteristic of a predominantly sapropelic facies. Microscopic examination confirms that such samples consist almost entirely of highly fluorescent, structureless sapropelic organic matter (SOM) of probable algal/bacterial origin. Similar intervals also occur in some wells on the basin margin, particularly in the Ryazanian, when the sapropelic facies extended across the whole basin. Samples from the flanks of the basin, however, are often leaner (2–5% TOC) and have mainly moderate hydrogen indices (200–400) which are suggestive of a predominantly dysaerobic environment, probably with some terrestrial influx. Microscopic examination confirms a 'mixed' organic facies where structureless organic matter can often be replaced by spores and inertinite as the dominant constituents. In addition, structureless organic matter deposited under such dysaerobic conditions often appears to be partially oxidized and degraded, is only weakly fluorescent and is probably of reduced generative potential. At least some of this hydrogen-poor structureless material is thought to be of terrestrial origin.

Some Kimmeridge samples from a deep basinal setting in the southern Viking Graben have very low hydrogen indices (*ca.* 100), although thermal maturities are moderate (see Fig. 7). This material is from black hemipelagic shale within a thick turbidite sequence, and microscopic examination reveals the presence of abundant inertinite and oxidized SOM, which was apparently carried into the basin along with coarse clastic material from source areas to the west. Hence, organic-rich shales of low source potential may also occur in the deep basin, and are often interbedded with other more oil-prone source rocks (see Fig. 6).

The calibration set thus confirms a wide range of organic facies within the Kimmeridge source interval, and provides a detailed correlation between optical and chemical typing methods. Above about VR/E = 0.7, hydrogen indices decline due to maturation effects, as will be discussed later.

Wireline log expression

With their unusual physical, as well as chemical, properties, rich source rocks have a distinctive expression on wireline logs (Meyer and Nederlof, 1984). Because of their high organic content, source rocks (and coals) have a low density (Schmoker, 1979) and a correspondingly high sonic transit time. As source rocks become mature, their resistivity often increases dramatically due to the presence of free hydrocarbons during the main phase of oil generation (Meissner, 1984). At high maturities, resistivity again declines as the remaining oil is progressively cracked to gas and expelled.

The most widely observed petrophysical property of rich marine source rocks is their high radioactivity, as indicated on gamma-ray logs (Bjørlykke *et al.*, 1975; Schmoker, 1981). The abnormal levels of radioactivity are largely due to the presence of uranium, which is preferentially enriched in organic matter deposited under stagnant, anoxic conditions. Studies made by the authors of Kimmeridge data indicate that there is often a close relation between *relative* gamma-ray count and total organic carbon content, but that *absolute* values of radioactivity on wireline logs may not be a direct indication of source potential. This arises because gamma-ray log response is influenced by both borehole size and mud weight (Schlumberger, 1979, p. 17), and in many deep, basinal wells the natural radioactivity of the Jurassic section is suppressed by the high mud weights used to combat high formation pressures. In addition, the ability of uranium to complex with organic matter under anoxic conditions also varies with organic type, such that the most oil-prone material may not have the highest gamma-ray response (Schlumberger, 1982).

Wireline logs are a poor indicator of source potential where source beds are too thin to be resolved by logging tools. This occurs particularly in turbidite sequences where thin but rich hemipelagic shales are interbedded with more massive sand bodies. Under these conditions, geochemical data from cuttings material may also be pessimistic, due to dilution effects. Pyrolysis data may be coupled with indications of source potential from wireline log character to provide an estimate of net source thickness, which can then form the basis of a quantitative estimate of hydrocarbon charge (Nederlof, 1979), see below.

A REFINEMENT OF THE 'OIL WINDOW' CONCEPT

The concept of the 'oil window' (Landes, 1967; Pusey, 1973) began with the definition of the oil 'floor' in terms of coal rank. White (1915) observed that in the Appalachian area oil no longer occured if associated coals were above about 65% fixed carbon, and thus formulated his 'carbon ratio theory'. The oil floor is essentially a cracking phenomenon, representing the lower (deepest) limit of thermal conditions suitable for the preservation of liquid hydrocarbons, and is reported by various authors to lie between VR = 1.2 and 1.35.

The onset of kerogen transformation to oil (the upper limit of the oil window) was originally determined by solvent extraction techniques (Philippi, 1965) and is thought to lie between VR = 0.5 and 0.7, depending to some extent on organic type. This broadly-defined oil window (VR = 0.5 to 1.35) therefore represents a generalized observation from many source rocks and in numerous basins. It embraces upper zones of generation and oil expulsion, and a zone beneath it where oil may exist after generation because thermal conditions (defined by the oil floor) are not sufficiently severe to cause large-scale cracking to gas. This lower zone of oil preservation is characterized by high levels of extractable hydrocarbons, but is not a site of major oil generation.

The Kimmeridge core calibration set therefore offers an opportunity to define more precisely the 'principal zone of oil generation' for a specific source rock. Such a calibration will allow more-precise use of thermal maturity maps in defining oil and gas kitchen areas in the North Sea.

Rock Eval pyrolysis data, being semi-quantitative, provide a potentially precise indicator of the generative process, as the relative pyrolysis yield (i.e. hydrogen index) of a source rock is seen to decline dramatically during the generation of oil (see Fig. 7). Taking the maturation path of Type II kerogen, little decline is seen in hydrogen index from its initial immature values of 600–700 until at least VR/E = 0.7. A major reduction in pyrolysis yield, presumably due to generation and expulsion of hydrocarbons, occurs between VR/E = 0.75 and 1.0, with the probable peak at around VR/E = 0.8. Electrical resistivity measurements of mature Kimmeridge source rocks also are highest at VR/E = 0.8 (Goff, 1983), tending to confirm this value for peak oil generation. By VR/E = 1.0, Kimmeridge source rocks have declined to a 'Pseudo-vitrinitic' composition and appear to be post-mature for oil generation. It is likely that by this stage active gas/condensate generation is already occurring.

A composite generative profile of Kimmeridge source rock extracts confirms the above model. Gas-chromatograms of solvent extracts from the calibra-

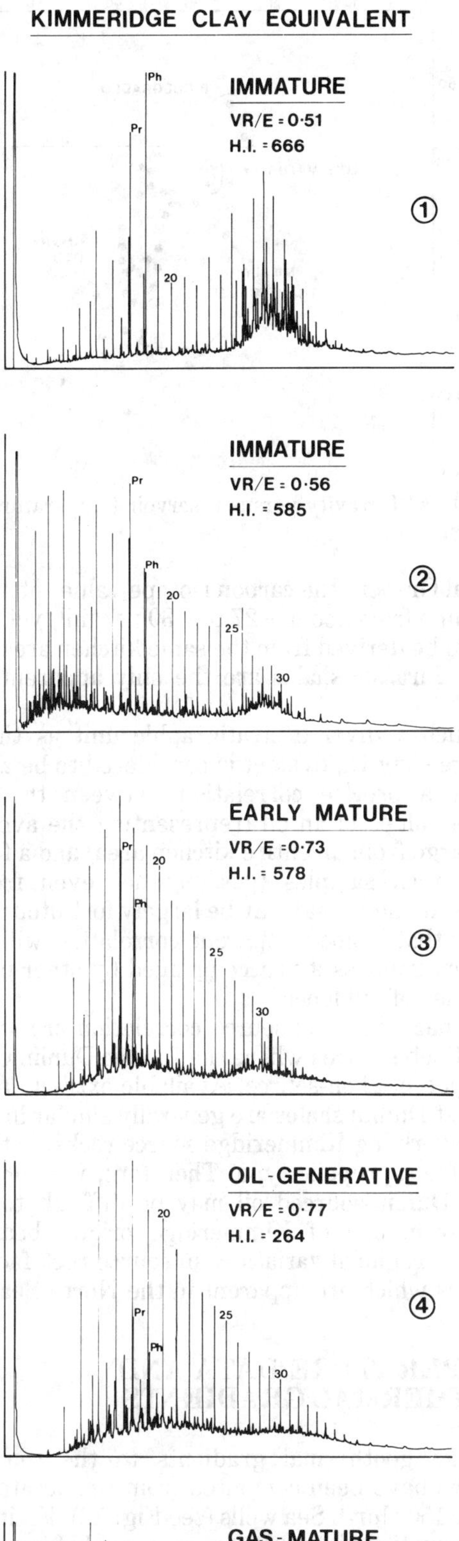

Fig. 8. Composite generative profile, Kimmeridge Clay Equivalent (Type II kerogen). Refer also to Fig. 7.

tion set indicate immaturity in samples of VR/E = 0.51 and 0.56 (Figs. 7 and 8, points 1 and 2). A sample at VR/E = 0.73 is largely 'oil-like' with some slight indications of immaturity (point 3). Extracts of source rocks at VR/E = 0.77 (point 4) and 1.0 (point 5) are fully mature and are closely comparable with many North Sea light crudes.

OIL GEOCHEMISTRY AND OIL-SOURCE CORRELATION

A number of typical oils from the Norwegian North Sea are illustrated in Fig. 9. The majority of North Sea crudes are medium to light (30–40 ° API) and paraffinic, with a low sulphur content (less than 0.5%), e.g. Brage, 30/6–5, Statfjord, etc. Another group of medium-to-heavy (24–28 ° API) naphthenic oils, which are deficient in normal alkanes, are believed to be biodegraded, e.g. Troll, Balder, Frigg and Gullfaks. Some relatively heavy (28–29 ° API) oils from the Egersund Sub-basin are not obviously biodegraded, e.g. Brisling, and may be low-maturity crudes. The lack of light ends and a high isoprenoid and sulphur content support this interpretation. The large paraffinic group of oils fits generally into the 'marine' type of Tissot and Roucache (1980), although a moderate wax content suggests a contribution of terrestrial plant material. Low pristane/phytane ratios are consistent with source rocks deposited in anoxic conditions.

In general, the API gravity of North Sea oils increases with reservoir depth, and the trend is even stronger for API versus formation temperature (see Fig. 10). The biodegraded oils are restricted to shallow reservoirs with present-day temperatures below 75 °C. A few oils which plot away from the dominant trend appear to be mildly transformed, perhaps by water-washing.

As was demonstrated earlier (Figs. 8 and 9) solvent extracts from thermally-mature shales of the sapropelic facies of the Kimmeridge Clay Equivalent compare closely in terms of normal alkane distribution with many North Sea paraffinic crudes. This observation is generally supported by more-detailed correlations using mass spectrometric techniques. Additional confirmation of the genetic link between Kimmeridge source rocks and North Sea oils is provided by artificial maturation experiments, where immature source rocks are heated under controlled laboratory conditions (see Fig. 11). A grossly-immature Kimmeridge source rock (Type II, hydrogen index = 666) has an extract GC-profile quite unlike any North Sea oil. After heating for 6 days at 330 °C (Lijmbach *et al.*, 1983), the resultant 'synthetic crude' is comparable with many naturally occurring North Sea oils.

Carbon isotopes are widely regarded as a powerful oil-source correlation tool, and have been used with apparent success in a number of basins (see Hunt, 1979). North Sea oils vary quite widely in carbon isotope values, from approximately −27 to −31 per mil, which may in isolation be taken as an indication of more than one source. However, carbon isotope data from solvent extracts of Kimmeridge cores in the calibration set cover a range from −26.8 to −31.1 per mil, and most of this regional range has also been observed in a single well section. Similar variations are described by Reitsema (1983) for Kimmeridge kerogens in the Brae area. In the Ekofisk area of the

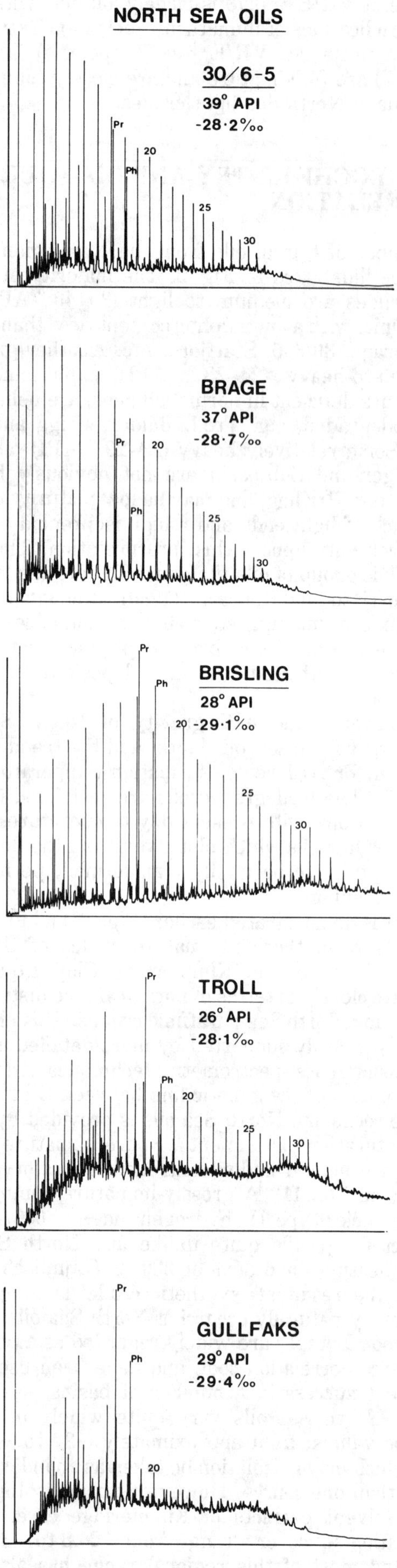

Fig. 9. Selected North Sea oils.

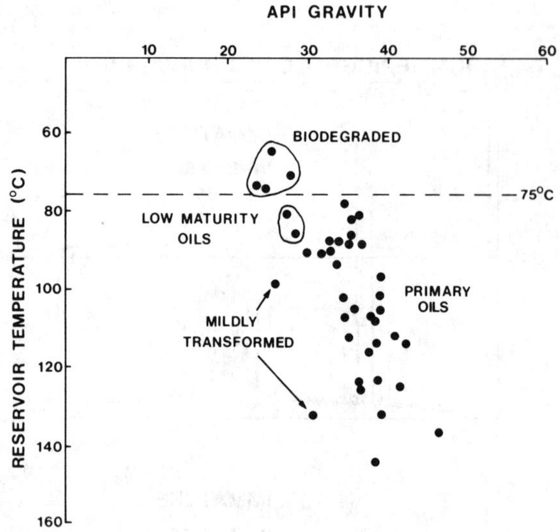

Fig. 10. API gravity versus reservoir temperature, North Sea oils.

Central Graben, the carbon isotope values of analysed oils range from about −27 to −30 per mil, yet they all seem to be derived from the same kitchen area, where Upper Jurassic shales are the only apparent source rocks.

If such a diverse stratigraphic unit as the Kimmeridge Clay Equivalent is considered to be a 'single' source, a precise correlation between the carbon isotope values of an oil (representing the average of the charge from an entire kitchen area) and a few spot source rock samples (possibly not even from the mature kitchen area) may be largely fortuitous. It also follows that a good apparent correlation will not be convincing unless it is accompanied by other independent lines of evidence.

The hazards in oil-source correlation are amplified in the Oseberg area where the Liassic Dunlin Group is also a potential oil-source. Available extract data indicate that Dunlin shales are generally similar in character to overlying Kimmeridge source rocks, both being of marine-sapropelic type. Therefore, it seems likely that a Dunlin-sourced oil may be difficult to distinguish from one of Kimmeridge origin, bearing in mind the regional variations in source rock facies and oil types which are apparent in the North Sea.

TEMPERATURE DATA AND GEOTHERMAL GRADIENTS

Average geothermal gradients to the top of the Jurassic have been computed from temperature data in over 150 North Sea wells (see Fig. 12). Estimates of true formation temperatures were made from wireline log data using a Horner-type plot, and have been calibrated with temperatures from drill stem tests where available.

Geothermal gradients within the Viking Graben are relatively low, ranging from 3.0 to 3.5 °C/100 m, whilst higher gradients occur along the flanks and on basement highs. A pronounced thermal anomaly in the Troll Field (5.0 °C/100 m) is attributed largely to the insultating effects of this giant gas accumulation. In contrast with the Viking Graben, the axis of the Central Graben is associated with high average gradients (more than 3.75 °C/100 m), although the data set may be biased since most wells in this area were drilled on salt-induced features.

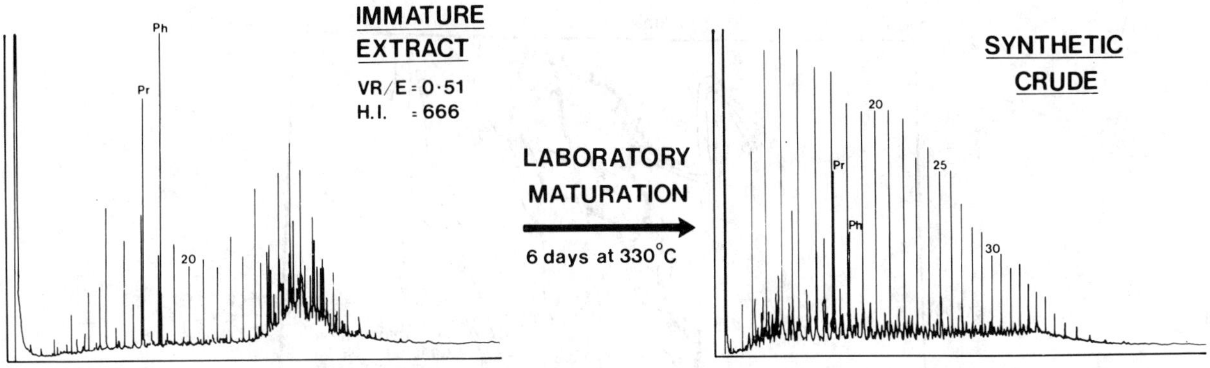

Fig. 11. Laboratory maturation experiment, Kimmeridge Clay Equivalent (type II kerogen).

THERMAL MATURITY TRENDS

Reliable vitrinite reflectance measurements are only obtained from properly identified teleocollinite, one of several species of vitrinite which may be distinguished microscopically (Buiskool-Toxopeus, 1983). For this reason the direct measurement of thermal maturity is difficult in many North Sea wells. Reflectance data from the Cretaceous and Tertiary are often spurious, and detailed maturity profiles of entire well sections (from which geothermal history may be reconstructed) are almost impossible to obtain. Vitrinite is virtually absent within the rich Kimmeridge source interval (a common feature of many Type II marine source rocks), and reliable measurements in many wells are obtained only from a few coaly intervals in the Middle to Lower Jurassic section.

In these studies, the thermal modelling method of Lopatin (1971) was used to estimate maturity levels from burial history and geothermal gradients. Temperature data were carefully evaluated, since the method of Lopatin is very sensitive to this parameter, and the results of modelling were calibrated in wells with reliable measurements to vitrinite reflectance. Present geothermal gradients were used (see Fig. 12), since they usually provide a good correlation between measured and estimated vitrinite reflectance.

Using seismic structure maps and well data, a regional thermal maturity map of the Northern North Sea was prepared for the top of the Kimmeridge source rock interval (i.e. top Humber Group), see Fig. 13. Mature 'kitchen' areas coincide closely with the Viking and Central Grabens, and deeper parts of the East Shetland Basin. In the east, source rocks of the Volgian–Kimmeridgian Tau Formation reach early oil maturity in the Egersund Sub-basin. Individual kitchen areas have been named to facilitate further discussion of hydrocarbon generation and migration.

HYDROCARBON MIGRATION AND ENTRAPMENT

The location of known hydrocarbon occurrences with respect to kitchen areas and stratigraphy provides an indication of likely hydrocarbon migration routes, which may then be confirmed by detailed mapping of drainage areas on regional seal horizons. Hydrocarbon distributions in the North Sea are complex, since reservoirs occur at several levels within the Mesozoic and Tertiary section. In some areas hydrocarbon migration patterns seem to be almost entirely vertical, and fields are coincident with mature kitchen areas, whilst in others large-scale lateral migration is apparent. On many occasions, migration has occurred to a stratigraphically lower level. Some parts of the basin appear to be predominantly gas-prone, e.g. the Frigg and Sleipner areas; others are prolific oil provinces, e.g. Ekofisk area and Tampen Spur.

Geochemical studies were coupled with wireline log indications of source rock thickness in the Humber Group for a number of key basinal wells (see Fig. 1), as follows:

Well 30/4-1 (Oseberg Kitchen)—200 m net source rock
Well 24/9-1 (Balder Kitchen) —420+ m net source rock
Well 15/3-1 (Sleipner Kitchen)—580 m net source rock
Well 2/11-1 (Ekofisk Kitchen) —570+ m net source rock

These data from the deep mature kitchen areas suggest that hydrocarbon charge vastly exceeds the available volume in adjacent traps. Many structures appear to be filled to spill-point and, in some areas, long chains of fields can be seen to be related to each other and a common kitchen area as a continuous fill-and-overspill sequence (Gussow, 1954).

Northern Viking Graben (60°30′ to 62°00′N)

Agat Kitchen

The most northerly occurrence of hydrocarbons in the North Sea is at 'Agat' (Myrland *et al.*, 1981), a small gas-condensate discovery in Lower Cretaceous submarine fan sands on the eastern flank of the Viking Graben (see Fig. 4). Drainage patterns indicate that the source of the gas is the deep Agat Kitchen about 20 km to the west, where presumed Humber Group source rocks are currently gas-mature (see Fig. 13).

Troll Kitchen

The giant Troll gas and oil field (Brekke *et al.*, 1981) and two smaller gas-condensate discoveries in block 35/8 are believed to be directly linked to the large gas-mature Troll Kitchen (see Figs. 13 and 14). The thermal origin of Troll gas is confirmed by carbon isotopic data (delta $^{13}C = -44.5$ per mil). The Troll Kitchen is a deep asymmetrical syncline, with a broad southeasterly flank (see Fig. 15). Migration of hydrocarbons from mature Jurassic source rocks is therefore focused preferentially east and southeast towards the Troll area, some 40 km distant. Although Humber Group source rocks are currently gas-mature in the undrilled central kitchen area, a significant prior oil charge is indicated by a biodegraded heavy-oil rim in

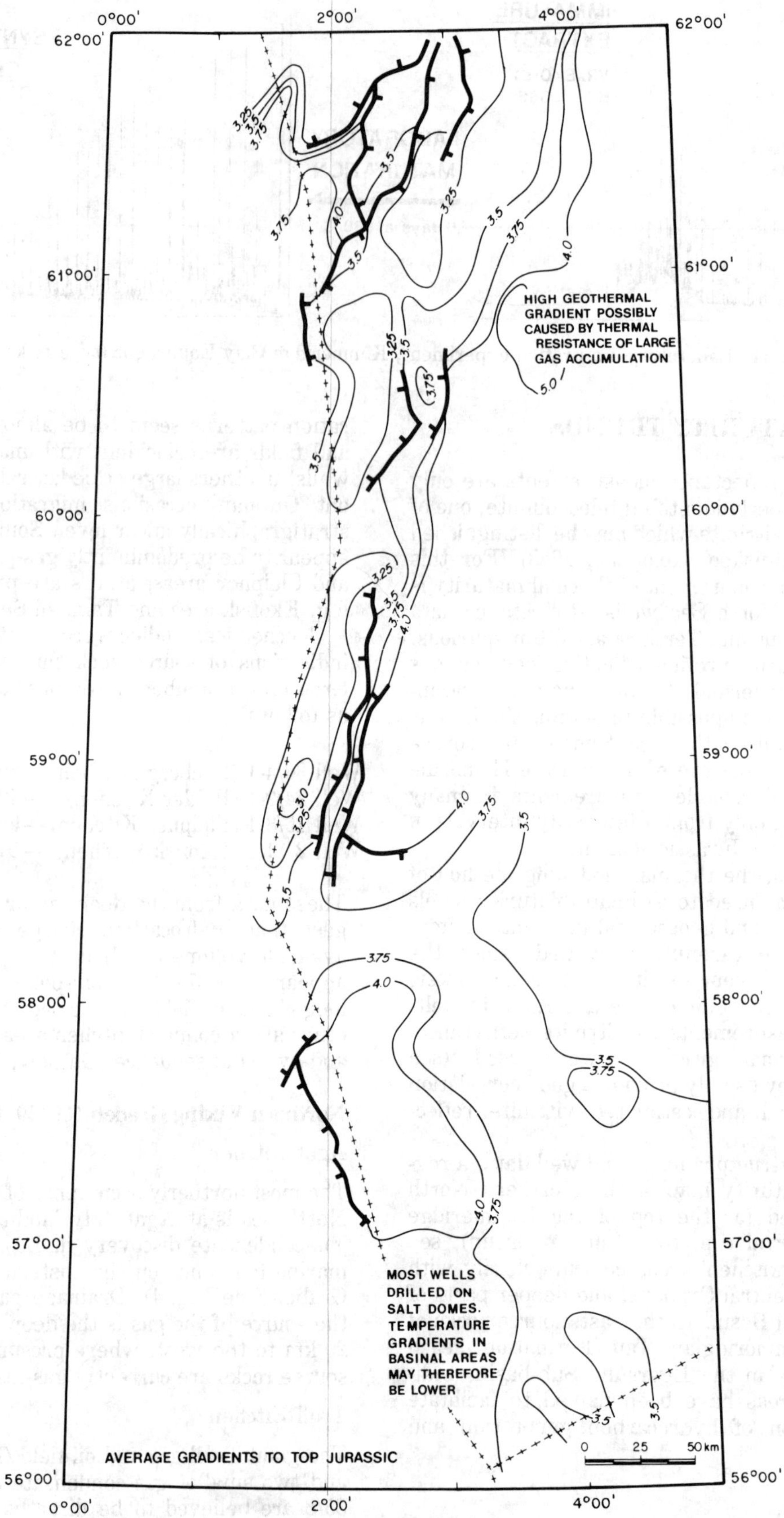

Fig. 12. Geothermal gradients (°C/100 m), Norwegian North Sea.

the Troll structure and evidence of oil-staining within the present Troll gas cap. It is estimated that oil generation from presumed Kimmeridge Clay Equivalent source rocks deep in the Troll Kitchen would have occurred mainly in the early Tertiary, whilst on its flanks the same sequence is currently still oil-mature (see Fig. 15). This is consistent with the character of Troll oil (see Fig. 9), which appears to be comprised of a biodegraded naphthenic residuum, possibly representing the early trapped oil, which has been diluted by a fresh paraffinic charge. Westerly migration of hydrocarbons out of the Troll Kitchen may be limited not only by basin geometry, but also by a major northeast trending fault system (see Fig. 14), and accumulations on the Tampen Spur are thought to be mainly related to another group of hydrocarbon kitchens.

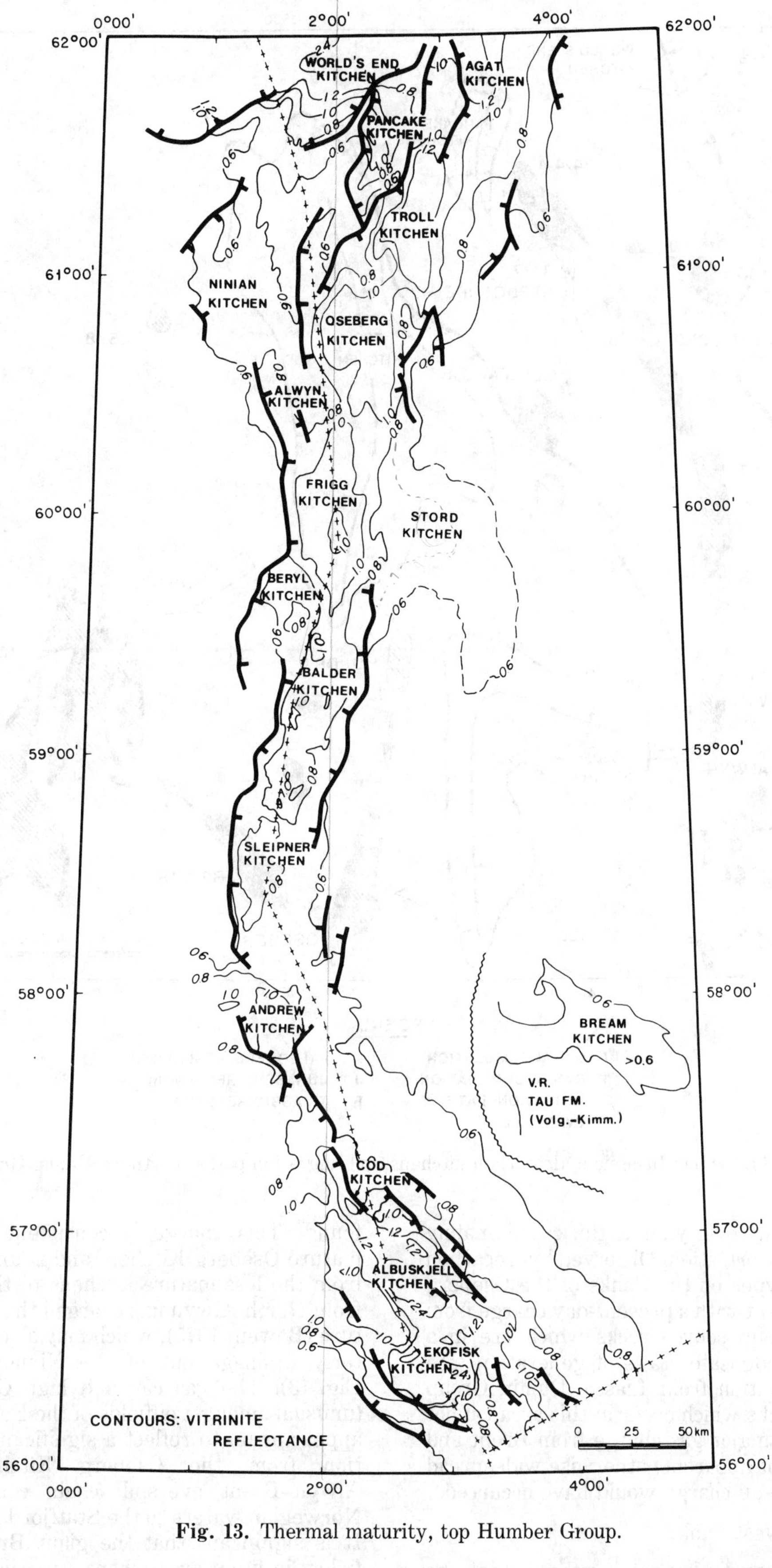

Fig. 13. Thermal maturity, top Humber Group.

Oseberg Kitchen (east flank)

Hydrocarbon migration patterns and spill relationships are complex in the Oseberg area. The Oseberg oil field (Larsen *et al.*, 1981) is charged directly from the Oseberg Kitchen only 10–15 km to the west. Overspill of oil from Oseberg appears to have fed the Brage structure to the east, which could also overspill into the western compartment of the Troll Field. The Huldra gas-condensate field appears to be charged by both the Oseberg and Troll Kitchens, and overspill may occur into the main Oseberg Field.

All hydrocarbons in the Oseberg–Troll area occur in Jurassic reservoirs. Deep synclinal well control suggests that the Humber Group in the Oseberg Kitchen consists of relatively thin oil-prone Kimmeridge Clay

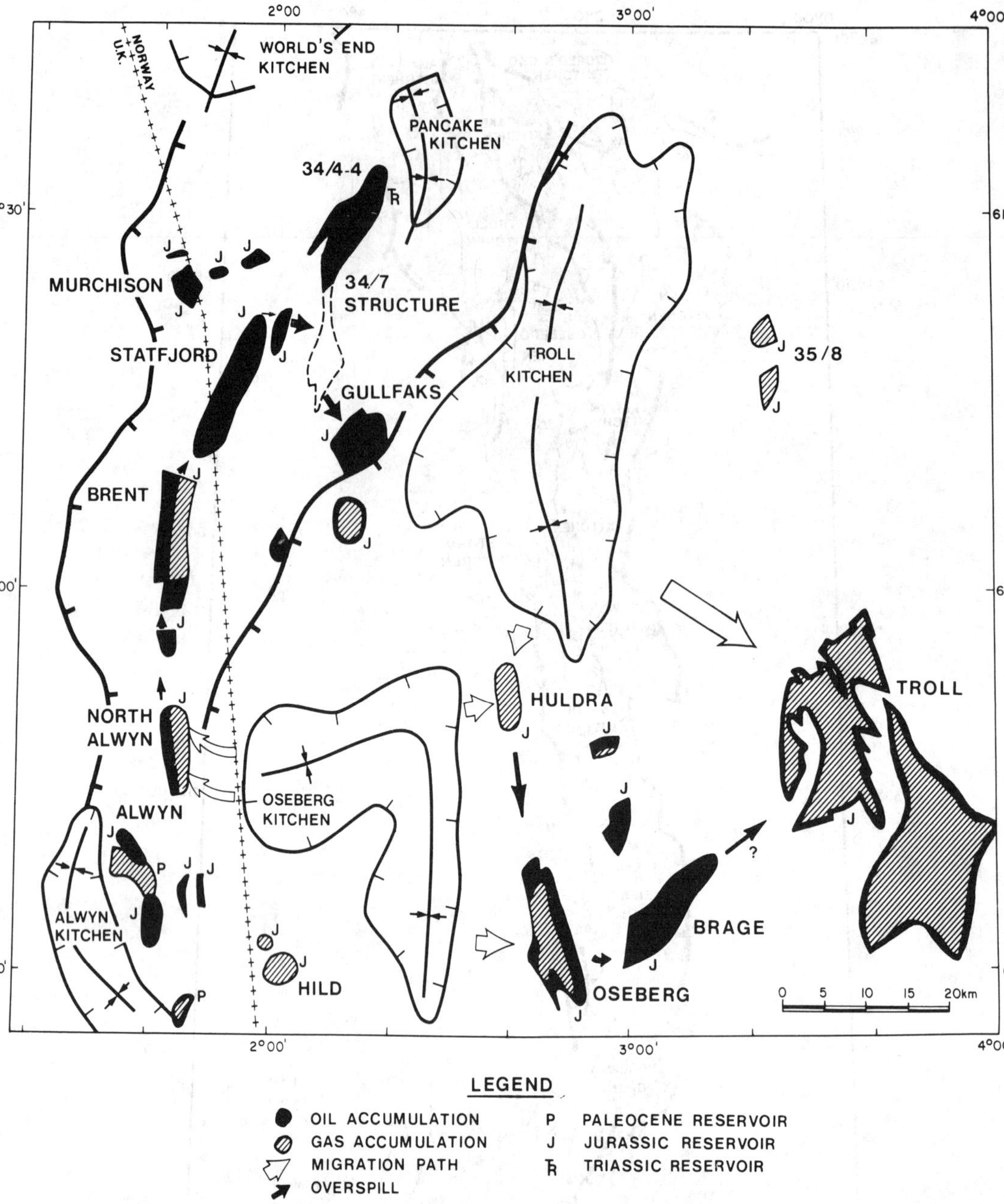

Fig. 14. Upper Jurassic hydrocarbon kitchens and migration paths, northern Viking Graben.

Equivalent (<100 m) overlying a thick and mainly gas-prone Heather Formation. Observed hydrocarbon distributions and types on the flanks of the Oseberg Kitchen are consistent with a present-day charge from upper Humber Group source rocks which are in a late-oil and gas-condensate stage of generation (see Fig. 13). A contribution from Liassic Dunlin Group oil-prone source rocks which occur in this area cannot be excluded, but a major gas charge from Brent and Statfjord coals is not apparent, otherwise widespread displacement of the oil charge would have occurred.

Oseberg Kitchen (west flank)

Hydrocarbons appear to have migrated out of the Oseberg Kitchen into a number of fields in UK waters on the west flank of the Viking Graben (see Fig. 14). Several accumulations in the Alwyn complex are well-situated to be charged by the Oseberg Kitchen, although the oil-mature Alwyn and Ninian Kitchens to the west also drain into the same structural trend. Confirmation of a major charge from the east is seen in North Alwyn, where high GOR oil and a gas cap are found. This charge is consistent with the highly-mature Oseberg Kitchen, and is unlikely to originate from the less-mature kitchens to the west. Overspill from North Alwyn has charged the giant Brent structure (Bowen, 1975), which may also benefit from westerly drainage out of the Ninian oil kitchen (see Fig. 13). The gas cap and high GOR oils in Brent (unusual amongst oilfields of the East Shetland Basin) appear again to reflect a significant charge contribution from the Oseberg Kitchen. The North Alwyn–Brent overspill chain continues back into Norwegian waters to the Statfjord Field (Kirk, 1980). It is significant that the giant Brent and Statfjord fields lie in an area where Jurassic source rocks are immature, and that these structures must rely entirely on lateral migration and overspill for their charge. North of Statfjord, a chain of small oil fields centred on Murchison appears to be charged from the northerly World's End Kitchen (see Fig. 14).

Structural studies indicate that the recent 34/4-4 oil discovery is on the northern extent of a very large rotated Jurassic–Triassic fault block which lies mainly

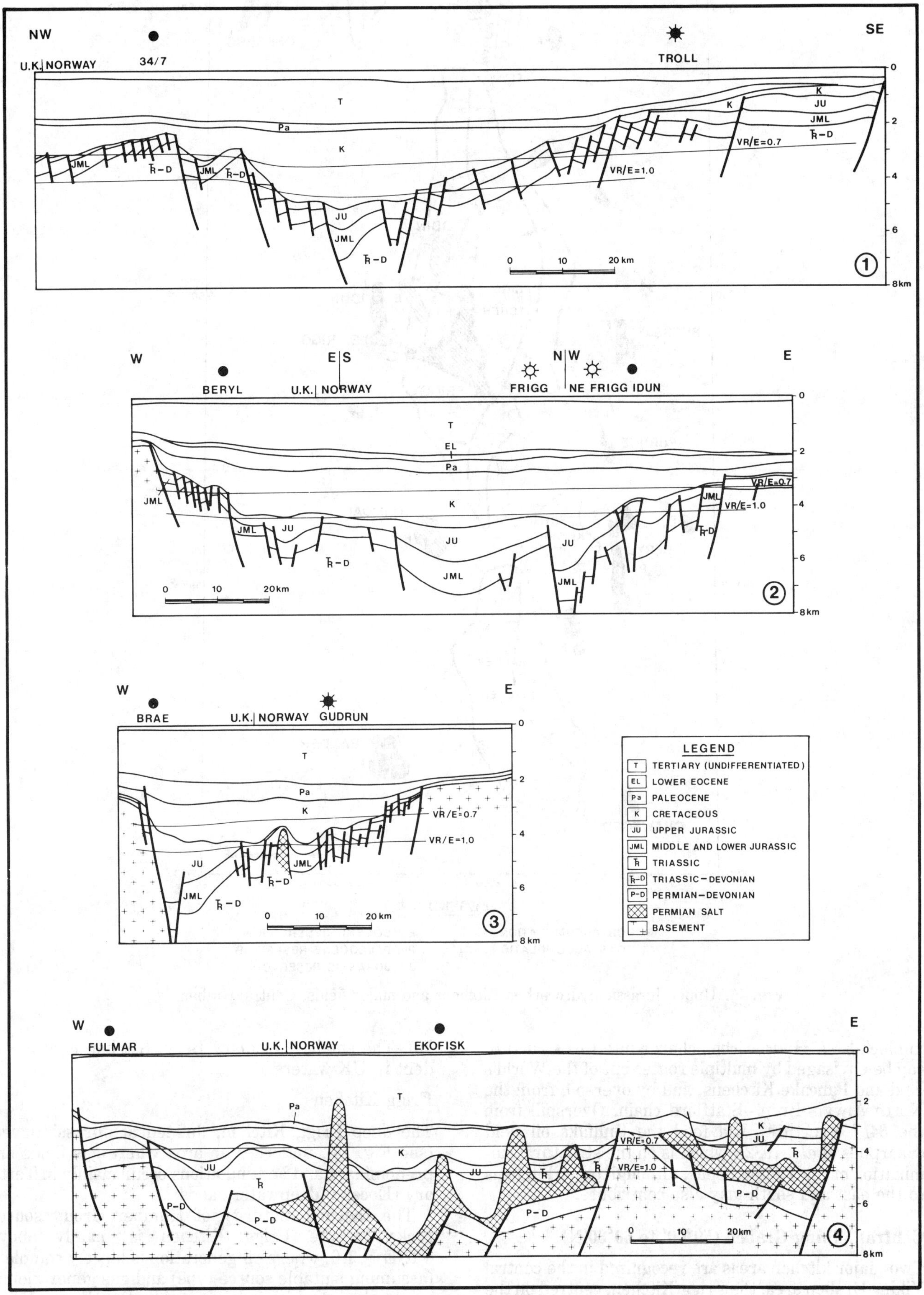

Fig. 15. Regional cross-sections, North Sea (Section 4, after Ziegler, 1982). For locations, see Fig. 1.

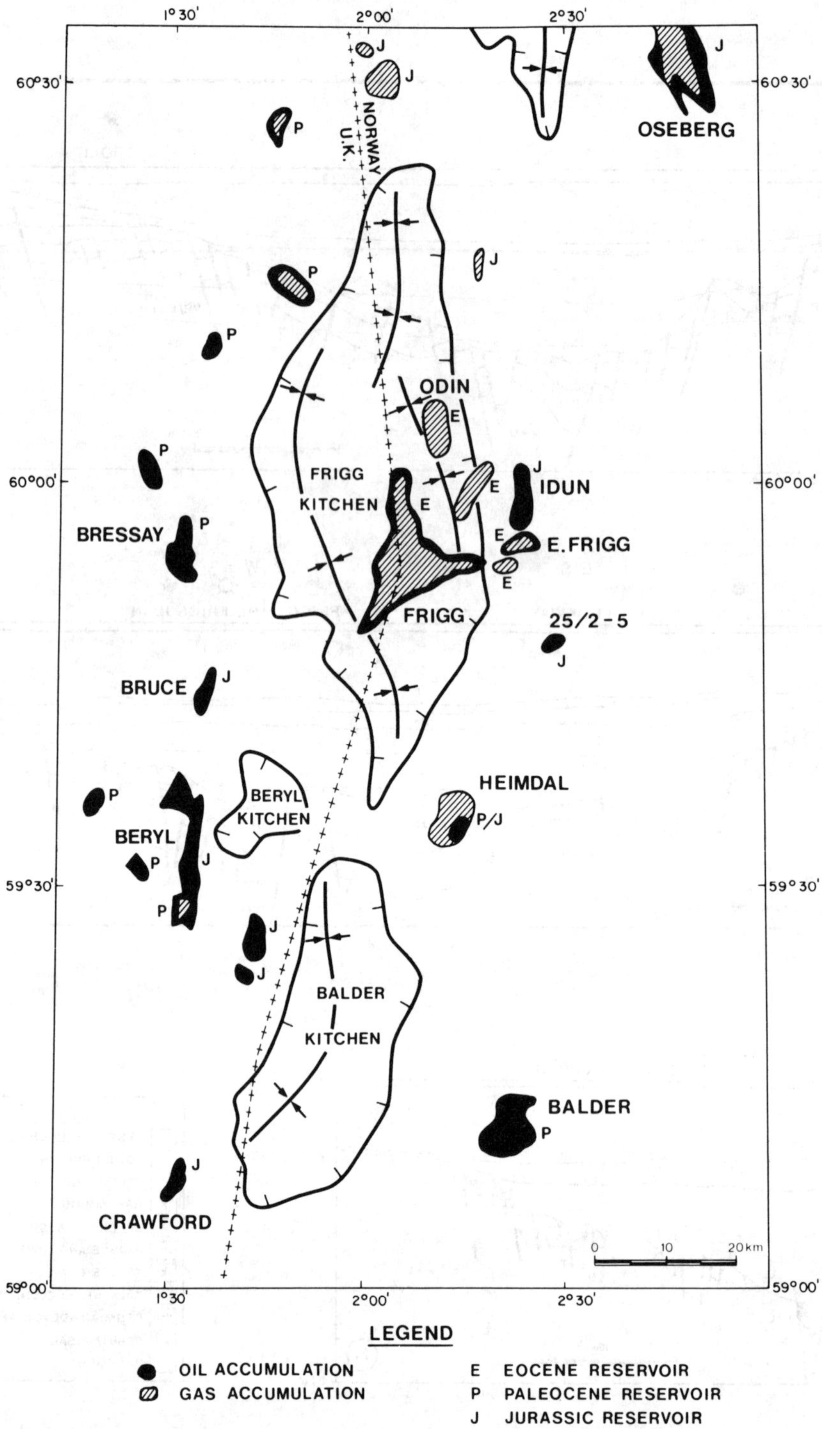

Fig. 16. Upper Jurassic hydrocarbon Kitchens and major fields, Central Graben.

in block 34/7. Hydrocarbon charge into this structure can be envisaged by multiple routes out of the World's End and Pancake Kitchens, and by overspill from the North Alwyn–Brent–Statfjord chain. Overspill from the 34/7 structure* will feed the Gullfaks oil field (Skarpnes *et al.*, 1982), which is on the structural culmination of the Tampen Spur, although other kitchens to the east and south may also contribute.

Central Viking Graben (59°00′ to 60°30′N)

Two major kitchen areas are recognized in the central Viking Graben area: the Frigg Kitchen, centred on the Frigg gas field, and the Balder Kitchen, to the immediate west of the Balder oil field (see Figs. 13 and 15). The small, subsidiary Beryl Kitchen is also evident in UK waters.

Frigg Kitchen

The deep Frigg Kitchen, defined at Jurassic levels (see Fig. 13), underlies an area where gas fields are predominant and accumulations occur mainly in Tertiary (Eocene) deepwater sands.

The thermal maturity of Humber Group source rocks in the Frigg Kitchen is mainly above VR/E = 1.0, where oil generation is largely complete (assuming suitable source type) and gas generation is predominant, which is confirmed by the Eocene Frigg Gas Field and its satellites (see Figs. 15 and 16). Oil has also clearly been generated in the Frigg Kitchen, as indicated by the heavy (biodegraded) oil at Frigg and East Frigg (Heritier *et al.*, 1980). Light oil was reco-

*Recently named the Snorre Field.

vered from Jurassic reservoirs on the eastern flank of the Frigg Kitchen at Idun and 25/2-5, where Humber Group source rocks are still within the oil window.

In the greater Frigg area, vertical migration of hydrocarbons from Jurassic source rocks has clearly been dominant, possibly facilitated by microfracturing of overlying abnormally pressured Cretaceous shales (Goff, 1983). The Eocene umbrella closure of the greater Frigg structure is largely coincident with the gas-mature Jurassic Frigg Kitchen. In the west, both vertical and lateral migration paths out of the Frigg Kitchen are indicated by a number of Paleocene heavy oil fields (e.g. Bressay) which occur in UK waters on the East Shetland Platform.

Apart from maturity considerations, trap timing may also play a role in the apparent gas-prone nature of the Frigg area. Since many of the shallow productive structures were not sealed until at least late Eocene times, thermal modelling studies suggest that these traps largely postdate the early oil charge from the Frigg Kitchen. Their content therefore reflects the later (and current) charge where gas predominates.

Balder Kitchen

A thick sequence of Humber Group source rocks is mature for oil and gas generation within the Balder Kitchen (see Figs. 13 and 16). The effective hydrocarbon charge from this area is predominantly oil, as is indicated by the Paleocene Balder oil field (Sarg and Skjold, 1982) and several Jurassic discoveries in UK waters. Oil in the Jurassic at Heimdal may be from either the Balder or Frigg Kitchens, although gas in Paleocene sands suggests that there is a significant contribution from the Frigg gas kitchen.

Beryl Kitchen

The small Beryl Kitchen, entirely within UK waters, appears to have charged a cluster of Jurassic and Paleocene oil fields in the Beryl–Bruce area (see Figs. 13 and 16).

Southern Viking Graben (58° to 59°N)

The southern Viking Graben differs significantly in structure and stratigraphy from its central and northern counterparts, and these differences have direct implications on hydrocarbon sources, migration paths and distributions. The southern Viking Graben is narrow and deep, and at Triassic and Jurassic levels is highly asymmetrical in structural profile (see regional cross-section 3, Fig. 15). At Middle Jurassic levels structure takes the form of a half-graben, with the deepest part of the basin in the west adjacent to the basin-bounding fault system. At top Jurassic level, the

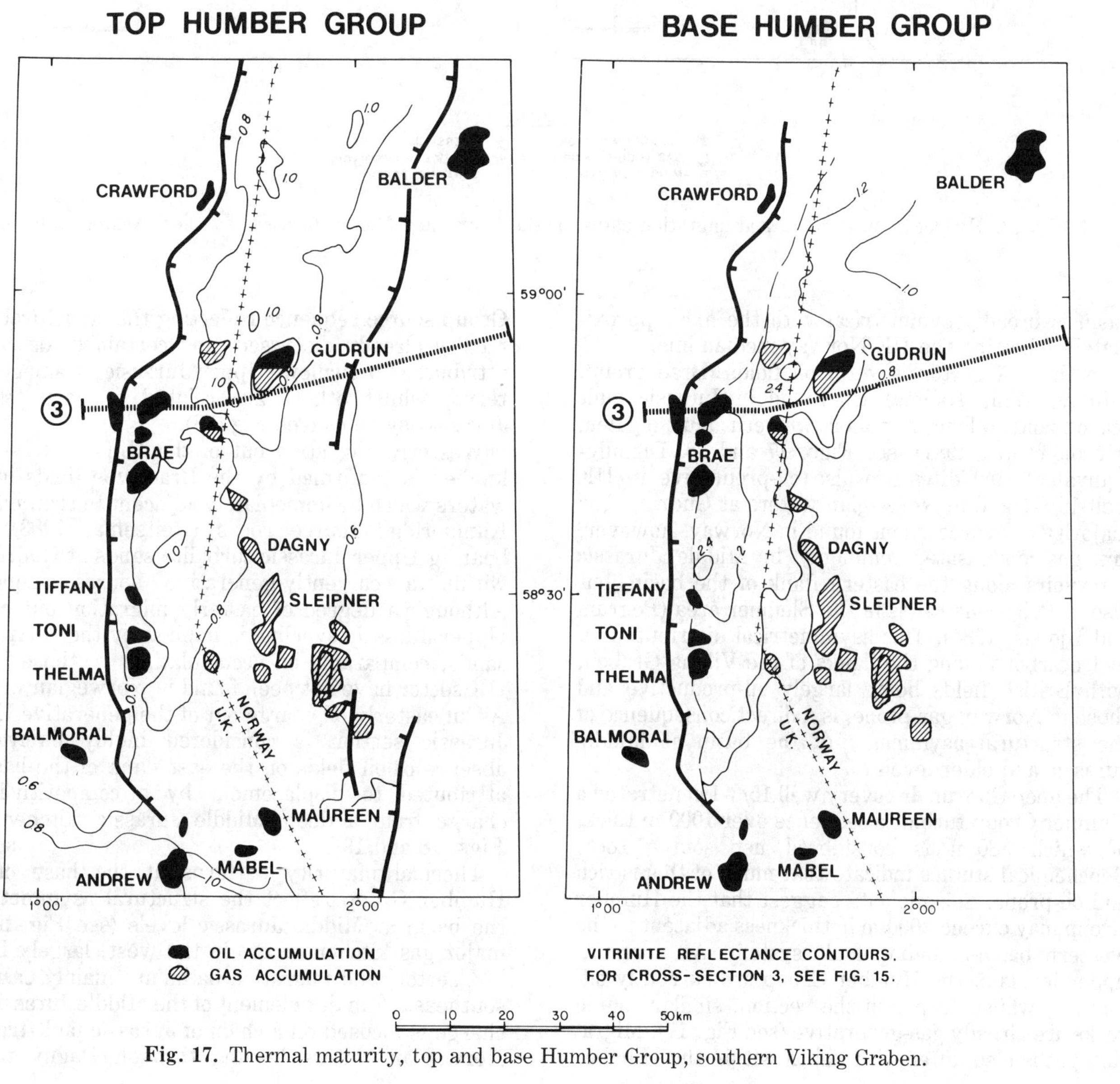

Fig. 17. Thermal maturity, top and base Humber Group, southern Viking Graben.

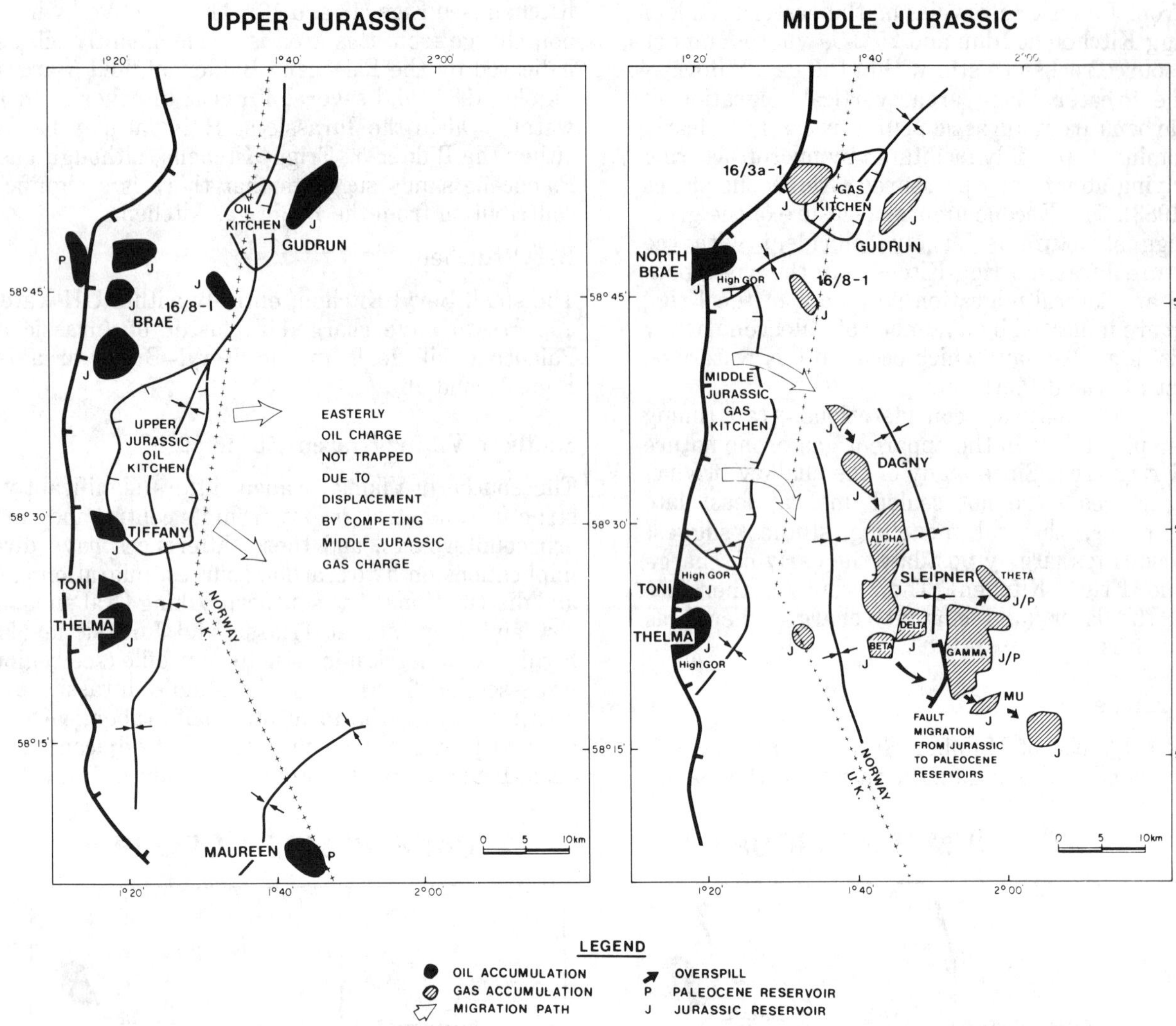

Fig. 18. Hydrocarbon kitchens and migration paths for the Upper and Middle Jurassic, southern Viking Graben.

basin is broadly symmetrical with the axis approximately following the UK/Norway median line.

In the UK sector, several oil fields (Brae group, Tiffany, Toni, Thelma) are found in Jurassic fault scarp sands along a major western fault system (Harms *et al.*, 1981), see Figs. 17 and 18. Distally-equivalent turbidites are also oil-productive in UK well 16/8-1 and in Norwegian waters at Gudrun. The majority of hydrocarbons found in Norway, however, are gas-condensate, commonly in Middle Jurassic reservoirs along the eastern flank of the basin, but also in Paleocene sands in the Sleipner area (Pegrum and Ljones, 1984). This asymmetrical distribution of hydrocarbons along the flanks of the Viking Graben, with British fields being largely oil-productive and those in Norway gas-prone, is a direct consequence of the structural asymmetry of the basin at Middle Jurassic and older levels.

The deep Gudrun discovery well 15/3-1 penetrated a Humber Group turbidite sequence over 1000 m thick, of which 580 m is considered net source rock. Geochemical studies indicate that much of this is rich and oil-prone. Seismic data suggest that the Humber Group may exceed 2000 m in thickness adjacent to the western basin-bounding fault system. As a result, upper levels of the Humber Group are currently oil-mature, whilst deeper in the section, similar source rocks are already gas-generative (see Fig. 17). Migration paths also differ with depth within the Humber Group source sequence reflecting the structural complexity already discussed, and certain fields can be attributed to shallow 'Upper Jurassic' drainage patterns, whilst others are related to deep 'Middle Jurassic' systems (see Fig. 18).

Westerly migration out of the Upper Jurassic oil kitchen is confirmed by the Brae-type fields in UK waters which lie immediately adjacent to rich, mature Kimmeridge source rocks (Reitsema, 1983). Oil-bearing Upper Jurassic turbidite sands at Gudrun lie within a currently-generative source sequence. Although a degree of easterly migration out of the Upper Jurassic syncline is implied by the Gudrun oil sands, comparable oil accumulations to those in the UK sector have not been found in Norwegian waters. As an easterly oil charge out of the generative Upper Jurassic section is considered highly likely, the absence of oil fields on the east flank of the basin is attributed to displacement by a competitive gas charge from a deep 'Middle Jurassic' kitchen (see Figs. 18 and 19).

Thermal maturity patterns at the base of the Humber Group reflect the structural asymmetry of the basin at Middle Jurassic levels (see Fig. 17). A major gas kitchen occurs in the west, largely in the UK sector, and migration paths are mainly east and southeast. A major element of the Middle Jurassic gas charge is focused on a chain of Jurassic fault-traps in Norwegian waters leading through Dagny to the

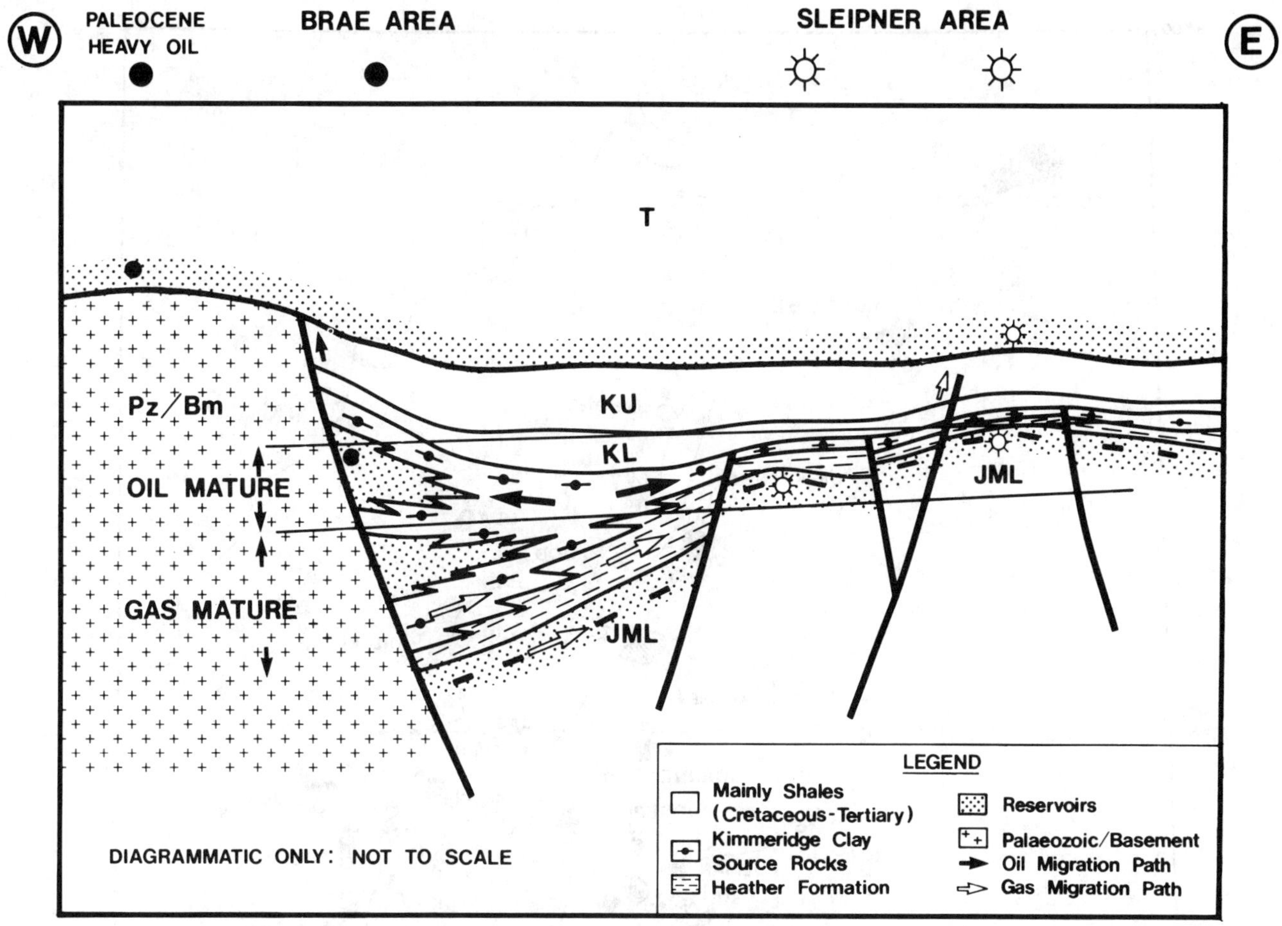

Fig. 19. Schematic cross-section, southern Viking Graben, showing kitchen areas and migration paths.

Sleipner complex (Larsen and Jaarvik, 1981; Fagerland, 1984). All of these structures contain gas-condensate (see Fig. 18). Gas is also found in the Paleocene at Gamma and Theta, which is attributed to vertical communication with the Jurassic migration route, possibly via a major fault along the western margin of the Gamma feature (Fagerland, 1984).

In the Gudrun area, Middle Jurassic migration paths are apparently radial, resulting in gas in deep reservoirs in UK 16/3a-1, UK 16/8-1 and Gudrun. The occurrence of high-GOR oil in North Brae (in contrast with Central and South Brae) is possible evidence of a contribution from deeper highly mature source rocks, due to an element of easterly dip at Middle Jurassic levels in this part of the kitchen. Similarly, Toni and Thelma, with Upper Jurassic reservoirs containing high GOR oils, lie in another area where the Middle Jurassic basin has a small western flank. The Maureen, Mabel and Andrew oil fields in Paleocene sands are thought to be related to other migration paths out of the southerly Andrew Kitchen (see Fig. 13).

A contribution to hydrocarbon accumulations in the southern Viking Graben from pre-Humber Group source rocks, particularly coals of the Vestland Group and Statfjord Formation, cannot be excluded. However, the wet gas accumulations of the Sleipner area are most compatible with an origin from post-mature oil-prone source rocks, such as are found in the lower Humber Group in this area.

Central Graben (56° to 58°N)

The Central Graben is structurally complex, since the primary pattern of the North Sea rift system is heavily modified with diapiric salt tectonics. As in the Viking Graben, the principal source unit is the Humber Group, which exceeds 1000 m in thickness in basinal areas. In the Norwegian sector, fields are commonly in fractured Danian–Maastrichtian chalk reservoirs (van den Bark and Thomas, 1980; Watts, 1983) which have been arched over deep intrusive Zechstein salt bodies (see Fig. 15). Paleocene sands are productive, particularly in UK waters (Carman and Young, 1981), and some hydrocarbons are also found in Jurassic and older reservoirs (Bailey *et al.*, 1981). Mapping thermal maturity at top Humber Group level (see Fig. 13), almost every field in the Central Graben is contained within a fairway defined by the VR/E = 0.6 contour, i.e. fields appear to be restricted closely to locations within or above the currently-generative Jurassic basin (see Fig. 20). Since in the Norwegian Sector fields lie mainly in Cretaceous reservoirs, a strong element of vertical migration is implied with little apparent lateral migration. Assuming an oil-floor of VR/E = 1.2, the central trend of the Cod–Albuskjell–Ekofisk gas kitchen is clearly defined. Almost every gas field in the Central Graben is either within or immediately adjacent to the limits of a gas-mature Jurassic kitchen, again confirming the dominance of vertical migration systems in this area.

Egersund Sub-basin

Three small oil discoveries have been made in the Egersund Sub-basin: Bream, Brisling and 18/10-1 (see Fig. 4). Reservoir sands are within the Middle Jurassic Vestland Group, and trapping structures are

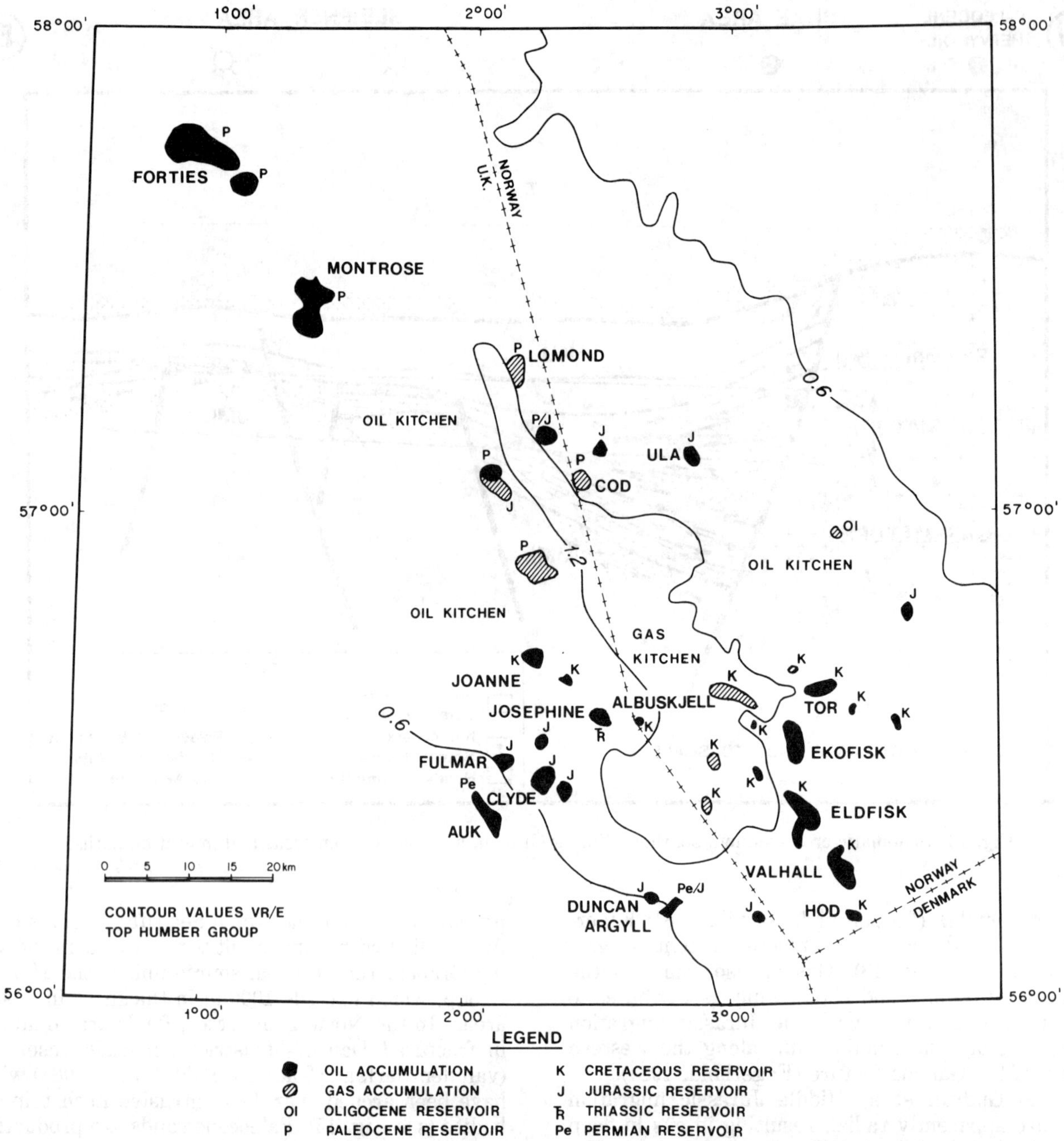

Fig. 20. Upper Jurassic hydrocarbon kitchens and major fields, Central Graben.

halokinetic anticlinal features in an area of major salt diapirism. Rich, oil-prone source rocks occur in the Volgian–Kimmeridgian Tau Formation, which is early mature for oil generation (VR/E = 0.6 to 0.8) over much of the Egersund Sub-basin (see Fig. 13). The three oil-bearing structures lie immediately adjacent to an arm of this Bream oil kitchen.

CONCLUSIONS

Hydrocarbon occurrences in the North Sea graben system can largely be explained in terms of the distribution and thermal maturity of source rocks within the Kimmeridge Clay Equivalent (Oxfordian–Ryazanian). Other source units, e.g. Dunlin Group, Brent Group and Heather Formation, may also make significant charge contributions locally. Although variations in organic facies occur within the dominant Kimmeridge source interval, these are thought to be, at best, only a secondary influence on the distribution of apparent oil- and gas-prone areas in the North Sea since, in general, Kimmeridge source rocks are present in their richest, most oil-prone facies in deep, mature kitchen areas of the Central and Viking Grabens. Other factors, such as thermal maturity of kitchen areas, timing relationships, migration paths and charge competition are considered to be more significant. The presence of excess charge in several areas has resulted in structures which are full to spill-point, and groups of fields can often be related to one another and a common kitchen as a continuous fill-and-overspill sequence.

ACKNOWLEDGEMENTS

The authors wish to acknowledge the assistance of Statoil, Norsk Hydro, Amoco Norway, Conoco Norway, Phillips Norway and the Norwegian Petroleum Directorate, who supplied oils and core material for analysis. Most of the oil and extract analysis were the work of geochemists at Koninklijke/Shell Exploratie en Produktie Laboratorium, the Netherlands; Rock Eval analyses were performed by Robertson Research International, Wales and IKU, Norway. Partners in production licence 054 kindly approved release of confidential data from the Troll Field. For helpful discussions and encouragement we thank J. R. Freake, R. J. Murris, P. A. Ziegler, J. W. A. Bodenhausen, J. M. A. Buiskool-Toxopeus and E. Talleraas. This paper is published with the permission of A/S Norske Shell, Forus, Norway.

REFERENCES

Bailey, C. C., Price, I. and Spencer, A. M. 1981. The Ula oil field, Block 7/12, Norway. In: *Norwegian Symp. on Exploration (NSE-81)*, Bergen. Norwegian Petroleum Society.

Barnard, P. C. and Cooper, B. S. 1981. Oils and source rocks of the North Sea area. In: Illing, L. V. and Hobson, G. D. (eds.), *Petroleum Geology of the Continental Shelf of Northwest Europe*. Institute of Petroleum, London, 521 pp.

Bjørlykke, K., Dypvik, H. and Finstad, K. G. 1975. The Kimmeridge Shale, its composition and radioactivity. In: Finstad, K. G. and Selley, R. C. (co-ordinators), *Proc. Jurassic Northern North Sea Symp. 1975*. Norwegian Petroleum Society.

Bowen, J. M. 1975. The Brent oil-field. In: Woodland, A. W. (ed.), *Petroleum and the Continental Shelf of North West Europe*, Vol. 1, *Geology*. Applied Science Publishers, Barking, 501 pp.

Brekke, T., Pegrum, R. M. and Watts, P. B. 1981. First exploration results in Block 31/2, offshore Norway. In: *Norwegian Symp. on Exploration (NSE-81)*, Bergen. Norwegian Petroleum Society.

Buiskool-Toxopeus, J. M. A. 1983. Selection criteria for the use of vitrinite reflectance as a maturity tool in petroleum exploration. In: Brooks, J. (ed.), *Petroleum Geochemistry and Exploration of Europe*. Blackwell Scientific Publications, Oxford, 379 pp.

Carman, G. J. and Young, R. 1981. Reservoir geology of the Forties oil-field. In: Illing, L. V. and Hobson, G. D. (eds.), *Petroleum Geology of the Continental Shelf of Northwest Europe*. Institute of Petroleum, London, 521 pp.

Clementz, D. M. 1979. Effect of oil and bitumen saturation on source rock pyrolysis. *AAPG Bull.*, **63**(12), 2227–2232.

Demaison, G. 1984. The generative basin concept. In: Demaison, G. and Murris, R. J. (eds.) *Petroleum Geochemistry and Basin Development. AAPG Memoir*, **35**, 1–14.

Demaison, G. J. and Moore, G. T. 1980. Anoxic environments and oil source bed genesis. *AAPG Bull.*, **64**(8), 1179–1209.

Dypvik, H., Rueslåtten, H. G. and Throndsen, T. 1979. Composition of organic matter from North Atlantic Kimmeridgian Shales. *AAPG Bull.*, **63**(12), 2222–2226.

Espitalie, J., Laporte, J. L., Madec, M., Marquis, F., Leplat, P. and Boutefeu, A. 1977. Méthode rapide de caracterisation des roches meres, de leur potential petrolier et de leur degré d'évolution. *Rev. Inst. Fr. Petrol*. **32**, 23–42.

Fagerland, N. 1984. A rift-related stratigraphic trap. *J. Petrol. Geol.*, **7**, 227–236.

Goff, J. C. 1983. Hydrocarbon generation and migration from Jurassic source rocks in the East Shetland Basin and Viking Graben of the northern North Sea. *J. Geol. Soc. London*, **140**, 445–474.

Gussow, W. C. 1954. Differential entrapment of oil and gas; a fundamental principle. *AAPG Bull.*, **38**(5), 816–853.

Harms, J. C., Tackenberg, P., Pickles, E. and Pollock, R. E. 1981. The Brae oil field area. In: Illing, L. V. and Hobson, G. D. (eds.), *Petroleum Geology of the Continental Shelf of Northwest Europe*. Institute of Petroleum, London, 521 pp.

Harwood, R. J. 1977. Oil and gas generation by laboratory pyrolysis of kerogen. *AAPG Bull.*, **61**(12), 2082–2102.

Heritier, F. E., Lossel, P. and Wathne, E. 1980. Frigg Field—large submarine-fan trap in Lower Eocene rocks of the Viking Graben, North Sea. In: Halbouty, M. T. (ed.), *Giant Oil and Gas Fields of the Decade 1968–1978. AAPG Memoir*, **30**, 59–79.

Hunt, J. M. 1979. *Petroleum Geochemistry and Geology*. W. H. Freeman and Company, San Francisco, 617 pp.

Kirk, R. H. 1980. Statfjord—a North Sea giant. in: Halbouty, M. T. (ed.), *Giant Oil and Gasfields of the Decade 1968–1978. AAPG Memoir*, **30**, 95–116.

Landes, K. K. 1967. Eometamorphism and oil and gas in time and space. *AAPG Bull.*, **51**(6), 828–841.

Larsen, R. M. and Jaarvik, L. J. 1981. The geology of the Sleipner Field complex. In: *Norwegian Symp. on Exploration (NSE-81)*, Bergen. Norwegian Petroleum Society.

Larsen, V., Aasheim, S. M. and Masset. J. M. 1981. 30/6-Alpha structure: a field case study in the silver block. In: *Norwegian Symp. on Exploration (NSE-81)*, Bergen. Norwegian Petroleum Society.

Lijmbach, G. W. M., van der Veen, F. M. and Engelhardt, E. D. 1983. Geochemical characterisation of crude oils and source rocks using field ionisation mass spectrometry. In: Bjorøy, M. *et al.* (eds.), *Advances in Organic Geochemistry 1981*. John Wiley and Sons, New York, 880 pp.

Lopatin, N. V. 1971. Temperature and geologic time as factors in Coalification. *Akad. Nauk. SSSR Izv. Ser. Geol.*, **3**, 95–106 (in Russian).

Meissner, F. F. 1984. Petroleum geology of the Bakken Formation, Williston Basin, North Dakota and Montana. In: Demaison, G. and Murris, R. J. (eds.), *Petroleum Geochemistry and Basin Evaluation*. AAPG Memoir, **35**, 159–179.

Meyer, B. L., and Nederlof, M. H. 1984. The identification of source rocks on wireline logs by density/resistivity cross plots. *AAPG Bull.*, **68**(2), 121–129.

Myrland, R., Messel, K. and Raestad, N. 1981. Gas discovery in 35/3: an exploration history. In: *Norwegian Symp. on Exploration (NSE-81)*, Bergen. Norwegian Petroleum Society.

Nederlof, M. H. 1979. The use of habitat of oil models in exploration prospect appraisal. *Proc. 10th World Petroleum Congr.*, Bucharest, September 1979.

Pegrum, R. M. and Ljones, T. E. 1984. 15/9 Gamma gas field offshore Norway, new trap type for North Sea Basin with regional structural implications. *AAPG Bull.*, **68**(7), 874–902.

Phillippi, G. T. 1965. On the depth, time and mechanism of petroleum generation. *Geochim. Cosmochim. Acta*, **29**, 1021–1049.

Pusey, W. C. 1973. The ESR-kerogen method, a new technique of estimating the organic maturity of sedimentary rocks. *Petroleum Times*, 12 January.

Reitsema, R. H. 1983. Geochemistry of the North and South Brae areas, North Sea. In: Brooks, J. (ed.), *Petroleum Geochemistry and Exploration of Europe*. Blackwell Scientific Publications, Oxford, 379 pp.

Robertson Research International Limited 1979. Southern Offshore Norway: the stratigraphy and petroleum geochemistry of the Jurassic to Tertiary sediments. Unpublished multi-client report.

Robertson Research International Limited 1981. Southern Offshore Norway Phase II Study: the stratigraphy and petroleum geochemistry of the Jurassic to Tertiary sediments. Unpublished multi-client report.

Sarg, J. F. and Skjold, L. J. 1982. Stratigraphic traps in Paleocene sands in the Balder area, North Sea. In: Halbouty, M. T. (ed.), *The Deliberate Search for the Subtle Trap. AAPG Memoir*, **32**, 197–206.

Schlumberger 1979. Log interpretation charts, 1979 Edition.

Schlumberger 1982. Natural gamma ray spectrometry, essentials of N. G. S. interpretation.

Schmoker, J. W. 1979. Determination of organic content of Appalachian Devonian shales from formation-density logs. *AAPG Bull.*, **63**(9), 1504–1537.

Schmoker, J. W. 1981. Determination of organic-matter content of Appalachian Devonian shales from gamma-ray logs. *AAPG Bull.*, **65**(7), 1285–1298.

Skarpnes, O., Briseid, E. and Milton, D. I. 1982. The 34/10 Delta prospect of the Norwegian North Sea: exploration study of an unconformity trap. In: Halbouty, M. T. (ed.), *The Deliberate Search for the Subtle Trap. AAPG Memoir*, **32**, 207–216.

Taylor, J. C. M. 1983. Bit-metamorphism, illustrated by lithological data from German North Sea wells. *Geol. Mijnbouw* **62**, 211–219.

Tissot, B. P. and Roucache, J. 1980. Principles of a genetic classification of crude oils. In: Miall, A. D. (ed.), *Facts and Principles of World Petroleum Occurrence. Can. Soc. Petrol. Geol. Memoir*, **6**, 1003 pp.

Tissot, B., Durand, B., Espitalié, J. and Combaz, A. 1974. Influence of the nature and diagenesis of organic matter in formation of petroleum. *AAPG Bull.*, **58**, 499–506.

van den Bark, E. and Thomas, O. D. 1980. Ekofisk: first of the giant oil fields in Western Europe. In: Halbouty, M. T. (ed.), *Giant Oil and Gas Fields of the Decade 1968–1978. AAPG Memoir*, **30**, 195–224.

Vollset, J. and Doré, A. G. (eds) (1984). A revised Triassic and Jurassic lithostratigraphic nomenclature for the Norwegian North Sea. *NPD Bull.* **3**.

Watts, N. L. 1983. Microfractures in chalks of Albuskjell field, Norwegian sector, North Sea: possible origin and distribution. *AAPG Bull.*, **67**(2), 201–234.

White, D. 1915. Some relations in origin between coal and petroleum. *Wash. Acad. Sci. J.*, **5**(6), 189–212.

Ziegler, P. A. 1980. Northwest European Basin: Geology and hydrocarbon provinces. In: Miall, A. D. (ed.), *Facts and Principles of World Petroleum Occurrence. Can. Soc. Petrol. Geol. Memoir*, **6**, 1003 pp.

Ziegler, P. A. 1981. Evolution of sedimentary basisn in North-West Europe. In: Illing, L. V. and Hobson, G. D. (eds.), *Petroleum Geology of the Continental Shelf of Northwest Europe*. Institute of Petroleum, London, 521 pp.

Ziegler, P. A. 1982. *Geological Atlas of Western and Central Europe*. Shell Internationale Petroleum Maatschappij B. V., 130 pp. + enclosures.

2 Correlation of the offshore sequences referred to the Kimmeridge Clay Formation—relevance to the Norwegian sector

A. G. Doré, J. Vollset† and G. P. Hamar‡*
*Conoco Norway Inc., Stavanger, Norway
†Norwegian Petroleum Directorate, Stavanger, Norway
‡Statoil, Stavanger, Norway

The lithologies referred to the Kimmeridge Clay Formation are acknowledged to include the most important petroleum source rocks known in the North Sea. However, the correlations and historical review presented herein demonstrate that widespread use of the term 'Kimmeridge Clay Formation' is a misleading piece of stratigraphic nomenclature which stems from conflicting definitions of the unit both offshore and onshore. This has led to striking differences in the way the unit is perceived in different basins of the North Sea area.

The base of a unit is regarded as a fundamental criterion in stratigraphy, and yet several quite different definitions for the base of the Kimmeridge Clay Formation exist in the literature. In eastern England the base is placed at a chronostratigraphic marker (the Oxfordian–Kimmeridgian boundary) even when present within a homogeneous claystone sequence. Immediately offshore this entire claystone sequence is included in the Kimmeridge Clay and an older, but more lithostratigraphically significant, horizon (the contact with Corallian limestones) is taken as the base. In total contrast, the base in the northern North Sea is viewed as the contact between bituminous claystones and underlying claystones of more normal character. The basic criteria on which the northern Kimmeridge Clay Formation was defined (anomalously high organic content and associated log responses) were never documented in the offshore type and reference sections of the southern UK North Sea. Similarly, bituminous shale developments are not typical of the Kimmeridge Clay as defined in the Netherlands sector.

Adoption of a local formation nomenclature for the Norwegian sector is urged as a means to more-precise correlation. A scheme proposed by the Norwegian Lithostratigraphic Nomenclature Committee is outlined. This classification is based on the recognition that the economically important organic shales are discrete and readily mappable units, but that they are not always synchronous or continuous between basins. A review of the role of the Kimmeridge Clay Formation in other national sectors is also suggested.

INTRODUCTION

In the recent work of the Norwegian Jurassic Lithostratigraphic Nomenclature Committee (1980–1984) much time was devoted to assessing the place of the Upper Jurassic–lowermost Cretaceous Kimmeridge Clay Formation in the Norwegian Lithostratigraphic scheme. During the course of this work it became clear that the term was no longer easy to apply in the Norwegian sector. The unit previously defined as the Kimmeridge Clay Formation was considered to be too broad, to contain too many facies types, and to be of limited use in correlation or in the understanding of basin evolution. These problems were compounded by imprecise and variable definitions of the unit in previous attempts at offshore lithostratigraphic classification. A decision was therefore taken to replace the existing nomenclature with a number of locally-defined units, and these proposals were published by Vollset and Doré (1984).

Unfortunately, the term 'Kimmeridge Clay Formation' is strongly entrenched in the literature and parlance of North Sea explorationists, in all involved national areas, with the possible exception of the Danish sector. Normally this consideration alone would be grounds for extreme caution in modifying the nomenclature. However, it is precisely this entrenchment which leads to misunderstanding and problems of correlation; although the term is widely used, there is no consensus as to its meaning. The problem is essen-

Petroleum Geochemistry in Exploration of the Norwegian Shelf © Norwegian Petroleum Society (Graham & Trotman, 1985) pp. 27–37.

tially a linguistic one. To illustrate this point, some popularly held beliefs (not necessarily published, but frequently heard) on the Kimmeridge Clay Formation are listed below.

The Kimmeridge Clay is variously thought to comprise the following.

(i) All shales of late Jurassic age, regardless of log character (particularly in the Norway–UK Central Graben).

(ii) A sequence of bituminous shales distinguished by very high gamma-ray readings and anomalously low seismic velocities (widely accepted in areas north of the Mid-North Sea High).

(iii) A sequence of shales defined by the particular class of kerogen found in its organic fraction.

(iv) Shales of Kimmeridgian age. Surprisingly, terms such as 'Kimmeridgian shale' still occur in the literature, where clearly referring to sequences of much broader chronostratigraphic range, or even to non-Kimmeridgian sequences. This probably results from the fact that the Kimmeridge Clay Formation of the English mainland is equated with the Kimmeridgian stage *sensu anglico* (see the following section).

Brookfield (1973), in referring to onshore British stratigraphy, wrote: 'The lithostratigraphic terminology of the Oxfordian and Kimmeridgian is truly a mess' (p. 137). In this brief account we attempt to show that some of these problems have been transferred to the offshore realm. In examining these problems we will expand upon our proposal to reclassify locally the Upper Jurassic–lowermost Cretaceous shale units, which include the richest source rocks known in the North Sea, and will draw attention to some of the conceptual problems that exist in Norway and in other national sectors.

THE 'SPREAD' OF THE KIMMERIDGE CLAY FORMATION FROM ITS TYPE AREA

Onshore England

The name 'Kimmeridge Clay' has been in use since the dawn of British stratigraphy in the early 19th Century (e.g. Fitton, 1836). The type area for the Kimmeridge Clay Formation is Dorset on the south coast of England (e.g. Arkell, 1947), where the unit consists of dark claystones with numerous limestone/dolomite stringers and phosphatic horizons. The lowermost part of the unit is developed in a calcareous, shelly facies but higher in the sequence bituminous clays and oil shales are frequent. The Kimmeridge Clay of the English mainland is essentially synonymous with the Kimmeridgian stage *sensu anglico* (e.g. Arkell, 1933; Cox and Gallois, 1981; Cope, 1980, fig. 14), or Kimmeridgian–early Middle Volgian in Boreal terminology.

Over much of its northeastern outcrop in England (Fig. 1) the Kimmeridge Clay overlies more claystones of Late Oxfordian age, the Ampthill Clay. The latter unit in turn overlies, and is in part equivalent to, limestones and calcareous sandstones of the Corallian Formation. The relationship between the Kimmeridge Clay and Ampthill Clay is generally obscure due to lack of exposure, and is mainly documented in early literature, summarized by Arkell (1933). It is alluded to here because the contact between these two units is of significance offshore.

The Ampthill Clay extends as far south as Oxfordshire. In the southern part of its outcrop there is a non-sequence between the Ampthill and Kimmeridge Clays, commonly expressed by a phosphate nodule band (Arkell, 1933). Elsewhere, however such as on Humberside, there is very little or no hiatus. Key sections at South Ferriby (Kelly and Rawson, 1983, p. 66) and Worlaby (Richardson, 1979, p. 13) show the Ampthill Clay–Kimmeridge Clay transition to be one of continuous deposition, with no perceptible lithological difference above and below the 'contact'. In fact, the facies of the basal Kimmeridge Clay and the underlying Ampthill Clay—dark, calcareous, shelly, non-bituminous claystone—are everywhere remarkably similar. In boreholes the 'contact' has almost invariably been determined on palaeontological grounds alone (e.g. Arkell, 1933, p. 418; Wright, 1980, p. 74). The base of the Kimmeridge Clay Formation occurs at the base of the lowermost (*baylei*) zone of the Kimmeridgian stage, but this marker has little significance in lithostratigraphic terms. Brookfield (1973, p. 141) writes: 'Despite contrary opinions . . . there is no major lithological or faunal break between the Oxfordian and Kimmeridgian stages in Britain'.

The Kimmeridge and Ampthill Clays thin northwards towards the Market Weighton High (Fig. 1). North of the high in the Yorkshire Basin boreholes have proved a thick Kimmeridge Clay sequence and limited exposures indicate, for the uppermost zones at least, strong similarities to the Dorset type area (Cope, 1974). The Kimmeridge Clay was long considered to rest directly on Corallian beds in this area, but recent boreholes have proven the existence of intervening Ampthill Clay. Once again, however, this identification appears to be based solely on late Oxfordian datings and not on any lithostratigraphic criterion (I.G.S., 1974).

Southern North Sea, UK sector

The first application of the name Kimmeridge Clay Formation to the offshore realm was by Rhys (1974) in the recommendations of a joint Oil Industry–I.G.S. committee on North Sea nomenclature. The formation was defined as the uppermost unit of the Humber Group, a primarily claystone sequence of Oxfordian—Kimmeridgian age found in the Southern North Sea Basin. Extension of the name offshore was intended to express correspondence with the deposits documented in Norfolk, Lincolnshire and Yorkshire, and hence with the Dorset type area. The type well for the offshore Kimmeridge Clay was 47/15–1 (Phillips) with 47/9–1 (Gulf) as a reference section. Both wells are approximately 65 km east of the Humber estuary. Well 48/22–2 (Burmah) also serves to illustrate the Kimmeridge Clay Formation and lies about 65 km ENE of the Wash (Fig. 1).

The Humber Group as defined by Rhys (1974) also includes the Corallian and the underlying Oxford Clay Formation. In the type well the Kimmeridge Clay Formation is only 34 m thick, and is shown to rest directly on the Corallian. The comparatively minor thickness of Kimmeridge Clay may indicate contemporaneous thinning towards an offshore extension of the Market Weighton High (Fig. 1). However, it is also clear from the wells that there is a major unconformity between the Kimmeridge Clay and overlying Volgian–Lower Cretaceous sediments (Rhys, 1974, figs. 6 and 7). It seems likely that, as on Humberside

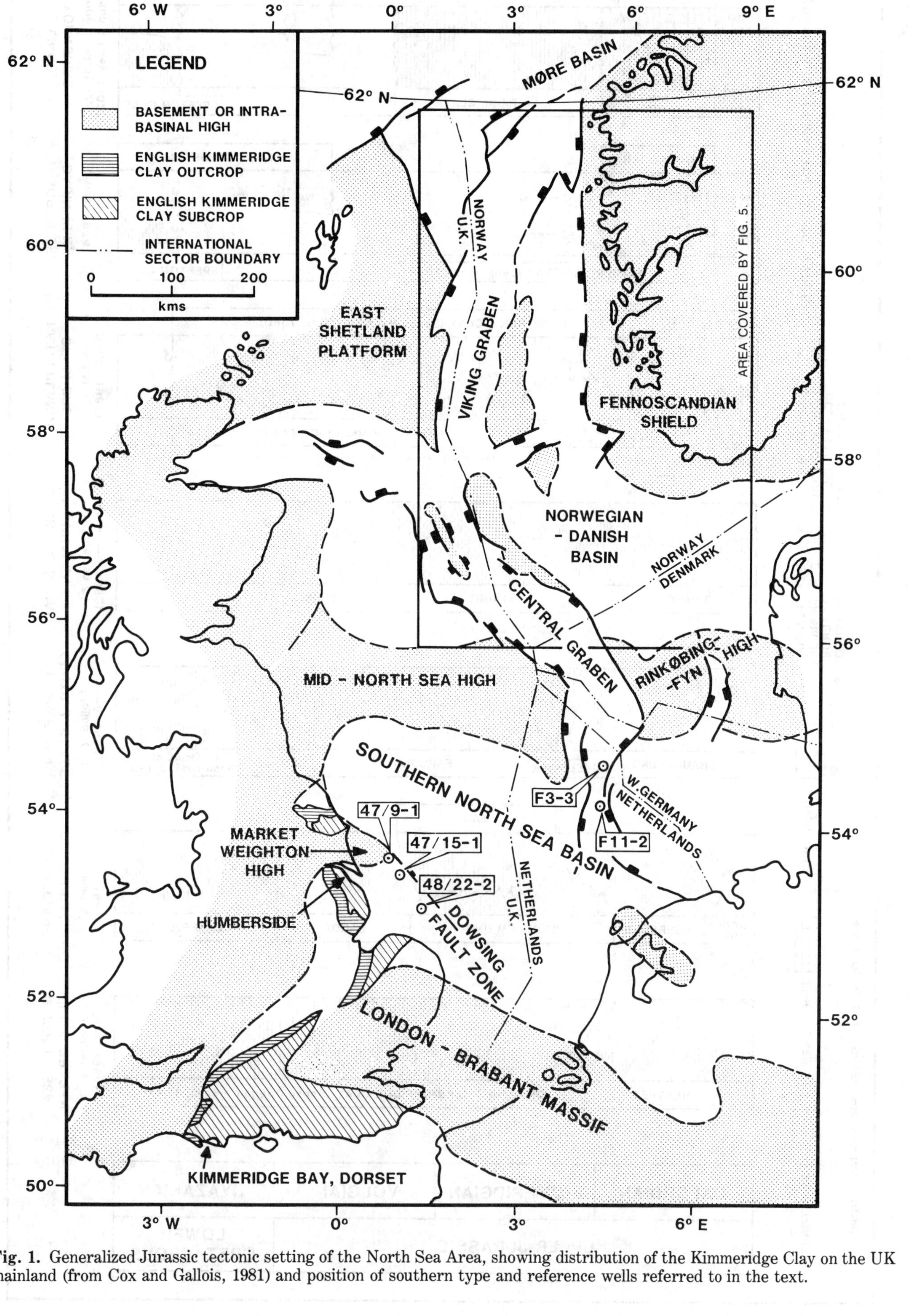

Fig. 1. Generalized Jurassic tectonic setting of the North Sea Area, showing distribution of the Kimmeridge Clay on the UK mainland (from Cox and Gallois, 1981) and position of southern type and reference wells referred to in the text.

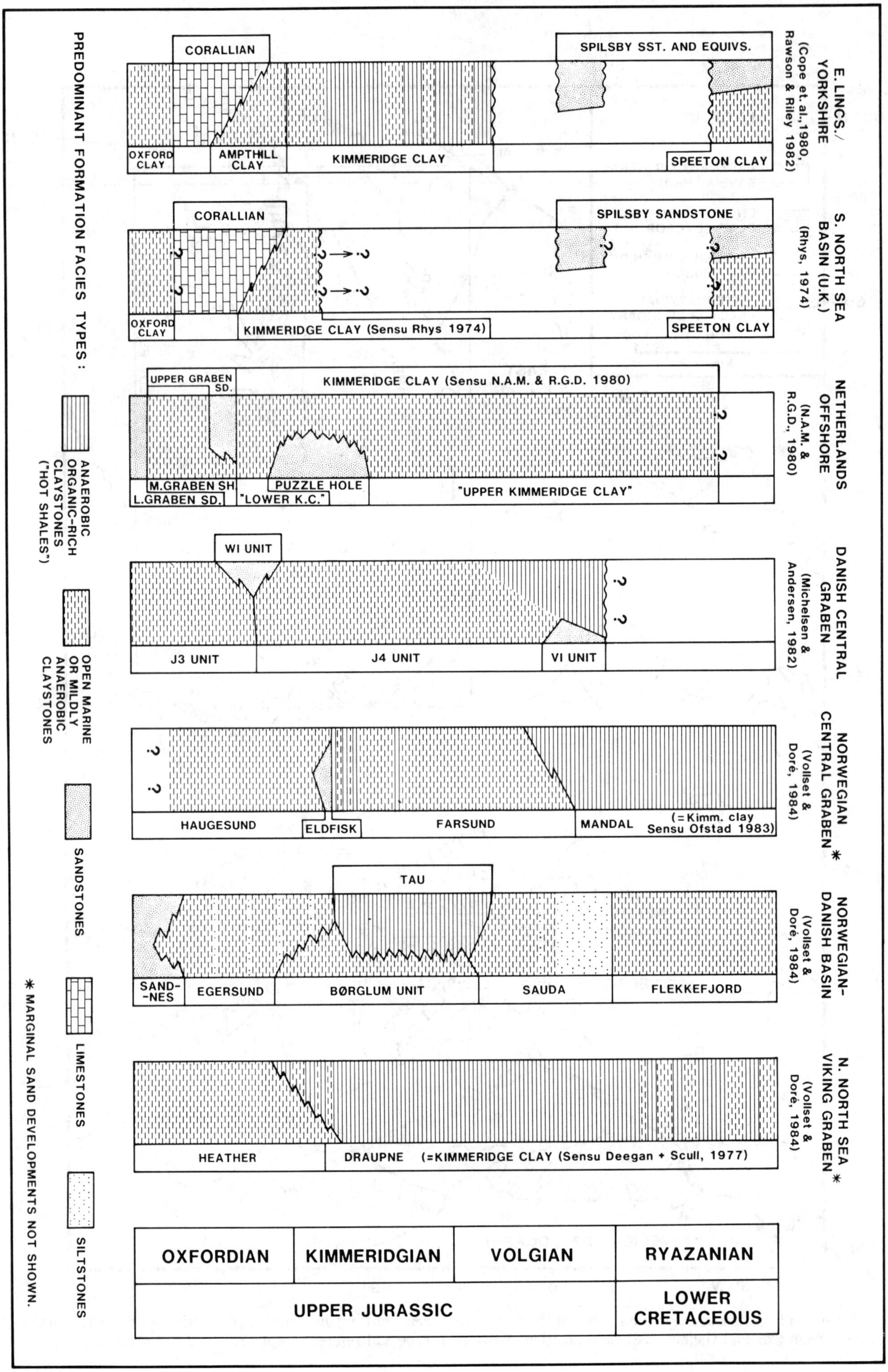

Fig. 2. Correlation chart for the Kimmeridge Clay and associated sequences in the North Sea. Unit nomenclature is indicated at formation level.

(e.g. Richardson, 1979) all but the lowermost zones of the Kimmeridge Clay have been removed by erosion. The Ampthill Clay is not delineated offshore, but Rhys (1974, p. 7) writes: 'The Kimmeridge Clay Formation in the type well section probably includes at its base a few feet of shale which on a *chronostratigraphic* basis would belong to the Ampthill Clay, but which cannot be identified as such without detailed paleontological evidence' and 'Elsewhere the Humber Group consists entirely of mudstones or shales, equivalent to the land area sequence of Oxford, Ampthill and Kimmeridge Clays, *which are indistinguishable on the available lithological evidence* (italics added).

The inclusion which has occurred of offshore correlatives of the Ampthill Clay in the Kimmeridge Clay makes sense in lithostratigraphic terms; offshore, as onshore, there is no obvious lithological criterion to separate the two units. Nevertheless, in using the name Kimmeridge Clay Formation offshore the stratigraphic range of the unit has been broadened, and its definition altered, compared to the 'chronostratigraphically defined' onshore unit (see Fig. 2).

Northern North Sea, UK and Norway sectors

Extension of the nomenclature to this area was carried out by Deegan and Scull (1977), who described well reference sections some 500 km to the north of the offshore type well. The Kimmeridge Clay Formation *sensu* Deegan and Scull occupies a much broader chronostratigraphic range (Oxfordian–Ryazanian) than the southern formation. It is defined by its high organic content, high gamma-ray readings (greater than 100 API units) and anomalously low sonic velocities. Deegan and Scull remarked that the formation in the southern North Sea is considerably less radioactive (less than 60 API units) but 'still has the typical low velocity'.

Examination of the southern Kimmeridge Clay type and reference sections confirms that the gamma-ray log shows a normal shale response, both in an absolute sense (i.e. when corrected for borehole and mud density effects) and in a relative sense (i.e. compared with other shales in the sequence). Furthermore, the velocities of the southern unit are not anomalously low for a claystone unit. The sonic log of Cretaceous type well 48/22–2 (Rhys, 1974) shows that the interval transit time gradient of the formation is more-or-less on trend with that of the overlying Speeton Clay Formation of the Lower Cretaceous (see Fig. 3(a)). Similar results are obtained by comparing the Kimmeridge Clay *sensu* Rhys with other argillaceous units (the Oxford Clay and Lias) in Jurassic type well 47/15–1. Interval transit times for the Kimmeridge Clay Formation *sensu* Rhys, when compared with normal compaction trends for northern North Sea shales (e.g. Chiarelli and Duffaud, 1980), appear, in fact, to be somewhat 'fast' (see Fig. 3(b)). This is perhaps not surprising in view of the position of the type and reference wells close to the Market Weighton High and Dowsing Fault Zone (Fig. 1), where burial and subsequent uplift may be expected to have occurred. The southern Kimmeridge Clay Formation only appears 'slow' when bounded by units of completely different lithology; e.g. where overlain by the Spilsby Sandstone (Volgian–Ryazanian) or Red Chalk (Albian), and underlain by Corallian limestones. Where the unit is bounded by argillaceous sequences no obvious distinction can be made.

In essence, therefore, the basic criteria on which the northern 'Kimmeridge Clay Formation' was defined were never documented in the southern type and reference sections.

Southern North Sea, Netherlands sector

The publication of a comprehensive Netherlands lithostratigraphic scheme by Nederlandse Aardolie Maatschappij and Rijks Geologische Dienst (1980) further extended the Kimmeridge Clay Formation eastwards into the Netherlands sector (see Figs. 1 and 2), again with an increased age range (Oxfordian–?Ryazanian) compared with the UK sector. The unit is described as medium to dark olive-grey, generally silty shales with numerous thin dolomite streaks. It is not defined by anomalous gamma-ray and sonic log responses, and an examination of the reference well F3–3 (N.A.M. and R.G.D., 1980, enc. 22) shows overall similarity with the formation in the UK sector of the southern North Sea, albeit a much thicker section. The base of the Kimmeridge Clay *sensu* N.A.M. and R.G.D. is within the Late Oxfordian, and it is clear that, as in the UK sector, shales which are time-equivalent and lithologically similar to the Ampthill Clay have been absorbed into the Kimmeridge Clay Formation (see Fig. 2).

In the southern part of the Dutch Central Graben the formation is divided by a marginal marine sand (Puzzle Hole Formation) into Lower and Upper Kimmeridge Clay Members. The lower unit is of Late Oxfordian age, while the upper unit is of Kimmeridgian–?Ryazanian age. These units do not, however, relate in any obvious way to the Upper and Lower divisions of the type Kimmeridge Clay, which correspond to the Upper and Lower Kimmeridgian stage *sensu anglico* (Arkell, 1947, p. 67; Cope, 1980, p. 77). This is a potential source of further confusion.

Interestingly, N.A.M. and R.G.D. (1980) do not extend the Humber Group across the median line into the Netherlands sector, instead referring the Kimmeridge Clay Formation to the locally-defined Scruff Group (p. 38). This perhaps reflects the inadequacy of the attenuated type and reference sections in the southern UK sector in providing a definition of the Humber Group (and its subdivisions) which can be confidently extended to other offshore areas.

Summary

1. The Kimmeridge Clay in the Yorkshire Basin and East Midlands Shelf can usually only be distinguished from the lithologically similar Ampthill Clay on palaeontological grounds—i.e. by the presence of Kimmeridgian fossils.

2. In the Southern North Sea Basin there is similarly no obvious stratigraphic marker at Base Kimmeridgian, but here 'Ampthill Clay equivalents' are included in the Kimmeridge Clay.

3. The 'hot' organic shales of the northern Kimmeridge Clay Formation *sensu* Deegan and Scull are lithologically distinct from the formation in the southern North Sea. Neither the formation *sensu* Rhys in the southern UK sector, nor the formation *sensu* N.A.M. and R.G.D. in the Netherlands sector, are distinguished by particularly high organic content.

4. Succeeding offshore extensions of the unit have

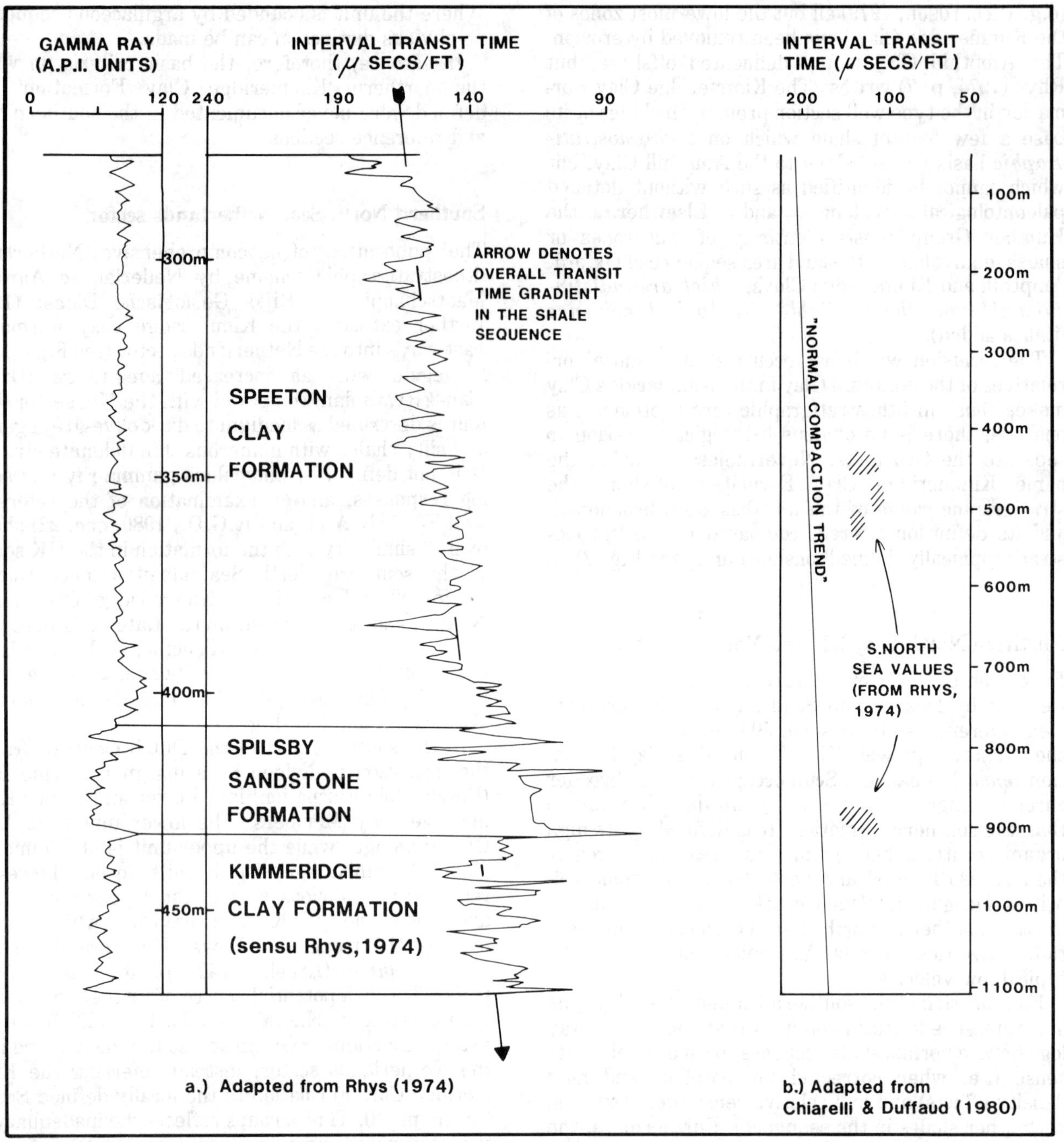

Fig. 3. Velocity characteristics of the Kimmeridge Clay in the southern North Sea. (a) GR-Sonic log of Cretaceous type well 48/22–2 (Rhys, 1974) showing similarity of Kimmeridge Clay velocity gradient with overlying Speeton Clay. (b) Comparison of southern Kimmeridge Clay transit times with normally compacted Jurassic shales in northern North Sea.

therefore resulted in significantly different definitions, while maintaining the same name.

RELEVANCE OF ANOXIC ENVIRONMENTS TO UPPER JURASSIC–LOWERMOST CRETACEOUS LITHOSTRATIGRAPHY

Because of its importance to source rock studies, recognition of anoxic environments is regarded as an important basis for stratigraphic classification in the Northern North Sea. Dominance of organic-rich oil-shale lithology was the essential criterion in the Deegan and Scull (1977) definition of the Kimmeridge Clay Formation. The character, distribution and regional significance of anoxic intervals is therefore critical in understanding the lithostratigraphic modifications which have subsequently been proposed.

Shale sequences deposited in a basin with anoxic bottom conditions are usually easily recognizable in the subsurface, from cuttings and on wireline logs such as gamma-ray, sonic, density and resistivity. In marine shales there is generally a good correlation between gamma-ray response and total organic carbon content, particularly when the gamma-ray is corrected for borehole and mud density effects. This is due to the tendency of uranium to form complexes with organic material under reducing conditions (e.g. Serra *et al.*, 1980). Usually high gamma ray response corresponds with depressed density/velocity values and high resistivities, the highest resistivities occurring in thermally mature sections. Note, however, that none of these log responses is a direct reflection of

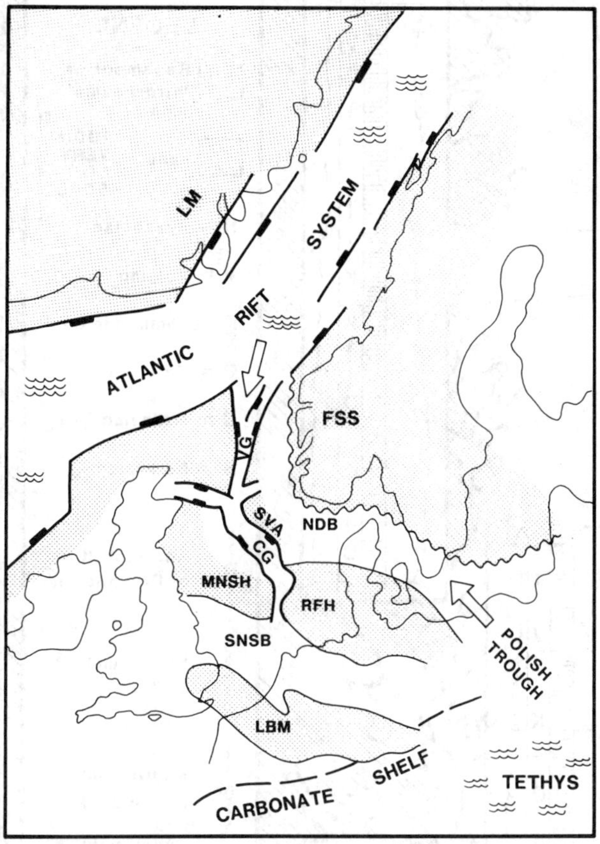

Fig. 4. Schematic palaeogeography of NW Europe–Greenland in late Jurassic to Ryazanian times. Shaded areas represent basement or intra-basinal highs. Arrows indicate connections of North Sea basin complex to major marine systems. LM = Laurentian Massif, FSS = Fennoscandian Shield, VG = Viking Graben, SVA = Southern Vestland Arch, NDB = Norwegian–Danish Basin, MNSH = Mid-North Sea High, CG = Central Graben, RFH = Rinkøbing–Finn High, SNSB = Southern North Sea Basin, LBM = London Brabant Massif.

the quality of organic matter (kerogen type) present in the shale. The term 'hot shale' is used herein as a shorthand to denote highly radioactive shales, which by implication are of high organic richness.

Models for the formation of anoxic shale sequences were discussed extensively by Demaison and Moore (1980). In one such model they demonstrate from studies of the Black and Baltic Seas how anoxic conditions are able to develop in a 'silled basin' setting, during times of worldwide transgression (p. 1204). Tyson *et al.* (1979) apply a similar model to explain facies assemblages in the type Kimmeridge Clay. They suggest that the major transgressive episodes of the Late Jurassic, which occurred over a shelf with limited connection to the Tethyan Ocean in southern Europe (see Fig. 4), resulted in a stratified water column with a stagnant, frequently anaerobic bottom layer. Deposition of anoxic or more 'normal' marine shales depended upon cyclical fluctuations within the water column. Seasonal convection of nutrients from the lower anoxic layers caused coccolith blooms, depositing the thin carbonate stringers which are typical of the type Kimmeridge Clay and some of the offshore shale sequences.

These models provide an overall regional framework within which anoxic bottom conditions, and consequent preservation of organic material, can occur. Given this tendency, however, the preferential occurrence of major anoxic shale developments was probably a function of more localized control. Studies of the northern North Sea area show that timing and duration of the organic shale sequences can vary from basin to basin (Rawson and Riley, 1982; Hamar *et al.*, 1983). Restriction of circulation within a particular structural province probably related to its overall bathymetric and geometric configuration, to its position on the NW European shelf, and to local tectonic effects. Detailed palynofacies studies also support this observation (pers. comm., M. F. Whitaker). This is nowhere more obvious than in the Norwegian sector. In the Norwegian Central Graben 'hot' shales of predominantly Ryazanian age occur (Mandal Formation of Hamar *et al.*, 1983). To the northeast in the Norwegian–Danish Basin a quite separate event deposited 'hot' shales in the Kimmeridgian–Volgian (Tau Formation of Olsen and Strass, 1982). There is no suggestion of a diachronous communication between these units across the Southern Vestland Arch. To the north in the Viking Graben the organic shales occupy a broader chronostratigraphic range, from Oxfordian to Ryazanian (Draupne Formation of Vollset and Doré, 1984) although 'warm' shales of lesser organic carbon content frequently typify the uppermost and lowermost parts of the sequence. Correlation of these units is shown in Fig. 2, and is addressed in more detail in the following section.

There are some hints at synchronicity of certain anoxic events; in particular, the onset of radioactive shales at the base of the Tau Formation probably has regional significance (see next section). In general, however, the anoxic intervals of the Northern North Sea are best recognized as locally developed lithostratigraphic units, which should only be extended from basin to basin with extreme caution.

NOMENCLATURE AND CORRELATION IN THE NORWEGIAN SECTOR

Figure 2 summarizes the current lithostratigraphic scheme for the Oxfordian–Ryazanian rocks of the Norwegian sector, and compares it with other areas considered in this account. A fuller and more descriptive lithostratigraphy for the Norwegian sector is given in Vollset and Doré (1984). It is only discussed here insofar as it relates to the central issue, which is the desirability of establishing a local nomenclature system for Norwegian waters.

Norwegian Central Graben

The Upper Jurassic of this area was not considered in detail by Deegan and Scull (1977), who concentrated mainly on type and reference sections in the Viking Graben and East Shetland Basin. Consequently, there has been some uncertainty in identifying the base of their 'Kimmeridge Clay Formation'. In some of the older well reports the contact is placed at the base of what is now called the Farsund Formation. This unit, which is 200 m thick or more in the graben centre, is a Kimmeridgian to Volgian shale sequence of moderately high organic content (see Fig. 2).

More commonly, the 'Kimmeridge Clay' has been identified with the overlying 'hot' shale sequence of late Volgian–Ryazanian age, now known as the Mandal Formation. This unit (type well 7/12–3A) is up to 70 m thick and thins over structural highs. It is fre-

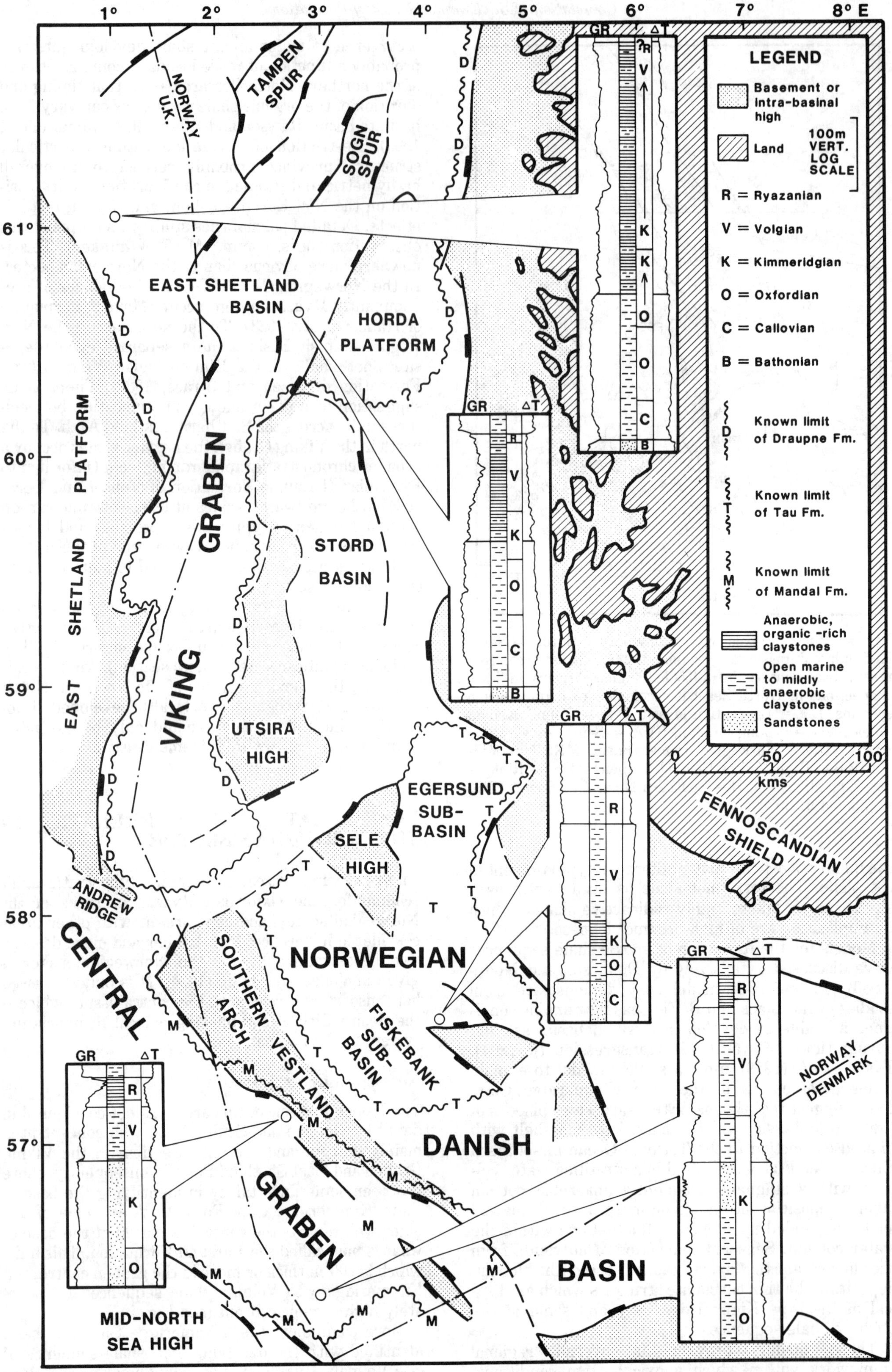

Fig. 5. Known distribution of organic shale intervals (Upper Jurassic–lowermost Cretaceous) in the Norwegian sector, with specimen log profiles. Gamma ray scale: 0–200 A.P.I. units. Sonic log scale 140–40 μs/ft.

quently typified by an unusually high proportion of algal kerogen. Its approximate distribution area is shown in Fig. 5. The unit was named the 'Kimmeridge Clay Formation' by Ofstad (1983), informally redesignated the Mandal Formation by Hamar *et al.* (1983) and formally defined by Vollset and Doré (1984). It appears to correlate southwards with highly radioactive shales in the upper part of 'Unit J4' in the Danish part of the Central Graben (Michelsen and Andersen, 1983, fig. 4) but is not seen further south in the published wells of the Dutch Sector (N.A.M. and R.G.D. 1980, Encl. 2).

The Mandal Formation appears to represent a predominantly Ryazanian anoxic event peculiar to the Central Graben, which is not matched by similar log responses elsewhere. It is possible that the establishment of open marine circulation between Tethys and the proto-Atlantic rift system in latest Ryazanian times was preceded by the incoming of slightly more oxygenated waters in the Norwegian–Danish Basin and East Shetland Basin (see Fig. 4). The central, more-protected position of the Central Graben may, in contrast, have allowed progressive stagnation to occur as transgression continued, resulting in the anoxic maximum of the Ryazanian.

Norwegian–Danish Basin

The Tau Formation—'hot' shales of Kimmeridgian to early Volgian age—is widely distributed in the Egersund and Fiskebank Sub-Basins of the Norwegian–Danish Basin (see Fig. 5) and is usually about 50 m thick. The type well for the formation is 9/4–3. It overlies a sequence of silty marine shales of Callovian–Kimmeridgian age, the Egersund Formation. The Tau Formation was formerly equated with the Børglum Unit, a dark (but less radioactive) homogeneous marine shale unit occurring in the Danish sector (Deegan and Scull, 1977, p. 22; Michelsen, 1978, pp. 21–22). Hamar *et al.* (1983) acknowledged the distinctive character of the Tau unit by separating it out as a member of the 'Børglum Formation', but the unit is now considered widespread enough to merit formation status (Vollset and Doré, 1984; see also Olsen and Strass, 1982). The Tau Formation and its bounding silty lithologies grade into the more homogeneous Børglum Unit in the southeastern part of the Fiskebank Sub-Basin (see Figs. 2 and 5).

Rawson and Riley (1982) pointed out that the middle Kimmeridgian appearance of 'hot shales' at the base of the Tau Formation corresponds to the first appearance of oil-shale facies in the *eudoxus* zone over the entire English Kimmeridge Clay outcrop. This event can also be recognized in the Norwegian Central Graben, where 'hot' streaks are not infrequent in the basal part of the Farsund Formation around the Southern Vestland Arch, and in the Viking Graben where it is often observed as an upward transition to 'hotter' shales within the Draupne sequence. This event, which clearly represents a widespread middle Kimmeridgian transgression, has potential use as a lithostratigraphic boundary in areas other than the Norwegian–Danish Basin (see Fig. 2).

The Tau Formation is overlain by a siltstone–claystone unit of Volgian age, the Sauda Formation. The succeeding finer grained deposits of the Flekkefjord Formation represent a major Ryazanian transgression (Rawson and Riley, 1982), and correspond in time with the much more radioactive shales of the Mandal Formation in the Central Graben (see Fig. 2).

Viking Graben and East Shetland Basin

The predominantly anaerobic Draupne Formation (type well 30/6–5) is present throughout the Viking Graben, East Shetland Basin and over part of the Horda Platform (see Fig. 5). It essentially corresponds to the northern 'Kimmeridge Clay Formation' as defined by Deegan and Scull (1977). It is of Oxfordian to Ryazanian age, approximately time-equivalent to the Farsund–Mandal sequence of the Central Graben, and overlies the predominantly open marine shales of the Heather Formation. The Draupne–Heather contact is diachronous between the Oxfordian and Kimmeridgian, probably in part a result of anoxic bottom conditions reaching progressively higher parts of the basin complex during the course of the Late Oxfordian–Middle Kimmeridgian transgressive episodes.

The thickness of the Draupne Formation is extremely variable, due to onlap of the basin margins, but probably reaches over 1000 m in axial parts of the Viking Graben. As defined in Vollset and Doré (1984), it encompasses a fairly wide range of lithotypes. Organic content varies (albeit against a generally high 'background') with the 'hottest' shales of all frequently occurring in the middle (Kimmeridgian–Volgian) portion of the sequence. A tripartite subdivision of the Draupne Formation on this basis may eventually be desirable. Minor sandstone and siltstone interbeds, probably turbiditic, are frequently present. They occasionally dominate the sequence, reducing overall gamma-ray response. Larger, more massive sand bodies are found enclosed within the Draupne shales close to the basin margins and intra-basinal highs of the UK sector; for example, the Magnus sand (De'Ath and Schuyleman, 1981). The 'Kimmeridge Clay Formation' of the Outer Moray Firth basin (Turner *et al.*, 1984), also Oxfordian–Ryazanian, has a similarly varied lithology with a major intercalated sand, the Claymore Sandstone Member. The sequence in this area appears to match well with the newly defined Draupne Formation. However, the further question of correlation with the Inner Moray Firth and North-East Scotland is not addressed here.

DISCUSSION

Organic shale sequences of approximately Late Jurassic age are a notable example of the 'persistence of facies' concept—the deposition of very similar lithologies over immense areas at certain points in the geological record (e.g. Ager, 1981). Widespread distribution of the organic shales can be explained by an overall sea level rise combined with a configuration favourable for restricted marine circulation in the megastructure of northern Europe (see earlier sections; also Ziegler, 1982, enc. 20). Roughly contemporaneous deposits of this type exist northwards along the Norwegian shelf into the Arctic regions and are found as far afield as Andøya, Tromsøflaket, Svalbard, East Greenland and Western Siberia (see Table 1).

Although in the past there has been a tendency to express the conceptual unity of such intervals in common lithostratigraphic terms (Arkell (1933, p. 613)

TABLE 1. Organic shale intervals roughly contemporaneous with 'Kimmeridge Clay Formation' in the Arctic region

Location	*Name of lithostratigraphic unit*	*Age*	*Authority*
Andøya (N. Norway)	Ratjønna Member (of Dragneset Fm)	?Kimmeridgian to Volgian	Dalland (1979)
Tromsøflaket	Under discussion	Kimmeridgian to Hauterivian	Westre (1948) and unpubl. well data
East Greenland	Hareelv Formation and Bjernberg Formation	Oxfordian to Kimmeridgian	e.g. Surlyk (1975)
Svalbard	Agardhfjellet Member (of Janusfjellet Fm)	Callovian to Volgian	e.g. Steel and Worsley (1984)
Western Siberia	'Bazhenov Deposit' and others	Various intervals within Callovian to Ryazanian	e.g. B.E.I.C.P. and Robertson Research (1980)

noted 'Kimmeridge Clay' in the Moscow region), this is not in accord with modern lithostratigraphic principles, since it presupposes a connection which has not been proven. A prime criterion for identifying a lithostratigraphic unit is continuity (or original continuity) with the type area (Hedberg, 1976). This is frequently impossible to demonstrate in a new area, and this is why local nomenclature systems evolve. Later, detailed correlation work between areas may make it desirable that certain units should be unified under a single name. However, in identifying the Kimmeridge Clay Formation over wide offshore areas this correlation has been taken for granted. It has been assumed that lithological similarity and continuity are self-evident, whereas in reality they have not been demonstrated.

We have remarked that the defining lithological characteristics of the Kimmeridge Clay in the northern North Sea do not correspond with those of the offshore type area. Even accepting that a fairly wide lateral variation in lithology may be acceptable within a formation, there remains a fundamental problem concerning the base of the unit. The recognition of a firmly defined base is becoming increasingly important in modern stratigraphy (e.g. Ager, 1981, pp. 67–79). In the case of the Kimmeridge Clay Formation there is no correspondence between the base of the unit onshore, where it is essentially a chronostratigraphic marker separating similar. Oxfordian and Kimmeridgian claystones, and offshore, where Oxfordian claystones have been included in the Kimmeridge Clay (see the second section). We are thus left with no clear directive about the base of the unit; there is no 'golden spike' for the Kimmeridge Clay Formation. It is this lack of definition which is partly responsible for the progressively broader and more disparate view of the Kimmeridge Clay formation we find as we move further afield in the North Sea area.

The reclassified units of the Norwegian sector are therefore advocated in preference to a diverse Kimmeridge Clay Formation of limited lithostratigraphic value. This has the advantage that correlation of sequences and events can be discussed free of the false sense of equivalence which the name Kimmeridge Clay Formation imposes. It should be remembered that lithostratigraphic units are 'localized in occurrence' and are defined 'to provide an observable framework of rock for reference in subsequent correlation and interpretation' (Holland *et al.*, 1978, pp. 7–8). Formation nomenclature is not intended as a vehicle for broad unifying concepts on a continental scale. This is the realm of 'event stratigraphy', biostratigraphy and seismic stratigraphy.

CONCLUSIONS

1. There are significant differences in the way the 'Kimmeridge Clay Formation' is defined—both in terms of lithotype and boundaries—between the onshore type area, the southern North Sea and the northern North Sea. This has made regional correlation difficult and led to lack of communication between individual workers.

2. Anoxic shale units of the northern North Sea are frequently readily mappable units which should be classed as formations; however, they are not necessarily synchronous and are of limited lateral continuity. The 'hot shale' definition of the Kimmeridge Clay Formation provided by Deegan and Scull (1977) has led to major problems of correlation, particularly in the Norwegian sector. Adoption of local nomenclature is urged for these intervals. This, at least, provides a lithostratigraphic scheme which is locally workable. Extension of these units between adjacent structural provinces may well be possible, but should be carried out with caution. The concept of a unified 'Upper Jurassic transgressive unit', containing anoxic shale intervals as well as 'normal' shales and other lithologies, could be expressed at group or supergroup level if desired.

3. Use of the term Kimmeridge Clay Formation in the southern UK and Netherlands sectors is questioned, and a review suggested. It may at least be desirable to rename the 'Upper' and 'Lower Kimmeridge Clay Members' of the Netherlands to avoid confusion with the Upper and Lower divisions of the type Kimmeridge Clay.

4. The Upper Jurassic–lowermost Cretaceous shale sequences of the North Sea in general are important enough to merit a thorough review. Many current problems of regional correlation and nomenclature could be resolved by a Correlation Group including representatives from each of the national sectors involved.

REFERENCES

Ager, D. V. 1981. *The Nature of the Stratigraphical Record* (2nd Edn). Macmillan Press.

Arkell, W. J. 1933. *The Jurassic System in Great Britain*. Clarendon Press, Oxford.

Arkell, W. J. 1947. The geology of the country around Weymouth, Swanage, Corfe and Lulworth. *Mem. Geol. Surv. G.B.*

B.E.I.C.P. (Bureau d'Etudes Industrielles et de Coopération de l'Institut Français du Pétrole) and Robertson Research International Ltd 1980. *The Petroleum Industry in the U.S.S.R.* Non-exclusive industry report.

Brookfield, M. 1973. Palaeogeography of the Upper Oxfordian and Lower Kimmeridgian (Jurassic) in Britain. *Palaeogeogr., Palaeoclimatol., Palaeoecol.*, **14**, 137–167.

Chiarelli, A. and Duffaud, F. 1980. Pressure origin and distribution of Jurassic of Viking Basin (United Kingdom–Norway). *Bull. Am. Ass. Petrol. Geol.*, **64**(8). 1245–1266.

Cope, J. C. W. 1974. New information on the Kimmeridge Clay of Yorkshire. *Proc. Geol. Ass.* **85**(2), 211–221.

Cope, J. C. W. 1980. Kimmeridgian correlation chart. In: Cope, J. C. W. *et al.* (ed.), *A Correlation of Jurassic Rocks in the British Isles. Part 2. Geol. Soc. Lond. Spec. Rep.* **15**.

Cox, B. M. and Gallois, R. W. 1981. The stratigraphy of the Kimmeridge Clay of the Dorset type area, and its correlation with some other Kimmeridgian sequences. *Rep. Inst. Geol. Sci.*, **80/4**.

Dalland, A. 1979. The sedimentary sequence of Andøya, Northern Norway. Depositional and Structural History. In: *Norwegian Sea Symp.* Norwegian Petroleum Society. Tromsø.

De'Ath, N. G. and Schuyleman, S. F. 1980. The geology of the Magnus oil field. In: Illing, L. V. and Hobson, G. D. (eds.), *Petroleum Geology of the Continental Shelf of N. W. Europe*. Institute of Petroleum, London, pp. 342–351.

Deegan, C. E. and Scull, B. J., (compilers) 1977. A proposed standard lithostratigraphic nomenclature for the Central and Northern North Sea. *Rep. Inst. Geol. Sci.* **77/25**; *Bull. Norw. Petrol. Direct.*, **1**. H.M.S.O., London.

Demaison, G. J. and Moore, G. T. 1980. Anoxic environments and oil source bed genesis. *AAPG Bull.*, **64**(8), 1179–1209.

Fitton, W. H. 1836. Observations on some of the strata between the Chalk and the Oxford Oolite in the south-east of England. *Trans. Geol. Soc. Lond.*, **2**(3), 103–388.

Hamar, G. P., Fjæran, T. and Hesjedal, A. 1983. Jurassic stratigraphy and tectonics of the south-eastern Norwegian offshore. In: *Petroleum Geology of the South-eastern North Sea and Adjacent Onshore Areas. Geol. en. Mijnb, The Hague.*

Hedberg, H. D. (ed.) 1976. *International Stratigraphic Guide: Guide to Stratigraphic Classification, Terminology and Procedure.* International Subcommission on Stratigraphic Classification of I.U.G.S. Commission on Stratigraphy. Wiley–Interscience, New York.

Holland, C. H. *et al.* 1978. A guide to stratigraphical procedure. *Geol. Soc. Lond. Spec. Rep. No. 10.*

I.G.S. (Institute of Geological Sciences) 1974. *Annual Report for 1973*. London.

Kelly, S. R. A. and Rawson, P. F. 1983. Some late Jurassic to mid-Cretaceous sections, as demonstrated on a field meeting, 18–20 May 1979. *Proc. Geol. Ass.*, **94**(1), 65–73.

Michelsen, O. 1978. Stratigraphy and distribution of Jurassic deposits of the Norwegian–Danish Basin. *Dan. Geol. Unders. Ser. B*, **2**.

Michelsen, O. and Andersen, C. 1983. Mesozoic structural and sedimentary development of the Danish Central Graben. In: *Petroleum Geology of the South-eastern North Sea and Adjacent Onshore Areas. Geol. Mijnb, The Hague.*

Nederlandse Aardolie Maatschappij B.V. and Rijks Geologische Dienst (N.A.M. and R.G.D.) 1980. *Stratigraphic Nomenclature of the Netherlands.* Verlog K. Nea. Geol. Mijnb. Genoot.

Ofstad, K. 1983. Geology of the southernmost part of the Norwegian section of the Central Trough. *N.P.D. Paper No. 32.*

Olsen, R. C. and Strass, I. F. 1982. The Norwegian–Danish Basin. *N.P.D. Paper No. 31.*

Rawson, P. F. and Riley, L. A. 1982. Latest Jurassic–Early Cretaceous events and the 'Late Cimmerian Unconformity' in the North Sea area. *Bull. Am. Ass. Petrol. Geol.*, **66**(12), 2628–2648.

Rhys, G. H. (compiler) 1974. A proposed standard lithostratigraphic nomenclature for the southern North Sea and an outline structural nomenclature for the whole of the (UK) North Sea. *Rep. Inst. Geol. Sci., No. 74/8.*

Richardson, G. 1979. The Mesozoic stratigraphy of two boreholes near Worlaby, Humberside. *Bull. Geol. Surv. G.B.* **58**.

Serra, O., Baldwin, J. and Quirein, J. 1980. Theory, interpretation and practical application of natural Gamma Ray Spectroscopy. *SPWLA, 21st A. Log. Symp. Trans.—Paper Q.*

Steel, J. R. and Worsley, D. 1984. Svalbard's post-Caledonian strata—an atlas of sedimentological patterns and paleogeographic evoluion. In: Spencer, A. M. *et al.* (eds.), *Petroleum Geology of the North European Margin.* Graham and Trotman, London, for Norwegian Petroleum Society, 436 pp.

Surlyk, F. 1975. Block faulting and associated marine sedimentation at the Jurassic–Cretaceous boundary, East Greenland. *Jurassic Northern North Sea Symp. Norwegian Petroleum Soc.*, Stavanger.

Turner, C. C., Richards, P. C., Swallow, J. L. and Grimshaw, S. P. 1984. Upper Jurassic stratigraphy and sedimentary facies in the Central Moray Firth Basin, North Sea. *Mar. Petrol. Geol.*, **1**, 105–117.

Tyson, R. V., Wilson, R. C. L. and Downie, C. 1979. A stratified water column environment model for the type Kimmeridge Clay. *Nature*, **277** (5695), 377–380.

Vollset, J. and Doré, A. G. (eds.) 1984. A revised Triassic and Jurassic lithostratigraphic nomenclature for the Norwegian North Sea. *N.P.D. Bull. No. 3.*

Westre, S. 1984. The Askeladden gas find—Troms 1. In: Spencer, A. M. *et al.* (eds.), *Petroleum Geology of the North European Margin.* Graham and Trotman, London, for Norwegian Petroleum Society, 436pp.

Wright, J. K. 1980. Oxfordian correlation chart. In: Cope, J. C. W. *et al.* (eds), *A Correlation of Jurassic Rocks in the British Isles. Part 2. Geol. Soc. Lond. Spec. Rep.* **15**.

Ziegler, P. A. 1982. *Geological Atlas of Western and Central Europe.* Shell Internasjonale Petroleum Maatschappij B. V., The Hague.

3 Organic geochemistry in exploration of the Northern North Sea

J. D. Field
Tenneco United Kingdom Inc., 1 Castle Yard, Richmond, Surrey, UK

A subsurface geochemical study has been conducted in the Northern North Sea with the following objectives:

(1) to identify potential hydrocarbon-source horizons;
(2) to determine their regional maturity, and hence the most likely type of hydrocarbons expelled into carrier beds and
(3) to identify secondary migration routes and to evaluate the probable gravity of accumulated hydrocarbons.

In the study, a combination of sedimentary facies, organic facies, regional structure maps and isopachs, and modelling techniques were used to evaluate the distribution and quality of mainly Jurassic source horizons. The Upper Jurassic Kimmeridge Clay Formation is the most important oil source horizon on the Tampen Spur and in the Northern North Sea graben system. Shales of the Heather Formation and Dunlin Group, and coals of the Brent Group and Statfjord Formation also have source potential, but mainly for gas. The state of maturation at top Brent Group and top Humber Group levels and likely hydrocarbon products from these horizons were assessed with well data and burial modelling in undrilled areas. The timing of expulsion and migration routes of hydrocarbons were determined based upon present and reconstructed structural configurations on the Tampen Spur (Alwyn–Statfjord trend, Brent Field), in the South Viking Graben (Brae, Sleipner, Thelma, etc., complexes) and in the North Viking Graben (Huldra, Oseberg, Brage, Troll fields). The overall source–migration–accumulation model can be tested against both known fields and barren structures, and results in a predictive ability for estimating hydrocarbons in untested prospects.

INTRODUCTION

A vital function of the petroleum geologist is to predict the type, or gravity, of hydrocarbons accumulated in an exploration area of interest. The probabilities of a structure containing oil or gas can have considerable influence on the economic perception of the prospect. In the Northern North Sea, the distribution of oil and gas fields and their related source rocks is becoming increasingly well-documented. This area therefore provides a natural laboratory to identify the subsurface processes which take place during the generation, migration and accumulation of hydrocarbons. The models developed are being applied in frontier basins with similar geology, particularly those bordering the North Atlantic rift zone in North-West Europe.

The areas of study are outlined in Fig. 1. A regional survey has been conducted over the area bounded by 58° and 62°N, and 0° and 6°E. Two sub-regional surveys in the North and South Viking Grabens are also described. Data presented in this paper are drawn from published material, in-house analyses of released well data, and modelling studies.

SOURCE POTENTIAL

The Kimmeridge Clay Formation and its lateral equivalents have been identified as the major source intervals in the Northern North Sea (Fuller, 1975; Barnard and Cooper, 1981; Goff, 1983). This formation is the source of the majority of oil and gas fields in the study area, including Brae (Reitsema, 1983), Brent (Bowen, 1975), Magnus (De'Ath and Schuyleman, 1981), Ninian (Albright *et al.*, 1980), Piper (Maher, 1981), Statfjord (Kirk, 1980) and Troll (Brekke *et al.*, 1981). Exceptions include the Frigg complex, where Middle and Lower Jurassic strata have been suggested as possible oil sources (Heritier *et al.*, 1980), and the Sleipner complex, where Middle Jurassic coals and basal Upper Jurassic shales of the Heather Formation may have sourced gas and condensate, respectively (Larsen and Jaarvik, 1981). Brent Group and Statfjord Formation coals are reported to be possible contributors of gas to the Brent and Troll fields. Middle Jurassic shales are a possible source of oil in the Alwyn area (Oudin, 1976), while unspecified Jurassic sources are invoked for Heimdal gas and condensate

Petroleum Geochemistry in Exploration of the Norwegian Shelf © Norwegian Petroleum Society (Graham & Trotman, 1985) pp. 39–57.

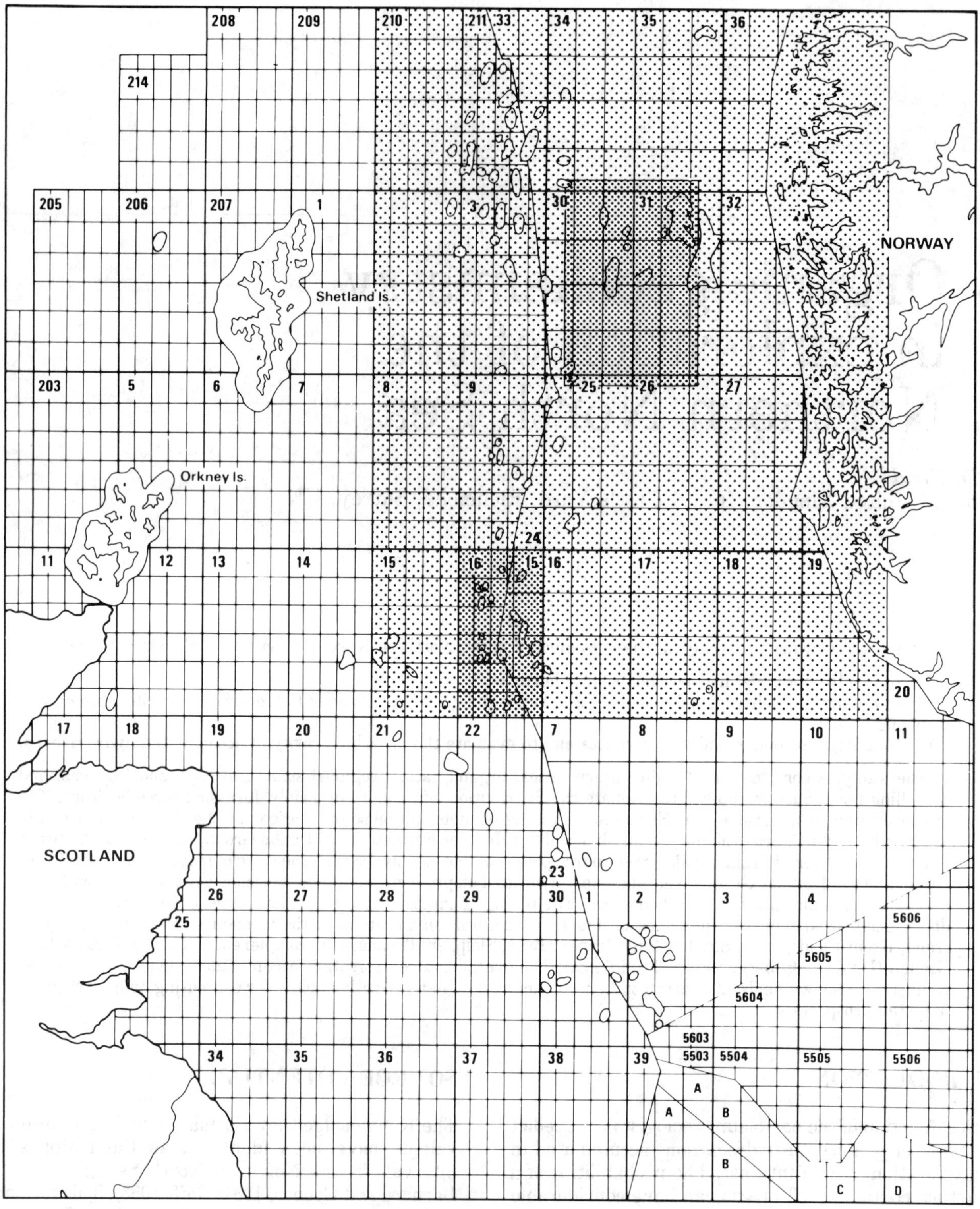

Fig. 1. Northern North Sea, location of study areas.

(Heritier *et al.*, 1980). The relative source potential of strata in the Northern North Sea is discussed below.

Palaeozoic source rocks are predicted to have virtually no economic potential in the study area. Carboniferous strata are absent north of the Outer Moray Firth, and the Permian Kupferschiefer and overlying lagoonal carbonates are restricted to the South Viking Graben south of 60°N (Ziegler, 1982). In the Mesozoic and Cenozoic section, major source horizons are developed in the Jurassic System (Fig. 2). Shales of the Humber Group, which is subdivided into the Kimmeridge Clay and Heather Formations, and Dunlin Group, and coals of the Brent Group and Statfjord Formation are apparent as potential source horizons. Figure 2 demonstrates the average potential of each horizon, and local enhancement of source quality is shown in a lighter tone. Total organic carbon (TOC) content is expressed in weight percent, and the hydrogen index values are representative of immature rocks.

Statfjord Formation

The Statfjord Formation consists of fluvial sandstones, siltstones, claystones and coals overlain by marginal–marine sandstones and shales. The formation has poor average source potential with TOC less than 1%. However, more than 10 m of coal is

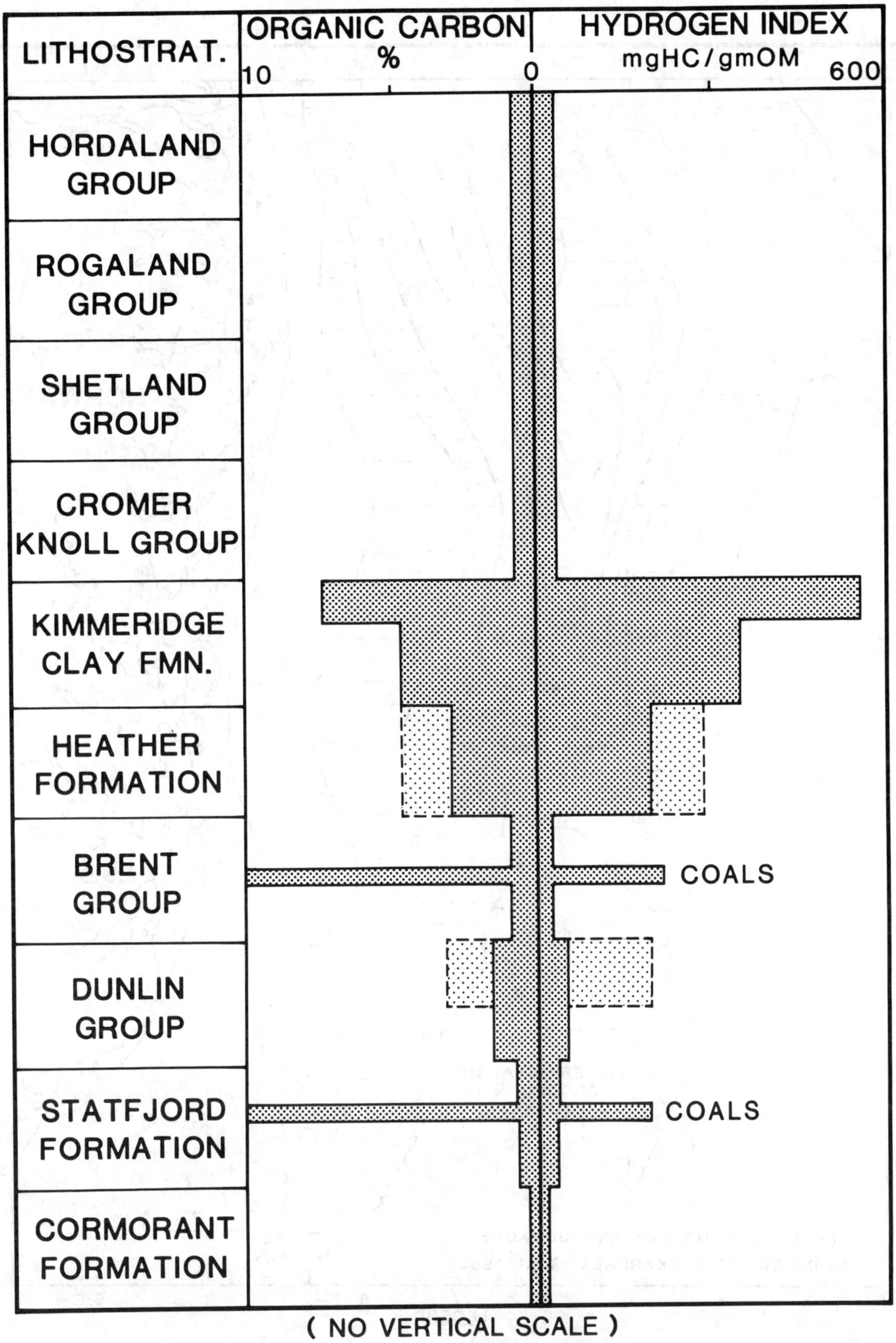

Fig. 2. Northern North Sea, source potential of Mesozoic and Cenozoic strata.

developed in the Raude and Eiriksson Members on the Horda Platform (Fig. 2) and it is a proposed secondary source of gas for the Troll accumulation (Brekke *et al.*, 1981).

Dunlin Group

The siltstones, shales and offshore-bar sandstones of the Dunlin Group represent deposits of the Early Jurassic marine transgression, which reached its maximum extent in Toarcian–Aalenian times with deposition of the Drake Formation (Skarpnes *et al.*, 1980) see Fig. 3. The source quality of the Dunlin Group, and particularly the Drake Formation, varies across the study area. On the Tampen Spur, the Dunlin Group has only limited source potential with an average TOC of 1.5% characterized by inertinitic kerogen (type IV). In the North Viking Graben area, however,

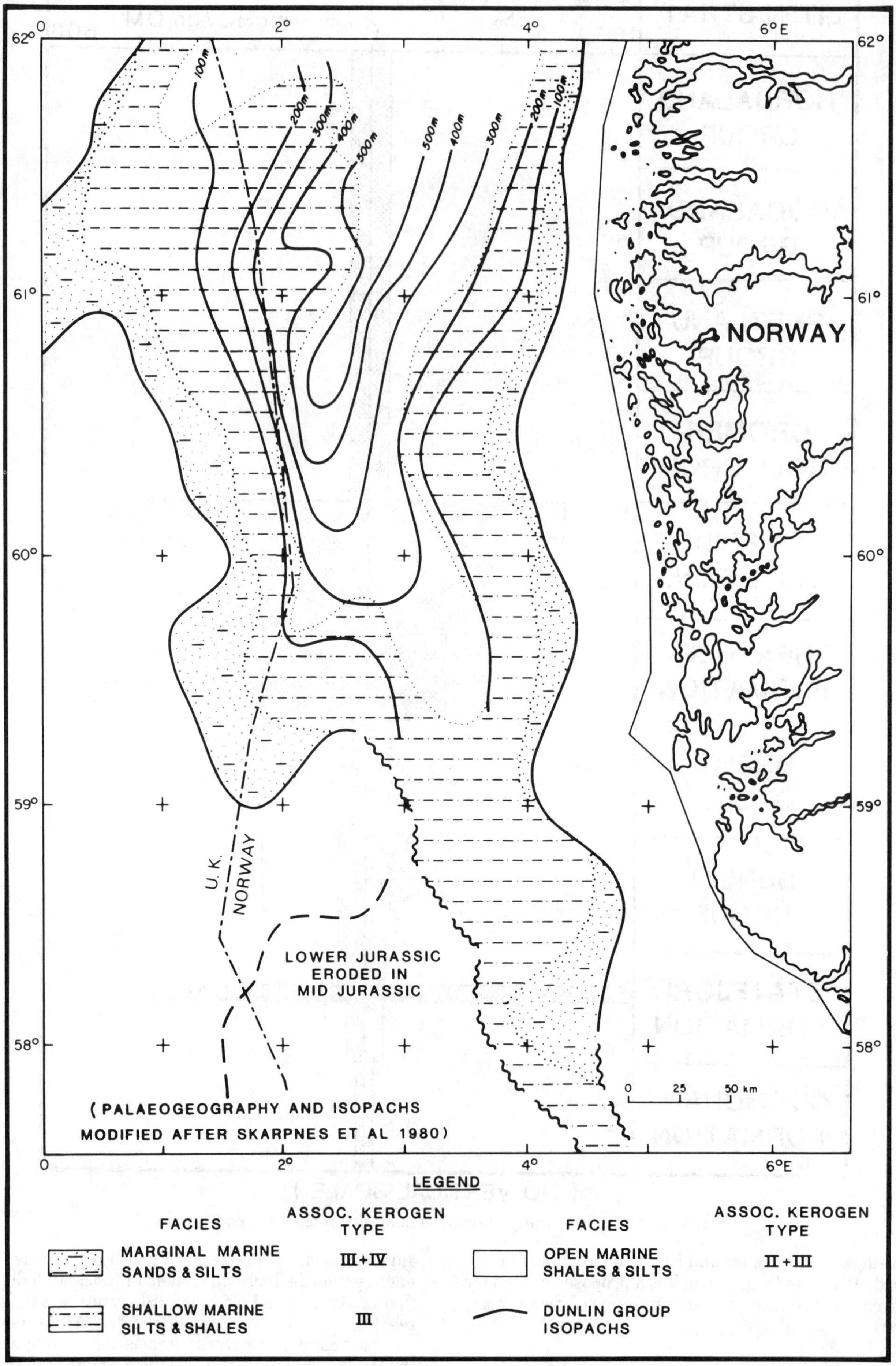

Fig. 3. Northern North Sea, Dunlin Group, isopachs, palaeogeography and kerogen types at time of maximum transgression.

average TOC ranges up to 3% with amorphous sapropel, exinite and vitrinite more prevalent (types II and III), resulting in hydrogen index values up to 200 (Ungerer *et al.*, 1984), see Fig. 2. The limited data available indicate that organic facies are related to the sedimentary facies belts outlined in Fig. 3. Marginal–marine and shallow-marine sands, silts and shales are dominated by inertinite and vitrinite (types IV–III), and the open-marine shales in the basin centre are characterized by vitrinite, exinite and amorphous sapropelic kerogens (types III–II). The basin–margin shales have only minor gas source potential. The basin-centre shales have gas and oil source potential in the Møre Basin, the North Viking and Sogn grabens, and possibly in the undrilled Stord Basin.

Brent Group

The Brent Group was deposited during the major Middle Jurassic regressive pulse associated with the volcanic updoming of the Outer Moray Firth (Eynon, 1981). The average source potential of these dominantly arenaceous, deltaic deposits is low, but associated delta-top coals and coal-bearing shales have considerable source potential. The coal-bearing sequences exhibit TOC up to 45 per cent (Oudin, 1976) and high hydrogen index values in the range of 200–250 (Durand and Paratte, 1983; Ungerer *et al.*, 1984). It has yet to be demonstrated that coals have contributed significantly to an oil field in the area of study, but these values indicate a considerable generative capability for liquid hydrocarbons and gas.

Oudin (1976) reports that the heavy, alkane-deficient oil of UK 3/15–2 exhibits good correlation with extracts of Middle Jurassic carbonaceous beds based on mass-spectrometry and gas-chromatography techniques. Heritier *et al.* (1980) indicate, based upon pristane/nC^{17} : phytane/nC^{18} observations, that the naphthenic oil of East Frigg originates from Lower and Middle Jurassic source rocks. However, in both cases the oils are altered, thus the correlations may be debatable. Goff (1983) argues for Frigg gas originating from sapropelic Upper Jurassic source horizons in the Kimmeridge Clay Formation, based on stable carbon isotope data. By association, it might be inferred that Frigg oil is also, at least partially, sourced from the Kimmeridge Clay Formation.

Heather Formation

The Heather Formation and its lateral equivalents have an average TOC of 2–2.5% and hydrogen index values ranging up to 200. It is characterized by vitrinite, with minor exinite and amorphous sapropel, and is capable of sourcing gas and associated light liquids. Larsen and Jaarvik (1981) consider the Heather Formation may have sourced condensate in the Sleipner complex. The Heather Formation exhibits strong local areal variations in source potential. Several restricted-circulation sub-basins appear to have developed in the East Shetland Basin as the structural highs developed in the late Callovian to early Kimmeridgian. As indicated in Fig. 2, source quality in these basins is enhanced with TOC up to 5% and hydrogen index values up to 300. In the Outer Moray Firth, Bissada (1983) reports an average TOC of 3.3%, with amorphous kerogen dominant. Woody kerogens, however, are preferentially developed in the south-eastern corner of UK Quadrant 15, a trend which continues into the South Viking Graben, where the Heather may be a partial source of Sleipner condensate and gas.

Kimmeridge Clay Formation

The Kimmeridge Clay Formation and its lateral equivalents have an average TOC of 6% and hydrogen index values ranging up to 600. As indicated in Fig. 2, optimum source quality is generally developed in the upper part of the formation. The shales are dominated by oil-prone, amorphous sapropel and exinite, but exhibit considerable facies-related, lateral variations as discussed by Barnard *et al.* (1981), Barnard and Cooper (1981) and Cooper and Barnard (1984). The richest source rocks were deposited in the graben centre and restricted sub-basins in the East Shetland Basin, far removed from terrestrially-derived, higher plant debris. On the basin margins and contemporaneous highs, agitated oxygenated conditions prevailed, which favoured the preservation of inertinite and vitrinite. As already stated, the Kimmeridge Clay Formation is believed to be the prime source of the majority of hydrocarbon accumulations in the Northern North Sea. Oil-to-source correlations by Fuller (1975), Oudin (1976), Grantham *et al.* (1980), Van den Bark and Thomas (1980), Bissada (1983) and Reitsema (1983) confirm this observation.

Cretaceous and Tertiary

Average TOC less than 1%, dominated by inertinite, and hydrogen index values less than 50 (see Fig. 2), indicate that only poor to fair source potential is present in these systems. Locally minor gas- and oil-prone kerogen is developed, but the source potential of this section is minor in comparison to the underlying Jurassic.

MATURATION

Maturation data based upon spore colouration and vitrinite reflectance studies are available from a number of structural provinces in the Northern North Sea (Goff, 1983; Pearson *et al.*, 1983; Bissada, 1983). These studies generally indicate that maturity gradients vary between structural provinces. The most marked difference is between the Viking Graben and its flanking highs: the Vestland Arch, the Horda Platform, the East Shetland Platform and the Tampen Spur. Table 1 summarizes the differences in the maturation gradients for these two classes of structural province. The greater depths of burial required in the graben axis to achieve a certain level of maturity compared with the flanking highs reflect its lower average geothermal gradient. Figure 4 summarizes the mean geothermal gradients in the Northern North Sea modified after the work of Carstens and Finstad (1981) and

TABLE 1. Maturation gradients in the Viking Graben and on its flanking highs

	R_0	*Viking Graben (m)*	*Graben Margin highs (m)*
Onset oil generation	0.55	3600	3100
Peak oil generation	0.75	4200	3600
Floor of oil generation	1.30	4900	4400

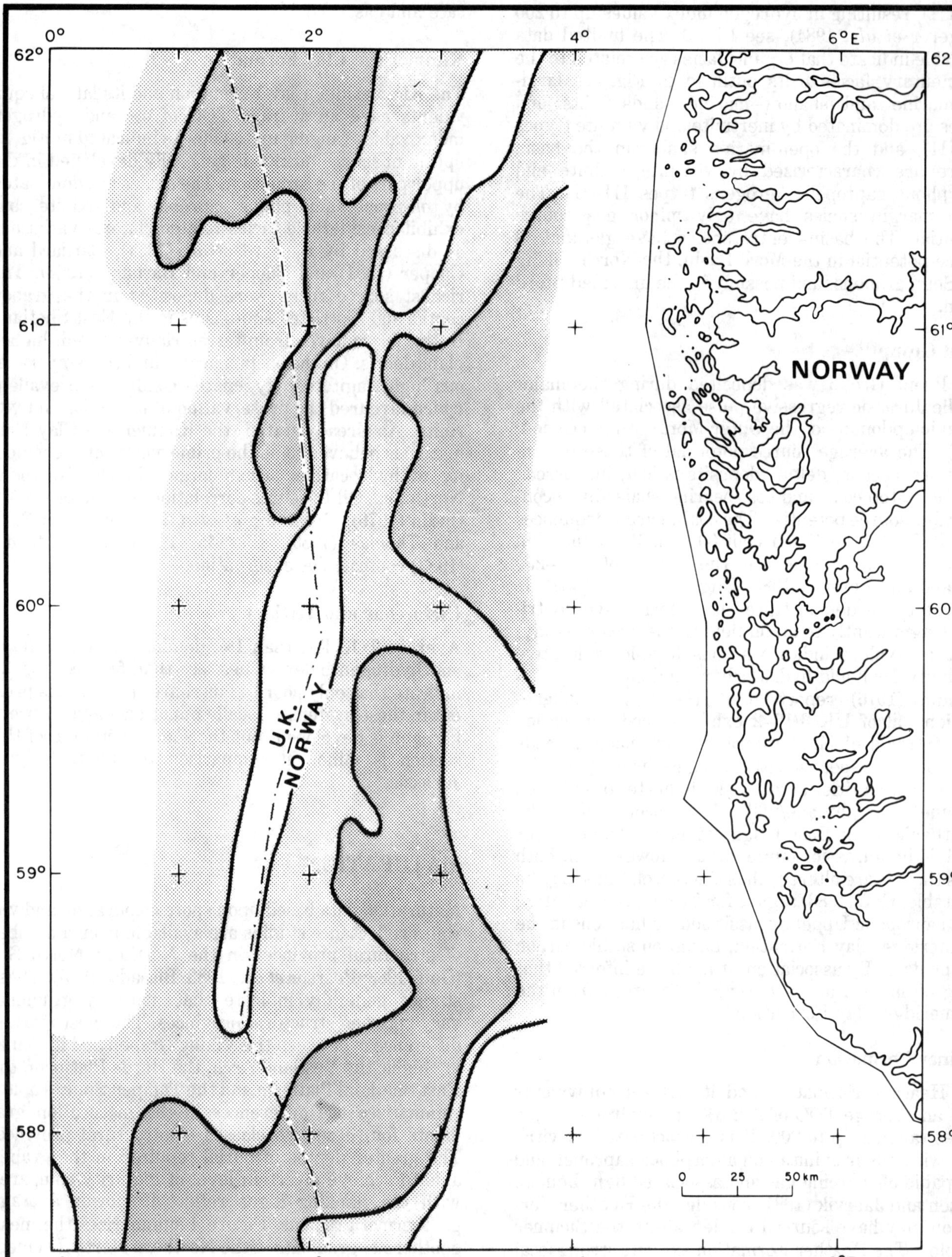

Fig. 4. Northern North Sea, mean geothermal gradients.

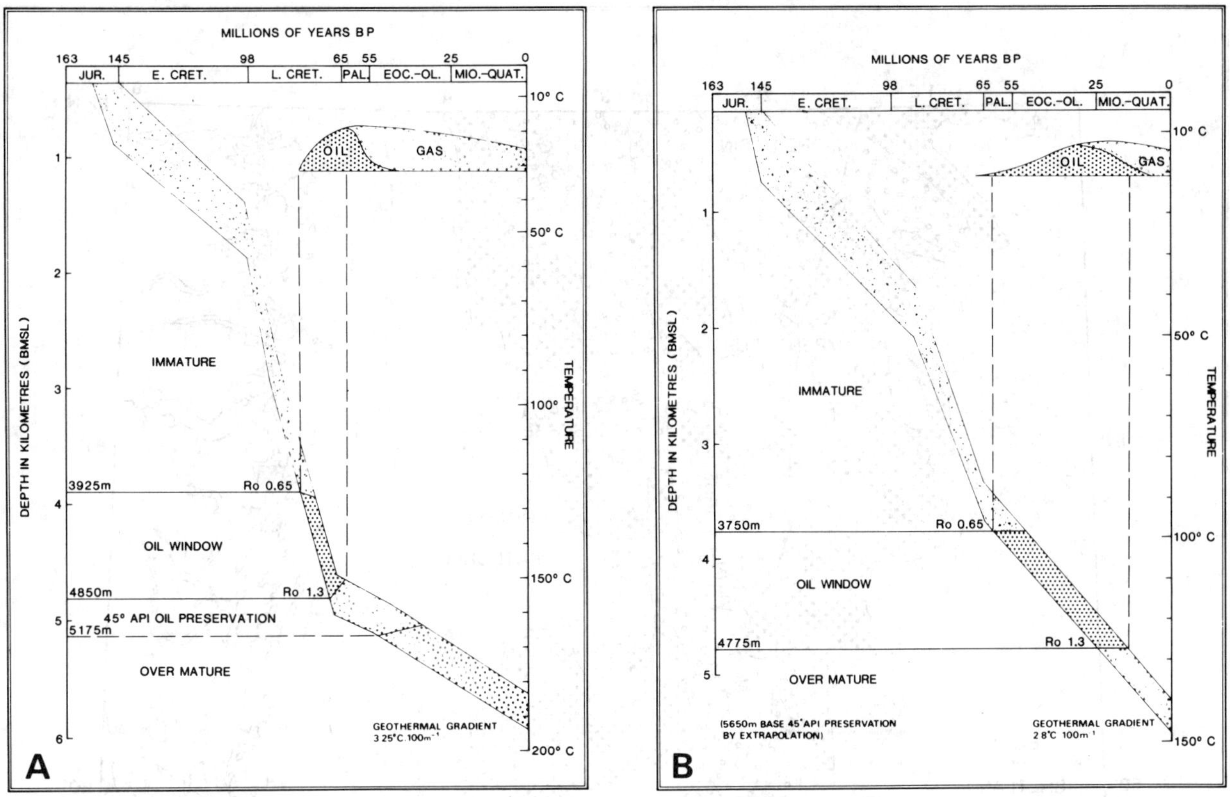

Fig. 5. Northern North Sea, maturation modelling profiles. (A) Blocks 34/9–35/2; (B) Blocks 30/2–30/5.

Evans and Coleman (1974). In contrast to the Central North Sea, the lowest geothermal gradients are located along the graben axis and the highest gradients on the flanking highs. These elevated gradients probably result from a combination of connate water flowing from the basinal areas to the flanking highs, and higher heat conductivities on the flanking highs.

In the deepest sections of the graben system, only limited well data are available and maturation modelling using the Lopatin Method, as outlined by Waples (1980), has been utilized. Two model profiles are presented for the graben axis in the area of blocks 34/9 and 35/2, and blocks 30/2 and 30/5 (see Fig. 5). In each instance, the model simulates burying a 300 m section of Upper Jurassic strata using the present geothermal gradient. Allowance for decompaction was incorporated using the porosity–depth relationships of Sclater and Christie (1980), but water depths were held constant. Depths and timing of each maturity stage are noted on the graphs.

Depth–structure maps were combined with maturation data in each structural province to derive level-of-maturation maps for the tops of the Brent and Humber Groups (see Figs. 6 and 7). Within the Middle and Lower Jurassic gas-zone maturities (R_0 greater than 1.3) are achieved in the Viking and Sogn graben systems and in the Møre Basin (see Fig. 6). Oil-window maturity levels are predicted on the flanks of the graben system, on the Tampen Spur, and in the Stord Basin. In the Upper Jurassic, gas-zone maturity levels are relatively more restricted within the graben system and in the Møre Basin (see Fig. 7). Oil generation is taking place within the graben system, along its flanks, and on the Tampen Spur. The northwestern part of the Stord Basin may be of sufficient depth for early oil generation, if the high geothermal gradients of the Horda Platform and Utsira High extend across the area.

MIGRATION

Primary migration of hydrocarbons from the predominantly fine-grained source rocks to coarser-grained carrier and reservoir rocks is now generally considered to take place in a separate hydrocarbon phase (Jones, 1981). The limited solubility of oil-gravity hydrocarbons in water and the reduced availability of movable water in the hydrocarbon-generating zone, argue against transportation in solution (Durand, 1983). In biphasic flow, to overcome capillary pressures of the normally water-wet shales, a continuous oil-wet phase of oil (Meissner, 1978), or kerogen (McAuliffe, 1980) is believed to be developed (Dickey, 1976). Expulsion of the hydrocarbon phase is aided by microfracturing, achieved by aquathermal pressuring (Barker, 1972) and expansion of the organic phase upon hydrocarbon generation (Momper, 1978). Solution in water of light aromatics and methane, and diffusion of low molecular weight n-alkanes are two alternative primary migration mechanisms for lower-weight hydrocarbons (Tissot, 1984; Leythaeuser *et al.*, 1983). Whatever the primary-migration mechanism, the process is aided by interdigitation or juxtaposition of source and carrier beds, and is impeded by a thick poorly drained source horizon.

Secondary migration of oil and gas is generally believed to occur as a discrete hydrocarbon phase (Tissot and Welte, 1978). Flow of gas-bearing formation waters may be an important migration mechanism for light hydrocarbons (Maximov, 1981), but on release from solution the hydrocarbons act as a discrete phase. The main processes acting to migrate the hydrocarbon phase are buoyant rise and hydrodynamic flow, while its movement is resisted by capillary pressures. Secondary migration routes due to buoyancy can be derived from structure maps combined with sand-distribution maps of the probable

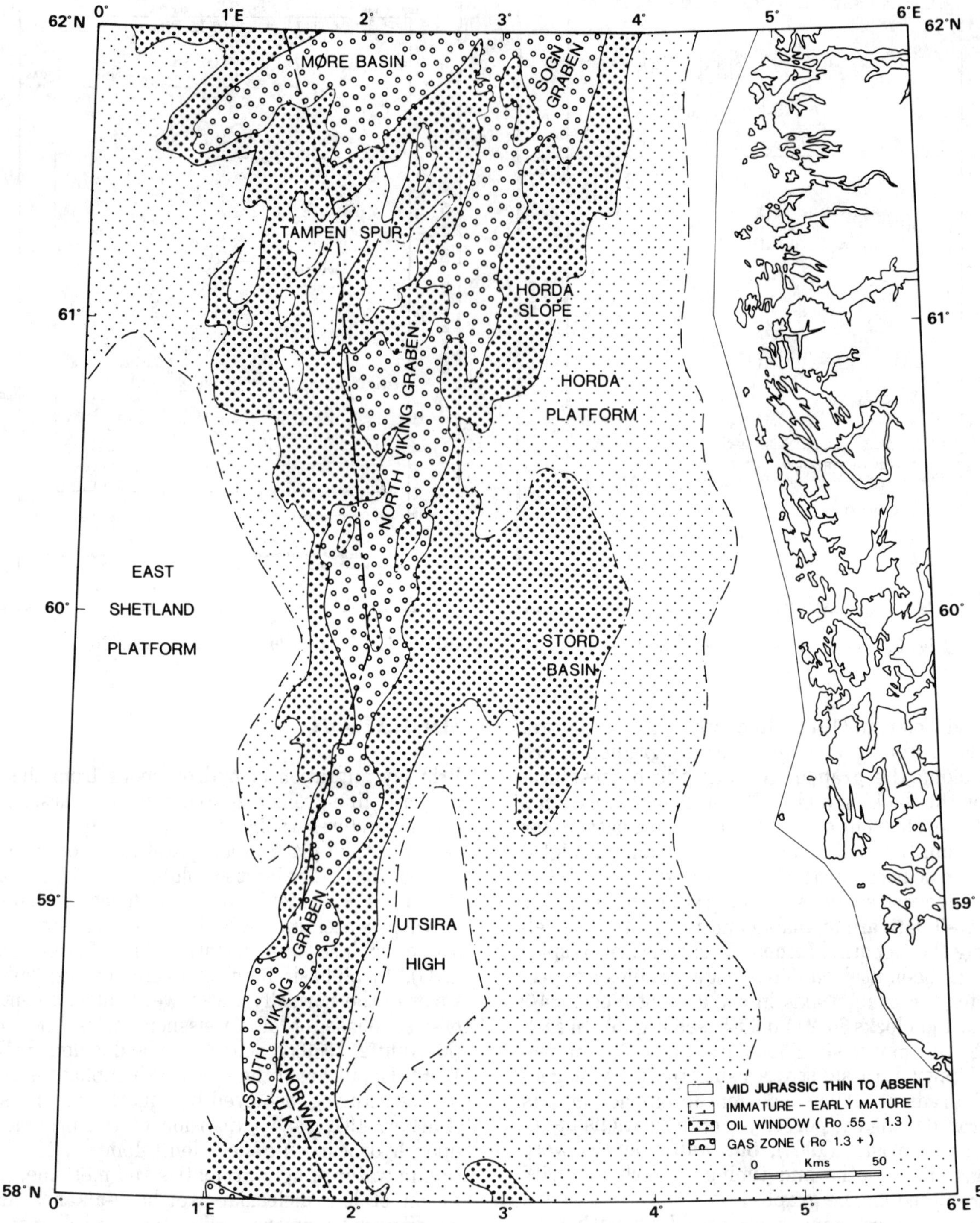

Fig. 6. Northern North Sea, level of maturation, top Brent Group.

conduit. Modification of these routes due to hydrodynamic flow can be assessed through the construction of potentiometric surface maps and subsequent resolution of the buoyancy and hydrodynamic force vectors (Dahlberg, 1982). For regional mapping, however, construction of detailed potentiometric maps is limited by the presence of flow barriers not detected by the sparse pressure data available.

Chiarelli and Duffaud (1980) presented an equipotentiometric map of the Jurassic in the Northern North Sea. They identified a zone of abnormal pressures centred on the graben, which was divided into a number of pressure compartments. Barriers between the compartments do not always follow obvious tectonic discontinuities. Detailed potentiometric mapping within each pressure compartment is required to define the hydrodynamics of each area. Secondary migration routes on Figs. 9, 10, 13, and 14, below, are based on buoyancy principles only. The flow of fluids from overpressured graben areas to flanking highs is anticipated to complement and reinforce the buoyancy vector.

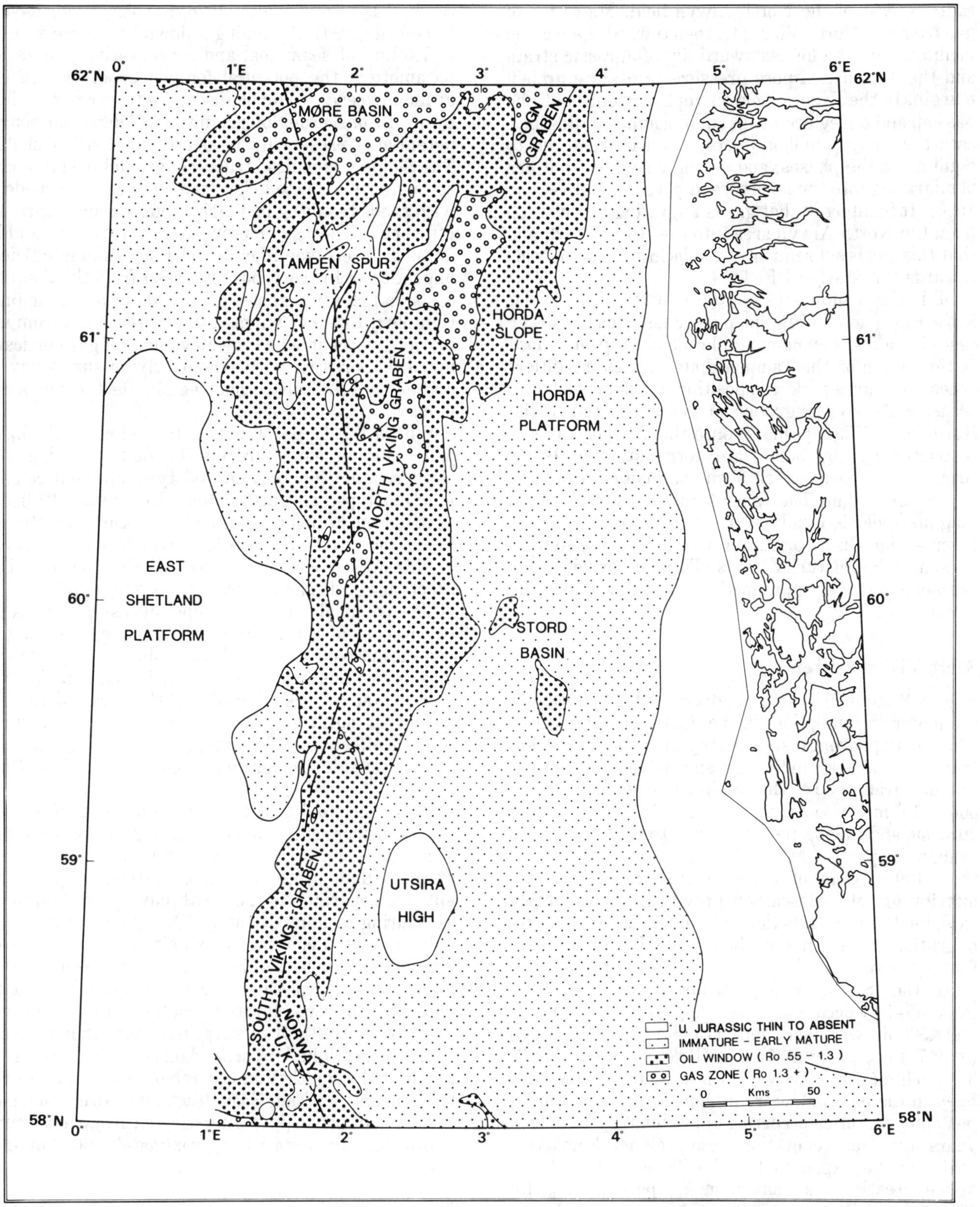

Fig. 7. Northern North Sea, level of maturation, top Humber Group.

PREDICTION OF HYDROCARBON ACCUMULATION PROPERTIES

Tampen Spur

The properties of accumulated hydrocarbons result from the combination of kerogen types, maturation states, and migration pathways in the drainage area. On the Tampen Spur, fields west of the Alwyn–Statfjord trend have low gas contents with no free gas caps and gas–oil ratios in the range of 20–150 m^3 gas/m^3 oil (110–850 scf/barrel). This results from the Upper Jurassic source envelope being predominantly in the middle maturation zone and deeper buried, potential gas-source rocks, such as Dunlin shales, being dominated by inertinite. The eastern margin of the Tampen Spur has relatively higher gas–oil ratios of 65–500 (365–2800), and free gas caps are developed on the North Alwyn and Brent fields.

In the Alwyn–Statfjord structural trend, the gas–oil ratio of each accumulation is proportional to its proximity to the deepest part of the graben in Quadrant 29, east of the North Alwyn field. Migration of gas from the North Viking Graben toward the west is facilitated by the low eastward dip of Jurassic strata and the decrease in pore pressure across the graben margin. In the Brent reservoirs of the Brent field, the gas cap and oil leg are not in equilibrium. Studies indicate that only 1 million years would be required to equilibrate the phases, and strongly suggest that gas is migrating into the structure at present (Bath *et al.*, 1980). It is inferred that gas is migrating northwards from the North Alwyn area into the Brent Field, and that this gas is subsequently displacing oil northwards towards the Statfjord Field.

Oil in fields supplied from the southern flank of the Møre Basin show different properties (such as lower gas–oil ratios for comparable depths) from those farther south on the Tampen Spur, suggesting differences in source-rock composition between the two areas. Extensive migration of gas generated in the Møre Basin (Figs. 6 and 7) onto the Tampen Spur is restricted by the northward termination of Brent Group sandstones, which form the major secondary-migration conduit. TheUpper Jurassic reservoir of the Magnus Field is locally sourced from the enveloping Kimmeridge Clay Formation, which lies close to peak oil-generation maturity levels. These reservoir sands are isolated from gas-productive shales downdip to the north.

South Viking Graben

Figure 8 summarizes the hydrocarbon properties of the major accumulations in the South Viking Graben. The most apparent distribution pattern is the concentration of gas fields in the east and oil fields in the west of the graben. In the Sleipner complex, up to 300×10^9 m^3 of gas is recoverable from Middle Jurassic and Paleocene reservoirs. In the west of the graben, Upper and Middle Jurassic reservoirs contain up to 100×10^6 m^3 of oil and 50×10^9 m^3 of gas. This distribution of hydrocarbon gravities can be largely explained as a function of different secondary-migration routes in the Middle and Upper Jurassic fluid systems.

In the South Viking Graben, oil generation (R_0 0.55–1.3) occurs between approximately 3500 m and 4900 m, with peak generation in type II kerogen (R_0 0.75) occurring at 4100 m (Pearson *et al.*, 1983). The rich oil-prone Upper Jurassic source rocks have been in the oil window from Miocene times onwards, peak oil generation occurring approximately 10 million years ago. Gas generation from Sleipner Formation coals can be expected to take place at R_0 0.7 and above, reaching a maximum at perhaps R_0 1.5 (Durand and Paratte, 1983), corresponding to depths of 4100 m and below 5000 m, respectively. The zones of maturation at top Brent Group and top Humber Group levels are mapped in Figs. 9 and 10, together with secondary-migration pathways.

Sleipner Field gas is considered to have predominantly originated from Brent Group (Sleipner Formation) coals (Larsen and Jaarvik, 1981). The isopach of these coals is presented in Fig. 11, and demonstrates that more than 30 m of seam coal is developed in the axis of the graben. As noted previously, this coal has a high hydrogen content and thus a high hydrocarbon-generating potential. Gas potential at an R_0 of 1.5 varies between 1.2 and 14×10^9 m^3 gas/km^3 of 1% TOC source rock, for type III and type II kerogens respectively (Tissot and Welte, 1978; Brooks *et al.*, 1984 in Cornford, 1984). Assuming a downdip source volume of 7.5 km^3 of seam coal and conservative generation parameters, the potential for gas generation in the graben exceeds volumes discovered in accumulations by up to one order of magnitude. Additional source potential is present in the overlying Heather shales. The Sleipner complex structures are full to spill-point, as might be expected since the coals are interbedded with sands resulting in a high expulsion and migration efficiency. The asymmetry of the graben at Middle Jurassic levels (see Figs. 9 and 12) results in efficient migration of hydrocarbons released from the Sleipner coals and Heather shales eastwards into the Sleipner Complex. Little or no gas travels westward from the graben depocentre and the movement of gas vertically upwards is restricted by the overlying shales having higher pore pressures than the Middle Jurassic sandstones.

Within the Upper Jurassic, the source-rock shales and reservoir sands are interbedded, resulting in a close relationship between oil type and source-rock kerogen type, and maturation. Reitsema (1983) has detailed the kerogen types and hydrocarbon matching in the Brae area. The depths to which oil is preserved in this area, considerably greater than 4000 m, is a function of this close source–reservoir coupling.

The synclinal axis of the Upper Jurassic is displaced to the east of the Middle Jurassic axis, resulting in significant volumes of Upper Jurassic sediments draining to the west (see Figs. 9, 10 and 12). The volume of source rocks downdip of the accumulations is more than sufficient to fill graben-margin structures to their effective spill-points, as indicated by accumulations outside the graben such as the West Brae Field.

It is concluded that the gross distribution of hydrocarbon properties can be explained by the preferred secondary migration routes in the Middle and Upper Jurassic fluid systems. These systems may not be entirely separate, as they both have pore fluid overpressuring to approximately 1.28 g/cm^3. Thin, probably discontinuous sands exhibit higher pressures. The lighter gravities of North Brae crude compared to South and Central Brae (see Fig. 8) have been attributed to longer migration routes into the former (Reitsema, 1983). It is suggested here that the concentration of lighter-hydrocarbon accumulations in the north of the South Viking Graben is associated with the syncline between East Brae and Gudrun, in which the base of the Upper Jurassic reaches a maximum depth for the area of approximately 6000 m (see Fig. 8).

North Viking Graben

As previously noted, the Humber Group is the prime potential source-rock horizon in the North Viking Graben. However, Brent Group coals with high extractabilities and hydrogen indexes, have good gas and light-liquid potential (see Fig. 2). Maturation states in the graben have been estimated from limited well data and maturation modelling (see Fig. 5). This modelling results in an oil window between approximately 3600 m and 4800 m, and peak oil generation at 4000–4200 m. Oil generation commenced in the Paleocene, and gas generation in the Miocene (see Fig. 5).

The areal distribution of these maturation zones has

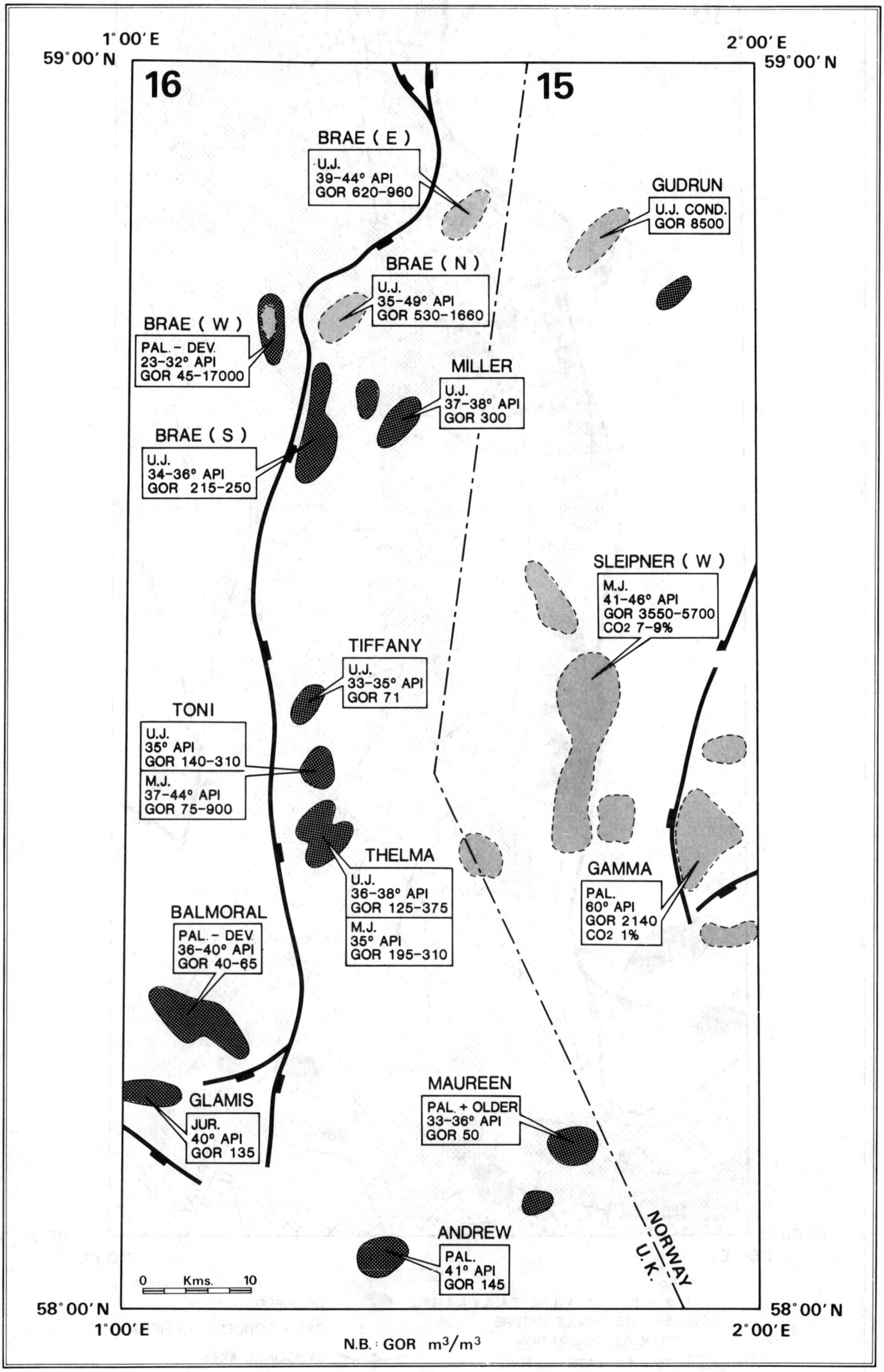

Fig. 8. South Viking Graben, hydrocarbon properties in major accumulations.

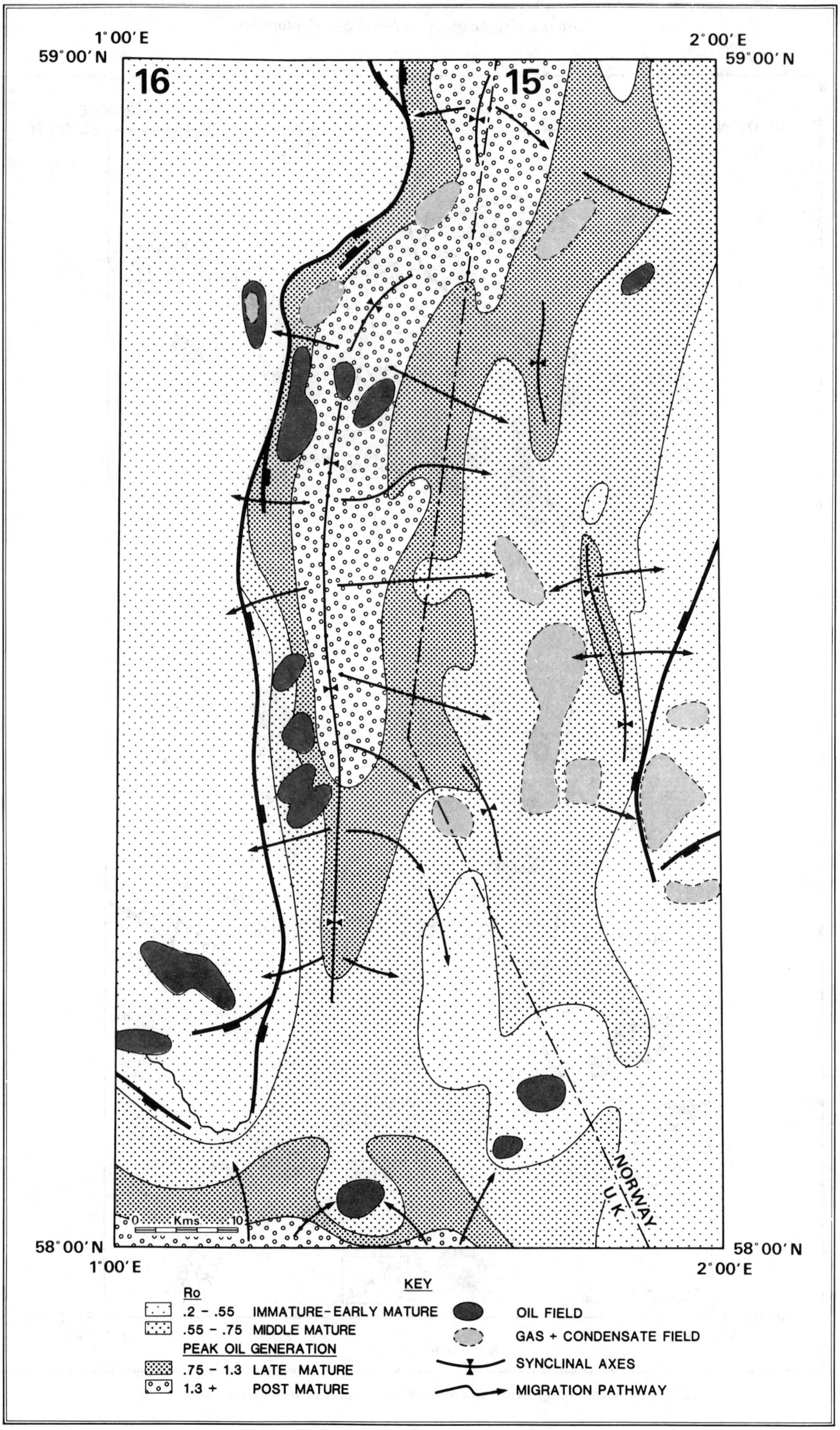

Fig. 9. South Viking Graben, top Brent Group, level of maturation and migration pathways.

Fig. 10. South Viking Graben, top Humber Group, level of maturation and migration pathways.

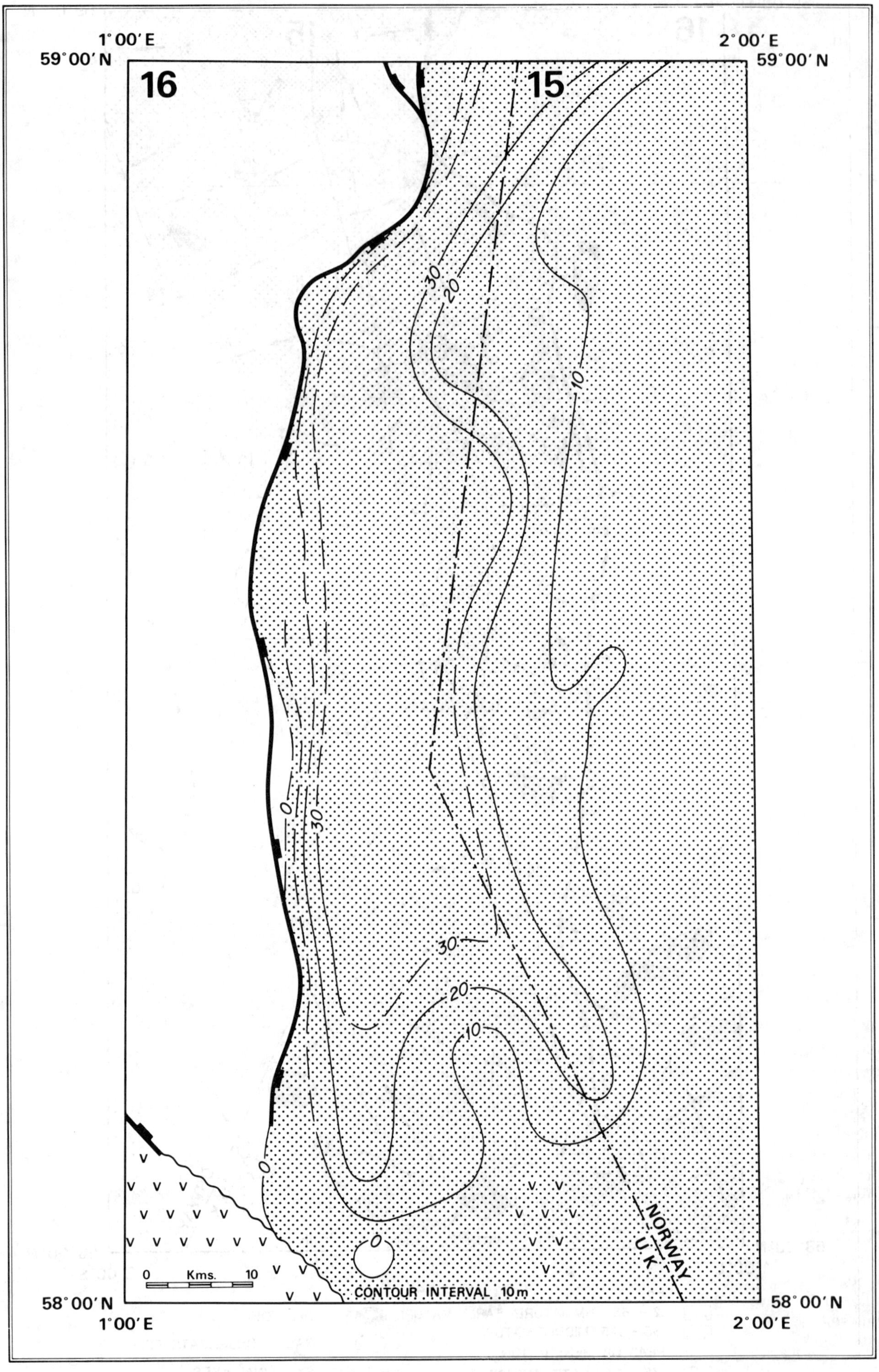

Fig. 11. South Viking Graben, Brent Group, coal isopach.

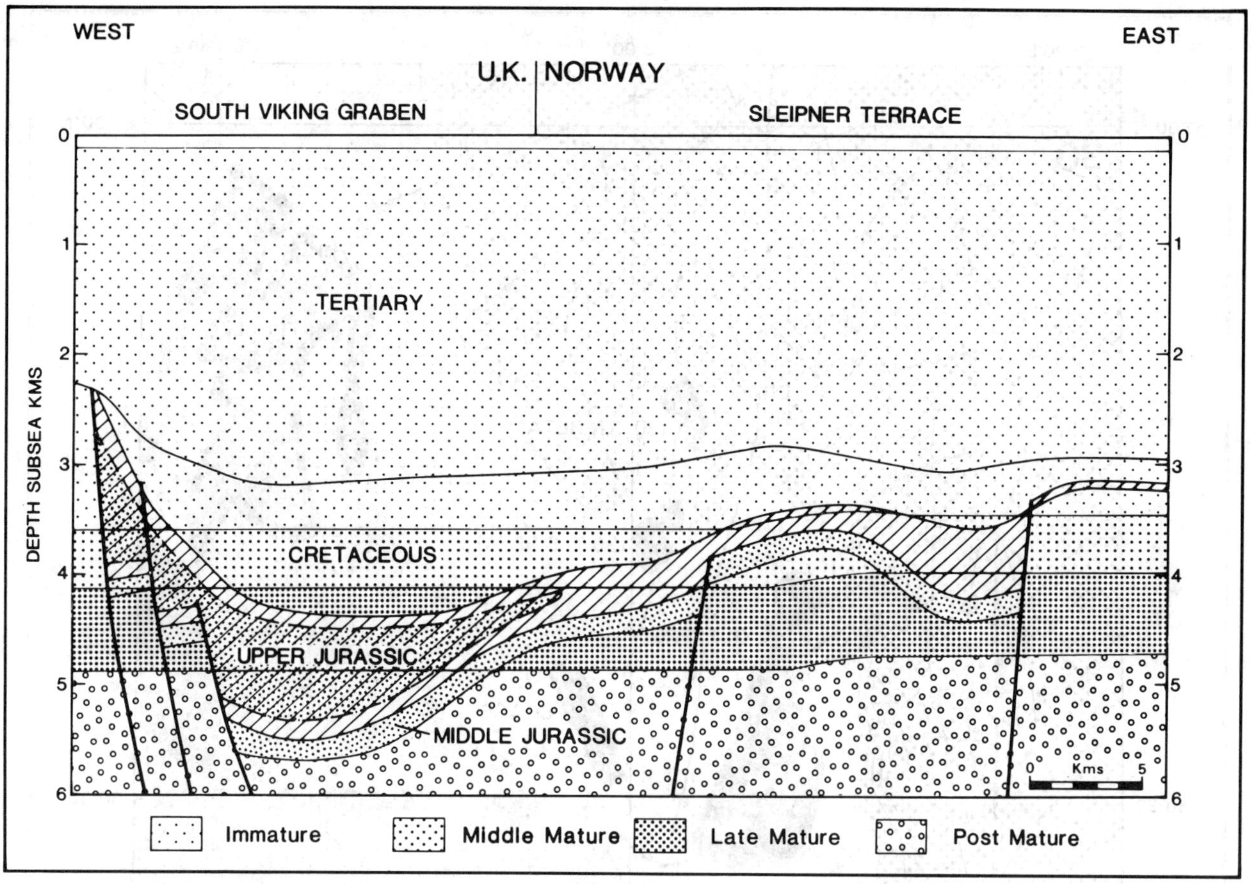

Fig. 12. South Viking Graben, E–W cross section at 58°30′N.

been mapped at top Middle and top Upper Jurassic horizons. Using regional isopachs and the maturation depths calculated above, isopachs of oil-mature and gas-mature Humber Group sediments are outlined on Figs. 13 and 14, respectively. When combined with probable secondary-migration routes, it is possible to estimate the relative proportions of oil or gas entering a particular structure or accumulation. The distribution of oil and gas reserves in the study area can be largely accounted for by these maps.

Significant oil reserves are present in Troll, Brage and Oseberg Fields and in the North and South Beta prospects in 30/3 and 30/6. Oil in the Troll accumulation has been largely sourced from the Horda Slope, north of the study area (see Fig. 7), with possible additional contributions through spillage of the Brage Field. Brage is possibly sourced from oil migrating through North and South Beta prospects. An alternative source of oil for Brage is from spillage of oil from the southeastern closure of Oseberg, due to displacement of early oil by later-stage gas. Another alternative is that oil migrates from the basin areas to the southeast (see Fig. 13). Oseberg oil is probably derived both from the Bergen Sub-basin to the northeast, and from the Viking Graben to the west.

Huldra, Troll and Oseberg Fields contain significant reserves of free gas. Gas in Troll has migrated from the Sogn Graben southwards up the Horda Slope into Troll, which lies on the culmination of the Horda Platform (see Fig. 7). Huldra gas and condensate can be derived in significant quantities from basins to the west, north and east. This field represents the first major structure on the migration routes from these source basins, and any early-accumulated oil has been, at least partially, displaced southwards. Oseberg has potential gas-source areas to the immediate west and possibly to the north, if sourced by spillage from the Huldra area. The areas to the west of Oseberg are probably sourcing gas into the structure, as is indicated by the presence of gas caps on the westernmost fault blocks.

The drainage basins of Huldra and Oseberg can be treated quantitatively to calculate the volumes of oil and gas generated and the entrapment efficiency. Table 2 presents approximate volumes of Humber and Brent Group source rocks in the maturity zones of the drainage areas of these structures, based on Figs. 13 and 14. A representative ultimate yield of oil per unit volume (km^3) of 1% TOC source rock characterised by type II kerogen is taken as 6×10^6 m^3 oil (Goff, 1983; Sluijk and Nederlof, 1984; Brooks *et al.*, 1984, in Cornford, 1984). For gas an equivalent yield of 14×10^9 m^3 gas is adopted. Brent Group coals, with relatively high hydrogen indexes, are estimated to have an ultimate yield of 275×10^9 m^3 gas/km^3 seam coal. Representative TOC values as indicated in Fig. 2 were utilized in the generation calculations.

The efficiency of entrapment can be calculated from the ratio of volumes of trapped oil and gas to those generated. The entrapment efficiency of Oseberg oil and gas is approximately 3%; Huldra gas has a lower value of approximately 1%. These values compare with the 20–30% efficiences noted by Goff (1983) for fields on the Tampen Spur. The overall low values of efficiency suggest restricted drainage of oil from the source envelope, escape or retention of oil and gas on migration routes, and escape of gas and oil from the accumulations through the caprock or by spillage at the closure. Inefficient primary and secondary migration from the Viking Graben and spillage from the

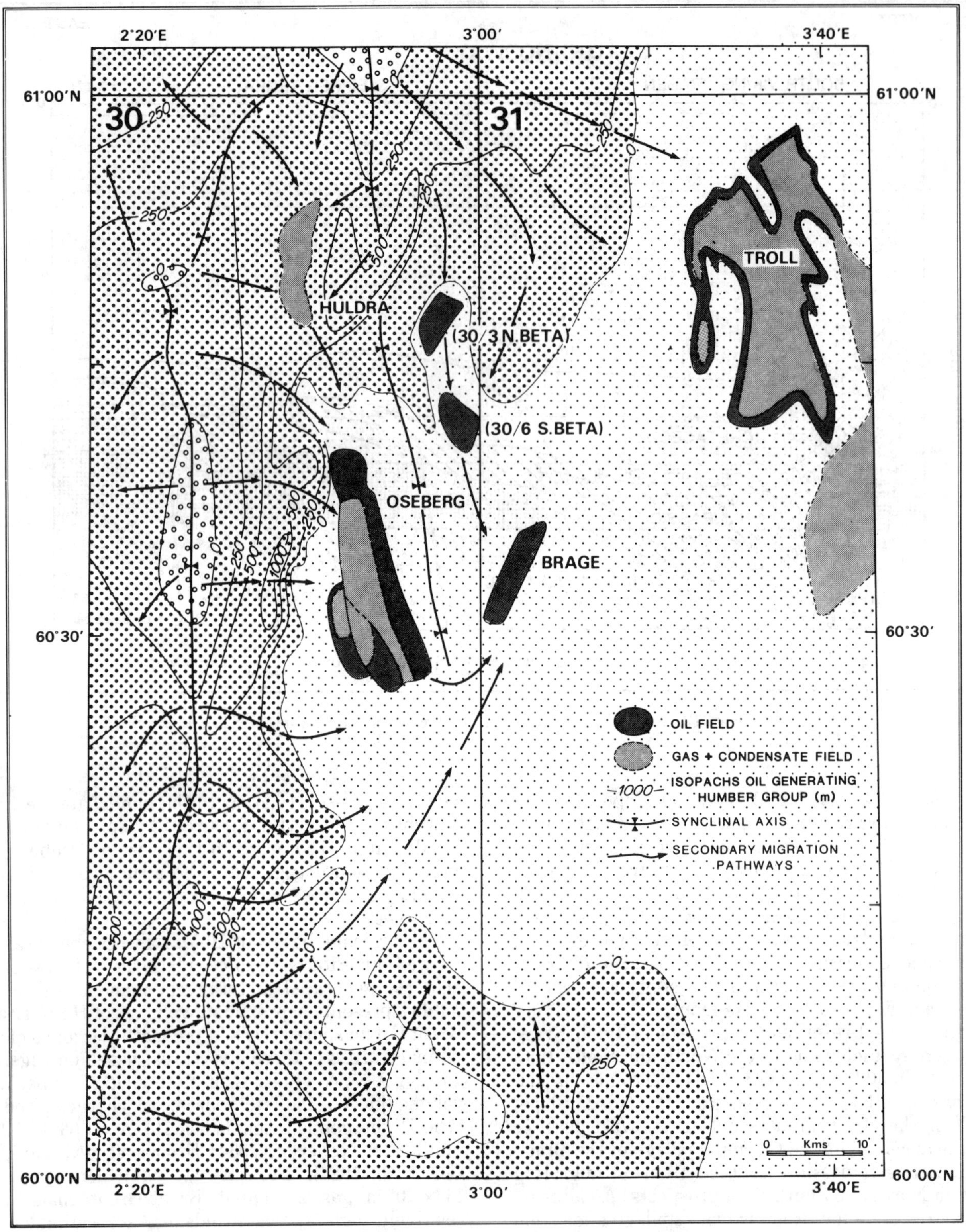

Fig. 13. North Viking Graben, Humber Group, isopach of oil-prone source rocks.

TABLE 2. North Viking Graben: drainage area volumes

	Source rock volume (km^3)		
	Oil-prone Humber Gp shale	*Gas-prone Humber Gp shale*	*Gas-prone Brent Gp coal*
Oseberg	190	55	0.5
Huldra	150	120	1.1

TABLE 3. North Viking Graben: gas-oil ratios (m^3/m^3)

	Source volume (m^3 gas-prone rock/m^3 oil-prone rock)	*Source generation (m^3 gas generated/m^3 oil generated)*
Oseberg	0.35	0.4×10^3
Huldra	1.0	1.2×10^3

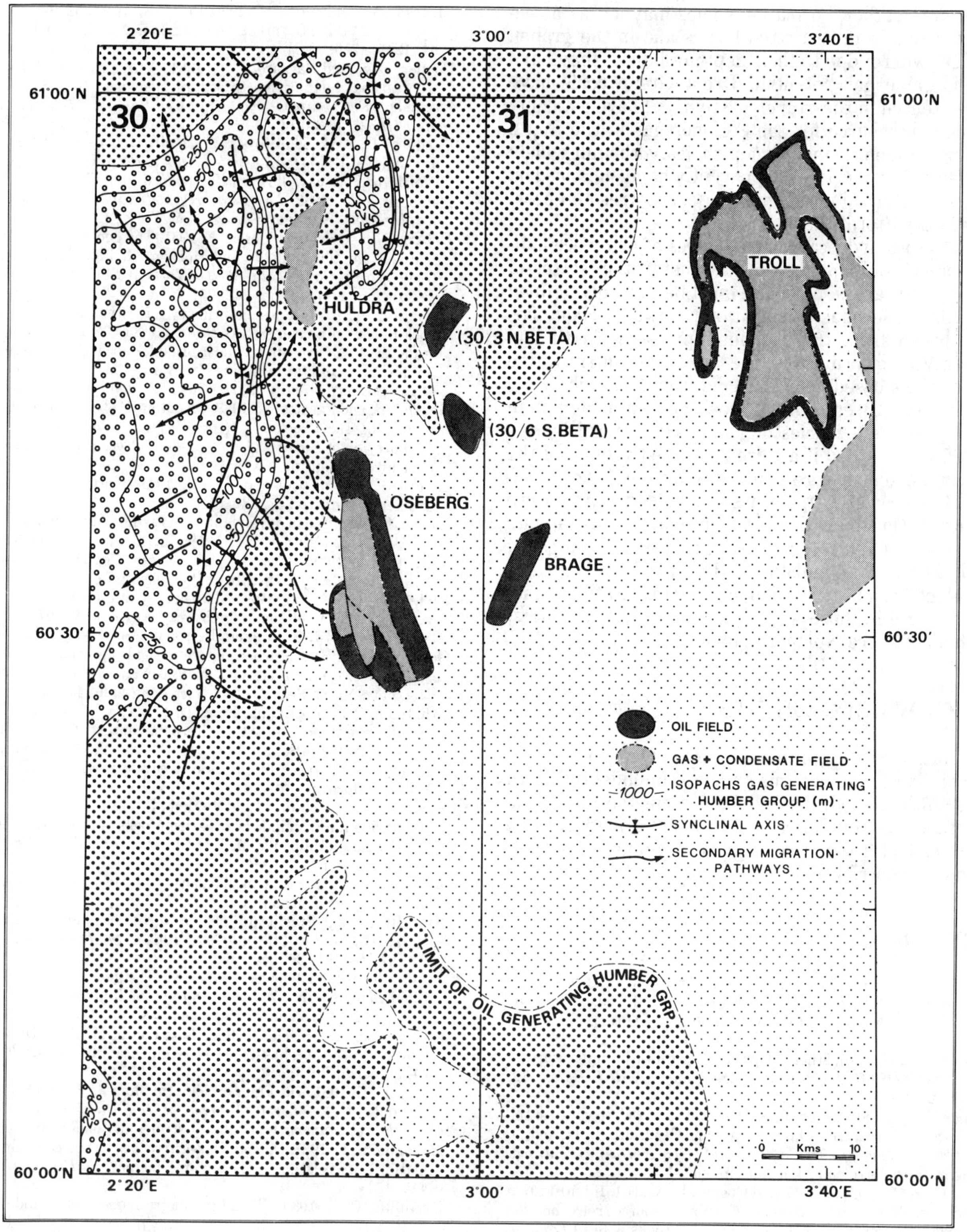

Fig. 14. North Viking Graben, Humber Group, isopach of gas-prone source rocks.

structures eastwards onto the Horda Platform are conceived to be the prime reasons for the low entrapment efficiency.

The gas–oil ratios of the drainage areas in units of rock and hydrocarbon volume are presented in Table 3. For source productivity, it appears that ratios of gas generation to oil generation greater than 1 ($\times 10^3$) may lead to gas-dominated accumulations such as Huldra. Ratios less than 0.5 ($\times 10^3$) may result in oil-dominated accumulations such as Oseberg. Although the models require further refinement, particularly with respect to migration efficiencies, such determinations result in a useful predictive capability in untested areas.

CONCLUSIONS

Potential source horizons are identified in shales of the Kimmeridge Clay Formation, Heather Formation and Dunlin Group, and coals of the Brent Group and Statfjord Formation. The Kimmeridge Clay Formation is widely developed and is the most important source horizon on the Tampen Spur and in the graben system. Heather Formation shales may act as a secondary source in restricted basins and in the graben system where, due to their depth of burial, they have considerable gas potential. Brent Group coals have gas potential in the graben system and have probably formed the source for the Sleipner gas in the South Viking Graben. Dunlin Group shales, although widely dominated by inertinite, apparently have enhanced oil and gas potential in the graben depocentres and possibly in the Stord Basin.

The gross hydrocarbon properties of many accumulations can be explained by the maturation states of their source areas, but these must be carefully defined in terms of secondary-migration pathways. Examples are drawn from the Tampen Spur and South and North Viking Grabens. In the latter area, quantitative calculations based upon sourcing from Humber Group shales and Brent Group coals indicate minimum entrapment efficiencies of 1–5%. Higher efficiencies would be recorded if a more restricted volume of the source envelope were being drained, or if the additional reserves in updip associated structures were included. Gas–oil ratios of the drainage areas in units of rock and hydrocarbon volume can be compared with known accumulations. The predictive models developed can be utilized in frontier exploration areas, particularly in those with similar geological histories, such as offshore Northern Norway.

ACKNOWLEDGEMENTS

The author wishes to thank Tenneco Oil Company for permission to publish this paper. I am indebted to D. J. Griffiths for editing and C. S. Smeeton for typing the text, and R. Salter and staff for draughting the figures. I also thank colleagues in Tenneco for their technical contributions.

REFERENCES

Albright, W. A., Turner, W. L. and Williamson, K. R. 1980. Ninian Field, U. Sector, North Sea. In: Halbouty, M. T. (ed.), *Giant Oil and Gas Fields of the Decade 1968–1978. AAPG Memoir*, **30**, 173–193.

Barker, C. 1972. Aquathermal pressuring-role of temperature in development of abnormal-pressure zones. *AAPG Bull.*, **56**, 2068–2071.

Barnard, P. C. and Cooper, B. S. 1981. Oils and source rocks of the North Sea area. In: Illing, L. V. and Hobson, G. D. (eds.), *Petroleum Geology of the Continental Shelf of North-West Europe.* Institute of Petroleum, London, pp. 169–175.

Barnard, P. C., Collins, A. G. and Cooper, B. S. 1981. Identification and distribution of kerogen facies in a source rock horizon —examples from the North Sea Basin. In: Brooks, J. (ed.), *Organic Maturation Studies and Fossil Fuels Exploration.* Academic Press, London, pp. 271–282.

Bath, P. G., Fowler, W. N. and Russel, M. P. 1980. The Brent Field, a reservoir engineering review. *Eur. Offshore Conf. and Exhibition*, London. EUR 164, pp. 179–192.

Bissada, K. 1983. Petroleum Generation in Mesozoic Sediments of the Moray Firth Basin, British North Sea Area. In: Bjorøy, M. (ed.), *Advances in Organic Geochemistry 1981.* Wiley, London, pp. 7–15.

Bowen, J. M. 1975. The Brent oil-field. In: Woodland, A. W. (ed.), *Petroleum and the Continental Shelf of North-West Europe*, Vol. 1., *Geology.* Applied Science Publishers, London, pp. 353–361.

Brekke, T., Pegrum, R. M. and Watts, P. B. 1981. First exploration results in Block 31/2. In: *Norwegian Symp. on Exploration (NSE-81).* Norwegian Petroleum Society, pp. NSE/16, 1–34.

Carstens, H. and Finstad, K. G. 1981. Geothermal gradients of the Northern North Sea Basin 59–62°N. In: Illing, L. V. and Hobson, G. D. (eds.), *Petroleum Geology of the Continental Shelf of North-West Europe.* Institute of Petroleum, London, pp. 152–161.

Chiarelli, A. and Duffaud, F. 1980. Pressure origin and distribution in Jurassic of Viking Basin (United Kingdom–Norway). *AAPG Bull.*, **64**, 1245–1250.

Cooper, B. S. and Barnard, P. C. 1984. Source rocks and oils of the Central and Northern North Sea. In: Demaison, G. and Murris, R. J. (eds.), *Petroleum Geochemistry and Basin Evaluation. AAPG Memoir*, **35**, 303–314.

Cornford, C. 1984. Source rocks and hydrocarbons of the North Sea. In: Glennie, K. W. (ed.), *Introduction to Petroleum Geology of the North Sea.* Blackwell Scientific, Oxford, pp. 171–204.

Dahlberg, E. 1982. *Applied Hydrodynamics in Petroleum Exploration.* Springer-Verlag, New York.

De'Ath, N. and Schuyleman, S. F. 1981. The geology of the Magnus Oilfield. In: Illing, L. V. and Hobson, G. D. (eds.), *Petroleum Geology of the Continental Shelf of North-West Europe.* Institute of Petroleum, London, pp. 342–351.

Dickey, P. A. 1976. 'Possible primary migration of oil from source rock in the oil phase. *AAPG Bull.*, **59**, 337–345.

Durand, B. 1983. Present trends in organic geochemistry in research on migration of hydrocarbons. In: Bjorøy, M. (ed.), *Advances in Organic Geochemistry 1981.* Wiley, London, pp. 117–128.

Durand, B. and Paratte, M. 1983. Oil potential of coals: a geochemical approach. In: Brooks, J. (ed.), *Petroleum Geochemistry and Exploration of Europe. Geol. Soc. Lond. Spec. Publ.* **12**, 255–265. (Blackwell Scientific, Oxford.)

Evans, T. R. and Coleman, N. C. 1974. North Sea geothermal gradients. *Nature* **247**, 28–30.

Eynon, G. 1981. Basin development and sedimentation in the Middle Jurassic of the Northern North Sea. In: Illing, L. V. and Hobson, G. D. (eds.), *Petroleum Geology of the Continental Shelf of North-West Europe.* Institute of Petroleum, London, pp. 196–204.

Fuller, J. G. C. M. 1975. Jurassic source-rock potential and hydrocarbon correlation, North Sea. In: *Jurassic Northern North Sea Symp. 1975.* Norwegian Petroleum Society, pp. 11/1–18.

Goff, J. C. 1983. Hydrocarbon generation and migration from Jurassic source rocks in the East Shetland Basin and Viking Graben of the Northern North Sea. *J. Geol. Soc. Lond.* **140**, 445–474.

Grantham, P. J., Posthuma, J. and De Groot, K. 1980. Variation and significance of the C_{27} and C_{28} triterpane content of a North Sea core and various North Sea crude oils. In: Douglas, A. G. and Maxwell, J. R. (eds.), *Advances in Organic Geochemistry, 1979.* Pergamon Press, Oxford, pp. 29–38.

Heritier, F. E., Lossel, P. and Wathne, E. 1980. Frigg Field—large submarine-fan trap in Lower Eocene rocks of the Viking Graben, North Sea. In: Halbouty, M. T. (ed.), *Giant Oil and Gas Fields of the Decade 1968–1978. AAPG Memoir*, **30**, 59–79.

Jones, R. W. 1981. Some mass balance and geological constraints on migration mechanisms. *AAPG Bull.*, **65**(1), 103–122.

Kirk, R. H. 1980. Statfjord Field—a North Sea giant. In: Halbouty, M. T. (ed.), *Giant Oil and Gas Fields of the Decade 1968–1978. AAPG Memoir*, **30**, 95–116.

Larsen, R. M. and Jaarvik, L. J. 1981. The geology of the Sleipner Field complex. In: *Norwegian Symp. on Exploration (NSE-81).* Norwegian Petroleum Society, pp. NSE/15, 1–31.

Leythaeuser, D., Schaefer, R. G. and Pooch, H. 1983. Diffusion of light hydrocarbons in subsurface sedimentary rocks. *AAPG Bull.*, **67**, 889–895.

Maher, C. E. 1981. The Piper oilfield. In: Illing, L. V. and Hobson, G. D. (eds.), *Petroleum Geology of the Continental Shelf of North-West Europe.* Institute of Petroleum, London, pp. 358–370.

Maximov, S. P. 1981. Oil and gas occurrences on the East-European platform as related to its tectonic history. *J. Petrol. Geol.*, **3**, 445–453.

McAuliffe, C. D. 1980. Oil and gas migration: chemical and physical constraints. *AAPG Studies in Geology, No. 10*, pp. 89–107.

Meissner, F. F. 1978. Petroleum geology of the Bakken Formation Williston Basin, North Dakota and Montana. In: Demaison, G. and Murris, R. J. (eds.), *Petroleum Geochemistry and Basin Evaluation. AAPG Memoir*, **35**, 159–179.

Momper, J. A. 1978. Oil migration limitations suggested by geological and geochemical considerations. *AAPG Course Notes Series No. 8*, pp. , 1–60.

Oudin, J. L. 1976. Étude géochimique du bassin de la Mer du Nord. *Bull. Centre Rech. Pau*, **10**, 339–358.

Pearson, M. J., Watkins, D., Pittion, J. L., Caston, D. and Small, J. S. 1983. Aspects of burial diagenesis, organic maturation and palaeothermal history of an area in the South Viking Graben, North Sea. In: Brooks, J. (ed.), *Petroleum Geochemistry and Exploration of Europe. Geol. Soc. Lond. Spec. Publ.* **12**, 161–173. (Blackwell Scientific, Oxford.)

Reitsema, R. H. 1983. Geochemistry of North and South Brae Areas, North Sea. In: Brooks, J. (ed.), *Petroleum Geochemistry and Exploration of Europe. Geol. Soc. Lond. Spec. Publ.* **12**, 203–212, (Blackwell Scientific, Oxford.)

Sclater, J. G. and Christie, P. A. F. 1980. Continental stretching: an explanation of the post mid-Cretaceous subsidence of the Central North Sea Basin. *Geophys. Res.*, **85**, 3711–3739.

Skarpnes, O., Hamar, G. P., Jakobsson, K. H. and Ormaasen, D. E. 1980. Regional Jurassic setting of the North Sea, north of the Central Highs. In: *The Sedimentation of the North Sea Reservoir Rocks*. Norwegian Petroleum Society, pp. XIII, 1–8.

Sluijk, D. and Nederlof, M. H. 1984. Worldwide geological experience as a systematic basis for prospect appraisal. In: Demaison, G. and Murris, R. J. (eds.), *Petroleum Geochemistry and Basin Evaluation. AAPG Memoir*, **35**, 15–26.

Tissot, B. P. 1984. Recent advances in petroleum geochemistry applied to hydrocarbon exploration. *AAPG Bull.*, **68**, 545–563.

Tissot, B. P. and Welte, D. H. 1978. *Petroleum Formation and Occurrence*. Springer-Verlag, Berlin.

Ungerer, P., Bessis, F., Chenet, P. Y., Durand, B., Nogaret, E., Chiarelli, A., Oudin, J. L. and Perrin, J. F. 1984. Geological and geochemical models in oil exploration; principles and practical examples. In: Demaison, G. and Murris, R. J. (eds.), *Petroleum Geochemistry and Basin Evaluation. AAPG Memoir*, **35**, 53–77.

van den Bark, E. and Thomas, O. D. 1980. Ekofisk: first of the giant oil fields in Western Europe. In: Halbouty, M. T. (ed.), *Giant Oil and Gas Fields of the Decade 1968–1978. AAPG Memoir*, **30**, 195–224.

Waples, D. 1980. Time and temperature in petroleum formation; application of Lopatin's method to petroleum exploration. *AAPG Bull.*, **64**, 916–926.

Ziegler, P. A. 1982. *Geological Atlas of Western and Central Europe*. Elsevier, Amsterdam.

4 Oil-generating coals

S. Thompson, B. S. Cooper†, R. J. Morley* and P.C. Barnard**
*Robertson Research International Ltd, Ty'n-y-Coed, Llanrhos, Llandudno, Gwynedd LL30 1SA, UK
†BSC Associates, The Beeches, Byfield Road, Chipping Warden, Nr. Banbury, Oxford OX17 1LD, UK.

Coals are oil source rocks in many of the Tertiary basins of Southeast Asia. The precursors of these hydrogen-rich and oxygen-poor coals are coastal plain peats which have mainly developed in an everwet and tropical climate. In these environments water flow and reworking can concentrate liptinitic kerogen in preference to vitrinitic kerogen. The distribution, petrography and chemistry of the coaly Miocene source rocks present in the Kutai Basin are described. Studies of Tertiary and contemporaneous coastal plain deposits can be applied to trends in source rock distribution in offshore Norway. Comments on the environment of deposition of coaly sediments in the basins of the Norwegian Sea are discussed with reference to their possible oil and/or gas sourcing potential. The Triassic–Jurassic coals of the Haltenbanken area may become more oil-prone towards the delta margins, and facies mapping could aid oil exploration in this area.

INTRODUCTION

The most productive oil source rocks of Indonesia are drift coals and associated shales deposited within tidal dominated coastal plain environments during the Tertiary. Present analogues for environments in which such sediments accumulate are to be seen in many parts of Southeast Asia where the processes of the sedimentation of organic matter have been studied. These studies of Tertiary and contemporaneous coastal plain deposits can be used as a guide to trends in source rock distribution in offshore Norway where coastal plain deposits occur in the Triassic and Jurassic.

The possibility that coals may act as oil source rocks has been widely discussed over the last few years (Thomas, 1982; Durand and Paratte, 1983), those of Indonesia having been evaluated by Roe and Polita (1979) and Durand and Oudin (1979). It is doubtful that significant quantities of oil are generated from vitrinite-rich coals; Boudou *et al.* (1984) suggest that when kerogen is rich in oxygen, carbon dioxide and water will be evolved in preference to hydrocarbons. However, it is clear that many coals, such as some of the liptinite-rich examples described by Teichmüller and Durand (1983), are hydrogen-rich and oxygen-poor at relatively low levels of thermal maturity.

Hydrogen-rich coals frequently occur in the subsurface in Indonesian basins, and accumulate as a result of everwet and tropical conditions which have persisted through the Tertiary to the present over much of Indonesia. In Kalimantan, everwet peat deposits extend over several thousand square miles, and their environment and development, including the more restricted areas likely to form very good quality oil source rocks, are a suitable analogue for discussion.

Triassic–Jurassic coals are relatively common in the Haltenbanken area of northern offshore Norway, and an area of delta plain deposits containing thick coals has been described (Rønnevik *et al.*, 1983); it is probable that coal-bearing deposits of this age and facies are extensive in the Norwegian Atlantic offshore province. In this paper we try to compare the coals of this area with those of the more extensively examined Indonesian swamp environments. For the purpose of demonstration, pyrolysis data from several Indonesian coals are presented, and the oil potential of such coals is discussed.

PRESENT DAY COASTAL PLAIN PEAT-FORMING ENVIRONMENTS IN SOUTHEAST ASIA

The distribution of peats in Southeast Asia is controlled essentially by climate (see Fig. 1). Areas which yield peats greater than 1 m in thickness are more-or-less wholly confined to everwet regions or areas exhibiting minor seasonality of climate (Morley, 1981). Outside these areas, humic organic matter is broken down by desiccation and oxidation during dry seasons, and true peats do not form. Peat formation occurs

Petroleum Geochemistry in Exploration of the Norwegian Shelf © Norwegian Petroleum Society (Graham & Trotman, 1985) pp. 59–73.

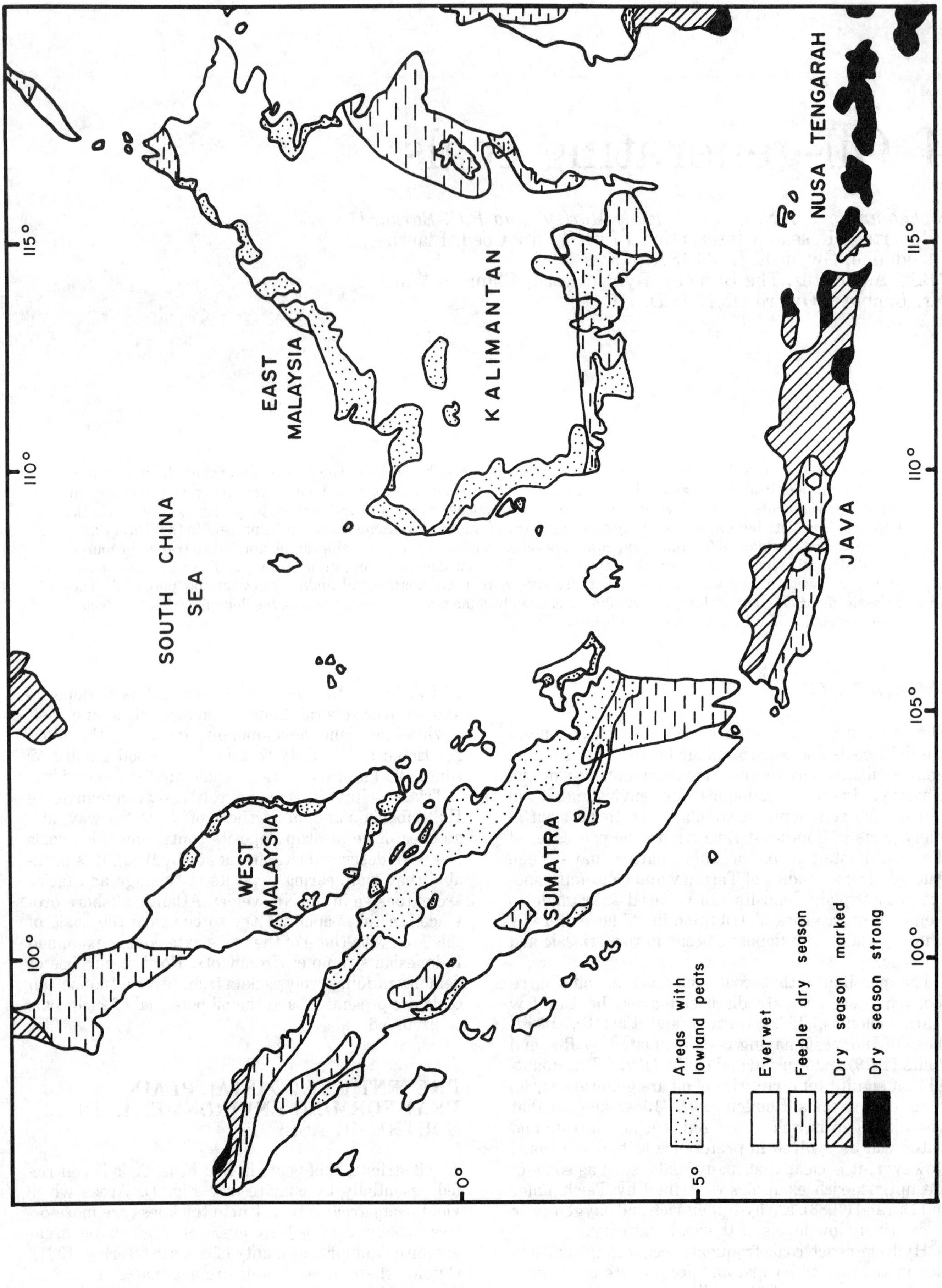

Fig. 1. Distribution of lowland peats in the Sunda region of Southeast Asia (after Morley, 1981).

over the upper coastal plain as far as the intertidal zone, where they are either bounded by beach ridges or give way to mangrove swamp. They are domed (ombrogenous), their water supply being maintained by rainfall only, and may reach up to 20 m in thickness (Anderson, 1964). They bear a raised water table and may be compared with blanket bogs of temperate areas.

The flora of Southeast Asian peat swamps consists predominantly of trees. The vegetation is strongly zoned and may be highly diverse on thin peats, but species-poor over thick parts (Fig. 2). Many of the trees, such as members of the angiosperm family Dipterocarpaceae, and the conifer *Agathis* (Araucariaceae) produce highly resinous wood. As peats accumulate, bacterial activity occurs throughout the peat, resulting in the breakdown of most woody material to a water-soluble semi-liquid humic acid gel.

Peat accumulation within prograding coastal plain environments has a profound effect on the manner of sediment accumulation. Peat accumulation takes place much more rapidly than levee development, and hence any depressions within the upper coastal plain rapidly become filled. Lakes, lagoons or crevasse splay deposits within interdistributary bays are thus rare. The peats are, however, highly unstable and are easily eroded by the scouring action of tidal channels and shifting water courses in the upper coastal plain. In strongly tidal deltas, autochthonous peats are rarely preserved as sediments. Once eroded, humic acid gels go into solution, whereas water-insoluble fragments, such as plant cuticle, plant waxes, resins and humified woody fragments become part of the suspension or traction load. Particulate organic matter is deposited in intertidal or inner shelf shallow water depths as streams lose energy when they reach the sea, whereas the soluble humic acid component will generally be carried to greater water depths before precipitation (as the precursor to amorphous or unstructured vitrinite) or oxidation. The process of separation of soluble and insoluble organic matter results in the accumulation of liptinite-enriched allochthonous (drift) peats and carbonaceous shales in estuarine or shallow marine environments along the coastal margin. This process can currently be observed in the modern Mahakam Delta, where allochthonous peats are deposited adjacent to beach ridges (Allen *et al.*, 1979). These liptinite-enriched peats may be further enriched by fresh liptinitic material derived from vegetation growing behind the coastal plain, of which resin-producing trees may form a large component.

Peats preserved within upper coastal plain deposits tend to retain their initial high proportion of humic acid gels, and are more likely to form vitrinite-rich lignites during burial. Such sediments are potentially gas source, rather than oil source, rocks. However, unusually high rates of water flow may flush out humic acid gels from these *in situ* peats, and in this way also give rise to enrichment in liptinitic kerogen.

TERTIARY COASTAL PLAIN DEPOSITS IN INDONESIA

Most of the sedimentary basins of western Indonesia contain coastal plain deposits which have significant oil source potential. Of these basins, the foremost oil producers are the Central and South Sumatra, Sunda, Northwest Java, and Kutai basins (see Fig. 3). The Oligocene or Miocene coastal plain deposits of these basins frequently comprise interbedded shales, sandstones and coals. The coals and shales are often good-quality oil source rocks, and not infrequently the coals have better oil generating capacity than the shales. Although the samples described in this paper come from several of these basins, for the purposes of descriptive convenience, and because of the availability of published information, we have concentrated on the Kutai Basin, which underlies the present Mahakam Delta of Kalimantan.

The basic geochemical data of the Mahakam Delta have been reported by Combaz and de Matharel (1978), Durand and Oudin (1979) and Verdier *et al.* (1980). A correlation of the oils to the coaly kerogens in the Kutai Basin (Mahakam Delta) in general, and the Balikpapan Group in the Handil field in particular, has been published by Schoell *et al.* (1983).

DISTRIBUTION OF BALIKPAPAN GROUP SOURCE ROCKS

The Balikpapan Group, which is of Middle Miocene, but locally also Late Miocene, age, represents a phase of delta progradation which has continued, with transgressive interruptions, to the present day Mahakam Delta phase of deposition (Oudin and Picard, 1982). Sediments have been deposited from one dominant eastward-flowing river system into a subsiding basin with its north–south axis parallel and close to the delta front, the axis migrating eastwards with time after an early basin inversion. A varied suite of facies representing upper delta plain to delta front environments with occasional transgressive limestones is present. The distribution of sediments and their contained organic matter has been strongly influenced by the interaction of tides and rivers which has concentrated the deposition of organic material within tidal flats and interdistributary areas of the lower coastal plain, and has resulted in the deposition of drift coals and black shales. The Balikpapan Group source rock sequence at its richest includes some 175 m (cumulative) of coal and nearly 1750 m (cumulative) of shale. The source rock sequence is in a zone trending north–south parallel to the lithofacies trends, and probably thins at the northern and southern margins of the main area of deltaic sedimentation. There are also rapid variations both to the east and to the west of the main north–south source rock zone, the shales becoming thinner and vitrinitic, and the coals becoming thinner and inertinitic (see Fig. 4). The coaly sequences of the Balikpapan Group of the Kutai Basin extend some 230 km NE–SW and about 180 km NW–SE underneath the Mahakam Delta. However, the area where coals (and shales) of good oil source quality occur in this group, extends about 175 km NE–SW and about 55 km NW–SE, and is broadly equivalent to the lower delta plain back swamp and intertidal areas. In other basins where a seasonal influence is more marked, black (and often algal-rich) shales deposited in semi-permanent lakes and lagoons, occur within the coal-bearing sequence.

EXAMPLES OF INDONESIAN COALS

Coal samples have been selected from the principal oil source rock formations of the Kutai Basin (three

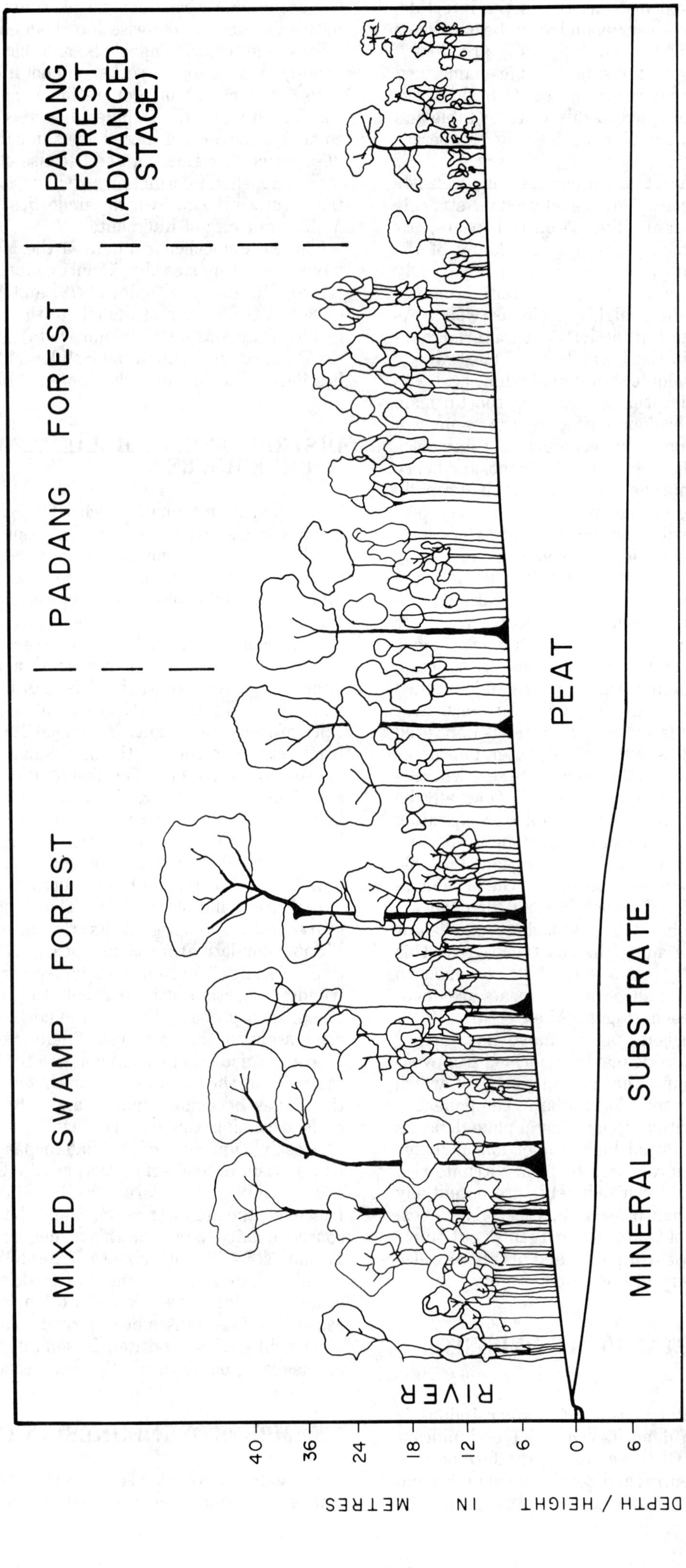

Fig. 2. Idealized cross-section of a domed peat swamp. Mixed swamp forest—high, species-rich forests on thin (1–3 m) peats. Padang forest—low, species-poor forests on thick (> 3 m) peats.

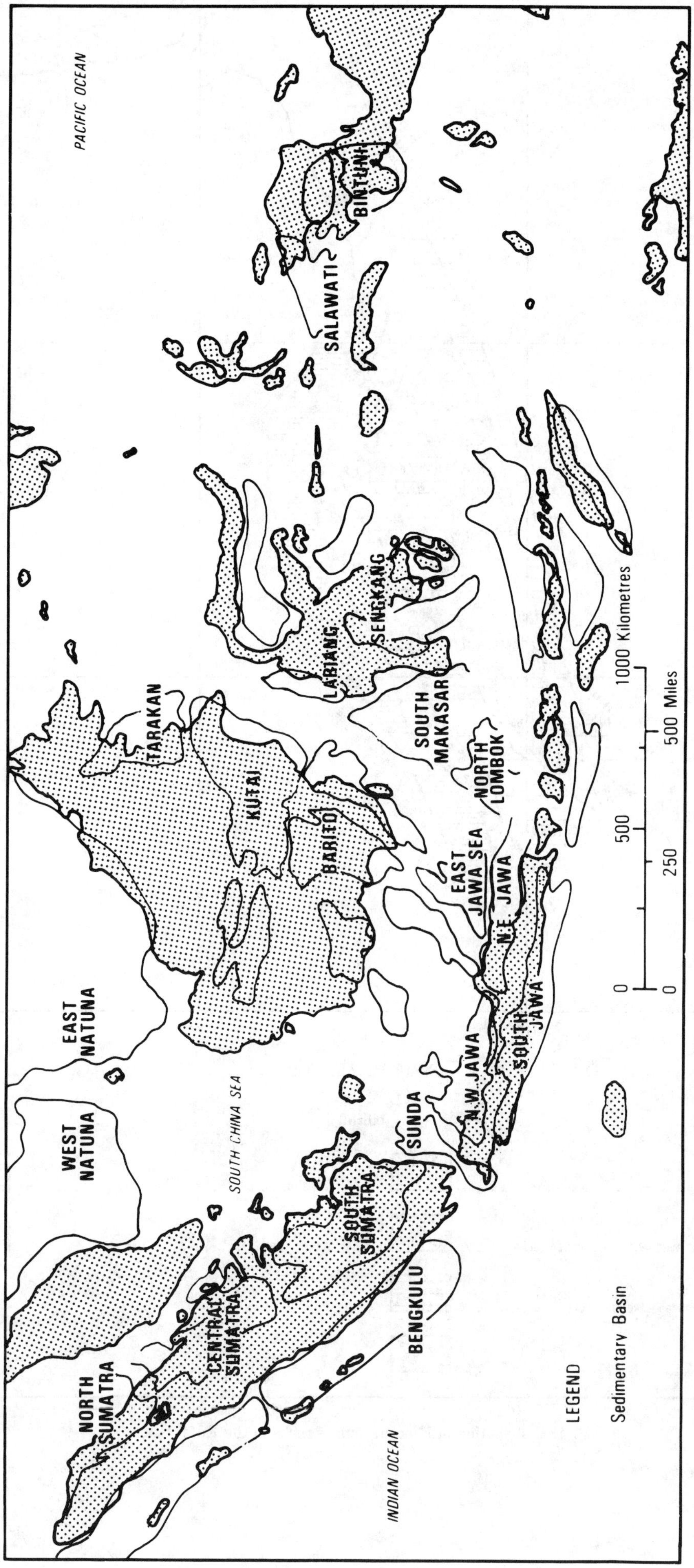

Fig. 3. Sedimentary basins of the Indonesian area.

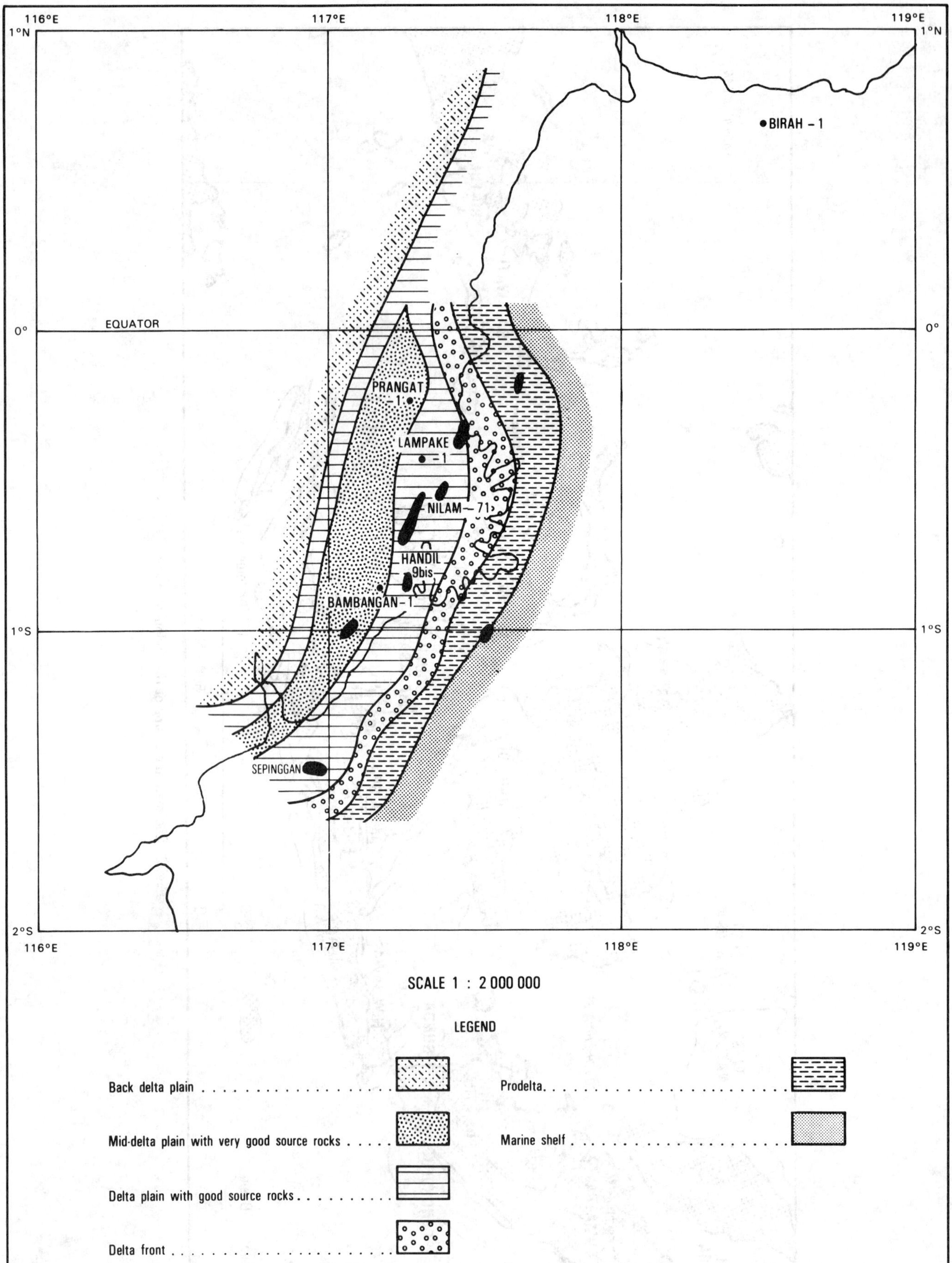

Fig. 4. Distribution of Miocene source rocks in the Kutai Basin.

wells), Northwest Java Basin (two wells) and Sunda Basin (one well) of western Indonesia. A key to the samples, which comprise 12 coals and two mixed lithologies, from Oligocene and Middle Miocene formations, is presented in Table 1. Sufficient published work exists for some comparisons to be made between the coals of the Haltenbanken area and the Indonesian coals discussed here. The chemical data for the samples are presented in Table 2, and the maturity and interpreted kerogen composition data are presented in Table 3.

METHODS

Organic carbon analysis was conducted on a Leco carbon analyser, and is inclusive of ash content. An IFP-Fina 'Rock-Eval' apparatus was used for

TABLE 1. Key to samples

Sample code	*Indonesian Basin*	*Depth (ft)*	*Formation*	*Age*
B-1	Kutai	3000	Balikpapan (Gp)	Middle Miocene
B-2	Kutai	4780	Balikpapan (Gp)	Middle Miocene
B-3	Kutai	7710	Balikpapan (Gp)	Middle Miocene
H-1	Kutai	5950	Balikpapan (Gp)	Middle Miocene
H-2	Kutai	8230	Balikpapan (Gp)	Middle Miocene
H-3	Kutai	8660	Balikpapan (Gp)	Middle Miocene
P-1	Kutai	3800	Balikpapan (Gp)	Middle Miocene
P-2	Kutai	8800	Balikpapan (Gp)	Middle Miocene
C-1	Northwest Java	7260	Lower Cibulakan	Oligocene
C-2	Northwest Java	7410	Lower Cibulakan	Oligocene
C-3	Northwest Java	7520	Lower Cibulakan	Oligocene
S-1	Northwest Java	8560	Lower Cibulakan	Oligocene
S-2	Northwest Java	8920	Lower Cibulakan	Oligocene
Y-1	Sunda	6420	Talang Akar	Oligocene

TABLE 2. Chemical data

			Rock—Eval pyrolysis data				
Sample code	*Lithology analysed*	*Total organic carbon (%)*	T_{max} (°C)	*HI*	*OI*	*Production index*	*Potential yield* (S_2) *(ppm)*
B-1	Coal	51.7	415	246	21	0.02	127 300
B-2	Coal	62.8	425	263	10	0.03	165 400
B-3	Brn-blk claystone + 10% Coal	12.8	434	360	12	0.08	46 280
H-1	Coal	58.6	422	347	10	0.12	203 650
H-2	Coal	69.6	442	354	5	0.06	247 050
H-3	Coal	72.1	437	355	21	0.11	256 280
P-1	Coal + 20% brn-blk shale	21.8	417	348	8	0.08	75 870
P-2	Coal	43.1	427	293	5	0.07	126 490
C-1	Coal	66.1	444	400	13	0.09	265 000
C-2	Coal	51.9	445	446	21	0.04	232 000
C-3	Coal	26.9	443	389	1	0.09	104 700
S-1	Coal	63.6	449	351	6	0.08	223 800
S-2	Coal	37.6	447	305	9	0.07	114 750
Y-1	Coal	49.7	439	461	12	0.02	229 600

TABLE 3. Maturity and kerogen composition data

				Interpreted kerogen composition			
Sample code	*Lithology analysed*	*Vitrinite reflectivity*	*Spore colour index*	*Intertinite*	*Vitrinite*	*Algal sapropel*	*Waxy (I + II) kerogen*
B-1	Coal	0.42(21)	2.5	60	15	0	25
B-2	Coal	0.45(22)	3.5–4.0	55	15	5	25
B-3	Claystone, brn-blk + 10% coal	0.65(23)[b]	5.0[a]	50	35	0	15
H-1	Coal	0.43([a])	4.5–5.0	40	15	20	25
H-2	Coal	0.40([a])	5.0–5.5[a]	35	5	0	60
H-3	Coal	0.49(28)[c]	5.0–5.5[c]	20	25	10	45
P-1	Coal + 20% brn-blk shale	0.52(22)	3.5[a]	40	25	15	20
P-2	Coal	0.62(23)	5.0–5.5[a]	30	35	15	20
C-1	Coal	0.59(30)	4.0–4.5[c]	30	20	0	50
C-2	Coal	0.60(30)	4.5–5.0[a]	10	40	0	50
C-3	Coal	0.63(29)	4.5–5.0[c]	25	25	0	50
S-1	Coal	0.63(22)	6.0[a]	25	5	5	65
S-2	Coal	0.65(22)	6.0[a]	30	15	0	55
Y-1	Coal	0.48([a])	4.0[a]	30	15	5	50

[a]Estimated from gradient. [b]Measured on coal. [c]Measured on kerogen of mixed cuttings.

pyrolysis analysis using 20 mg aliquots of crushed sample. Vitrinite reflectivity was measured on a Leitz Orthoplan microscope against glass standards. Spore colour was measured under transmitted light on unoxidized kerogen fractions using a standard palynological microscope with incident u.v. light attachments.

PETROGRAPHY OF THE ANALYSED SAMPLES

Under incident light, the coals appear to be composed mainly of vitrinite, although layers of liptinitic material are not uncommon. The areas of coal with lower reflectance and lower relief than semi-fusinite, and higher reflectance than liptinite (e.g. sporinite) are usually given the generic name vitrinite, and their reflectance is measured. However, the term vitrinite may include up to six or more sub-type (maceral types) of humic kerogen, all with different morphology, reflectance, and chemical composition (Stach, 1975).

Apart from normal vitrinite, two other types are frequent or even dominant. The reflectance of vitrinite in the samples analysed is often relatively high, particularly at low levels of maturity, with respect to the spore colours. Benedict *et al.* (1968) described anomalously high reflecting vitrinite in Appalachian coals and called it pseudo-vitrinite. They consider that it results from alteration, probably by partial oxidation, either before or during the peat stage of coalification. Thompson and Benedict (1974) noted that this material did not behave like vitrinite in terms of coking properties, being more inert. The morphology of pseudo-vitrinite is similar to that of the high reflecting vitrinite noted in the analysed samples.

In many cases the vitrinite of the Indonesian samples resembles the coal maceral desmocollinite, described by Teichmüller in Stach (1975), which has a relatively low reflectance. Electron micrographic studies by Taylor (1966) have suggested that this maceral is vitrinite impregnated with submicroscopic liptinite, possibly resinite, intercalations. In some cases the groundmass of the 'vitrinite' exhibits fluorescence, and the inclusion of amorphous type II (liptinitic) kerogen in a non-fluorescent matrix derived from humic acid gel (the precursor of vitrinite *s.s.*) would seem a probable explanation.

Hence, although apparently vitrinite-rich, the pyrolysis analysis of these coals in terms of types I, II and III kerogens may indicate a different composition.

CHEMICAL CHARACTERISTICS OF THE ANALYSED SAMPLES

Chemically, the coals of coastal plain environments of Indonesia are different from the Carboniferous and Tertiary coals of North-West Europe, in that the bulk of their kerogen, which has low relief and is grey under reflected light, is hydrogen-rich and oxygen-poor. These properties are clearly demonstrated by the pyrolysis data (see Table 2) in that the hydrogen indices of these coals are often above 300, and the oxygen indices are usually below 15. By comparison, vitrinitic Carboniferous coals of North America and Europe at similar levels of maturity may be expected to have hydrogen indices of about 150, and oxygen indices of about 35. These Tertiary coals also contain relatively high contents of clays. Calculation of kerogen composition with respect to the hydrogen and oxygen indices of theoretical type I, II, III and IV kerogen end members (Barnard *et al.*, 1981) indicates that vitrinite is subordinate to oil-prone (type I and II) kerogens. The calculated liptinitic (I and II) component of these coals ranges between 45% and 65% and the vitrinite contents are generally below 15%. However, incident u.v. light microscopy does not reveal anything like 45% of fluorescent components, although yellow fluorescent exsudatinite and orange-yellow fluorescent algal, spore and cuticle debris are common. Again, this shows that the groundmass contains a mixed vitrinite and liptinite composition. The liptinite of such kerogen compositions either in coals or disseminated in shales, acts as an oil source (Snowdon and Powell, 1982).

CHARACTERISTICS OF THE OILS GENERATED FROM COALS AND ASSOCIATED SHALES

The oils of the Indonesian basins discussed here are often waxy and associated with condensates and gas. They have been described by Sutton (1979), Roe and Polito (1979), Seifert and Moldowan (1981) and Schoell *et al.* (1983). Those generated at a relatively low level of thermal maturity have very high pristane contents and a marked odd-over-even preference in the generally pronounced C_{27} to C_{33} waxy component (see Fig. 5). At higher levels of maturity these features tend to become less pronounced or absent. In most of the basins, the carbon isotope ratios of the C_{15+} fractions of these oils are around $-30‰$ PDB, suggestive of derivation from higher plant kerogen. In the few Indonesian basins in which Tertiary coastal plain deposits have accumulated under the influence of a more marked seasonal climate, organic-rich lacustrine and lagoonal shales are developed as source rocks in the mid-coastal plain. They yield oils with carbon isotope ratios of around $-22‰$ PDB, which Sutton (1979) attributed to salt marsh biota.

Analysis of the cyclic hydrocarbons by gas chromatography–mass spectrometry reveals a number of wax and resin-derived components in oils from Indonesian basins (Grantham *et al.*, 1983). The presence of these hydrocarbons, which are considered to be the maturation products of resins, is best detected using mass fragmentography. The mass fragmentograms of (a) an Indonesian oil generated from a coaly source rock, (b) an Indonesian oil generated from a non-coaly source rock and (c) a coaly source rock, are presented in Figs. 6, 7 and 8, respectively. The mass spectra of these resin-derived hydrocarbons have base peaks at m/e 163, and subdominant peaks at m/e 259, 217 and 191. Hence the presence of these hydrocarbons is clearly shown by the m/e 163 fragmentogram, which in their absence only detects the hopane series, as in Fig. 7.

OIL SOURCE ROCKS IN THE NORWEGIAN OFFSHORE AREA

It has previously been demonstrated by a number of workers (Cooper and Barnard, 1984; Goff, 1984) that one of the most important oil source rocks in the Central and Northern North Sea is the Late

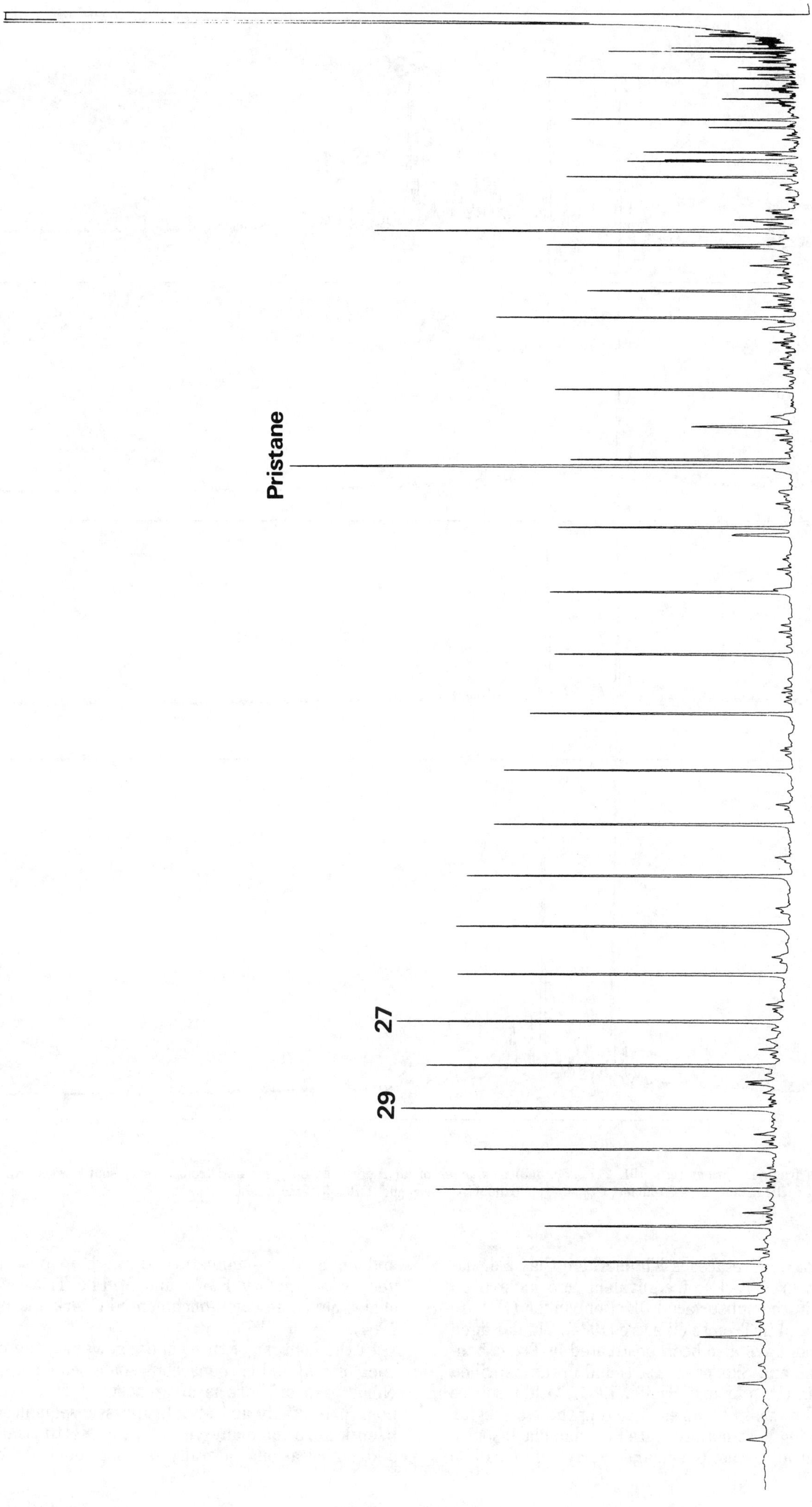

Fig. 5. Gas chromatogram of a waxy oil (34°API gravity) from Indonesia.

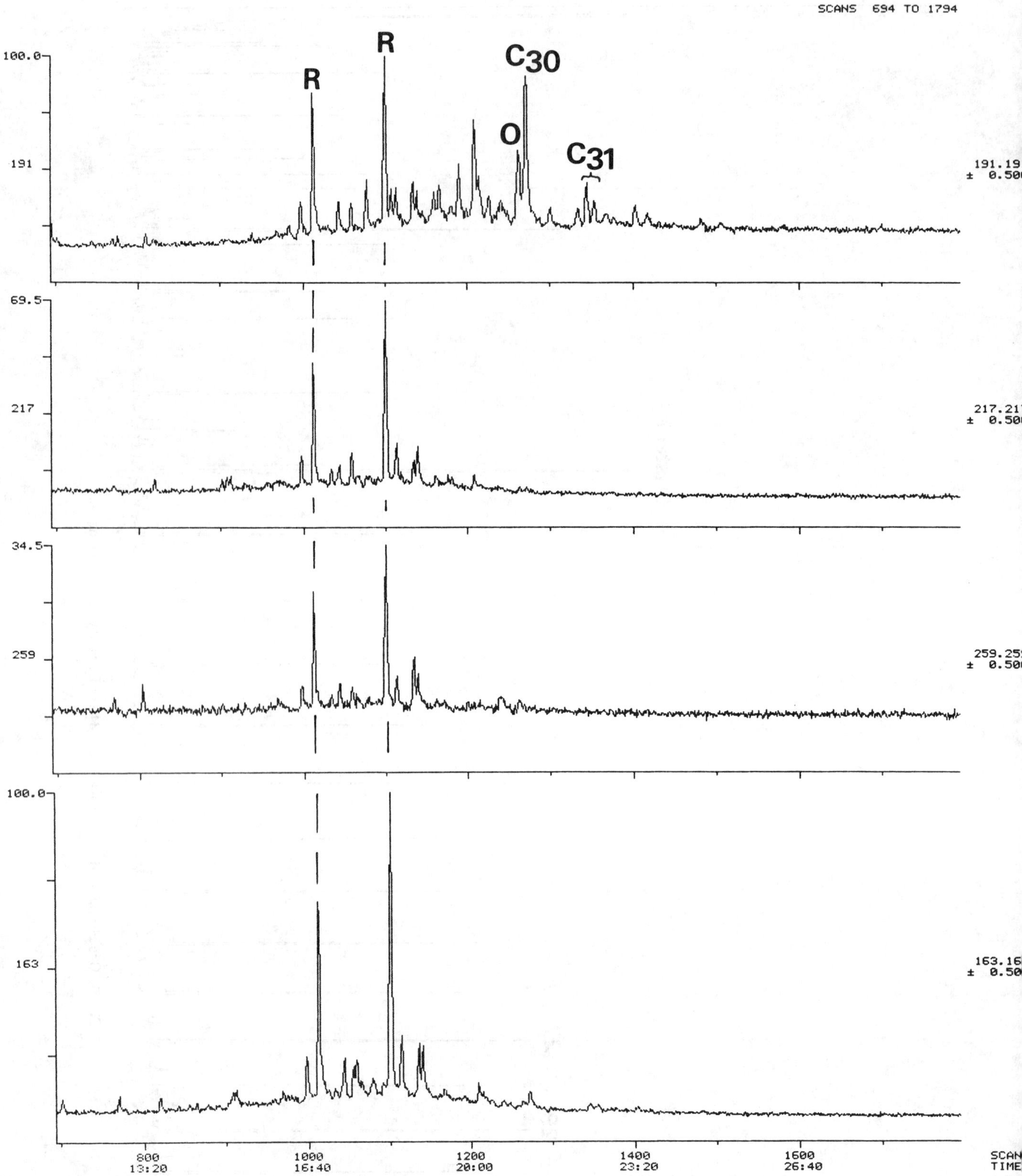

Fig. 6. Mass fragmentograms (*m/e* 191, 217, 259, 163) of alkanes of an Indonesian oil generated from a coaly source rock. Key: R = resin-derived cyclic alkane; O = Oleanane; C_{30} = C_{30} hopane; C_{31} = 22S and 22R C_{31} homohopane.

Jurassic–Early Cretaceous Kimmeridge Clay Formation. Oil source rocks of equivalent age have been described in the subsurface in Haltenbanken (Hollander, 1984) and in Tromsø (Westre, 1984). Similar aged source rocks have also been postulated in Trænabanken (Larsen and Skarpnes, 1984) and in the undrilled Møre Basin (Hamar and Hjelle, 1984). Other source rocks are known in the areas fringing the Norwegian Sea, including post-mature Late Permian black shales which outcrop in East Greenland (Surlyk *et al.*, 1984) and on Svalbard immature through to post-mature very good quality Early and Middle Triassic black shales have been well-documented (Mørk and Bjorøy, 1984).

In the context of this paper, however, the occurrence of coal and carbonaceous shale sequences in the Norwegian offshore is of greaest interest. The carbonaceous Early and Middle Jurassic sequences have been known for many years in the North Sea Basin, and on occasions laterally discontinuous drift coals

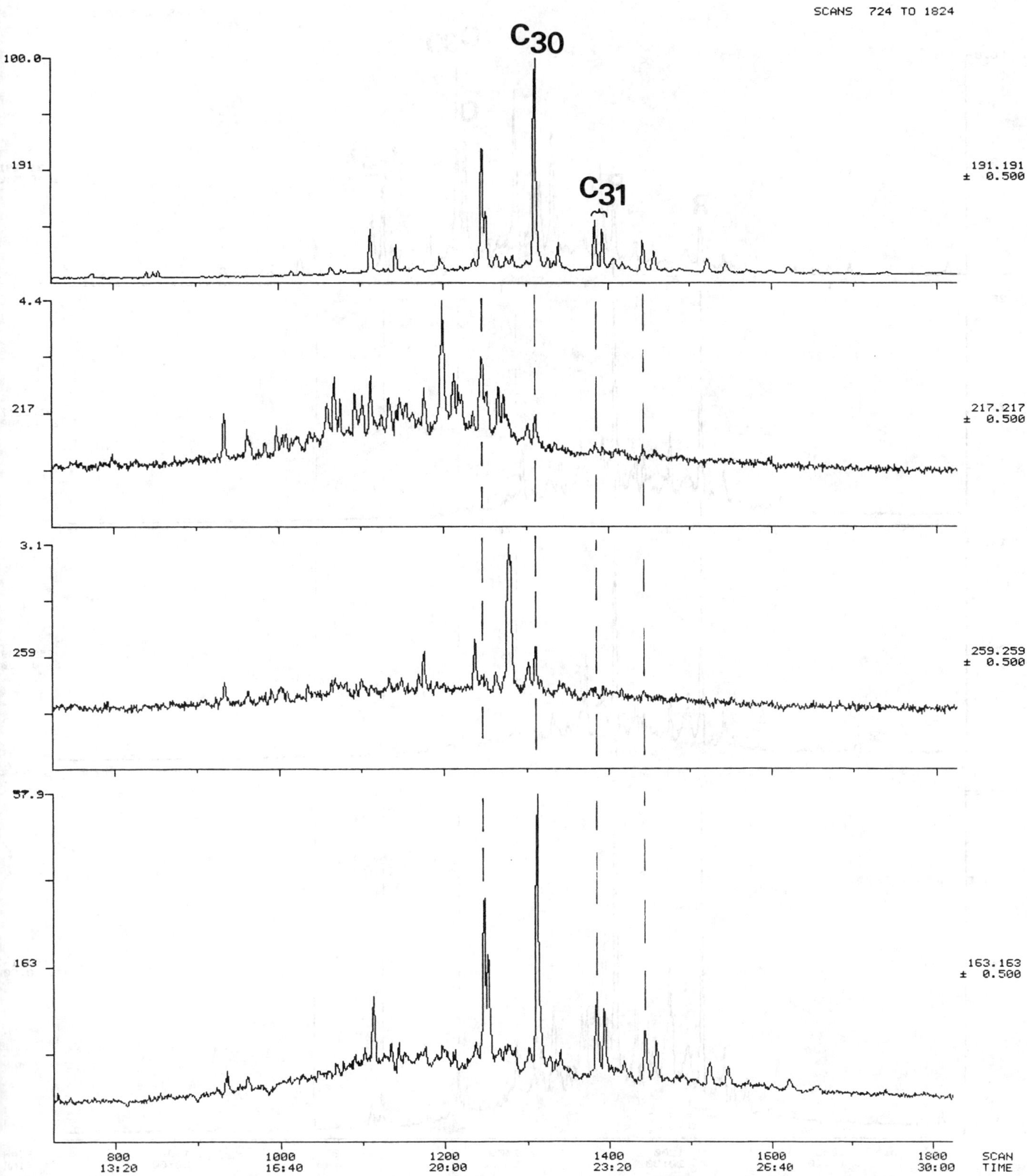

Fig. 7. Mass fragmentograms (*m/e* 191, 217, 259, 163) of alkanes of an Indonesian oil generated from a non-coaly source rock. Key: C_{30} = C_{30} hopane; C_{31} = 22S and 22R C_{31} homohopane.

have been found to contain significant enrichments of liptinitic material. The contribution of these source rocks to discovered hydrocarbon accumulations in this area is not thought to be very great, although several of the larger gas with heavy oil accumulations in the Norwegian sector have not yet been attributed to the usual Kimmeridge Clay Formation source rock.

In the more northerly drilled areas of Haltenbanken and Tromsø, the Kimmeridge Clay Formation equivalent appears to be regionally rather thin and insufficiently buried to generate major amounts of oil. By contrast the Late-Triassic–Early Jurassic coaly sequence is both thick (greater than 500 m) and deeply buried (greater than 3000 m).

A postulated reconstruction of the Late Triassic–Early Jurassic basin is shown in Fig. 9. On the eastern margins of the basin along the coast of Norway red sandstones and shales with several intervals of halite were deposited during the Triassic. In the Rhaetic, the climate appears to have dramatically

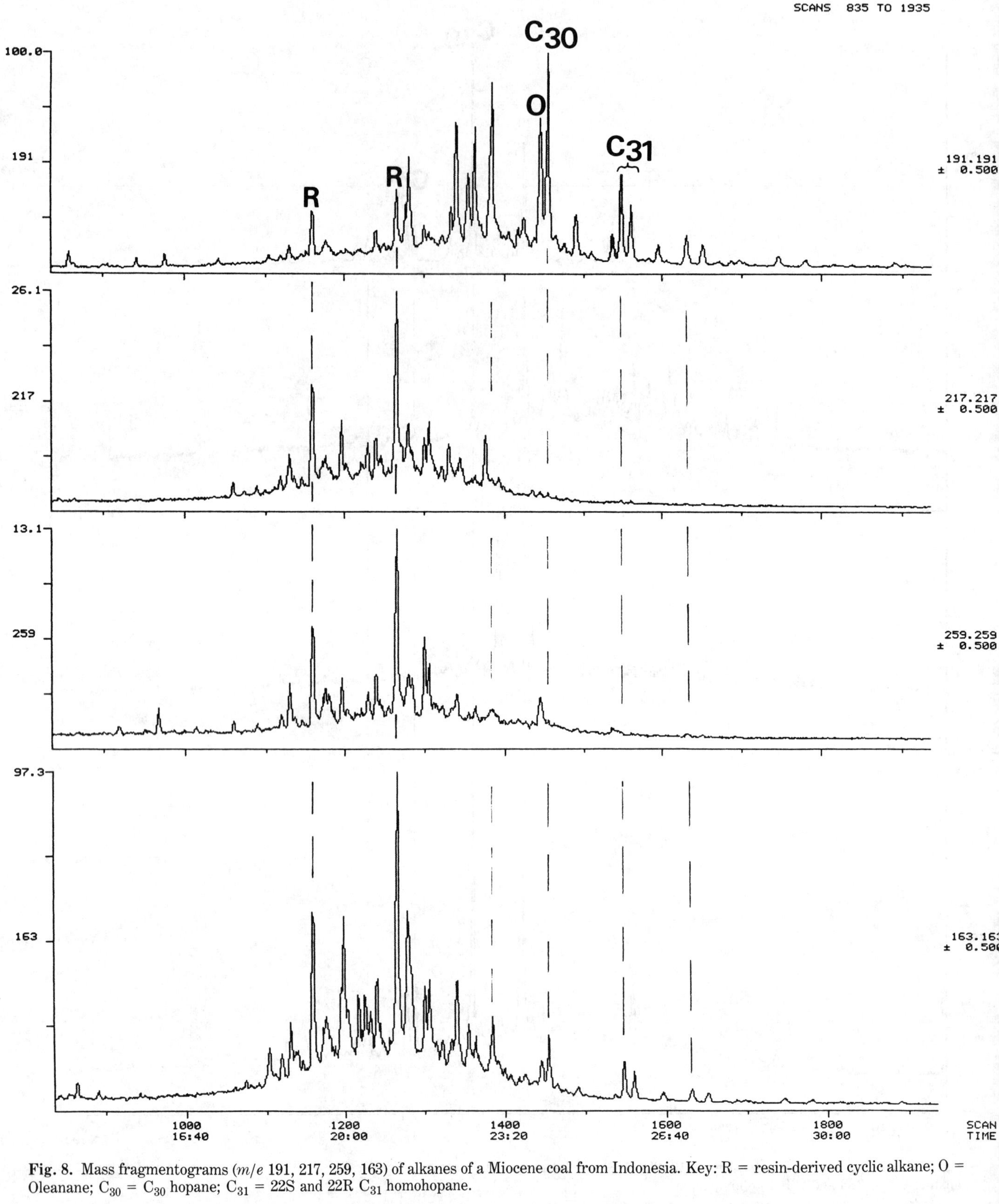

Fig. 8. Mass fragmentograms (m/e 191, 217, 259, 163) of alkanes of a Miocene coal from Indonesia. Key: R = resin-derived cyclic alkane; O = Oleanane; C_{30} = C_{30} hopane; C_{31} = 22S and 22R C_{31} homohopane.

changed (Jacobsen and van Veen, 1984), and this change may be associated with the progressive northward drift of the continental mass. At this time Central and Southern Norway are reported to have been at a latitude of about 30°N with a subtropical climate, while the northern province of Svalbard was in a temperate regime at a latitude of about 45°N.

The change in climate from an arid to an apparently very wet climate in the Norian–Rhaetian, resulted in the establishment of significant fluvial and deltaic sediment transport and depositional systems on the basin margins at the same time as a major marine transgression began. The marine margins of the coastal plain environments are not recognized in the latest Triassic sediments in the North Sea Basin or the west of Scotland. In these areas, only non-marine fluvial and upper delta deposits are seen. Their coastal plain and marine equivalents are presumed to be present in the subsurface in the Møre Basin and in the Faeroes Trough.

From Late Triassic through to Cretaceous time, the area presently covered by the Norwegian Sea was a

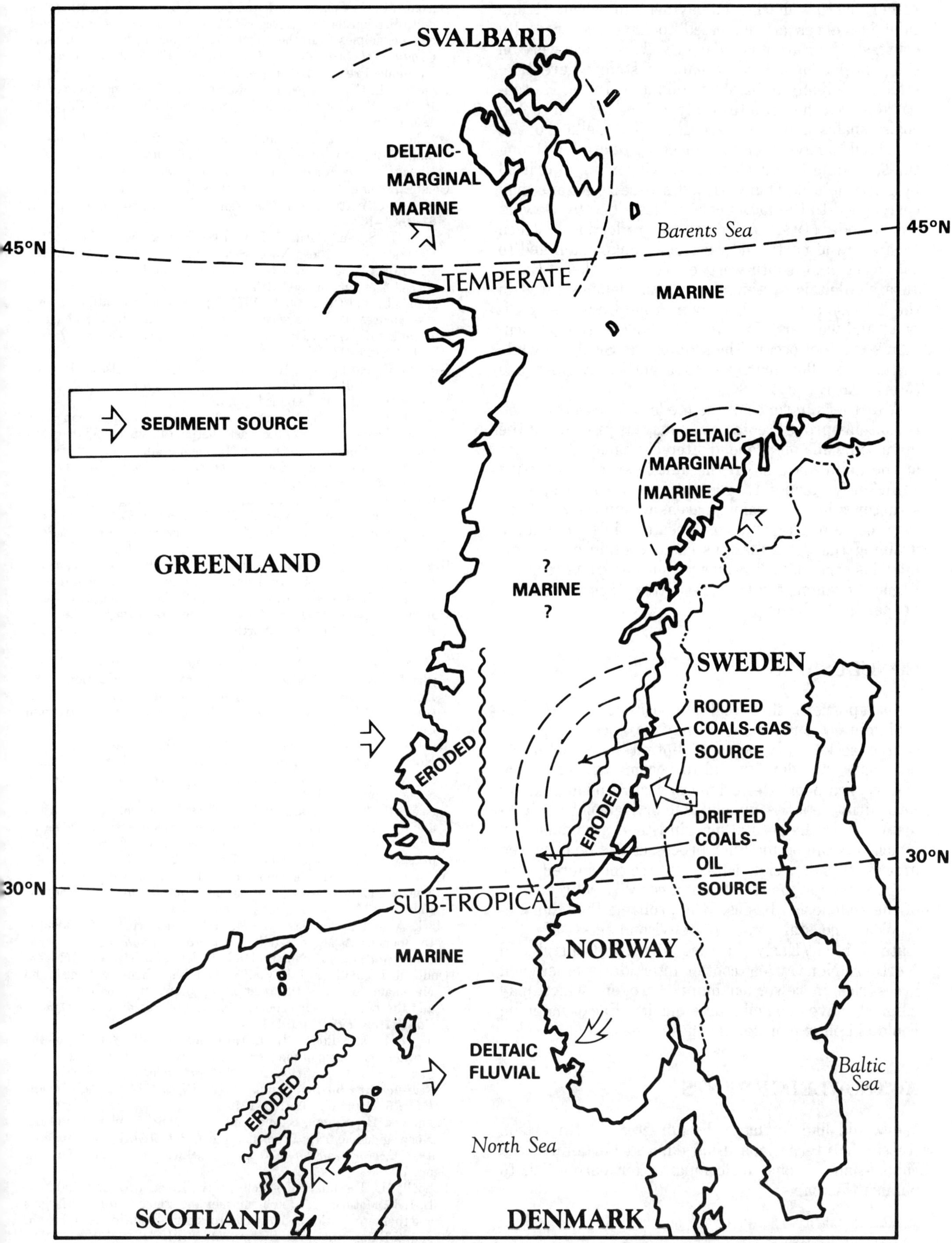

Fig. 9. Palaeolatitudes and palaeoclimates of the Norwegian area during the Triassic and Jurassic.

rather narrow marine embayment in which strong tidal influences are expected to have occurred. By contrast, the Barents Sea and Svalbard areas were on the margins of a large oceanic system where tidal effects are likely to have been limited. In the Haltenbanken area the delta top coaly sediments with rooted coals, shales and sands are seen. The delta front is expected to have been influenced by relatively strong tides, leading to enrichment by winnowing of drifted liptinitic debris. This enriched source rock zone may correspond to the facies change described by Gowers and Lunde (1984) from seismic reflections, which occurs to the north and west of Trænabanken and to the west and south-west of Haltenbanken. Early Jurassic deltaic systems have been demonstrated in the Tromsø area but here with a more northerly climate and weaker tidal effect similar oil-prone enrichments may not occur. The sequence of Svalbard while deltaic to shallow marine is apparently only gas-prone (Mørk and Bjorøy, 1984).

A further inference of the model we have proposed is that gas-prone, rooted coals may be present on the delta top while oil-prone, drifted coals may be present at the delta margins. If the deltaic system is buried sufficiently deeply to generate hydrocarbons, then structures having drainage areas within the delta top may contain gas and condensate while structures draining the delta margins could contain oil. If this model is correct then in an area such as Haltenbanken it may be suggested that the better oil prospects are deeper in the basin.

CONCLUSIONS

It is apparent that coals deposited in particular sedimentary environments may be good quality oil source rocks. A process of liptinite enrichment is necessary in order for coal to become hydrogen-rich and oxygen-poor. Reworking of coastal plain peats to form drift deposits in tidal flat or lagoonal environments along the coastal margin is one mechanism by which this can occur. Such processes can be observed in present Indonesian deltas, and are considered to be responsible for the formation of coaly oil source rocks of the Indonesian basins. It is probable that the subtropical coastal swamp environments of the Late Triassic/Early Jurassic offshore Central and Northern Norway also contain intertidal or inner shelf areas rich in reworked humic kerogen, which may similarly have good oil source quality. Facies mapping could aid oil exploration in these areas.

ACKNOWLEDGEMENTS

We would like to thank the Directors of Robertson Research International Limited, and Pertamina, the Indonesian national oil company, for permission to publish this paper.

REFERENCES

Allen, G. P., Laurier, D. and Thouvenin, J. 1979. Etude sédimentologique du Delta de la Mahakam. *Notes et Mémoires No. 15.* Compagnie Française des Pétroles, Paris, pp. 1–156.

Anderson, J. A. R. 1964. The structure and development of the peat-swamps of Sarawak and Brunei. *J. Trop. Geogl.*, **18**, 7–16.

Barnard, P. C., Collins, A. G. and Cooper, B. S. 1981. Identification and distribution of kerogen facies in a source rock horizon—examples from the North Sea Basin. In: Brooks, J. (ed.), *Organic Maturation Studies and Fossil Fuel Exploration.* Academic Press, London, pp. 271–282.

Benedict, L. G., Thompson, R. R., Shigo, J. J. III and Aikman, R. P. 1968. Pseudovitrinite in Appalachian coking coals. *Fuel*, **47**, 125–143.

Boudou, J. P., Pelet, R. and Letolle, R. 1984. A model of the diagenetic evolution of coaly sedimentary organic matter. *Geochim. Cosmochim. Acta*, **48**, 1357–1362.

Combaz, A. and De Matharel, M. 1978. Organic sedimentation and genesis of petroleum in Mahakam Delta Borneo. *AAPG Bull.*, **62**, 1684–1695.

Cooper, B. S., and Barnard, P. C. 1984. Source rock and oils of the Central and Northern North Sea. In: Demaison, G. and Murris, R. J. (eds.), *Petroleum Geochemistry and Basin Evaluation. AAPG Memoir*, **35**, 303–315.

Durand, B. and Oudin, J. L. 1979. Exemple de migration des hydrocarbures dans une série deltaique: le delta de la Mahakam, Kalimantan, Indonesie. *10th World Petroleum Congr., Proc.*, Vol. 2, pp. 3–11.

Durand, B. and Paratta, M. 1983. Oil potential of coals. In: Brooks, J. (ed.) *Petroleum Geochemistry and Exploration of Europe.* Blackwell Scientific, Oxford, pp. 285–292.

Goff, J. C. 1984. Hydrocarbon generation and migration from Jurassic source rocks in the East Shetland Basin and Viking Graben of the Northern North Sea. In: Demaison, G. and Murris, R. J. (eds.), *Petroleum Geochemistry and Basin Evaluation. AAPG Memoir*, **35**, 273–303.

Gowers, M. B. and Lunde, G. 1984. The geological history of Trænabanken. In: Spencer, A. M. *et al.* (eds.), *Petroleum Geology of the North European Margin.* Norwegian Petroleum Society, pp. 237–251.

Grantham, P. J., Posthuma, J. and Baak, A. 1983. Triterpanes of Far-eastern crude oils. In: Bjorøy, M. *et al.* (eds.), *Advances in Organic Geochemistry 1981.* Wiley, Chichester, pp. 675–683.

Hamar, G. P. and Hjelle, K. 1984. Tectonic framework of the Møre Basin and the Northern North Sea. In: Spencer, A. M. *et al.* (eds.), *Petroleum Geology of the North European Margin.* Norwegian Petroleum Society, pp. 349–359.

Hollander, N. 1984. Geohistory and hydrocarbon evaluation of the Haltenbank area. In: Spencer, A. M. *et al.* (eds.), *Petroleum Geology of the North European Margin.* Norwegian Petroleum Society, pp. 383–388.

Jacobsen, V. and van Veen P. 1984. The Triassic offshore Norway North of 62 °N. In: Spencer, A. M. *et al.* (eds.), *Petroleum Geology of the North European Margin.* Norwegian Petroleum Society, pp. 317–329.

Larsen, R. M. and Skarpnes, O. S. 1984. Regional interpretation and hydrocarbon potential of the Trænabanken area. In: Spencer *et al.* (eds.), *Petroleum Geology of the North European Margin.* Norwegian Petroleum Society, pp. 217–236.

Morley, R. J. 1981. Development and vegetation dynamics of a lowland ombrogenous peat swamp in Kalimantan Tengah, Indonesia. *J. Biogeog.*, **8**, 383–404.

Mørk, A. and Bjorøy, M. 1984. Mesozoic source rocks on Svalbard. In: Spencer, A. M. *et al.* (eds.), *Petroleum Geology of the North European Margin.* Norwegian Petroleum Society, pp. 371–383.

Oudin, J. L. and Picard, P. F. 1982. Genesis of hydrocarbons in the Mahakam delta and the relationship between their distribution and the over-pressure zones. *Indonesian Petrol. Ass., 11th A. Conv. 1982, Proc.*, pp. 181–202.

Roe, G. D. and Polito, L. J. 1979. Source rocks for oils in the Ardjuna sub-basin of the Northwest Java basin, Indonesia. *Proc. Seminar on Generation and Maturation of Hydrocarbons in Sedimentary Basins*, Manila, 1977. *UN/ESCAP, CCOP Techn. Publ.* **6**, Bangkok, pp. 180–194.

Rønnevik, H., Eggen, S. and Vollset, J. 1983. Exploration of the Norwegian shelf. In: Brooks, J. (ed.), *Petroleum Geochemistry and Exploration of Europe.* Blackwell Scientific, Oxford. pp. 71–93.

Schoell, M., Teschner, M., Wehner, H., Durand, B. and Oudin, J. L. 1983. Maturity related biomarker and stable isotope variations and their application to oil/source rock correlation in the Mahakam delta, Kalimantan. In: Bjorøy, M. *et al.* (eds.), *Advances in Organic Geochemistry 1981.* Wiley, Chichester, pp. 156–163.

Seifert, W. K. and Moldowan, J. M. 1981. Paleoreconstruction by biological markers. *Geochim. Cosmochim. Acta*, **45**, 783–794.

Snowdon, L. R. and Powell, T. G. 1982. Immature oil and condensate—modification of hydrocarbon generation for terrestrial organic matter. *AAPG Bull.*, **66**, 775–788.

Stach, E. 1975. *Textbook of Coal Petrology.* Borntraeger, Berlin, 428 pp.

Surlyk, F., Piasecki, S., Rolle, F., Stemmerik, L., Thomsen, E. and Wrang, P., 1984. The Permian Basin of East Greenland. In: Spencer, A. M. *et al.* (eds.), *Petroleum Geology of the North European Margin*. Norwegian Petroleum Society, pp. 303–317.

Sutton, C. 1979. Depositional environments and their relation to chemical composition of Java Sea crude oils. *Proc. Seminar on Generation and Maturation of Hydrocarbons in Sedimentary Basins*, Manila, 1977. *UN/ESCAP, CCOP Techn. Publ.* **6**, Bangkok, pp. 163–179.

Taylor, G. H. 1966. The electron microscopy of vitrinites. *Coal Sci., Adv. Chem., Ser.* **55**, Washington, DC, pp. 274–283.

Teichmüller, M. and Durand, B. 1983. Fluorescence microscopical rank studies on liptinites and vitrinites in peat and coals, and comparison with results of Rock–Eval pyrolysis. *Int. J. Coal Geol.*, **2**, 197–230.

Thomas, B. M. 1982. Land-plant source rocks for oil and their significance in Australian basins. *APEA Journal*, **22**(1), 164–178.

Thompson, R. R. and Benedict, L. G. 1974. Vitrinite reflectance as an indicator of coal metamorphism for cokemaking. *Geol. Soc. Am. Spec. Paper No. 153*, pp. 95–108.

Verdier, A. C., Oki, T. and Suardy, A. 1980. Geology of the Handil field (East Kalimantan, Indonesia). In: Halbouty, M. T. (ed.), *Giant Oil and Gas Fields of the Decade 1968–1978. AAPG Memoir*, **30**, 399–421.

Westre, S. 1984. The Askeladdan gas find—Troms 1. In: Spencer, A. M. *et al.* (eds.), *Petroleum Geology of the North European Margin*. Norwegian Petroleum Society, pp. 33–41.

5 Geochemistry of greater Ekofisk crude oils

W. B. Hughes, A. G. Holba, D. E.Miller and J. S. Richardson
Phillips Petroleum Company, Bartlesville, Oklahoma 74004, USA

Detailed analyses of bulk properties and of molecular composition by gas chromatography and gas chromatography–mass spectrometry of 30 oils from the greater Ekofisk area including Hod, Valhall, Eldfisk, Edda, Ekofisk, West Ekofisk, Albuskjell, Northwest Tor, and Tor fields show the following.

The major differences among the oils are due to maturity, which decreases systematically by field along a NW–SE trend as follows: Albuskjell > Northwest Tor, Tor, West Ekofisk > Ekofisk, Edda, Eldfisk > Valhall, Hod. These maturity variations could be due to differences in the degree of thermal stress experienced by the source rocks, to thermal alteration in the reservoirs, or to some combination of these two factors.

Minor, but definite, source differences exist which are not distributed systematically by field. The oils fall in the following order of increasing terrestrial input and oxicity: Eldfisk (oil 27) < Eldfisk (oil 26) < Ekofisk, Tor, Northwest Tor < Eldfisk, West Ekofisk, Hod, Edda (oil 17) < Albuskjell, Valhall, Ekofisk (oil 12), Edda (oil 18). These variations are the result of locally different source rock subfacies within the Upper Jurassic. Within the Ekofisk, Valhall, Edda, West Ekofisk and Eldfisk structures the Danian and/or Cretaceous reservoirs appear to be differentially filled from these source subfacies.

INTRODUCTION

The discovery of the Ekofisk Field in the Norwegian sector of the North Sea has been cited as one of the major finds of the decade 1968–1978, and the first of the giant oil fields in Western Europe (Van den Bark and Thomas, 1980). Since Ekofisk, many additional fields, including a number of giants, have been discovered in the North Sea, with present estimated recoverable reserves of greater than 40 billion barrels of oil and 130 trillion ft^3 of gas (Dolan, 1983).

Organic geochemistry has clarified the source of Ekofisk oils (Van den Bark and Thomas, 1980), the source in broad terms of central and northern North Sea oils (Oudin, 1976; Barnard and Cooper, 1981, 1984), and the source/maturity relationships among oils from the major grabens (Grantham *et al.*, 1980; Mackenzie *et al.*, 1983). The economic and historical significance of the Ekofisk area merits a comprehensive geochemical study of greater Ekofisk crude oils. The objective of this study was to infer the source (organic input and depositional environment) and thermal maturity relationships of the oils. This was accomplished by obtaining detailed geochemical analyses, including bulk properties, saturated and aromatic molecular fossils and carbon isotopes, of 30 greater Ekofisk crude oils.

SAMPLES AND METHODS OF ANALYSIS

The 30 greater Ekofisk crude oils included in the study are listed in Table 1 and located by well and field of origin on the index map in Fig. 1.

After precipitation of asphaltenes with 20× volumes of *n*-pentane, and filtration, the oils were separated by medium-pressure liquid chromatography on the combination of a cyanophase pre-column and an activated silica main column. The saturate and aromatic fractions were eluted with *n*-hexane from the main column, and the polar fraction was backflushed with dichloromethane from the pre-column. The aromatic fraction was analyzed directly by gas chromatography using a flame photometric detector (GC–FPD) and by gas chromatography–mass spectrometry (GC–MS). The total saturate fraction was analyzed by gas chromatography using a flame ionization detector (GC–FID) and, following removal of *n*-alkanes by molecular sieve adduction, by GC–MS. The *n*-alkanes liberated by dissolution of the molecular sieves with hydrofluoric acid were analyzed by GC–FID.

GC–FID analysis of the total saturate fraction was made on a 30 m SE-54 fused silica capillary column heated from 120°C to 320°C at 4°C/min. The *n*-alkane fraction GC–FID analysis was made on the same column heated from 150°C to 320°C at 8°C/min.

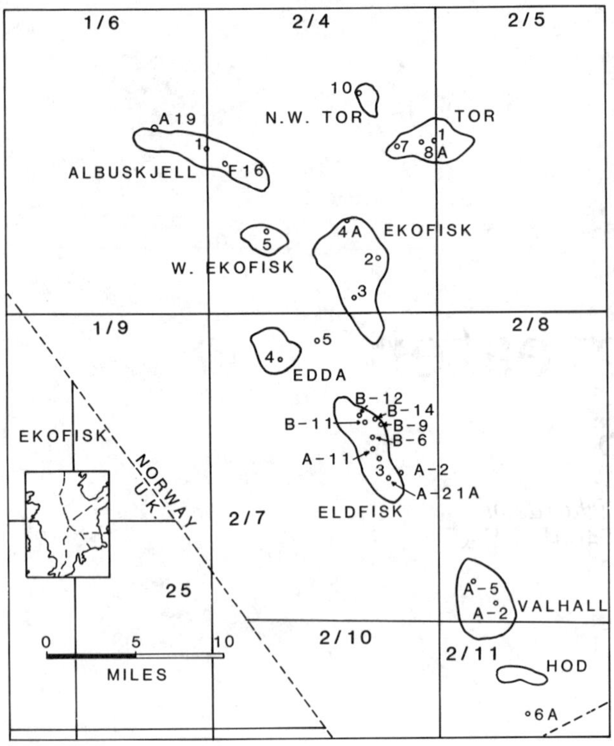

Fig. 1. Location of crude oils analyzed by well and field of origin.

GC–FPD analysis of the aromatic fraction was made on a 60 m DB-1 fused silica capillary column heated from 148°C to 225°C at 5°C/min. followed by heating to 280°C at 10°C/min.

GC–MS analysis of the non-adduct portion of the saturate fraction was made using a 60 m DB-5 fused silica capillary column heated from 150°C to 310°C at 5°C/min. The column was coupled directly to the source of a Hewlett–Packard mass-selective detector operated in the selected ion mode. The aromatic fraction GC–MS analysis was made using a 90 m SE–54 fused silica capillary column heated from 40°C to 310°C at 6°C/min. The column was connected via a heated (300°C) line to a Kratos MS25 mass spectrometer operated in the scan mode (50–500 a.m.u.).

The carbon isotopic compositions of the total crude oils were determined using a Craig combustion system and a MM903 isotope ratio mass spectrometer.

Multivariate factor analysis was accomplished with ARTHUR 81, a statistical software package obtained from Infometrix, Inc.

GEOLOGICAL SETTING OF THE GREATER EKOFISK AREA

The eight fields included in the study lie in a general NW–SE geographic trend within the Central Graben in the southern part of the Norwegian sector of the North Sea (see Fig. 1). Most of the oils are produced from either the Ekofisk Formation of Danian age or the Tor Formation of Maastrichtian age (see Table 1 for reservoir formations and depths). Three oils are from the Hod Formation of Campanian–Turonian age. The remaining oil is from the Jurassic in the Eldfisk field. The Ekofisk, Tor, and Hod reservoirs are fractured chalks located on structural highs, which often overlie salt diapirs. Overpressured Paleocene age shales generally act as seals for the greater Ekofisk reservoirs.

Rapid subsidence, coupled with a eustatic rise in sea level, resulted in the deposition of organic-rich mudstones during the late Jurassic. Reduced circulation within the Central Graben, caused by differential subsidence resulted in anoxic conditions at the sediment–water interface with resulting preservation and burial of the organic matter (Van den Bark and Thomas, 1980; MacKenzie *et al.*, 1983a). These organic-rich Upper Jurassic shales are generally believed to be the source of most crude oils in the northern and central North Sea (Barnard and Cooper, 1981, 1984).

RESULTS

Bulk properties

The geochemical properties we define as bulk properties are listed in Table 2 for the greater Ekofisk oils. The crudes are medium to high API gravity and low in sulfur nitrogen. As observed for many crude oils (Tissot and Welte, 1978), there is an inverse correlation between sulfur and API gravity (see Fig. 2). In addition, this correlation follows a general geographical trend by field with West Ekofisk/Albuskjell having the highest API gravities and lowest sulfur contents and Valhall/Hod having the lowest API gravities and highest sulfur contents (see Fig. 2)*. The nitrogen content shows a linear correlation with the sulfur content ($R^2 = 0.73$).

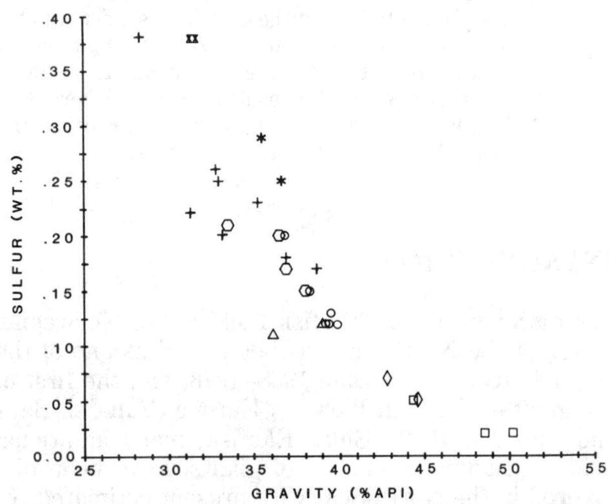

Fig. 2. Variation in gravity with wt% sulfur for greater Ekofisk crude oils (see Table 1 for field symbol key).

The oils are very low in asphaltenes, with the notable exception of the Eldfisk Jurassic oil. The other chemical class compositional data (saturate, aromatic, and polar fractions) are plotted on the triangular diagram in Fig. 3. A trend with field is evident with the Albuskjell oils highest in saturates and lowest in polars, and the Hod oil next to the lowest in saturates and highest in polars. The Eldfisk Jurassic oil is significantly different from the other Eldfisk oils and similar in its chemical class composition to the Hod oil.

*The Northwest Tor oil is very similar to the Tor oils, and is plotted with them in this and all subsequent figures.

TABLE 1. Greater Ekofisk crude oils

Sample number	*Field symbol*	*Field*	*Well*	*Depth (m)*[a]	*Formation*
1	○	Tor	2/4-8A	3194.3–3203.4	Tor
2	○	Tor	2/4-8A	3014.5–3041.9	Ekofisk
3	○	Tor	2/5-1	3154.1–3190.6	Tor
4	○	Tor	2/5-1	3224.8–3233.9	Tor
5	○	Tor	2/4-7	3304.0–3328.4	Tor
6	○	Northwest Tor	2/4-10	3304.0–3322.3	Tor
7	□	Albuskjell	1/6-1	3270.5–3279.6	Tor
8	□	Albuskjell	2/4-F16	3283.9–3302.8	Tor
9	□	Albuskjell	1/6A-19	3266.8–3285.7	Tor
10	◇	West Ekofisk	2/4-5	3124.2–3151.6	Ekofisk
11	◇	West Ekofisk	2/4-5	3270.5–3310.1	Tor
12	⬡	Ekofisk	2/4-2	3169.0–3199.5	Ekofisk
13	⬡	Ekofisk	2/4-3	3218.7–3224.8	Ekofisk
14	⬡	Ekofisk	2/4-3	3127.2–3200.4	Ekofisk
15	⬡	Ekofisk	2/4-4A	3163.8–3203.4	Tor
16	⬡	Ekofisk	2/4-4A	3041.9–3075.4	Ekofisk
17	△	Edda	2/7-4	3105.9–3113.5	Ekofisk
18	△	Edda	2/7-5	3209.5–3218.7	Tor
19	+	Eldfisk	2/7B-12	NA	Ekofisk
20	+	Eldfisk	2/7B-14	NA	Tor
21	+	Eldfisk	2/7B-6	2933.1–2991.0	Tor
22	+	Eldfisk	2/7B-11	NA	Ekofisk + Tor
23	+	Eldfisk	2/7B-9	NA	Ekofisk + Tor
24	+	Eldfisk	2/7A-11	2812.4–2863.6	Ekofisk
25	+	Eidfisk	2/7A-2	NA	Tor
26	+	Eldfisk	2/7A-21A	2816.0–2844.7	Hod
27	+	Eldfisk	2/7-3	3502.2–3517.4	(Jurassic)
28	*	Valhall	2/8-A5	2790.1	Hod
29	*	Valhall	2/8-A2	3293.1–3304.0	Tor
30	XX	Hod	2/11-6A	NA	Hod

NA = Not available. [a]True vertical depth.

TABLE 2. Bulk properties of greater Ekofisk crude oils

Sample number	*Gravity (°API)*	*Sulfur (wt%)*	*Nitrogen (wt%)*	*Class composition (wt%)*				$\delta^{13}C$ *TOT*[a] (‰)
				Saturate	*Aromatic*	*Polar*	*Asphaltene*	
1	39.9	0.12	0.05	70.4	19.4	9.2	0.9	−28.0
2	39.5	0.13	0.05	70.0	19.7	9.0	1.0	−27.8
3	39.4	0.12	0.04	68.7	18.8	11.6	0.8	−27.9
4	39.2	0.12	0.08	70.0	17.9	11.1	1.0	−27.8
5	31.0	0.11	0.05	71.5	18.5	9.7	0.3	−27.3
6	38.2	0.15	0.06	64.6	22.6	12.1	0.6	−28.1
7	44.4	0.05	0.04	78.6	16.2	4.9	0.3	−28.3
8	48.6	0.02	0.01	86.8	12.0	1.2	0.0	−28.2
9	50.3	0.02	0.01	89.0	9.8	1.2	0.0	−28.3
10	42.8	0.07	0.02	58.2	35.8	5.5	0.5	−28.2
11	44.6	0.05	0.04	73.9	18.4	7.2	0.5	−28.2
12	33.5	0.21	0.13	57.7	23.8	16.9	1.6	−27.3
13	36.5	0.20	0.15	56.3	24.5	18.5	0.7	−27.4
14	36.8	0.20	0.13	55.7	22.9	20.6	0.8	−27.4
15	36.9	0.17	0.08	58.4	20.6	19.6	1.4	−27.4
16	38.0	0.15	0.07	54.9	23.8	17.2	0.3	−27.6
17	36.2	0.11	0.6	67.4	19.7	11.7	1.1	−27.4
18	39.0	0.12	0.07	68.3	20.5	11.4	NA	−28.7
19	38.7	0.17	0.15	59.5	24.2	15.5	0.3	−28.0
20	31.4	0.22	0.22	58.9	23.3	17.5	0.2	−27.6
21	36.7	0.20	0.14	57.4	22.9	17.8	1.9	−27.2
22	37.0	0.18	0.13	58.1	24.4	17.2	0.3	−27.2
23	33.0	0.25	0.14	57.2	24.9	16.9	1.0	−27.4
24	35.3	0.23	0.18	56.7	23.0	20.2	0.1	−27.2
25	33.2	0.20	0.12	59.7	22.7	17.2	0.4	−27.2
26	36.6	0.20	0.12	62.0	20.0	17.4	0.6	−27.2
27	28.3	0.38	0.29	42.4	21.2	22.9	13.6	−26.8
28	36.6	0.25	0.17	57.5	22.5	19.6	0.4	−27.3
29	35.5	0.29	0.13	54.5	23.4	21.8	0.3	−27.3
30	31.5	0.38	0.18	49.1	25.1	24.7	1.1	−27.1

NA = Not available. [a]Total crude relative to PDB.

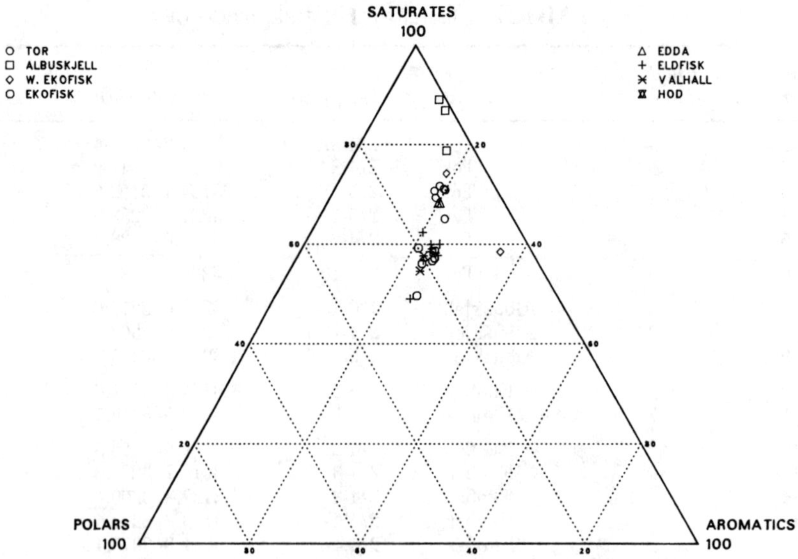

Fig. 3. Distribution by wt% of saturate, aromatic and polar fractions for greater Ekofisk crude oils (see Table 1 for field symbol key).

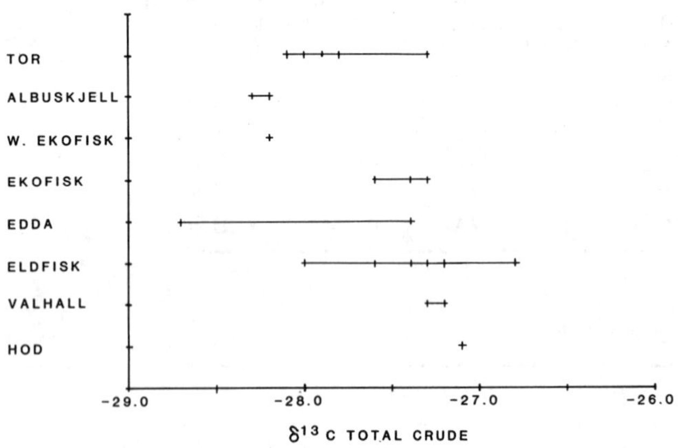

Fig. 4. Variation in $\delta^{13}C$ of total crude with field of origin for greater Ekofisk crude oils.

The $\delta^{13}C$ values for the total crudes span the relatively narrow range of −26.8 to −28.7‰ (see Table 2). A plot of $\delta^{13}C$ values by field (Fig. 4) indicates that, in general, the oils become isotopically lighter in going from Hod to Albuskjell, although there is considerable overlap of the ranges for one field with another.

Molecular properties

The molecular properties of the greater Ekofisk oils are various ratios of the relative concentrations of discrete molecules as determined by GC or GC–MS analysis. For convenience, the molecular properties have been divided into two general categories: source input/depositional environment indicators (Table 3) and maturity indicators (Table 4). No doubt some overlap of these two categories is possible. In the discussions which follow, the molecular properties will be grouped by chemical family.

n-Alkanes/isoprenoids

All of the oils have *n*-alkane distributions biased toward low molecular weight (maximum at C_{16} or C_{17}) and an average odd–even predominance (Scalan and Smith, 1970) close to unity. There is no evidence for biodegradation of the oils.

The pristane/phytane ratios for the oils (see Table 3) determined by GC–FID analysis of the saturate fraction fall in the very narrow range of 1.25–1.50 (excluding the Eldfisk Jurassic oil (pristane/phytane = 0.98) which appears anomalous compared with the Danian–Cretaceous oils). The pristane/phytane ratios do not appear to follow any trend with field of origin, implying that the redox potential of the depositional environments were similar for the source rocks of all the oils. Values of <2 for the pristane/phytane ratio fall within the range associated with anoxic conditions (Didyk *et al.*, 1978).

Steranes

Molecular properties based on the sterane family of chemical fossils were determined by GC–MS analysis of the molecular sieve non-adduct portion of the saturate fraction. The ions monitored for the steranes were m/z 217, 218, and 259.

Representative m/z 217 sterane mass fragmentograms for the greater Ekofisk oils are shown in Fig. 5. The thermal maturities of the oils are reflected in the relative proportions of 24-ethylcholestane isomers (Mackenzie *et al.*, 1983*b*). The C_{29} biological sterol precursor has the 5α, 14α, 17α–20R stereochemistry which is inherited by the C_{29} sterane (peak 8 in Fig. 5) formed during diagenesis. This isomer is converted to the 5α, 14α, 17α–20S configuration (peak 6 in Fig. 5) during catagenesis. Both 20R and 20S 5α, 14α, 17α isomers are converted to the more stable 5α, 14β, 17β isomers (peaks 7 in Fig. 5) with increasing maturation (Petrov *et al.*, 1976; Mackenzie *et al.*, 1980, 1982; Seifert and Moldowan, 1981; van Graas *et al.*, 1982).

The 20S to 20R + 20S ratio of 5α, 14α, 17α, 24-ethylcholestane isomers ranges from 0.25 to 0.52 (see Table 4) for the oils and decreases fairly systematically from the Tor to the Hod field (Fig. 6(A)). The most mature oils (Tor, Albuskjell and oil 11 from West Ekofisk) fall in or near the proposed equilibrium range of 0.50–0.55 (Mackenzie *et al.*, 1980, 1981*a*). The ratio of 5α, 14β, 17β to 5α, 14α, 17α plus 5α, 14β, 17β (20R + 20S) ethylcholestane isomers range from 0.33 to 0.63 (see Table 4) and also show a reasonably sys-

TABLE 3. Molecular properties of greater Ekofisk crude oils: organic input and depositional environment indicators

Sample number	*Pristane phytane*	*DBT/ phen.*	*Diasterane/ sterane*	*Norhopane/ hopane*	*26S/ 28S*	$C_{27}\beta\beta$ / $C_{28}\beta\beta$	*Aquatic/ total*
1	1.40	0.06	1.31	0.43	0.39	1.28	0.69
2	1.40	0.08	1.32	0.48	0.49	1.27	0.69
3	1.40	0.08	1.44	0.47	0.44	1.33	0.71
4	1.50	0.10	1.31	0.44	0.46	1.31	0.68
5	1.50	0.06	1.19	0.48	0.41	1.14	0.68
6	1.36	0.09	1.19	0.44	0.48	1.21	0.70
7	1.38	NA	1.32	0.49	NA	1.09	0.68
8	1.35	0.08	1.28	0.49	0.59	1.26	0.67
9	1.54	0.06	1.41	0.53	0.62	1.20	0.68
10	1.44	0.10	0.91	0.55	0.57	1.12	0.67
11	1.44	0.11	1.41	0.51	0.78	1.42	0.70
12	1.47	0.10	0.37	0.46	0.63	0.92	0.63
13	1.39	0.14	0.55	0.53	0.60	1.28	0.70
14	1.37	0.12	0.54	0.51	NA	1.20	0.68
15	1.45	0.11	0.66	0.51	0.52	1.26	0.70
16	1.38	NA	0.64	0.53	NA	1.27	0.69
17	1.48	0.18	0.73	0.50	0.67	1.39	0.70
18	1.52	0.19	0.52	0.43	0.58	1.13	0.66
19	1.42	0.16	0.38	0.48	0.47	1.01	0.61
20	1.41	0.20	0.41	0.48	0.60	1.22	0.69
21	1.30	0.20	0.45	0.56	0.52	1.29	0.65
22	1.39	0.17	0.46	0.48	0.53	1.28	0.69
23	1.32	0.20	0.40	0.50	0.56	1.23	0.68
24	1.33	0.21	0.37	0.48	0.61	1.01	0.61
25	1.33	0.20	0.42	0.48	0.51	1.12	0.64
26	1.25	0.19	0.57	0.61	0.48	1.21	0.70
27	0.98	0.25	0.44	0.53	0.47	1.37	0.71
28	1.44	0.26	0.17	0.44	0.67	0.86	0.56
29	1.39	0.27	0.22	0.52	0.69	1.02	0.63
30	1.34	0.25	0.40	0.62	0.67	1.11	0.67

Pristane/phytane—GC–FID peak areas from total saturates fraction.
BDT/phen—GC–MS areas of dibenzothiophenes (*m/z* 184)/phenanthrene (*m/z* 178).
Diasterane/sterane—GC–MS (*m/z* 217) areas of peaks (1 + 2)/(3 + 4 + 5) as in Fig. 5.
Norhopane/hopane—GC–MS (*m/z* 191) areas of peaks 27/29 as in Fig. 11.
26S/28S—GC–MS (*m/* 231) areas of peaks 35/37 as in Fig. 14.
$C_{27}\beta\beta/C_{28}\beta\beta$—GC–MS (*m/z* 218) areas of peaks 9/10 as in Fig. 8(A).
Aquatic/total—GC– (*m/z* 218) areas of peaks (9 + 10)/(9 + 10 + 11) as in Fig. 8(A).
See also Table 8 for peak identification.
NA = Not available.

tematic increase with field of origin (see Fig. 6(B)) from Tor/Albuskjell (most mature) to Valhall/Hod (least mature). Based on intensities from the *m/z* 217 mass fragmentograms, the equilibrium value for the 14β, 17β/14α, 17α + 14β, 17β ratio appears to be around 0.63. This is slightly lower than the range of 0.65–0.75 given by Mackenzie *et al.* (1980, 1981*a*). However, this is not surprising since Mackenzie *et al.* used the *m/z* 218 trace to obtain the 14β,17β peak areas, which would give larger numbers for these isomers, hence higher 14β, 17β/14α, 17α + 14β, 17β ratios.

Several significant conclusions about the maturities of the greater Ekofisk oils may be drawn from the sterane isomer distributions. First, there is a major N–S and minor W–E trend from more to less mature. Second, the Tor and Albuskjell oils and oil 11 from West Ekofisk are of advanced maturity, having reached essentially the thermodynamic equilibrium distribution of C_{29} isomers. Third, the West Ekofisk oils are significantly more mature than the Ekofisk oils. Fourth, the range of maturities spanned by the Eldfisk and Ekofisk oils are very similar. Fifth, relative to oil generation, the oils range from early to late stage oil window maturity.

Information about the chemical environment during early diagenesis may be obtained from the *m/z* 217 mass fragmentogram. The conversion of stanols to diasterenes, which are ultimately reduced to diasteranes, in sediments is believed to be catalyzed by acidic sites on certain types of clays (Rubenstein *et al.*, 1975; Sieskind *et al.*, 1979). Thus, the ratio of diasteranes to regular steranes in an oil reflects the activity of unbuffered acidic clays during diagenesis of the source rock organic matter. This ratio was determined for the oils using the C_{28} diasteranes (peaks 1 and 2 in Fig. 5) and C_{28} steranes (peaks 3–5 in Fig. 5). The ratios are given in Table 3 and plotted by field in Fig. 7. The diasteranes make up from 15% (oil 28 from Valhall) to 59% (oil 3 from Tor) of the total steranes. The diasterane/sterane ratio increases very systematically by field from Valhall to Tor; the Hod oil appears to have an anomalously high ratio relative to the Valhall oils. Again, a significant difference between the two West Ekofisk oils is observed. The Ekofisk oils have values similar to the Eldfisk oils, but significantly different from West Ekofisk.

The relative abundances of C_{27}, C_{28}, and C_{29} steranes in oils reflects the carbon number distribution of the sterols present at the time of deposition and thus reflects the type of organic matter input to the source rocks of the oils (Huang and Meinschein, 1979). The

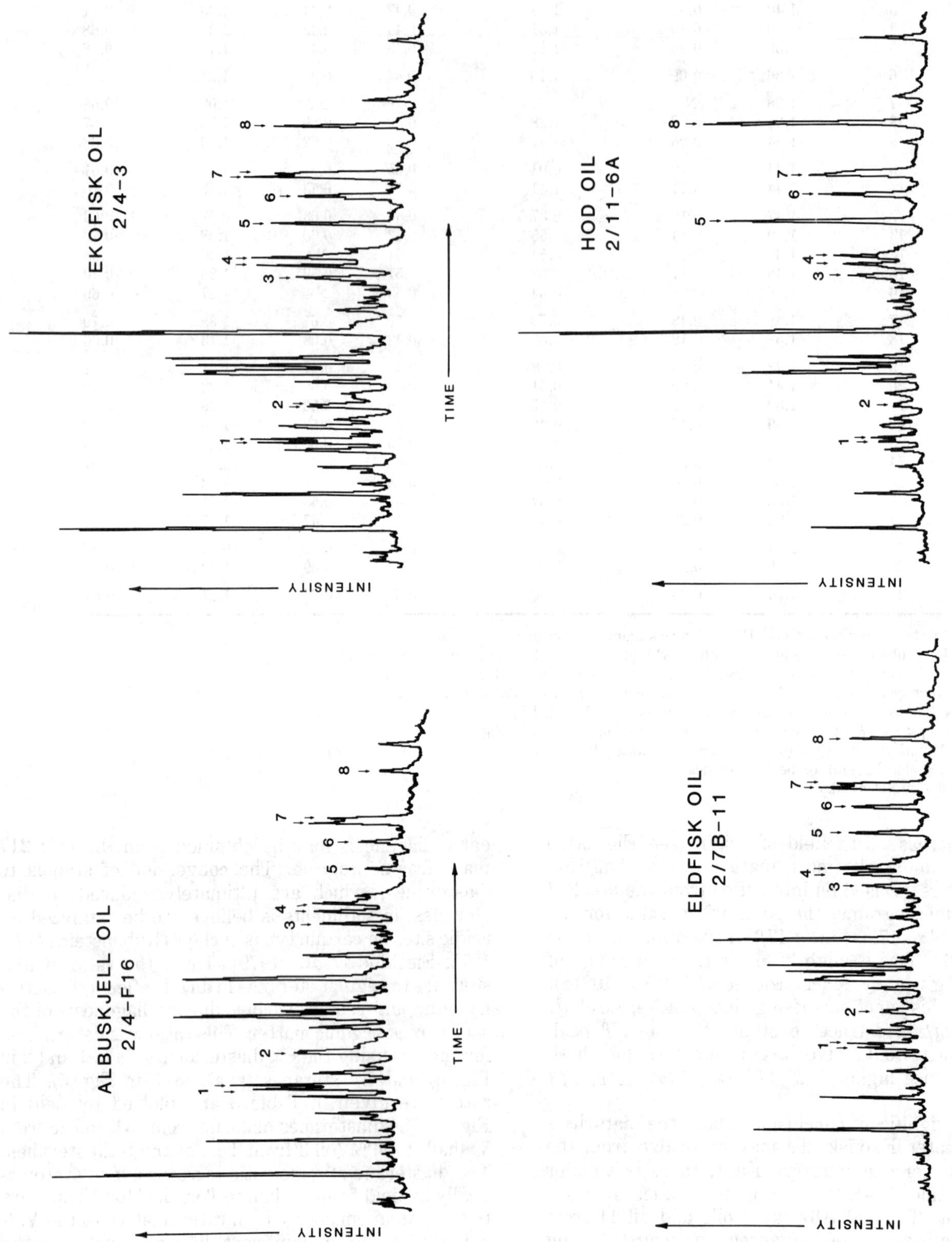

Fig. 5. Representative *m/z* 217 mass fragmentograms for greater Ekofisk crude oils (see Table 8 for peak identification key).

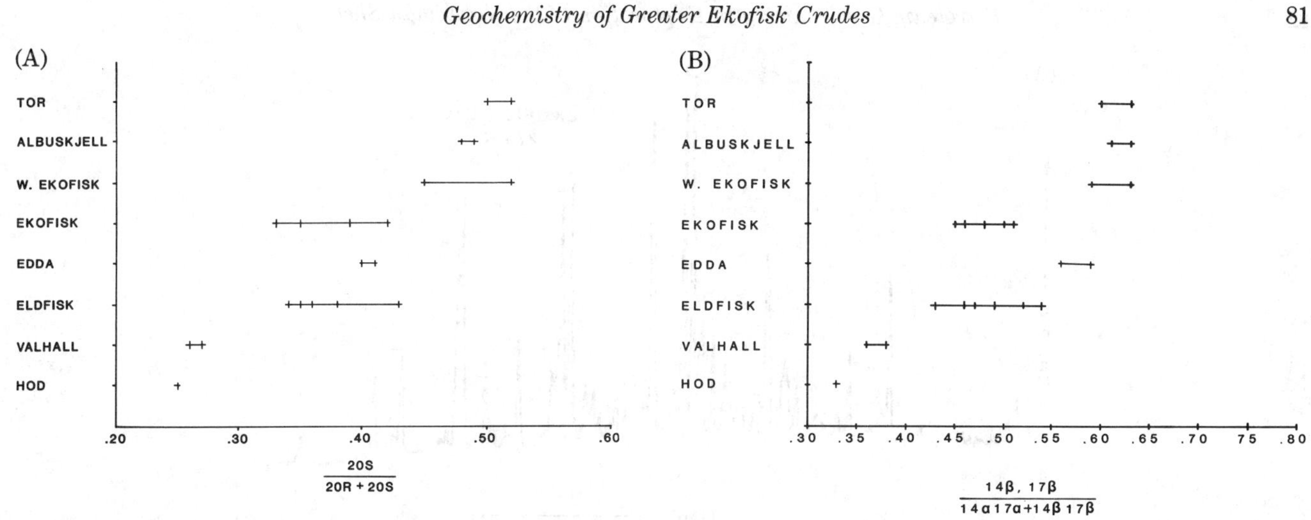

Fig. 6. (A) Variation in 20S/20R + 20S 5α, 14α, 17α-24-ethylcholestane ratio with field of origin for greater Ekofisk crude oils. (B) Variation in ββ/αα + ββ (20R + 20S) 24-ethylcholestane ratio with field of origin for greater Ekofisk crude oil.

sterane carbon number distributions were determined from the relative intensities of C_{27} (peaks 9), C_{28} (peaks 10), and C_{29} (peaks 11) 5α, 14β, 17β (20R + 20S) steranes in the *m/z* 218 mass fragmentogram (a typical *m/z* 218 trace is shown in Fig. 8(A)). The sterane carbon number distributions are plotted on the triangular diagram in Fig. 9. Very similar distributions are found for all the oils, and there is no trend with field of origin.

TABLE 4. Molecular properties of greater Ekofisk crude oils: maturity indicators

Sample number	*20S (20R + 20S)*	*ββ/ (ββ + αα)*	*Ts/Tm*	$C_{20+21}/C_{26+27+28}$	*MPI-1*
1	0.50	0.63	2.49	1.21	0.76
2	0.52	0.60	2.60	1.53	0.78
3	0.50	0.63	2.99	1.47	0.77
4	0.52	0.63	3.07	1.96	0.74
5	0.52	0.63	2.33	1.11	0.75
6	0.50	0.63	2.69	1.17	0.64
7	0.48	0.61	1.88	NA	NA
8	0.48	0.63	2.61	2.54	0.72
9	0.49	0.61	3.06	2.54	0.72
10	0.45	0.59	1.95	1.07	0.74
11	0.52	0.63	2.98	1.53	0.74
12	0.33	0.46	1.09	0.55	0.62
13	0.35	0.48	1.54	0.42	0.60
14	0.33	0.45	1.50	0.48	0.62
15	0.42	0.51	1.69	0.49	0.63
16	0.39	0.50	1.43	NA	NA
17	0.40	0.56	2.25	0.69	0.49
18	0.41	0.59	2.08	0.59	0.49
19	0.38	0.49	1.12	0.36	0.57
20	0.34	0.43	1.29	0.34	0.57
21	0.38	0.46	1.01	0.28	0.56
22	0.35	0.47	1.53	0.42	0.59
23	0.34	0.46	1.19	0.31	0.58
24	0.38	0.49	1.12	0.37	0.54
25	0.36	0.49	1.19	0.27	0.59
26	0.43	0.54	1.07	0.39	0.57
27	0.35	0.52	1.02	0.16	0.58
28	0.27	0.36	0.96	0.21	0.54
29	0.26	0.38	0.98	0.22	0.55
30	0.25	0.33	0.93	0.15	0.56

20S/20R + 20S—GC–MS (*m/z* 217) areas of peaks 8/(6 + 8) as in Fig. 5.
ββ/ββ + αα—GC–MS (*m/z* 217) areas of peaks 7/(6 + 7 + 8) as in Fig. 5.
Ts/Tm—GC–MS (*m/z* 191) areas of peaks 24/26 as in Fig. 11.
$C_{20+21}/C_{26+27+28}$—GC–MS (*m/z* 231) areas of peaks (33 + 34)/(35 + 36 + 37 + 38 + 39) as in Fig. 14.
MPI-1—GC–MS (*m/z* 178 and 192) peak areas where MPI-1 = 1.5(3 − +2-methylphenanthrenes)/(phenanthrene + 9- + 1-methylphenanthrenes).
Also see Table 8 for peak identifications.
NA = Not available.

The ratio of C_{27} and C_{28} to the total of C_{27}, C_{28}, and C_{29}ββ steranes generally reflects the ratio of organic matter derived from aquatic organisms to the sum of aquatic- and terrestrially-derived organic matter. The aquatic/total ratios for the greater Ekofisk oils (see Table 3) range from 0.60 to 0.72, suggesting a predominantly aquatic input with a significant terrestrial contribution.

However, a recent example indicates that oils from pre-Ordovician carbonate sources, which predate the appearance of extensive land vegetation, have high C_{29} sterane contents (McKirdy *et al.*, 1983). In the present case, significant terrestrial input is supported by our observation of conifer diterpenoids in three of the oils (see below) and by reports of land plant-derived kerogen in the presumed Upper Jurassic (Kimmeridge Clay Formation and equivalents) source rocks (Van den Bark and Thomas, 1980; Barnard and Cooper, 1984).

The sterane carbon number distribution was also determined from the *m/z* 259 mass fragmentogram (Fig. 8(B)) using the relative intensities of the C_{27} (peaks 12 and 13), C_{28} (peaks 14 and 15), and C_{29} (peaks 16 and 17) diasteranes. The diasterane carbon number distributions are also plotted on the triangular diagram in Fig. 9 for comparison with the ββ steranes. The distributions are quite similar, and the six oils which fall outside the main cluster (toward the C_{29} vertex)

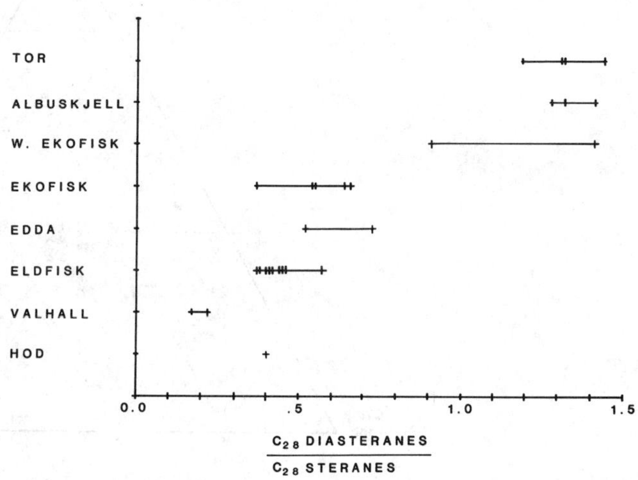

Fig. 7. Variation in C_{28} diasterane/C_{28} sterane ratio with field of origin for greater Ekofisk crude oils.

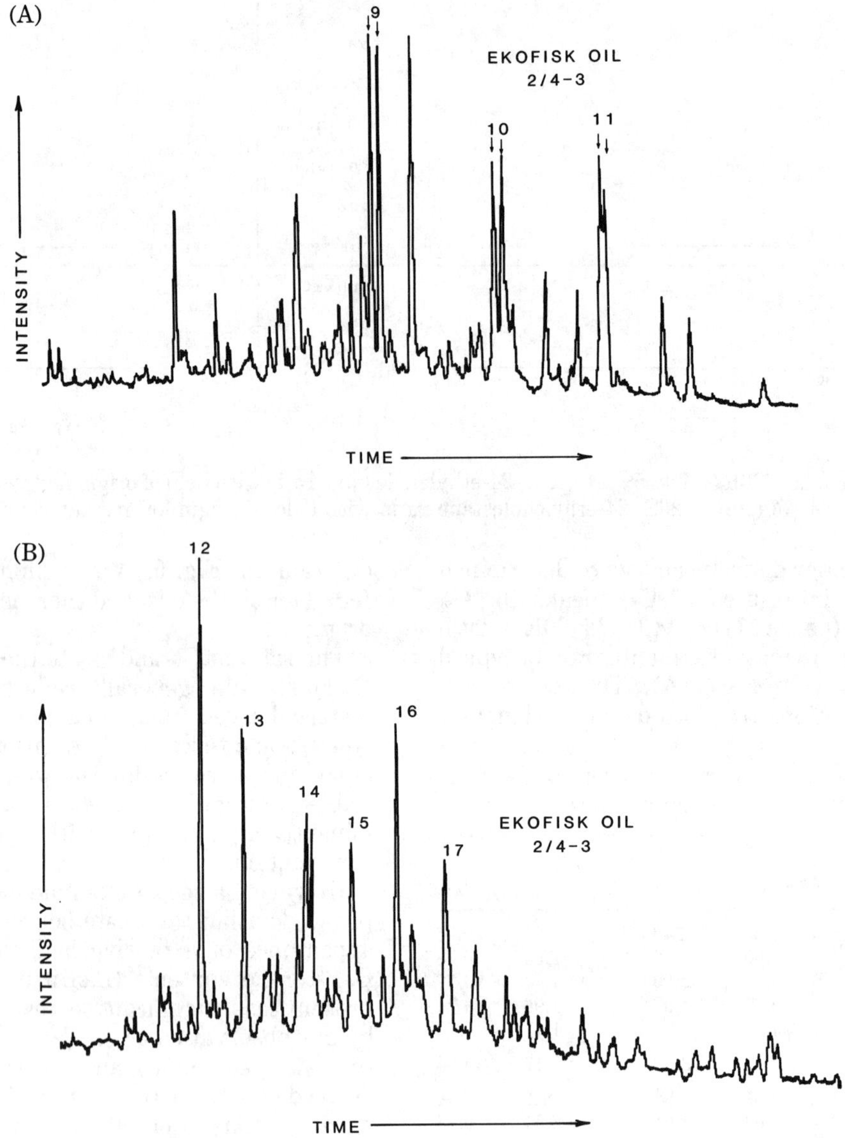

Fig. 8. (A) m/z 218 mass fragmentogram for 2/4–3 Ekofisk oil (see Table 8 for peak identification key). (B) m/z 259 mass fragmentogram for 2/4–3 Ekofisk oil (see Table 8 for peak identification key).

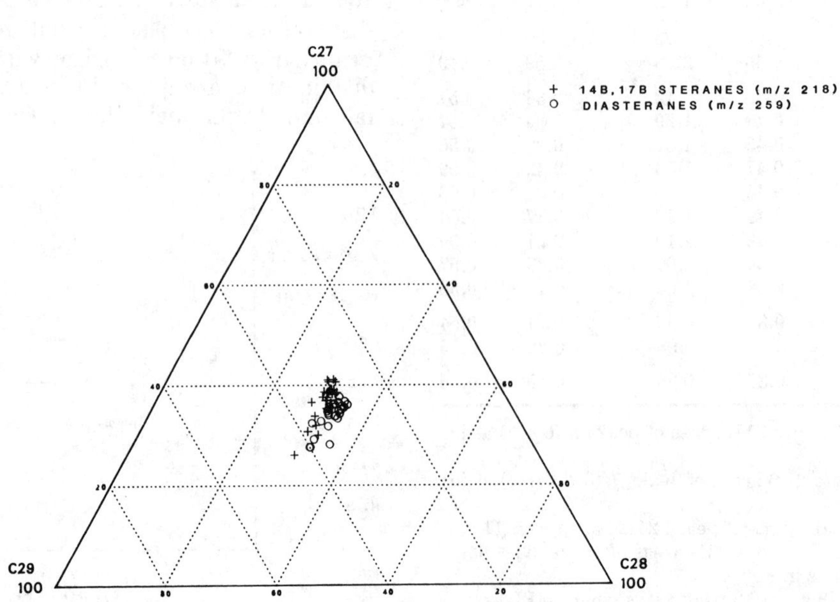

Fig. 9. Carbon number (C_{27}, C_{28}, C_{29}) distributions of 15 α,14β,17β, (20R + 20S) steranes (+) and of 13β,17 α–20 (R + S) diasteranes (O).

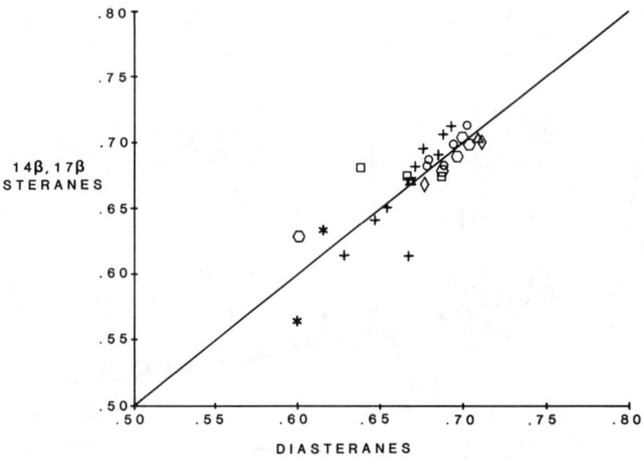

Fig. 10. Relationship between aquatic/total ratio (C_{27} + C_{28}/C_{28} + C_{28} + C_{29}) calculated from 5α, 14β, 17β (20R + 20S) steranes and 13β, 17α (20R + 20S) diasteranes.

are the same in both cases. These samples are oil 12 from Ekofisk, oils 19, 24, and 25 from Eldfisk, and oils 28 and 29 from Valhall (see Table 1).

The aquatic/total sterane ratios calculated from the ββ sterane and diasterane distributions are compared graphically in Fig. 10; the solid line in the figure represents a 1 : 1 equivalence. As this plot shows, the two ratios give essentially the same result. This indicates no preferential conversion by carbon number of sterols to diasteranes during diagenesis.

Terpanes

The representative *m/z* 191 mass fragmentograms shown in Fig. 11 contain significant amounts of the hopane series of pentacyclic triterpenoids (peaks 24, 26, 27, 29, 31 + 32) commonly found in crude oils and rock extracts. Present in variable amounts are a series of C_{23}–C_{26} and C_{28}–C_{30} tricyclic terpenoids (peaks 18–23 and 25). Of low abundance in all the oils are moretane (peak 30) and normoretane (peak 28). In addition, several other compounds detected in the *m/z* 191 mass fragmentogram are of interest. These are bisnorhopane (peak Y), previously reported (Grantham *et al.*, 1979) in North Sea oils; a 24 carbon tetracyclic terpenoid (peak X) whose structure has recently been established (Rinaldi, 1984); two 30 carbon pentacyclic triterpenoids (peaks A and B) of unknown structure (molecular ions *m/z* 412, base peaks *m/z* 191); and a 29 carbon pentacyclic triterpenoid (peak Z) of unknown structure (molecular ion *m/z* 398, base peak *m/z* 191).

Overall, the *m/z* 191 mass fragmentograms for the oils are fairly similar. Features in common include a predominance of hopanes over tricyclics, hopane > norhopane, and low concentrations of moretanes. However, definite trends exist in the *m/z* 191 fingerprint which are related to the geographical location of the field of origin of the oil. Although the range of values within some of the fields is fairly large, the ratio of 18 α- to 17 α-trisnorhopane (Ts/Tm) increases in going north from Hod to Albuskjell and Tor (see Table 4 and Fig. 12(A)). The Ts/Tm ratios for West Ekofisk and Edda are higher than for Ekofisk. There is also a considerable difference in this ratio for the two West Ekofisk oils. The Ts/Tm ratio has been proposed as a maturity indicator (Seifert and Moldowan, 1978) with higher values indicating greater maturity. The major N–S and minor W–E geographic trend toward higher maturity in Ts/Tm is the same trend observed in the 20R/20R + 20S and ββ/αα + ββ sterane maturity indicators.

Several other terpane molecular ratios show similar geographic trends as Ts/Tm with field of origin (see Table 5). These are the ratio of the C_{23}–C_{26} tricyclic diterpenoids (peaks 18–21 in Fig. 11) to hopane (peak 29), the ratio of the unknown C_{30} pentacyclic triterpenoid (peak B) to hopane, and the ratio of the unknown C_{29} pentacyclic triterpenoid (peak Z) to norhopane (peak 27). In all cases, these ratios generally decrease from north to south, as does Ts/Tm. In contrast to Ts/Tm, however, Albuskjell shows somewhat higher values than Tor and West Ekofisk in the tricyclic/hopane and compound B/hopane ratios. Also in contrast to Ts/Tm, the values for the other three ratios are similar for the Edda and Ekofisk oils. In general, however, the ratios of compounds B and Z and of the C_{23}–C_{26} tricyclics to hopane behave very similarly to Ts/Tm; these three ratios versus Ts/Tm have linear correlation coefficients (R^2) of 0.59–0.77.

Contrary to the other molecular terpane ratios (see Tables 4 and 5), the ratio of norhopane (peak 27) to hopane (peak 29) remains relatively constant from field to field (see Table 3 and Fig. 12(B)). A similar constancy by field was observed in the carbon number distribution of steranes of the same configurations.

The bisnorhopane (peak Y) to hopane (peak 29) ratio (see Table 5) shows an intermediate behavior. No systematic geographic trend by field of origin is observed; most of the oils have values in the range 0.07–0.23, and the ratio for the Hod oil is in the same range as those of the Tor oils. The Albuskjell and West Ekofisk oils show significantly higher values (0.29–0.44). Only a poor linear correlation is found with Ts/Tm ($R^2 = 0.19$). However, the most mature oils based on the Ts/Tm ratio tend to have the highest bisnorhopane /hopane ratios.

Aromatics

The aromatic fractions of the greater Ekofisk crude oils were analyzed by GC–FPD and by GC–MS in the full scan mode. The fingerprint ions used in this study were *m/z* 178, 184, 192, 198 and 231.

Representative gas chromatograms of the aromatic fraction using a flame photometric (sulfur-selective) detector are shown in Fig. 13. The major organosulfur components are benzothiophenes (peaks 40); dibenzothiophene (peak 41); and a series of methyl- (peaks 42–44), dimethyl- (peaks 45), and trimethyl- (peaks 46) dibenzothiophenes. In going from Hod to Eldfisk to Ekofisk there is a disappearance of the benzothiophenes, a decrease in dibenzothiophene relative to methyl derivatives, and a redistribution of isomers among the methyl-, dimethyl-, and trimethyldibenzothiophenes. The last change was quantitated from the *m/z* 198 mass fragmentogram, and is reflected in the ratio of the 1-methyl isomer to the sum of the 1-, 2-, 3-, and 4-methyl isomers (MDBTR in Table 5). It was observed previously (Hughes, 1984) that oils with the distribution of methyldibenzothiophene isomers found in the Hod oil, i.e. peak 42 > peak 44 > peak 43, are from carbonate or calcareous shale sources or sources deposited in highly reducing environments. It was also found that redistribution of the methyldibenzothiophenes to the pattern in the Albuskjell oil, i.e. peak 42 >> peak 43 > peak 44, results from advanced maturation of such oils. Changes in the distribution of dibenzothiophenes have also been observed by Radke *et al.* (1982*a*).

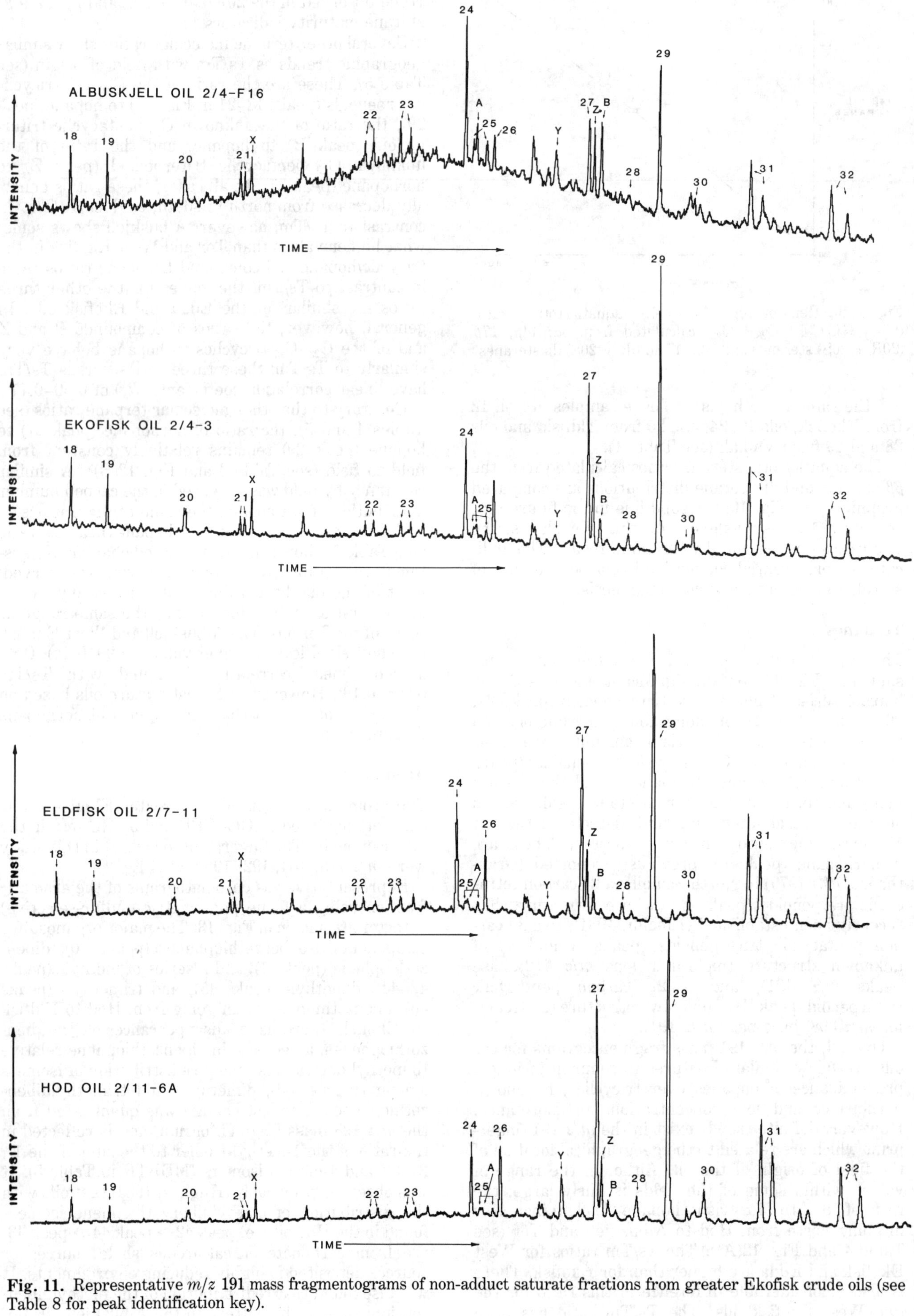

Fig. 11. Representative m/z 191 mass fragmentograms of non-adduct saturate fractions from greater Ekofisk crude oils (see Table 8 for peak identification key).

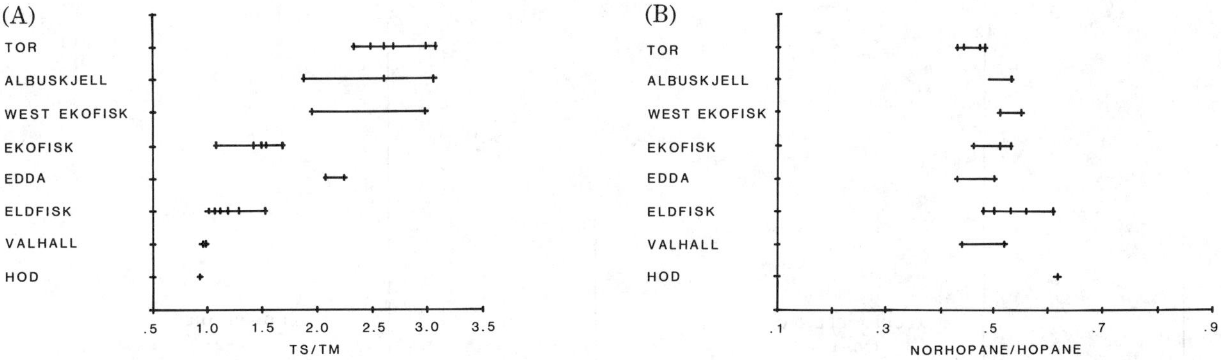

Fig. 12. (A) Variation in 18α,21β (Ts) to 17α,21β (Tm) 22,29,30-trisnorhopane ratio with field of origin for greater Ekofisk crude oils. (B) Variation in 17α,21β-30-norhopane to 17α,21β-hopane ratio with field of origin for greater Ekofisk crude oils.

The relative distribution of phenanthrene and its 1-, 2-, 3-, and 9-methyl derivatives has been proposed as a maturity indicator, termed the methylphenanthrene index (MPI-1), by Radke *et al.* (1982a). The MPI-1 values for the greater Ekofisk oils (see Table 4) are highest in the northern (Tor, Albuskjell) and West Ekofisk oils; intermediate in Ekofisk; and lowest in Eldfisk, Valhall, and Hod. The Edda and Northwest Tor oils have anomalously low values. The MPI-1 indicator does not differentiate Valhall and Hod oils from Eldfisk as do the sterane and triterpane maturity indicators in Table 4.

The ratio of C_{20} + C_{21} to C_{26} + C_{27} + C_{28} triaromatic steranes from the *m/z* 231 mass fragmentograms (peaks 33 + 34/35–39 in Fig. 14) decreases systematically from Albuskjell to Hod (see Table 4). This molecular ratio is similar to the C_{20}/C_{28} + C_{20} aromatic sterane ratio used by Mackenzie *et al.* (1981*b*, *c*) as a maturity index operative in the early to late oil window range; both indicators increase with maturity. Based on the C_{20-21}/C_{26-28} ratio, Ekofisk oils are more mature than Eldfisk oils and West Ekofisk and Edda oils are more mature than Ekofisk oils. This maturity trend matches that observed for the 20R/20R + 20S

TABLE 5. Selected molecular properties of greater Ekofisk crude oils

Sample number	*MDBTR*	*Bisnorhopane/ hopane*	*29 Pentacyclic/ Norhopane*	*Triterpane B/ hopane*	*Tricyclics/ hopane*
1	0.10	0.23	1.12	0.39	1.48
2	0.09	0.17	1.02	0.37	0.86
3	0.13	0.16	0.96	0.37	1.41
4	0.10	0.13	0.98	0.35	1.25
5	0.11	0.23	1.04	0.41	1.42
6	0.12	0.16	0.94	0.31	1.27
7	NA	0.44	1.02	0.38	NA
8	0.10	0.43	0.91	0.52	1.58
9	0.11	NA	0.90	0.59	2.28
10	0.13	0.29	0.54	0.39	0.89
11	0.09	0.31	0.95	0.41	1.85
12	0.15	0.11	0.37	0.19	NA
13	0.17	0.09	0.47	0.11	0.66
14	0.16	0.08	0.44	0.11	0.38
15	0.16	0.08	0.49	0.12	0.61
16	NA	0.09	0.46	0.12	0.85
17	0.18	0.12	0.47	0.13	0.85
18	0.18	0.10	0.45	0.13	0.25
19	0.18	NA	0.51	0.19	NA
20	0.19	0.08	0.36	0.08	0.17
21	0.21	0.19	0.52	0.20	NA
22	0.19	0.07	0.38	0.08	0.17
23	0.20	0.07	0.34	0.07	0.14
24	0.19	NA	0.51	0.19	NA
25	0.20	NA	0.56	0.20	NA
26	0.20	NA	0.67	0.30	NA
27	0.26	0.12	0.31	0.04	0.27
28	0.23	0.07	0.33	0.07	0.08
29	0.25	0.10	0.32	0.06	0.08
30	0.25	0.17	0.28	0.05	0.19

MDBTR—GC–MS (*m/z* 198) areas of peaks 1-methyldibenzothiophene/(4 + 2 + 3 + 1-methyldibenzothiophenes).
Bisnorhopane/hopane—GC–MS (*m/z* 191) areas of peaks Y/29 as in Fig. 11.
29 Pentacyclic/norhopane—GC–MS (*m/z* 191) areas of peaks Z/27 as in Fig. 11.
Triterpane B/hopane—GC–MS (*m/z* 191) areas of peaks B/29 as in Fig. 11.
Tricyclics/hopane—GC–MS (*m/z* 191) areas of peaks (18 + 19 + 20 + 21)/29 as in Fig. 11.
Also see Table 8 for peak identifications.
NA = Not available.

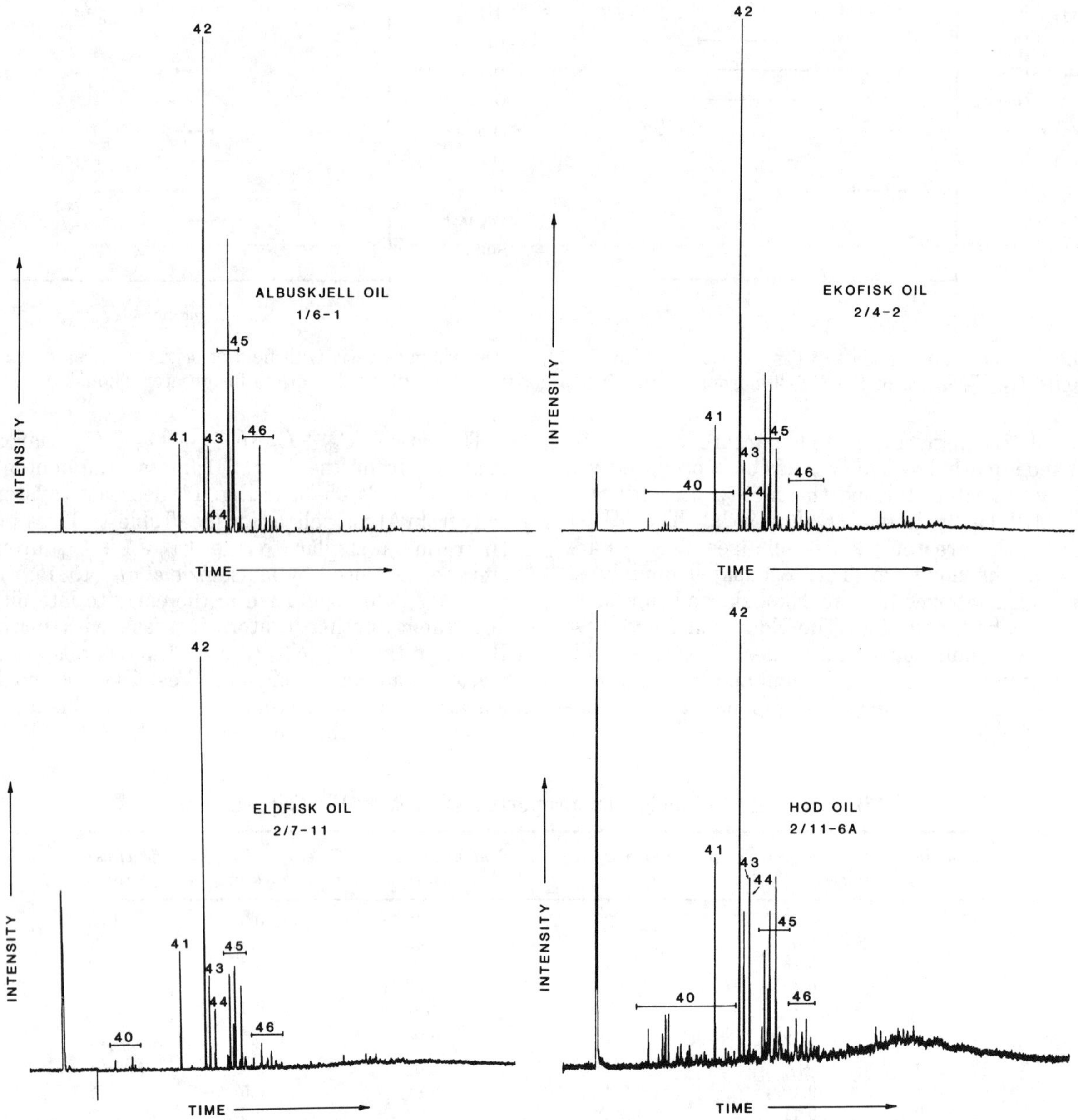

Fig. 13. Representative GC–FPD chromatograms of aromatic fractions from greater Ekofisk crude oils (see Table 8 for peak identification key).

and $\beta\beta/\alpha\alpha + \beta\beta$ sterane ratios and the Ts/Tm trisnorhopane ratio.

The relative distribution of C_{26}–C_{28} triaromatic steranes (peaks 35–39 in Fig. 14) is similar for all the oils, suggesting a similar source organic input (Mackenzie *et al.*, 1983*a*). Some variation in internal ratios is observed, however. For example, the ratio of the C_{26}, 20S (peak 35) to the C_{28}, 20S (peak 37) triaromatic steranes (see Table 3) varies from 0.39 (Tor, sample 1) to 0.78 (West Ekofisk, sample 11). The variation is not systematic by field, or with maturity indicators such as $\beta\beta/\alpha\alpha + \beta\beta$ steranes and Ts/Tm, or with other source indicators, such as the ratio of C_{27}/C_{29} $\beta\beta$ steranes (peaks 9 and 11, respectively, in Fig. 8(A)). It appears that different factors may control the carbon number distribution of triaromatic and saturated steranes.

The Valhall and Hod oils have significant levels of the conifer-derived (Thomas, 1969) diterpenoid retene and, in low concentrations, its precursors tetrahydroretene and simonellite. The absence of these diterpenoids in the other greater Ekofisk oils could be due to a source (organic input) difference, to the destruction of these compounds in the other oils, which are more mature than the Valhall/Hod oils, or to some combination of source and maturity.

The ratio of dibenzothiophene to phenanthrene is thought to be an indicator of source rock lithology (Holba, unpubl. data). The low values for the greater Ekofisk oils (see Table 3) are in the range for shaly source rocks. The dibenzothiophene/phenanthrene ratio increases systematically from Albuskjell and Tor to Valhall and Hod, being similar in this regard to the maturity indicators. This ratio correlated highly ($R^2 = 0.89$) with the methyldibenzothiophene isomer ratio (see Table 5).

DISCUSSION

Source versus maturity effects

A classical problem in organic geochemistry is to disentangle the effects of source (which we take to include both organic input and depositional environment) and maturity on bulk and molecular properties

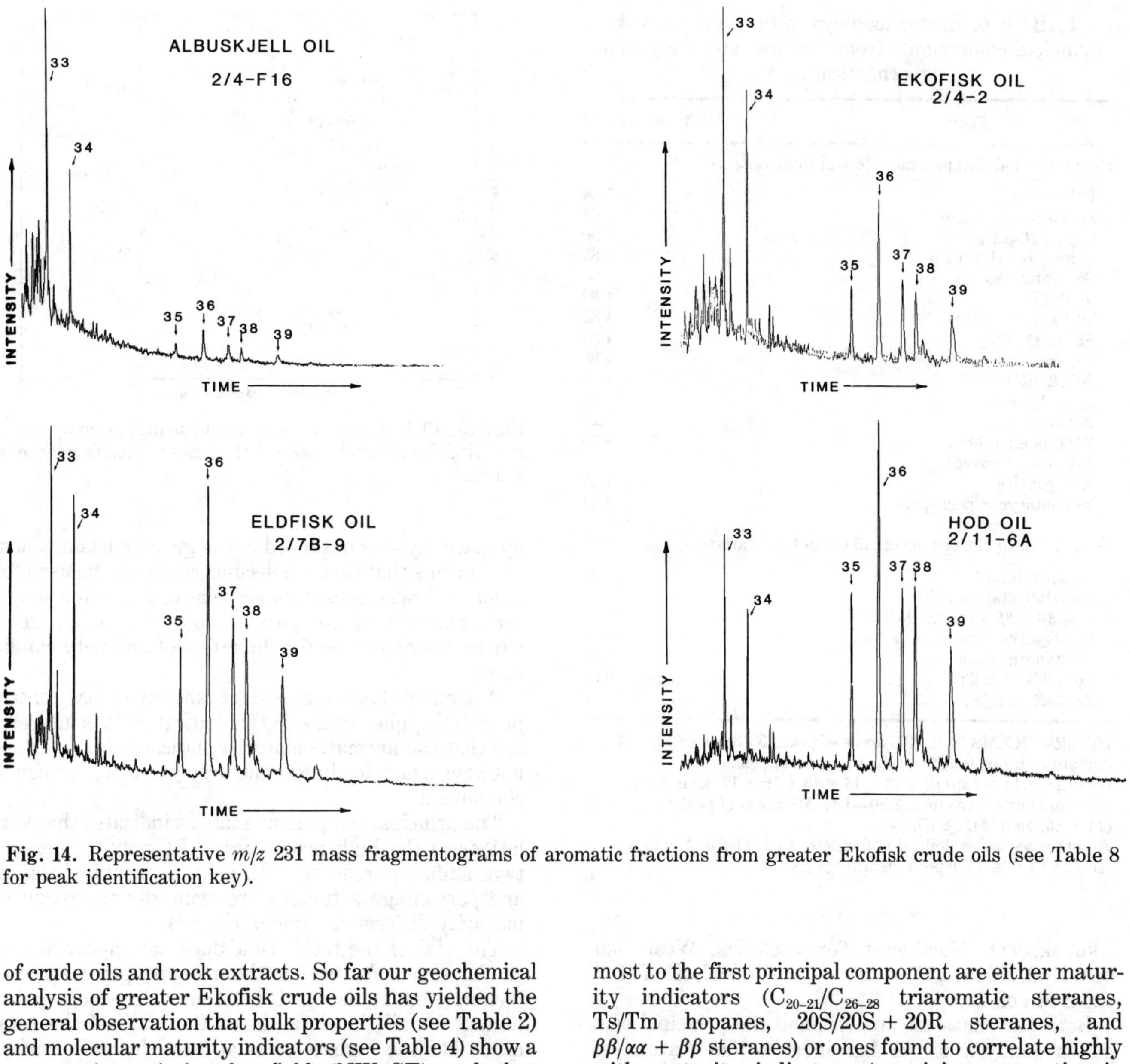

Fig. 14. Representative m/z 231 mass fragmentograms of aromatic fractions from greater Ekofisk crude oils (see Table 8 for peak identification key).

of crude oils and rock extracts. So far our geochemical analysis of greater Ekofisk crude oils has yielded the general observation that bulk properties (see Table 2) and molecular maturity indicators (see Table 4) show a systematic variation by field (NW–SE) and that molecular source indicators (see Table 3) generally show very little or no systematic change (exceptions are the dibenzothiophene/phenanthrene and diasterane/sterane ratios). This observation suggests that maturity, and not source, is the major factor for the differences in geochemical properties of the oils.

To assist in interpreting the large amount of geochemical data in this study and in determining what are the most important factors in controlling the variation among the samples we have used principal component (eigenvector) analysis (Harper *et al.*, 1984). This statistical technique reduces the dimensionality of the data by extracting the minimum number of factors (eigenvectors) which best span the original data. The original data in this case consisted of all the bulk and molecular properties listed in Tables 2–5 plus nine additional molecular ratios.

Two eigenvectors were found to account for 64.4% of the total variance. Most (31.1%) of the remaining variance was accounted for by nine additional eigenvectors with individual contributions of 1.3–7.1%.

Table 6 lists the percentage of the total variance accounted for by each of the first two principal components (eigenvectors) and the individual contribution of the most important molecular or bulk properties to each eigenvector. The properties contributing the most to the first principal component are either maturity indicators (C_{20-21}/C_{26-28} triaromatic steranes, Ts/Tm hopanes, 20S/20S + 20R steranes, and $\beta\beta/\alpha\alpha + \beta\beta$ steranes) or ones found to correlate highly with maturity indicators (remaining properties in Table 6). The fact that two properties are important in the same eigenvector means that they are correlated. The properties contributing the most to the second principal component are source-related (organic input or depositional environment) indicators.

Based on the principal component analysis of all the geochemical data, we conclude that the differences in bulk and molecular properties among the oils are due predominantly to maturity (51% of total variance), and, to a significantly lesser extent, to source (13% of total variance).

A cross-plot (see Fig. 15) of the first and second principal components shows the distribution of the oils with respect to the major factors of maturity and source. The samples are distributed along the first principal component in the order (with increasing maturity): Hod, Eldfisk oil 27, Valhall < Eldfisk, Ekofisk, Edda < Tor, Northwest Tor, West Ekofisk, Albuskjell oil 7 < Albuskjell. The distribution of oils by field along the maturity principal component corresponds very well to the NW–SE trend previously observed in maturity indicators (see Table 4 and Figs. 6 and 12(A)).

The samples are distributed along the second principal component in the order (with increasing terrestrial input and oxicity): Eldfisk oil 27 < Eldfisk oil 26 <

TABLE 6. Factor loadings in first and second principal components from multivariate analysis of geochemical data

Feature	*Variance (%)*
First Principal Component (51.4% of total variance)	
Polars (%)	5.04
Diasterane/sterane	5.03
$C_{20+21}/C_{26+27+28}$	4.90
Triterpane B/hopane	4.87
Saturates (%)	4.85
MDBTR	4.82
20/28 R	4.82
Sulfur (%)	4.81
Ts/Tm	4.80
20S/(20R + 20S)	4.75
$\beta\beta/(\beta\beta + \alpha\alpha)$	4.55
Nitrogen (%)	4.29
DBT/phenanthrene	4.29
Tricyclics/hopane	4.28
API gravity	4.27
29 pentacyclic/norhopane	4.01
Second Principal Component (13.0% of total variance)	
Aquatic/total (m/z 218)	13.41
Aquatic/total (m/z 259)	12.44
$C_{27}\beta\beta/C_{28}\beta\beta$ steranes (m/z 218)	11.37
C_{27}/C_{28} Diasteranes (m/z 259)	10.47
Pristane/phytane	6.94
26S/28S (m/z 231)	6.35
Asphaltenes (%)	5.16

20/28R—GC–MS (m/z 231) areas of peaks 33/39 as in Fig. 14.
Aquatic/total (m/z 259)—GC–MS areas of peaks (12 + 13 + 14 + 15)/(12 + 13 + 14 + 15 + 16 + 17) as in Fig. 8(B).
C_{27}/C_{28} Diasteranes (m/z 259)—GC–MS areas of peaks (12 + 13)/(14 + 15) as in Fig. 8(B).
All other variables defined in footnotes to Tables 3, 4, and 5.
Also see Table 8 for peak identifications.

Ekofisk, Tor, Northwest Tor < Eldfisk, West Ekofisk, Hod, Edda oil 17 < Albuskjell, Valhall, Ekofisk oil 12, Edda oil 18.

Both our 'classical' analysis and the principal component analysis of the data indicate that maturity is the most important factor controlling the geochemical features of the greater Ekofisk oils. Thus, this group of oils represents a case in which source effects are reasonably constant, permitting the effect of maturity alone to be assessed.

The diasterane/sterane ratio (see Table 3), which is taken to be a depositional environment indicator of lithology, appears to be, for the greater Ekofisk oils, primarily maturity controlled. This conclusion is substantiated by the high loading of this ratio in the first (maturity) principal component (see Table 6) and by the smooth increase in the ratio with the $\beta\beta/\alpha\alpha + \beta\beta$ C_{29} sterane ratio (see Fig. 16). The deviation of the Hod oil from the general trend may be due to source (high clastic/terrestrial input?). For similar reasons, the variations in the dibenzothiophene/phenanthrene ratio (see Table 3), the methyldibenzothiophene isomer ratio and the triterpane-derived ratios in Table 5 are largely due to maturity differences among the oils.

The diasterane/sterane, dibenzothiophene/phenanthrene, and methyldibenzothiophene isomer molecular ratios discussed above are examples of geochemical parameters which respond to source, lithology and maturity. Probably the most important control is source lithology, which determines the initial values for the ratios. Subsequently, the ratios are modified

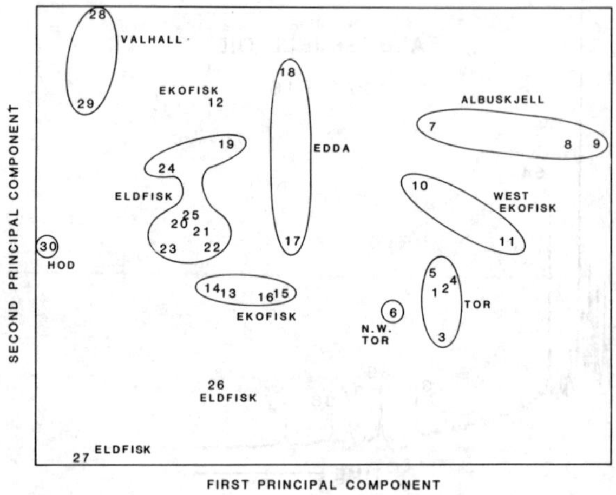

Fig. 15. Plot of first versus second principal components resulting from factor analysis of greater Ekofisk geochemical data.

by maturity—as observed in the greater Ekofisk oils. This means that these molecular ratios can be used for oil/oil and oil/source rock correlations only for oils or rock extracts of comparable maturity or in cases where compensation for the effect of maturity can be made.

A similar duality of source and maturity control probably applies to the Ts/Tm indicator. It is interesting that the aromatic maturity indicator MPI-1, does not have a high loading to the first (maturity) principal component.

The principal component analysis indicates that variations in the bulk properties, API gravity, percentage sulfur, percentage nitrogen, percentage polars, and percentage saturates are primarily the result of maturity differences among the oils.

The $\delta^{13}C$ of the total crude does not appear among the important features in the maturity principal component (see Table 6). Inspection of the correlation matrix for all the data shows that $\delta^{13}C$ of the total crude has the highest correlation with the maturity-related features API gravity ($R^2 = 0.62$), percentage sulfur ($R^2 = 0.64$) and $14\beta,17\beta/14\alpha,17\alpha + 14\beta,17\beta$ sterane ratio ($R^2 = 0.44$) and a low correlation with the source-related features m/z 218 aquatic/total ratio ($R^2 = 0.02$) and pristane/phytane ratio ($R^2 = 0.28$).

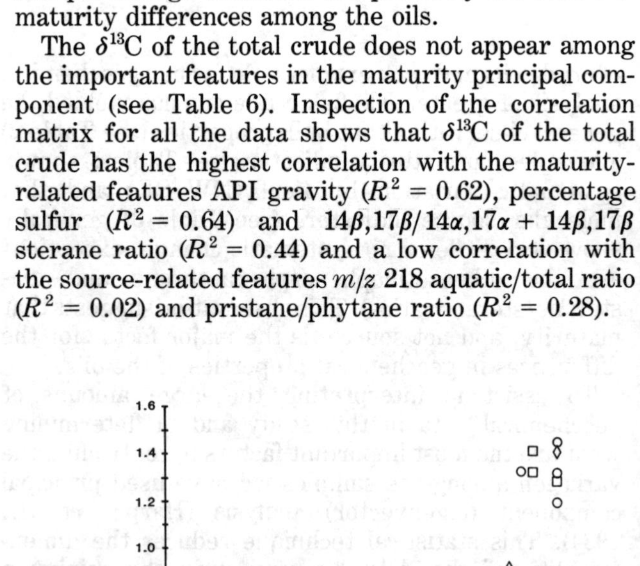

Fig. 16. Variation in C_{28} diasterane/sterane ratio with $\beta\beta/\alpha\alpha + \beta\beta$ (20R + 20S) 24-ethylscholestane ratio for greater Ekofisk crude oils (see Table 1 for field symbol key.

Disregarding the outlying points, namely oil 18 from Edda and oil 27 from Eldfisk, Fig. 4 indicates a NW–SE, hence maturity, trend, with the oils becoming isotopically lighter with increasing maturity by approximately 1‰. This trend is opposite to conventional ideas that oils become isotopically heavier with maturation (Silverman, 1967; Sofer, 1984).

A possible explanation is thermal alteration (cracking) of the oil, either before expulsion in the source rock or after accumulation in the reservoir. Such a cracking process would tend toward the formation of a low molecular weight, condensate-like fraction and a high molecular weight, coke-like fraction, with the former isotopically lighter and the latter isotopically heavier on average than the original whole oil. The lower molecular weight fraction would be preferentially recovered during production, hence the measured whole-oil $\delta^{13}C$ value would decrease as this process progresses, i.e. as the oils become more mature.

A number of the molecular indicators involving members of homologous series are maturity independent and appear as significant features in the second (source) principal component. In this group are those ratios based on the carbon number distribution in the saturated steranes and diasteranes and the triaromatic steranes. The carbon number distribution of $\beta\beta$ 20(S + R) steranes has been cited as reliable (unaffected by maturity) source indicator (Mackenzie *et al.*, 1982). Of interest is the occurrence of pristane/phytane in this group of indicators. Our data show that over the range of maturities found in the greater Ekofisk oils, pristane/phytane is unaffected. It is also interesting that the principal component analysis shows the percentage asphaltenes in the oils to be source controlled (see Table 6).

The distribution of the oils along the second principal component gives important information about the source of the oils. The Jurassic oil from Eldfisk (sample 27) falls at one extreme of the source component (see Fig. 15). Its geochemical characteristics include the highest aquatic/total ratio, lowest pristane/phytane ratio and high sulfur content/low gravity for its maturity. We conclude that oil 27 is a locally sourced oil from a subfacies of the Upper Jurassic deposited in a marine and very anoxic environment. Oil 26 from Eldfisk falls significantly outside the other Eldfisk oils and near the Eldfisk Jurassic oil. This oil is from the deepest of the Cretaceous reservoirs (Hod Formation) and may be a mixture of the Jurassic type oil and the type represented by the main group of Eldfisk oils. Excluding these two oils, the second (source) principal component indicates that the Tor, Northwest Tor, and Ekofisk oils are from the most marine and anoxic source relative to the other oils.

The distribution along the source component axis in Fig. 15 also shows that oil 12 falls significantly outside the cluster of the other Ekofisk oils, and that the two Valhall oils, the two Edda oils and, to a lesser extent, the two West Ekofisk oils, are separated. In each case, the pairs of oils from Valhall, Edda, and West Ekofisk are from different reservoir formations. Within Eldfisk, oils 19 and 24 are separated from the main group; it is interesting that these two oils are the only unmixed Ekofisk Formation productions. A plausible interpretation is that the Ekofisk, Valhall, Edda, West Ekofisk, and Eldfisk structures have received oil from different source subfacies of the Late Jurassic and that the Danian and/or Cretaceous reservoirs may be differently filled from these source subfacies.

Inspection of the molecular ratios used in this study shows that they fall into two general categories (see Table 7): those based on homologous molecular pairs and those based on isomeric molecular pairs. Based on the results of this study, the following generalizations can be made. Over the range of maturities exemplified by the greater Ekofisk oils, which we estimate to be approximately equivalent to 0.55–1.00 ± 0.05 vitrinite reflectance, molecular ratios based on homologous pairs are unaffected by maturity, and may be used unequivocally as source indicators and as oil/oil and oil/rock correlation parameters (examples of these ratios are listed in column A of Table 7). Over the range of maturities of the greater Ekofisk oils, molecular ratios based on isomeric pairs change systematically with maturity (examples of these ratios are listed in column B of Table 7). Those isomeric ratios where initial value is not source dependent, such as $\beta\beta/\alpha\alpha + \beta\beta$ steranes, may be used universally as maturity indicators. Those isomeric ratios whose initial value varies with source may be used as comparative maturity indices only when source is constant.

Bisnorhopane

Since the first report of 17α(H), 18α(H), 21β(H)-28,30-bisnorhopane (Seifert and Moldowan, 1978) in the Monterey shale, the factors controlling its rather enigmatic occurrence in oils and the biological source of this triterpane have been of interest. Grantham *et al.* (1980) have previously reported the presence of bisnorhopane in North Sea oils. They also found the ratio of bisnorhopane to hopane tended to increase with increasing sulfur content. In this study, the opposite is true; bisnorhopane/hopane generally increases with decreasing sulfur content (see Fig. 17). The two samples with the greatest deviation from the

TABLE 7. Ratios based on homologous and isomeric molecular pairs

A *Homologous Molecular Pairs*	*B* *Isomeric Molecular Pairs*
Pristane Phytane	5α,14α,17α,20(R + S)-24-ethylcholestane 5α,14β,17β,20(R + S)-24-ethylcholestane
5α,14β,17β,20R(or 20S)-cholestane 5α,14β,17β,20R(or 20S)-24-methylcholestane	5α,14α,17α,20R(or 20S)-24-methylcholestane 13β,17α,20R(or 20S)-24-methyldiacholestane
17α(H),21β(H)-30-norhopane 17α(H),21β(H)-hopane	17α(H),21β(H)-22,29,30-trisnorhopane 18α(H)-21β(H)-22,29,30-trisnorhopane
C_{23} tricyclic terpane C_{24} tricyclic terpane	1-methyldibenzothiophene 4-methyldibenzothiophene

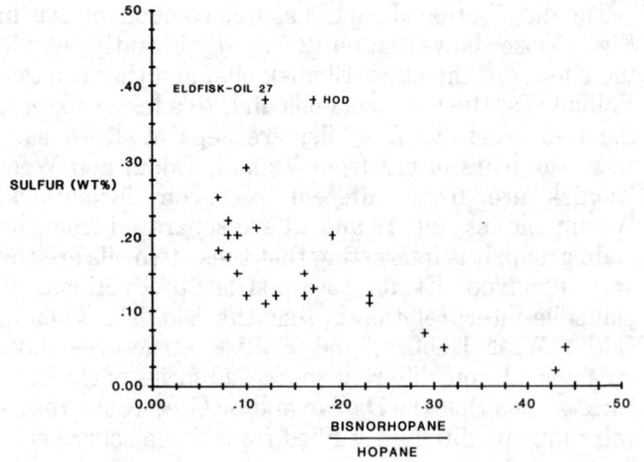

Fig. 17. Variation in wt% sulfur with bisnorhopane/hopane ratio for greater Ekofisk crude oils.

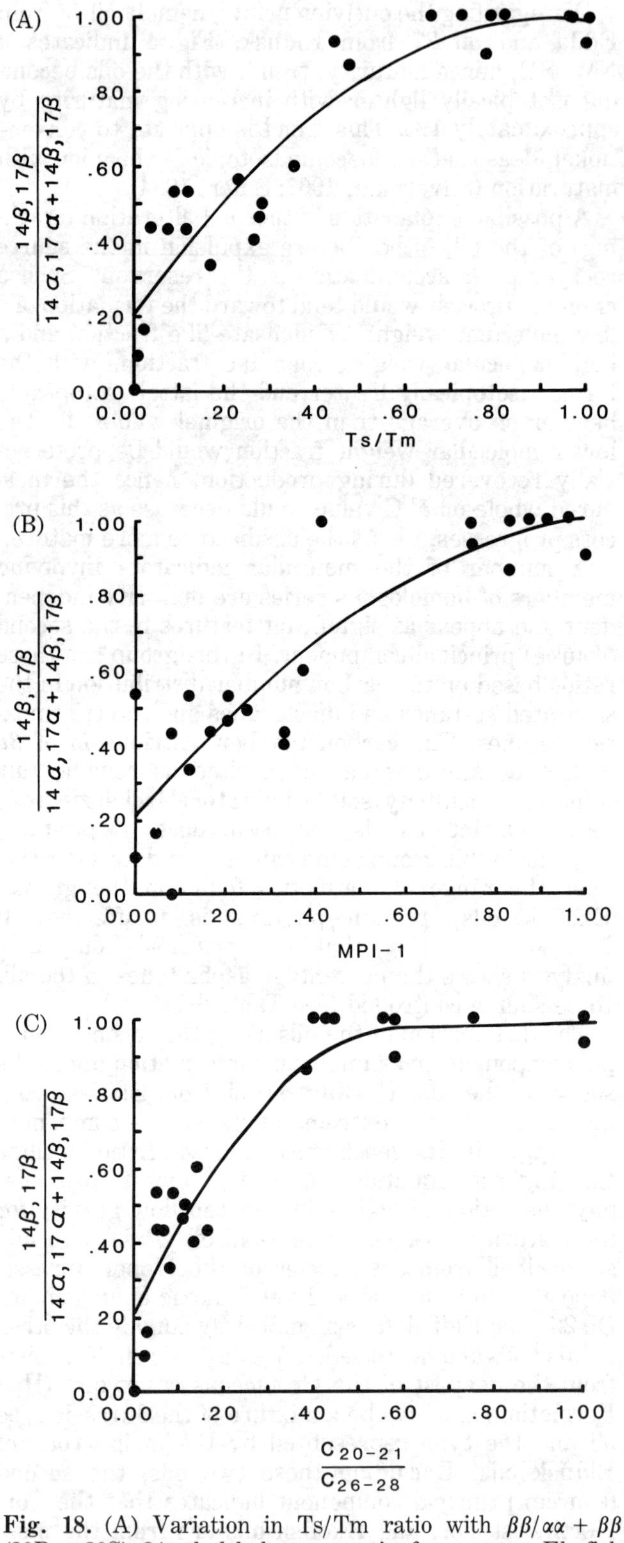

Fig. 18. (A) Variation in Ts/Tm ratio with $\beta\beta/\alpha\alpha + \beta\beta$ (20R + 20S) 24-ethylcholestane ratio for greater Ekofisk crude oils. (B) Variation in MPI-1 with $\beta\beta/\alpha\alpha + \beta\beta$ (20R + 20S) 24-ethylcholestane ratio for greater Ekofisk crude oils. (C) Variation in C_{20-21}/C_{26-28} triaromatic sterane ratio with $\beta\beta/\alpha\alpha + \beta\beta$ (20R + 20S) 24-ethylcholestane ratio for greater Ekofisk crude oils.

general trend are the Eldfisk Jurassic oil (sample 27) and the Hod oil. We have already seen that the Eldfisk oil is sourced by a different subfacies than the other oils. Why the Hod oil deviates from the trend is not clear. We have also already established that sulfur content is controlled by maturity.

We believe that our data and that of Grantham *et al.* (1980) can be explained by the following argument. The major control on bisnorhopane/hopane ratio is source. However, among samples from a similar source, and perhaps when bisnorhopane/hopane is initially rather low, maturity is the controlling factor and an inverse relationship exists between this ratio and sulfur content. This relationship is not direct but indirect, i.e. they are both responding to a common factor—thermal stress. Moreover, the relationship need not be linear, since the decrease in sulfur and increase in bisnorhopane/hopane ratio can respond to thermal stress at different rates. Our data show the process responsible for increasing the bisnorhopane/hopane ratio has a higher activation energy than the process of desulfurization of the oil.

If this hypothesis is correct, then the decrease in bisnorhopane/hopane ratio with decreasing sulfur content observed by Grantham *et al.* (1980) is source- rather than maturity-controlled. Also, in this scenario, the bisnorhopane/hopane ratio in the Hod oil and possibly the Eldfisk Jurassic oil are source-controlled. Since the Hod oil is very similar to the other oils in source indicators (see Table 3), it is not clear why the bisnorhopane/hopane ratio should be in such high (relative) concentration in this sample. Evidently, whatever is controlling the bisnorhopane content is not reflected in other commonly measured geochemical properties.

Comparison of maturity indicators

Intercomparison of the various thermal indicators used in this study can give information about their sensitivity to thermal stress. To make these comparisons we have plotted Ts/Tm, MPI-1, and C_{20-21}/C_{26-28} triaromatic sterane ratios versus $\beta\beta/\alpha\alpha + \beta\beta$ sterane ratio. Since we are interested in assessing maturity effects only, we have eliminated those samples which appear to have a strong source component. These samples are the Eldfisk Jurassic oil (sample 27) and the Eldfisk oil which appears to contain a significant amount of the Jurassic-type oil (sample 26). The two Edda oils were also eliminated as they appear to have very anomalous MPI-1 values. To remove the effects of differing ranges in which each indicator is operative, the indicators were scaled by subtracting the lowest measured value from each observation and dividing the result by the observed range. The resulting plots are shown in Fig. 18(A)–(C). The solid curves were calculated by a polynomial regression analysis of the data.

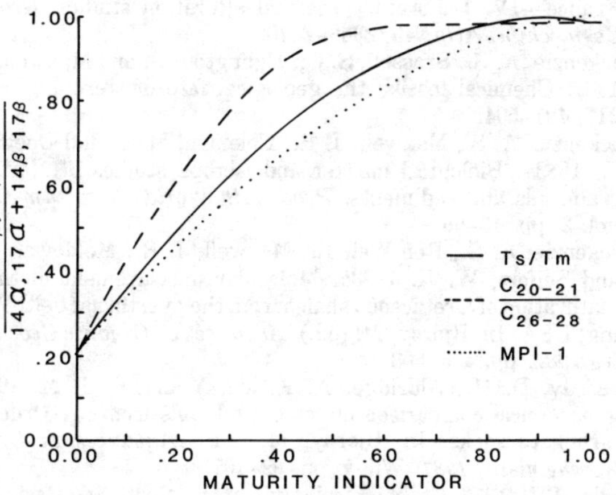

Fig. 19. Comparison of regression curves for Ts/Tm ratio, MPI-1, and C_{20-21}/C_{26-28} triaromatic sterane ratio with $\beta\beta/\alpha\alpha + \beta\beta$ (20R + 20S) 24-ethylcholestane sterane ratio for greater Ekofisk crude oils.

The general shape of the three plots is similar, and indicates that the aromatic sterane and triterpane indicators are less sensitive to thermal stress (this assumes that heating rates were not drastically different for the sources of the oils). The best-fit curves for all three indicators are plotted versus the $\beta\beta/\alpha\alpha + \beta\beta$ sterane indicator on the same axes in Fig. 19. Relative to sterane isomerization and for heating rates comparable with the greater Ekofisk area, sensitivity to thermal stress increases in the order: C_{20-21}/C_{26-28} triaromatic sterane ratio < Ts/Tm ≈ MPI-1. The rates of the processes responsible for the changes in these thermal indicators decrease in the same order.

CONCLUSIONS

The results of bulk property, carbon isotope, and saturated and aromatic molecular fossil analyses of 30 oils from the greater Ekofisk area of the Norwegian sector of the North Sea suggest the following.

The major differences among the oils are due to maturity, which decreases systematically by field along a NW–SE trend as follows: Albuskjell > Northwest Tor, Tor, West Ekofisk > Ekofisk, Edda, Eldfisk > Valhall, Hod. Whether the maturity variations are due to differences in the degree of thermal stress experienced by the source rocks, to thermal alteration in the reservoirs, or to some combination of these two factors is not known.

Minor, but definite, source differences exist which are not distributed systematically by field. The oils fall in the following order of increasing terrestrial input and oxicity: Eldfisk (oil 27) < Eldfisk (oil 26) < Ekofisk, Tor, West Tor < Eldfisk, West Ekofisk, Hod, Edda (oil 17) < Albuskjell, Valhall, Ekofisk (oil 12), Edda (oil 18). These variations are the result of locally different source rock subfacies within the Upper Jurassic. Within the Ekofisk, Valhall, Edda, West Ekofisk, and Eldfisk structures the Danian and/or Cretaceous reservoirs appear to be differentially filled from these source subfacies.

Over the maturity range of the greater Ekofisk oils (early to late oil window; approximately 0.55–1.00 R_0) molecular ratios based on homologous pairs are relatively unaffected by maturity and may be used unequivocally as source indicators and in correlations. These include the pristane/phytane and norhopane/hopane ratios and the ratios of steranes of different carbon numbers with the same stereochemistry. Molecular ratios based on isomeric pairs change systematically with maturity, and may be used as comparative maturity indicators only when source is constant or the initial value is not source-dependent. These include, in the former category, the Ts/Tm hopane ratio and, in the latter category, the C_{29} sterane isomer ratios. Also included is the source-related indicator, the diasterane/sterane ratio.

TABLE 8. Peak identification key for figures

Peak	*Compound(s)*
m/z 217 GC–MS sterane trace (Fig. 5)	
1	13β,17α,20S-24-methyl-diacholestane
2	13β,17α,20R-24-methyl-diacholestane
3	5α,14α,17α,20S-24-methyl-cholestane
4	14β,17β(20S + 20R)-24-methyl-cholestanes
5	5α,14α,17α,20R-24-methyl-cholestane
6	5α,14α,17α,20S-24-ethyl-cholestane
7	14β,17β,(20S + 20R)-24-ethyl-cholestanes
8	5α,14α,17α,20R-24-ethyl-cholestane
m/z 218 GC–MS sterane trace (Fig. 8(A))	
9	14β,17β(20S + 20R)-cholestanes
10	14β,17β(20S + 20R)-24-methyl-cholestanes
11	14β,17β(20S + 20R)-24-ethyl-cholestanes
m/z 259 GC–MS diasterane trace (Fig. 8(B))	
12	13β,17α,20S-diacholestane
13	13β,17α,20R-diacholestane
14	13β,17α,20S-24-methyl-diacholestanes
15	13β,17α,20R-24-methyl-diachloestanes
16	13β,17α,20S-24-ethyl-diacholestane
17	13β,17α,20R-24-ethyl-diacholestane
m/z 191 GC–MS terpane trace (Fig. 11)	
18	23 carbon tricyclic terpenoid
19	24 carbon tricyclic terpenoid
20	25 carbon tricyclic terpenoids
21	26 carbon tricyclic terpenoids
22	28 carbon tricyclic terpenoids
23	29 carbon tricyclic terpenoids
24	18α(H),21β(H)-22,29,30-trisnorhopane (Ts)
25	30 carbon tricyclic terpenoids
26	17α(H),21β(H)-22,29–30-trisnohopane (Tm)
27	17α(H),21β(H)-30-norhopane
28	17β(H),21α(H)-30-nomoretane
29	17α(H),21β(H)-hopane
30	17β(H),21α(H)-moretane
31	17α(H),21β(H),22(S + R)30-homohopanes
32	17α(H),21β(H),22(S + R)-30,31-bishomohopanes
A	30 carbon pentacyclic triterpenoid (unidentified)
B	30 carbon pentacyclic triterpenoid (unidentified)
X	24 carbon tetracyclic terpenoid
Y	17α(H),18α(H),21β(H)-28,30-bisnorhopane
Z	29 carbon pentacyclic triterpenoid (unidentified)
m/z 231 GC–MS triaromatic sterane trace (Fig. 14)	
33	20 carbon triaromatic sterane
34	21 carbon triaromatic sterane
35	26 carbon (20S) triaromatic sterane
36	26 carbon (20R) + 27 carbon (20S) triaromatic steranes
37	28 carbon (20S) triaromatic sterane
38	27 carbon (20R) triaromatic sterane
39	28 carbon (20R) triaromatic sterane
Thiophenic sulfur GC–FPD trace (Fig. 13)	
40	C_2 + C_3-benzothiophenes
41	dibenzothiophene
42	4-methyldibenzothiophene
43	2- + 3-methyldibenzothiophene
44	1-methyldibenzothiophene
45	dimethyldibenzothiophenes
46	trimethyldibenzothiophenes

Comparison of maturity indicators shows that within the suite of greater Ekofisk oils they have the following sensitivity to thermal stress: sterane isomerization > methylphenanthrene index ≈ Ts/Tm hopane ratio > C_{26-21}/C_{26-28} triaromatic sterane ratio.

ACKNOWLEDGEMENTS

The authors wish to thank the following individuals for helpful discussions or assistance in obtaining information for this study: D. C. Boatwright, M. A. Sloan, J. S. Shveima, and M. E. Smith, Exploration and Production Group, Phillips Petroleum Company; W. G. Lyon, Research and Development, Phillips Petroleum Company; and J. Brewster, Phillips Petroleum Company Norway. The experimental assistance of J. A. Lytle, J. S. Spears, and T. Smith is also acknowledged.

REFERENCES

Barnard, P. C. and Cooper, B. S. 1981. In Illing, L. V. and Hobson, G. D. (eds.), *Petroleum Geology of the Continental Shelf of North West Europe*, Institute of Petroleum, London, pp. 169–175.

Barnard, P. and Cooper, B. S. 1984. Source rocks and oils of the central and northern North Sea. In Demaison, G. and Murris, R. J. (eds.), *Petroleum Geochemistry and Basin Evaluation. AAPG Memoir*, **35**, pp. 303–314.

Didyk, B. M., Simoneit, B. R. T., Brassell, S. C. and Eglington, G. 1978. Organic geochemical indicators of paleoenvironmental conditions of sedimentation. *Nature*, **272**, 216–222.

Dolan, P. 1983. North Sea oil province: a failed rift basin. *Oil Gas J.*, 185–196 (31 January).

Grantham, P. J., Posthuma, J. and de Groot, K. 1980. Variation and significance of the C_{27} and C_{28} triterpane content of a North Sea core and various North Sea crude oils. In Douglas, A. G. and Maxwell, J. R. (eds.), *Advances in Organic Geochemistry 1979*. Pergamon Press, Oxford, pp. 29–48.

Harper, A. M., Meuzelaar, H. L. C. and Given, P. H. 1984. Correlations between pyrolysis mass spectra of Rocky Mountain coals and conventional coal parameters. *Fuel*, **63**, 793–802.

Huang, W-Y. and Meinschein, W. G. 1979. Sterols as ecological indicators. *Geochim. Cosmochim. Acta*, **43**, 739–745.

Hughes, W. B. 1984. The use of thiophenic organosulfur compounds in characterizing crude oils derived from carbonate versus siliciclastic sources. In: Palacas, J. G. (ed.), *Petroleum Geochemistry and Source-Rock Potential of Carbonate Rocks. AAPG Studies in Geology* **18**.

Mackenzie, A. S., Patience, R. L., Maxwell, J. R., Vandenbroucke, M. and durand, B. 1980. Molecular parameters of maturation in the Toarcian shales, Paris Basin, France—I. Changes in the configurations of a cyclic isoprenoid alkanes, steranes and triterpanes. *Geochim. Cosmochim. Acta*. **44**, 1709–1723.

Mackenzie, A. S. and Maxwell, J. R. 1981*a*. Assessment of thermal maturation in sedimentary rocks by molecular measurements. In: Brooks, J. (ed.), *Organic Maturation Studies and Fossil Fuel Exploration*, Academic Press, London, pp. 239–254.

Mackenzie, A. S., Hoffmann, C. F. and Maxwell, J. R. 1981*b*. Molecular parameters of maturation in the Toarcian shales, Paris Basin, France—III. Changes in aromatic steroid hydrocarbons. *Geochim. Cosmochim. Acta*, **45**, 1345–1355.

Mackenzie, A. S., Lewis, C. A. and Maxwell, J. R. 1981*c*. Molecular parameters of maturation in the Toarcian shales, Paris Basin, France—IV. Laboratory thermal alteration studies. *Geochim. Cosmochim. Acta*, **45**, 2369–2376.

Mackenzie, A. S., Brassell, S. C., Eglington, G. and Maxwell, J. R. 1982. Chemical fossils: the geological fate of steroids. *Science*, **217**, 491–504.

Mackenzie, A. S., Maxwell, J. R., Coleman, M. L. and Deegan, C. E. 1983*a*. Biological marker and isotope studies of North Sea crude oils and sediments. *Proc. 11th World Petroleum Congr.* vol. 2, pp. 45–56.

Mackenzie, A. S., Ren-Wei, L., Maxwell, J. R., Moldowan, J. M. and Seifert, W. K. 1983*b*. Molecular measurement of thermal maturation of Cretaceous shales from the overthrust belt, Wyoming, USA. In: Bjorøy, M. (ed.), *Advances in Organic Geochemistry 1981*. pp. 496–503.

McKirdy, D. M., Aldridge, A. K. and Ypma, P. J. M. 1983. A geochemical comparison of some crude oils from pre-Ordovician carbonate rocks. In: Bjorøy, M. (ed.), *Advances in Organic Geochemistry 1981*. Wiley, pp. 99–107.

Oudin, J. L. 1976. Geochemical study of the North Sea Basin. *Bull. Centre Rech. Pau–SNPA*, **10**, 339–358.

Petrov, A. A., Pustil'nikova, S. D., Abrjutina, N. N. and Kagramonova, G. E. 1976. Petroleum steranes and triterpanes. *Neftekhimiya*, **16**, 411–427.

Radke, M., Welte, D. H. and Willsch, H. 1982*a*. Geochemical study on a well in the western Canada basin: relation of the aromatic distribution pattern to maturity of organic matter. *Geochim. Cosmochim. Acta*, **46**, 1–10.

Radke, M., Willsch, H., Leythhaeuser, D. and Teichmüller, M. 1982*b*. Aromatic components of coal: relation of distribution pattern to rank. *Geochim. Cosmochim. Acta*, **46**, 1831–1848.

Rubenstein, I., Sieskind, O. and Albrecht, P. 1975. Rearranged sterenes in a shale: occurrence and simulated formation. *J. Chem. Soc. Perkin Trans.*, **I**, 1833–1836.

Rinaldi, G. G. L. 1984. Novel series of tetracyclic terpanes in crude oils and rock extracts. *Am. Chem. Soc. Abstr. Papers, 187th Natn. Mtg;* GEOC-57.

Scalan, R. S. and Smith, J. S. 1970. An improved measure of the odd–even predominance in the normal alkanes of sediment extracts and petroleum. *Geochim. Cosmochim. Acta*, **34**, 611–620.

Seifert, W. K. and Moldowan, J. M. 1978. Applications of steranes, terpanes, and monoaromatics to the maturation, migration, and source of crude oils. *Geochim. Cosmochim. Acta*, **42**, 77–95.

Seifert, W. K. and Moldowan, J. M. 1981. Paleoreconstruction by biological markers. *Geochim. Cosmochim. Acta*, **45**, 783–795.

Seifert, W. K., Carlson, R. M. K. and Moldowan, J. M. 1981. Geomimetric synthesis, structure assignment, and geochemical correlation application of monoaromatized petroleum steroids. In: Bjorøy, M. (ed.), *Advances in Organic Geochemistry 1981*. pp. 710–724.

Sieskind, O., Joly, G. and Albrecht, P. 1979. Simulation of the geochemical transformations of sterols: superacid effect of clay minerals. *Geochim. Cosmochim. Acta*, **43**, 1675–1679.

Silverman, S. R. 1967. Carbon isotopic evidence for the role of lipids in petroleum. *J. Am. Oil Chemists Soc.*, **44**, 691–695.

Sofer, Z. 1984. Stable carbon isotope compositions of crude oils: applications to source depositional environments and petroleum alterations. *AAPG Bull.*, **68**, 31–49.

Thomas, B. R. 1969. Kauri resins—modern and fossil. In: Eglington, G. and Murphy, M. T. J. (eds.), *Organic Geochemistry*. Springer-Verlag, New York, pp. 599–619.

Tissot, B. and Welte, D. H. 1978. *Petroleum Formation and Occurrence*. Springer-Verlag, New York, 538 pp.

van Graas, G., Baas, J. M. A., van de Graaf, B. and de Leeuw, J. 1982. Theoretical organic geochemistry. I. The thermodynamic stability of several cholestane isomers calculated by molecular mechanics. *Geochim. Cosmochim. Acta*, **46**, 2399–2402.

Van den Bark, E. and Thomas, O. D. 1980. Ekofisk: first of the giant oil fields in western Europe. in: Halbouty, M. T. (ed.), *Giant Oil and Gas Fields of the Decade 1968–1978. AAPG Memoir*, **30**, 195–224.

6 Correlation of Northern North Sea oils: the different facies of their Jurassic source

Mark A. Northam, Mobil Research and Development Corporation, Dallas Research Laboratory, Dallas, Texas 75881, USA

Fifty-four crude oils from the Norwegian Shelf and the UK sector of the North Sea have been compared geochemically to determine their genetic relationships. All are considered to have been sourced from the Jurassic Kimmeridge Clay. Although most compositional differences among the oils are the result of post-accumulation processes, some of the parameters reflect organic facies changes in the source. These facies changes are a result of the varying depositional environments present in the Viking and Central Grabens during the Jurassic. Evidence for the facies changes can be found in their stable carbon isotope profiles. Correlation of North Sea crude oils allows identification of Kimmeridge-sourced sub-families, which may be useful in delineating drainage areas and migration pathways. Comparison of the shape of the isotope profiles within one of these oil sub-families is helpful in explaining the maturity trend of oils in the Beryl Field, and may aid in determining timing of migration of hydrocarbons.

INTRODUCTION

The North Sea has been of great interest to geochemists because it forms a natural laboratory in which to study the processes of generation and migration in a large family of oils. The source of these oils is generally accepted to be the organic-rich shales of the Kimmeridge Clay Formation. Because the oils are reservoired at different levels in fields which are spread over a relatively large area, there is a strong likelihood that depositional environment, organic facies, and thermal maturity changes are present in their source rocks. The wide range of compositional variations which are present in the North Sea oil set have, however, been attributed mainly to thermal maturity differences in their otherwise homogeneous source rocks.

Cornford *et al.* (1983) came to this conclusion for a large suite of UK North Sea oils. They compared maturity-dependent biomarker ratios determined for the oils and extracts of source rocks at known maturity levels in order to explain the origin of the different oil types. As expected, post-generation processes such as biodegradation and asphaltene precipitation were found to be responsible for the formation of some of the heavy oils. They also found that certain biomarker ratios were somewhat useful in correlating oils with regional facies of the source rock.

A suite of oils from 12 fields in the Viking and Central Grabens of the North Sea (see Fig. 1) have been analyzed at Mobil over the past years. Our conventional geochemical data lead to the same conclusions arrived at by other groups. The oils were also examined using a technique called isotope profiling, a parameter which is sensitive to depositional environment as well as thermal maturity. Results of our study of these oils show that there are, indeed, organic facies differences which can be recognized in the Jurassic source rocks.

NORTH SEA OILS—GEOCHEMICAL DATA

The 54 oils in this study were obtained from reservoirs at depths between 6000 and 13 700 ft (1830–4175 m). All have been subjected to compositional analyses and oil–oil correlation techniques, which are standard geochemical procedures and need no special explanation in a paper of this scope. Results show that the oils have ranges of API gravities, sulfur contents, and hydrocarbon compositions which fit within the ranges published by Cornford *et al.* (1983), who showed that these parameters vary with and are dependent upon the thermal maturity of North Sea oils.

Biomarker analysis, the determination of the relative distribution of chemical fossils present in oils and rock extracts, is generally considered to be a sensitive indicator of the type of organic material which was deposited and incorporated into the source rock

Petroleum Geochemistry in Exploration of the Norwegian Shelf © Norwegian Petroleum Society (Graham & Trotman, 1985) pp. 93–99.

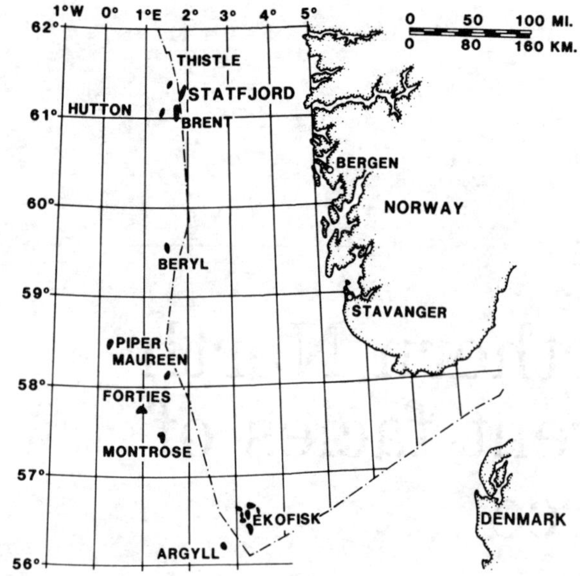

Fig. 1. Northern North Sea oil fields.

kerogen. The families of compounds most frequently used for biomarker analysis are the steranes and terpanes, because of their resistance to degradation. The distribution of biogenic steranes (20R-5α, 14α,17α-C_{27}, -C_{28}, and -C_{29} steranes) in the oils is dependent almost entirely upon the input to the source rock of sterols from indigenous organisms. Oils generated in the same source inherit the same biomarker fingerprints. The mass chromatograms of steranes (m/e = 217) and terpanes (m/e = 191) are very similar for all of the North Sea oils, excepting maturity related differences. That the oils in this study fall into a tight group on a plot of regular sterane distributions (Fig. (2a)) leads to the conclusion that the oils were generated in related source rocks and that the contribution of biological material to the Kimmeridge Clay was uniform over the entire area of this study.

The relative thermal maturities of oils from a given family can be determined by measuring the extent of the isomerization of several thermally sensitive molecules (Seifert and Moldowan, 1981; MacKenzie *et al.*, 1982). For this study, two transformations of the C_{29} sterane were graphically compared after the method of Seifert and Moldowan (1981). The first is the transformation from the biogenic $\alpha\alpha$ 20R isomer (see above) to the more thermally stable $\beta\beta$ isomer (5α,14β,17β (20R)-C_{29} sterane). The second is the transformation from the biogenic $\alpha\alpha$ 20R isomer to the more thermally stable $\alpha\alpha$ 20S isomer. The plot of $\alpha\alpha$ 20S/$\alpha\alpha$ 20R versus $\beta\beta$ 20R/$\alpha\alpha$ 20R (20S/20R versus $\beta\beta/\alpha\alpha$) for this data set (Fig. 2(b)) shows the broad range of maturities covered by the oils from the 12 represented fields.

Not all the variations in the composition of these oils can be ascribed to maturation effects. For instance, the whole oil stable carbon isotope ratios of this sample set range from −27.4 to −30.2 per mil versus PDB. The distribution of isotope ratios, shown in Fig. 3(a), shows that most of the oils group around a value of −29 per mil. The nearly 3 per mil spread in values suggests differences in the isotope ratios of the source rock kerogen and, hence, in the original organic input. Similarly, the variation in the ratio of pristane to phytane (Pr/Ph) for this sample set (see Fig. 3(b)) suggests subtle changes in the depositional environment of separate source facies. Although these values span the range published by Cornford *et al.* (1983), the Pr/Ph values do not show any trend with maturity. The Pr/Ph data suggest that the source rock depositional environments vary from slightly reducing to slightly oxidizing. There is no obvious relationship between the stable isotope ratio and Pr/Ph ratios for these oils, since the former is dependent entirely upon the nature of the organic input and the latter is dependent primarily upon the oxidation state of the depositional environment. Since neither of these two parameters is sensitive to thermal exposure, it is difficult to use standard geochemical analyses to describe in detail the variations which exist in the source of North Sea oils without an extensive study of the rock samples themselves.

CARBON ISOTOPE PROFILES

A carbon isotope profile of a crude oil is constructed by measuring the stable carbon isotope ratio ($\delta^{13}C$) of the whole oil, 11 distillation cuts, and the over 300 °C residuum of the distillation. The distillation cuts are taken at 25°C intervals from 50°C to 300°C. This range represents hydrocarbons with volatilities between those of n-C_5 and n-C_{17}. The $\delta^{13}C$ values of the cuts and residuum are plotted versus the maximum distillation temperature of the cut on a diagram as illustrated in Fig. 4. The shape of the profile and its relationship to the whole oil $\delta^{13}C$ values are important for oil–oil correlation purposes.

Carbon isotope profiles have been utilized in internal reports at Mobil Research and Development Corporation for more than 20 years. Over 800 oils from worldwide sources have been profiled, mainly for oil–oil correlation studies. Isotope profiles are only warranted in special cases. They have been found to be most useful when used to detect subtle differences in large families of oils generated in an apparently homogeneous source rock, as is the case in the North Sea. Their interpretation can be complicated, however, because the shape of a given profile depends on several factors and can be changed by any process which alters the composition of an oil. Isotope profile shapes vary greatly for different oils, just as biomarker fingerprints vary due to differences in the make-up of the source kerogens. Biodegradation and water-washing, when severe enough, can render an isotope profile useless for correlation purposes. It is helpful, therefore, to understand as much about a family of oils as possible before interpreting isotope profiles.

Thermal exposure is another of the natural processes which can alter the composition of a crude oil. Oil composition changes with increasing maturity of the source rock during generation, and with thermal exposure after expulsion. Several chemical parameters can be used to determine maturity levels of oils, but cannot make the distinction between thermal exposure before or after expulsion from the source rock. The shape of an isotope profile is sensitive to the effects of thermally induced change in oils and can often be used to distinguish between increasing maturity in the source and thermal exposure after expulsion for *families* of oils. Making this distinction is important when timing of generation is considered for exploration purposes.

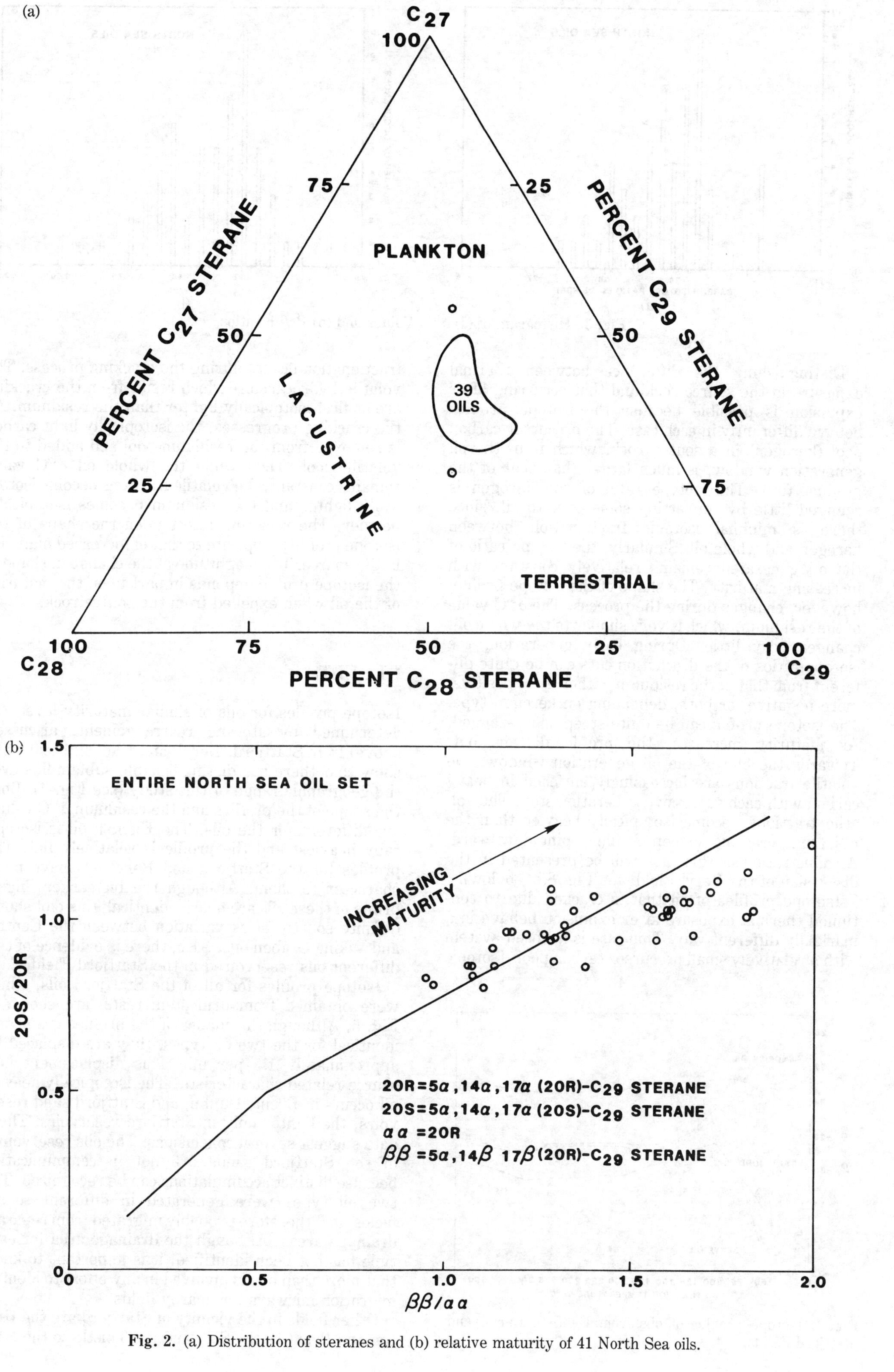

Fig. 2. (a) Distribution of steranes and (b) relative maturity of 41 North Sea oils.

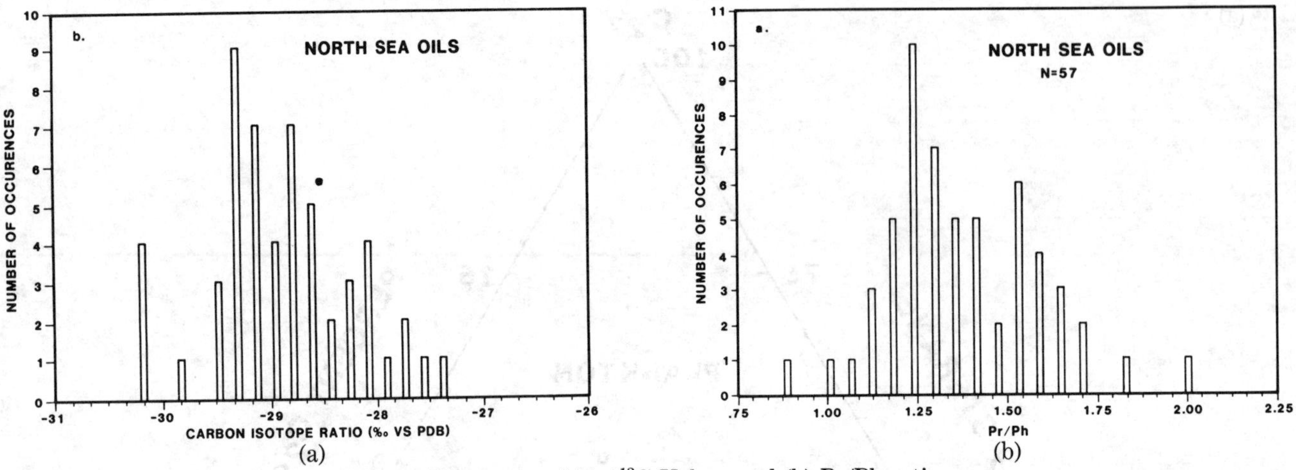

Fig. 3. Histogram of (a) $\delta^{13}C$ Values and (b) Pr/Ph ratios.

Distinguishing the difference between thermal exposure in the source rock and that occurring after expulsion is possible because the isotope profiles behave differently in each case. The precursor carbon pool (kerogen) in a source rock, which is in the oil generation window, is much larger than that of the product (oil). The isotope ratio of the kerogen is changed little by converting some of it to oil, since there is minimal isotopic fractionation between kerogen and whole oil. Similarly, the isotope ratio of the oil generated remains relatively constant with increasing maturity. The shape of the isotope profile, however, changes during the process. The $\delta^{13}C$ value of the residuum, which is very similar to the whole oil, changes very little. During early generation, the isotope ratios of the distillation cuts can be quite different from that of the residuum, either more positive, more negative, or both, depending on kerogen type. The isotope profile can be quite steep or zig-zagged. As maturity increases, the profile flattens out. Towards the end of the oil generation window, the volatile fractions are increasingly enriched in heavy carbon with each successive generation step. The volatile fractions become isotopically heavier than the residuum, creating a steep profile sloping downward. An illustration of this case will be presented in the discussion of the Beryl Field (see Fig. 8(b), below).

Isotope profiles of oil that is cracked due to continued thermal exposure after expulsion behave in a markedly different way. This case is a closed system with a relatively small precursor carbon pool. Isotopic fractionation occurs during the cracking process. The volatile hydrocarbons which result from the cracking are at first isotopically lighter than the residuum. As the cracking progresses, the isotopically light carbon is removed from the residuum pool and added to the volatile pool. Thus while the whole oil $\delta^{13}C$ value remains constant, the volatile fractions become isotopically lighter and the residuum becomes isotopically heavier. The resulting effect upon the shape of the isotope profile is opposite to that of increased maturity in the source. The magnitude of the change in shape of the isotope profile depends in part upon the maturity of the oil when expelled from the source rock.

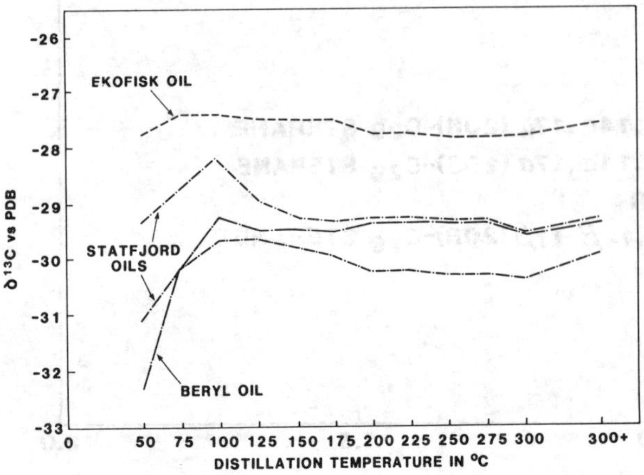

Fig. 4. Isotope profiles of oils from Ekofisk, Beryl, and Statfjord Fields.

RESULTS

Isotope profiles for oils of similar maturity levels (as determined by sterane rearrangement, discussed above) from Statfjord, Beryl, and Ekofisk Field wells show that there are distinguishable subfamilies over the geographic range of this study (see Fig. 4). Both the shape of the profiles and the residuum $\delta^{13}C$ values are different for the oils. The Ekofisk oil is isotopically heaviest and the profile is relatively flat. The profiles for the Statfjord and Beryl oils have more character to them. Although the biomarker fingerprints of these oils are nearly identical, this plot shows definite source facies variation between the Central and Viking Graben oils. Also, there is evidence of two different oils reservoired in the Statfjord Field.

Isotope profiles for all of the Statfjord oils, which were obtained from drill stem tests, are shown in Fig. 5. Although the shapes of the profiles are nearly identical for the two oil types, they are displaced by approximately 1.5 per mil. This displacement is a source-related characteristic. The isotopically heavier oil occurs in Brent, Dunlin, and Statfjord sand reservoirs, the lighter only in Statfjord reservoirs. These data suggest several conclusions. The oils reservoired in the Statfjord sands are not in communication because distinct accumulations can be recognized. The two oil types were generated in different source facies, and therefore probably migrated from separate drainage areas. Although the drainage area for each type has not been identified, it is important to know that more than one is involved in any effort to identify migration fairways for nearby fields.

Other fields in the vicinity of Statfjord are the Hutton and Brent to the SW, and the Thistle to the NW.

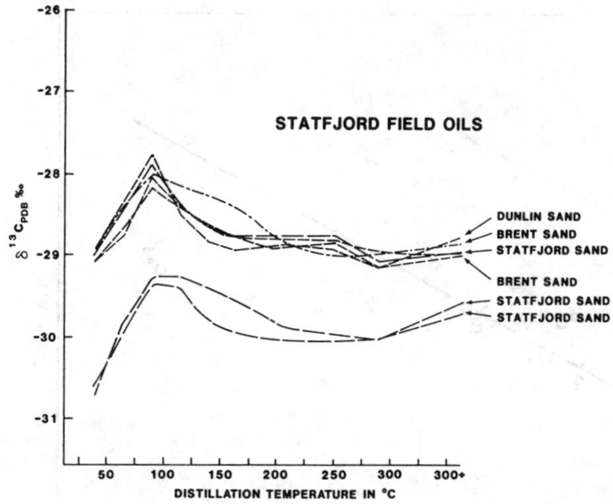

Fig. 5. Isotope profiles of Statfjord Field oils.

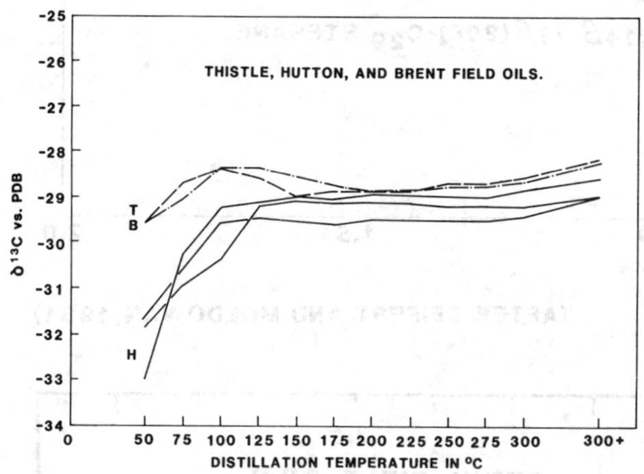

Fig. 6. Isotope profiles of Thistle, Hutton, and Brent Field oils.

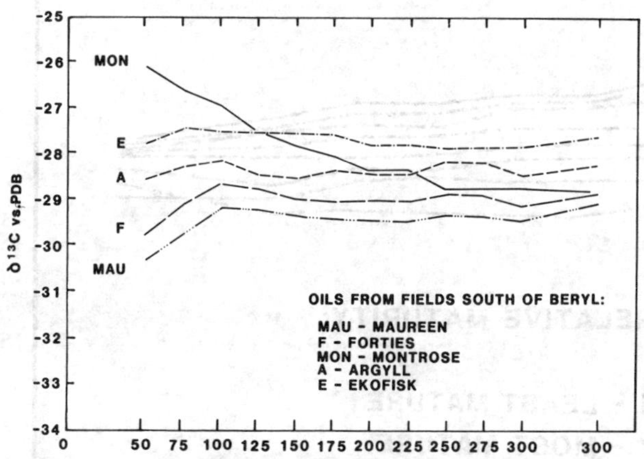

Fig. 7. Isotope profiles of oils from fields south of Beryl.

Isotope profiles for oils in these fields are very similar to the major Statfjord oil type in shape, but the $\delta^{13}C$ values for the residua are approximately 1 per mil heavier (see Fig. 6). These profiles have all the characteristics that would be expected of a Statfjord oil which has received thermal exposure after expulsion. In fact, in a laboratory experiment, one of the Statfjord oils was heated in a bomb for 15 d at 350 °C to see the effect on the shape of its isotope profile. The profile after heating was nearly identical to those from the Hutton field (R. M. Squires, pers. comm.). It appears that these oils may have been derived from the same source facies as the Statfjord oils, but have been thermally altered after expulsion from that source. This information may be useful in conjunction with structure maps and temperature gradient data for mapping the drainage area of Statfjord, Hutton, Brent, and Thistle Fields.

As stated above, the isotope profile for the oil from the Ekofisk Field in the Central Graben is markedly different, both in shape and $\delta^{13}C$ values, from those for the Statfjord Field. Going generally north from Ekofisk to Beryl, there is a regular progression from the Ekofisk to Statfjord type of profile (see Fig. 7) indicating a systematic change in the source rocks as one moves from the Viking Graben into the Central Graben. Interestingly, the representative from the Montrose Field is a high gravity oil. It has an isotope profile characteristic of an oil generated near the end of the oil window. Yet its rearranged steranes indicate that it is less mature than the others in this group. The steranes occur in low abundance in the Montrose oil. It appears that this is a high maturity oil, depleted in high molecular weight hydrocarbons, which has extracted biomarker compounds from less-mature rock somewhere along its migration pathway.

Finally, the Beryl Field presents an interesting case. Eleven oils covering a broad range of maturities (see Fig. 8(a)) were studied. The oils were taken from reservoirs between 9000 and 11 000 ft (2700–3300 m) deep. There is no consistent correlation between maturity and reservoir depth. Although stacked reservoirs in one well have oils whose maturities increase with depth, the difference in depth is not sufficient to invoke thermal alteration in the reservoir. In another case, the shallowest oil in the well is the most mature.

Seven successive maturity levels have been numbered, with 1 as the least mature, for comparison of their isotope profiles, shown in Fig. 8(b). It is apparent that there is a systematic change in shape from the steep immature profiles to the flatter mature profiles. There is very little change in the $\delta^{13}C$ value of the residua. This is an excellent example of successive episodes of generation and expulsion from an increasingly more mature source rock within one drainage area. The timing of generation can be determined in a relative sense. It may be possible to determine pathways of migration for these oils by integrating the timing data with paleo-reconstructed structure maps of the Beryl Field.

CONCLUSIONS

Several overall conclusions can be drawn from this study. Biomarker analyses show that the North Sea oils have been generated in one common source. Isotope profiles indicate that different organic facies of this source can be recognized by identifying variations in the isotopic composition of the oils. This information can be used for determining the continuity of reservoir sands and for identifying accumulations which have migrated from different drainage areas. Isotope profiles are also useful for determining the relative thermal maturities of oils in the subfamilies of the North Sea, and whether the maturity was due to pre- or post-expulsion thermal exposure. The geochemical interpretations, when combined with geological data, can be helpful in determining timing

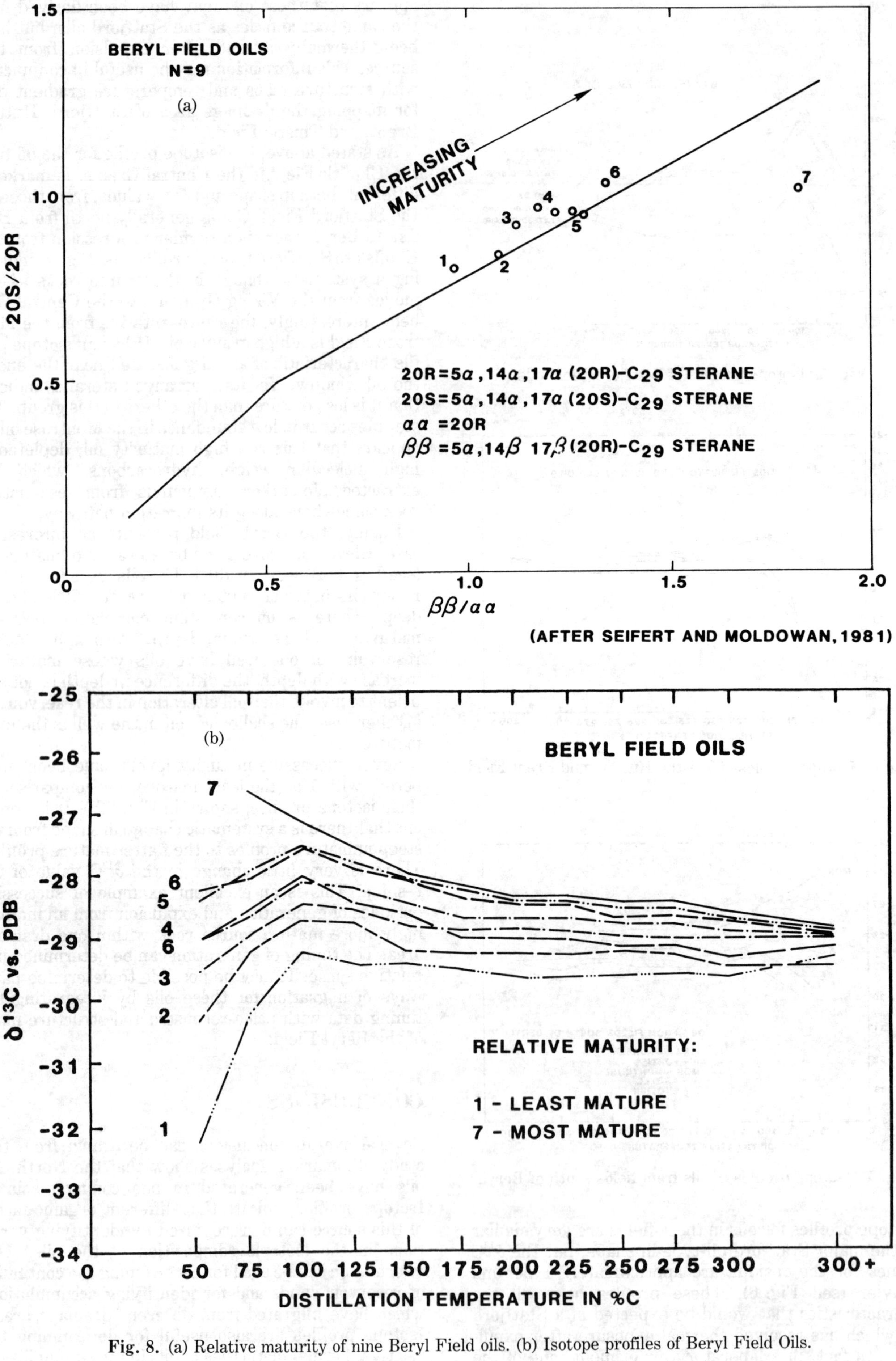

Fig. 8. (a) Relative maturity of nine Beryl Field oils. (b) Isotope profiles of Beryl Field Oils.

of migration and the history of structural development.

ACKNOWLEDGEMENTS

R. M. Squires of Mobil Research and Development Corporation constructed the isotope profiles used in this report. His efforts and discussions are greatly appreciated. W. L. Orr, also of MRDC, added valuable insight to the interpretation of the isotope data. I thank the management of Mobil Research and Development Corporation for releasing this work for outside publication.

REFERENCES

Cornford, C., Morrow, J. A., Turrington, A., Miles, J. A. and Brooks, J. 1983. Some geological controls on oil composition in the UK North Sea. In: Brooks, J. (ed.), *Petroleum Geochemistry and Exploration of Europe*. Blackwell Scientific, Oxford, pp. 175–194.

MacKenzie, A. S., Lamb, N. A. and Maxwell, J. R. 1982. Steroid hydrocarbons and the thermal history of sediments. *Nature*, **295**, 223–226.

Seifert, W. K. and Moldowan, J. M. 1981. Paleoreconstruction by biological markers. *Geochim. Cosmochim. Acta*, **45**, 783–794.

7 Oil–oil and oil–source rock correlation, Northern North Sea

L. Schou, S. Eggen† and M. Schoell‡*
*Continental Shelf Institute (IKU), trondheim, Norway
†Statoil, Stavanger, Norway
‡Chevron Oil Field Research Company, La Habra, California, USA

Nine oils from seven different discoveries in the Northern North Sea area and 10 Jurassic shale samples from three wells in the same area have been analysed and correlated using molecular marker and isotopic properties. The oils were divided into two groups, the main difference between the groups being the maturity. Highest maturity is encountered in the two samples from blocks 30/4 and 35/8. The most likely source kerogens for the oils are different maturity stages of the Draupne and Heather Formations such as in 30/7-7, possibly with some input from the Drake Formation, represented by the maturity stage in 30/3-1.

INTRODUCTION

Nine oil samples from seven different discoveries in the Northern North Sea area and ten Jurassic shale samples from three wells in the same area (Fig. 1, Table 1) have been geochemically analysed and compared. The main objective was to study the relationship between oils trapped in different structures and potential source rocks in the same area.

GEOLOGICAL SETTING

The main tectonic framework of the area studied is shown in Fig. 1.

The Northern North Sea Basin is an area that underwent rifting and crustal stretching during the 'Kimmerian' tectonic episode. Crustal thinning during Jurassic to early Cretaceous times resulted in strong thermal subsidence in the Cretaceous. The depocentre during the Upper Jurassic to Cretaceous was centred along an axial graben trend (see Fig. 2).

The Northern North Sea area is a prolific oil-producing basin. The combination of rich and mature source rocks, large structures and excellent reservoir quality have resulted in oil or gas being found in many structures in the area.

The main source rocks in the area are believed to be Jurassic organic-rich shales (Rønnevik *et al.*, 1983). Shales of Jurassic age are thermally immature on the Horda Platform, whereas in the graben area oil maturity is reached at approximately 3.5 km. The Jurassic in the central part of the graben is at present in the gas window (deeper than 5 km) (Rønnevik *et al.*, 1983).

Potential source rocks are found in several formations within the Jurassic. Both the Upper Jurassic (Draupne and Heather Formations), and the Lower Jurassic (Drake and Burton/Amundsen Formations) have the characteristics of potentially rich source rocks for both oil and gas.

Based on organic richness and kerogen type, the Draupne Formation is by far the 'richest' source rock in the area. In terms of migration efficiency, the Lower Jurassic and intra-Brent shales may be more effective source rocks due to close juxtaposition with carrier/reservoir beds.

SAMPLES AND EXPERIMENTAL METHODS

Oil samples

Oil samples were obtained from the operators for the different discoveries that were to be studied. The samples were stored in glass bottles for a short time before they were sent for analysis.

Rock samples

A major problem in correlation studies is to obtain representative samples from a source rock interval. An attempt was made to solve this problem by picking a large number of samples throughout the formations studied, and then mixing them before extraction. One analysed rock sample in this study therefore repre-

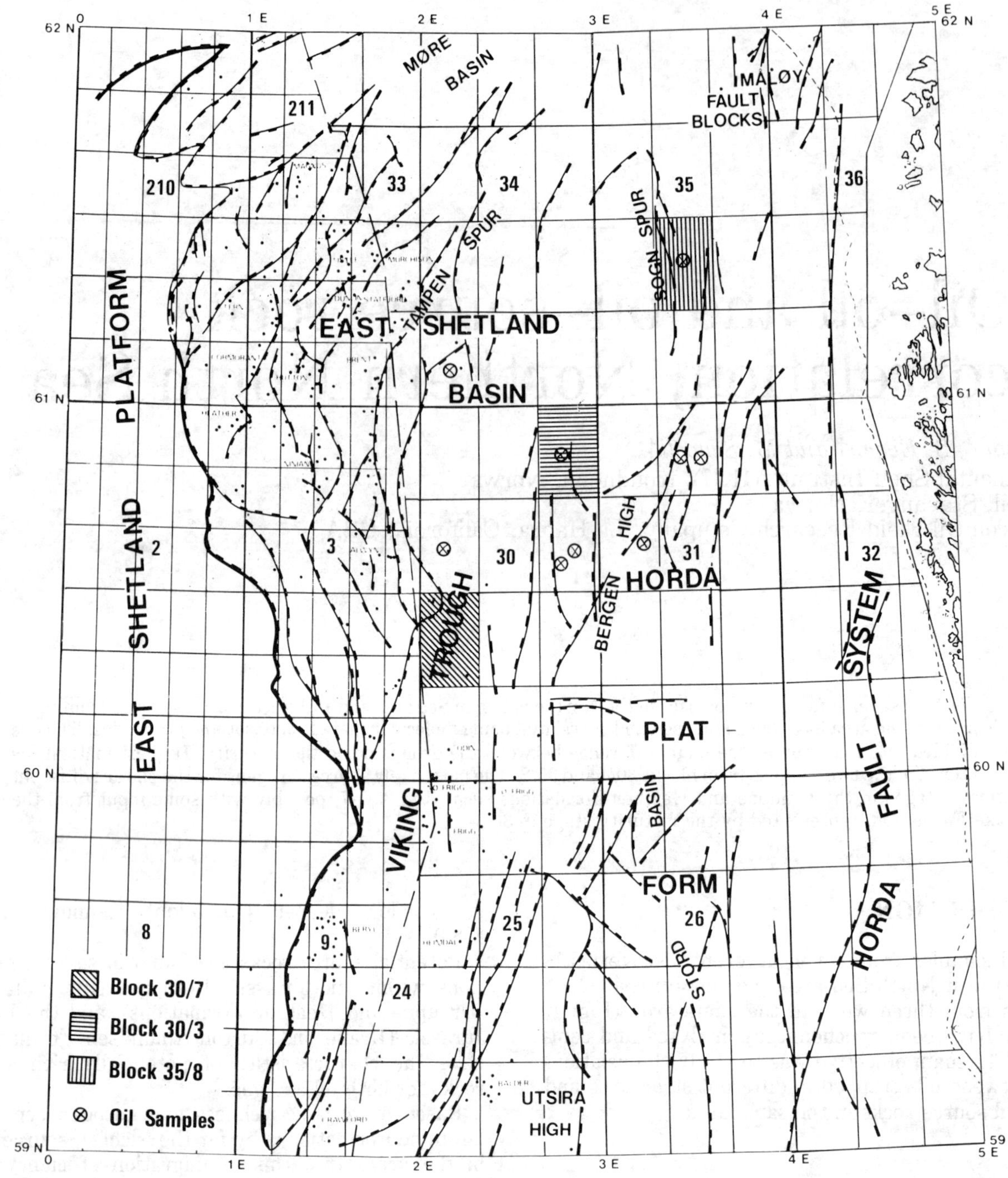

Fig. 1. Main structural elements, top Jurassic.

sents the average over a several hundred metre thick interval (see Table 1).

Experimental methods

A flow diagram (Fig. 3) illustrates the different analytical techniques which were employed.

Total organic carbon (TOC) analyses were run on samples pretreated with HCl for carbonate removal. The determinations were made on a Leco CR12 carbon analyser.

Rock–Eval pyrolysis: crushed samples (100 mg) were analysed on a Rock–Eval pyrolyser.

Extractable organic matter (EOM) and chromatographic separation: approximately 50 g of powdered rock was extracted by flowblending for 3 min using dichloromethane (DCM) as solvent. Aliquots of the oil samples were diluted with n-pentane (40 ml/mg), left at 4 °C for 24 h, then filtered to separate the asphaltene fractions. The filtrates and the EOM from the extracts were separated into a saturated, an aromatic and a non-hydrocarbon fraction using a MPLC system with n-hexane as eluant (Radke *et al.*, 1980).

Gas chromatographic (GC) analyses: all of the GC analyses were performed on polar glass capillary columns (OV-1 or OV-101). Hydrogen was used as the carrier gas for the whole oils and the saturated hydrocarbons, while helium was used for the aromatic hydrocarbons. Medium length columns (25 and 30 m) were used for the whole oils and the saturated compounds while a 100 m column with a specially designed backflushing system was used for the separation of the aromatic hydrocarbons. The analyses were performed on three different GCs, an HP 5730 (whole oils), a Varian 3700 (aromatic hydrocarbons) and a Carlo Erba Fractovap 2150 (saturated hydrocarbons). The

TABLE 1. Source rock and oil samples analysed

Figure Sample No.	*Code*	*Well*	*Depth interval (m)*	*Formation*
Source rocks				
M-9138(31)		30/3-1	3873–3950	Drake
M-9139(32)		30/3-1	4128–4227	Burton/Amundsen
M-9028(11)		30/7-7	3900–3925	Draupne
M-9029(12)		30/7-7	3925–4200	Heather
M-9030(13)		30/7-7	4200–4500	Heather
M-9031(14)		30/7-7	4500–4730	Drake
M-9032(15)		30/7-7	4790–4910	Burton/Amundsen
M-9235(21)		35/8-1	3180–3350	Draupne
M-9136(22)		35/8-1	3350–3500	Heather
M-9137(23)		35/8-1	3700–3800	Drake
Oils				
M-49		31/2-2	1512	Upper Jurassic
M-50		30/4-2	3870	Middle Jurassic
M-51		31/2-5	1503	Upper Jurassic
M-52		31/4-5	2135	Middle Jurassic
M-53		35/8-1	3565	Middle Jurassic
M-54		30/6-4	2630	Middle Jurassic
M-55		34/10-2	3355	Middle Jurassic
M-56		30/6-3	2430	Middle Jurassic
M-57		30/3-2	2870	Middle Jurassic

temperature programs applied were 80 °C (2 min) to 260 °C at 4 °C/min for the hydrocarbon fractions and −50 °C (2 min) to 260 °C at 4 °C/min for the whole oils. The data processing for all the GC analyses was performed on a VG Multichrom system.

GC–MS analysis: the analyses were performed on a VG 70–70H GC–MS–DS equipped with a Varian 3700 GC. Helium was used as the carrier gas, but apart from this the other GC conditions were similar to the single GC analyses. Multiple ion detection (MID) was applied for the saturated steranes and terpanes, and the ratios from the mass chromatograms were calculated from peak heights in the appropriate chromatogram.

Isotope analysis: the methods for isotopic analysis are as described in Schoell *et al.* (1983). The internal standard NBS 22 lubricating oil ($\delta^{13}C$) was run eight times during the period of analysis, and revealed a value of -29.73 ± 0.05.

RESULTS AND DISCUSSION

Source rocks

Data on the source rock intervals studied are presented in Tables 1 and 2. Samples have been picked from wells where the Jurassic formations are within the oil window (3–5 km) at present. However, few wells have been drilled in the deeper part of the graben, thus limiting the choice of suitable locations in this area.

The organic content of the studied source rocks varies from 1.6% to 4.32%, the richest being in the Draupne and Heather Formations of wells 30/7-7 and 35/8-1. The Lower Jurassic shales have less than 2% organic carbon content. It should be borne in mind that these TOC values represent *averages*, and as such they fit well with the natural spread in values published by Ronnevik *et al.* (1983).

The kerogens were separated out and analysed, both for elemental and for isotopic composition. In a modified van Krevelen diagram the samples from 35/8–1 plot in the transition area between immature/mature (Fig. 4). The 30/3–1 samples plot in the oil window together with samples from 30/7–7.

The high maturity of the samples from both 30/3–1 and 30/7–7 makes their kerogen classification dubious. However, type II or mixed type II/III can be inferred from Fig. 4. The deepest sample from 35/8–1 (Lower Jurassic) appears to be the only pure type III kerogen (see Tables 1 and 2). The elemental analysis and pyrolysis data are in good agreement.

Isotope data on kerogens and extracts as well as elemental analyses are plotted in Fig. 5 and summarized in Table 3. It is evident that no dramatic change in isotopic composition takes place with increasing depth of burial. Neither is there a relationship between geographic position and isotopic composition of the kerogens. Only the samples from 30/7–7 show an increase in negative numbers versus depth. Six out of 10 kerogens have an almost identical isotopic composition, with a mean value of −25.4 ± 0.1% (three samples from 30/7–7 and all samples from 35/8–1). This could be an effect of the way the samples were averaged before analysis.

The parallel change in the isotopic composition of the kerogens and the extracts suggests that most of the extracts are indigenous. The two lowermost samples of 30/7–7 are strongly depleted in heavy C-isotopes in the NSO and aromatic fractions (Fig. 5), although this may be a result of incomplete compound class separation.

The general distribution of the terpane and sterane biomarkers confirms that there are only relatively small variations between the various formations in the three wells. The main differences are thought to be due to differences in maturity. The steranes in the Draupne and Heather Formations show a very similar distribution; the main difference seen is the maturity-related increase in $14\beta,17\beta$-steranes and in %20S $14\alpha,17\alpha$-steranes with increasing burial depth. A similar picture is seen in the terpane distribution (shown in Fig. 6). The higher maturity of 30/7–7 is seen both in the absolute amounts of C30 hopane and in the maturity ratios. The Tm/Ts ratio (Seifert and

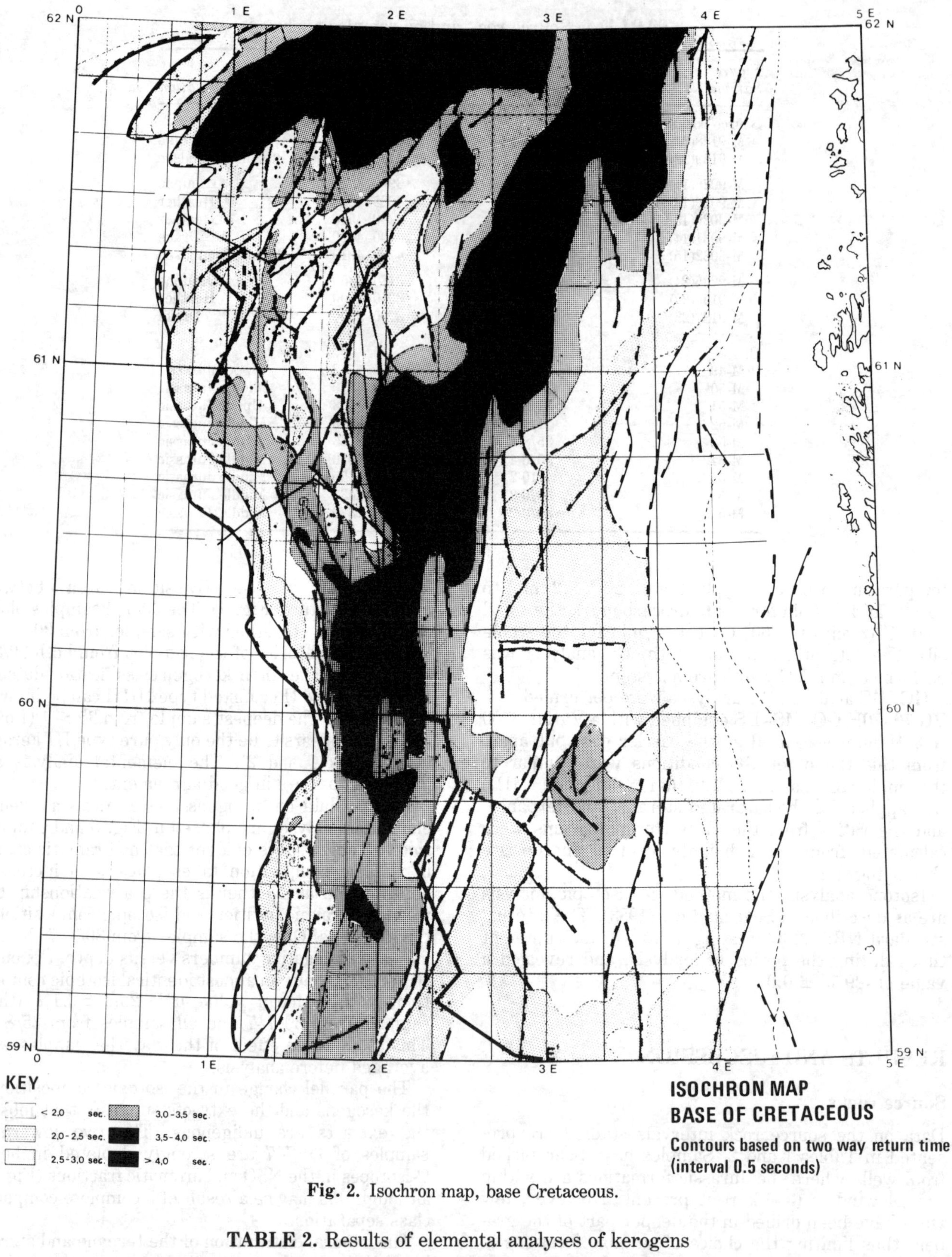

Fig. 2. Isochron map, base Cretaceous.

TABLE 2. Results of elemental analyses of kerogens

Figure code	*Sample no.*	*Well*	*Depth (m)*	*C (%)*	*H (%)*	*O (%)*	*Ash (%)*	*H/C*	*O/C*
11	M-9028	30/7-7	3900–3925	26.86	1.31	2.17	40.0	0.585	0.061
12	M-9029	30/7-7	3925–4200	30.41	1.50	2.38	42.3	0.592	0.059
13	M-9030	30/7-7	4200–4500	27.78	1.50	2.48	42.0	0.648	0.077
14	M-9031	30/7-7	4500–4730	27.88	1.11	3.11	42.6	0.478	0.084
15	M-9032	30/7-7	4790–4910	29.37	1.23	3.34	41.8	0.502	0.085
21	M-9135	35/8-1	3180–3350	33.14	2.49	4.36	37.7	0.902	0.099
22	M-9136	35/8-1	3350–3500	24.08	1.85	3.24	45.4	0.922	0.101
23	M-9137	35/8-1	3700–3800	15.20	1.00	3.40	52.6	0.789	0.168
31	M-9138	30/3-1	3875–3950	29.28	1.77	3.46	42.3	0.725	0.089
32	M-9139	30/3-1	4128–4227	34.94	1.88	4.71	37.5	0.646	0.101

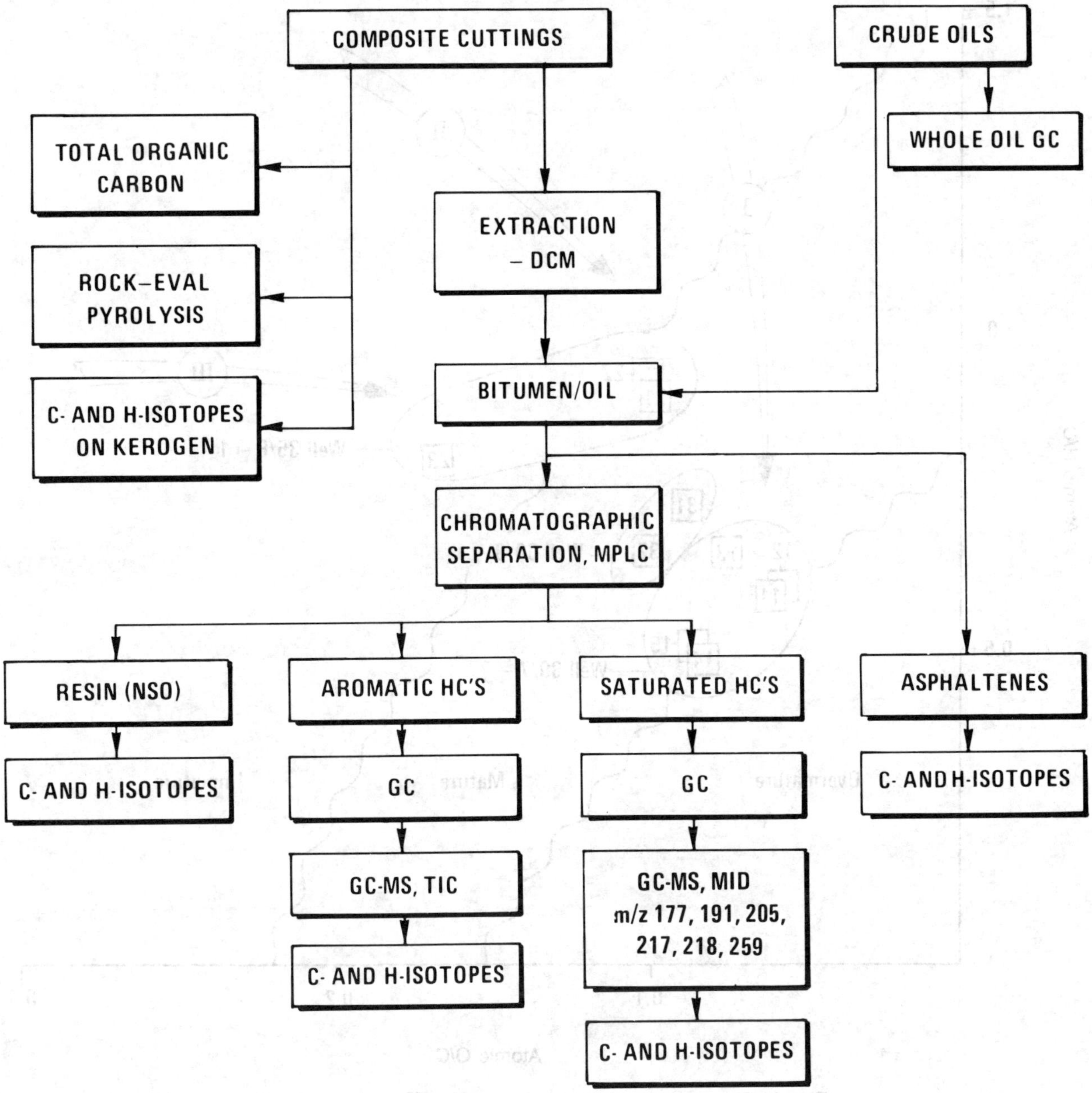

Fig. 3. Flow diagram, analytical procedure.

TABLE 3. Results of isotopic analysis of kerogens and extracts

				SHC		AHC		NSO		Kerogen	
Figure code	*Sample no.*	*Depth (m)*	*Well*	$\delta^{13}C$	δD	$\delta^{13}C$	δD	$\delta^{13}C$	δD	$\delta^{13}C$[a]	δD[b]
11	M-9028	3900–3925	30/7-7	−27.7	−113	−26.0	−116	−25.8	−110	−25.4	−105
12	M-9029	3925–4200	30/7-7	−27.7	−114	−26.0	−109	−25.6	−105	−24.4	−106
13	M-9030	4200–4500	30/7-7	−27.9	−120	−26.4	−105	−26.1	−104	−24.6	−105
14	M-9031	4500–4730	30/7-7	−28.3	−109	−27.6	−103	−27.9	−105	−25.3	−103
15	M-9032	4790–4910	30/7-7	−28.4	−117	−27.4	−105	−27.7	−103	−25.5	−130
21	M-9135	3180–3350	35/8-1	−28.4	−138	−26.8	−124	−26.7	−124	−25.6	−119
22	M-9136	3350–3500	35/8-1	−28.0	−131	−26.6	−116	−28.8	−125	−25.5	−123
23	M-9137	3700–3800	35/8-1	−27.9	−121	−26.3	−118	−26.4	−112	−25.3	—
31	M-9138	3873–3950	30/3-1	−28.7	−124	−27.2	−121	—	−119	−26.3	−111
32	M-9139	4128–4227	30/3-1	−28.5	−123	−27.3	−115	−27.5	−112	−27.0	−106

[a]Decarbonated.
[b]Demineralized.

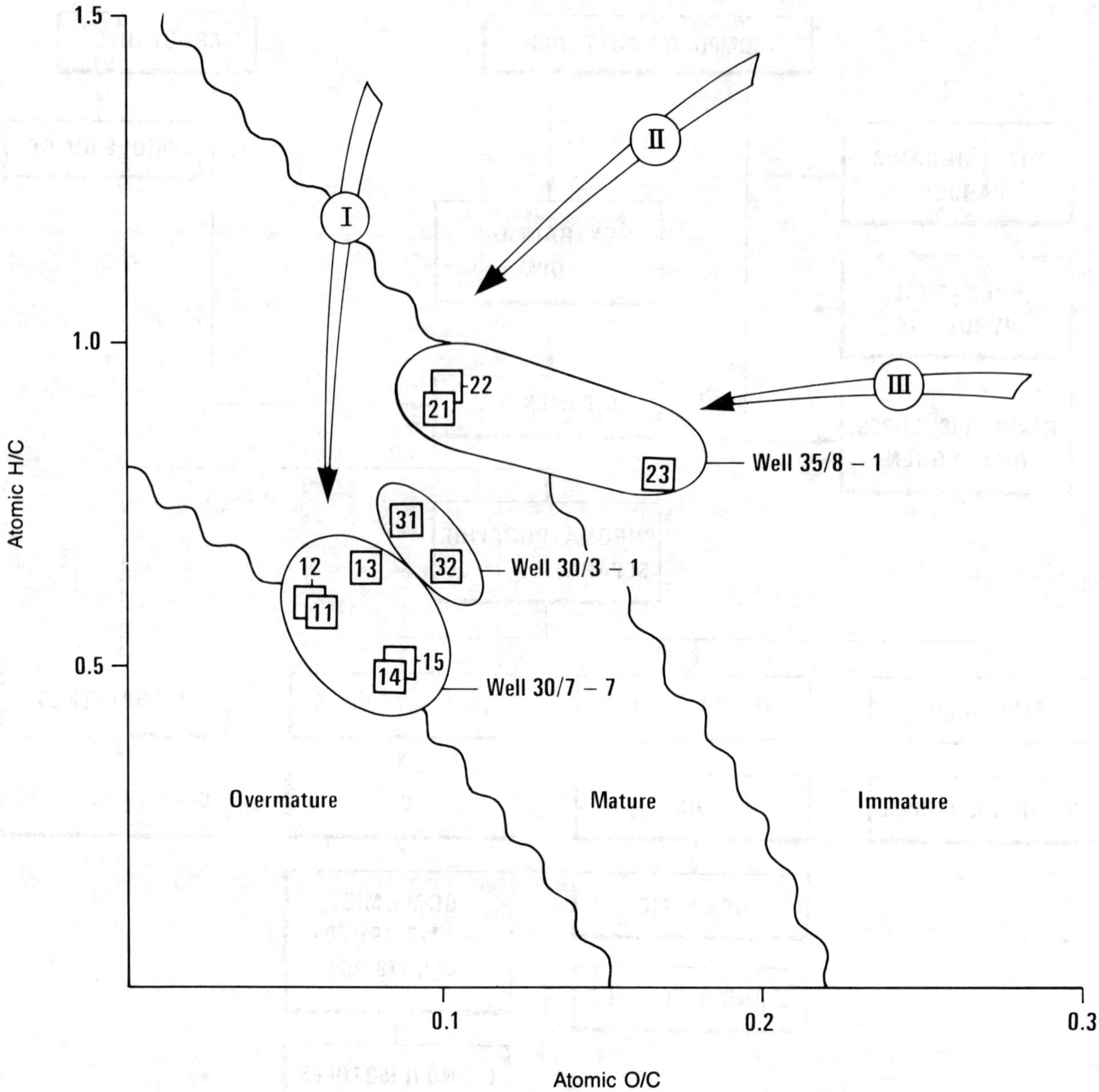

Fig. 4. Elemental analysis of kerogen (Van Krevelen diagram).

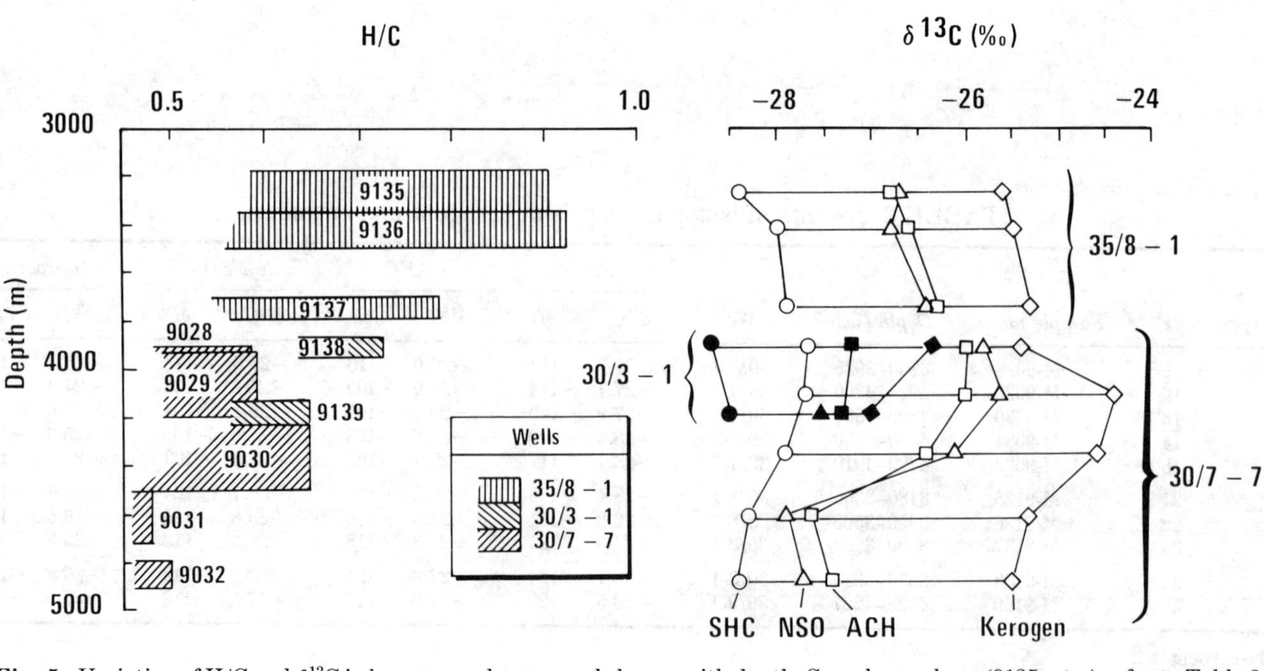

Fig. 5. Variation of H/C and δ^{13}C in kerogen and compound classes with depth. Sample numbers (9135, etc.) refer to Table 2.

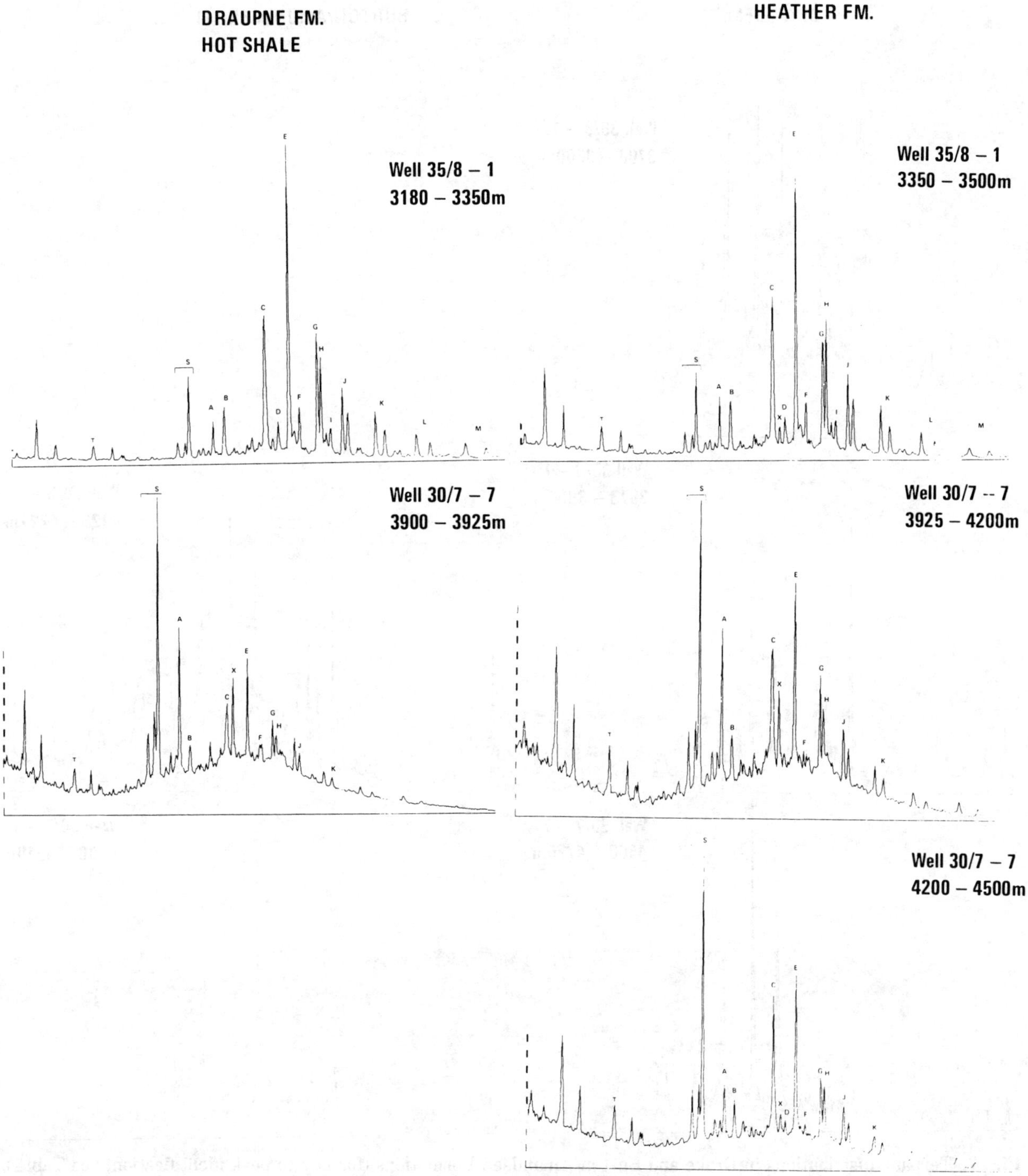

Fig. 6. Terpane distributions in Draupne and Heather Formations (for key to peak identification, see Table 6).

Moldowan, 1978) has a very low value in the top two samples in 30/7–7, compared with the other three samples. These uppermost two samples also contain a high relative amount of an unknown compound (X), that is very often seen in samples where the general concentration of terpanes is low.

The difference between the two Heather samples in 30/7–7 may be an effect of the composite samples that were used. From the terpane distribution, the uppermost sample of the two seems to be a mixture of the Draupne one above and the one below. This could be due to caving of Draupne material into the uppermost Heather interval, or it may reflect a more-or-less continuous change in character from Draupne down through the Heather.

The lowermost Heather sample in 30/7–7 is similar to the Draupne and Heather samples in 35/8–1, apart from the lower concentration of terpanes in 30/7–7. This concentration difference is probably caused by the difference in maturity of the two wells.

The Drake and Burton/Amundsen samples in the three wells show the same increase in the sterane maturity parameters with increasing depth, as was seen for the Draupne and Heather samples. The main difference seems to be a higher input of C29 steranes in the 30/7–7 sample. In the terpane distribution the picture is somewhat more complex (Fig. 7). In 30/7–7, the absolute content of hopanes in both the Drake and the Burton/Amundsen Formations is increased by a factor of 10 compared with the Heather samples in this well. The Burton/Amundsen sample in 30/3–1 is different from the others, in that the lower molecular weight tricyclic terpanes are more abundant in 30/3–1.

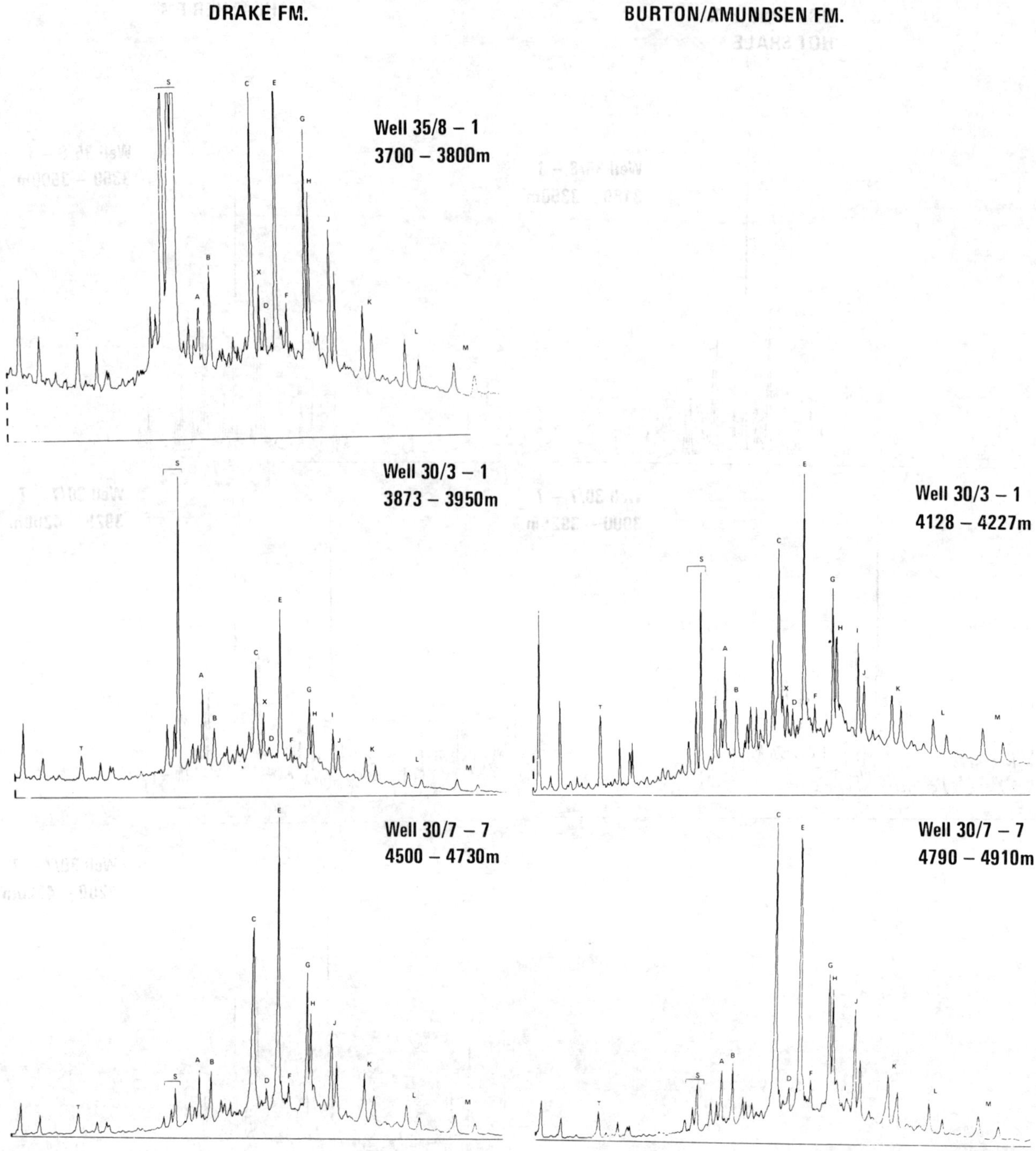

Fig. 7. Terpane distributions in Drake and Burton/Amundsen Formations (for key to peak identification, see Table 6).

Thus, most of the maturity ratios show fairly good depth relationships, the only ratio that does not is the Tm/Ts ratio, which has no clear trend. This may be due to variations in organic matter input. It is apparent that the Drake and Burton/Amundsen samples from 30/7–7 and the Burton/Amundsen sample in 30/3–1 are significantly different from the other samples.

In summary, all of the intervals, apart from the deepest Drake in 35/8–1, may have potential for generating oil. The Draupne and Heather intervals in 35/8–1 are immature, while all the other samples are from within the oil window.

Oils

The majority of the oils analysed were taken from oil and oil/gas discoveries on the Horda Platform (Fig. 1). Two liquid samples from an oil/condensate field (35/8) and a gas/condensate field (30/4 and 30/7) situated on the flanks of the Viking Trough were also analysed, as was one sample from the Tampen Spur area (34/10). The oils were all from Jurassic reservoirs at depths ranging from 1500 m to nearly 4000 m (see Table 1). A decrease in density (API gravity) related to depth is evident in Fig. 8. The two deepest samples may be classified as condensates.

The isotopic data show that within the six shallowest oil samples there is a trend towards more negative C- and D-isotope values with increasing depth (Fig. 8 and Table 4). The two deepest samples (from 30/4–2 and 35/8–1) seem to fall on a different trend.

The gross composition, determined by chromatographic separation, is presented in Table 5. All of the oils analysed have more than 60% light components. Three of the oils (31/2–5, 30/6–4, 30/3–2) have high

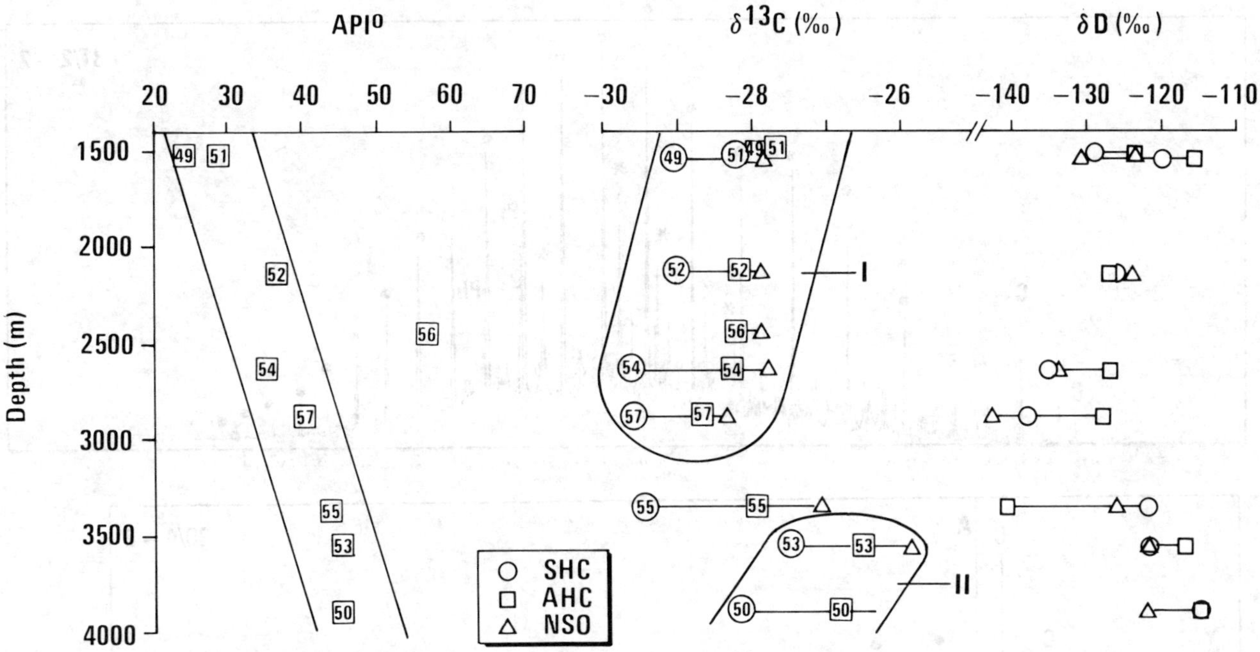

Fig. 8. Variation of gravity of oils and isotopic composition of compound classes with depth. Sample numbers (49, etc.) refer to Table 4. I and II are proposed oil groups.

TABLE 4. Reservoir data and isotopic composition of the oil fractions

Sample no.	*Depth (m)*	*API (degrees)*	*GOR (m³/m³)*	*SAT* δ^{13}	*SAT* δD	*ARO* $\delta^{13}C$	*ARO* δD	*HET* $\delta^{13}C$	*HET* δD
M-51	1503	27.6	66	−28.8	−129	−27.8	−124	−27.8	−124
M-49	1512	24.0	59	−29.0	−120	−28.0	−116	−27.9	−131
M-52	2135	35.5	82	−28.0	−126	−28.2	−127	−27.9	−124
M-56	2430	54.6	3230	—	—	−28.2	—	−27.9	—
M-54	2630	32.2	100	−29.6	−135	−28.3	−127	−27.8	−134
M-57	2870	38.9	84	−29.6	−138	−28.7	−128	−28.3	−143
M-55	3355	43.8	200	−29.4	−122	−28.0	−141	−27.1	−126
M-53	3565	44.4	3953	−27.5	−122	−26.6	−117	−25.9	−122
M-50	3870	42.7	4715	−28.2	−115	−26.9	−115	−26.3	−122

TABLE 5. Gross composition

Sample no.	*Amount of oil (mg)*	*Sat HC (mg)*	*Sat HC (%)*	*Aro HC (mg)*	*Aro HC (%)*	*HC (mg)*	*HC (%)*	*NSO (mg)*	*NSO (%)*	*Asph. (mg)*	*Asph. (%)*	*Light HC (mg)*	*Light HC (%)*
M-49	178.7	48.1	27	18.7	10	66.8	37	5.0	3	0.2	0.1	106.7	60
M-50	219.1	49.1	22	9.8	4	57.9	26	3.9	2	0.9	0.4	156.4	71
M-51	202.2	37.1	18	19.3	10	56.4	28	13.9	7	3.2	1.6	128.7	64
M-52	205.2	44.4	22	16.3	8	60.7	30	9.2	4	1.5	0.7	134.0	65
M-53	202.4	45.8	23	12.9	6	58.7	29	3.8	2	0.3	0.2	139.6	70
M-54	195.4	42.5	22	19.3	10	61.8	32	12.4	6	3.5	1.8	177.7	60
M-55	208.6	41.0	20	11.1	5	32.1	25	1.9	1	0.01	0.00	154.6	74
M-56	215.3	20.5	10	7.2	3	27.1	13	0.8	0.3	0.1	0.05	187.3	87
M-57	219.6	49.6	23	14.4	7	64.0	29	7.9	4	3.8	1.7	143.9	66

asphaltene content (more than 1%). The hydrocarbon content varies from 17% to 37%.

Gas chromatographic separation of the oils indicates that there are significant differences in some of the samples. Whole oil gas chromatograms (nine oils) reveal that the general distribution of hydrocarbons is similar for seven of the oils (represented by the oil from 30/4 in Fig. 9). The exceptions are the two oils from 31/2, which seem to have undergone different degrees of biodegradation: one shows a significant loss of light compounds, while the other has lost all n-alkanes. Figure 9 shows the GC traces representative of the three different types of oil. GC analyses of the saturated hydrocarbon fractions distinguish only two of the groups, since the C15+ hydrocarbons have the same general distribution in all samples apart from the most heavily biodegraded one. Aromatic hydrocarbons exhibit only small variations betwen the oils.

GC–MS analyses show that the sterane distribution is very similar in all the oils, although some differences are seen within the terpanes (see Fig. 10). Varying amounts of 28,30-bisnorhopane and tricyclic terpanes are seen, in addition to more maturity-related variations. From these data there seems to be two

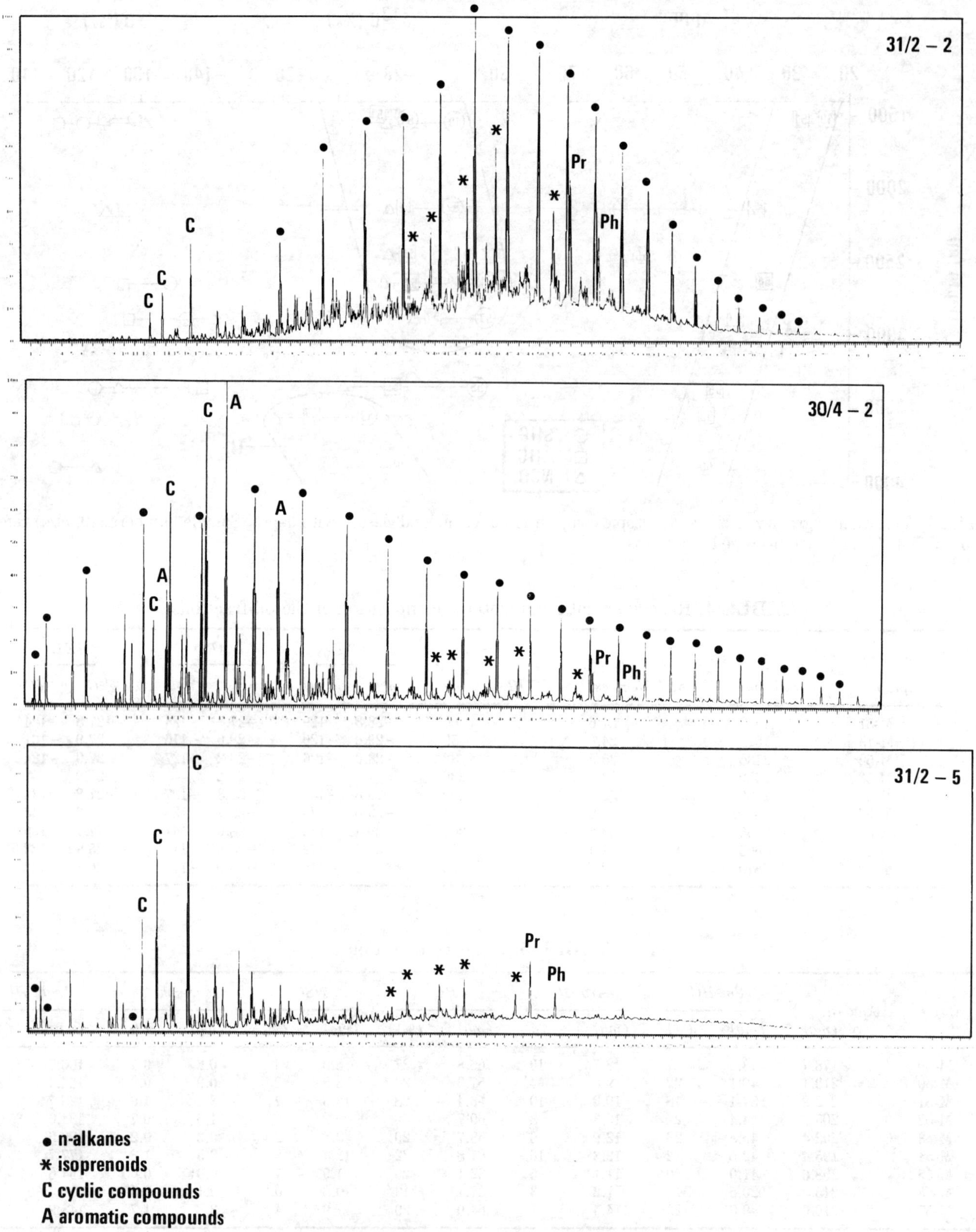

Fig. 9. Whole oil gas chromatograms.

main groups of oils, which is in good agreement with the isotopic data. The oils from 30/4 and 35/8 form a different group (group II) from the rest of the oils (group I), see Fig. 8 and Table 4.

Maturity of oils and source rocks

In an attempt to compare the maturity of the oils with source rock maturity, plots of some maturity ratios against depth for both oils and source rock samples were made. The plot representing the relative amount of 14β,17β C29 steranes is shown in Fig. 11. A fairly good linear increase occurs with depth for source rock values up to 70–75%. Cornford *et al.* (1983) state that the upper part of the curve with the biggest changes in the values is in the early mature range, while the deeper source rocks are moderately mature. All of the oils fall within a narrow range with values between 70% and 80%. These data would suggest that source rocks of maturities similar to those below approximately 3800–4000 m have sourced these oils.

A similar relationship was seen for the relative

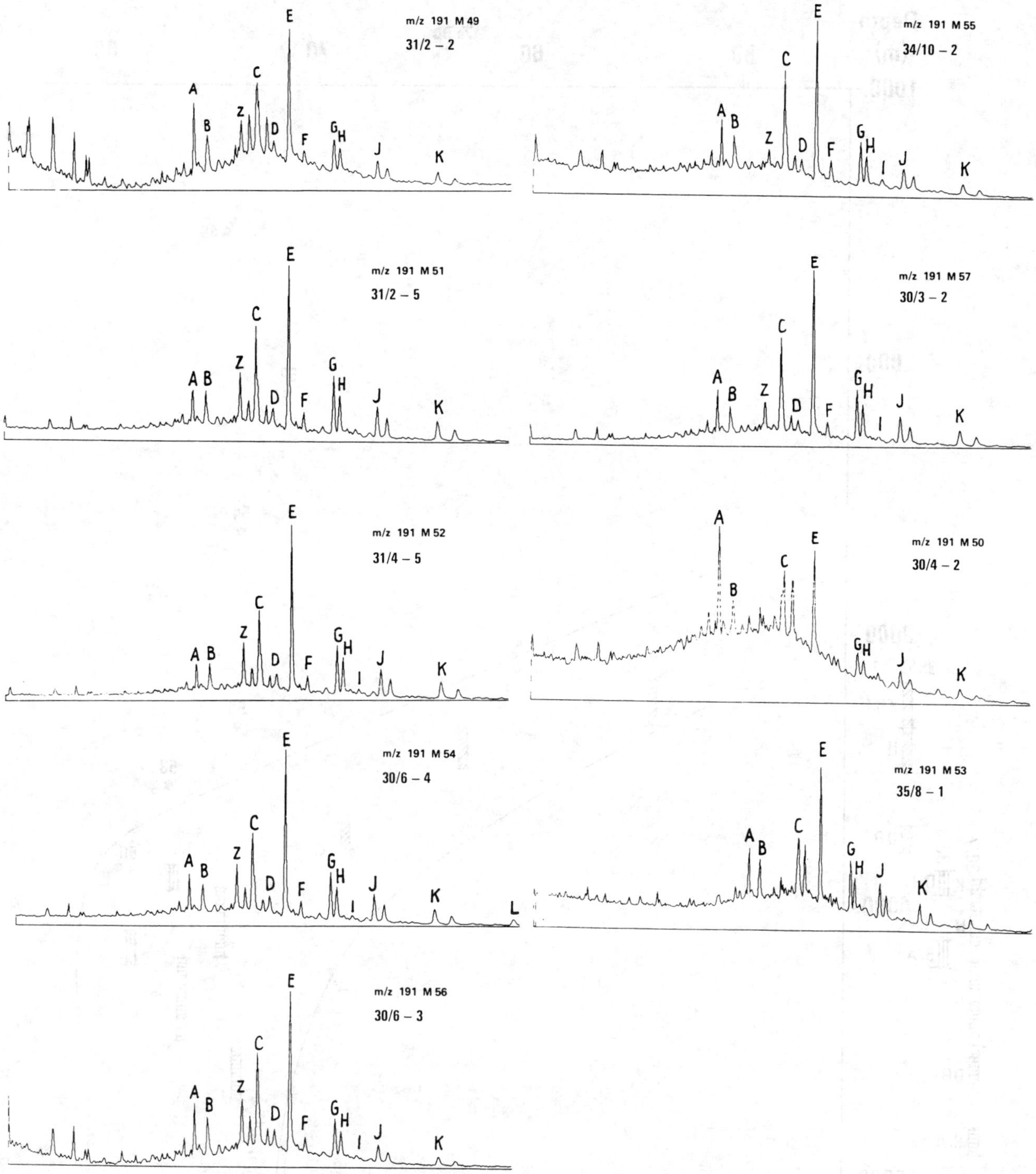

Fig. 10. Terpane distribution in oil samples (for key to peak identification, see Table 6).

TABLE 6. Mass chromatograms representing terpanes (*m/z* 191)

A	Ts, 18α(H)-trisnorneohopane	$C_{27}H_{46}$	K	17α(H)-trishomohopane (22S,22R)	$C_{33}H_{58}$
B	Tm, 17α(H)-trisnorhopane	$C_{27}H_{46}$	L	17α(H)-tetrakishomohopane (22S,22R)	$C_{34}H_{60}$
C	17α(H)-norhopane	$C_{29}H_{50}$	M	17α(H)-pentakishomohopane (22S,22R)	$C_{35}H_{62}$
D	17β(H)-normoretane	$C_{29}H_{50}$	Z	bisnorhopane	$C_{28}H_{48}$
E	17α(H)-hopane	$C_{30}H_{52}$	X	unknown triterpane	$C_{30}H_{52}$
F	17β(H)-moretane	$C_{30}H_{52}$	P	tricyclic terpane	$C_{23}H_{42}$
G	17α(H)-homohopane (22S)	$C_{31}H_{54}$	Q	tricyclic terpane	$C_{24}H_{44}$
H	17α(H)-homohopane (22R) + unknown triterpane (gammecerane?)	$C_{31}H_{54}$	R	tricyclic terpane (17R,17S)	$C_{25}H_{46}$
			S	tetracyclic terpane	$C_{24}H_{42}$
I	17β(H)-homomoretane	$C_{31}H_{54}$	T	tricyclic terpane (17R,17S)	$C_{26}H_{48}$
J	17α(H)-bishomohopane (22S,22R)	$C_{32}H_{56}$			

Letters refer to Figs. 6, 7 and 10.

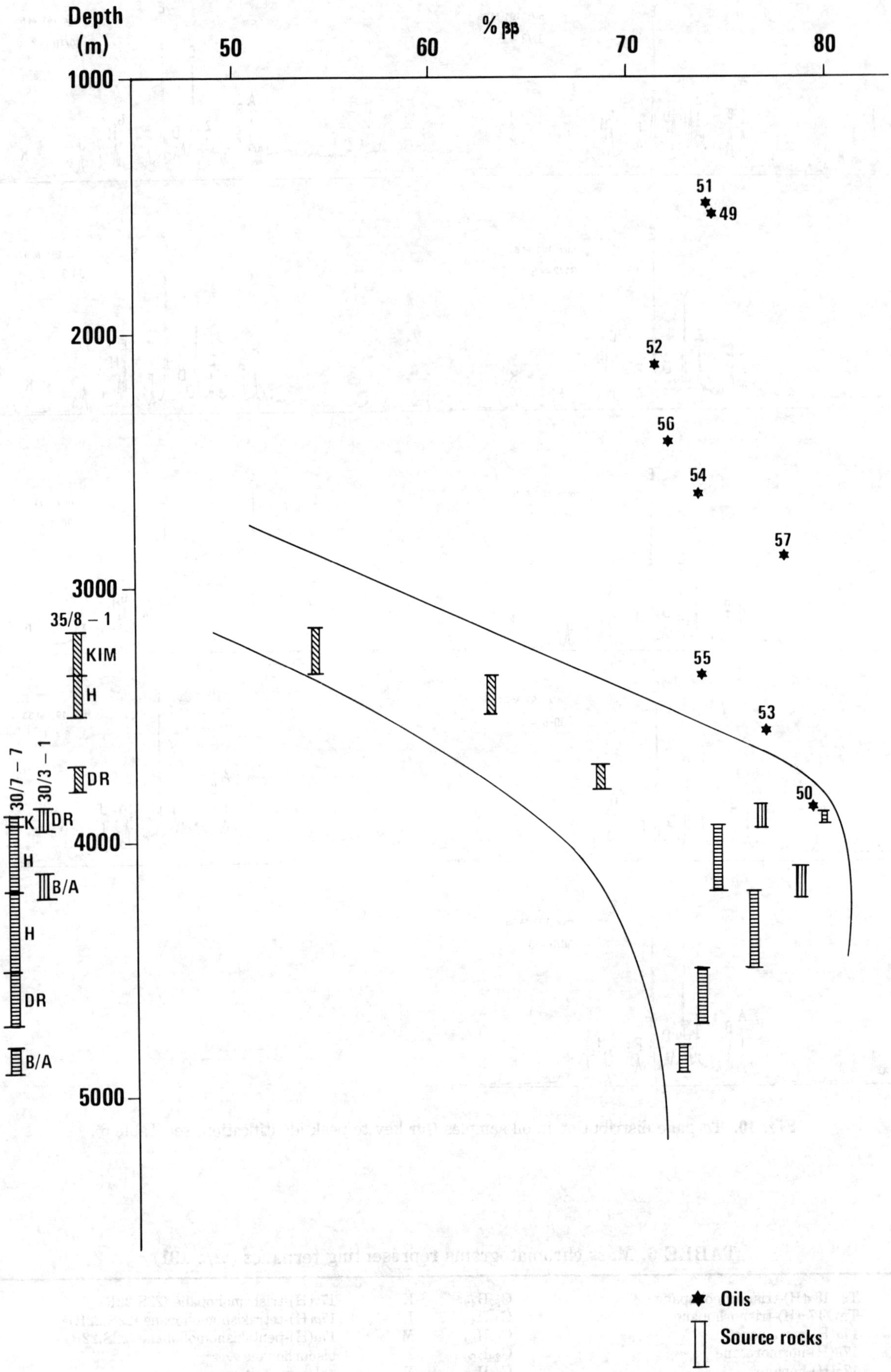

Fig. 11. Depth plot of percentage $\beta\beta$ C_{29} steranes. Sample numbers (49, etc.) and formations refer to Table 1.

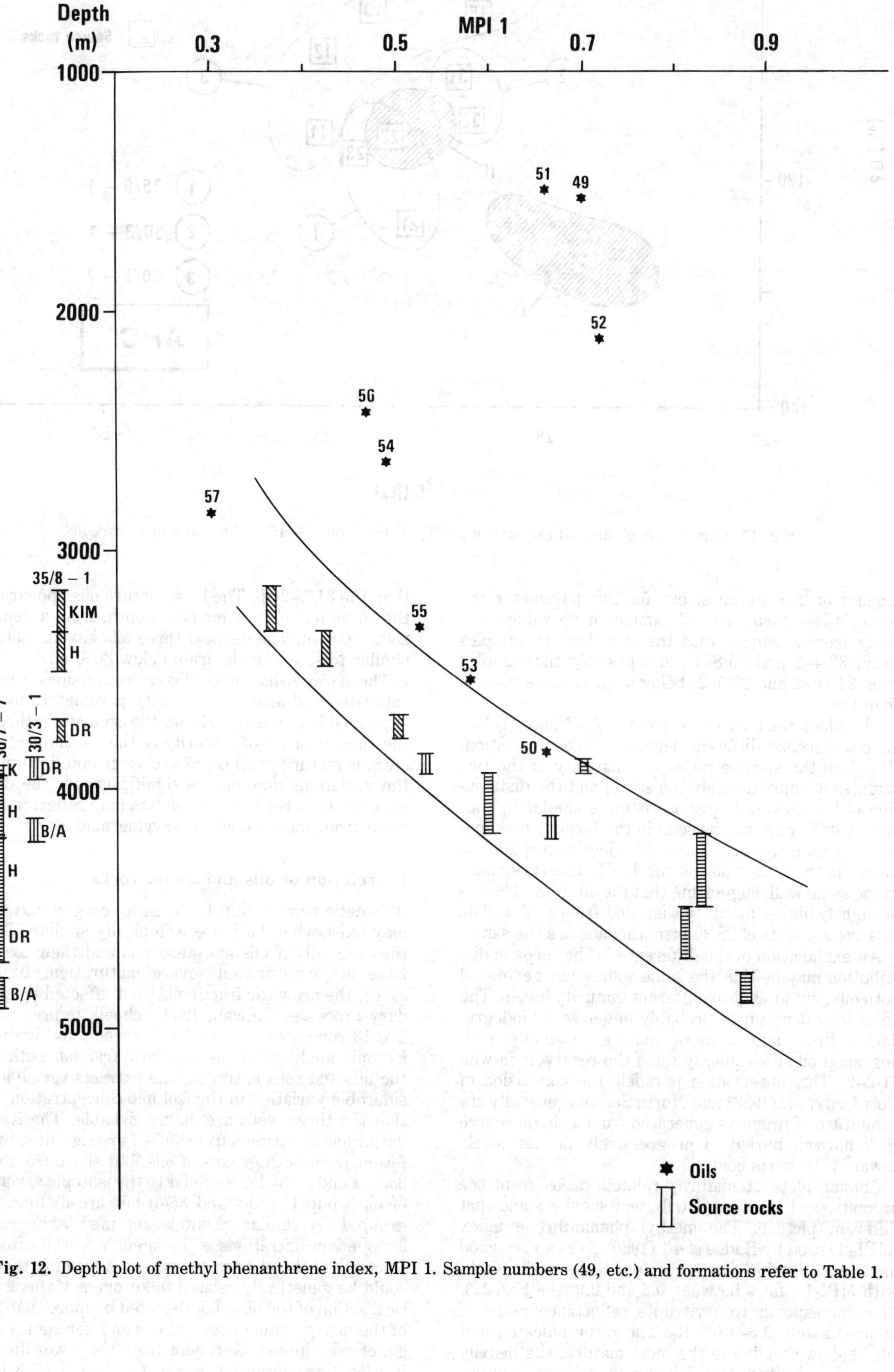

Fig. 12. Depth plot of methyl phenanthrene index, MPI 1. Sample numbers (49, etc.) and formations refer to Table 1.

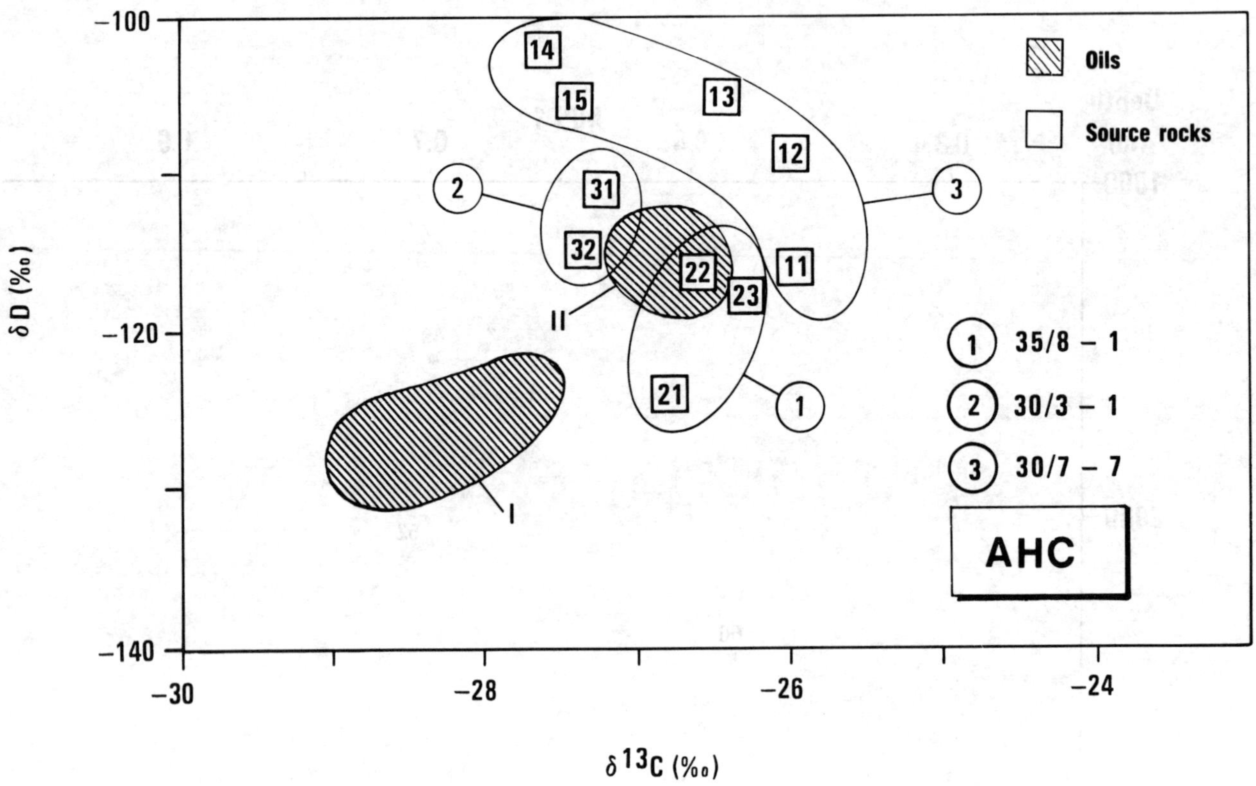

Fig. 13. Comparison of oils and extracts in a $\delta^{13}C/\delta D$ cross-plot. AHC = Aromatic hydrocarbons.

amount of 20S steranes, but for this parameter the trend is less clear. The oil samples have values in a fairly narrow range, with the two deepest samples (from 30/4–2 and 35/8–1), and possibly the samples from 34/10–2 and 30/3–2, being slightly more mature than the rest.

The shallowest two oils, from 31/2–2 and 31/2–5, have undergone different degrees of biodegradation. Based on the sterane ratios the maturity of the two samples is approximately the same, and the distribution of biomarkers in general is quite similar in both oils. A difference can be seen in the terpane distribution, in that the abundance of tricyclic terpanes is higher in the 31/2–2 oil and the Tm/Ts ratio is smaller in the same well, suggesting that the oil from 31/2–2 is of slightly higher maturity than that from 31/2–5. The relative amounts of 28,30-bisnorhopane are the same.

An explanation of the difference in the terpane distribution may be that the same source has generated both oils, but at slightly different maturity levels. The oil of lowest maturity probably underwent biodegradation first, and a more mature originally non-degraded oil subsequently filled the reservoir in well 31/2–2. This observation parallels the conclusion of Cornford *et al*. (1983) that North Sea oils generally are 'a mixture of products generated from a single source rock horizon buried to progressively deeper levels towards the basin centre'.

Similar plots of maturity-related ratios from the aromatic hydrocarbon distribution show a somewhat different picture. The methyl phenanthrene index (MPI–1) used by Radke *et al*. (1980) gives a very good linear relationship for the source rocks with depth, with MPI-1 values between 0.3 and 0.9 (see Fig. 12). This corresponds to a vitrinite reflectance range of approximately 0.5–1.0% R_0. The ratio indicates that the shallowest oils are the most mature, the heavily biodegraded oil in 31/2–5 being slightly less mature than the 31/2–2 oil. The least mature oils appear to be the three oils of intermediate depth, from a depth of 2500–3000 m. The deepest three oils have a maturity similar to source rocks from below 3600 m.

These somewhat contradictory conclusions from the saturated and aromatic maturity parameters may be explained in two ways. Either the aromatic ratio gives the correct order of maturity of the oil samples—the sterane maturity ratios are so close to equilibrium that the variations may not be significant—or the difference between the two sets of data may reflect a mixed input from source rocks of varying maturity.

Correlation of oils and source rocks

A genetic relationship between a kerogen and an oil may exist when both are isotopically similar. This is the case only if oils and kerogens and their extracts have not experienced severe maturation. In most cases, the aromatic fraction is least affected by secondary processes (Lewan, 1982; Schoell, in press). Figure 13 compares the results of carbon and hydrogen isotopic analyses of the aromatic fraction both from the oils and rock extracts. The extracts reveal a considerable variation in the isotopic concentration, such that the three wells are distinguishable. The isotopic variations in extracts from 30/7–7 are significantly different from both groups of oils, but those from wells 35/8–1 and 30/3–1 are similar to the isotopic variations of oil group II (30/4 and 35/8) but are distinct from group I. A similar result is obtained when carbon isotope concentrations of kerogens and the aromatic fractions of the oils are compared. Assuming that oils could be genetically related to kerogens if the aromatic fraction of the oil is not depleted by more than 1.5% of the heavy isotope, several oils and kerogens can be genetically linked. Kerogens from the lower Jurassic in 30/3–1 are potential sources for the oils of group I,

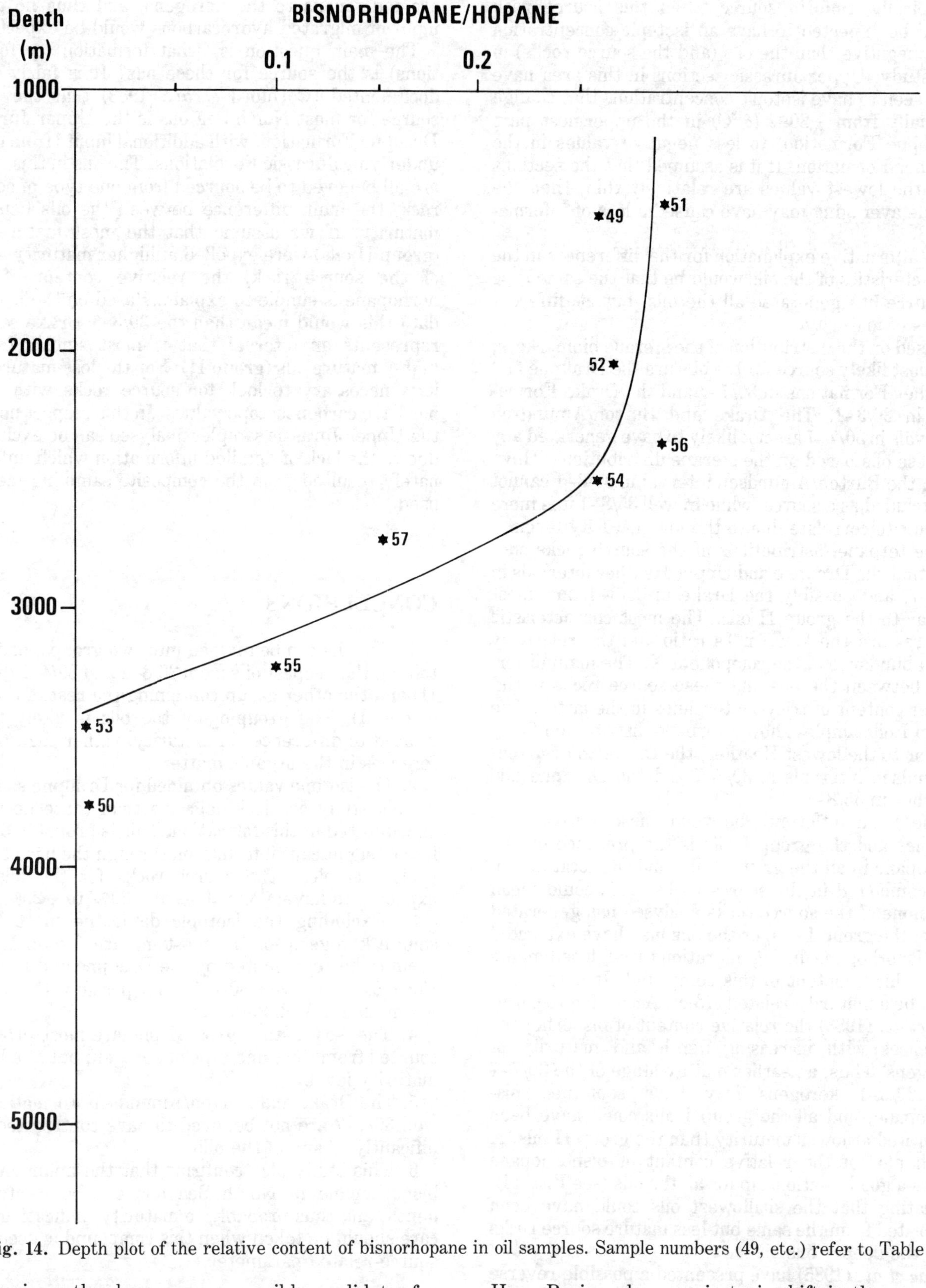

Fig. 14. Depth plot of the relative content of bisnorhopane in oil samples. Sample numbers (49, etc.) refer to Table 1.

and various other kerogens are possible candidates for group II.

The genetic significance of the grouping of the oils is equivocal. The isotopic differences between the two groups could be due to the primary isotopic differences in the source rocks. The clear relationship between the isotope data and depth and gravity suggests that the differences found may be due to thermal maturation of the oils in the reservoirs. This could mean that the oils of group II formed from oils of group I through thermal alteration. An almost identical change in isotopic composition was observed in the Big Horn Basin, where an isotopic shift in the aromatic and NSO-fraction of the oils was attributed to thermal alteration (Chung *et al.*, 1981).

The oils from 35/8–1 and 30/4–2 are the deepest of all the oils studied, and are characterized by the highest gas/oil ratios. This may suggest that these oils have undergone considerable thermal alteration. Thus the isotopic concentrations of these oils may not be representative of their source. Since the causes of the isotopic shift are not known, any further inference would be speculative.

Assuming that all of the oils were derived from an

isotopically uniform source, then this source rock would be expected to have an isotopic concentration more negative than the oils (and the source rocks) in this study. Upper Jurassic sections in this area have been seen to have isotopic concentrations that change gradually from −30‰ ($\delta^{13}C$) in the uppermost part (Draupne Formation) to less negative values in the Heather Formation. If it is assumed that the sections with the lowest values are relatively thin, then the sample averaging may have caused a loss of information.

An alternative explanation for the differences in the characteristics of the oils would be that the same type of source has generated all the oils, but at different stages of maturity.

Based on the distribution of the sterane biomarkers, the most likely source for the oils are the Draupne and Heather Formations in 30/7–7 and the Drake Formation in 30/3–1. The Drake and Burton/Amundsen intervals in 30/7–7 are not likely to have generated any of these oils based on the sterane distributions. However, the Burton/Amundsen interval in 30/3–1 cannot be excluded as a source, while in well 35/8–1 it is more difficult to correlate due to the low maturity levels.

The terpane distributions of the source rocks suggest that the Draupne and Upper Heather intervals in 30/7–7, and possibly the Drake in 30/3–1, are most similar to the group II oils. The most characteristic features are the low Tm/Ts ratio and the relatively high abundance of the component X. The main difference between the oils and these source rocks is the higher content of tricyclic terpanes in the latter. The group I oil samples show a terpane distribution fairly similar to the lowest Heather, the Drake and Burton/Amundsen intervals in 30/7–7 and the Draupne and Heather in 35/8–1.

The main difference between these source rock samples and the group I oils is the presence of bisnorhopane in all the group I oils, and the total lack of this compound in the source rocks. This could mean that none of the source rocks analysed has generated any of the group I oils, or the oils may have extracted the bisnorhopane during migration through sediments with a high content of this compound. It may, however, be a maturity-related effect. According to Cornford *et al.* (1983) the relative content of bisnorhopane decreases with increasing depth and maturity of kerogens. Thus, an earlier mature stage of the 30/7–7 and 30/3–1 kerogens may have contained bisnorhopane, and all the group I oils could have been generated at lower maturity than the group II oils. A depth plot of the relative content of bisnorhopane shows a good relationship for all the oils (see Fig. 14), indicating that the shallowest oils could have been generated from the same but less mature source rocks than the deeper oils.

Huc *et al.* (1985) have presented a possible reverse trend for the bisnorhopane concentration. Over a fairly thin Upper Jurassic section they found an increase in the bisnorhopane content with increasing depth. The maturity of this section was, however, very uniform, and the effect of increased bisnorhopane content is probably due to changes in organic matter, rather than maturity variations.

The Burton/Amundsen interval in 30/3–1 is very different from all the oils due to the presence of a very high abundance of tri- and tetracyclic terpanes. This could be due to migrated hydrocarbons in this sample, although the isotopic data suggest that the extracts are indigenous to the kerogens, and thus no great input of migrated hydrocarbons would be expected.

The main question is, what formation (or formations) is the source for these oils? It is fairly well-documented (Cornford *et al.*, 1983) that the main source for most North Sea oils is the Upper Jurassic Draupne Formation, with additional inputs from other underlying Jurassic Formations. The oils in this study are all believed to be sourced from one type of source rock, the main difference between the oils being in maturity. If we assume that the most mature oils (group II oils) were expelled at higher maturity levels of the source rock, the relative content of bisnorhopane is simple to explain. Based on the isotopic data this would mean than the 30/3–1 Drake sample represents an interval that is most similar to the highly mature oils (group II). For the less mature oils it is necessary to look for source rocks with more negative carbon isotope values. In this respect none of the Upper Jurassic samples analysed can be excluded, due to the lack of detailed information which unfortunately resulted from the composite sampling method used.

CONCLUSIONS

1. The oils can be divided into two groups, one containing the deepest oils from 35/8–1 and 30/4–2 (group II) and the other group containing the rest of the oils (group I). The grouping of the oils is likely to be related to differences in maturity, rather than to differences in the organic matter.
2. The isotope values obtained for Draupne samples in this study are less negative than those normally encountered in this formation. This is probably due to loss of significant information through the use of composite samples. The source rocks for the oils are expected to have $\delta^{13}C$ values of −30‰ to −29‰.
3. Excluding the isotopic data, the most likely source kerogens for the most mature (group II) oils seem to be represented by the Draupne and Heather Formations in well 30/7–7, and possibly the Drake Formation in well 30/3–1.
4. The less mature group I oils are most probably sourced from the same type of kerogen, but at a lower maturity level.
5. The Drake and Burton/Amundsen Formations in well 30/7–7 are not believed to have contributed significantly to any of the oils.
6. This study also confirms that the abundance of bisnorhopane in North Sea oils can be related to depth, and thus probably to maturity, indicating that care should be taken when this compound is used as a source-related parameter.

ACKNOWLEDGEMENTS

We would like to acknowledge the following oil companies for giving permission to present these data: BP Norge A/S; Gulf Oil corporation, Norway; Norsk Hydro a.s.; A/S Norske Shell and Statoil.

In addition, we would like to thank the technical staff at IKU for the analytical work, and also Tone Aas who made all the figures.

REFERENCES

Chung, H. M., Brand, S. W. and Crizzle, P. L. 1981. Carbon isotope geochemistry of Palaeozoic oils from the Big Horn Basin. *Geochim. Cosmochim. Acta*, **45**, 1803–1815.

Cornford, C., Morrow, J. A., Turrington, A., Miles, J. A. and Brooks, J. 1983. Some geological controls on oil composition in the U.K. North Sea. In: Brooks. J. (ed.), *Petroleum Geochemistry and Exploration of Europe. Geol. Soc. Spec. Publ.* **12**, 35–50. (Blackwell Scientific, Oxford)

Huc, A., Irwin, H. and Schoell, M. 1985. Organic matter quality changes in an Upper Jurassic Shale section from the Viking Graben, (this volume).

Lewan, M. D. 1982. Effects of thermal maturation on stable organic carbon isotopes as determined by hydrous pyrolysis of Woodford Shale. Paper presented at *Norwegian Petrol. Soc. Mtg 6409 'Stable Isotopes in Hydrocarbon Exploration'*, Stavanger, October 1982.

Radke, M., Willsch, H. and Welte, D. H. 1980. Preparative hydrocarbon group type determination by automated MPLC. *Analyt. Chem.*, **52**, 406–411.

Rønnevik, H., Eggen, S. and Vollset, J. 1983. Exploration of the Norwegian Shelf. In: Brooks, J. (ed.), *Petroleum Geochemistry and Exploration of Europe. Geol. Soc. Spec. Publ.* **12**, 71–93. (Blackwell Scientific, Oxford.)

Schoell, M., Faber, E. and Coleman, M. L. 1983. Carbon and hydrogen isotopic compositions of NBS 22 and NBS 21 stable isotope reference materials: an inter-laboratory comparison. *Org. Geochem.*, **5**, 3–6.

Schoell, M. in press. Paper submitted to *Geol. Jb.*

Seifert, W. K. and Moldowan, J.M. 1978. Applications of steranes, terpanes and monoaromatics to the maturation, migration and source of crude oils. *Geochim. Cosmochim. Acta*, **42**, 77–95.

Part II
Modelling

8 Modelling of petroleum genesis and migration with a bidimensional computer model in the Frigg sector, Viking Graben

P. Ungerer, A. Chiarelli† and J. L. Oudin‡*
*IFP, France
†SNEA(P)-ELF, France
‡CFP-TOTAL, France

Expulsion of oil from source rocks and subsequent migration towards traps are assumed to take place as a discrete phase distinct from water. The main cause of expulsion is compaction of sediments during burial, while secondary migration is ruled by buoyancy. They are described by a diphasic and bidimensional computer model based on Darcy's law and the relative permeability concept. The extent of formation of oil at depth is provided by a kinetic model, using the knowledge of basin thermal history and distribution of organic sedimentary facies.

The example presented is a typical tilted block structure of the Frigg sector. According to this diphasic model, hydrocarbon expulsion began around 70 Ma from the Brent coals, and a few million years later from the Kimmeridge Clay and Heather Formations, leading readily to an accumulation in the Brent sandstones at the top of the block. This accumulation developed up to the present time, and partial dysmigration occurred through the caprocks at the top of the block.

Although gas migration should be computed by a triphasic model, it has been attempted to simulate it by the same diphasic model, using low density and viscosity values for the hydrocarbon phase. The model then found an extensive dysmigration, leading additionally to an accumulation in the Paleocene sands, as is actually found. The heavy oil ring in the Frigg field is thought to be the result of a migration in the gas phase, the dissolved oil being condensed during the upwards migration of the gas because of the important pressure drop, and later being biodegraded.

INTRODUCTION

The final aim of migration modelling—which implies necessarily oil generation modelling—is to give predictions about the location and volume of oil accumulations in a given basin. However, work is still in progress, and the present step is to simulate the migration processes in well-documented cases with numerous control parameters: pressures, volumes of proven accumulations, and so on. Once the model has proved to be satisfactory in such a case, its predictions may become reliable for other locations in the basin, provided a minimum of data is available. In any case, the use of the migration model needs some geological and geochemical data from neighbouring wells. If the burial history, the lithology of the series or the richness of the source rocks are unknown, it is impossible to quantify the amount of hydrocarbons generated, and any study of migration is impossible.

The Viking Graben, North Sea, is a typical area where the processes of oil genesis and migration may be reasonably reconstructed. Most of the wells have been drilled since 1970: thus a lot of reliable geochemical data has been collected. These data enable us to assess the quality of the source rocks, which are of Jurassic age. Among these the richest are: the Kimmeridge Clay Formation, which contains organic matter of generally good quality; the Heather Formation, generally less rich but thicker; and Brent coals, which may produce oil (Durand *et al.*, 1983), with their low yield compensated by a high organic content.

The geological features of the North Sea are well-documented in terms of the stratigraphy, the sedimentology and the tectonic history. The present structure of the Viking Graben is the result of a rifting period that occurred mainly during Middle and Late Jurassic. The rifting declined progressively during Late Cretaceous and was followed by regional subsidence until the present time. The Triassic and Jurassic sediments were extensively faulted during the rifting period, which results in tilted block structures, as seen on E–W cross-sections (Fig. 1). The bottom part of these blocks is commonly 5000–8000 m deep in the central part of the graben, whereas their top is often truncated by the Cimmerian unconformity, due to a phase of erosion or non-deposition in the Late Jurassic. The Jurassic and older sediments are capped by a continuous Cretaceous and Tertiary series, com-

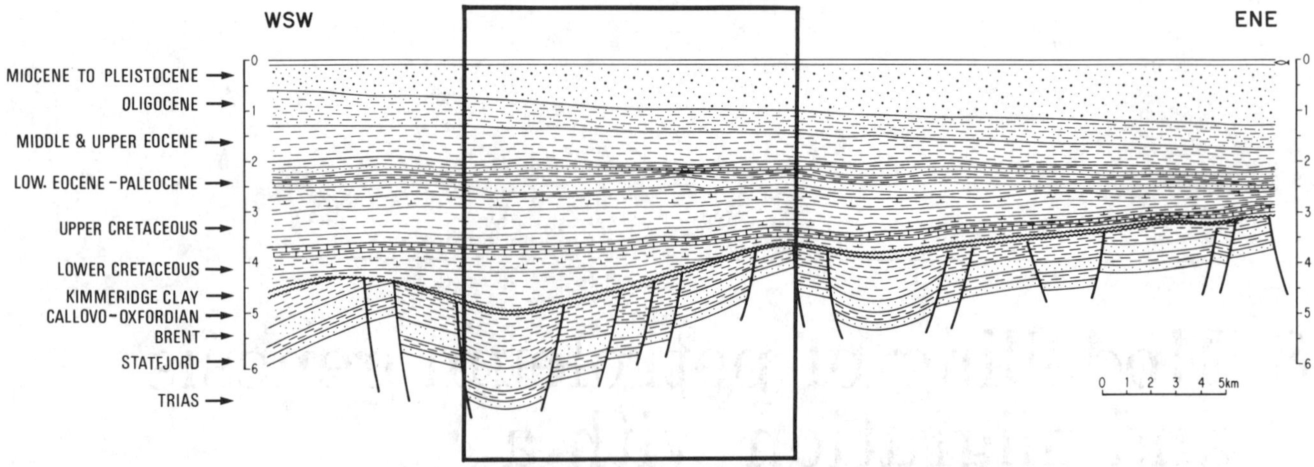

Fig. 1. Schematic section of part of the Viking Graben in the North Sea. The framed zone is the one selected for application of the model. It is bounded by the two largest faults, which define a tilted block below the major unconformity.

monly 2500–4000 m thick in the central part of the graben. Since the thickness of eroded sediments at the top of the blocks is quite limited, the burial of sediments may be reconstructed with good accuracy.

The Frigg sector, which has been selected for the application of the model, is representative of the hydrocarbon deposits of the Viking Graben: oil and gas are found in Lower and Middle Jurassic sandstones (Statfjord and Brent) at the top of the tilted blocks, and in the Paleocene sands where the accumulation of gas is particularly large. As described by Chiarelli and Duffaud (1980) and Carstens and Dypvik (1981), abnormal fluid pressures commonly occur in the Jurassic, whereas the Paleocene is at hydrostatic pressure (Heritier *et al.*, 1979).

Coupled to the organic matter maturation data, these features are the principal control parameters for the model. It could be claimed that such a deterministic model has numerous uncontrolled parameters, such as typical permeabilities, kinetic parameters, viscosities, etc. These parameters could always be adjusted to account for the available control parameters. Actually, we keep exactly the same set of uncontrolled parameters when switching from one basin to another. Of course, the geochemical and geological data differ. Thus a satisfactory match with field observations is really significant.

PHYSICAL AND CHEMICAL PROCESSES INVOLVED

It is generally accepted that the generation of petroleum results from maturation of the sedimentary organic matter, as controlled by temperature and time. There is no evidence that pressure plays an important role (Tissot and Welte, 1978, p. 183). The modelling of maturation has been described by several authors: Huck and Karweil (1955), Tissot (1969), Lopatin and Bostick (1973), Tissot and Espitalie (1975), among others.

On the other hand, migration of oil, and especially expulsion from the source rocks, is still not well-known, although a wide range of mechanisms have been proposed:

1. migration in aqueous solution;
2. migration by diffusion and
3. migration of hydrocarbons and water in separate phases.

Although the first two monophasic processes have been proposed by various authors in the past, current evidence, and data (Durand, 1981) strongly favour the third, biphasic process, which has consequently been adopted for the present migration model.

The main driving force for expulsion is considered to be the compaction of the source rock under the sediment load. In the first stage of maturation, the oil saturation within the pore system of the source rock is low, so that the oil or gas phase is discontinuous, and migration is prevented by capillary forces. Later, provided the petroleum potential is sufficiently high, the hydrocarbon phase becomes continuous and expulsion can start. Once the hydrocarbons have reached a permeable bed, buoyancy becomes the most important driving force.

This scheme is valid if the hydrocarbon phase is unique, which is probably the case for most source rocks within the oil-generating zone (Ungerer *et al.*, 1981). However, it may not be the case at shallower depth, where the decrease in temperature and pressure can cause separation into a gas phase and an oil phase.

GENERATION AND MIGRATION MODELLING

Based on the above considerations, the modelling of migration should include the computation of:

1. oil and gas generation;
2. compaction;
3. tri-phasic fluid flow and
4. phase behaviour of the hydrocarbon mixtures.

These four processes are interrelated and, as a consequence, the migration model should integrate all four into a unique system. It should also be at least two-dimensional, since lateral migration of petroleum is generally needed for its accumulation in traps.

This problem is even more complex than the triphasic and three-dimensional computer models that have been developed in reservoir engineering. Effectively, the porous medium is not static, but changing as a result of burial and compaction, and the process takes place on a million year time-scale. This high complexity prevents us from accounting for the process 4, above, in the present migration model. Therefore, the hydrocarbon mixtures are treated as monophasic.

The generation of oil is computed from the kinetic model published by Tissot and Espitalie (1975), with the difference that the present model makes no provision for secondary degradation of hydrocarbons (dismutation into gas and carbonaceous residue). Gas formation can also be simulated, modifying it by a coefficient which is assumed to account for the secondary cracking yield.

Hydrocarbon expulsion is influenced by pressure in the compaction process. Thus it is not possible simply to use porosity versus depth relationships, as this method does not allow for possible undercompaction. Following Smith (1971), compaction is represented in the present model by the law of effective stress, whereby effective stress is considered to be a function of porosity. The function differs from one lithology to another, reflecting the fact that different rocks compact in different ways.

Darcy's law, extended to polyphasic flows through the relative permeability concept, has been adopted for the model. According to Matheron (1967), application of Darcy's law to very slow monophasic flows is perfectly justified. Its extension to polyphasic flows by the relative permeability concept has, on the other hand, no theoretical justification and it must be considered as a pragmatic solution only. This formulation accounts for the buoyancy of oil versus water.

Pressures in the hydrocarbon and water phases are assumed to be linked by the capillarity equation. The intrinsic permeability of the porous medium is computed from the porosity and the average grain size of the rock. Source rocks or permeable beds through which fluids pass in the course of secondary migration are treated similarly, taking anisotropy into account. Furthermore, in the source rock kerogen is converted into oil, ultimately leading to an increased hydrocarbon pressure within its pores (Du Rouchet, 1981).

When pressure (P) increases in a porous rock, rupture occurs when one of the effective stress components becomes negative. This is evidenced either by creep (diapirism), or by the opening of fissures. These fissures, through which fluids may flow, close as soon as pressure drops. To account for this process, it is assumed here that if pressure exceeds the minimum stress S_3, rock permeability increases in proportion to the square of the difference ($P - S_3$).

All of the basic equations of the model are detailed in the Appendix.

The data required by the model are mainly the geometry of the cross-section, the age and lithology of the stratigraphic levels, the subsidence and temperature history and, finally, the initial petroleum potential of the various sediments. The results consist of a reconstruction of compaction (porosity, pressure), of oil generation (maturation level) and of migration (oil saturations in source rocks and reservoirs).

APPLICATION OF THE MODEL TO A TILTED BLOCK STRUCTURE OF THE FRIGG SECTOR

Geological and geochemical data

The E–W cross-section considered (see Fig. 1) has been derived from the interpretation of a seismic profile typical of the central part of the Viking Graben in the Frigg area. Several wells on the cross-section or in the vicinity serve as control points for the stratigraphy and the sedimentology of the various units.

The geometry and the lithology of the sediments is not very well ascertained in the deepest parts (i.e. under the unconformity in the deepest parts of the tilted blocks). Indeed, the seismic resolution is poor and there are no control wells in synclinal locations. In this study, the role of the Paleozoic sediments, which have never been reached by drilling in the area, and of the Triassic sediments, which never show notable properties, either as reservoirs or as source rocks, were neglected.

In the application of the present model, one is limited to an 8 × 20 elements grid, and this prevents representation of the whole cross-section of Fig. 1. Consequently, a zone limited by two major faults under the unconformity, which includes the deepest source rocks as well as possible traps in the Brent reservoirs at the top of the block, was considered. The minor faults inside the block are assumed not to act as impermeable barriers to fluid migration, and consequently the Statfjord and Brent sandstones were represented as continuous beds in the grid (see Fig. 2).

The selection of boundary conditions has been made as follows:

1. at the bottom of the grid, impermeable conditions (no fluid flow);
2. laterally, it was assumed that the faults located at the boundaries of the block form impermeable barriers under the unconformity (no fluid flow);
3. above the unconformity, it is assumed that water flow through the lateral boundaries is limited by hydraulic resistance which decreases upward. Effectively, the relatively permeable units of the Upper Cretaceous appear on seismic cross-sections to be in connection with hydrostatic zones. For instance, this lateral resistance was assumed to be negligible for the Paleocene sands boundaries, which enabled representation of their hydrostatic regime.

Six classes of lithology have been used for the elaboration of the grid: clay, sand, sandy silt, silty clay, marl and coal, with the lithological and geochemical characteristics shown in Table 1.

TABLE 1. Lithological and geochemical characteristics of the six classes of lithology

	Formation	*Lithology*	*Mean organic carbon content (wt %)*	*Mean initial oil potential (mg/g OC)*
	Eocene–Paleocene	Sand	0	0
		Silty sand	0.5	210
		Silty clay	0.7	210
		Clay	1	
	Senonian	Marl	0.7	70
		Clay	0.7	70
	Turonian	Marl	1.5	70
	Cenomanian	Marl	1.5	70
	Lower Cretaceous	Clay	1.5	70
Upper Jurassic	Kimmeridge	Clay	7	310
	Heather	Clay	6.5	210
Middle Jurassic	Brent	Sand	0	0
		Coal	30	220
Lower Jurassic	Dunlin	Clay	3.5	210
	Statfjord	Sand	0	0

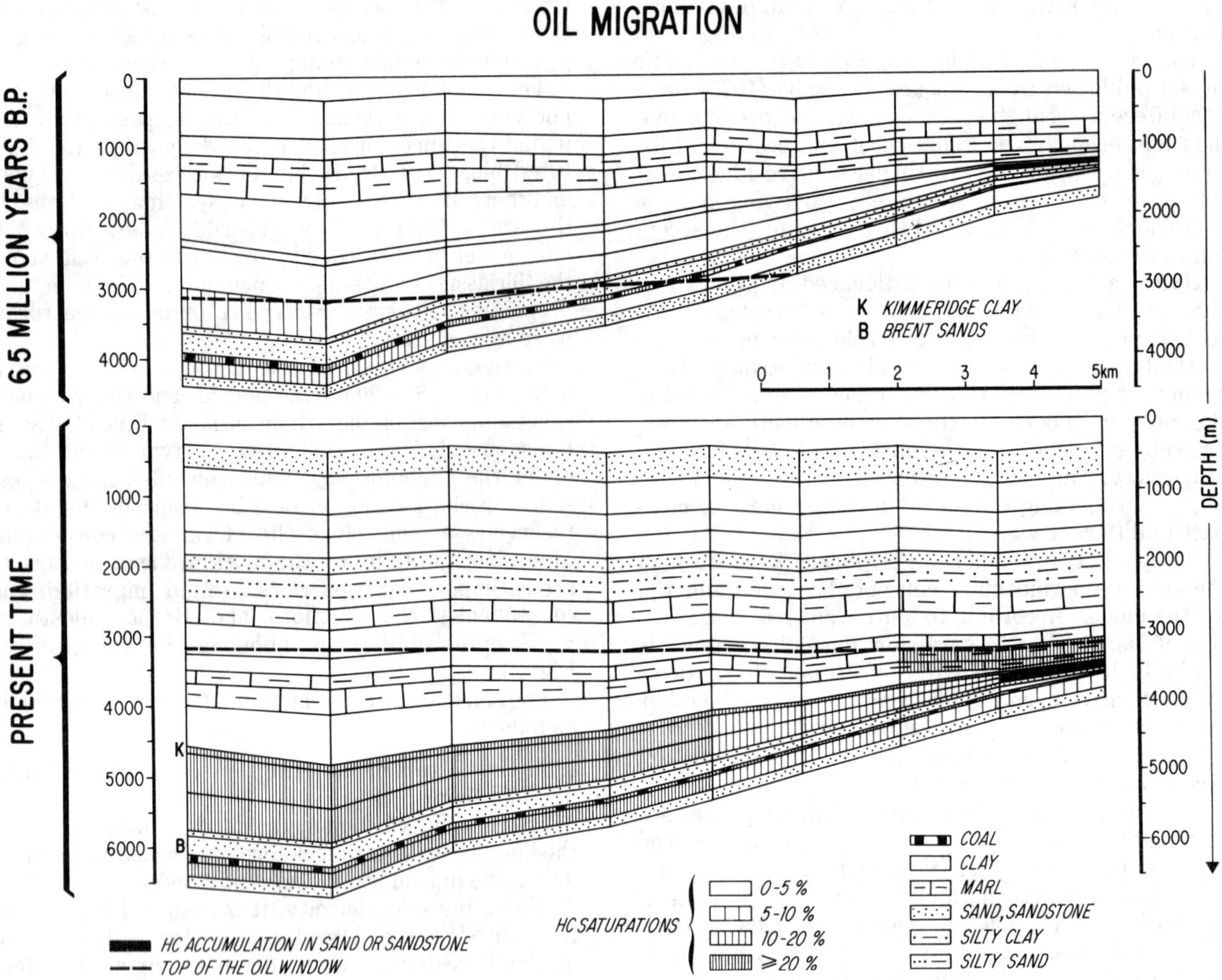

Fig. 2. Representation of the computed oil saturations for two successive stages of the cross section. The first stage corresponds to the beginning of oil accumulation (Maastrichtian, 65 Ma B.P.). At the present stage, a pool has formed in the permeable levels of the Lower Jurassic, and dysmigration has occurred in the Upper Jurassic and Lower Cretaceous at the top of the block. The top of the oil formation zone corresponds to a 10% conversion ratio of the potential.

It may be noted that a relatively high oil potential has been assigned to Brent coals. An elementary analysis of them does, in fact, indicate that they are richer in hydrogen than the kerogens which serve as a reference for type III of Tissot *et al.* (1974). In addition, the oil potentials assigned to the Kimmeridge Clay, and especially to the Heather Formation, are lower than for good quality, marine organic matter, which is suggested by their elementary analysis and pyrolysis results in adjacent wells.

Paleotemperatures have been derived from burial reconstruction, assuming a geothermal gradient of 30 °C/km and a surface temperature of 5 °C.

Computed history of oil formation and migration

Before considering the results of the model in a particular case such as the block used here, it should be noted that they can be very sensitive to the input data. Since these data are uncertain, all the figures (timing of the events, volumes accumulated, pressure, etc.) must be regarded as estimations only.

As computed by the model, the zone of important oil formation (10% conversion of the oil potential) corresponds to a depth of about 3000 m. This result is in good agreement with present maturation measurements. The first source rocks that reached the maturation zone were the Dunlin shales and Brent coals in the bottom part of the block, 110–100 Ma ago. At this time, the Dunlin had still a very low hydrocarbon saturation (5%) and therefore a low relative permeability to oil (see Fig. 5, below) that did not permit expulsion. The Brent coals, however, had a high saturation by the same time, as a consequence of their good potential, high organic carbon content and low porosity. Thus they started to expel hydrocarbons into the Brent sandstone, although in very small amounts since only a small part of the block was sufficiently mature. In these sandstones of high intrinsic permeability, secondary migration is found to occur with low residual saturations (1–2%) which is not possible before significant amounts of hydrocarbons have been expelled. As a consequence, the formation of an oil pool is found to start only when the expulsion is important, i.e. 75–65 Ma ago. At this time, the Heather source rocks entered the oil formation zone, but the Kimmeridge Clay did not. However, the oil saturation of the Heather (<20%) was not sufficient for oil expulsion. The saturations in the cross-section, as found 65 Ma ago, are represented by Fig. 2. Four classes of saturations are represented: 0–5%, 5–10%, 10–20% and 20% or more.

Subsequently, the Kimmeridge Clay, Heather and Dunlin source rocks in their turn, supplied the reservoirs: for instance, the Heather was found to expel at 55–50 Ma in the deepest parts of the section. Because

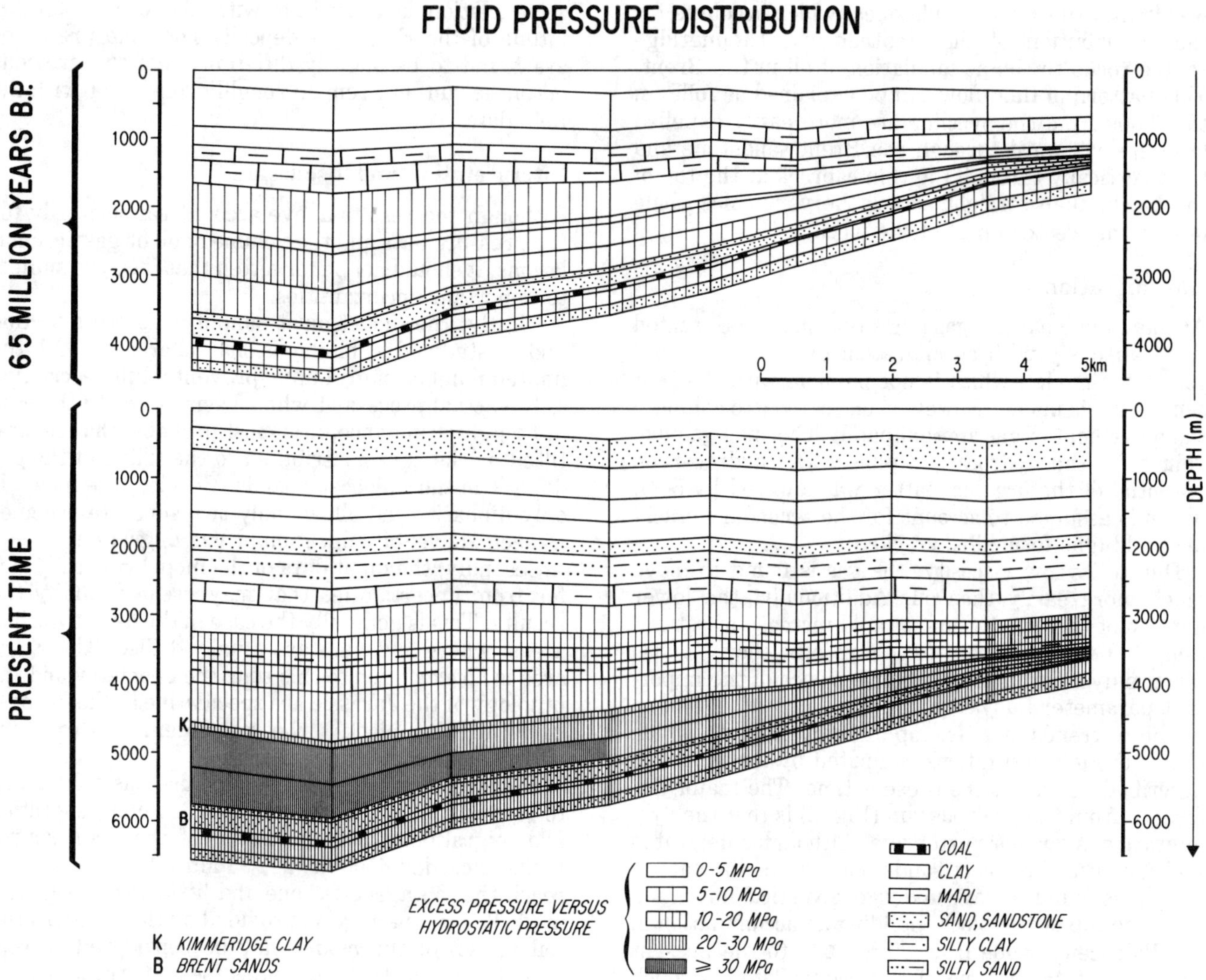

Fig. 3. Representation of the computed excess pressures versus hydrostatic, for the same stages as Fig. 2. The area of greatest excess pressures corresponds to the Heather and Kimmeridge Clay Formations in the deepest part of the block. The Upper Cretaceous and Tertiary sediments are quasi-hydrostatic.

of their lower organic carbon content, the Kimmeridge and Heather source rocks are found to have a lower expulsion efficiency than the Brent coals. However, their greater thickness compensates for this, and they are the most productive source rocks of the block if the whole geological history is considered.

Dysmigration is also found to have occurred through the Upper Jurassic and Lower Cretaceous caprocks at the top of the block from about 55 Ma to the present time. This was due to the high pressures in the Lower Jurassic sandstones, as discussed below. As a consequence, the hydrocarbon saturation increases in the Lower Cretaceous units, just above the oil deposit, as seen in Fig. 2. This result matches the oil shows that have been encountered in these units during drilling.

The calculated quantities of accumulated hydrocarbons in the Brent, extrapolated from the two-dimensional model to the three-dimensional real world, are of the same order of magnitude as those observed. In the case presented, the ratio of accumulated to generated hydrocarbons was found to be 0–10%, depending on the stage considered in the history of the pool. The major part of the hydrocarbons generated is thus found to have remained in the source and carrier rocks or to have undergone dysmigration—the latter being far from negligible.

Computed history of fluid pressure and compaction

As soon as the Upper Jurassic caprock was deposited, it provided a barrier to fluid migration which resulted in undercompaction and high pressures in the Brent and Statfjord sandstones, as predicted by the model from 135 Ma to the present time. The pressure regime is illustrated by Fig. 3, which shows the computed excess pressures versus hydrostatic for the successive stages (65 Ma and the present time). Five classes of excess pressure have been established: 0–5 MPa, 5–10 MPa, 10–20 MPa, 20–30 MPa and 30 MPa or more. Since water flows from areas of high excess pressure towards areas of lower excess pressure, an idea of the directions of water flow can be obtained.

The order of magnitude of the excess pressure computed by the model at the top of the Brent sandstone is quite comparable with the one reported by pressure tests. In addition, the hydrostatic regime of the Paleocene sands is also reproduced. The pressure profile is thus very similar to the one reported by Carstens and Dypvik (1981): the excess pressure increases progressively from the hydrostatic Upper Cretaceous to the overpressured Lower Jurassic, but does not reach, or even approach, the geostatic pressure.

The greatest excess pressures are found in the Jurassic shales, in a synclinal location. As a consequence, the fluids are considered to be expelled downwards into the Brent sandstones, which accounts for the contribution of the Heather and Kimmeridge source rocks to the accumulation of oil in the Brent. This pattern of fluid flow can be explained as follows: the fluids from the syncline are more easily expelled laterally eastward through the Brent sandstone and then vertically through the thin caprock at the top of the block, than straight across the very thick shale unit in the western part of the block.

Gas migration

As mentioned earlier, gas migration should be treated simultaneously with oil migration by a triphasic oil/gas/water model, which is not possible with the present diphasic model. However, one can try to compute the movement of gas approximately, choosing gas-like values for the hydrocarbon phase. The hydrocarbon potential of the organic matter was lowered by 50%, which is assumed to account for the secondary cracking yield (gas formed/oil cracked).

Due to its low viscosity, the gas is found to move much more easily than oil. Additionally, the lower density of the gas favours a rapid increase of saturation, and thus a higher expulsion efficiency. It also favours dysmigration, as a consequence of the important parameter buoyancy.

These trends are exemplified by Fig. 4, which shows the gas saturations computed by the model as described above, at the present time. The major difference from the previous run (Fig. 2) is that the dysmigration is far more extensive. Although a deposit is still found in the Brent sandstone, the major part of the gas is found to have undergone vertical migration from the top of the block, building an accumulation in the Paleocene sands from 35 Ma B.P. to the present time. The total volume of gas accumulated is much larger than the previously computed volume of oil accumulated. When extrapolated to the three-dimensional real world, the volume of the Paleocene accumulation is consistent with the order of magnitude of the Frigg gas deposit. The fluid pressures are found to be slightly different from the previous oil-phase run, but remain roughly consistent with the field data.

Interpretation and discussion

Although the results above account satisfactorily for the pressure data and the volume of oil or gas accumulations, it should not be forgotten that they are subject to important uncertainities.

Firstly, the model must simplify the cross-section under study. Its bidimensional character and the limited number of grid units prevent a full description of the actual process at whole basin scale. In the case of Frigg, for instance, it is most probable that several tilted blocks have contributed to the filling of the gas deposit in the Paleocene sands. Besides, the model is only diphasic, and allows only two separate schemes as seen above: oil migration or gas migration.

Secondly, the input data of the model generally suffer from uncertainties that may greatly modify the results. This is generally the case in the deep synclinal areas, where few wells are usually drilled. The Frigg area is no exception to this rule: the geometry and the lithology of the Jurassic are speculative in the deepest parts of the blocks, the palaeotemperatures are roughly estimated, and so on.

However, these results may be used as a guideline to reconstruct the history of hydrocarbon generation and migration. Until the burial of the main source rocks exceeded 4500 m, i.e. 65 Ma B.P., they did not reach the metagenesis zone and little gas was generated. The reconstruction provided by the standard run (oil phase) of the model may be considered as representative of the real processes at these times. Later, increasing amounts of gas were produced. Although the oil/gas mixture is most probably monophasic in the high pressure zone, the density and vis-

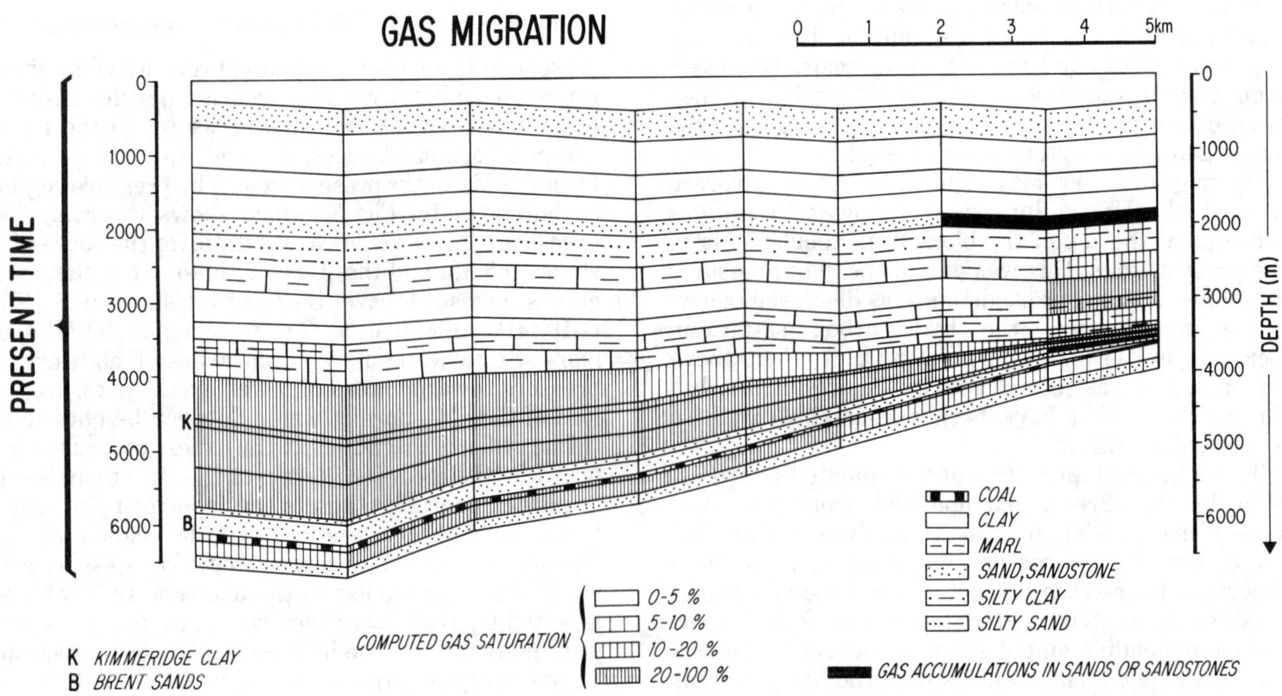

Fig. 4. Representation of the computed gas satgurations at the present time, assuming gas-like density and viscosity values for the hydrocarbon phase. The petroleum potentials have been lowered by 50% in order to account for the secondary cracking yield. The model finds an extensive dysmigration through the Cretaceous sediments, and a subsequent accumulation in the Palaeocene sands.

cosity of the hydrocarbon phase are much lower than for a conventional oil. In the present-day cross-section, a major part of the source sequence is in the gas generation zone. It is not surprising, then, that the present real state is nearer to the 'gas phase' run than the 'oil phase' run. Indeed, the Paleocene gas field is reproduced only by the 'gas phase' run.

It could be claimed that the oil phase run should account for an oil accumulation in the Paleocene, since a heavy oil ring is reported under the Frigg gas (Heritier *et al.*, 1979). In the opinion of the authors, this oil has migrated in gaseous solution, during the upward migration of the gas from the Brent accumulation through the Cretaceous units.

In this context, Nogaret (1983) proved that an initial hydrocarbon mixture, which is monophasic in the overpressured zone, separates out into an oil phase and a gas phase when the pressure and temperature decrease. Upwards, successively lighter oils condense. Their amount is far from negligible: assuming an initial oil : gas ratio of 1 : 3 in weight, 40% of the initial oil is condensed between 3000 m and 2500 m, and an additional 7% between 2500 m and 2000 m. These figures would be slightly different in the case of Frigg, but they prove that the process is quantitatively important. Further biodegradation of the condensed oil explains the present heavy character of the Paleocene oil, as mentioned by Heritier *et al.* (1979), who found that it contained virtually no normal alkanes.

Sensitivity of the model to geological and geochemical data

A knowledge of the model sensitivity is very useful for two main purposes.

(1) It gives an estimate of the possible variations in results that are caused by the uncertainty of the data. This provides a range of results concerning maturations, pressures, accumulations, etc. If the control parameters fall within this range, the model may be considered as satisfactory.

(2) It makes it possible to assess the relative influence of the parameters. This can serve as a guide for further exploration, searching for favourable locations in the area by investigation of the key parameters for oil accumulation.

In the case of Frigg, the influence of several parameters has been investigated. For instance, the geothermal gradient was increased and both generation and migration of oil were found to have occurred earlier in the geological history, and accumulations were found to be larger. Other important parameters are: petroleum potential of the source rocks, presence of permeable beds interbedded with source rock units, and maximum depth of the basement. The values at control points have been found to fall within the range caused by possible uncertainties, which proves the reliability of the model.

CONCLUSION

The present oil formation and migration model is still limited by computational difficulties, since it cannot represent the whole basin nor the oil/gas thermodynamic equilibrium. However, it has proved efficient in predicting the main control parameters, which are the depth of the oil generation zone, the pressure regime and the approximate size of hydrocarbon accumulations.

It is found by the model that successively, the Brent coals and the Heather and Kimmeridge Clay Formations expelled oil towards the Brent reservoir, leading to an oil accumulation at the top of the block between 75 and 65 Ma B.P.

The present situation is best reproduced by the model if gas-like density and viscosity values are selected for the hydrocarbon phase. The model then predicts an extensive dysmigration through the Upper Jurassic and Lower Cretaceous caprocks, and reproduces the gas accumulation in the Paleocene sands. The heavy oil ring beneath the gas field is thought to be the result of the retrograde condensation of the oil from the upward migrating gas phase, followed by biodegradation.

The satisfactory reconstruction of the block considered is, however, very sensitive to such data as the geothermal gradient, the petroleum potential of the source rocks, and so on. For this reason, it is thought that this approach can lead only to rough quantitative estimates of the accumulations in undrilled areas. However, it helps the explorationist to understand the process of oil accumulation and its key parameters. It gives a consistent frame for the reconstruction of oil formation and migration through geological time, and enables comparison of different prospects. Thus it is a very useful tool for hydrocarbon exploration.

APPENDIX: OIL FORMATION AND MIGRATION MODEL*

Hydrocarbon formation

The degradation of organic matter is described in the Tissot and Espitalie model (1975) by a series of six parallel chemical reactions obeying first-order kinetics and the Arrhenius law:

$$Q = \sum_{k=1}^{6} (q_{k0} - q_k),$$

$$\frac{dq_k}{q_k} = A_k e^{-E_k/RT} dt,$$

where:

Q = quantity of hydrocarbons formed (mg/g C_{org} initial),
q_{k0} = initial oil potential relative to reaction k (mg/g C_{org} initial),
q_k = residual oil potential relative to reaction k (mg/g C_{org} initial),
A_k = reaction coefficient (s^{-1}),
E_k = activation energy (cal/mole),
R = ideal gas constant ($R \approx 2$ cal/mole K),
t = time (s) and
T = temperature (K).

The values of parameters q_{k0}, A_k and E_k are fixed for each type of organic matter whose degrada-

*The equations used in the model have been taken from Tissot and Espitalie (1975) for oil generation, and from Marle (1972) and Scheidegger (1960) for the fluid flow calculations. The authors were also inspired by Smith (1971) for the description of compaction.

tion is to be described, on the basis of the classification of Tissot *et al.* (1974).

Representation of fluid flows (water and oil)

The formulation of fluid flows in porous environments is based on Darcy's law adapted to the diphasic case by the use of relative permeabilities:

For water:

$$V_e = \frac{-KK_{re}\,\rho_e g}{\mu_e}\ \text{grad}\ H_e;$$

For the hydrocarbon phase, assumed to be unique:

$$V_h = \frac{-KK_{rh}\,\rho_h g}{\mu_h}\ \text{grad}\ H_h;$$

where:
K = intrinsic permeability (m^2),
V_e, V_h = filtration velocity of water and hydrocarbons (m/s),
μ_e, μ_h = dynamic viscosities of water and hydrocarbons (Pa.s),
K_{re}, K_{rh} = relative permeabilities of water and hydrocarbons (dimensionless),
ρ_e, ρ_h = densities of water and hydrocarbons (kg/m^3),
g = acceleration due to gravity ($\approx$9.81 m/s^2),

H_e = $(P/\rho_e g) - Z$ = head (water),
H_h = $(P_h/\rho_h g) - Z$ = head (hydrocarbons),
P, P_h = pressure of water and oil (Pa) and
Z = depth (m).

Characteristic intrinsic permeabilities of the porous environment are calculated according to porosity by the Kozeny–Carman formula, introducing an anisotropy coefficient:

$$K_x = \frac{0.2\ \phi^3}{(1-\phi)^2 S_0^2},$$

$$K_y = \lambda K_x,$$

where:
K_x, K_y = horizontal and vertical intrinsic permeabilities (m^2),
ϕ = porosity (dimensionless),
S_0 = specific area of rock matrix (m^2/m^3) and
λ = anisotropy coefficient, less than unity (dimensionless).

Once the permeability coefficients for each element of the grid have been calculated, those applicable to each boundary are derived by interpolation.

Relative permeabilities K_{re} and K_{rh} are assumed to depend on oil saturation by curves similar to those used for modelling fluid movement in reservoirs (see Fig. 5).

Capillarity equation

$$P_h - P = P_c,$$

where:
P_h = pressure in the oil phase (Pa),
P = pressure in the water phase (Pa) and
P_c = capillary pressure (Pa).

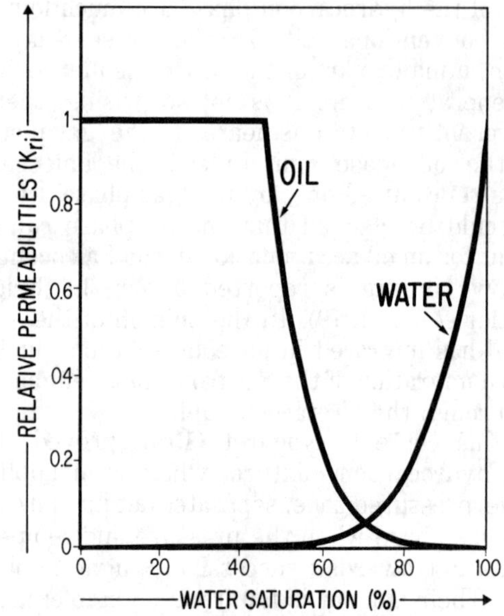

Fig. 5. Form of relative permeability curves used for modelling. For low saturation values, relative permeability is very low, but not nil.

The value of P_c is computed from a mean radius of access to pores R such that capillary pressure in the clays is approximately 2 MPa.

Representation of compaction

The equation of effective stress is used to express stresses between the solid matrix and fluids:

$$S_g = \sigma + P,$$

where:
S_g = total stress (Pa),
σ = effective stress (Pa) and
P = pressure (Pa).

As it was not possible to do otherwise, only the vertical component of stress was considered. This component is given by the following relationship, which expresses the weight of overlying sediments:

$$S_g = \int_0^z \rho g\ dz,$$

where:
z = depth of point considered (m),
ρ = total density of rock (kg/m^3) and
g = acceleration due to gravity.

Total stress, S_g, is then termed geostatic load. To describe sediment compaction it is assumed, following Smith (1971), that effective stress, σ, is a function of porosity:

$$\sigma = s_k(\phi).$$

The index k signifies that the function s_k differs from one lithology to another (see Fig. 6). The s_k functions were determined from porosity measurements on well logs in series with a hydrostatic system of fluid pressure, as it is then easy to calculate the effective stress corresponding to each measurement.

The concept of compaction irreversibility is introduced by linking σ, s_k and ϕ as soon as values

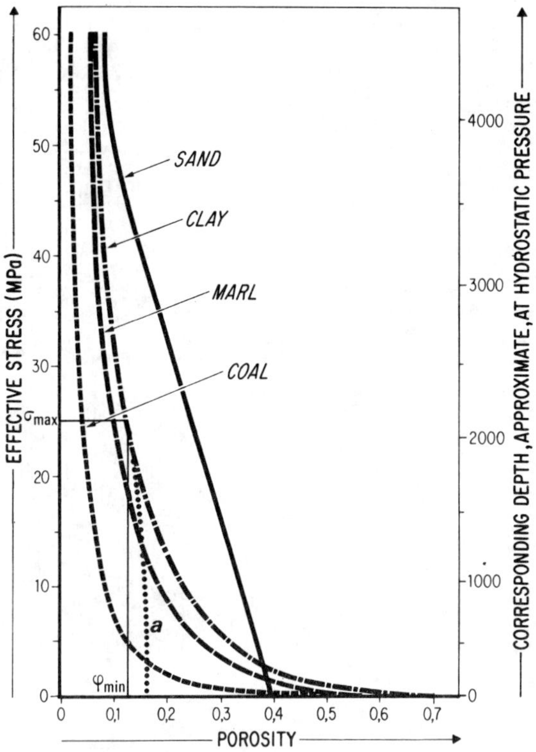

Fig. 6. Effective stress versus porosity curves used in studying a North Sea block. If effective stress declines, the curve followed (a) is different from the initial one and joins it at the point of maximum compaction reached (ϕ_{min}, σ_{max}).

of ϕ stop decreasing continuously in a grid unit and tend to increase above the ϕ_{min} value then reached, using an empirical relationship which is easy to calculate. This relationship expresses the fact that once values of ϕ increase, they no longer follow the initial $\sigma = s_k$ curves but, rather, much steeper ones which meet the $\sigma = s_k$ curves at a tangent at the point where the ϕ_{min} value is reached (see Fig. 6).

The density for the water (ρ_e) and that for oil (ρ_h) have been assumed to be constant.

The density and viscosity values used correspond to a low viscosity oil containing dissolved gas:

$\rho_h = 750$ kg/m^3 (density) and
$\mu_h = a\,e^{b/T}$ (dynamic viscosity according to Andrade formula in Latil (1975)).

T is the temperature, a and b are selected so as to obtain a viscosity of 10×10^{-3} Pa s at 40°C and 1.15×10^{-3} Pa s at 160°C.

Water viscosity according to temperature is calculated by the standard Bingham formula (Latil, 1975).

Algorithm and boundary conditions

The geological section under investigation from the beginning of subsidence up to the present time is represented by a finite-element grid with conditions at the limits as follows: (1) at the base of the grid, which it is assumed represents an impermeable bedrock affected by subsidence, zero flow and displacement specified; (2) at the upper boundary of the grid, hydrostatic pressure and overburden by sediments of uniform 2000 kg/m^3 density, free of vertical displacements and (3) at the right and left limits of the grid, flow limitation by hydraulic resistance, varying with examples treated, free of vertical displacements.

The number of grid units in the upward direction increases with evolution. An extra row is added when the depth of the top of the last grid level exceeds the initial thickness of the following level. The number of grid units is constant breadthwise. Using the equations set forth earlier, water and hydrocarbon flow velocities can be calculated at each moment; these velocities are used in each iteration to produce the equations of mass balance of water, hydrocarbons and solid matrix for each element of the grid, which in turn enables computation of the pressures, porosities, saturations and node depths through a finite-difference method.

REFERENCES

Carstens, H. and Dypvik, H. 1981. Abnormal formation pressure and shale porosity. *AAPG Bull.*, **65**, 344–350.

Chiarelli, A. and Duffaud, F. 1980. Pressure origin and distribution in Jurassic of Viking basin (United Kingdom–Norway). *AAPG Bull.*, **64**(8), 1245–1266.

Durand, B. 1981. Present trends in organic geochemistry in research on migration of hydrocarbons. Paper presented at the *10th Int. Organic Geochem. Congr.*, Bergen, September 1981.

Durand, B., Parratte, M. and Bertrand, P. 1983. Le potentiel en huile des charbons: une approche géochimique. *Revue IFP*, **38**(6), 713–721.

Du Rouchet, J. 1981. Stress fields, a key to oil migration. *AAPG Bull.*, **65**, 74–85.

Heritier, F. E., Lossel, P. and Wathne, E. 1979. Frigg field—Large submarine fan trap in Lower Eocene rocks of North Sea, Viking Graben. *AAPG Bull.*, **63**(11), 1999–2020.

Huck, G. and Karweil, J. 1955. Physikalishe Problem der Inkohlung. *Brennst.-Chem.*, **36**, 1–11.

Latil (1975). Récupération assistée. Technip, 275 pp.

Lopatin, N. V. and Bostick, N. H. 1973. Geology in coal catagenesis. In: *Nature of Organic Matter in Recent and Ancient Sediments*. Symp. Nauka, Moscow, pp. 79–90 (in Russian).

Marle 1972. Les écoulements polyphasiques en milieux poreux. *Cours de production*, tome IV. Technip.

Matheron, G. 1967. *Eléments pour une Théorie des Milieux poreux*. Masson, Paris, 164 pp.

Nogaret, E. 1983. Solubilité des hydrocarbures dans les gaz comprimés; application à la migration du pétrole dans les bassins sédimentaires. *Thesis*, Ecole des Mines, Paris (May).

Scheidegger. 1960. The Physics of flow through Porous Media. University Press, Toronto.

Smith, J. E. 1971. The dynamics of shale compaction and evolution of pore fluid pressure. *Math. Geol.*, **3**(3), 239–269.

Tissot, B. P. 1969. Premières données sur les mécanismes et la cinétique de la formation du pétrole dans les sédiments. Simulation d'un schéma réactionnel sur ordinateur. *Revue IFP*, **XXIV**(4), 470–501.

Tissot, B. and Espitalie, J. 1975. L'évolution thermique de la matière organique des sédiments: applications d'une simulation mathématique. *Rev. Inst. Fr. Pétrol.*, **30**, 743–777.

Tissot, B. and Welte, D. H. 1978. *Petroleum Formation and Occurrence*. Springer-Verlag, Berlin.

Tissot, B., Durand, B., Espitalie, J. and Combaz, A. 1974. Influence of nature and diagenesis of organic matter in formation of petroleum. *Am. Ass. Petrol. Geol. Bull.*, **58**(3), 499–506.

Ungerer, P., Behar, E. and Discamps, D. 1981. Tentative calculation of the overall volume expansion of organic matter during hydrocarbon genesis from geochemistry data. Implications for primary migration. Paper presented at the *10th Int. Organic Geochem. Congr.*, Bergen, September 1981.

9 Heat flow, geothermal gradients and maturation modelling on the Norwegian continental shelf using computer methods

R. H. Leadholm, Thomas T. Y. Ho† and S. K. Sahai†*
*Exploration Department, Conoco Norway Inc., Stavanger, Norway
†Regional Research, Conoco Exploration Research Division, Ponca City, Oklahoma, USA

Two computer programs have been developed to model maturation parameters. The first program (referred to as K-MOD) relates thermal conductivity to interval velocity and then computes either geothermal gradient or heat flow, depending on which is known. The program can be run using interval velocities obtained from either well or seismic data. The second program (referred to as M-MOD) is a maturation modelling program based on time–temperature, theoretical kinetic equations. M-MOD requires burial history, interval geothermal gradients and unconformity data as input. The program then computes present-day equivalent vitrinite reflectance (R_e) values for up to 21 horizons along with palaeo-temperature, palaeo-depth and palaeo-R_e values for designated time-stratigraphic horizons.

Sensitivity studies were conducted, and show that reasonable interval velocity errors do not greatly affect M-MOD R_e output. The studies also show that both programs are sensitive to errors in heat flow estimates.

Using these programs, heat flow, geothermal gradient and estimated R_0 (R_e) maps were constructed for the North Sea, Haltenbanken and Troms areas. Heat flow was first mapped using K-MOD and corrected temperature data from available wells. These data were extrapolated and mapped over each basin. Interval geothermal gradients were then computed for selected seismic shotpoints, again using K-MOD. Finally, M-MOD was run and present-day R_e values were mapped to delimit oil and gas windows.

Approximately 100 seismic and well data points were modelled in each basin. Before running the bulk of the data, cross-plots were constructed to ensure data reliability. A good correlation was established between laboratory-measured R_0 values and M-MOD-computed R_e values. For the North Sea, the Upper Jurassic 'hot shale' was modelled, and R_e maps show it to be oil-mature on the flanks of the Viking and Central Grabens, becoming gas/condensate-mature towards the axes. Both the 'hot shale' and the lower Jurassic 'Coaly Unit' were modelled in Haltenbanken, with resulting R_e maps indicating a mature oil and a mature gas window, respectively. In Troms, the top and base of the 'hot shale' were modelled, and a mature oil window was delimited on the north and west flanks of the Hammerfest Basin.

Results from the three basins suggest that utilization of the K-MOD and M-MOD programs together with interval velocity information provides a rapid means to map maturation trends over large basin-wide areas. The programs are especially useful in basins with limited well data.

INTRODUCTION

In recent years the search for oil and gas has become more risky and more selective because of political instability in producing-countries, tighter exploration budgets, and the high cost of exploring for elusive and remote plays. This problem is augmented by decreases in oil and gas prices. To cope with these business challenges it is necessary to develop new methods, such as computerized basin analysis, to find suitable prospects quickly and precisely.

One of these computerized techniques was proposed by Welte and Yükler (1981), who published a modelling method that integrated geological, hydrodynamic, thermodynamic and geochemical data. For maturation estimates they used borehole-based time and temperature data to calculate maturation level following the method of Lopatin (1971). This approach is adequate for a mature basin with sufficient well control. However, it is difficult or impossible to use it in a frontier basin where access to well samples is limited or non-existent. For this reason, it is obvious that a method based on surface geophysical data is needed to generate the time and temperature input required for modelling maturation in an unexplored basin.

This paper discusses a seismic–geochemical technique for estimating time/temperature data and offers a brief treatment of a kinetic method developed by Con-

Petroleum Geochemistry in Exploration of the Norwegian Shelf © Norwegian Petroleum Society (Graham & Trotman, 1985) pp. 131–143

oco for modelling organic maturation. In Norway, this technique has been applied to the regional maturation mapping of the North Sea, Haltenbanken and Troms areas.

THEORETICAL BACKGROUND

Heat flow

Heat flow is the amount of heat that flows out of the Earth in a unit time, and it has a rather uniform vertical distribution in a geological column. Heat flow cannot be measured directly. It is always calculated from measured thermal conductivity (k) of the rocks and the geothermal gradient (dt/dZ) by using the expression

$$\text{heat flow } (Q) = K\ (dT/dZ). \tag{1}$$

The average heat flow for the Earth is about 1.5 heat flow units (HFUs), where 1 HFU represents 1 μcal of heat that flows out of a 1-cm square in 1 s.

Generally, the heat flow for an area can be directly obtained or can be extrapolated from the measurements found in published information that covers worldwide heat flow distribution (e.g. Lee, 1963; Lee and Uyeda, 1965; Jessop *et al.*, 1976; Cermak and Ryback, 1979). The heat flow distribution is controlled by basin history. It is low in basins associated with Precambrian shields (<1.0 HFU) and with Paleozoic or older orogenies (e.g. Anadarko–Ardmore Basin with a heat flow of <1.2 HFU). However, it is somewhat higher in basins with Mesozoic or Tertiary orogenies (e.g. West Natuna Basin, South China Sea, with a heat flow of 1.9 HFU) and even higher in basins with Tertiary volcanism (e.g. >3 HFU in South Sumatra).

The heat flow in an area is also affected by local geological factors such as differential rates of sedimentation and erosion, intrusions, hydrothermal activity, radioactivity, etc. For example, faster sedimentation decreases the heat flow more than slower sedimentation. Conversely, faster uplift and erosion increase the heat flow more than do slower uplift and erosion. Volcanic intrusions can drastically alter the geothermal regime (although temporarily), thereby greatly affecting organic maturation in the vicinity of the intrusive body.

Sometimes, heat flow in an area may be determined from the corrected bottom-hole temperature measurements in a nearby well. The corrected temperatures are used to calculate the geothermal gradient from the expression

$$(dT/dZ) = [T_c(Z) - T_s]/Z, \tag{2}$$

where $T_c(Z)$ is the corrected at depth Z, and T_s is the mean annual surface temperature. The thermal conductivity of the geological column to depth Z is determined by the procedure discussed in the following section. The heat flow is then calculated from equation (1).

Thermal conductivity

The implication of the heat flow equation (1) is that the temperature gradient changes with variations in the thermal conductivity of the rocks. Our main interest is in subsurface temperatures, because they control the formation and preservation of oil and gas. Therefore,

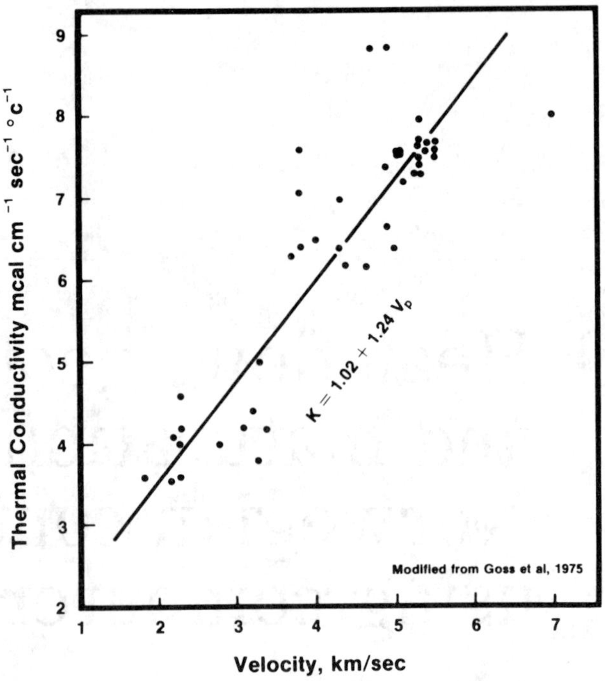

Fig. 1. Thermal conductivity versus sonic velocity. Modified from Goss *et al.* (1975).

it is necessary to understand how thermal conductivity changes with depth. This is accomplished by understanding the effect of various geological factors on the thermal conductivity of common rocks.

There are several physical parameters that affect the thermal conductivity of rocks. The most important of these for the present purposes are porosity, density and fluid saturation. Pressure and temperature also affect thermal conductivity to some extent. For example, there is a general decrease in conductivity with an increase in temperature. Also, the thermal conductivity increases with an increase in pressure.

The factors that affect the thermal conductivity of a rock similarly affect the velocity of seismic waves in the rock. It has been shown previously (e.g. Gross *et al.*, 1975; Houbolt and Wells, 1980) that thermal conductivity (K) is proportional to sonic velocity (see Fig. 1). This study reviews a computer program (referred to as K-MOD) that is based on a similar relationship. The program solves the basic heat flow equation for either heat flow or geothermal gradient, depending on which is known. The geothermal gradient data are then input together with burial history, unconformity, intrusion and surface temperature data into a maturation modelling program (referred to as M-MOD). The M-MOD program is based on time–temperature theoretical Arrhenius-type kinetic equations. The general form below refers to a crossplot:

$$\ln R_E = m\int_0^t e^{-E/RT(t)}dt + C, \tag{3}$$

$$T(t) = T_0 + rt,$$

where R_E is the equivalent vitrinite reflectance value, m is the slope, E is activation energy, c is intercept of the regression line, R is the ideal gas constant, and T is subsurface temperature (in K), which is a function of heating time (t). Heating time (t) is calculated from the difference in geological ages between the top and base of a stratigraphic unit.

This program differs from the Lopatin method (Lopatin, 1971; Waples, 1980) in two important respects. First, it calculates maturity in terms of equivalent vitrinite reflectance, R_e, rather than a time–temperature index, TTI. Second, the method is based on integration of the time–temperature kinetic equations rather than a graphical summation of time–temperature data. The use of these equations treats maturation rate as a variable, rather than assuming a constant doubling for every 10 °C as in the Lopatin method. In this respect it is similar to the work of Tissot and Espitalie (1975). The M-MOD program is also capable of modelling uplift and erosion.

The greatest advantage in using the K-MOD and M-MOD programs to map maturation is that they can both be run on data obtained directly from seismic sections. A seismic–geochemical analysis of remote basins is thus possible. For example, an early version of this program predicted the first oil discovery by the Norwegian industry north of latitude 62°N at Tyrihans (Ho and Sahai, 1982). Any available well data can be tied in to confirm results. A further advantage of the K-MOD program is that, because it relates interval velocity to thermal conductivity, the effects of overpressure on geothermal gradient are also taken into consideration.

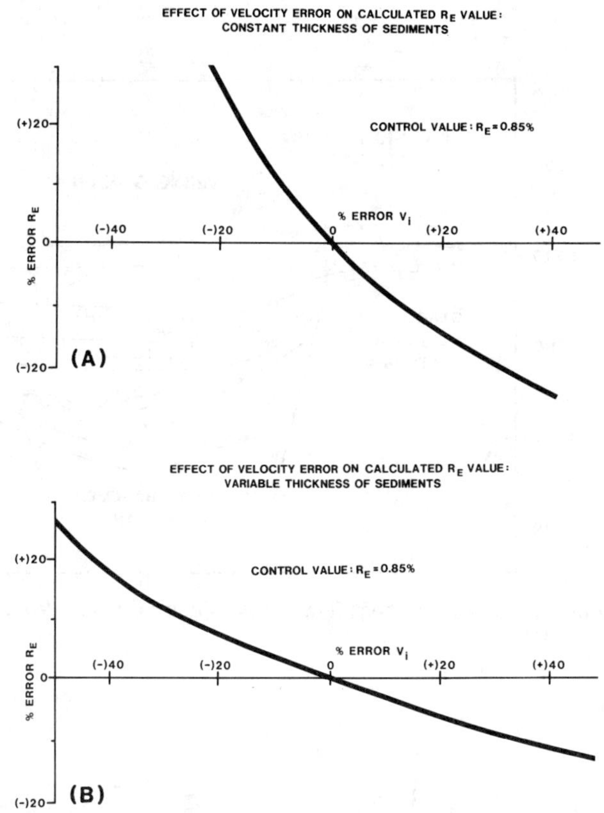

Fig. 2. Effect of velocity error on calculated R_0 values. (A) Stratigraphic sequence thicknesses are held constant. This is representative of a case using well data to pick tops and erroneous sonic logs or check shot data to calculate interval velocity. (B) Variable thickness of sediments. This is representative of a case using DIX corrected stacking velocities. Note that velocity error essentially cancels out thickness error.

APPLICATION

Heat flow, geothermal gradient and maturation R_e (equivalent R_0) maps were constructed for the North Sea, Haltenbanken and Troms areas of the Norwegian continental shelf using the K-MOD and M-MOD programs. Both seismic and well data were utilized, and results showed a good correlation with existing geochemical data. More than 400 seismic and well data points were modelled. Before running the bulk of the data, program sensitivity to variations in interval velocity and heat flow was first checked.

Program sensitivity

Errors in interval velocity determination affect both K-MOD and M-MOD output. Therefore, sensitivity studies, such as shown in Fig. 2, were conducted for a test point where the actual R_0 of the source horizon is 0.85%. Figure 2(A) shows one representative example where well data were used to pick interval thicknesses and erroneous sonic data were used to calculate velocities. A 20% increase in interval velocity causes a significant (14%) decrease in predicted R_0. However, Fig. 2(B) illustrates that if we vary the interval velocities *and use the same erroneous velocities* to calculate interval thicknesses, a 20% velocity input error causes less than a 7% error in predicted R_0 values. This latter example is representative of a case where erroneous DIX-corrected stacking velocities are used, and suggests that in remote areas stacking velocities are acceptable for K-MOD input because interval velocity errors effectively cancel out thickness errors.

Figure 3 is a representative profile of heat flow calculated from well data in the Viking Graben. It shows that in the Viking Graben heat flow varies from 1.2 HFU on the East Shetlands Platform to more than 1.7 HFU on the Utsira High. These variations are interpreted to be primarily due to timing of tectonic movements. Previous work reported in the literature supports a similar variation of heat flow in this region (Toth *et al.*, 1981).

Figure 4 is based on a M-MOD run for a specific data point where all parameters, except heat flow, are held constant. Varying the heat flow from 1.1 to 1.8 HFU moves the predicted R_0 for a specific source rock interval in this example all the way from the marginally mature window (0.6–0.8% R_0), completely through the bulk generation window (0.8–1.2% R_0) and into the late generation window (>1.2% R_0). This high sensitivity suggests that it is important to estimate and map heat flow as accurately as possible before modelling maturation.

Area of study, data base, modelled horizons

Figure 5 delimits the areas studied in the North Sea, Haltenbanken and Troms regions. More than 400 seismic and well data points were modelled and the distribution of this data is detailed in Fig. 6. The study began in the North Sea, where there are numerous wells and where geochemical data are available to check modelling results. It then moved north to the more-sparsely explored basins, where there is essentially only seismic data with which to work. The data points for modelling were drawn from a seismic base of 38 000 km (23 612 miles).

In all three areas the computed maturation state of the Upper Jurassic 'hot shale' was mapped (see Fig. 7). This is an organic-rich, radioactive marine shale whose prolific source potential for both oil and gas is well-documented in the literature (Demaison,

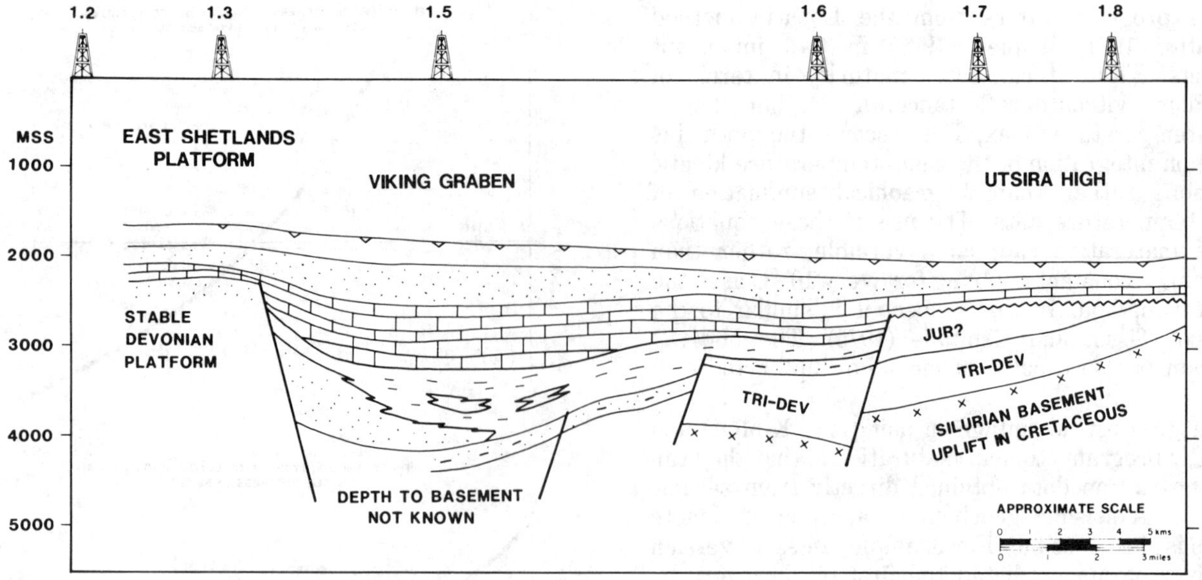

Fig. 3. Variance in heat flow across Viking Graben. Values where calculated from corrected bottom hole temperatures and K-MOD.

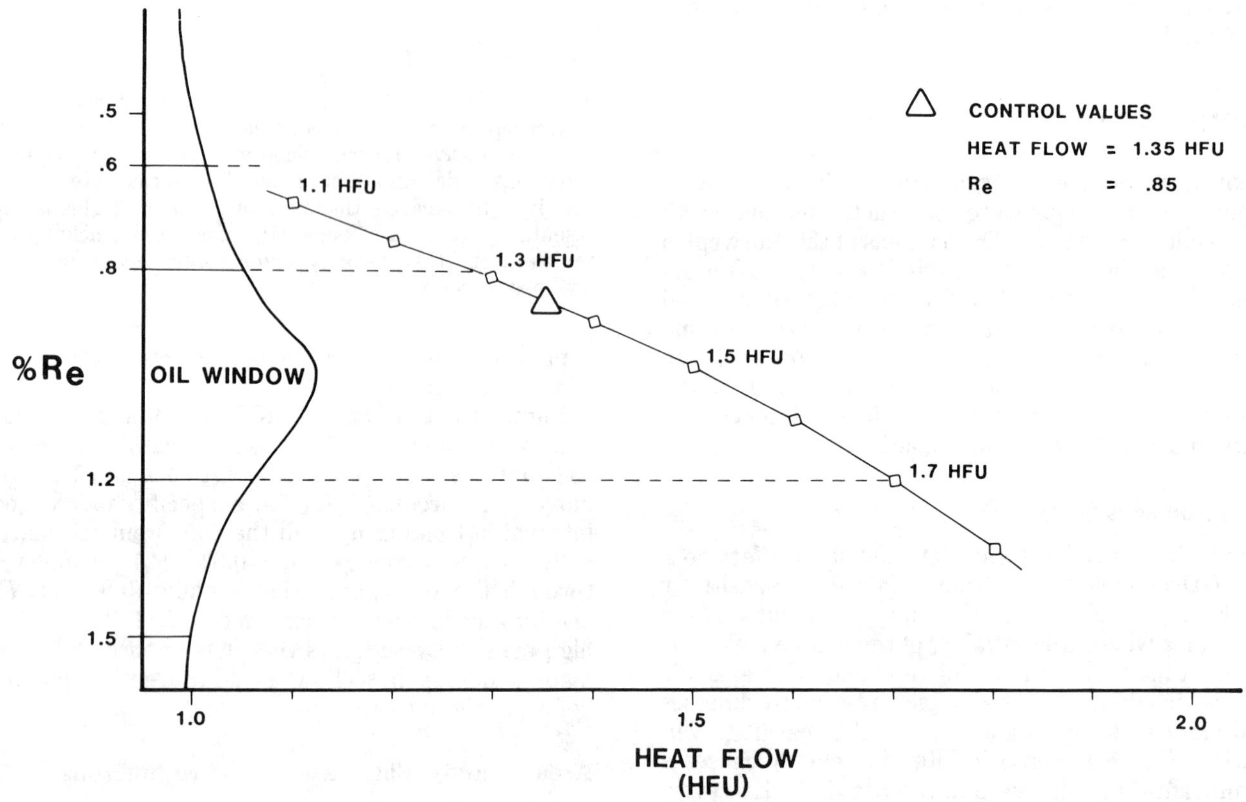

Fig. 4. Sensitivity of M-MOD to heat flow error. Varying the heat flow from 1.1 to 1.8 HFU moves the calculated R_0 value completely through the oil window.

1984; Pearson *et al.*, 1983; Rønnevik *et al.*, 1983). The name of this unit is presently being revised by the Norwegian Lithostratigraphic Nomenclature Committee. It has been known variously as the 'hot shale', the 'Kimmeridge Clay', and, most recently, as the Draupne Formation or Mandal Formation (Vollset and Doré, 1984). For simplification, this source horizon is refered to only as the 'hot shale' herein. In addition to the 'hot shale' the Lower Jurassic 'coaly unit' was mapped as a gas-prone horizon in the Haltenbanken area (see Fig. 7).

Assumptions

Geothermal gradient and heat flow are considered constant through time. Goff (1983) has pointed out that such an assumption introduces an optimistic error into maturation estimates. However, the present data cross-plots (see Fig. 8) show a good correlation between laboratory measured R_0 values and M-MOD modelled values, thus demonstrating that this error, if present, is not significant. The data were not examined statistically.

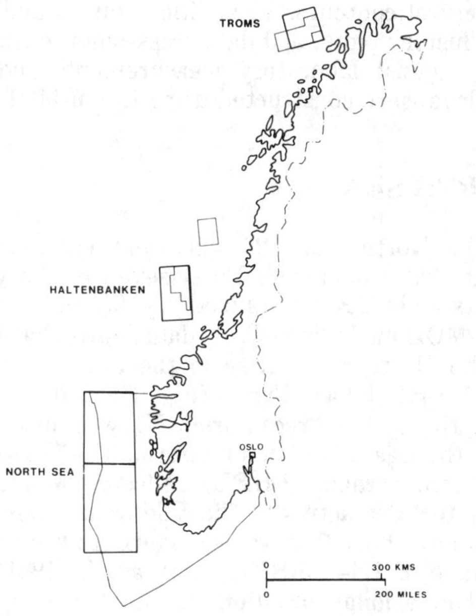

Fig. 5. Location plot showing areas of study.

Surface temperatures were estimated using a mean annual water bottom temperature chart published by Evans and Coleman (1974). An attempt was made to use extrapolated surface temperatures based on well data; however, this method resulted in M-MOD output that did not correlate with available geochemical data.

Recently, several authors have considered the effects of fluid flow on the observed vertical variations of goethermal gradients and heat flows in well-bores (Carstens and Finstad, 1981; Oxburgh, 1984). In this paper, it has been assumed that fluids are static and average heat flow values calculated from the well data have been used.

METHODOLOGY

To summarize the methodology (see Fig. 9): the first goal was to estimate and map heat flow using either published values and/or well data run through K-MOD. After the heat flow maps were constructed, interval geothermal gradients were determined for

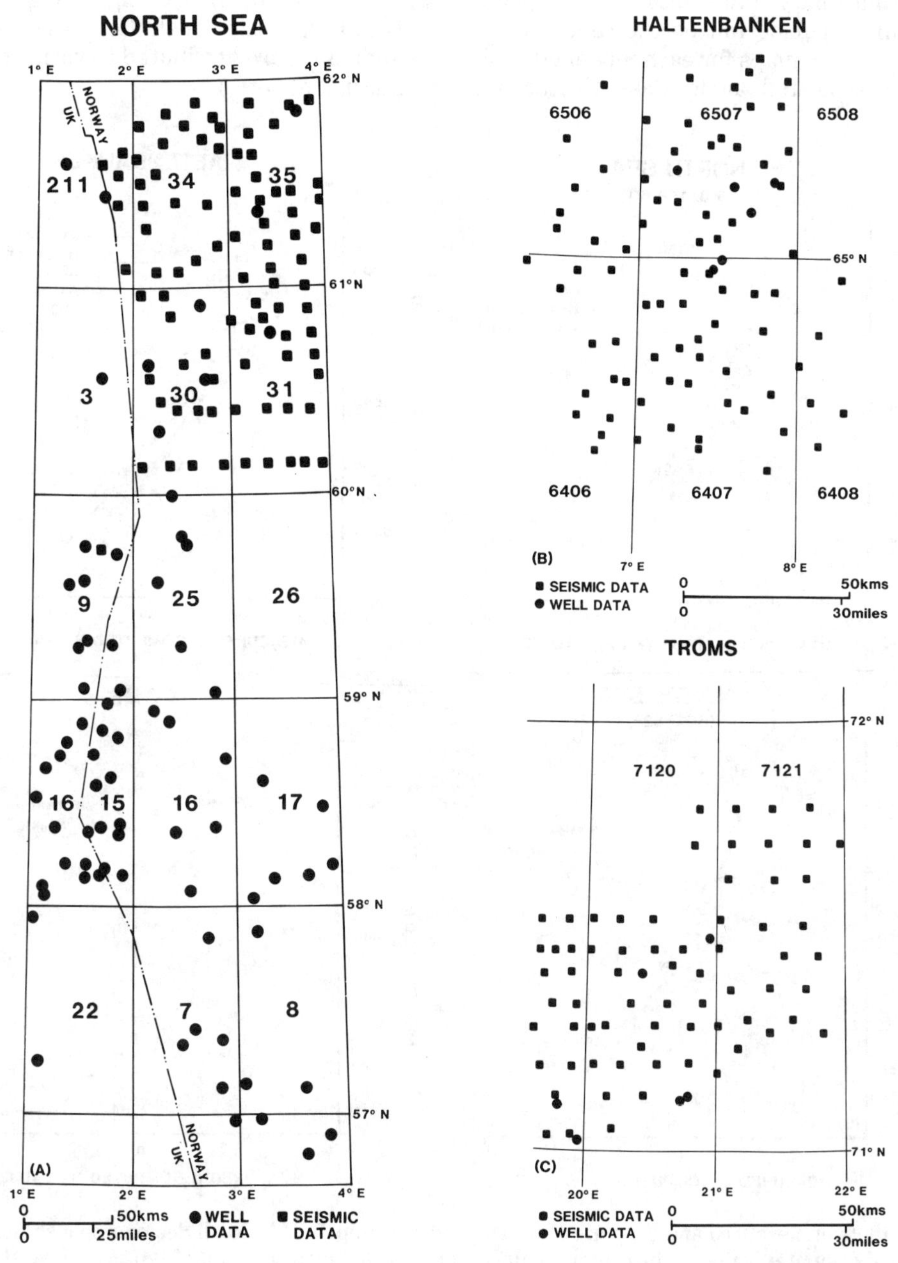

Fig. 6. Distribution of modelled points: (A) North Sea, (B) Haltenbanken and (C) Troms. A total of 277 seismic data points and 132 wells were used in the study.

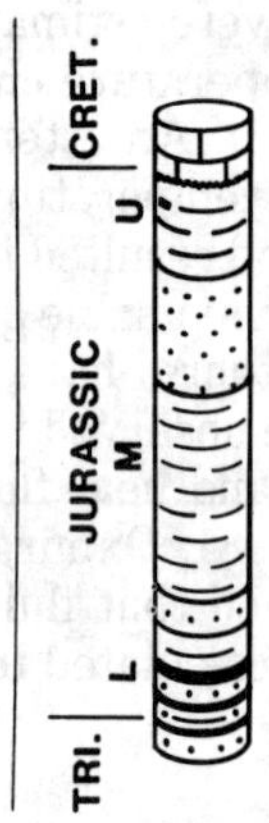

Fig. 7. Representative stratigraphic column. The 'hot shale' was modelled in the North Sea. Haltenbanken and Troms areas. The 'coaly unit' was modelled only in Haltenbanken.

each data point using either well data and/or seismic data and the program K-MOD once again. The geothermal gradients were then input together with burial history and unconformity data into the program M-MOD. Program output permitted the construction of basin-wide maturation maps for each modelled horizon. Maps are interpolated along Base Cretaceous structural contours. In addition, burial and maturation history plots, and data cross-plots to check accuracy against laboratory measurements and seismic displays were constructed using the M-MOD data.

NORTH SEA

In the North Sea, 120 wells and 119 seismic data points were modelled. Forty-seven of the well data points lie in UK waters (see Fig. 6(A)).

K-MOD modelling of well data shows that heat flow in the North Sea varies in the approximate range 1.2–1.8 HFU (see Fig. 10(A)). Heat flow is lowest over the stable Precambrian Norwegian Shield and over the East Shetlands Platform. The Utsira High is hot (greater than 1.7 HFU), reflecting Mesozoic–Tertiary tectonic activity. The Viking and Central Grabens have heat flow values similar to the worldwide average of 1.4–1.5 HFU. Toth *et al.* (1981) showed similar regional variations in the UK sector of the Central Graben. Geothermal gradients computed from both well and seismic data vary from approximately 30 to 45 °C/km, increasing along the same regional trends as the heat flow (see Fig. 10(B)). Local anomalies are attributed to variations in lithology (salt intrusions, etc.).

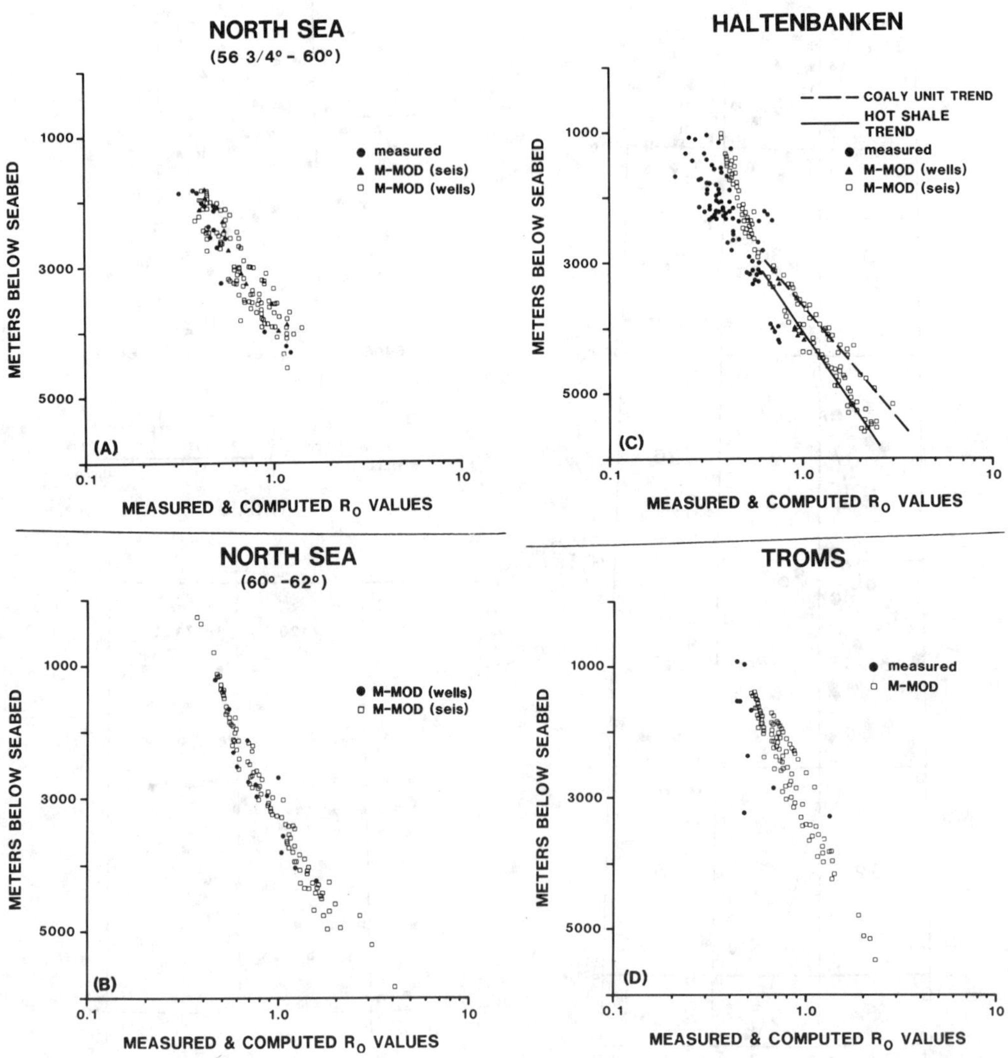

Fig. 8. Data cross-plots of measured and computed R_0 values versus depth. (A) North Sea example shows good correlation of laboratory values and computed values. The computed data were divided by a constant (1.08) to achieve this fit. (B) North Sea example shows that similar results are achieved using either well or seismic data to run K-MOD and M-MOD. (C) Haltenbanken example shows overall good correlation. Divergence in trends below 3000 m is attributed to variance in thermal conductivity with depth. (D) The limited measured data in Troms shows a fair correlation with computed values.

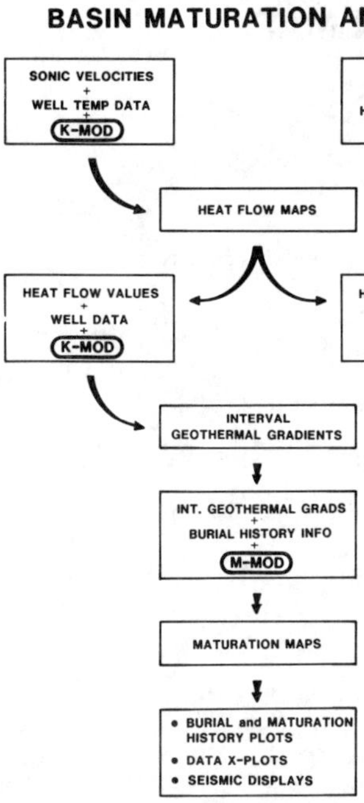

Fig. 9. This flowchart shows systematic approach to mapping and documenting basin maturation.

Figures 8(A) and (B) are cross-plots constructed to demonstrate quality control of M-MOD data output. R_0 values obtained from laboratory measurements are plotted, along with M-MOD data against depth. Figure 8(A) is based on the abundant well data south of latitude 60 °N. The M-MOD data were divided by a constant (1.08) to achieve this good fit. The data were not adjusted in any of the other areas. Figure 8(B) compares R_e values based on well data with R_e values based on seismic data for the area between 60 °N and 62 °N and demonstrates that the data sources are compatible.

In the North Sea, R_e (equivalent R_0) maps were only constructed for the top 'hot shale' horizon. Figure 11 is a sketch of the complete map and Fig. 12 is a detail of that map. The map shows that the 'hot shale' is oil-mature along the flanks of the Viking and Central Grabens becoming more thermally mature towards the axes. All of the major platforms and highs, as well as the Stord Basin and Norwegian–Danish Basin are immature at this level. These results thus confirm previous published work by Rønnevik *et al.* (1983) and Goff (1983). Cornford *et al.* (1983) showed from geochemical observations that the onset

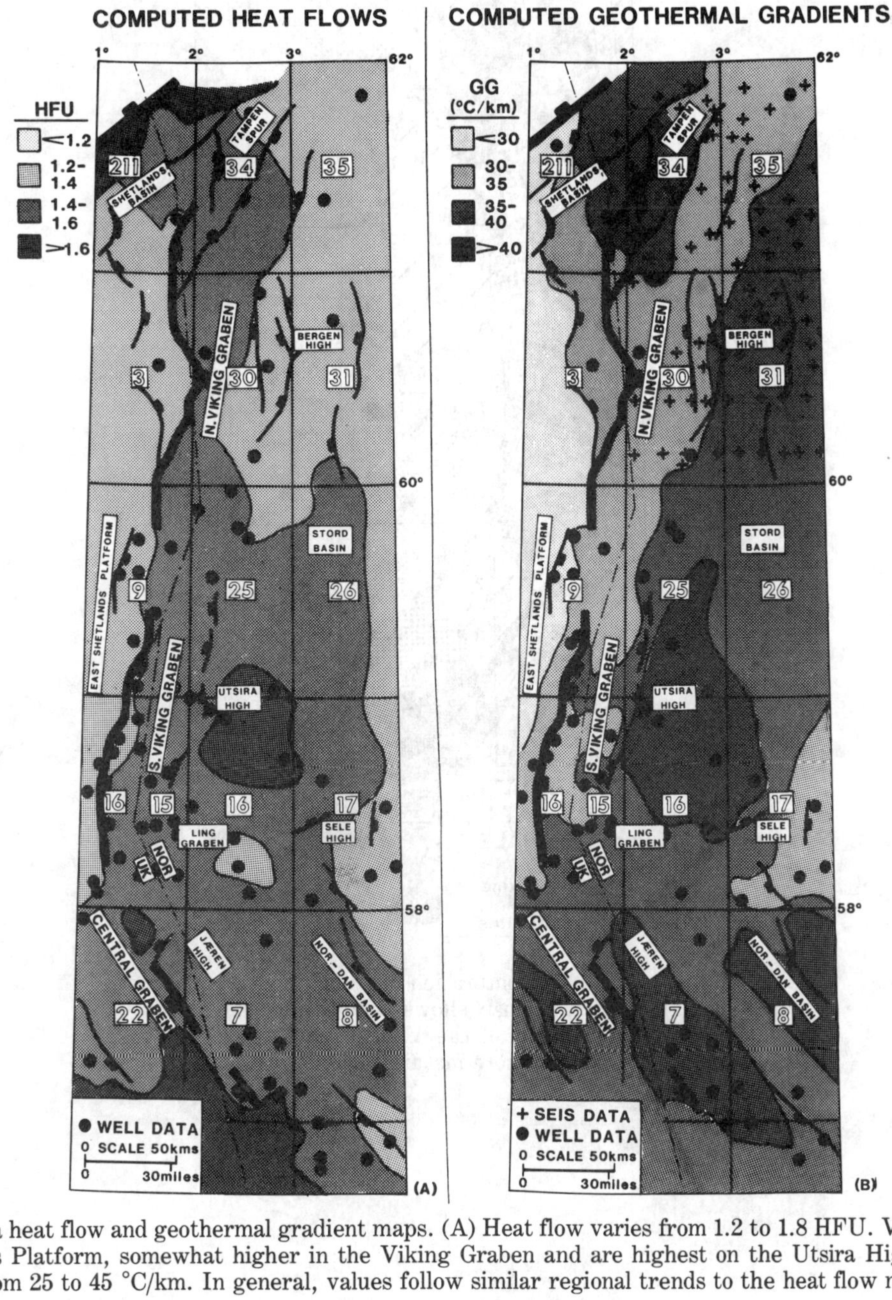

Fig. 10. North Sea heat flow and geothermal gradient maps. (A) Heat flow varies from 1.2 to 1.8 HFU. Values are lowest on the East Shetlands Platform, somewhat higher in the Viking Graben and are highest on the Utsira High. (B) Geothermal gradients range from 25 to 45 °C/km. In general, values follow similar regional trends to the heat flow map.

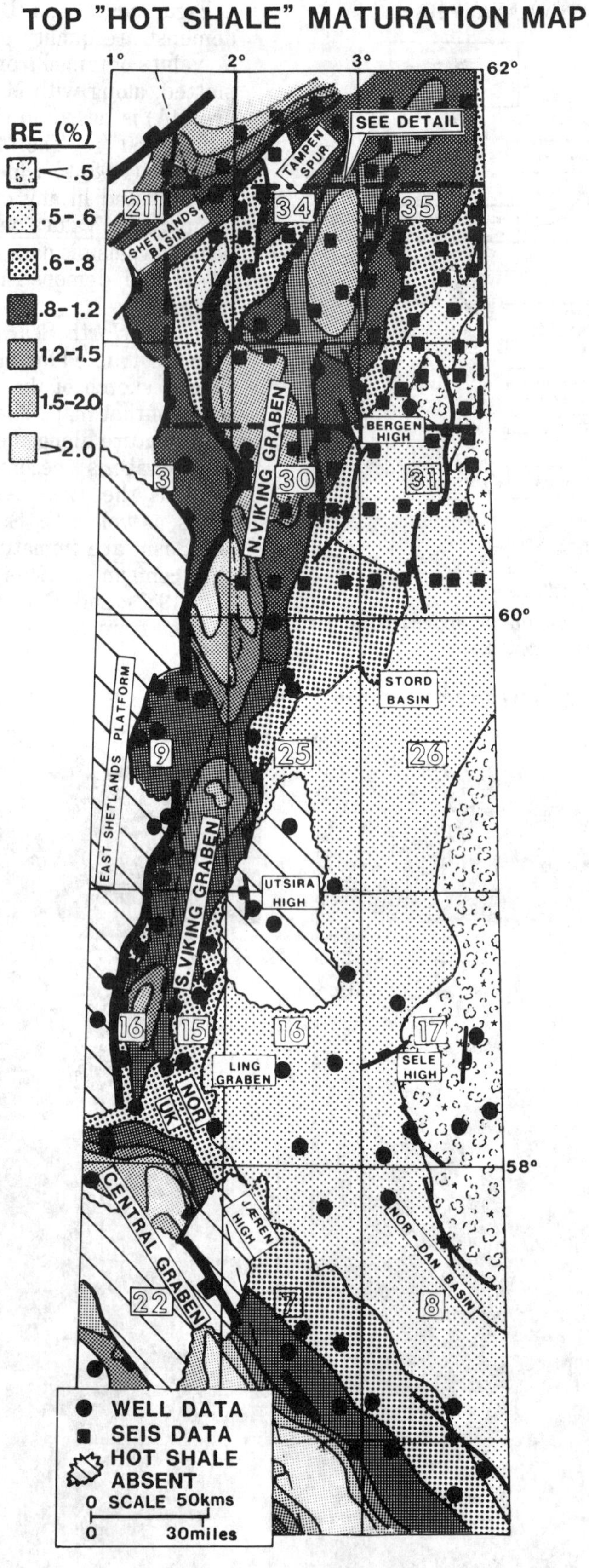

Fig. 11. North Sea maturation map for the top of the 'hot shale' horizon. The map shows that the formation is oil-mature on the flanks of the Viking and Central Grabens becoming thermally more mature towards the axes.

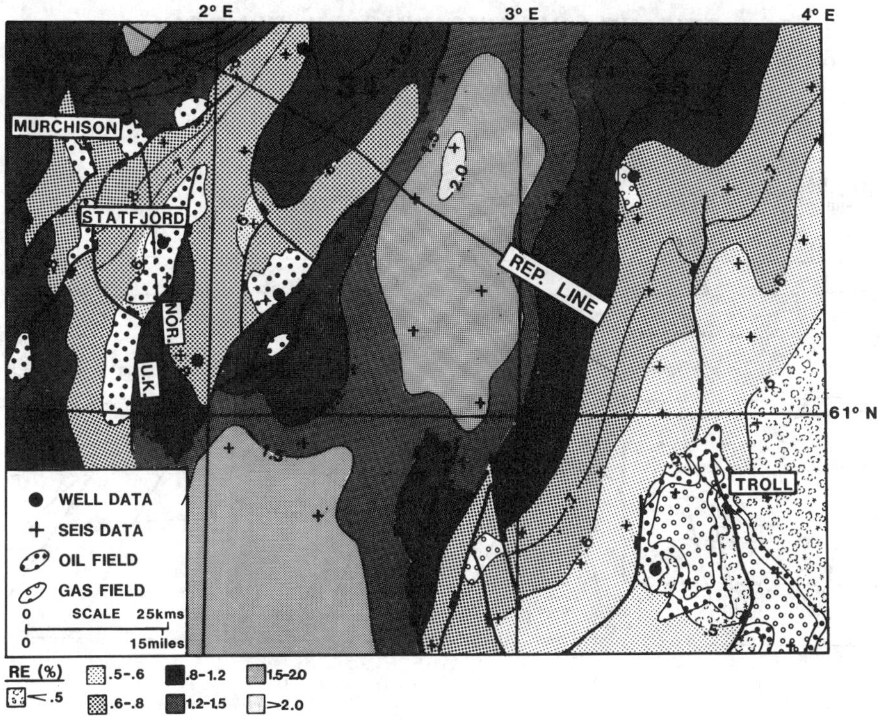

Fig. 12. Detail of North Sea maturation map. Note that some of the large fields (Troll, Statfjord, etc.) lie outside the mature windows for their respective hydrocarbon types.

of oil generation in the 'hot shale' occurs at 170–200 °F in this area. The temperatures predicted by K-MOD are compatible with these findings.

The present detailed mapping (see Fig. 12), indicates that some of the large fields lie outside the peak generation windows for their respective hydrocarbon types. This implies that significant migration has occurred or an alternative source horizon exists. Troll is the only large field that lies completely within an immature area. Fields in the East Shetlands Basin are predominantly oil-prone, while some fields on the east flank of the Viking Graben contain large amounts of gas. The maturation map for the 'hot shale' suggests that the East Shetlands Basin oil fields are sourced from within adjacent individual fault compartments, while the fields on the east flank of the Viking Graben drain more thermally mature shales in the graben centre.

Using M-MOD data it is possible to construct a geoseismic display to give a cross-sectional view of the basin maturation. Figure 13(A) is a sketch of a complete line display and Fig. 13(B) is a detail of that sketch. The lines are coded to show present hydrocarbon windows for the 'hot shale' horizon. The displays are constructed by using M-MOD to model depths to nominated R_e values at selected shotpoints along the line. These depths are converted to two-way times and the windows are then plotted. Across the top of the line the present R_e values for the top of the 'hot shale' are annotated.

The M-MOD program can also generate a burial and maturation history display for each data point modelled (see Fig. 14). These displays show both the present depth to various significant R_e values and the geological age at which these values were reached. The plots also show the burial history of the source horizon. Perhaps the most important use for this type of display is to study the timing of hydrocarbon generation and migration. For example, Fig. 14(A) shows that at the location modelled (see Fig. 13(A)), oil generation in the 'hot shale' occurred from early Miocene to middle Pliocene times. Rønnevik *et al.* (1983) showed similar results in their study of the Viking Graben. In order to gain a better understanding of the timing of migration, it is necessary to extrapolate these data and construct regional palaeo-R_e maps.

HALTENBANKEN

After determining that M-MOD R_e values agree with laboratory-measured R_0 values in areas with dense well control (see Figs. 8(A) and (B)), the study was extended north to more frontier basins. In Haltenbanken 98 seismic data points and six wells were modelled (see Fig. 6(B)).

The heat flow map for Haltenbanken is based on limited well data and the understanding of local tectonic history (see Fig. 15). The heat flow varies from 1.1 to 1.7 HFU, with the highest values found to the west. This is opposite to the trend expected from sedimentary thickness, and is attributed to Mesozoic–Tertiary volcanic activity in the Vøring Basin.

A good correlation between M-MOD R_e data and measured R_0 values was established (see Fig. 8(C)). Two horizons were modelled, and the cross-plot shows that below 3000 m, the 'Coaly Unit' matures with a less steep gradient than the 'hot shale'. This is attributed to variations in the thermal conductivity of the intervening sediments with depth.

Figures 16(A) and (B) are sketch maps of the maturation state for both the top 'hot shale' and the top Coaly Unit'. The Haltenbanken 'hot shale' maturation map shows that the formation is oil-mature over nearly half of the mapped area. On the Trøndelag Plat-

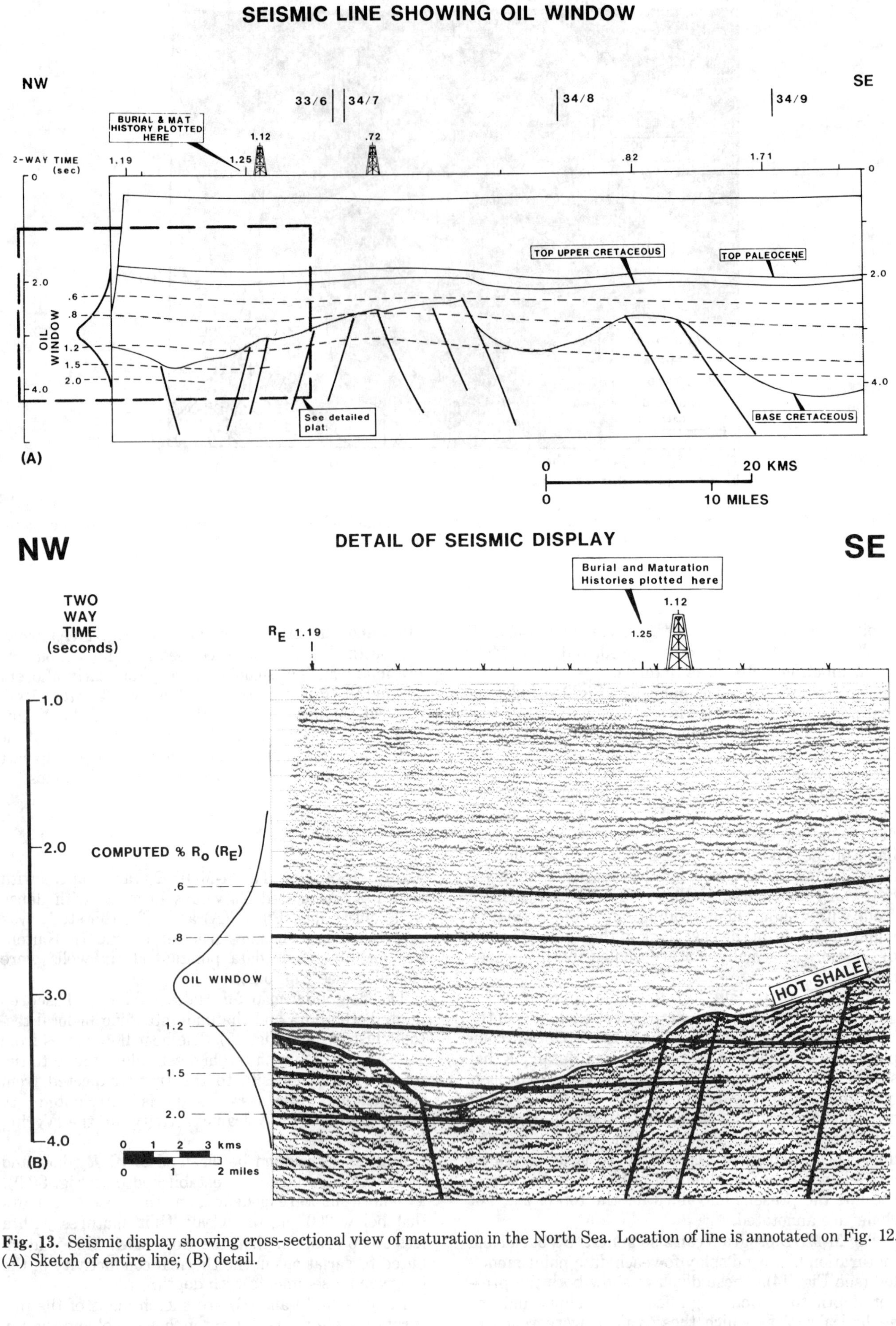

Fig. 13. Seismic display showing cross-sectional view of maturation in the North Sea. Location of line is annotated on Fig. 12. (A) Sketch of entire line; (B) detail.

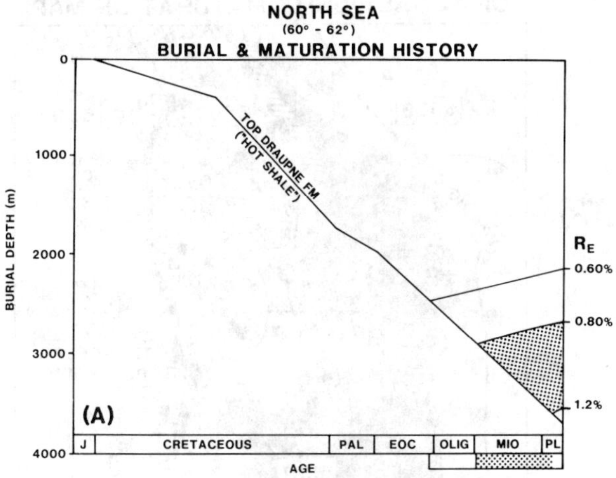

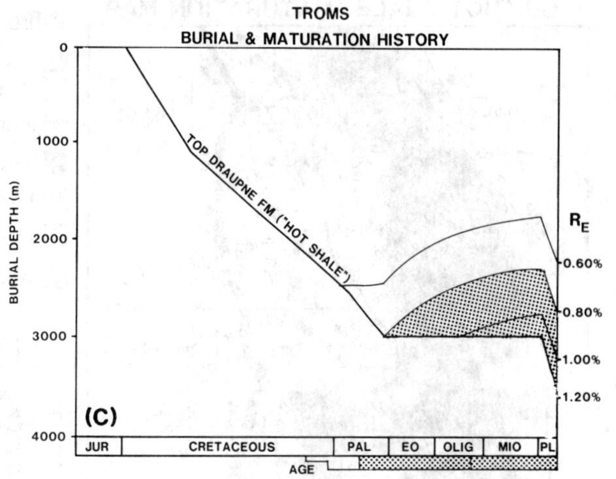

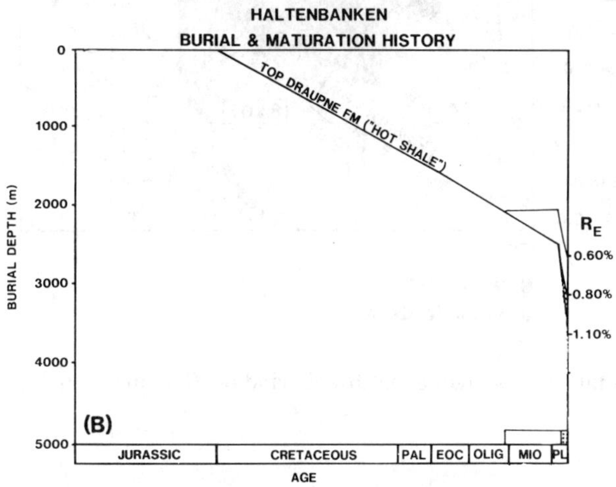

Fig. 14. Burial and maturation histories plotted for: (A) North Sea, (B) Haltenbanken and (C) Troms. The plots demonstrate that sedimentation rate is a controlling factor in oil generation timing.

form the 'hot shale' is immature ($R_e < 0.5$), but moving out into the Trøndelag and Vøring Basins the formation rapidly becomes thermal gas-mature to overmature ($R_e > 1.5\%$). This is attributed to both depth of burial and the high heat flows associated with Mesozoic–Tertiary volcanics in these basins.

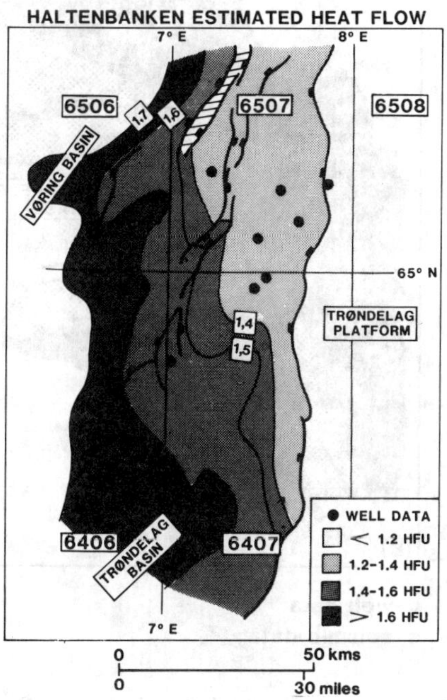

Fig. 15. Heat flow map for Haltenbanken based on well data and local tectonic history.

The Lower Jurassic 'Coaly Unit' maturation map delimits a large region of peak gas generation potential. This area almost directly underlies the 'hot shale' peak oil generation window. As a result, additional high GOR oil discoveries, similar to the results of wells drilled in the Tyrihans Field, are anticipated in Haltenbanken.

Figure 14(B) is an example of a burial and maturation history modelled for the location indicated on the sketch maps (Fig. 16). It shows that at this position the 'hot shale' reached peak oil generation in the Pliocene. This suggests that for this area even the most recent structures are prospective. Once again, regional palaeo-R_e maps are needed before generalized assumptions are made.

TROMS

Further north in the Troms areas, 80 seismic data points and six wells were modelled (see Fig. 6(C)). The limited laboratory R_0 data available show a fair correlation with M-MOD R_e output (see Fig. 8(D)). In Troms, maturation maps were generated for both the top and the base of the 'hot shale' (see Fig. 17), which is oil-mature, particularly along the western and northern margins of the Hammerfest Basin. Although no commercial oil has yet been discovered in Troms, several wells in the basin have encountered oil shows, and there is still hope for finding commercial oil in as yet unlicensed acreage.

Figure 14 compares burial and maturation history plots from areas where the 'hot shale' lies at similar

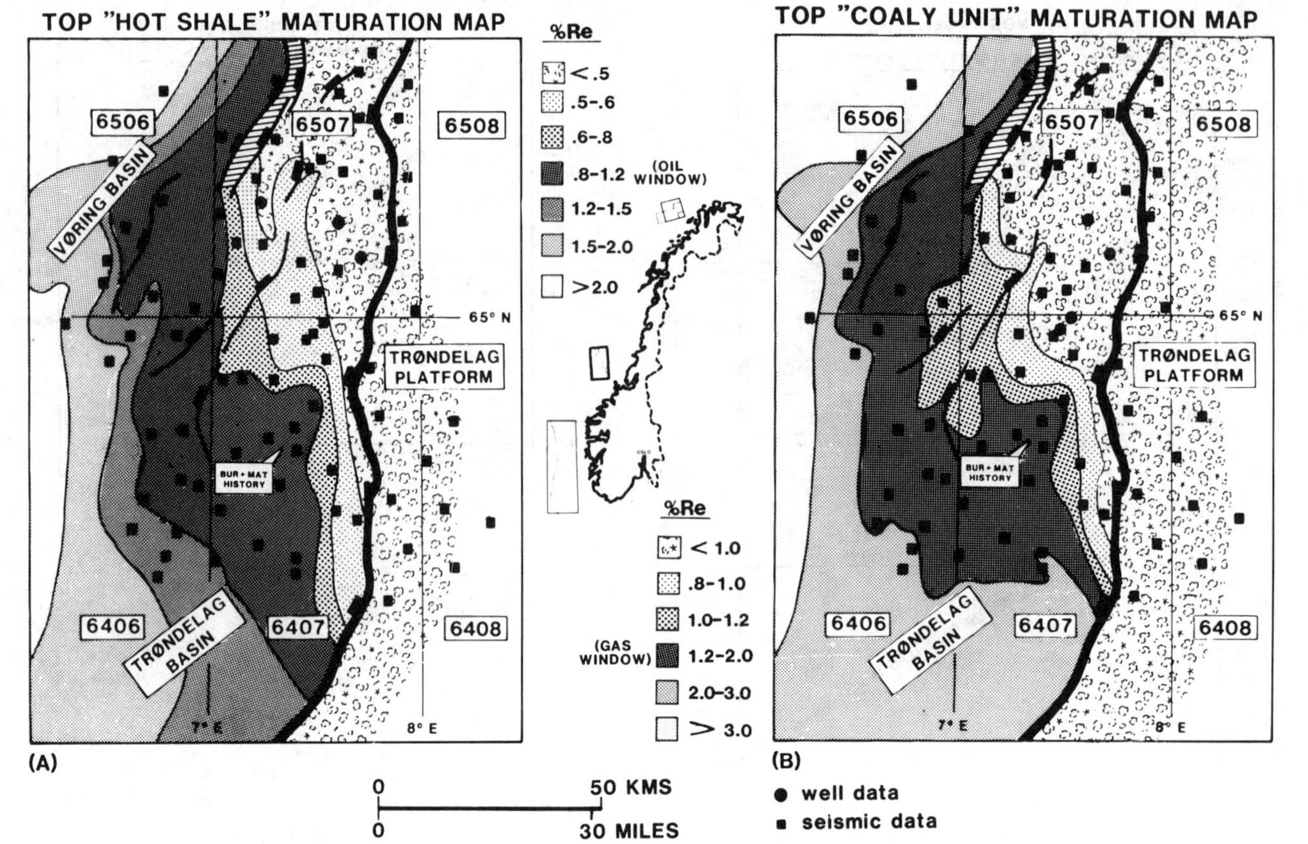

Fig. 16. Maturation maps for the Haltenbanken area. (A) Top 'hot shale' map showing mature oil window; (B) top 'Coaly Unit' map showing mature gas window.

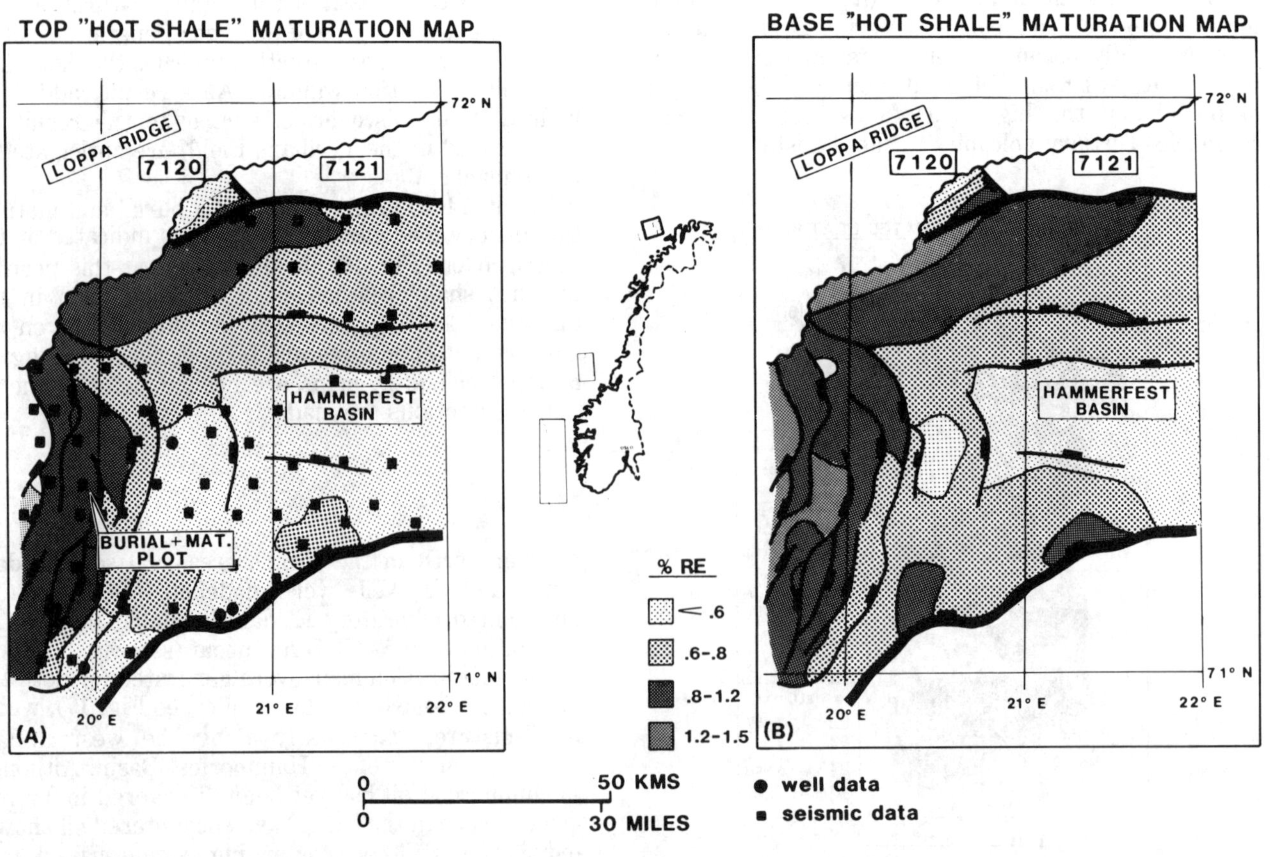

Fig. 17. Maturation maps for the Troms area. (A) Top 'hot shale' map showing mature oil window in the northwestern and western portions of Hammerfest Basin; (B) base 'hot shale' map showing that a greater area lies within the oil window at this level.

depths in Troms, Haltenbanken and the North Sea. A comparison of the shape of burial curves in these three areas suggests that sedimentation history is a major controlling factor in timing of oil generation and migration from the 'hot shale'. As is shown in the figures, time of onset of peak oil generation and migration is earliest (Paleocene) in the Troms, latest (Pliocene) in Haltenbanken, and intermediate (early Miocene) in the North Sea.

CONCLUSIONS

Sensitivity studies have shown that reasonable interval velocity errors will not greatly affect M-MOD R_e output. It has also been shown that heat flow maps should be constructed before maturation is estimated.

Computer-estimated R_e values generated for the North Sea, Haltenbanken and Troms areas, show good correlation with laboratory-measured R_0 values. In the North Sea, the 'hot shale' is at peak oil generation on the flanks of the Viking and Central Grabens, becoming gas/condensate-mature towards the axes. Major platform areas as well as the smaller flanking basins are all undermature. Maturation was mapped on two horizons in the Haltenbanken area. The 'hot shale' is oil-mature over nearly half of the basin, and the 'coaly unit' is gas-mature over approximately the same area. Additional high GOR discoveries similar to the Tyrihans Field are expected. In the Hammerfest Basin the 'hot shale' is oil-mature along the western and northern margins, and becomes rapidly overmature as the formation plunges into the Tromsø Basin.

The R_e maps demonstrate that utilization of K-MOD and M-MOD programs along with interval velocity information provides a rapid means of mapping regional maturation trends over large basin areas. Cross-plots have established that the results are relatively accurate. The method is especially useful in basins with limited well data. Even in basins with abundant well data this seismic–geochemical technique can be useful considering the uncertainties associated with direct vitrinite reflectance measurement (Toxopeus, 1983).

Additional information (such as migration and timing) can be gained in a basin by using the burial and thermal histories to reconstruct the maturity and structure at a specific geological time and by extrapolating these data to construct regional palaeo-R_e maps.

The M-MOD and K-MOD programs used in this study are in a constant state of modification. Future versions will eventually allow for the incorporation in more-complex models such factors as vertical anomalies in heat flow, basin fluid flux, decompaction, temporal variations in heat flow and/or geothermal gradients, 3-dimensional modelling, etc. Such a system is never complete but, hopefully, this discussion has demonstrated that a relatively simple, 2-dimensional modelling method is now of practical use.

ACKNOWLEDGEMENTS

The authors wish to thank the entire Conoco Norway staff, and in particular Emin Kabadayi, Peter Glagola and John Adams for their efforts in helping to compile and interpret the masses of data that were used in this study.

REFERENCES

Bjorøy, M. and Hall, P. B. 1983. A rich Middle Triassic source rock in the Barents Sea area. *15th A. Offshore Technol. Conf.*, Houston, 379–392.

Carstens, H. and Finstad, K. G. 1981. Geothermal gradients of the Northern North Sea Basin, 59–62°N. In: Illing, L. V. and Hobson, G. D. (eds.), *Petroleum Geology of the Continental Shelf of North-West Europe.* Institute of Petroleum, London, pp. 152–161.

Cermack, V. and Rybach, L. (eds.) 1979. *Terrestrial Heat Flow in Europe.* Springer-Verlag, New York, 328 pp.

Cornford, C., Morrow, J. A., Turrington, A., Miles, J. A. and Brooks, J. 1983. Some geological controls on oil composition in the UK North Sea. In: Brooks, J. (ed.), *Petroleum Geochemistry and Exploration of Europe.* Blackwell Scientific, Oxford, pp. 175–194.

Demaison, G. 1984. The generation basin concept. In: Demaison, G. and Murris, R. J. (eds.), *Petroleum Geochemistry and Basin Evaluation. AAPG Memoir*, **35**, 1–14.

Evans, T. R. and Coleman, N. C. 1974. North Sea geothermal gradients. *Nature*, **247**, 28–30.

Goff, J. C. 1983. Hydrocarbon generation and migration from Jurassic source rocks in the East Shetlands Basin and Viking Graben of the Northern North Sea. *Geol. Soc. Lond.* **140**, 445–474.

Goss, R., Coumbs, J. and Timur, A. 1975. Prediction of thermal conductivity in rocks from other physical parameters and from standard geophysical well logs. *SPWLA 16th A. Logging Symp.*, pp. 1–21.

Ho, Thomas, T. Y. and Sahai, S. K. 1982. Estimation of organic maturity from seismic and heat flow data (abstract). *AAPG Bull.*, **66**(5), 581–582.

Houbolt, J. J. H. C. and Wells, P. R. A. 1980. Estimation of heat flow in oil wells based on a relation between heat conductivity and sound velocity. *Geol. Mijnbouw*, **59**(3), 215–214.

Jessop, A. M., Hobart, M. A. and Sclater, J. G. 1976. *The World Heat Flow Data Collection—1975.* Earth Physics Branch, Geothermal Service of Canada, Geothermal Series **5**, 125 pp.

Lee, W. H. K. 1963. Heat flow data analysis. *Rev. Geophys.*, **1**, 449–479.

Lee, W. H. K. and Uyeda, S. 1965. Review of heat flow data: in terrestrial heat flow. *Am. Geophys. Union Monogr.* **8**, 87–190.

Lopatin, N. V. 1971. Temperature and geologic time as factors in coalification. *Akad. Nauk. SSSR, Izv. Ser. Geol.*, **3**, 95–106 (in Russian).

Oxburgh, E. R. 1984. Thermal gradients in sedimentary basins (abstract). *AAPG Bull.*, **68**(6), 797.

Pearson, M. J., Watkins, D., Pittion, J. L., Caston D. and Small J. S. 1983. Aspects of burial diagenesis, organic maturation and palaeothermal history of an area in the South Viking Graben, North Sea. In Brooks, J. (ed.), *Petroleum Geochemistry and Exploration of Europe.* Blackwell Scientific, Oxford, pp. 161–173.

Rønnevik, C. H., Eggen S. and Vollset J. 1983. Exploration of the Norwegian Shelf. *NPD-Contribution no. 5.* Also in: Brooks, J. (ed.), *Petroleum Geochemistry and Exploration of Europe.* Blackwell Scientific, Oxford, pp. 71–94.

Tissot, B., and Espitalie, J. 1975. L'evolution thermique de la matier organique des sédiments: applications d'une simulation mathématique. *Revue IFP*, **30**, 943–999.

Toth, D. J., Lerche, I., Petroy, D. E., Meyer, R. J. and Kendall, C. G. St C. 1981. Vitrinite reflectance and the derivation of heat flow changes with time. In: *Advances in Organic Geochemistry.* Wiley, New York, pp. 588–596.

Toxopeus, J. M. A. 1983. Selection criteria for the use of vitrinite reflectance as a maturity tool. In: Brooks, J. (ed.), *Petroleum Geochemistry and Exploration of Europe.* Blackwell Scientific, Oxford, pp. 295–307.

Vollset, J. and Doré, A. G. (eds.) 1984. A revised Triassic and Jurassic lithostratigraphy for the Norwegian North Sea. *NPD Bull.*, **3**.

Waples, D. W. 1980. Time and temperature in petroleum formation—application of Lopatin's method to petroleum exploration. *AAPG Bull.*, **64**(6), 916–926.

Welte, D. H. and Yükler, M. A. 1981. Petroleum origin and accumulation in basin evolution—a quantitative model. *AAPG Bull.*, **65**(8), 1387–1396.

10 Modelling burial history, temperature and maturation

Peter K. Jensen, Lise Holm† and Erik Thomsen†*
*Risø National Laboratory, Department of Energy Technology, DK-4000, Roskilde, Denmark
†The Geological Survey of Denmark, Thoravej 31, DK-2400, Copenhagen NV, Denmark

A model that simulates burial history, temperature and maturation is described. The subsidence of a surface layer, including compaction, is determined as a function of time by an analytical expression. Compaction of underlying layers and erosion are also considered. Formation temperatures are calculated analytically, with conductivity being a function of porosity. Maturity modelling is based on an empirical expression relating vitrinite reflectance, temperature and time.

The model is tested against vitrinite reflectance profiles from the Danish part of the Central Graben and the Danish Sub-basin. Past heat flow conditions are also evaluated, and a petroleum geological interpretation is given.

INTRODUCTION

The mathematical model which is presented in this paper can be used to simulate maturity conditions of organic matter in sediments and has been developed to support the evaluation of hydrocarbon potential in the Danish area.

Alteration of organic matter is dependent on formation temperature and time (Tissot and Welte, 1978). Therefore the temperature must be modelled through time. The formation temperatures are determined from burial depth, heat conductivities, surface temperatures and heat flow.

The mathematical formulation given here for the burial history and for the temperature calculations is based on analytical expressions, in contrast to other similar models that are numerical procedures. A brief presentation of the mathematical model is described below.

It is assumed that the depths of the formations are known at a given time. At a later time the depth to the base of the surface layer is calculated from an analytical expression. This new depth corresponds to the new depth of the top of the underlying formation. The new depth of the base of this layer is then calculated from another analytical expression. The next to deepest layer is handled similarly.

During uplift the compaction of the sediments is considered to be unchanged; that is, the compaction of a layer is equal to the compaction stage at the maximum burial depth.

The formation temperatures are calculated at the base of each of the layers. The conductivity is determined from parameters characterizing the rock type and the degree of compaction.

A simple expression relating vitrinite reflectance (maturity), time and temperature has been chosen since geochemical data are currently sparse for the Danish area.

A fast-working computer program has been developed which determines the burial history, the formation temperatures and the maturity. The amount of input data may be limited.

The use of the model is demonstrated by vitrinite reflectance simulations for a well in the Danish Central Graben and a well in the Danish Sub-basin (see Fig. 1). The simulations are then compared with actual measurements and a petroleum geological interpretation is given for the two wells.

THE MODEL

Burial history

The burial history may be sketched ideally as in Fig. 2. The sequences are divided into a number of units, where each unit is characterized by a certain porosity–depth relationship:

$$\phi(z) = \phi_0 e^{-az}, \qquad (1)$$

where ϕ is the porosity, z is the depth, ϕ_0 is the sur-

Petroleum Geochemistry in Exploration of the Norwegian Shelf © Norwegian Petroleum Society (Graham & Trotman, 1985) pp. 145–152.

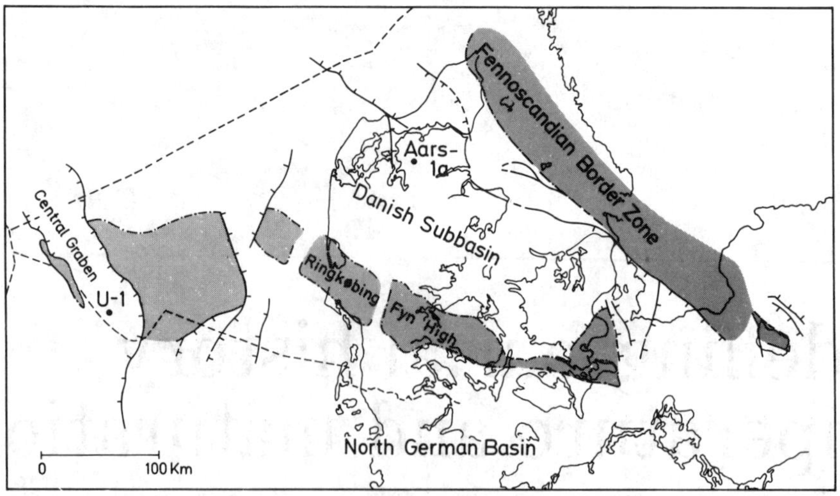

Fig. 1. Location of the wells Aars-1a and U-1, and the main structural elements of the study area. The shaded areas are structural highs.

face porosity and a is a compaction coefficient. The values of ϕ_0 and a may be different for each unit.

Geological time (see Fig. 2) is divided into time spans each of which is characterized by a constant sedimentation rate of the rock mass, or by a constant erosion rate.

The development of the mathematical formulation of the burial is treated in two stages. First the accumulation of sediments in a surface layer is considered, then the compaction of underlying layers is handled. Modelling of erosion is treated separately.

During the development of the surface layer, a constant mass of particles or grains is added to the surface. The grains are buried and move towards greater depths with a certain velocity. In the following, an expression will be deduced from which the particle (or grain) velocity can be found as a function of time. Knowing the velocity, the time necessary for a particle to reach a given depth can be determined. Thus, the depth of the base of the surface layer can also be determined as a function of time.

In the following, a co-ordinate system is used which is fixed relative to the surface of the sediments.

The flow of mass during the sedimentation of the surface is a physical analogue to flow of a compressible fluid. Therefore the equation of continuity known from hydrodynamics is applicable:

$$\frac{\partial(\rho v)}{\partial z} = -\frac{\partial \rho}{\partial t} \tag{2}$$

where v is the particle velocity, t is the time and ρ is the skeletal density defined by the equation

$$\rho = \rho_m(1 - \phi), \tag{3}$$

where ρ_m is the matrix density.

Equation (2) shows that the change of mass per unit volume, fixed in space, equals the flow of mass into and out of the volume. The unknown velocity, v, is to be found.

The pore fluid is assumed to escape towards the surface.

The right-hand side of equation (2) is zero, since the density and thus also the porosity is constant with time for a given depth (according to equation (1)). The boundary condition to equation (2) is the mass rate of rock matrix which is added to the surface. The initial condition is the density distribution.

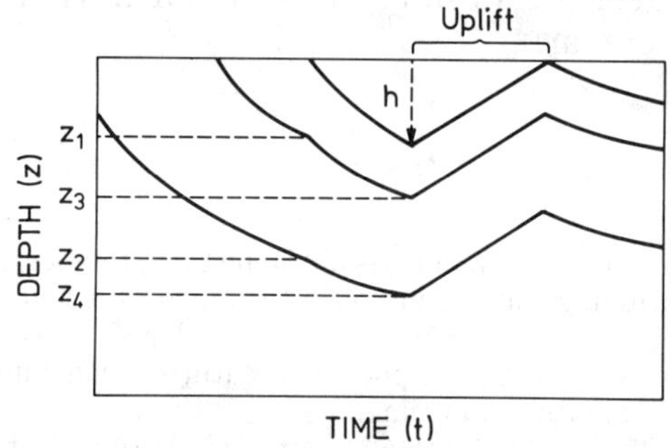

Fig. 2. Sketch of the physical model for simulating burial history.

Combination of equations (1)–(3) results in a first-order linear differential equation:

$$\frac{\partial v(z)}{\partial z} + \frac{\phi_0\, a e^{-az}}{1 - \phi_0 e^{-az}}\, v(z) = 0 \tag{4}$$

A solution to this equation has been found following Bronstein and Semendjajew (1963, p. 378):

$$v(z) = v_0(1 - \phi_0)(1 - \phi_0 e^{-az})^{-1}, \tag{5}$$

where v_0 is the constant particle velocity at the surface.

With equation (5), an important result is obtained. For a given depth, z, the velocity of the base of the surface layer can be calculated.

It is seen from the equation that the velocity will decrease with depth, because a is a positive number.

The time necessary for the surface layer to reach depth z can be calculated from

$$t = \int_0^z \frac{1}{v(z')}\, dz', \tag{6}$$

and the following equation is obtained:

$$t = \frac{1}{v_0(1 - \phi_0)}\left[z + \frac{\phi_0}{a}\,(e^{-az} - 1)\right]. \tag{7}$$

Unfortunately, the depth z cannot be separated on one side of the equation, and z must be found by an iterative procedure. It is seen from equation (7) that the depth to the base of the surface layer can be calculated as a function of time without subdividing this layer.

The parameters v_0, ϕ_0 and a must be known. The parameters ϕ_0 and a have already been discussed. v_0 is calculated when a dating, T_b, is known for the base of the surface layer using equation (7):

$$V_0 = \frac{1}{(1-\phi_0)T_b}\left[z + \frac{\phi_0}{a}(e^{-az} - 1)\right], \tag{8}$$

where z is the thickness of the layer.

With respect to the compaction of layers underlying the surface layer, a useful expression was derived by Sclater and Christie (1980). They used the mass conservation of the rock matrix and an exponential decrease of porosity with depth. Only the result will be given here.

z_1 and z_2 are the initial depths to, respectively, the top and the bottom of a unit (see Fig. 2). During burial, the top of the layer is moved to the depth z_3. The new depth, z_4, of the base of the layer can be calculated from

$$z_4 + \frac{\phi_0}{a} e^{-az_4} = z_2 - z_1 + \frac{\phi_0}{a}(e^{-az_2} - e^{-az_1} + e^{-az_3}) + z_3. \tag{9}$$

Again, the equation is solved by an iterative procedure.

The burial history of several layers can now be calculated. The base of the surface layer is calculated through time by use of equation (7) and the bases of the deeper layers are calculated using equation (9).

Compaction is considered to be an irreversible process. During uplift, each of the unit thicknesses is unchanged (see Fig. 2), but the porosity–depth relation (equation (1)) for a formation is changed during uplift. That is, if the porosity–depth relation for a unit is identical with equation (1), then the porosity–depth relation after uplift, h, will be

$$\phi_h(z) = \phi_0 e^{-a(z+h)}, \tag{10}$$

or

$$\phi_h(z) = \phi_0 e^{-ah} e^{-az}. \tag{11}$$

In practice the porosity function $\phi_h(z)$ may be determined from log data, and equation (10) is used to determine the original porosity function $\phi(z)$ before uplift.

It can be seen from equation (11) that the compaction coefficient a is unchanged during uplift, but the surface porosity $\phi_0 e^{-ah}$ is changed.

It should be noted that the surface velocity v_0 in equation (8) for a particular unit must be determined for parameters corresponding to the situation where the unit is at the surface, before uplift occurred.

Formation temperatures

For a given sedimentary unit the temperature T at the base at depth z_2 is calculated according to the expression

$$T(z_2) = T(z_1) + W \int_{z_1}^{z_2} \frac{1}{K(z)} \, dz, \tag{12}$$

where $T(z_1)$ is the temperature at the top at depth z_1, W is the heat flow and $K(z)$ is the conductivity of the unit. K is a function of the porosity (Robertson, 1979):

$$K = K_F + \gamma^2(K_S + pC - K_F), \tag{13}$$

where K_F is the pore fluid conductivity, $\gamma = 1 - \phi$, K_S is the solid matrix conductivity, p is the quartz content and C is a constant.

The integral in equation (12) can be evaluated analytically, and equation (12) becomes

$$T(z_2) = T(z_1) + W \frac{1}{aA} \frac{1}{(K_F/A) + 1}$$

$$\left[-\ln(1-\gamma_2) + \tfrac{1}{2}\ln\left(\frac{K_F}{A} + \gamma_2^2\right) + \frac{1}{K_F/A}\,\mathrm{arctg}\left(\frac{\gamma_2}{K_F/A}\right) + \ln(1-\gamma_1) - \tfrac{1}{2}\ln\left(\frac{K_F}{A} + \gamma_1^2\right) - \frac{1}{K_F/A}\,\mathrm{arctg}\left(\frac{\gamma_1}{K_F/A}\right)\right], \tag{14}$$

where $A = K_S + pC - K_F$, $\gamma_2 = 1 - \phi_0 e^{-az_2}$ and $\gamma_1 = 1 - \gamma_0 e^{-az_1}$.

Vitrinite reflectance

An empirical relation evaluated by Middleton and Falvey (1983) relating the vitrinite reflectance R_0 to temperature T and time t has been used here:

$$(R_0)^{a'} = b \int_0^t \exp\,[cT(t')]dt', \tag{15}$$

where a', b and c are constants.

By empirical studies, Middleton and Falvey found that $a' = 5.635$, $b = 2.7 \times 10^{-6}$/Ma and $c = 0.068/°C$. The value of c is determined from a parameter of organic alteration introduced by Royden *et al.* (1980) which doubles for each 10.2°C increase in temperature.

In the present study the constants a' and b were modified to satisfy the measured R_0 values. The modification procedure will be discussed below. It was found that $a' = 13.63$, $b = 6.52 \times 10^{-9}$/Ma and $c = 0.068/°C$.

The integral in equation (15) is solved numerically.

The simulator

Having outlined the equations, we now turn to the construction of the computer code.

An idealized picture of a burial history is shown in Fig. 2. Examples of input data are given in Tables 1–4. These examples are discussed in a following section. A flow diagram for the computer code is given in Fig. 3.

The computer code first reads the necessary input data for the burial history calculations (see Fig. 3). These are first the time span divisions, which are

TABLE 1. Time-span diversions for the burial history at the well Aars-1a

No.	*Time-span* (*×10⁶ years*)	*Present depth* (*mb. KB*)	*Erosion* (*m*)
1	0–1	128	—
2	1–2	—	220
3	2–60	—	—
4	60–73	815	—
5	73–83	1450	—
6	83–98	1792	—
7	98–138	2194	—
8	138–150	2374	—
9	150–188	2500	—
10	188–210	3205	—

periods of constant sedimentation or erosion rate. Then the computer program reads the preserved formation depths and the erosion depths. The surface porosities and the compaction coefficients (ϕ_0 and a in equation (1)) are read for preserved as well as eroded formations. On the basis of these data, the burial history is determined in a later subroutine.

The data necessary for the temperature calculations are the thermal rock constants, the heat flux and the surface temperature. The thermal constants are read for each formation. The surface temperature, as well as the assumed heat flux, is read as a function of time. The formation temperatures are calculated on the basis of the burial history.

TABLE 2. Time-span divisions for the burial history at the well U-1

No.	*Time-span (×10⁶ years)*	*Present depth (mb. surface)*	*Erosion (m)*
1	0–2	560	—
2	2–11	1413	—
3	11–60	2153	—
4	60–75	2425	—
5	75–144	—	no depn. or eros.
6	144–156	2853	—
7	156–178	3297	—
8	178–188	—	800
9	188–210	—	—
10	210–213	3344	—

TABLE 3. Rock properties for the well Aars-1a

Time-span no.	*Present surface porosity*	*Compaction coefficient (×10⁻³/m)*	*Thermal constants* K_F *(W m⁻¹ K⁻¹)*	K_S *(W m⁻¹ K⁻¹)*	*p (ratio)*	*C (×10⁻³ W m⁻¹ K⁻¹)*
1	0.62	0.594	0.60	8.1	0.65	39
2	—	—	—	—	—	—
3	0.62	0.594	0.60	8.1	0.20	125
4	0.55	0.681	0.60	4.3	0.05	0
5	0.50	0.739	0.60	3.8	0.05	0
6	0.42	0.446	0.60	3.3	0.05	0
7	0.32	0.345	0.60	8.1	0.45	−139
8	0.30	0.436	0.60	8.1	0.70	−81.8
9	0.45	0.324	0.60	8.1	0.90	−59.0
10	0.50	0.572	0.60	8.1	0.35	−188.0

K_F = Fluid conductivity, K_S = Solid matrix conductivity, p = Quatz content, C = constant.

TABLE 4. Rock properties for the well U-1

Time-span no.	*Present surface porosity*	*Compaction coefficient (×10⁻³/m)*	*Thermal constants* K_F *(W m⁻¹ K⁻¹)*	K_S *(W m⁻¹ K⁻¹)*	*p (ratio)*	*C (×10⁻³ W m⁻¹ K⁻¹)*
1	0.32	0.405	0.60	1.52	0.90	44.0
2	0.60	0.505	0.60	1.52	0.40	83.4
3	0.50	0.125	0.60	1.52	0.30	105.0
4	0.60	0.354	0.60	3.0	0.05	0.0
5	—	—	—	—	—	—
6	0.60	0.366	0.60	1.52	0.35	13.3
7	0.60	0.274	0.60	1.52	0.35	62.0
8	—	—	—	—	—	—
9	0.60	0.367	0.60	1.52	0.35	62.0
10	0.50	0.367	0.60	1.52	0.35	62.0

The vitrinite reflectance parameters a', b and c (in equation (15)) are read for calculation of maturity.

When the input data are read, the burial history is computed backwards in time, beginning with present formation depth. New depths are calculated for each time step Δt. For every time step it is checked whether the time is within an erosion period or not. For erosion periods, the mean erosion rate for the particular time-span is first calculated. The new depths are then determined linearly with time. As mentioned in connection with equations (10) and (11), the porosity–depth relationship of the formations is changed during the uplift. For times within a sedimentation period, the surface sedimentation velocity is first calculated using equation (8) (constant for the particular time-span). Then the depth to the surface layer is calculated using equation (7). The depths of all underlying layers are determined from equation (9).

When the burial history is determined, the temperature history can be calculated. This is done by application of equation (14). The temperature is calculated at the base of each of the formations, beginning with the surface layer and going downwards. The calculations are backwards in time beginning with present temperatures.

Equation (15) is now used to determine the vitrinite reflectance as a function of time at the base of each layer. These calculations are done forwards in time ending at the present time, where the simulated values of vitrinite reflectance can be compared with measured values.

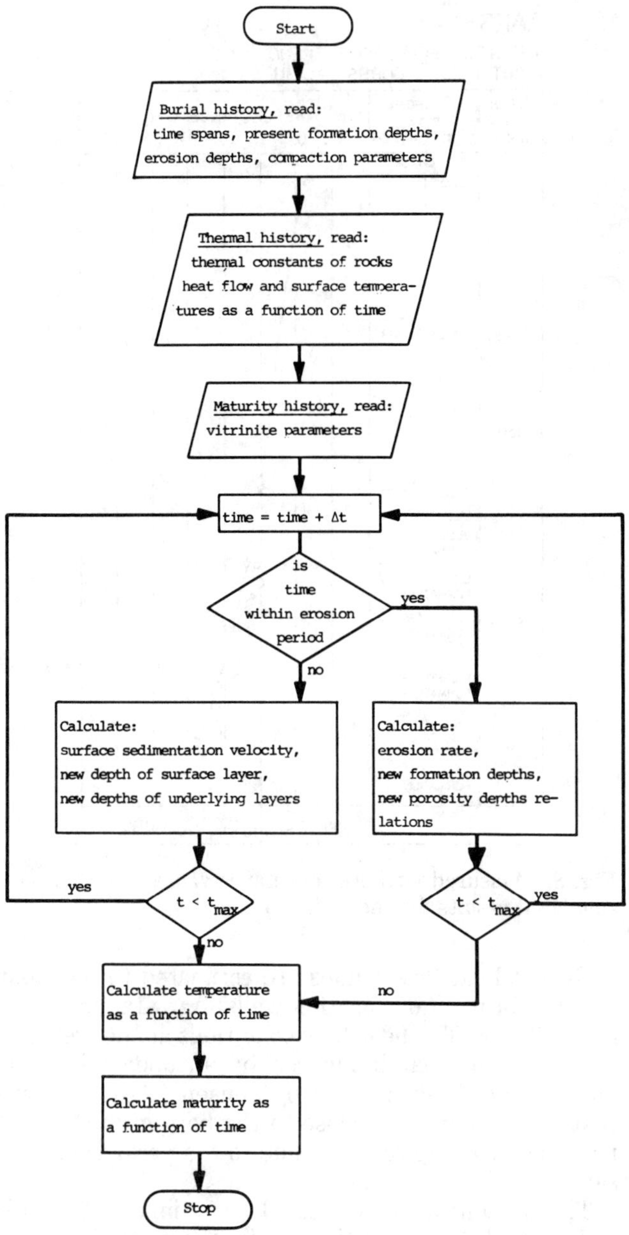

Fig. 3. Flow-diagram of the simulator.

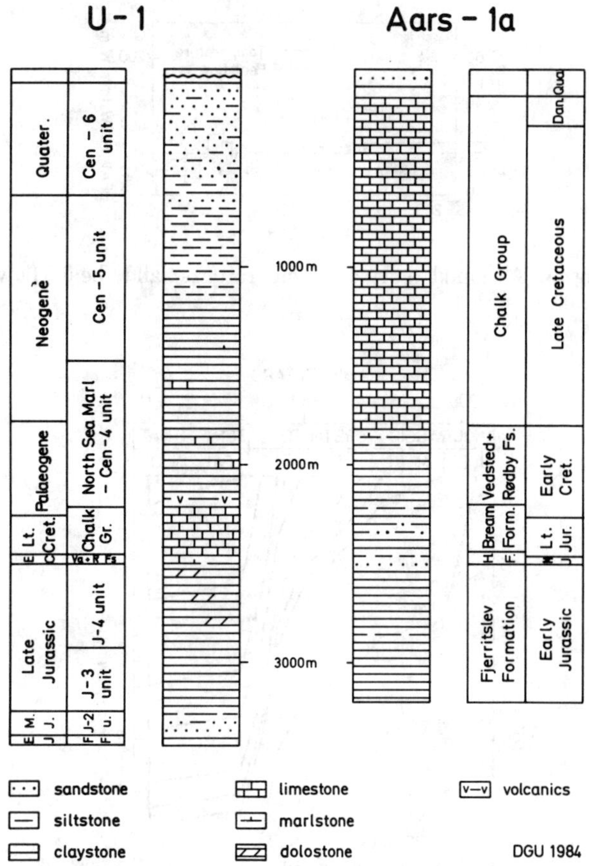

Fig. 4. Lithology, chrono- and lithostratigraphy of the wells Aars-1a and U-1. Va + R Fs = Valhall + Rødby Formations, FF = Fjerritslev Formation, HF = Haldager Formation.

SIMULATION EXAMPLES

Regional setting

The two wells investigated are situated in two different basins. The Aars-1a well was drilled in the Danish Sub-basin and the U-1 well in the Central Graben (see Fig. 1). The two basins are of different origin, the Danish Sub-basin being an intracratonic basin, and the Central Graben part of a rift system. Consequently, their subsidence histories are different (see Figs. 6 and 7, below). The main subsidence period of the Danish Sub-basin was the Triassic while the Central Graben suffered strong subsidence during the Late Jurassic and the Cenozoic, especially the Neogene and the Quaternary (Holm, 1983). The thermal histories are different, too, with higher heat flow during the Jurassic in the Central Graben due to active rifting. Both basins are strongly influenced by halokinesis.

The Aars-1a well was chosen because it has been thoroughly investigated and because it is situated centrally in the Sub-basin. The Fjerritslev Formation, the most promising source rock in this area (Lindgreen and Thomsen, 1982), is rather deeply buried here, about 3000 m. In addition, core material is available from this well. The Aars area is not influenced by halokinesis. The Mid-Cimmerian event which strongly affected the whole North Sea region did not cause erosion in this area. The only period with significant erosion is assumed to be the Quaternary, in relation to glaciation during which a few hundreds of metres of Tertiary sediments are supposed to have been removed.

The U-1 well was chosen because it is representative of the stratigraphic development of the Mesozoic and Cenozoic in the Danish part of the Central Graben. Furthermore, the stratigraphy is well-known. The Mid-Cimmerian event is here supposed to have caused erosion of 500 m (compacted value) of the Fjerritslev Formation. The U-1 well is mildly influenced by a deep-seated salt pillow which evolved during the Late Jurassic, giving rise to a slightly condensed Upper Jurassic sequence.

The lithologies, chrono- and lithostratigraphy of the two wells are shown in Fig. 4.

Input and simulation results

The time-spans corresponding to the drilled sequences of the wells U-1 and Aars-1a are listed in Tables 1 and 2. In the tables are shown a numbering of the time-spans, the present depth for each time-span, and the erosion. The value of the erosion is the total depth of the erosion in the particular time-span. Tables 3 and 4 show the estimated rock parameters. The present surface porosity is found by plotting depth versus logarithmic porosity for each lithological unit. The

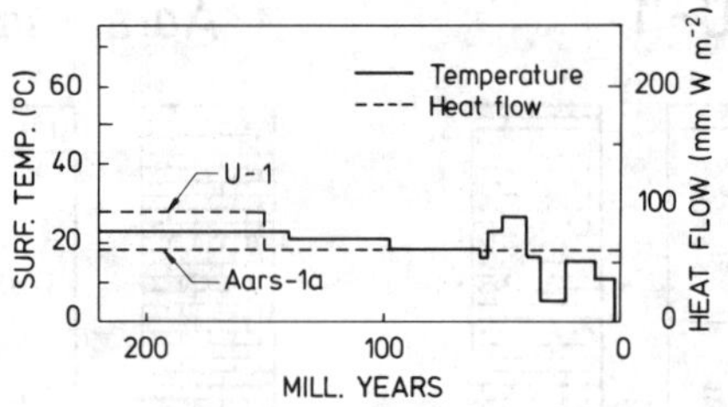

Fig. 5. Assumed surface temperatures and heat flow conditions.

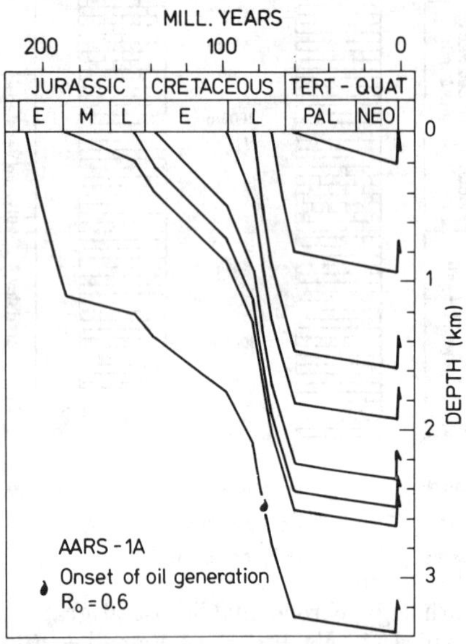

Fig. 6. Burial history corrected for compaction for the well Aars-1a, and time of onset of oil generation.

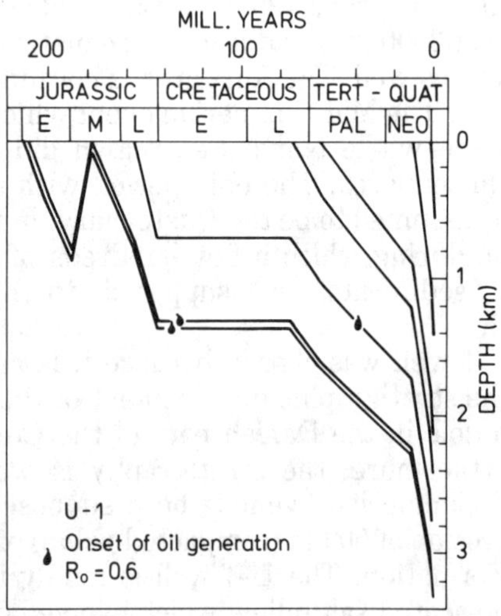

Fig. 7. Burial history corrected for compaction for the well U-1, and time of onset of oil generation.

porosities were determined from wireline log measurements. From the same plots the compaction coefficients were determined. Knowledge about surface porosities for the lithologies involved was also incorporated.

The thermal constants were initially determined from figures given by Robertson (1979) and from conductivity measurements in the case of the well Aars-1a (Balling, 1981). Later, the C values in equation (13) were adjusted to obtain calculated recent temperatures equal to the measured temperatures (Balling, pers. comm.; Jensen, 1983).

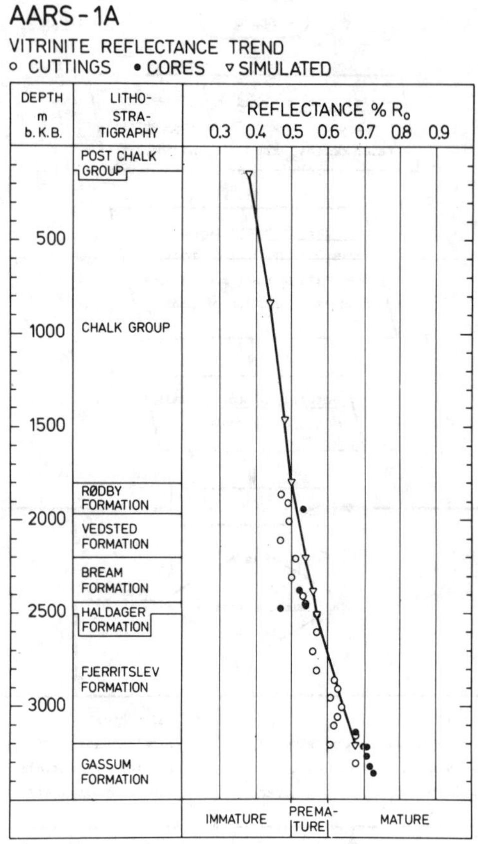

Fig. 8. Measured vitrinite reflectance values together with simulated results for the well Aars-1a.

Recent heat flow values are estimated to be about 60 and 74 mW/m^2 for the wells Aars-1a and U-1, respectively. The heat flow conditions at the well U-1 may be somewhat influenced by an underlying salt pillow. Calculations made by Jensen (1981) indicate that the temperature close to the bottom of the well may increased up to 15 °C due to the presence of the salt.

The present heat flow could be estimated for both wells as already mentioned. In the following it is explained how the past heat flow is estimated and how the parameters a' and b (in equation (15)) are adjusted to the available data. First, it can be seen that a double logarithmic plot of measured vitrinite reflectance and the calculated time–temperature integral should fall on a straight line, if the temperature and therefore the heat flow are estimated correctly. From the straight-line plot, the constants a' and b can be determined. It is assumed here that the thermal parameters, the surface temperature and the burial history are correctly determined. A plot was made for the two wells, assuming constant heat flow conditions equal to the recent heat flow. All of the data for the Aars-1a well and low R_0 values for the U-1 well fell on a straight line. It could then be concluded that the assumed constant heat flow condition for the Aars-1a well was reasonable. The heat flow in the case of the U-1 well had to be increased to about 95 mW/m^2 for ages greater than 150 Ma. This is in accordance with modelling of heat flow conditions during the subsidence of the Central North Sea Basin (Sclater and Christie, 1980). Figure 5 shows the assumed heat flow models.

Surface temperatures in the North Sea area have been evaluated by Buchardt (1978) for the Tertiary. Cretaceous and Jurassic temperatures were assumed

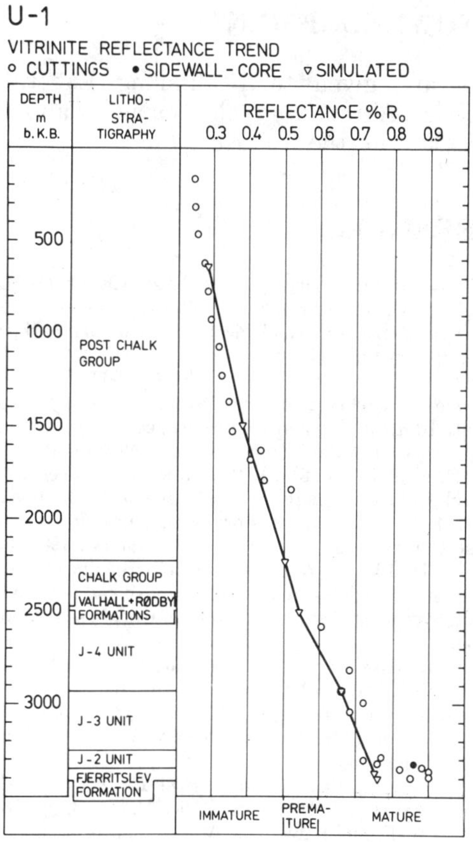

Fig. 9. Measured vitrinite reflectance values together with simulated results for the well U-1.

after personal communication with Buchardt. The surface temperatures used in the calculations are shown in Fig. 5.

The burial histories, corrected for compaction, are seen in Figs. 6 and 7 for the two wells. A simulated vitrinite reflectance equal to 0.6% R_0, which is regarded as a general threshold value for the onset of oil generation (Thomsen *et al.*, 1983), is marked on the figures. Simulated and measured vitrinite reflectance values are plotted as a function of depth for the two wells in Figs. 8 and 9. It is seen that the simulated values are close to the measured values.

DISCUSSION

The model

The model is one-dimensional, and therefore its use is restricted. It should be applied with caution in areas where large lateral tectonic movements have occurred and where the horizontal heat flux component is considerable. On the other hand, a one-dimensional model implies simpler equations, and analytical solutions can be found. Furthermore, the computing time and the amount of input data are strongly reduced.

Several of the so-called 'back stripping' models have been discussed in the literature (Perrier and Auiblier, 1974; Du Rouchet, 1980; Sclater and Christie, 1980; Falvey and Deighton, 1982; Middleton and Falvey, 1983). In these models, lithologically uniform sequences must be divided into a number of subunits, and the calculation accuracy increases with this number. The model which is presented here works with a restricted number of units.

Data quality

The burial histories of the two wells investigated have been compiled by means of biostratigraphical analyses carried out by the Geological Survey of Denmark. These studies have been supported by lithostratigraphical correlations to wells in the close vicinity. The reliability of the data presented is considered good, except for the Tertiary and the Quaternary of the Aars-1a well. Here, the magnitude of original deposition and subsequent erosion must be regarded as uncertain.

The porosity data are based on petrophysical log readings, mainly the acoustic sonic, supplemented by a few sample measurements. The reliability of the absolute porosity values are considered to be only fair, but the porosity trends are considered to be reliable.

The temperature measurements in the well Aars-1a are probably within 1–2 °C of true formation temperature. The present heat flow is estimated to lie between 50 and 80 mW/m^2. This range reflects the uncertainty of the conductivity estimate. The temperatures obtained in the well U-1 were measured a short time after cessation of the mud circulation. The temperatures have been corrected, but the accuracy is unknown (Jensen, 1983). Also, the conductivities and the heat flow estimates are uncertain since no heat conductivity measurements were performed.

From the Aars-1a well, core and cuttings samples were available for analysis of vitrinite reflectance. Lignite used as a drilling mud additive restricted the use of several of the cuttings samples. However, the rather high number of cores and the presence of organic-rich intervals and coal in samples from the Haldager and the Gassum Formations, permitted proper control of the rank conditions. Thus, the data quality is rated high.

From the U-1 well, cuttings samples were available for analysis, supplemented by a single sidewall-core from the J-2 unit. According to biostratigraphical studies, caving is significant in the Mesozoic intervals. This is clearly revealed in several of the samples investigated in this study. Fair amounts of vitrinite have been recorded in samples from the CEN-4 and the CEN-5 units and from the J-2 unit. The latter data comprise a few cuttings samples with coal fragments, the rank of which is confirmed by measurements from the one side-wall core from this unit. Thus, in general, the vitrinite reflectance data quality is rated fair in this well.

Results of analysis

Aars-1a

The Aars-1a well is of special interest, since it represents the most mature Mesozoic deposits that so far have been drilled in the Danish Sub-basin. The reflectance trend reaches about 0.73% R_0 at a depth of about 3350 m (see Fig. 8). This is significantly higher than the reflectance range of 0.36–0.58% R_0 for Upper Triassic–Jurassic–Lower Cretaceous deposits at depth intervals between 200 and 2800 m, that has previously been reported from this area (Thomsen *et al.*, 1983).

Modelling shows that the onset of oil maturity in the Aars-1a well took place during the late Cretaceous (see Fig. 6). This timing is regarded as favourable relative to structural development for the area in question.

However, none of the wells drilled through these deposits, including the Aars-1a well, has yielded commercial petroleum. The lack of oil is attributed to the unfavourable terrestrial type of organic matter in the Gassum Formation and the basal part of the Fjerritslev Formation. However, a more favourable type of organic matter has been encountered in the overlying F-III and F-IV Members of the Fjerritslev Formation, which, under proper burial conditions, could represent potential oil source intervals.

U-1

Oil and gas were encountered in the U-1 well. The J-4 unit is regarded as the principal source unit by comparison with other parts of the Danish Central Graben area (Thomsen *et al.*, 1983). The reflectance trend presented in this study (see Fig. 9) is an extension of a previous preliminary study published by Michelsen (1982). The higher number of samples investigated permits a more thorough interpretation of the maturity conditions and reveals a higher maturity gradient than previously recorded. The J-4 unit is within the zone of early oil generation, while the J-2 unit and the Fjerritslev Formation are within the zone of maximum oil generation according to the R_0 values suggested by Tissot and Welte (1978).

The modelling shows that the onset of the oil generation took place during the Palaeogene (see Fig. 7). This timing is regarded as favourable relative to structural development, and is in accordance with general hydrocarbon migration concepts in the study area.

CONCLUSIONS

It has been shown that the model can simulate satisfactorily the measured vitrinite reflectance trends for two wells in the Danish area with quite different burial histories and heat flow conditions. It is, furthermore, possible to estimate past heat flow conditions for the two wells in accordance with the observed tectonic development.

ACKNOWLEDGEMENTS

Thanks are given to N. Balling Laboratory of Geophysics, Aarhus who collected and analysed temperature data for the well Aars-1a.

REFERENCES

Balling, N. 1981. Principper for temperaturberegning. In: Michelsen, O. (ed.), *Kortlægning of Potentielle Geotermiske Reservoir i Danmark*. Geological Survey of Denmark, Series B, No. 5, pp. 13–19.

Bronstein, I. N. and Semendjajew, K. A. 1963. *Taschenbuch der Matermatik*. Teubner, Leibzig, 584 pp.

Buchardt, B. 1978. Oxygen isotope palaeotemperatures from the Tertiary period in the North Sea areas. *Nature*, **275**, 121–123.

Falvey, D. A. and Deighton, I. 1982. Recent advances in burial and thermal geohistory analysis. *APEA Journal*, **22**, 65–81.

Holm, L. 1983. Subsidence history of the Jurassic sequence in the Danish Central Graben. *Danm. Geol. Unders.*, **1982**, 39–51.

Jensen, P. K. 1981. Geotermiske modelberegninger ved endelige elementers metode og analytiske metoder specielt med henblick på saltstrukturer. *Thesis*, Institute of Geophysics, Copenhagen, 270 pp.

Jensen, P. K. 1983. Formation temperatures in the Danish Central Graben. *Danm. Geol. Unders.*, **1982**, 91–106.

Lindgreen, H. and Thomsen, E. 1982. *Investigation on the Source Rock Potential of the Danish Onshore Area*. Danish Geological Survey, Copenhagen, 91 pp.

Michelsen, O. (ed.) 1982. *Geology of the Danish Central Graben*. Geological Survey of Denmark, Series B, No. 8, pp. 73–86.

Middleton, M. F. and Falvey, D. A. 1983. Maturation modelling in Otway Basin, Australia. *AAPG Bull.*, **67**, 271–279.

Perrier, R. and Auiblier, J. 1974. Thickness changes in sedimentary layers during compaction history, methods for quantitative evaluation. *AAPG Bull.*, **58**, 507–520.

Robertson, E. C. 1979. *Thermal Conductivity of Rocks*. Open File Report, 79–356, U.S. Geological Survey, 31 pp.

Rouchet, J. Du 1980. Le programme diagen, deux procédures pour apprécier l'évolution chemique de la matière organique. *Bull. Cent. Rech. Exploration–Production Elf-Aquitaine*, **4**(2), 813–831.

Royden, L., Sclater, J. E. and Von Hertsen, R. P. 1980. Continental margin subsidence and heat flow: important parameters in formation of petroleum hydrocarbons. *AAPG Bull.*, **64**, 173–187.

Sclater, J. G. and Christie, P. A. F. 1980. Continental stretching: an explanation of the Post- Mid-Cretaceous subsidence of the Central North Sea Basin. *J. Geophys. Res.*, **85**(B7) 3711–3739.

Thomsen, E., Lindgreen, H. and Wrang, P. 1983. Investigation on the source rock potential of Denmark. In: Kaaschieter, J. P. H. and Reijers, T. J. A. (eds.), *Petroleum Geology of the Southeastern North Sea and the Adjacent Onshore Areas. Geol. Mijnbouw*, **62**, 221–239.

Tissot, B. and Welte, H. 1978. *Petroleum Formation and Occurrence*. Springer-Verlag, Berlin, 538 pp.

11 Basin analysis and maturation modelling, onshore Denmark; a case study from the Danish First Round

Steinar Sørensen
Rogalandsforskning/Rogaland Regional College, P.O. Box 2540, Ullandhaug, 4001 Stavanger, Norway

In the Danish Sub-basin exploration for hydrocarbons has continued for decades without any commercial success. At present the petroleum industry is again showing an increased interest in Danish areas. The axis of the Danish Sub-basin runs parallel to the Tornquist–Teissure Line and perpendicular to the Oslo–Rhine Graben System. Mesozoic tectonics of the area were dominated by subsidence and halokinetic movements, in contrast to the Cenozoic, when inversion and partial erosion took place in the eastern parts of the basin. In this study the Lower Jurassic Fjerritslev Formation is considered to be the principal source rock, but so far only immature sections have been encountered in the subsurface, possibly because most of the features drilled are salt structures where the source rocks have never reached a depth greater than 1500 m. In modelling thermal history it is assumed that source rock maturation is related exponentially to temperature and linearly to time. Temperature data for wells were obtained from open-file information in the Danish Geological Survey. An estimate of palaeo-heat flux was based on an interpretation of the tectonic and depositional history of the Danish Sub-basin, and is taken into account in the model. By applying heat flux changes through time and adjusting geothermal gradients for differences in lithological heat conductivity, porosity variations and age, the geothermal history of the Danish Sub-basin has been modelled more precisely. It is concluded that locally-mature Lower Jurassic source rocks may occur in the Danish Sub-basin, and have potential to generate enough oil to fill some untested structures already identified.

INTRODUCTION

In the Danish Sub-basin exploration for hydrocarbons has continued for decades without any commercial success. From an exploration viewpoint, interest in this area has been limited for two reasons: the petroleum potential of the basin has been regarded as only fair, and the whole area was held solely by the Dansk Undergrunds Consortium. However, blocks relinquished by the DUC in 1982 were recently offered for exploration to the entire industry. In April 1984, 22 companies were awarded licences in the Danish First Round, and at present there are more applicants for blocks in Denmark than in Norway, among them some of the major oil companies.

The present study was part of a technical evaluation prepared for a Danish First Round Application, and consequently techniques and concepts, rather than specific prospects will be described.

DATA BASE

Data available for this study included 43 released wells from the greater Danish area, as well as 59 000 km of seismic data, although locally the data base was more limited. In the Danish Sub-basin, some 20 wells were available. Pyrolysis data and other geochemical information were available from nine wells. A fairly detailed stratigraphy (see Fig. 1) of the Danish Sub-basin is available in numerous publications, mainly from the Danish Geological Survey (DGU) and, hence, formation ages are established with some confidence.

Accurate temperature information is often a problem in thermal history analysis, but fortunately a large amount of useful data is also available from the DGU. This information is part of a Danish project on geothermal energy and geothermal reservoirs in onshore Denmark (Michelsen *et al.*, 1980, 1981). In these publications heat flux, as well as lithological heat conductivity values, are listed. Variations in heat conductivity are considered as a function of lithology, geological age (for units with basically the same lithology) and porosity.

Based on this information, geothermal gradients from the Danish Sub-basin published by Madsen (1975, 1977) were modified and used as guidelines in this study. Heat flux values from several places in onshore Denmark are published by Michelsen *et al.*

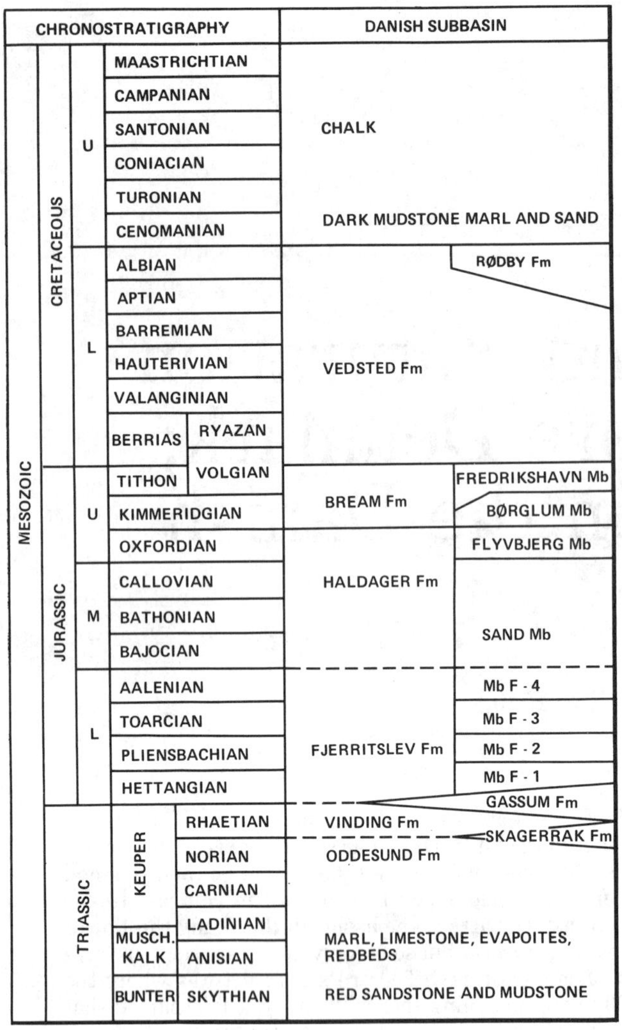

Fig. 1. Stratigraphic subdivision of the Danish Sub-basin.

(1980, 1981) and are calculated from the average geothermal gradients and the average lithological heat conductivities (Balling *et al.*, 1981).

STRUCTURAL OUTLINE

The Danish Sub-basin is part of the greater Norwegian–Danish Basin, and in this paper it refers to the north Jylland area and its adjacent waters (see Fig. 2). The axis of the Danish Sub-basin runs parallel to the Tornquist–Teissure Line and perpendicular to the trend of the Oslo Rift and Rhine Graben System.

Denmark, and the North Sea generally, lie close to the Y-shaped junction of three major plates, the Fennoscandian Shield, the North American–Greenland Plate and the South European–African Block (Zwart and Dornsiepen, 1978; Strømberg, 1981).

Five orogenic phases have affected Denmark and have played an important role in the evolution of the area. These phases are (a) Grenvillian, (b) Cadomian, (c) Caledonian, (d) Variscan and (e) the Alpine (Surlyk, 1980). From the analysis of tectonic elements, five major trends have been recognized (see below), of which the first three are of major importance in the structural and sedimentary evolution of the Danish Sub-basin.

(1) NW–SE (130–150°) trending faults, termed the Dinaric Trend (Lindstrøm, 1967). These trends are dominant in the Fennoscandian Border Zone.

(2) WNW–ESE (110–120°) trending faults, termed the Hercynian Trend, outlining parts of the Fennoscandian Border Zone and at an angle of about 25° to the Dinaric Trend.

(3) NNE–SSW (010–030°) trending faults, termed the Rhenian Trend and illustrated by the *en echelon* axis of the Oslo Rift and Rhine Graben System. Many of the salt walls in the German Basin follow this Trend (Pratsch, 1979).

(4) NNW–SSE (160–170°) trending faults, termed the Kattegat Trend. These faults seem to be arranged in a disorderly *en echelon* pattern in the Danish Sub-basin.

(5) ENE–WSW (040–070°) trending faults, termed the Skagerak Trend. This trend is represented by only a few faults in the Skagerak area.

The five structural trends listed above are also recognized from magnetic, gravimetric, Landsat and bathymetric data (Frost, 1977; Dikkers, 1977).

DEPOSITIONAL HISTORY

In the Danish Sub-basin, Paleozoic rocks are reported from only two wells, Rønde-1 (Rasmussen *et al.*, 1971) and Nøvling-1 (Rasmussen *et al.*, 1973). Hence the interpretation of the Paleozoic section is mainly based on seismic data. From seismic lines a relatively thick sedimentary sequence is recognized between a basement reflector and the base of the Mesozoic. The Paleozoic section includes thick Zechstein salt deposits, which have influenced the Mesozoic and Cenozoic evolution of the basin.

During the Triassic, significant subsidence took place in Denmark, particularly in the Danish Sub-basin. Sediments consist mainly of fluvial to continental deposits (Bertelsen, 1978; Michelsen *et al.*, 1981), with a thickness exceeding 1500 m locally (Mors-1 well). During Keuper time the sedimentary load on the underlying Zechstein salts reached more than 1000 m, and halokinesis was initiated. Halokinetic movements have been occurring spasmodically since Late Triassic times, creating extensive salt structures and rim synclines in the Danish Sub-basin.

Basin subsidence continued strongly throughout the Early Jurassic (Michelsen, 1978) declining through the Middle and Late Jurassic and into the Cretaceous.

Subsidence was renewed during the Tertiary, with the main depocentre in the Danish Sub-basin being located just north of the Holmsland Block (Nielsen *et al.*, 1984). The northeastern part of Jylland was affected by Laramide inversion during the Paleogene and by the Fennoscandian uplift during the Neogene (Ziegler, 1982).

In summary, the depositional history in the Mesozoic was dominated by basin subsidence and halokinetic movements, as opposed to the Cenozoic when inversion and partial erosion occurred (Nielsen *et al.*, 1984; Pegram, 1984).

LOWER JURASSIC FJERRITSLEV FORMATION

In this study the Lower Jurassic Fjerritslev Formation is regarded as the source rock, as it has a favourable kerogen composition and is buried deeply enough to be mature (Lindgren and Thomsen, 1982; Thomsen

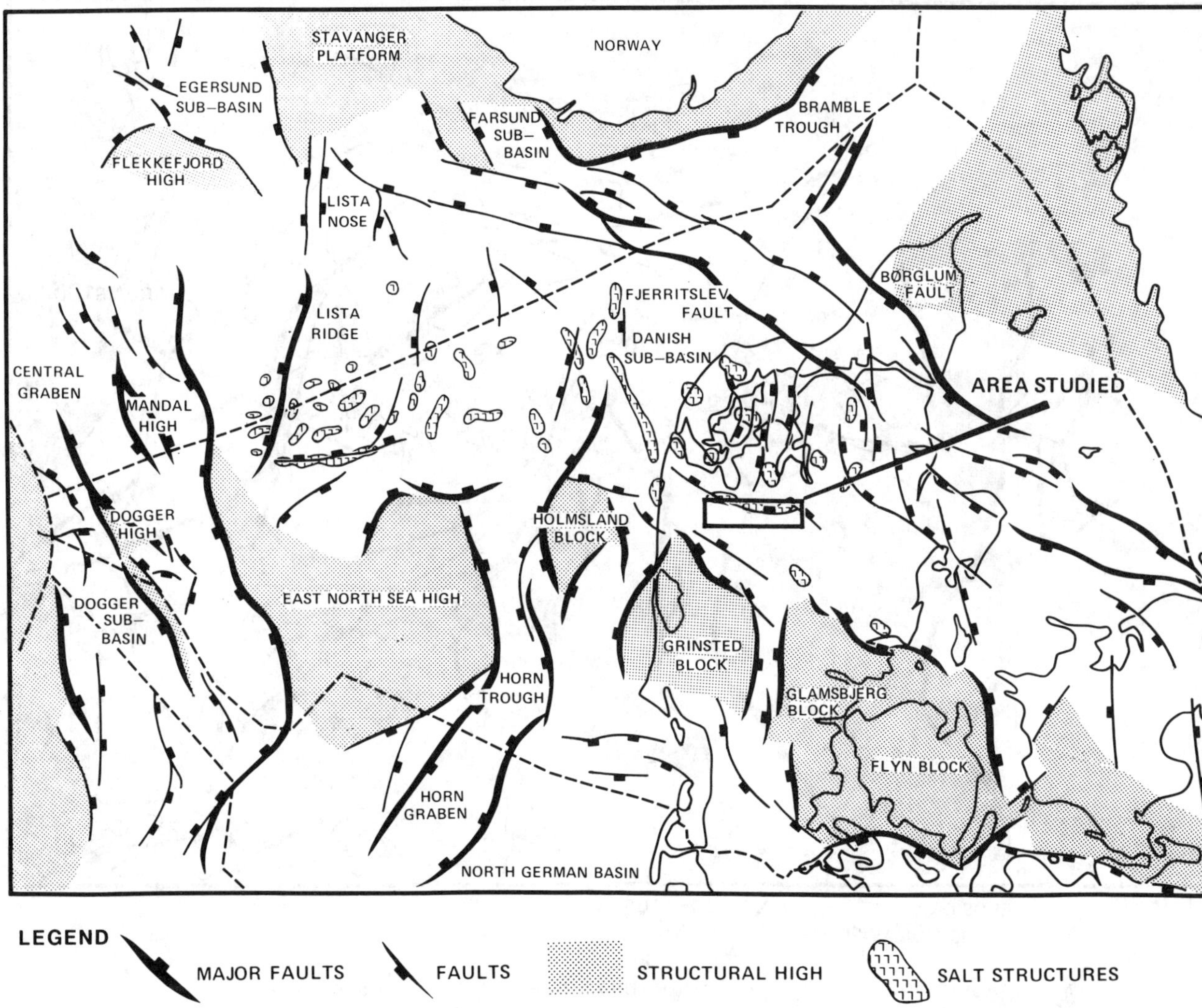

Fig. 2. Tectonic framework of Denmark.

et al., 1983). It is divided into four Members (Michelsen, 1978), as follows.

Member F-I (Hettangian–Pliensbachian) was deposited under marine conditions and consists of deposits grading from shallow to deep environments. The total organic carbon (TOC) content is of the order of 1–2%, and the organic matter consists of a mixture of humic and amorphous material.

Member F-II (Pliensbachian) was deposited in a well-oxygenated shallow marine environment. TOC ranges from 1% to 3%, and the organic matter consists also of a humic/amorphous mixture.

Member F-III (Toarcian) was deposited in a deep water marine environment at a time when sea level changes and basin isolation resulted in a restricted bottom water circulation. This unit is dominated by amorphous material with a TOC of the order of 1–6%.

Member F-IV (Aalenian) was deposited locally in a lagoonal environment with strongly reducing conditions. The organic content consists of a mixture of amorphous and humic material and TOC ranges from 1% to 3%.

Of the four members of the Fjerritslev Formation, Member III (of Toarcian age) is regarded as the best source rock because of its high TOC and favourable kerogen composition. Member III is comparable with the Drake Shale of the Northern North Sea (Cooper and Barnard, 1984), to the Posidonia Shale in the Netherlands (Bodenhausen and Ott, 1981) and the Toarcian source rocks of the Paris Basin (Tissot *et al.*, 1971).

The Lower Jurassic Fjerritslev Formation in the Danish Sub-basin reaches thicknesses in excess of 1000 m locally (see Fig. 3). All members of the formation have an organic carbon content which classifies them as potential source rocks. Substantial parts of the Formation have a favourable kerogen composition for oil generation. The principal question which this study addresses is the degree of thermal maturation. In order to determine thermal maturity, a temperature, depth and time relationship has been established in order to make semi-quantitative estimates, based on an assumed thermal history.

METHODS

In mapping the geothermal history of a basin, often only two temperature data points are known, i.e. the surface temperature (A) and a subsurface temperature (B) (in this model the temperature at the present depth of the source rock). A straight line joining point A to point B provides the average geothermal gradient.

Alternatively, the temperature profile may be divided into several segments with different gradients (Gretener, 1981), which represent reality more closely since heat conductivity varies with lithology. To estimate the geothermal gradient within a lithological unit

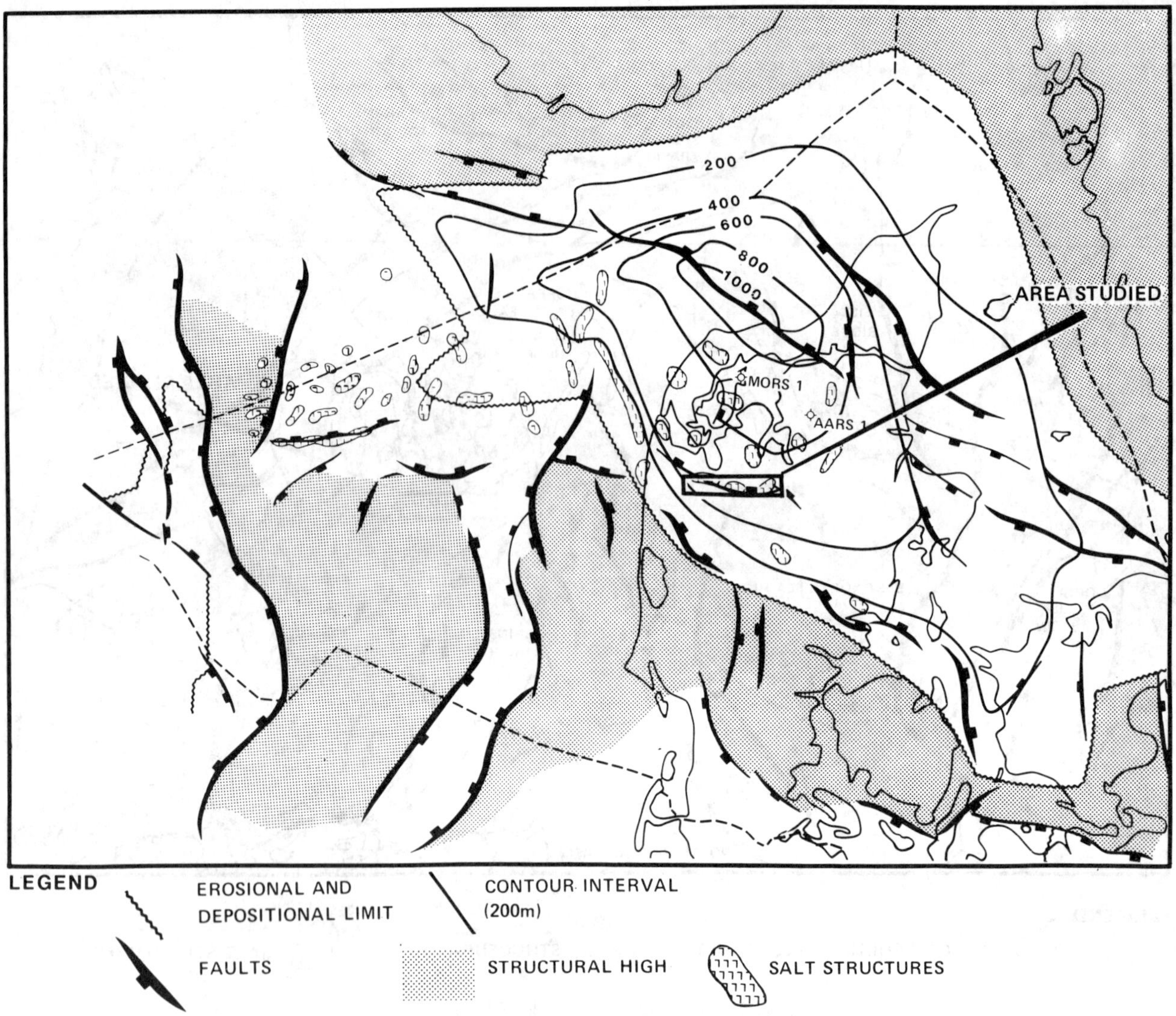

Fig. 3. Isopach map of Fjerritslev Formation.

one must know the heat flux in the area (q_0) as well as the heat conductivity (k) of the rocks. The geothermal gradient within each lithological unit can be expressed as the quotient of the two, as follows:

$$q_0/k = \text{gradient}. \tag{1}$$

The temperature calculations are derived from the analytical expression of Michelsen *et al.* (1981):

$$T_n = T_0 + \sum_{i=1}^{n}\left\{\frac{(Z_i - Z_{i-1})}{k_i}\left[q_0 - \sum_{j=1}^{i-1} A_j(Z_j - Z_{j-1})\right] - \frac{A_i(Z_i - Z_{i-1})^2}{2k_i}\right\}, \tag{2}$$

where:

T_n = temperature at the nth level,
T_0 = surface temperature,
q_0 = heat flow of the unit,
A_j = heat production within the unit,
k_i = lithological heat conductivity and
Z_i = depth to base of unit i.

The surface temperature of the area is set at 10 °C. The heat production (A_j) is very low, with an average of 1.1 μW/m^3, which is regarded as negligible for the actual temperature calculations (Balling *et al.*, 1981). This leaves the expression

$$T(z) = T_0 + \frac{q_{0_z}}{k} - \frac{1}{2}\frac{A}{k}z^2. \tag{3}$$

The open file information from the Danish Geological Survey on heat conductivity measurements as a function of lithology, age and porosity, provides useful data for heat conductivity corrections through various lithological sections. These measurements indicate heat conductivity variations from as low as 1.34 W m^{-1} K^{-1} in parts of the Paleocene clays of the Viborg-5 well to 4.20 W m^{-1} K^{-1} for the Middle–Late Jurassic sandstones of the Gassum-1 well (Balling *et al.*, 1981). It then becomes obvious that heat conductivity variations have to be accounted for in subsurface temperature estimation.

In equation (3), q_{0_z} represents the heat flux through the different units, and is regarded as a function of time. The present modelling of heat flux is based on the assumption that present-day heat flux values are probably lower than palaeo-heat flux values in the area because there has been little tectonic activity in the area throughout the Neogene. In contrast, five earlier orogenic phases have affected the Danish area, and hence it is assumed that increased heat flux occurred during such tectonically active periods and for a time after the main tectonic event (see McKenzie, 1978, 1981; Richardson and Oxburgh, 1978; Royden *et al.*, 1980; Turcotte and Ahrens, 1977).

In the following example the thermal history of one of the rim synclines just north of the Grinsted Block is analysed (see Fig. 2). In order to map the geothermal history of an area of interest, the burial history also

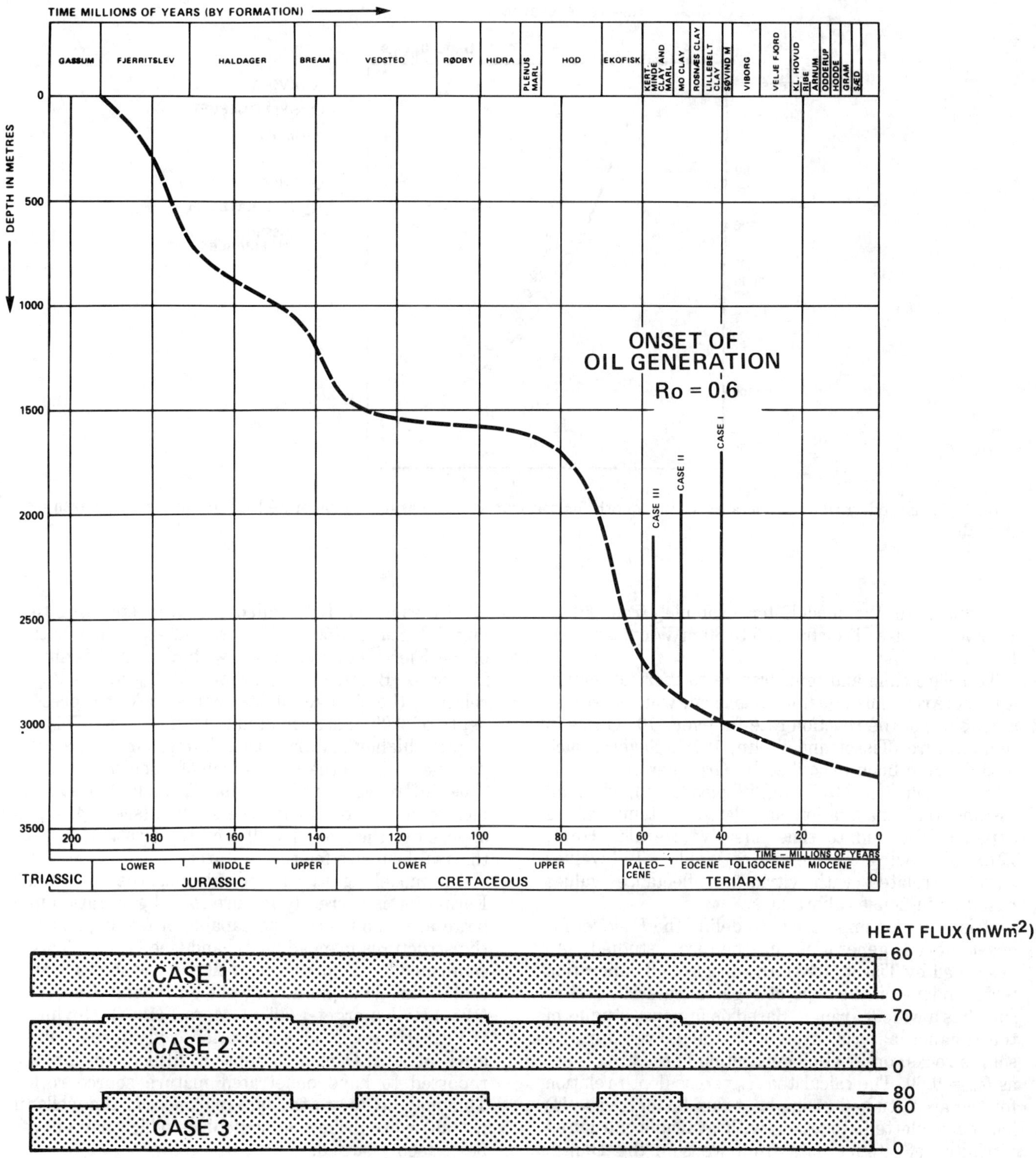

Fig. 4. Subsidence history of the area studied.

had to be determined, and was based on an interpretation of relevant seismic sections (see Fig. 4). The well-tie in the area is good, hence the stratigraphy can be reliably correlated. The section has been depth-converted by means of local velocity analysis. The lithology of the section is interpreted on the basis of seismic stratigraphy, and correlated with known lithologies from the nearby wells. The age, lithology, porosity and heat conductivity relations have been calibrated against empirical data published by the Danish Geological Survey (Michelsen *et al.*, 1980, 1981). The burial history plot of the study area (see Fig. 4) indicates that major subsidence occurred in three distinct phases—during the Early Jurassic as a continuation of the Triassic subsidence (Bertelsen, 1978), during the Late Jurassic and during the Late Cretaceous–Early Tertiary.

In the present geothermal modelling, three different heat flux histories were considered (see Fig. 4). A conservative estimate of the geothermal history is represented by temperatures adjusted for lithological heat conductivity, but keeping the heat flux in the area at the present value constant with time (case I, Fig. 4). The geothermal history becomes more favourable for maturation if an increased heat flux during the three periods of high tectonic activity is

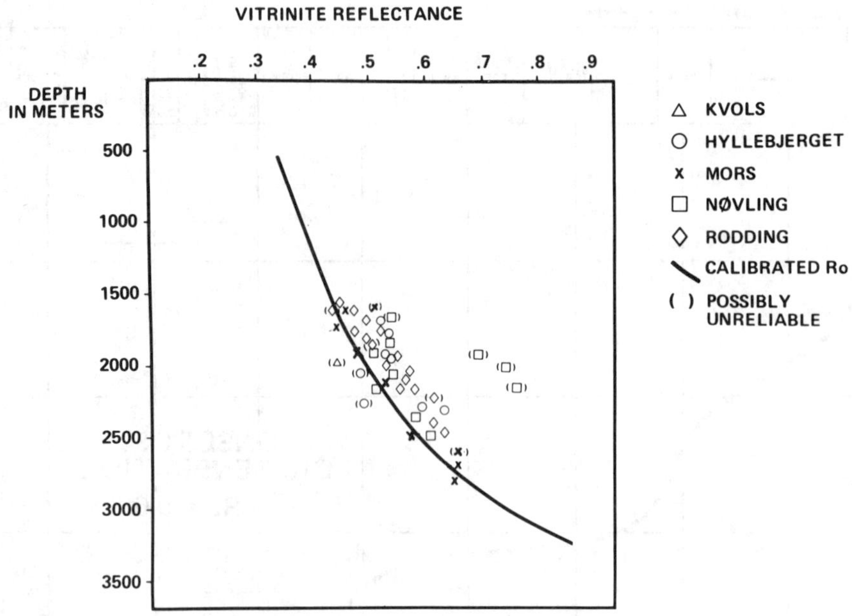

Fig. 5. Measured vitrinite reflectance versus depth for the Danish Sub-basin, and calibrated trend for the study area (see Fig. 2).

introduced to the model, for example, from 60 to 70 mW/m^2 (case II) or from 60 to 80 mW/m^2 (case III, Fig. 4).

To relate time and temperature to thermal maturity, the Arrhenius equation was used, which assumes a doubling of the reaction rate for every 10 °C rise in temperature (Tissot and Welte, 1978). Several such models have been described by Gretener and Curtis (1982), Hood *et al.* (1975) and Waples (1980), which all assume that maturation is related to temperature exponentially and to time linearly (see also Hunt, 1979). Following the method of Waples, TTI values were correlated with vitrinite reflectance values based on in-house calibrations.

It is of major importance to define the R_0-level for onset of oil generation in the area studied. As described by Tissot (1984), the onset of oil generation is dependent on the type of organic material, and as such has a wide R_0-range. Based on in-house studies of the organic material present in the Lower Jurassic source rocks, onset of oil generation in the area is set as $R_0 = 0.60$. The calculated R_0 versus depth relation for the area studied (using heat flux history, case II) has been plotted, together with available measured vitrinite reflectance data from wells in the Danish Sub-basin (see Fig. 5). Histograms of all vitrinite reflectance measurements were available, and the quality of the data used is considered to be good. The two trends are in general agreement, and observed discrepancies can be attributed to subtle differences in burial and thermal history between the location modelled and individual wells elsewhere in the basin where measurements were made.

RESULTS

The timing of the onset of oil generation, as deduced from the calibrated R_0–depth relation for cases I, II and III is shown in Fig. 4. In the most likely case (II), this is estimated to have occurred at the location modelled during the Late Paleocene, corresponding to a burial depth of 2750 m. The present areal distribution of the Fjerritslev Formation with $R_0 > 0.60$ is shown in Fig. 6. Based on the calibrated R_0 versus depth relation, the R_0 level of the source rock at a present depth of 3200 m is estimated to be $R_0 = 0.85$. This is a slightly higher maturity than is measured at a comparable depth in the Aars-1 well (Jensen *et al.*, 1985). This difference is probably due to the higher subsurface temperature in the area studied (see Fig. 3), as opposed to the Aars-1 well, and to variations in geothermal histories for the two areas.

The modelling studies indicate that the Fjerritslev Formation is currently mature for oil generation in a large area, and should be capable of feeding some of the structures mapped on Jylland (see Fig. 6). This is confirmed by the presence of oil and gas shows in several of the old exploration wells on Jylland (Gassum-1 (1951) Hyllebjerget-1 (1976), Mors-1 (1967), Nøvling-1 (1966)). However, no exploration wells in the Danish area of the Danish Sub-basin drilled so far are reported to have penetrated mature source rocks. This may be due to the fact that most of the wells (if not all) were drilled on structures which have never been deeply buried.

CONCLUSIONS

By means of thermal modelling techniques, Lower Jurassic source rocks are estimated to be oil-mature onshore in Jylland. The source rocks may have been generating oil since Late Paleocene times. It seems probable that a high percentage of the source rock kerogen has been converted to oil.

Possible explanations for the lack of significant success in the area may include limited drainage into the drilled structures, ineffective trapping mechanisms or lateral facies variations of the source rock. However, the author believes that hydrocarbons may be found in the area, perhaps trapped below the edge of mushroom-shaped salt structures.

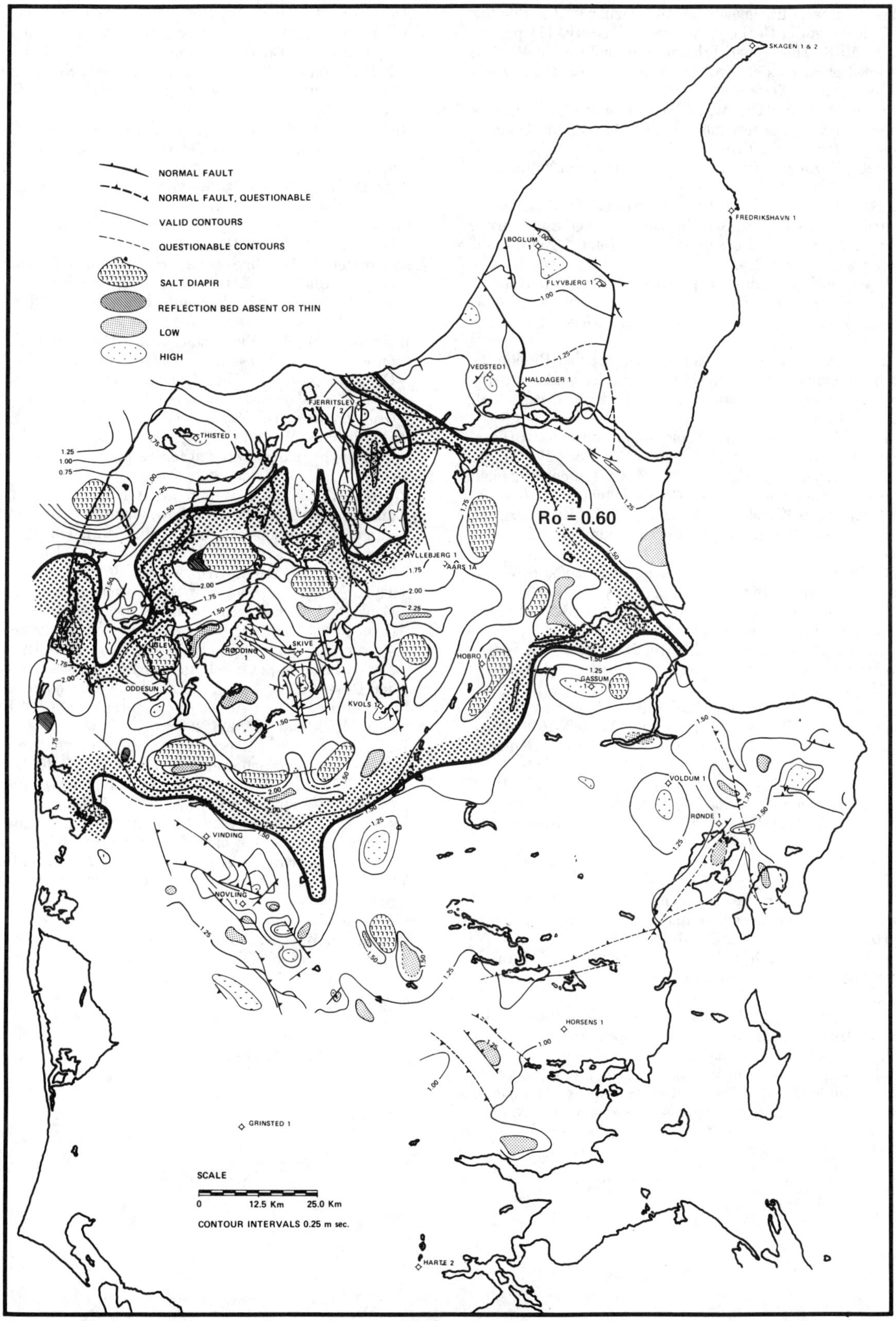

Fig. 6. Areal distribution of mature Lower Jurassic source rocks.

ACKNOWLEDGEMENTS

The author would like to thank Superior Oil Company for the permission to publish this paper. Thanks are also due to my colleagues, K. E. Trommestad and W. H. Edwards, for the seismic interpretation, to N. F. Peterson for great help with the geochemistry and to Dr O. H. J. Christie of Rogalandsforskning and P. Møller-Pedersen of A/S Norske Shell for fruitful discussions and suggestions on the manuscript.

REFERENCES

Balling, N., Kristiansen, J. I., Breiner, N., Poulsen, K. D., Rasmussen, R. and Saxov, S. 1981. Geothermal measurements and subsurface temperature modelling in Denmark. *GeoSkrifter No. 16*, Department of Geology, Aarhus University, 171 pp.

Bertelsen, F. 1978. The Upper Triassic–Lower Jurassic Vinding and Gassum Formations of the Norwegian–Danish Basin. *Danmarks Geol. Unders. Serie B*, **3**, 26 pp.

Bodenhausen, J. W. A. and Ott, W. F. 1981. Habitat of the Rijswijk oil province, onshore Netherlands. In: Illing, L. V. and Hobson, G. D. (eds.), *Petroleum Geology of the Continental Shelf of North-West Europe*. Institute of Petroleum, London, pp. 301–309.

Cooper, B. S. and Barnard, P. C. 1984. Source rocks and oils of the Central and Northern North Sea. In: Demaison, G. and Murris, R. J. (eds.), *Petroleum Geochemistry and Basin Evaluation. AAPG Memoir*, **35**, 303–314.

Dikkers, A. J. 1977. Sketch of a possible lineament pattern in North-West Europe. In: Frost, R. T. C. and Dikkers, A. J. (eds.), *Fault Tectonics in North-West Europe. Geol. Mijnbouw*, **56**, 275–285.

Frost, R. T. C. 1977. Tectonic patterns in the Danish Region (as deduced from comparative analysis of magnetic, landsat, bathymetric and gravity lineaments). *Geol. Mijnbouw*, **56**(4), 351–362.

Gretener, P. E. 1981. Geothermics: using temperature in hydrocarbon exploration. *AAPG Education Course Note Ser.*, **17**, 170 pp.

Gretener, P. E. and Curtis, C. D. 1982. Role of temperature and time on organic metamorphism. *AAPG Bull.*, **66**(8), 1124–1149.

Hood, A., Gutjahr, C. C. M. and Heacock, R. L. 1975. Organic metamorphism and the generation of petroleum. *AAPG Bull.*, **59**(6), 986–996.

Hunt, J. M. 1979. *Petroleum Geochemistry and Geology*, W. H. Freeman & Co., San Francisco, 617 pp.

Jensen, P. K., Holm, L. and Thomsen, E. 1985. Modelling burial history, temperature and maturation (this volume).

Lindgren, H. and Thomsen, E. 1982. Investigations on the source rock potential of the Danish onshore area. *Danm. Geol. Unders.* (June 1982) 91 pp.

Lindstrøm, M. 1967. Om den tektoniske utvecklingen av Skånes Paleozoikum. *Medd. Dansk Geol. Forening*, **17**, 163–165.

Madsen, L. 1975. Approximate geothermal gradients in Denmark and the Danish North Sea Sector. *Danm. Geol. Unders*, **1974**, 5–16.

Madsen, L. 1977. Geotermisk Energi i Danmark—en geologisk vurdering. *Dansk Geol. Foren.*, Årskrift for 1977, København, (January 1978), 29–40.

McKenzie, D. 1978. Some remarks on the development of sedimentary basins. *Earth Planet. Scie. Lett.*, **40**, 25–32.

McKenzie, D. 1981. The variation of temperature with time and hydrocarbon maturation in sedimentary basins formed by extension. *Earth Planet. Sci. Lett.*, **55**, 87–98.

Michelsen, O. 1978. Stratigraphy and distribution of Jurassic deposits of the Norwegian–Danish Basin. *Danm. Geol. Unders., Serie B*, **2**, 28 pp.

Michelsen, O., Saxov, S., Leth, J. A., Andersen, C., Balling, N., Breiner, N., Holm, L., Jensen, K., Kristiansen, J. I., Laier, T., Nygaard, E., Olsen, J. C., Poulsen, K. D., Priisholm, S., Raade, T. B., Sørensen, T. R. and Wurtz, J. 1980. *Landsdækkende Geotermisk Undersøgelser*. Del 1: *Kortlægning og Analyse of Reservoirer*. Dansk Olie & Naturgas A/S, København, 103 pp.

Michelsen, O., Saxov, S., Leth, J. A., Andersen, C., Balling, N., Breiner, N., Holm, L., Jensen, K., Kristiansen, J. I., Laier, T., Nygaard, E., Olsen, J. C., Poulsen, K. D., Priisholm, S., Raade, T. B., Sørensen, T. R. and Würtz, J. 1981. Kortlægning of potentielle geotermiske reservoirer i Danmark. *Danm. Geol. Unders., Serie B*, **5**, 96 pp.

Nielsen, O. B., Sørensen, S., Thiede, J. and Skarbø, O. 1984. Differential subsidence of the Cenozoic North Sea. *AAPG Bull.* (submitted).

Pegrum, R. M. 1984. The extension of the Tornquist Zone in the Norwegian North Sea. *Norw. Geol. Tidsskr.*, **64**(1), 39–68.

Pratsch, J. C. 1979. Regional structural elements in Northwest Germany. *J. Petrol. Geol.*, **2**(2), 159–180.

Rasmussen, L. B., Dinesen, A., Henriksen, S. E., Bang, I., Senestad, E., Buch, A., Christensen, O. B., Michelsen, O. and Jacobsen, F. L. 1971. Dybdeboringer Rønde nr. 1 på Djursland. *Danm. Geol. Unders. III Række, Serie no. 39*, 123 pp.

Rasmussen, L. B., Baartman, J. C., Henriksen, S. E., Kristoffersen, F. N., Dinesen, A., Bang, I., Stenestad, E., Buch, A., Christensen, O. B., Michelsen, O., Hansen, T. J. and Jacobsen, F. L. 1973. Dybdeboringen Nøvling nr. 1 i Midtjylland. *Danm. Geol. Unders., III Række, Serie No. 40*, 181 pp.

Richardson, S. W. and Oxburgh, E. R. 1978. Heat flow, radiogenic heat production and crustal temperatures in England and Wales. *Geol. Soc. Lond.*, **135**, 323–337.

Royden, L., Sclater, J. G. and von Herzen, R. P. 1980. Continental margin subsidence and heat flow: important parameters in formation of hydrocarbons. *AAPG Bull.*, **64**(2), 173–187.

Strømberg, A. G. B. 1981. The European Caledonides and the Tornquist Lineament. *Geol. Foren. Stockh.*, **103**, 393–410.

Surlyk, F. 1980. Geology of the European countries: Denmark. Geology of the European countries. *Comité National Français de Geologie (C.N.F.G.), 26th Int. Geological Congress.*

Thomsen, E., Lindgren, H. and Wrang, P. 1983. Investigations on the source rock potential of Denmark. In: Kaasschieter, J. P. H. and Rejers, T. J. A. (eds.), *Petroleum Geology of the Southwestern North Sea and Adjacent Onshore Areas* (The Hague, 1982). *Geol. Mijnbouw*, **62**, 221–239.

Tissot, B. P. 1984. Recent advances in Petroleum geochemistry applied to hydrocarbon exploration. *AAPG Bull.*, **68**(5), 545–563.

Tissot, B. P. and Welte, D. H. 1978. *Petroleum Formation and Occurrence*. Springer-Verlag, Berlin, 538 pp.

Tissot, B., Califet-Debyser, Y., Deroo, G. and Oudin, J. L. 1971. Origin and evolution of hydrocarbons in early Toarcian shales. Paris Basin, France. *AAPG Bull.*, **55**(12), 2177–2193.

Turcotte, D. L. and Ahren, J. L. 1977. On the thermal and subsidence history of sedimentary basins. *Geophys. Res.*, **82**(26), 3762–3766.

Waples, D. W. 1980. Time and temperature in Petroleum Formation, application of Lopatin's method to petroleum exploration. *AAPG Bull.*, **64**(6), 916–926.

Ziegler, P. A. 1982. *Geological Atlas of Western and Central Europe*. Elsevier Applied Science Publishers, Barking, 130 pp.

Zwart, H. J. and Dornsiepen, U. F. 1978. The tectonic framework of Central and Western Europe. *Geol. Mijnbouw*, **57**, 627–654.

12 Quantitative modelling of the sourcing of giant oil fields in the East Shetland Basin (Abstract)

J. C. Goff
Department of Energy, Petroleum Engineering Division, London, UK

In the East Shetland Basin, Late Jurassic rifting formed tilted fault blocks and unconformity traps, sealed by onlapping Cretaceous mudstones. Important oil-bearing reservoirs are Middle and Lower Jurassic marginal/non-marine sheet sandstones. Mature Upper Jurassic mudstone and Middle Jurassic coal source rocks, located in half grabens downdip of the major fields, and in the flanking western Viking Graben, generated oil in Eocene to Miocene times.

Study of the reservoir pressures, distribution and structure defines the major Jurassic aquifers in the East Shetland Basin. The geometrical relationship of these aquifers to the source rock kitchens defines the volumes of source rock drained by the major oil fields.

Total organic carbon, pyrolysis and soluble extract analyses were performed on thick source rock sequences, representative of the source rock kitchens, to determine their organic richness, oil yield and maturity. A model of oil expulsion is developed, based on the squeezing of oil out of the mature source rocks during progressive burial (compaction). Quantitative modelling of the volume of oil generated and expelled from the source rocks accounts for the in-place reserves of the fields studied.

Part III
Case Histories

13 The geological history and geochemical evaluation of Block 2/2, Offshore Norway

Roy H. Gabrielsen, Steinar Ulvøen, Arvid Elvsborg and Ole Fredrik Ekern
Saga Petroleum a.s. P.b. 9. 1322 Høvik, Norway.

Block 2/2 is situated on the northeastern flank of the Central Graben within the 'Ula Trend', which is characterized by the development of Upper Jurassic reservoir sandstones. The 2/2 area underwent a relatively quick subsidence along the WNW-trending Hidra Fault Zone in late Jurassic times. The tectonic activity was influenced by halokinesis, and the geometry along the fault zone varies significantly. Three wells have so far been drilled in block 2/2. Besides gas discoveries in Oligocene sands in wells 2/2–1 and 2/2–2, oil has also been found in Upper Jurassic sandstones in well 2/2–1.

This study includes an organic geochemical interpretation of mainly the Jurassic section in block 2/2, and a discussion of the Oligocene gas discoveries. Source rocks are identified in the Upper Jurassic Mandal and Farsund Formations and in the coal-bearing Middle Jurassic Bryne Formation. Indications of migrated hydrocarbons are evident in the Rogaland Group and older formations in all three wells. The Jurassic oil discovery correlates best with Farsund Formation source rocks.

INTRODUCTION

The present paper is a summary of an integrated study by Saga Petroleum a.s. of the Jurassic sequence in Block 2/2 in the Norwegian Sector of the North Sea. The study resulted in a model for the structural and sedimentological development within Block 2/2. This model, in turn, served as a base for geochemical evaluation of the block.

Block 2/2 was a part of the Norwegian 6th round, and was awarded in August 1981 to a group including Arco Norway Inc., Mobil Exploration, Norway, Norsk Hydro A.S., Saga Petroleum a.s. and Statoil, with Saga Petroleum as the operator. The first well (2/2–1) was drilled in 1982 in the southeastern part of the block, resulting in an oil discovery in an Upper Jurassic reservoir. Subsequently, two more wells have been drilled in the block.

In addition to the present authors, several people have been involved in the study. The structural model is to a large extent based upon a seismic interpretation carried out in the Permian part of the sequence by Atle Edvardsen and Gyrd Sundsbø, while computer work was done by Knut Aanstad and Odd Alstad. Stuart Gowland and Louis Macchi made significant contributions to the tectono-sedimentological model, and the geochemical study has benefitted from cooperation with Torbjørn Throndsen.

TECTONIC FRAMEWORK OF BLOCK 2/2

Block 2/2 is situated on the northeastern flank of the Central Graben and encompasses the southeastern part of the Hidra Fault Zone, which forms the transition between the Southern Vestland Arch (Rønnevik *et al.*, 1975; Hamar *et al.*, 1980) and the Central Graben (see Fig. 1).

The Hidra Fault Zone trends WNW–ESE to NW–SE, and consists of several fault segments, partly arranged in an *en échelon* pattern. The faults are clearly influenced by halokinesis, and have to a large degree been developed as growth faults separating rotational fault blocks. According to Skjerven *et al.* (1983) a northeastward rotation of the so-called Hidra Fault Blocks took place in late Jurassic times, thus post-dating a more regional middle Jurassic uplift and eastward tilting of the Vestland Arch. To the southwest, the Hidra Fault Zone is bounded by the master faults of the Central Graben. At an early stage the fault zone was influenced by halokinesis, when considerable volumes of salt flowed away from the vicinity of the Hidra Fault Zone in a southwesterly direction. The combination of halokinesis and normal faulting associated with the Hidra Fault Zone has defined an area not only characterized by a series of potential hydrocarbon traps, but also by an elongate zone of favourable reservoir sands, the 'Ula Trend' (Bailey *et*

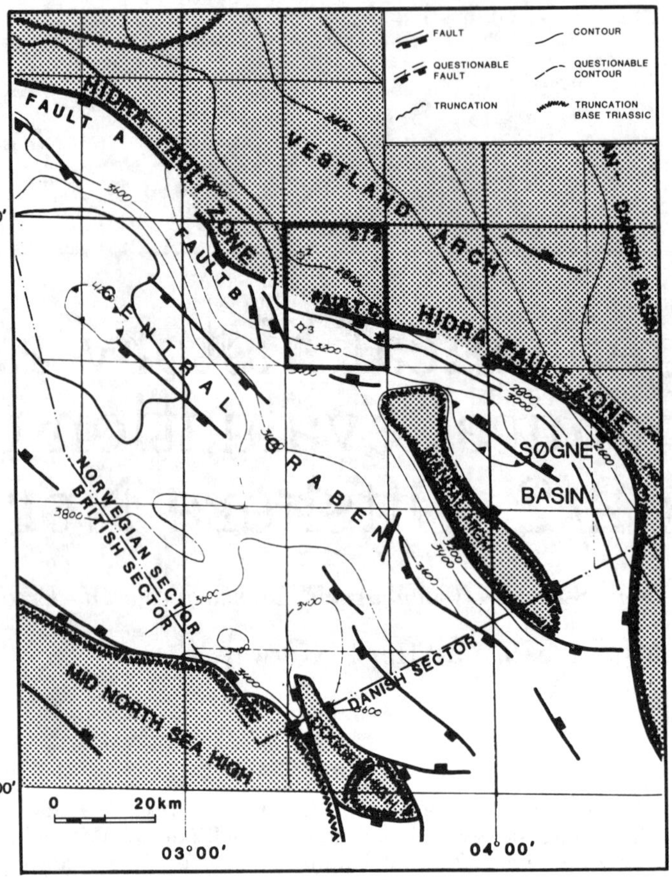

Fig. 1. Main structural elements of the northeastern flank of the Central Graben in the vicinity of Block 2/2. Contouring in two-way reflection time on Base Cretaceous level. Based on compilation by N. Fagerland.

al., 1981) within which the Ula Field is found in Block 7/12.

Roughly, three major fault segments may be identified in the Hidra Fault Zone, for which the notations A, B and C have been proposed (see Fig. 1). In detail, however, these three elements again may be subdivided into a series of minor fault segments. Also, a series of salt pillows are found on the downthrown (southwestern) side of the fault zone and may, together with a trend of rotated faults blocks, be said to define a discontinuous high between the major faults of the Hidra Fault Zone and the Central Graben.

STRATIGRAPHY

At present, several systems for stratigraphic nomenclature are in use for the Central North Sea (e.g. Deegan and Scull, 1977; Hamar *et al.*, 1983; Hesjedal and Hamar, 1983). However, a complete survey of the stratigraphy for the Norwegian sector south of 62 °N is in preparation by an official committee (Vollset and Doré, 1984), and in the present paper the preliminary version of this proposal has been used. Vollset and Doré (1984) discrimate between the Sleipner area, Norwegian/Danish Basin and the Central Graben area for the Jurassic part of the sequence, and a summary

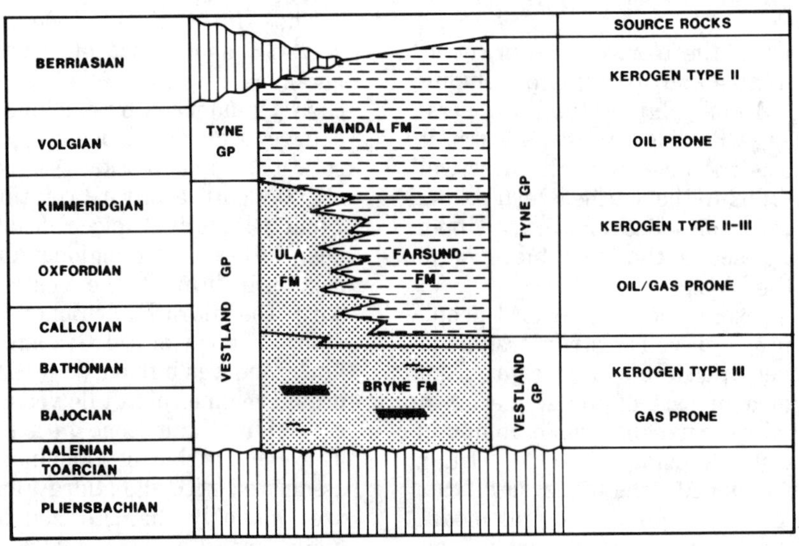

Fig. 2. General stratigraphy for the Jurassic interval in Block 2/2 with source rock units indicated.

of their scheme for the Central Graben is given in Fig. 2.

STRUCTURAL AND SEDIMENTOLOGICAL MODEL, BLOCK 2/2

In the structural evaluation of Block 2/2 all significant faults were classified and named according to geometry and relation to the main tectonic features in the area (Fig. 3). By careful reconstruction on migrated, depth-converted seismic sections, the geometric configuration as a function of time along strike of the Hidra Fault zone was restored. One example of this type of reconstruction is given in Fig. 4. These data were compared with conventional decompacted, back-stripped subsidence curves from wells 2/2–1 and 2/2–3. Even though the method does not allow corrections for periods of non-deposition or erosion, minimum rates of subsidence, axes of relative subsidence and differences in timing of structural events may be defined. The analysis, which incorporates several N–S and NE–SW trending lines, suggests that the structural development of the Hidra Fault Zone in Block 2/2 is not synchronous. The early fault activity within the Hidra Fault Zone was, furthermore, strongly influenced by halokinesis, and was restricted to a few faults within the zone. This situation is also reflected in the sedimentological record, and is clearly demonstrated in the differences in the late Jurassic sedimentological development between wells 2/2–1 and 2/2–3 (see Fig. 5).

In the following, a schematic structural and sedimentological history for Block 2/2 is given. The names and position of the structural elements, and the timing of activition of these features is seen in Figs. 3 and 6.

Pre-Triassic

The pre-Zechstein fault pattern is dominated by NNW-trending elements, and a considerable relief is indicated. When trends and styles are considered there is a striking difference between these faults and the faults mapped higher up in the sequence (see Fig. 4). However, a pronounced hinge-line at this level coincides with the trend of the later Hidra Fault Zone. The fault systems of this age are associated with the Permian sedimentary basins, and gave a surface relief which has controlled the thickness variations of the evaporites.

Triassic–Middle Jurassic

The fault pattern defined on Top Salt level is radically different from the pre-Zechstein system. The first WNW-trending fault blocks were activated in the Triassic, and continued to develop throughout the Late Triassic and Jurassic. The activating agent was

STRUCTURAL ELEMENTS, BLOCK 2/2

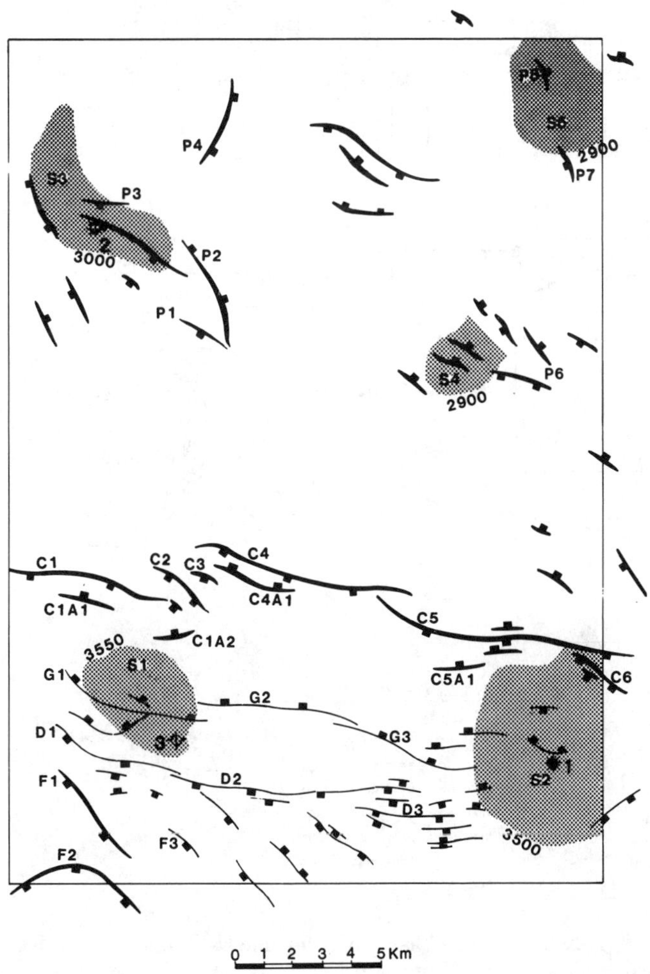

Fig. 3. Some structural elements and well locations in Block 2/2. Single-lined faults are defined on 'top Ula Formation' level, double-lined faults on Base Cretaceous level. Numbering refers to text and Fig. 7. S1–S5 indicate salt pillows.

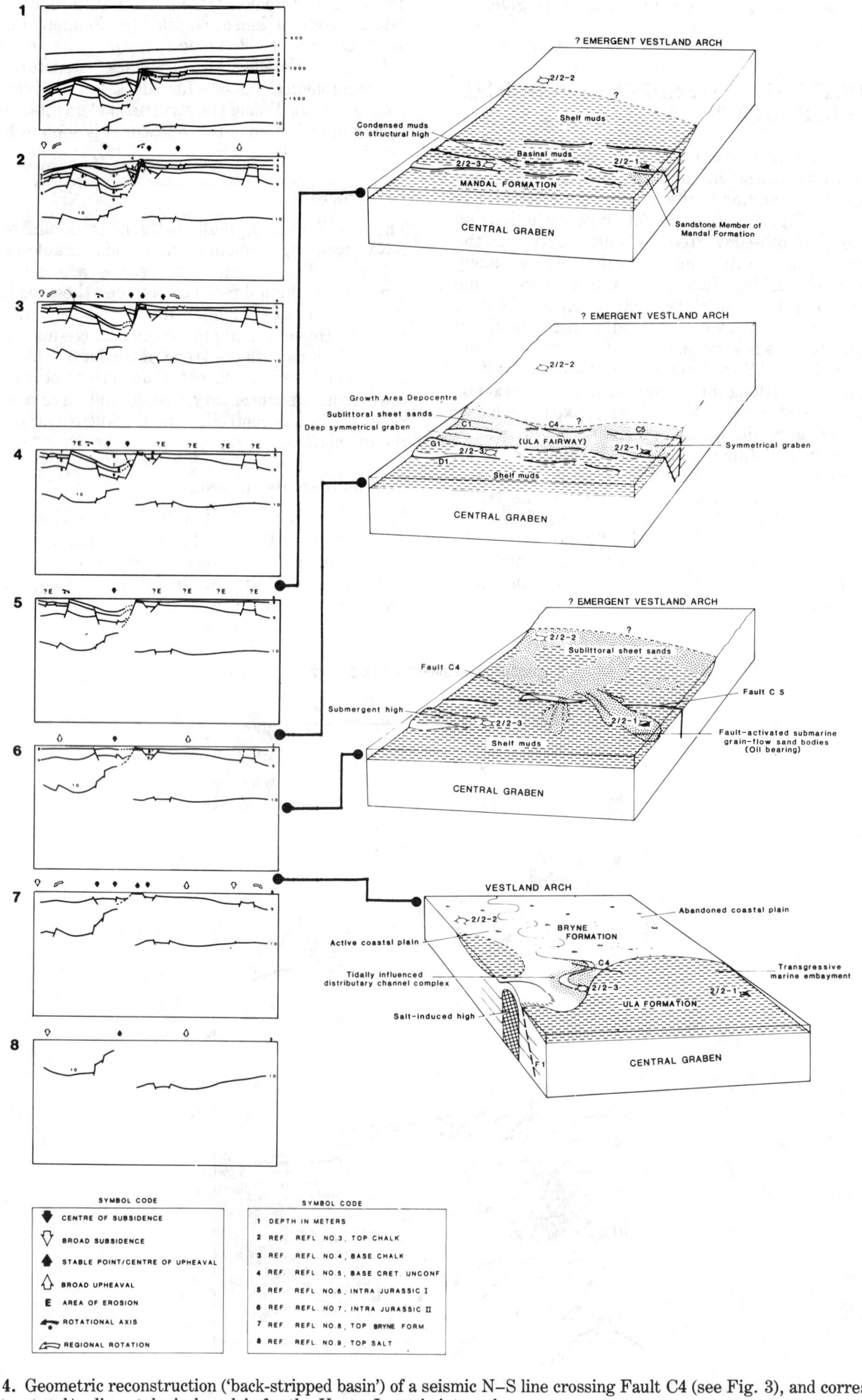

Fig. 4. Geometric reconstruction ('back-stripped basin') of a seismic N–S line crossing Fault C4 (see Fig. 3), and corresponding structural/sedimentological models for the Upper Jurassic interval.

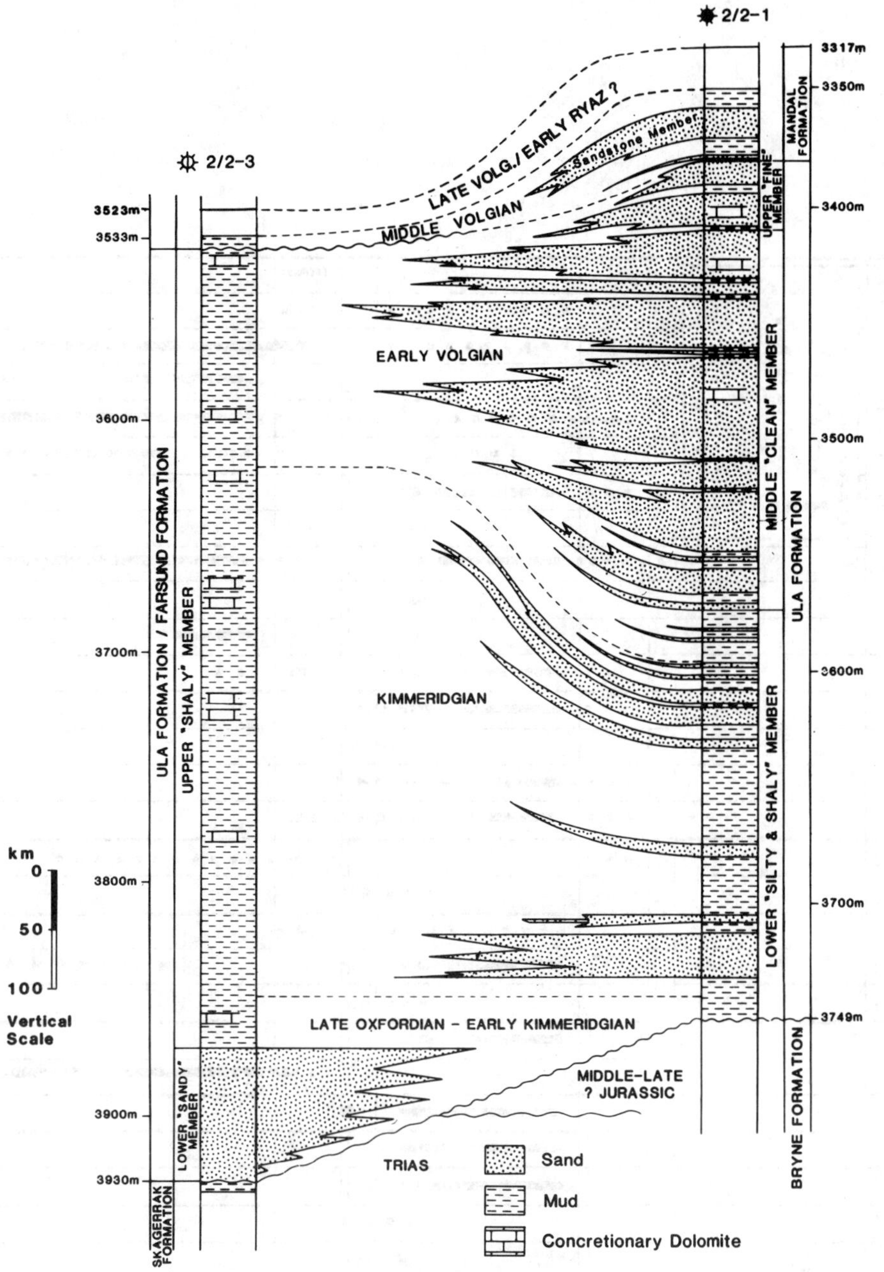

Fig. 5. Schematic lithofacies correlation within the reservoir interval of Wells 2/2-1 and 2/2-3. Note lateral relation between the Farsund Formation Source rocks and the Ula Formation sandstones in the two wells. Sedimentological correlation by Louis Macchi.

possibly a combination of a NNE–SSW-directed tensional stress system and halokinesis. The first activity was restricted to a few segments of the fault zone, and in Block 2/2 the earliest faults to develop were C4 and its antithetic counterpart C4A1 (see Figs. 3 and 6). Also, several minor faults (the P-faults of Figs. 3 and 6) mirror the early halokinetic processes. Obviously, a structural low was generated between the axis of the salt pillows S1 and S2 and the C-fault trend, and this had already started to act as a sedimentary trap in the Triassic. The Bryne Formation was deposited on a swampy coastal plain during the middle Jurassic.

Callovian–Kimmeridgian

In this period the area was tectonically relatively quiet. A eustatic sea level rise in early Kimmeridgian was followed by a slight relative emergence of the area towards the end of the period. Subsidence along most faults of the P-system as well as the C-system slowed down, and a slight south-southwesterly regional rotation took place.

The most important structural elements were the salt pillows in the northern part of Block 2/2 (S3 and S4, see Fig. 3) and the associated faults. These defined the structurally highest points, and may even have coincided with the basin margin in this time interval. Also, some activity was occurring along Fault System C, and again C4 was the most active element. Very slight flexuring accompanied by relatively high rates of deposition is indicated in the C1 and C2 segments, whereas faulting probably took place along the C4 and C5 segments (see Figs. 3 and 6). Even some of the faults of the G and F systems were active.

The deposition at this time took place in a shallow zone with a smooth relief (Fig. 4). Excepting the salt-induced high in the north, the only possible positive elements were a horst-like feature immediately north of the C4-fault and the S1-area south of the C1-fault.

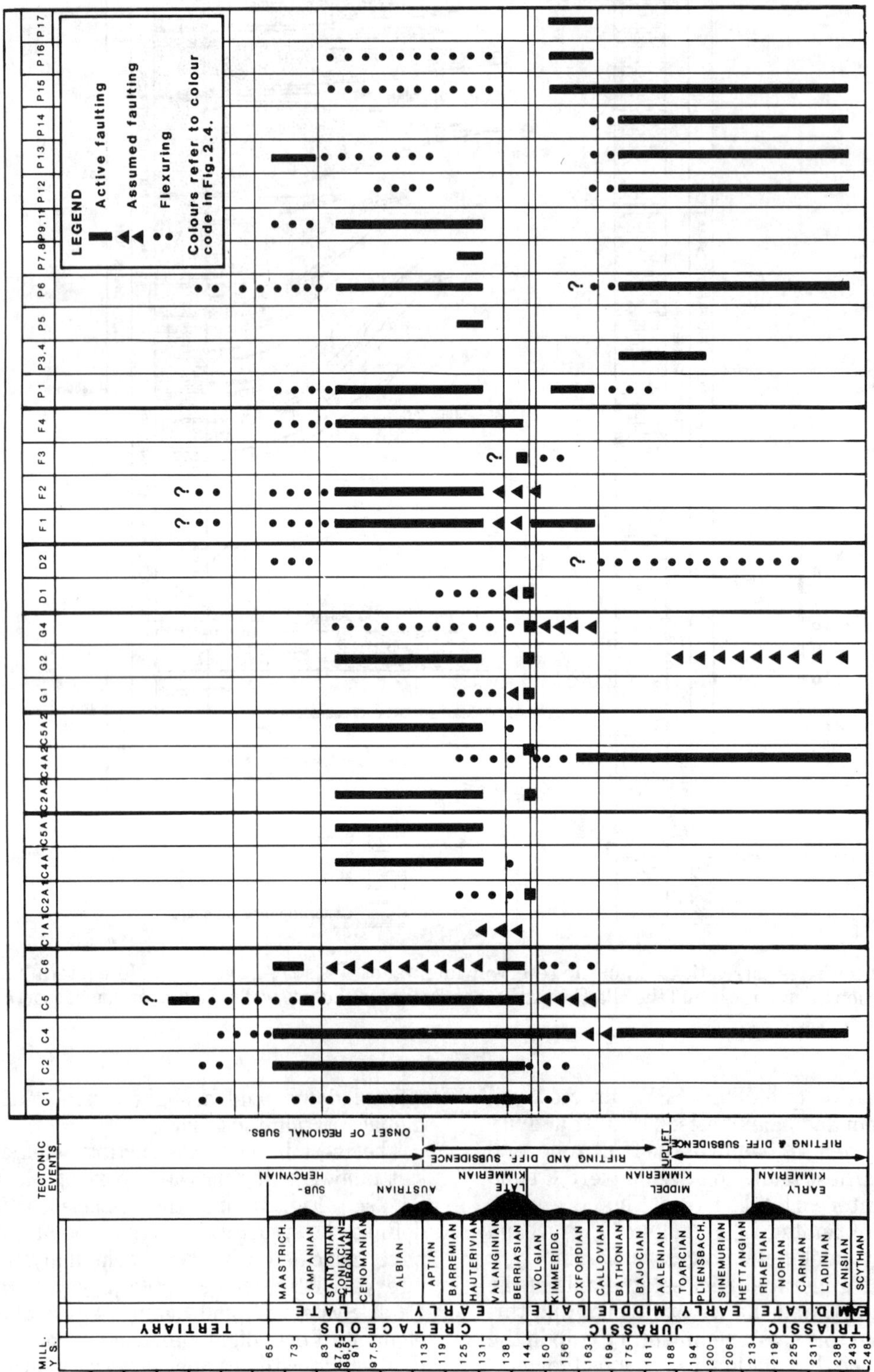

Fig. 6. Timing of activity of different structural elements in Block 2/2. Names (C1, C1A1, etc.) refer to code in Fig. 3.

The lack of obvious structural highs during this period may be surprising, considering that the sands of the Ula Formation were deposited in this time interval. However, the input of grain-flow deposits of the Ula Formation reflects the relative uplift of the northern parts of the area, possibly combined with tectonic activity along the C4 and C5 segments.

The influence of the different tectonic elements and differences in proximity to the source areas are clearly demonstrated by the varying stratigraphy seen in the different wells. Thus, the Oxfordian–early Kimmeridgian interval is represented by an 80 m thick sandy sequence in well 2/2–3, while in well 2/2-1 only an 8 m thick sandstone is present (see Fig. 5).

Towards the end of Kimmeridgian times, an increasing amount of fine sands were introduced into the 2/2-1 area (see Fig. 5). These are believed to have originated as shoals along the flanks of the Vestland Arch, and have been transported into the basin as grain flows, possibly triggered by earthquakes associated with the emergent activity of faults C4 and C5 (Fig. 4).

Early–middle Volgian

During the early Volgian, a considerable relief developed south of Fault System C. All of the structural elements were now active, either as flexures or faults. Also, the antithetic CA-faults were activated, defining a more-or-less continuous graben-like system which was filled in by very fine to coarse sands belonging to the Middle 'Clean' Member (see Fig. 5). This unit contains the best reservoir rocks in the area. The geometrical configuration varied considerably along strike of the graben structure. During this period the 2/2–3 area formed a structural high above a salt pillow (S1 in Fig. 3), and sand deposits were not preserved.

In middle Volgian times an increase in water depth is reflected in deposition of shales of the Mandal Formation which encompasses thin stringers of storm-emplaced sands (Middle Volgian Sandstone Member of Fig. 5, see also Fig. 4).

Late Volgian–early Berriasian

During the late Volgian and early Berriasian fault activity was restricted to Fault System C. Sediments of this age are only preserved in a shallow half-graben immediately to the south of this fault system, and very little structural information is accordingly available from this period. To the north and south of the graben-like structure non-deposition or erosion prevailed.

Early Cretaceous

Throughout most of the early Cretaceous several of the fault systems in the area were reactivated, resulting in a series of local structural highs and lows. Again, the geometrical configuration varied along strike of Fault System C as different antithetic faults developed. The point of maximum subsidence was associated with Fault System C, and rotation towards this feature persisted. The area to the south of the graben-like structure still underwent relative uplift, whereas the areas to the north were subject to a period of even subsidence.

Late Cretaceous

In late Cretaceous (Campanian) to early Tertiary times the configuration shifted dramatically from a situation of subsidence dominated by the developing Fault System C, to one where the rapid subsidence of the Central Graben started to control the depositional pattern. Even during this period Fault System C was important in acting as an axis of rotation, but the main movements along this structure had ceased.

Tertiary

From the Danian through the Tertiary the deformation in the 2/2 area was moderate. A southward rotation continued throughout this period, but although a slight flexuring above Fault System C may be traced, these faults no longer defined an important rotational axis. In addition to subsidence in the south, halokinesis affected the topography by slight doming of the strata immediately above the salt pillows.

SOURCE ROCKS

Two potential source rock units were deposited within the Block 2/2 area according to the history described above (see Fig. 2). The middle Jurassic Bryne Formation was deposited during a period of slow subsidence when the tectonic activity was restricted to a few faults, and the major movements were probably controlled by halokinesis. The depositional environment was a swampy coastal plain. The Bryne Formation contains gas-prone source rocks.

The late Jurassic Tyne Group consists of the Farsund and Mandal Formations. These units were deposited during a period of increasing tectonic activity associated with Callovian to Kimmeridgian and Volgian transgressive events.

Mandal Formation

The Mandal Formation consists of dark grey claystone, and is readily recognized from the 'hot shale' response on the gamma log. The unit is very thin in the wells drilled in the block, and only a few reliable data are available.

Average organic carbon content, pyrolysate yield (Rock–Eval S_2-peak), hydrogen index, kerogen type and likely products are listed in Table 1. The content of organic carbon is high, and the pyrolysate yield confirms a high potential for generation of hydrocarbons. Corresponding high hydrogen index values indicate oil-source capacity.

Although the unit is very thin in the wells and probably within the Block 2/2 area in general, it is well

TABLE 1.

Source rock unit	*Mandal Fm.*[a]	*Farsund Fm.*	*Bryne Fm.*[b]
Organic carbon (wt%)	5.72	3.67	7.5
Pyrolysate (S_2) yield (mg/g rock)	35.80	4.67	11.5
Hydrogen index	626	144	150
Kerogen type	II	II–III	III
Predominant hydrocarbon product	Oil	Oil/gas	Gas

[a]One sample only.
[b]Interpreted values, not measured.

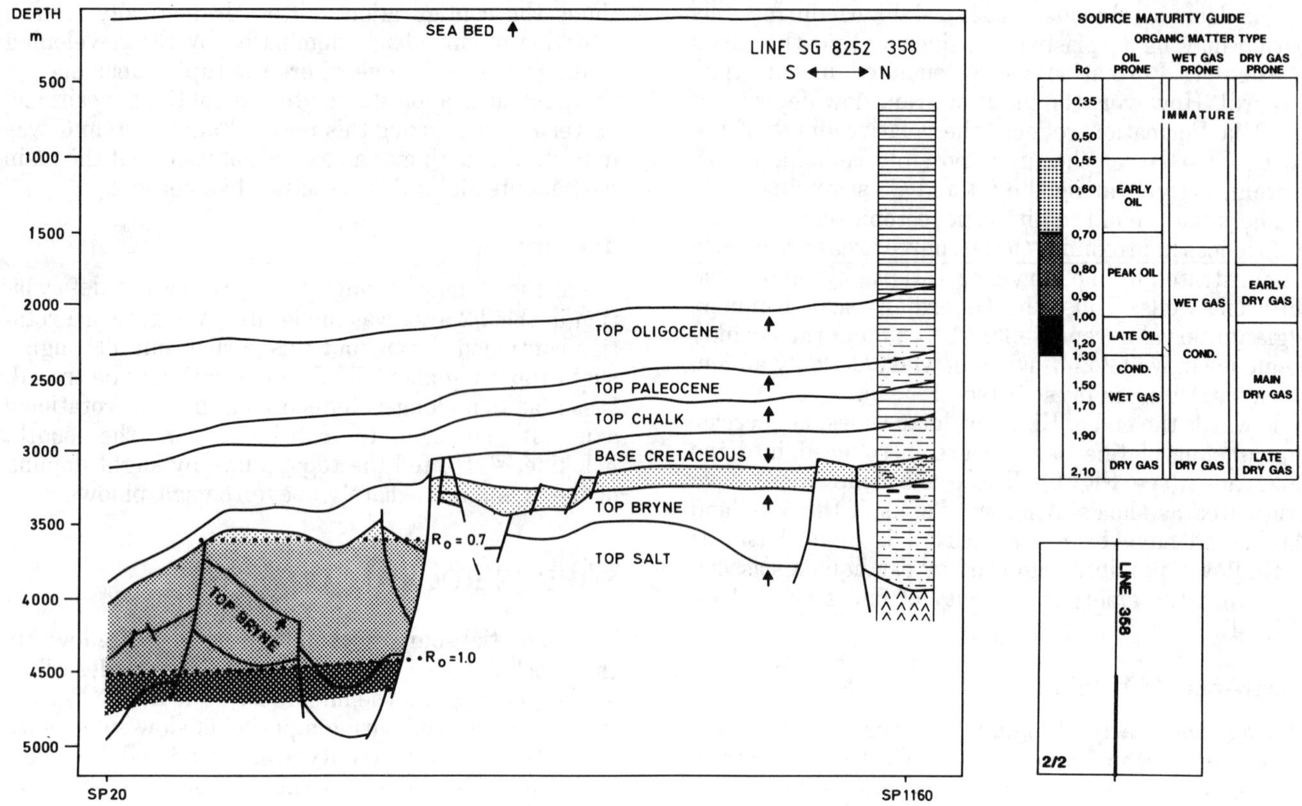

Fig. 7. Maturity cross-section N–S Block 2/2. For location see Figs. 4 and 5.

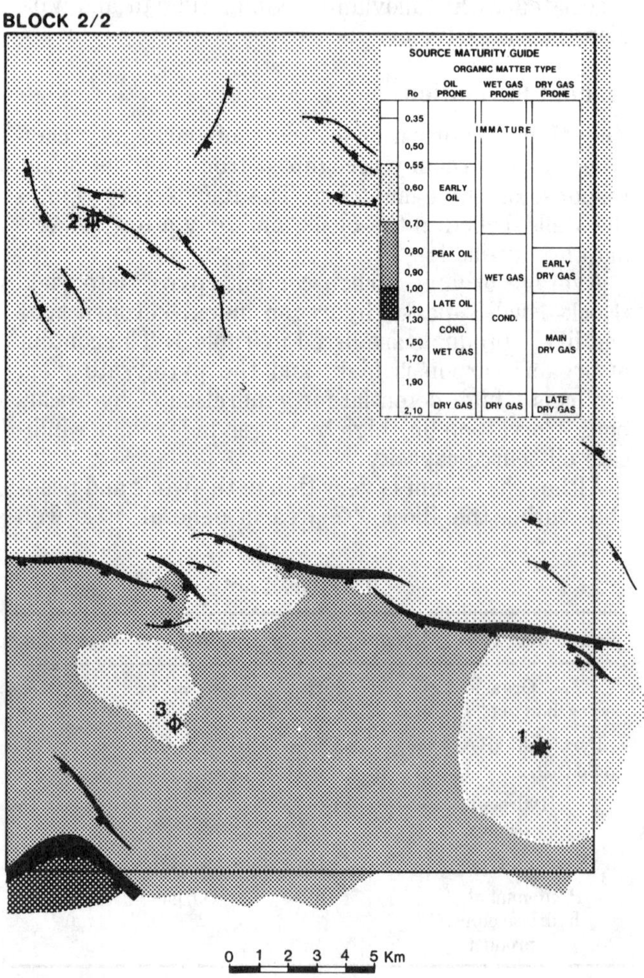

Fig. 8. Maturity map, Top Mandal Formation.

established that sufficient volumes are present in structurally deeper areas to make it the major oil source rock in the Central Graben. Its contribution to the 2/2–1 oil discovery, however, appears to be of minor importance.

A maturity cross-section is shown in Fig. 7, and a maturity map at the Base Cretaceous seismic reflector, representing the top Mandal Formation, is presented in Fig. 8. The maturity pattern within the block shows a variation from early mature to late mature, implying that the Mandal Formation is an active, although possibly small, oil source even within the block.

Farsund Formation

The bulk of the Tyne Group is made up of the Farsund Formation. This unit is particularly well-characterized in well 2/2–3, where it is penetrated with a total thickness of 353 m. The unit is relatively homogeneous, and consists of medium to dark grey claystones.

Average organic carbon content, pyrolysate yield (Rock–Eval S_2-peak), hydrogen index, kerogen type and likely products are listed in Table 1. Figure 9 shows a source rock profile of the unit. The content of organic carbon is moderate to high and ranges from 1.08 to 13.80 wt%, with an average of 3.67 wt%. The pyrolysate yield ranges from 0.68 to 17.52 mg/g rock, with corresponding hydrogen indices ranging from 45 to 273. These give averages of 4.67 mg/g rock and 144, respectively, which are moderate values and suggest a type III kerogen. It should, however, be taken into account that the samples analysed are early to peak mature, indicating that the pyrolysis values are depressed compared with similar samples from an immature setting.

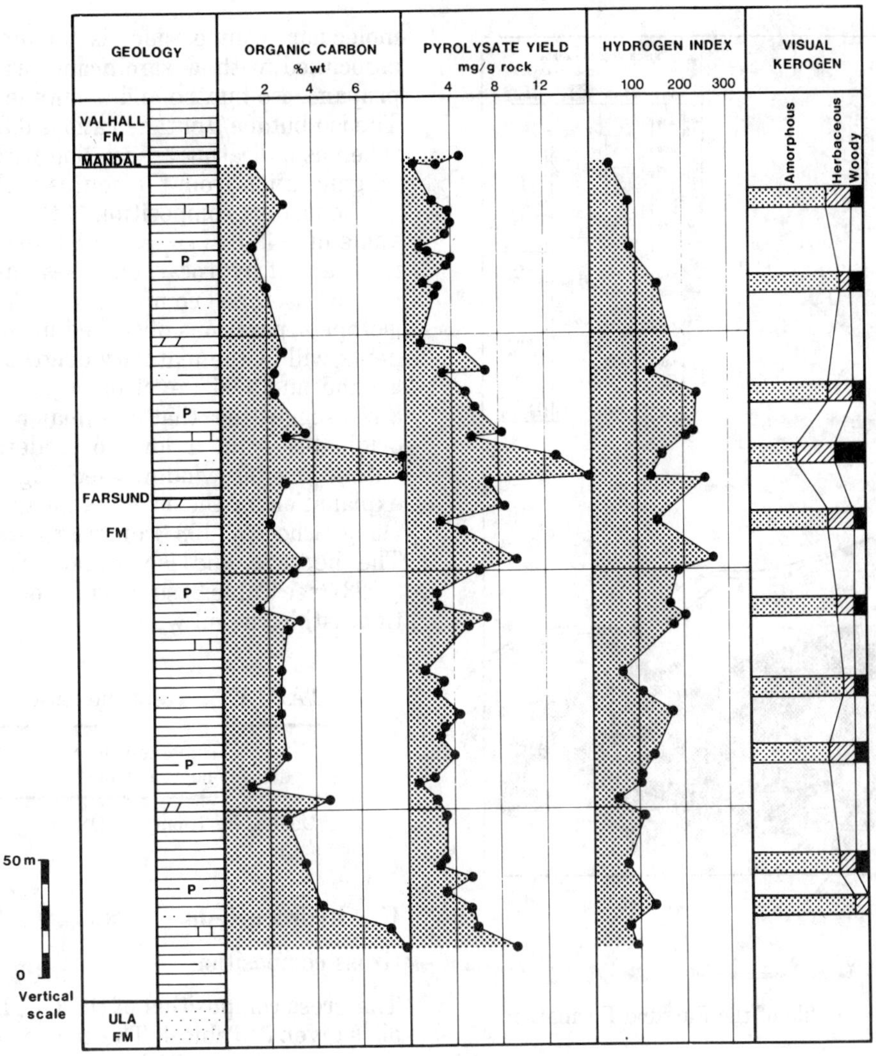

Fig. 9. Maturity map, Top Bryne Formation.

Pyrolysis–GC analysis, on the contrary, indicates kerogen type II. A pyrogram shows n-alkane/alkene homologues ranging beyond n-C_{20}, and the abundance of aromatic compounds is low relative to the aliphatics.

The organic matter composition is dominated by amorphous organic components associated with a significant proportion of herbaceous and minor amounts of woody/coaly material. The amorphous components show predominantly weak brownish fluorescence, in parts with more intense yellowish fluorescence, indicating liquid hydrocarbon potential. The hydrocarbon potential of the Farsund Formation can therefore presumably be rated as moderate for both oil and gas.

The maturity maps at base Cretaceous and the top Bryne Formation (Figs. 7, 8 and 10) together provide information on the maturity setting of the Farsund Formation. The patterns indicate lateral variation from early mature to late mature, implying that the Farsund Formation is an active and relatively voluminous source rock even within the block area. In addition, as is indicated in Fig. 5, the Farsund Formation in well 2/2–3 interacts favourably with the lower part of the Ula Formation in well 2/2–1. This type of setting is ideal with respect to lateral drainage from the source rock directly into the reservoir rock. This, combined with effective cap rocks, can explain the location of the 2/2–1 oil accumulation in the lowermost part of the Ula Formation.

Bryne Formation

The Bryne Formation is developed in wells 2/2–1 and 2/2–2, with thicknesses of 65 and 78 m, respectively. The sequence consists of alternating coal, shale and sandstone. The coal and carbonaceous shale, characterized by densities less than 2.0 g/cm^3, make up a cumulative thickness of only 14 and 5.5 m, respectively, in the two wells. This suggests that their source rock volume is restricted, and that the unit is only a secondary source for gas.

DETECTION OF MIGRATED HYDROCARBONS

The headspace gas composition over the pre-Chalk section in the three wells is shown in Fig. 11 and indicates the presence of thermogenic hydrocarbon gases.

In wells 2/2–1 and 2/2–3 the abundance of gas is low to moderate throughout the section. The wetness (C_{2+}/C_{1+}), however, is as high as 0.7, and the isobutane/n-butane ratio is mostly lower than 0.5. These characteristics can confidently be interpreted in terms of a thermogenic origin for the headspace gas.

In well 2/2–2, on the other hand, the wetness (C_{2+}/C_{1+}) is mostly below 0.2 and the iso-butane/n-butane ratio is above 1.0. This points towards a predominantly immature origin for this headspace gas.

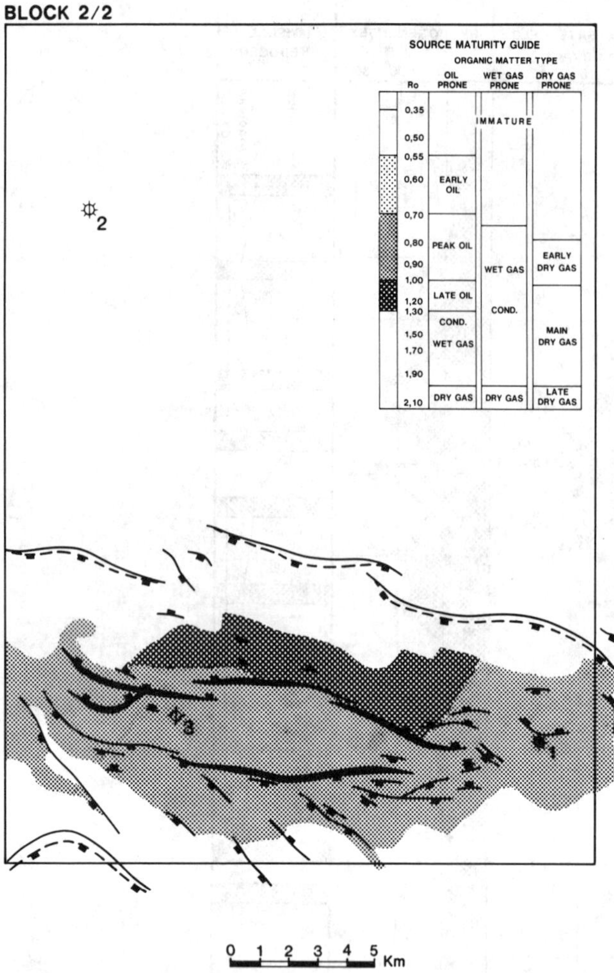

Fig. 10. Source rock profile of the Farsund Formation.

Moreover, it suggests that the Jurassic section in well 2/2–2 is devoid of thermogenic and migrated hydrocarbons.

GEOCHEMISTRY OF THE 2/2–1 OIL

Some bulk properties of the gaseous hydrocarbons from well 2/2–1 are given in Table 2. The oil from this well is light, having an API gravity of 44°, the sulphur content is low (0.15%) and the trace elements nickel and vanadium are present in very low quantities (1 mg/l).

TABLE 2. Gas composition, well 2/2–1

	Molecular composition (%)	*Carbon isotope composition* $\delta^{13}C$
Methane	68.9	−48.60
Ethane	16.8	−35.69
Propane	10.9	−31.29
Iso-butane	0.9	−29.34
N-Butane	2.4	−29.44

$(C_2 - C_4)/(C_1 - C_4) \times 100 = 31.1$.
Iso-butane/*N*-butane = 0.38.

Gaseous hydrocarbons

The molecular and carbon isotope composition of the hydrocarbon gases are also included in Table 2. The molecular composition is dominated by methane associated with a significant proportion of ethane, propane, iso-butane and *n*-butane, making up 31.1%. The iso-butane/*n*butane ratio is 0.38, a value normally taken as indicating association with the main phase of oil generation from the source rock.

The isotope composition of the methane with a $\delta^{13}C$ value of −48.6‰ supports a thermogenic origin. The full suite of hydrocarbon gases has been analysed for carbon isotope composition. The relatively wide isotope separations observed in the 2/2–1 gas components will fit the maturity diagram of James (1983) at around an LOM (level of organic metamorphism) of 8–9, suggesting that the major portion of the gas originates from a low to moderate maturity level within the oil window (see Fig. 12). An analogous explanation for the origin of the gas can be inferred by using Schoell's (1983) diagrams, as shown in Fig. 13. The heptane and iso-heptane values of Thompson (1983) (see Table 3) suggest a moderate thermal evolution within the oil window.

TABLE 3. Heptane ratios, well 2/2–1

Heptane-value[a]	22.6
Iso-heptane value[a]	1.28

[a]Defined by Thompson (1983).

C_{15+} hydrocarbons

Gross composition

The gross composition of the C_{15+} hydrocarbons of the oil is given in Table 4. The composition is highly paraffinic consisting of 79.0% saturates, 16.2% aromatics and 4.8% non-hydrocarbons.

Gas chromatography

A gas chromatogram of the saturated hydrocarbon is shown in Fig. 14. Carbon preference index and isoprenoid hydrocarbon ratios are given in Table 4.

The gas chromatogram shows a unimodal distribution with a maximum below n-C_{17}, a slight hump around n-C_{23} and extending out to n-C_{33}. There is a slight odd-over-even predominance beyond n-C_{21} (CPI = 1.1). Based on these features the oil appears to have a low to moderate maturity. On the other hand, the pristane/n-C_{17} ratio is only 0.4, and could indicate a higher maturity. The pristane/phytane ratio of 2.2 is relatively high for a North Sea oil, and is suggestive of a terrestrially-derived nature for the source rock organic matter.

TABLE 4. Well 2/2–1 oil

Gross composition (%)	
Saturates	79.0
Aromatics	16.2
Non-hydrocarbons	4.8
GC-ratios	
Carbon preference index	1.1
Pristane/phytane	2.2
Pristane/n-C_{17}	0.4
Carbon isotope composition ($\delta^{13}C$)	
Saturates	−27.4
Aromatics	−26.1

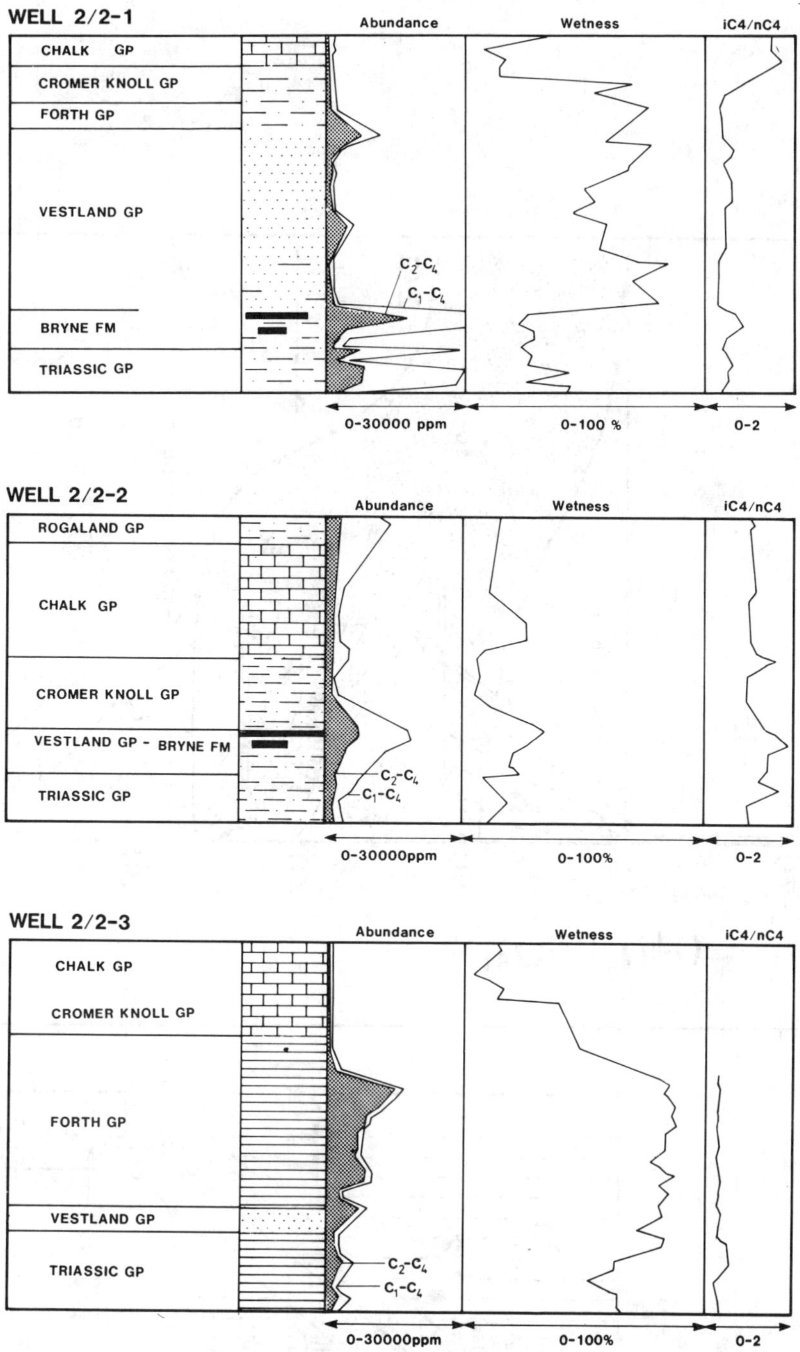

Fig. 11. Headspace gas composition over the pre-Chalk section in the wells 2/2–1, 2/2–2 and 2/2–3.

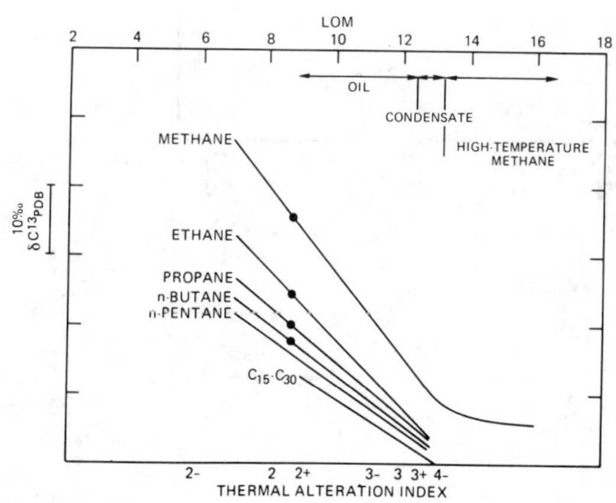

Fig. 12. Gaseous hydrocarbons from the 2/2–1 oil discovery plotted in James' (1983) maturity diagram.

Carbon isotope composition

The carbon isotope compositions for the C_{15+} fractions of the oil are included in Table 4. The $\delta^{13}C$ values of -27.4 and -26.1‰ respectively, for the saturates and aromatics are moderate for North Sea oils.

Biomarkers

In order to determine the source rock maturity for North Sea oils, Cornford *et al.* (1983) defined a set of ratio parameters based on the high carbon number biomarker molecules. The oils were grouped into three categories: 'early mature', 'fully mature' and 'condensate' as indicative of high maturity. A comparison of the 2/2-1 oil with this classification scheme is shown in Table 5. It is evident that the 2/2-1 oil fits into the early mature to fully mature category.

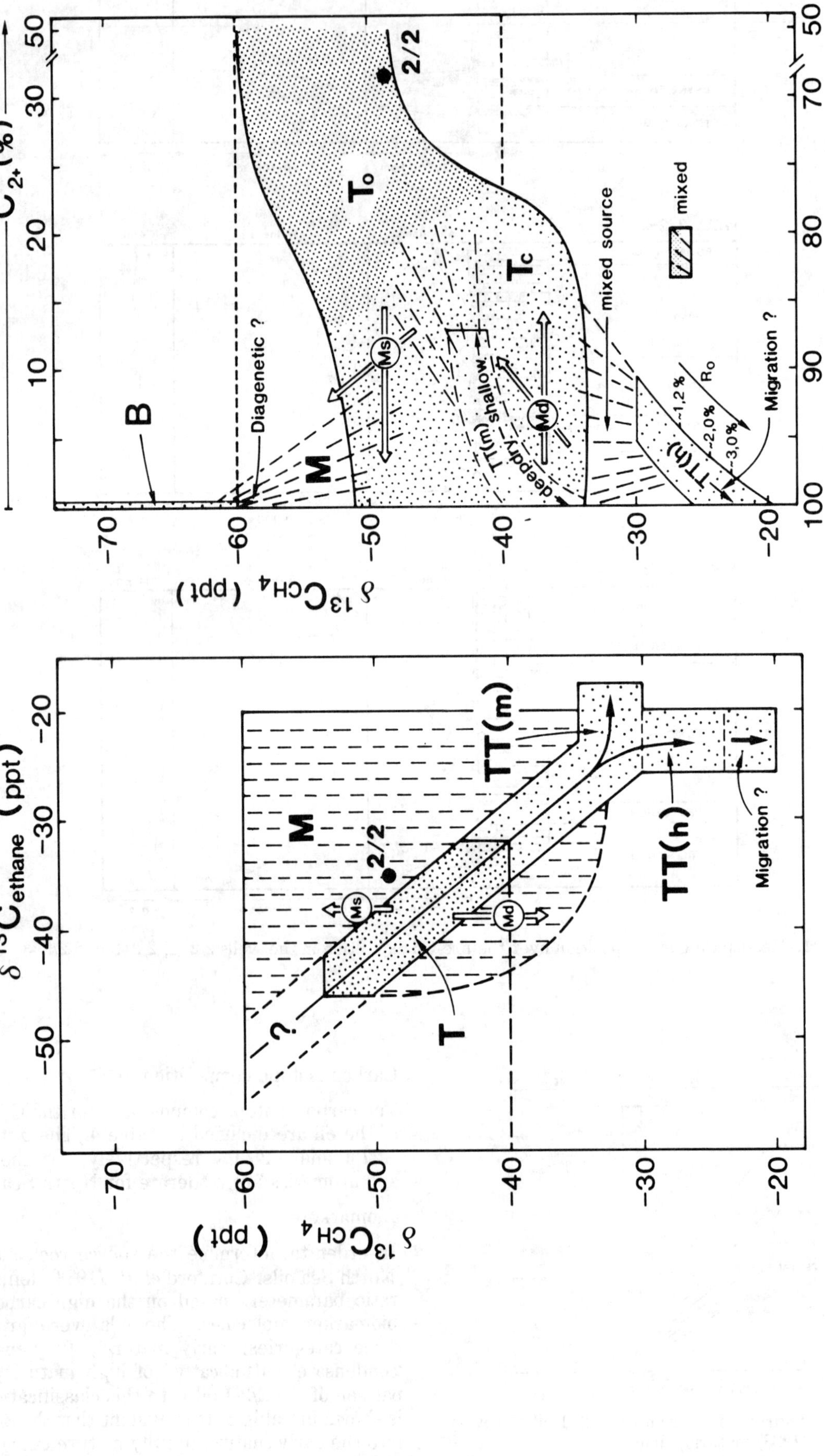

Fig. 13. Gaseous hydrocarbons from the 2/2–1 oil discovery plotted in Schoell's (1983) diagrams.

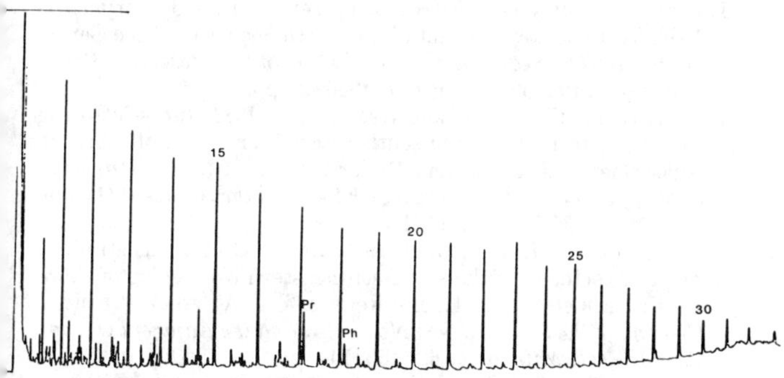

Fig. 14. Gas chromatogram of the saturated hydrocarbons in a stripped oil sample from the 2/2–1 discovery.

TABLE 5. Biomarker ratios, 2/2–1 oil

Biomarker ratio[a]	*Early mature*	*Fully mature*	*Condensate*	*2/2–1 oil*
Tt 1	*40*	57	90	25
Tt 4	48	33	*25*	8
Tt 5	*12*	8	0	14
Tt 8	*61*	62	66	56
St 9	38	*67*	80	54

[a]After Cornford *et al.* (1983).

CORRELATION

It has been established that the Farsund Formation is the prime candidate for the possible source rock for 2/2-1 oil. It is present in the area in a significant volume, it is mature and has a favourable setting with respect to effective drainage from source rock into the reservoir. In order to test this hypothesis, a correlation between the 2/2-1 oil and the Farsund Formation in well 2/2-3 was attempted.

Hydrocarbons extracted from the Farsund Formation show a front-biased n-alkane distribution, slight odd-over-even predominance beyond n-C_{21}, with pristane/n-C_{17} and pristane/phytane ratios of around 0.5 and 1.8, respectively. These features are also observed in the 2/2–1 oil, see Table 6.

One of the more reliable methods for oil-source correlation is the use of GC–MS 'fingerprints' of the high carbon number steranes and triterpanes. The present studies indicate that there is a high degree of similarity in biomarkers in the oil from well 2/2–1 and rock extracts from the Farsund Formation of well 2/2–3.

It may be argued that the extracted hydrocarbons are non-indigenous to the Farsund Formation, making the apparent correlation invalid. All of the samples have very high abundances of extractable hydrocarbons, averaging 2043 ppm of rock, and high hydrocarbon/organic matter and hydrocarbon/total extract ratios, averaging 0.082 and 0.69, respectively. These enhanced values can be interpreted in terms of good oil source capacity or, alternatively, in terms of non-indigenous hydrocarbons.

TABLE 6. Oil-source correlation

Parameter	*2/2–1 oil*	*Farsund Fm.*
Carbon preference index	1.1	1.1
Pristane/phytane	2.2	1.8
Pristane/n-C_{17}	0.4	0.5

In order to solve this problem a hydrous pyrolysis test was performed according to the method of Winters *et al.* (1983). In this experiment hydrocarbons are artificially produced from the source rock kerogen. A representative sample from the Farsund Formation was pyrolysed and the hydrocarbons produced were found to correlate fairly well with the previously extracted hydrocarbons, thus suggesting that the hydrocarbons are indigenous to the formation.

DISCUSSION AND CONCLUSIONS

Late Jurassic shales are generally considered to be the main source rock of the major oil discoveries in the North Sea, although the geographical spread between the fields are considerable (e.g. Barnard and Cooper, 1981). Early and middle Jurassic source rocks are also considered to be of importance in some areas (Fuller, 1975; Oudin, 1976). Despite general similarities in the history of hydrocarbon maturation and migration over large parts of the North Sea graben system, more detailed investigations often reveal subtle differences in timing and development between the various parts of the area.

In Block 2/2 the significance of local structural and sedimentological events is clearly demonstrated. Firstly, the distribution of source rocks and reservoir rocks is controlled by the structural development along the strike of the Hidra Fault Zone. The structuring was in part governed by halokinesis, which in turn was dependent on the configuration and sedimentary thickness variations in the Zechstein basin. Local variations in development of the Hidra Fault Zone caused considerable facies variations within the 'Ula Trend' throughout the Jurassic, when the Middle Jurassic Bryne Formation and the Late Jurassic Tyne Group (which includes the source rocks of the Farsund and Mandal Formations) were deposited. Simultaneously, the reservoir sands of the Ula Formation were developed adjacent to the most active parts of the fault zone. This has caused a geometrical juxtaposition of potential source rocks of the Farsund Formation and the reservoir rocks of the Via Formation which is very favourable for effective hydrocarbon drainage. On the other hand, communication between the upper part of the potential source sequence (the Mandal Formation) and the reservoir rocks is not favoured.

The Upper Jurassic Mandal and Farsund Formations classify as marine oil-prone source rocks, whereas the Middle Jurassic Bryne Formation is a coal-bearing gas-prone source rock. The Mandal and Bryne Formations are only developed as thin intervals within the block area, and significant hydrocarbon contribution from these units has to rely upon long-distance migration from outside the block area. The Farsund Formation, on the other hand, has a more favourable setting. It is present within the block area in a significant volume, it is mature and has a close relationship with Ula Formation sandstones which favours effective drainage into the reservoir.

The 2/2–1 oil is of low to moderate maturity and appears to originate from a source rock with a significant terrestrial input. Good indications are given of a positive correlation between the oil and the thick Farsund Formation in well 2/2–3. It should, however, be pointed out that this does not rule out the possibility of other source rocks, for instance a thicker development of the Mandal Formation.

ACKNOWLEDGEMENTS

The authors would like to thank Arco Norway Inc., Mobil Exploration, Norway, Norsk Hydro A. S., Statoil and Saga Petroleum a.s. for permission to publish this paper.

As indicated in the introduction, several people were involved in the study which formed the basis for the present paper, and we wish to thank them all for enthusiasm and useful discussions during the work. Jill-Renée Sonrier Mørk and Liv Ravdal typed and drafted several versions of the manuscript and the figures, respectively.

REFERENCES

Bailey, C. C., Price, I. and Spencer, A. M. 1981. The Ula Oil Field, Block 7/12, Norway. In: *Proc. Norwegian Symp. on Exploration*, Bergen, 1981. Norwegian Petroleum Society, NSE/18, 26 pp.

Barnard, P. C. and Cooper, B. S. 1981. Oils and source rocks of the North Sea Area. In: Illing, L. V. and Hobson, G. D. (eds.), *Petroleum Geology of the Continental Shelf of North-West Europe*. Institute of Petroleum, London, pp. 169–175.

Cornford, C., Morrow, J. A., Turrington, A., Miles, J. A. and Brooks, J. 1983. Some geological controls on oil composition in the U.K. North Sea. In: Brooks, J. (ed.), *Petroleum Geochemistry and Exploration of Europe. Geol. Soc. Spec. Publ. No. 12.* 71–93. (Blackwell Scientific, Oxford.)

Deegan, C. E. and Scull, B. J. 1977. A proposed standard lithostratigraphic nomenclature for the Mesozoic of the Central and Northern North Sea. *Rep. Inst. Geol. Sci., No. 77/25. Norwegian Petrol. Directorate Bull.*, **1**, 36 pp.

Fuller, J. G. C. M. 1975. Jurassic source rock potential and hydrocarbon correlation, North Sea. in: Finstad, K. G. and Selley, R. C. (eds.), *Proc. Jurassic Northern North Sea Symp.* Norwegian Petroleum Society JNNSS/11, 1–18.

Hamar, G. P., Jacobsen, K. H., Ormaasen, D. E. and Skarpnes, O. 1980. Tectonic development of the North Sea north of the Central Highs. In: *The Sedimentation of the North Sea Reservoir Rocks*. Norwegian Petroleum Society, 1980, 23 pp.

Hamar, G. P., Fjæran, T. and Hesjedal, A. 1983. Jurassic stratigraphy and tectonics of the southeastern Norwegian offshore. In: Kaasschieter, J. P. H. and Reijers, T. J. A. (eds.), *Petroleum Geology of the Southeastern North Sea and the Adjacent Onshore Area. Geol. Mijnbouw*, **62**, 103–114.

Hesjedal, A. and Hamar, G. P. 1983. Lower Cretaceous stratigraphy and tectonics of the south-southeastern Norwegian offshore. In: Kaasschieter, J. P. H. and Reijers, T. J. A. (eds.), *Petroleum Geology of the Southeastern North Sea and the Adjacent Onshore Areas. Geol. Mijnbouw*, **62**, 135–144.

James, A. T. 1983. Correlation of natural gas by use of carbon isotopic distribution between hydrocarbon components. *Am. Ass. Petrol. Geol. Bull.*, **67**, 1176–1191.

Oudin, J. 1976. Geochemical study of the North Sea basin. *Bull. Centre Rech., Pau.*, **10**, 339–358.

Rønnevik, H. C., van den Bosch, W. and Bandlien, E. H. 1975. A proposed nomenclature for the main structural features in the Norwegian North Sea. In: Finstad, K. G. and Selley, R. C. (eds.), *Proc., Jurassic Northern North Sea Symp.* Norwegian Petroleum Society, JNNS/18, 6 pp.

Schoell, M. 1983. Genetic characterization of natural gases. *Am. Ass. Petrol. Geol. Bull.*, **67**, 2225–2238.

Skjerven, J., Rijs, F. and Kalheim, J. E. 1983. Late Palaeozoic to Early Cenozoic structural development of the south-southeastern Norwegian North Sea. In: Kaasschieter, J. P. H. and Reijers, T. J. A. (eds.), *Petroleum Geology of the Southeastern North Sea and the Adjacent Onshore Areas. Geol. Mijnbouw*, **62**, 35–45.

Thompson, K. F. M. 1983. Classification and thermal history of petroleum based on light hydrocarbons. *Geochim. Cosmochim. Acta*, **47**, 303–316.

Vollset, J. and Doré, A. G. (eds.) 1984. A revised Triassic and Jurassic lithostratigraphic nomenclature for the Norwegian North Sea. *NPD-Bull. No. 3*, 53 pp.

Winters, J. C., Williams, J. A. and Lewan, M. D. 1983. A laboratory study of petroleum generation by hydrous pyrolysis. In: Bjorøy, M. *et al.* (eds.), *Advances in Organic Geochemistry 1981*. Wiley, Chichester, pp. 524–533.

14 Organic matter quality changes in an Upper Jurassic shale sequence from the Viking Graben

A. Y. Huc, H. Irwin† and M. Schoell‡*
*Institut Français du Pétrole, France
†Statoil, Stavanger, Norway
‡Aemeler Strasse 3, D3000 Hannover, FRG

Changes in the characteristics of the organic matter in a 120 m thick relatively immature (T_{max} = 428–436°C) section of Upper Jurassic shales from a borehole in the Norwegian sector of the Viking Graben are presented. Cuttings and sidewall cores have been analysed. Organic carbon concentration remains fairly high throughout the interval studied and averages 5–6%. However, the quality of the organic matter exhibits a dramatic change with depth, which is evident as a sharp drop in the hydrogen index (from 600 to 200) and H/C atomic ratio of the kerogens (from 1.25 to 0.80) across the boundary between the Draupne and Heather Formations. This change in kerogen type is associated with a decrease in bitumen content and modifications in the extractable hydrocarbons including biomarkers. In particular, 17α(H),18α(H),21β(H)-C_{28}-C_{30} bisnorhopane is absent in the 'hot' Draupne shales, is present in the underlying 'cold' shales and increases downwards within the Heather Formation. A large change in isotopic composition is clearly related to this organic quality pattern. The upper hot shales are depleted in ^{13}C ($\delta^{13}C = -30‰$) and in deuterium ($\delta D = -125‰$); this gradually changes downwards into the lower shales to a $\delta^{13}C$ of $-24‰$ and a D-value of $-100‰$, i.e. a comparative enrichment in carbon-13 and deuterium.

INTRODUCTION

The Kimmeridge Clay Formation in the Viking Graben of the North Sea consists generally of dark grey-brown to black carbonaceous, radioactive claystone. The formation varies considerably in thickness, exceeding several hundred metres in the central part of the graben, but may be missing due to non-deposition or erosion on the outer edges of the basin and on the crest of some structural blocks. These shales are often not differentiated in terms of organic type, and are frequently considered as homogeneous good source rock.

As regional geochemical studies progress it becomes obvious that such a statement can no longer be supported. Demaison *et al.* (1983) have shown, on the basis of over 2200 analyses by Rock–Eval pyrolysis and total organic carbon (TOC) content, that large variations in kerogen type can be observed laterally on a basinal scale. Moreover, they pointed out that these changes in kerogen types can also be recognized along a vertical dimension in individual boreholes. Such an observation cannot be ignored in future basin geochemical studies, particularly when prediction of regional source bed occurrence and organic facies variations is required by the explorationists.

So it is now evident that more information is needed on the time–space distribution of kerogen types and quantity. Moreover, such studies would permit a deeper insight into the relationship between depositional factors and organic matter composition. This study provides an example of vertical changes in characteristics of the organic matter in 120 m of an Upper Jurassic shale sequence penetrated by a borehole in the Viking Graben area.

THE SEQUENCE

The sequence examined (see Fig. 1) encompasses the Draupne Formation (i.e. Kimmeridge Clay equivalent) and the top of the Heather Formation. It is dominantly a sequence of interbedded marine shales and claystones, but changes to siltstone at the base, when limestone and sandstone interbeds also appear. The typical 'hot' shales occur at the top of the sequence but the gamma-ray response drops dramatically at 2625 m, further decreases gradually to 2655 m, and then remains fairly uniform down to the base of the sequence studied. The maturity of the sequence lies between 0.5 and 0.55% R_0 and can be termed immature to early mature.

Petroleum Geochemistry in Exploration of the Norwegian Shelf © Norwegian Petroleum Society (Graham & Trotman, 1985) pp. 179-183.

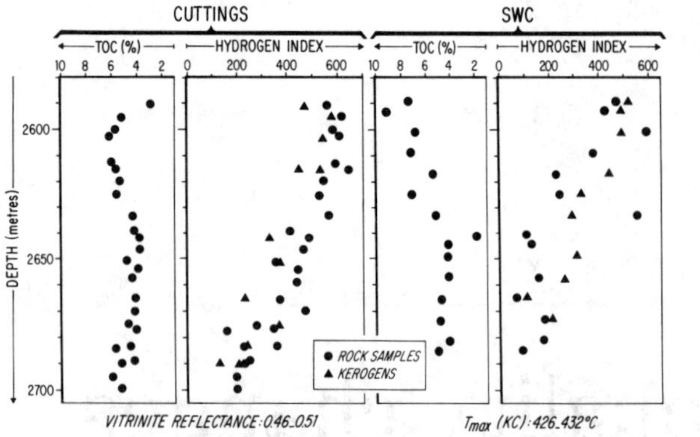

Fig. 1. Vertical distribution of total organic carbon (TOC) and hydrogen index and a comparison between cuttings and sidewall samples.

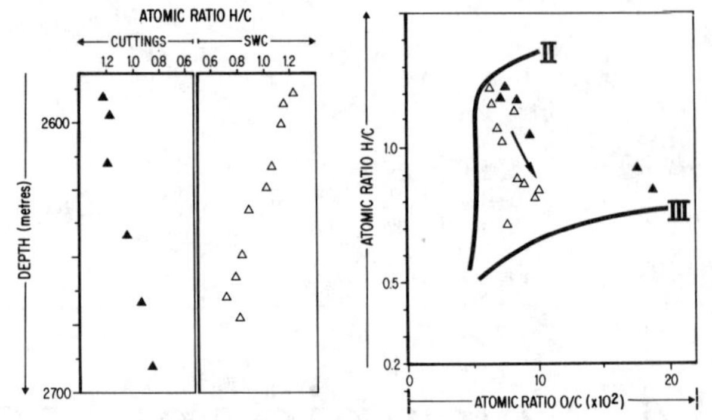

Fig. 2. Variation in the elemental composition of kerogens.

EXPERIMENTAL

The present study is based on analyses of cuttings and of sidewall cores. Organic characterization on whole rock samples has been carried out by Rock–Eval pyrolysis and total organic carbon (TOC) content measurements.

Kerogens have been isolated by destruction of minerals by acid attack then subjected to Rock–Eval pyrolysis, elemental analysis and stable isotope (^{13}C and D) analyses.

The powdered shales were extracted using CH_3Cl, then the extract was separated into saturated, aromatic and NSO fractions by thin-layer chromatography. The saturated fraction was analysed by GC and GC–MS techniques.

Clay mineralogy was carried out using standard X-ray diffraction procedures on oriented samples.

RESULTS AND DISCUSSION

TOC and Rock–Eval pyrolysis data on rock samples and kerogens are presented in Fig. 1. In the 120 m section selected, the total organic carbon is fairly high and lies in the range 3–8%. Hydrogen index values of the rock samples and the kerogens are close to each other and thus the mineral matrix effect (Espitalié *et al.*, 1980) is limited in this set of samples. A striking feature is the regular decrease of the hydrogen index, from 600 to 200, as a function of depth. It should be noted that the organic matter in this section is relatively immature, exhibiting Rock–Eval T_{max} values of 426–432°C. Thus, variations in the hydrogen index reflect changes in the nature of organic matter, and are only very slightly affected by thermal alteration. Moreover, the depth interval of only 120 m precludes any significant thermally-induced differences between the top and the bottom of the section.

When comparing the trends of decreasing hydrogen index between cutting samples and sidewall cores (see Fig. 1), there are significant differences. Both data sets are regularly decreasing with depth, but the decline is sharper for the sidewall core samples. This observation suggests that caving from the uppermost part of the section might be contaminating the downhole cuttings samples, causing a smoothing of the trend. This interpretation is consistent with indications of caving based on a caliper log examination. However, the caving effect does not obscure the general trend of the organic variation which is indicated by both sets of samples.

The kerogen elemental composition mirrors the hydrogen index trend: the atomic H/C ratio declines from 1.2 to 0.8 with increasing depth, and probably indicates that the deepest kerogen samples were deposited in a more oxygenated environment than their shallower counterparts (see Fig. 2). The elemental compositions indicate a transition from kerogens comparable with type II toward kerogens of type III, as defined by Tissot *et al.* (1974). A comparison of H/C ratio values over the depth interval studied confirms the differential behaviour between cuttings and sidewall core samples, as discussed earlier (see Fig. 2).

The isotopic composition of sidewall core kerogens provides additional evidence of a regular variation in the type of organic matter, as there is a progressive and significant enrichment in both carbon-13 ($\delta^{13}C$ from −30 to −24‰) and in deuterium (δD from −125 to −100‰) with increasing depth (see Fig. 3).

The ultimate causes of isotopic variation in kerogens remain speculative (Schoell, 1984). Whatever the reason for the changes, they reflect modifications in the organic material which are indicated by a technique which is totally independent of Rock–Eval pyrolysis and kerogen elemental analyses.

A reliable interpretation of organic facies from the study of a bitumen requires that neither thermal effects nor migration phenomena have altered the extract composition. The present example seems to fulfil such conditions. Firstly, T_{max} values supported by elemental analyses of kerogen indicate that the organic matter can be considered to be relatively immature, i.e. before the oil-generation zone. This situation minimizes the chance of thermal alteration of the initial organic molecules, their dilution by thermally generated hydrocarbons, and also the possibility

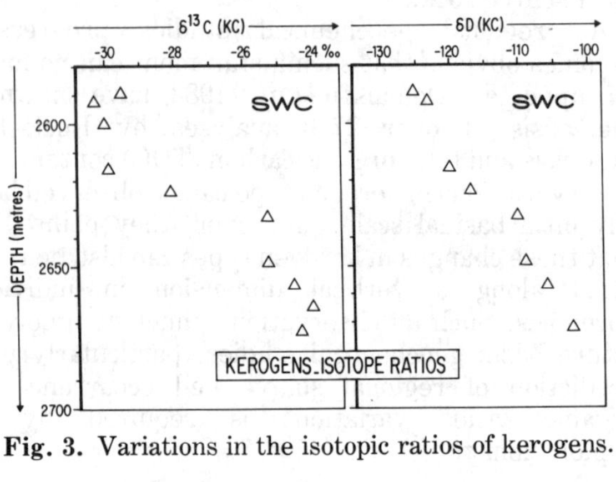

Fig. 3. Variations in the isotopic ratios of kerogens.

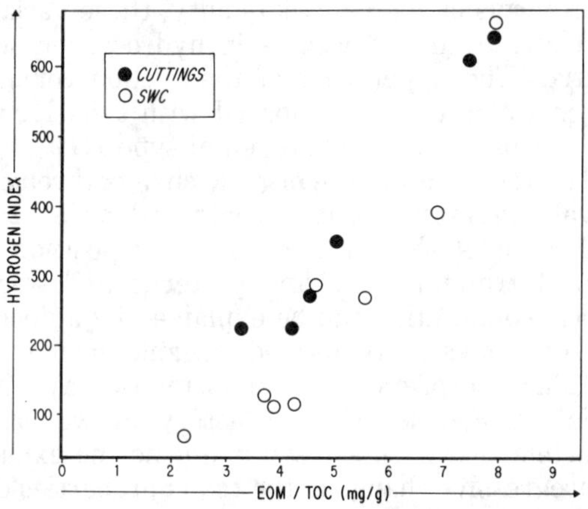

Fig. 4. Extractable organic matter (EOM) content versus hydrogen index.

of migration effects. Furthermore, a clear relationship is found between the hydrogen index of kerogens and the extractable organic matter (EOM) contents of the related sediments (see Fig. 4). Thus, there is no evidence of noticeable hydrocarbon accumulation of depletion due to migration, at least in the C_{15+} fraction.

Assuming an *in situ* origin for the bitumen, it should be noted that a regulated decrease in the hydrocarbon content mirrors the decrease of hydrocarbon potential of the kerogen as a function of depth in the studied section (see Fig. 5).

This quantity trend is reinforced by a quality trend: the *n*-alkane pattern exhibits a regular gradation between a distribution with a marked preference for low to medium molecular weight *n*-alkanes (*n*-C15 to *n*-C20) at the top, and a distribution with increased high molecular weight *n*-alkanes (*n*-C25 to *n*-C37) with a pronounced odd–even carbon predominance at the bottom of the section. The latter distribution is inherited from terrestrial higher plants while the former is usually associated with marine organisms. Thus, the changes in *n*-alkane 'fingerprints' may be interpreted as changes in organic input; the importance of the terrestrial contribution increases in the lower part of the Draupne and in the Heather Formation. This interpretation is consistent with the variations in the chemical composition of the kerogens. The parallel variations in kerogen properties and the extractable hydrocarbon patterns is further evidence for an absence of migration phenomena in the interval.

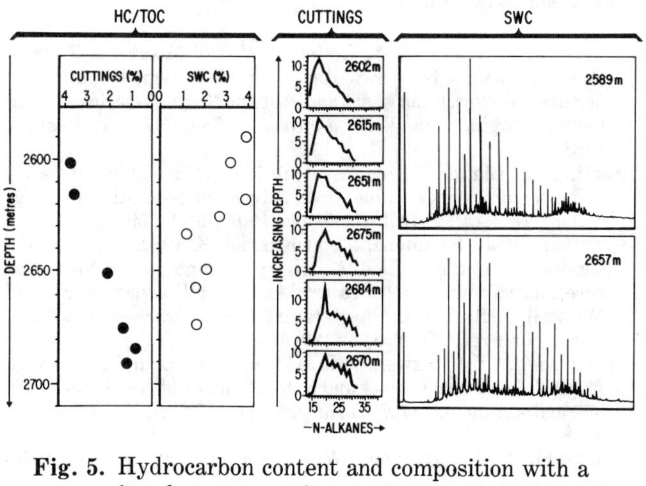

Fig. 5. Hydrocarbon content and composition with a comparison between cuttings and sidewall samples.

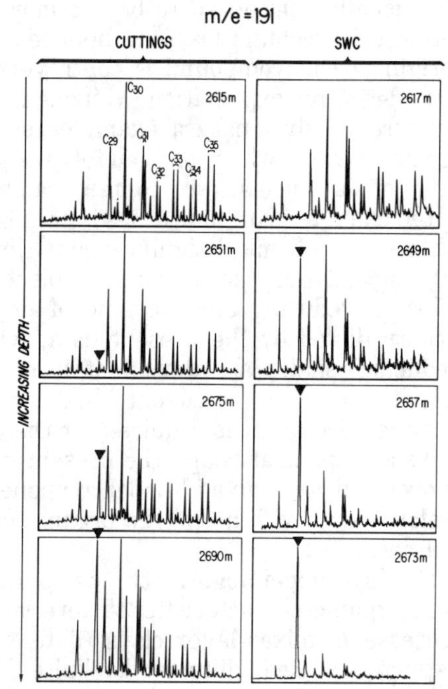

Fig. 6. Variations in the C_{28}-bisnorhopane relative content in mass chromatograms m/e = 191, with a comparison between cuttings and sidewall samples.

The parallel variations described above are also reflected in the occurrence of the C_{28} nuclear-demethylated triterpane: 17α(H),18α(H),21β(H)-28-30-bisnorhopane (Seifert *et al.*, 1978). Bisnorhopane is frequently observed in North Sea oils and sediments (Mackenzie *et al.*, 1983) and has been identified by GC–MS in most of the present samples (see Fig. 6). Because it is relatively uncommon, it is thought to be useful as an oil-source correlation parameter.

Depth-related variations in the relative proportions of this C_{28} hopane in m/e 191 terpane 'fingerprints' is quite spectacular. It is missing in the 'hot' Draupne shales, and becomes the major peak in the deepest levels of the Heather Formation where it even tends to become the only peak in the sidewall core samples (see Fig. 6). It has been documented that the relative abundance of bisnorhopane decreases with increasing maturity (Grantham *et al.*, 1980; Moldowan and Seifert, 1984). The fact that oils in the study area often contain a higher relative abundance of bisnorhopane than the immature source rocks is difficult to explain. It is even more puzzling that the Draupne shales,

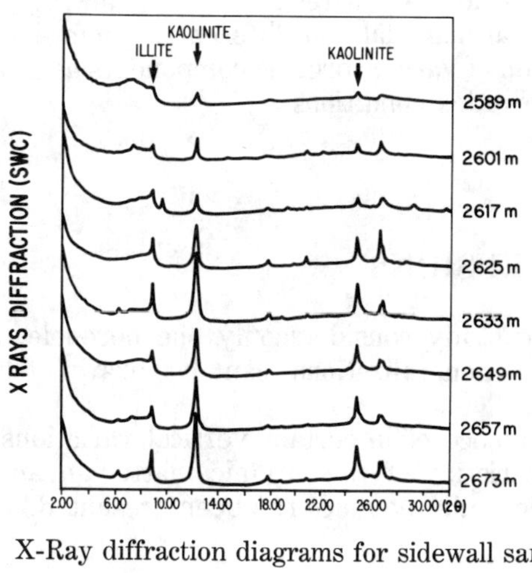

Fig. 7. X-Ray diffraction diagrams for sidewall samples.

which are usually considered to be the most favourable source rocks, contain no bisnorhopane at all.

The origin of this compound is controversial, and there is no definitive explanation available in the current literature. Study of natural examples, such as the one presented herein, where one can follow significant variations of a biomarker content along with documented changes in organic facies is possibly a way to decipher its geochemical significance. Grantham *et al.* (1980) suggest that the presence of bisnorhopane may reflect a reducing environment of deposition. This is contradicted by the present data, where the most 'reducing' shales are devoid of bisnorhopane. However, it cannot be ruled out that a particular depositional environment is required for the development of bisnorhopane although the present data suggest the prerequisite of an additional component, such as the relationship with a specific organic facies, as described here.

Vertical clay mineralogical trends parallel the changes in organic characteristics. There is a downward decrease in mixed-layer clays (illite–smectite) and a distinct increase in illite and kaolinite. Traces of chlorite appear at 2633 m and below. The progressive decline in illite–smectite and the appearance of chlorite can be explained by diagenesis, but the increase in illite and kaolinite is most readily explained by an increase in detrital components. Particularly, increased amounts of kaolinite can indicate proximity to land.

The clay mineral trend thus reflects the transgressive sequence of basin development from deltaic conditions in the underlying Brent Group, through a progressively deepening marine basin of the Heather Formation, leading finally to the widespread restricted environment represented by the Draupne Formation. The vertical clay mineral distribution seen here is expected to be expressed laterally, reflecting facies variations within the depositional basin. The mineralogical data support the interpretation that organic matter changes are related predominantly to variations in the proportions of terrestrial material, which increase downwards in the sequence. However some of the characteristics of the organic matter may have also been influenced by the redox properties of the environment of deposition. The Heather Formation was deposited in oxic to sub-oxic bottom waters, whereas the Draupne Formation is supposed to have been deposited in anoxic bottom waters. Differential preservation in these two contrasting environments could account for the variations seen.

It is very likely that the variations in organic matter observed here are a result of both varying input of terrestrial material and differential/preferential preservation of various organic components as a result of varying redox conditions.

CONCLUSIONS

1. The study considers only one borehole, and no regional generalizations can be drawn from the results.

2. Evidence of important vertical variations in the characteristics of the organic matter in an Upper Jurassic shale sequence has been presented.

3. In terms of source rock quality, these variations reflect significant differences in hydrocarbon source potential. The upper part of the section contains a kerogen which can be compared with type II, while the lower part contains kerogen of type III.

4. The regularity of the organic change through the interval suggests that it is controlled by a gradual modification of the environment of deposition. The observed trend in *n*-alkane patterns and kerogen chemical composition can be explained by a declining input of terrestrially-derived organic matter with time. This interpretation is supported by clay mineral associations and depositional history. However, it is not certain that such a trend might not be explained by a progressive change in the redox properties of the depositional medium, the marine organic matter being destroyed in an oxidizing environment and the proportion of terrestrial organic matter therefore being preferentially increased under such conditions.

5. It is apparent from comparison of geochemical data from cuttings and sidewall cores that there is a sample quality problem. To some extent contamination by cavings affects the lower part of the sequence when cuttings data are used. In this example it does not significantly change the general observations, but it indicates that care should be taken when interpreting data obtained from cuttings samples. Alternatively, sidewall cores are unambiguous in terms of depth and caving pollution, but one should be aware that they sample specific horizons and may not provide representative values for geochemical parameters at the formation scale.

6. The kind of sequence which has been studied is possibly favourable for understanding the significance of C_{28}-hopane in sediments, because the occurrence and quantity of this molecule can be matched with a gradual variation in the environment of deposition. Thus, further studies of the sedimentological conditions during the deposition of this sequence would be of interest.

ACKNOWLEDGEMENTS

The authors thank Statoil and partners for their permission to publish these data.

REFERENCES

Demaison, G., Holck, A. J. J., Jones, W. and Moore, G. T. 1983. Predictive source bed stratigraphy; a guide to regional petroleum occurrence North Sea basin and eastern North American Continental Margin. *11th World Petroleum Congr. PD 1*, London, 1983.

Espitalié, J., Madec, M. and Tissot, B. 1980. Role of mineral matrix in kerogen pyrolysis; influence on petroleum generation and migration. *Am. Ass. Petrol. Geologists Bull.*, **64**(1), 59–66.

Grantham, P. J., Posthuma, J. and de Groot, K. 1980. Variation and significance of the C_{27} and C_{28} triterpane content of a North Sea core and various North Sea crude oils. In: Douglas, A. G. and Maxwell, J. R. (eds.), *Advances in Organic Geochemistry 1979*. Pergamon Press, Oxford, pp. 29–48.

Mackenzie, A. S., Maxwell, J. R., Coleman, M. L. and Deegan, C. E. 1983. Biological marker and isotope studies of North Sea crude oils and sediments. *11th World Petroleum Congr. PD 1* London, 1983.

Moldowan, J. M. and Seifert, W. K. 1984. Structure proof and significance of stereoisomeric 28,30-bisnorhopanes in petroleum and petroleum source rocks. *Geochim. Cosmochim. Acta*, **48**, 1651–1661.

Schoell, M. 1984. Stable isotopes in petroleum research. In: *Advances in Petroleum Geochemistry*, Vol. 1. Academic Press, 215–245.

Seifert, W. K., Moldowan, J. M., Smith, G. W. and Whitehead, E. V. 1978. First proof of structure of a C_{28}-Pentacyclic triterpane in petroleum. *Nature*, **271**, 436–437.

Tissot, B., Durand, B., Espitalié, J. and Combaz, A. 1974. Influence of nature and diagenesis of organic matter in formation of petroleum. *Am. Ass. Petrol. Geologists Bull.*, **58**(3), 499–506.

15 Organic geochemistry of the Oseberg Field (I)

B. Dahl and G. C. Speers
Norsk Hydro a.s., Research Centre, Bergen, Norway

Oseberg crude oil is typical of the highly paraffinic, high quality North Sea crudes with low sulphur, asphaltenes and trace metal contents. A tar mat exists in part of the Oseberg Field, and is considered to result mainly from *in situ* de-asphaltening of the crude followed by gravity segregation and concentration near the oil–water contact.

Shales from the Humber Group are the most likely source of Oseberg oil, with the Kimmeridge Clay Formation being the richest source rock. Conventional geochemical parameters are unsatisfactory for maturity assessments in well 30/6–1, so molecular parameters were employed. The latter also provide better indications of migrated hydrocarbons and contamination.

The Humber Group is not considered mature enough in the Oseberg area to have generated the reservoir oil. The major source area is most likely to the west and north-west of the field area.

INTRODUCTION

The Oseberg Field lies mainly in Block 30/6 (formerly called the 'Silver Block') to the west of Bergen in the Norwegian sector of the North Sea. A portion of the field extends into the northern sector of Block 30/9 (see Figs. 1 and 2).

The licence for Block 30/6 was awarded in April 1979 to a group comprising Statoil (operator), Elf, Total, Norsk Hydro, Mobil and Saga. Operatorship was transferred to Norsk Hydro in April 1982. In August 1982, the northern part of Block 30/9 was awarded to a consortium of Norsk Hydro (operator), Statoil and Saga.

The first well was drilled in the summer of 1979, and found gas at approximately 2330 m in Brent Group sandstones. The two following wells also encountered gas in the same formation. A fourth well established the presence of oil as well as gas, and showed that the Oseberg Field had a large oil leg. The maximum height of the gas column is approximately 320 m, and the oil column is approximately 200 m. To date, 15 wells have been drilled in Block 30/6 and three wells in Block 30/9.

GEOLOGICAL SETTING

The Oseberg Field lies on a structural feature called the Bergen High, west of the Horda Platform (see Fig. 1). The structural characteristics of the area are the result of tensional rifting which took place mainly during the Mesozoic. Fault trends are predominantly N–S to NNW–SSE, with some NE–SW rifting also evident, the latter believed to be due to reactivation of older Caledonian structural trends.

The Oseberg Field is situated in a rotated, eastward-dipping fault block resulting from the main tectonically active rifting phase (see Figs. 1 and 3). The reservoir is Brent Group sandstone which is truncated on the crest of the fault block, and the unconformity is overlain by Lower Cretaceous marls and clays which act as a caprock for the accumulation.

PETROLEUM GEOCHEMISTRY

Potential source rocks

Source rock evaluations have been made on samples from well 30/6–1 and four other wells in the area, and are summarized in Fig. 4. The evaluations were produced by a number of service organisations as well as Norsk Hydro Laboratories in Bergen.

The results of Rock–Eval screening of cuttings samples from the first Oseberg Well, 30/6–1, are shown in Fig. 5. This well was drilled on the crest of the Alpha structure, and the Kimmeridge Clay is absent at this location (lithostratigraphic nomenclature used according to Deegan and Scull (1977)). Below the Cretaceous unconformity, only a thin (approximately 30 m) section of silty Heather Formation is

Petroleum Geochemistry in Exploration of the Norwegian Shelf © Norwegian Petroleum Society (Graham & Trotman, 1985) pp. 185–195.

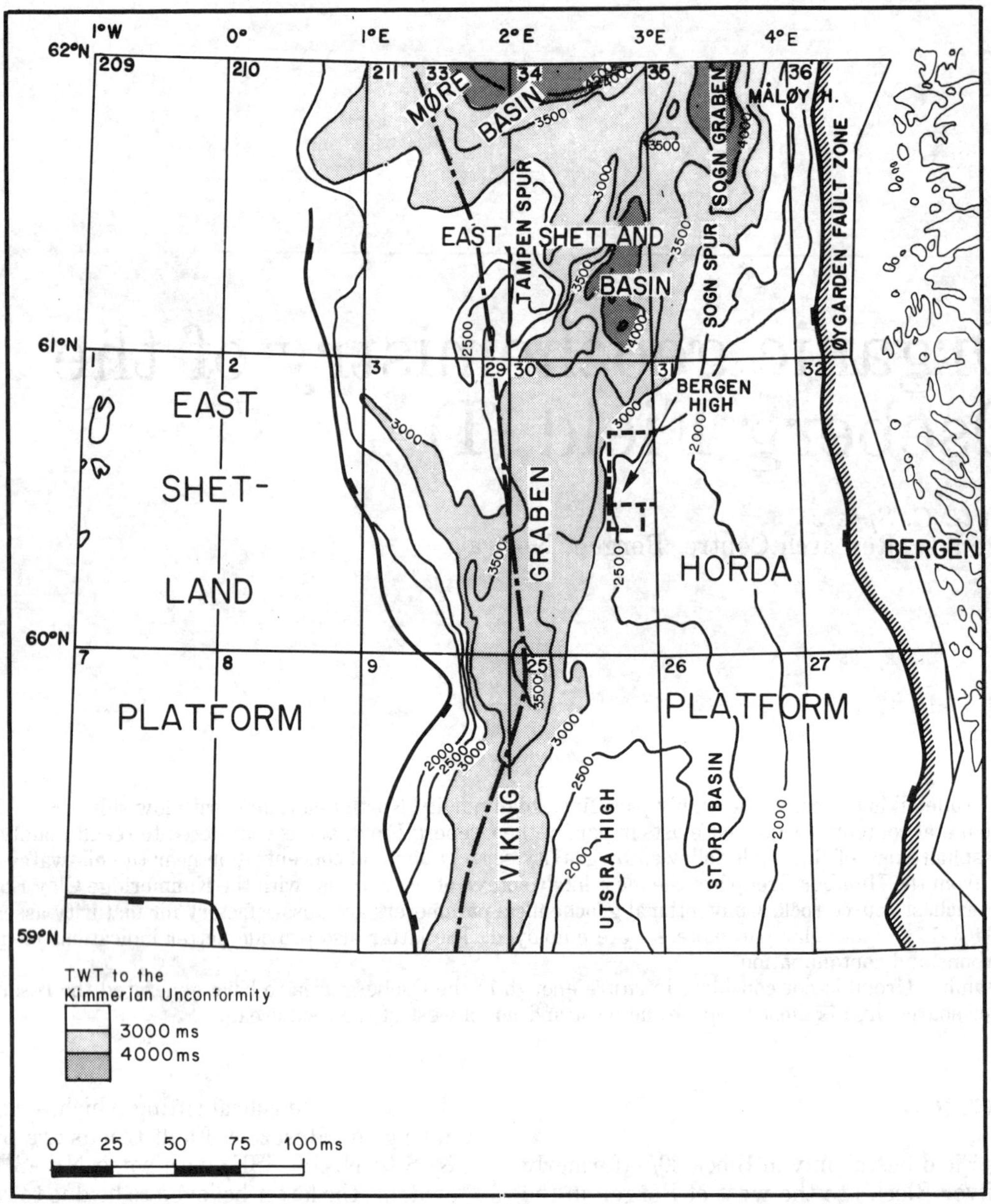

Fig. 1. Regional map of the Northern North Sea (59–62° N) showing location of the Bergen High (Oseberg Field).

present. Although the sampling density was rather low, the results shown in Fig. 5 indicate that only the Rogaland and Dunlin Groups of sediments have any significant hydrocarbon potential in this well.

Figures 4 and 5 show that, within the Paleocene Rogaland Group, the Balder and Sele Formations are of most interest. Total organic carbon contents (TOC) are in the range 1–1.5 wt%, and residual hydrocarbon potentials (S_2) are 3–3.7 kg/t. Rock–Eval hydrogen and oxygen indices for these formations, together with some results from the Lista Formation, are plotted on a modified Van Krevelen-type diagram in Fig. 6. The results suggest that these formations have little liquid hydrocarbon potential (probably gas/condensate), and that quality decreases with depth.

Data for the Lista Formation shown in Fig. 6 are linked with TOC values of <1.0 wt% and residual hydrocarbon potentials (S_2) < 1 kg/t. Kerogen from this formation would be classified as Type III only, but the scatter of oxygen index values within the samples is not considered to be due to maturity variations, but rather to the problems associated with determining this parameter in whole rock samples (e.g. Katz, 1983). The screening samples obtained from the Paleocene samples in 30/6–1 are consistent with results from the same formations in the Statfjord (Kirk, 1980) and Ekofisk areas (Van den Bark and Thomas, 1980).

Results from the Lower Jurassic Dunlin Group in 30/6–1 indicated some source potential in the Drake, Burton and Amundsen Formations. However, as the Burton and Amundsen Formations contain mainly siltstones and clays, the results shown in Fig. 5 probably give an unrealistic indication of overall formation source potential. Assessment of the source potential of the Burton and Amundsen Formations could be influenced by cavings from the overlying Drake Formation, but this was not checked by palaeontological studies.

The Drake Formation in 30/6–1 consists of silts and claystones interbedded with marls, and has a moderate hydrocarbon potential with S_2 values ranging from 1 to 5.8 kg/t and TOC contents in the range 1–2 wt%. The lower part of the Drake Formation has the highest hydrocarbon potential, hydrogen indices in this section indicating some liquid hydrocarbon potential

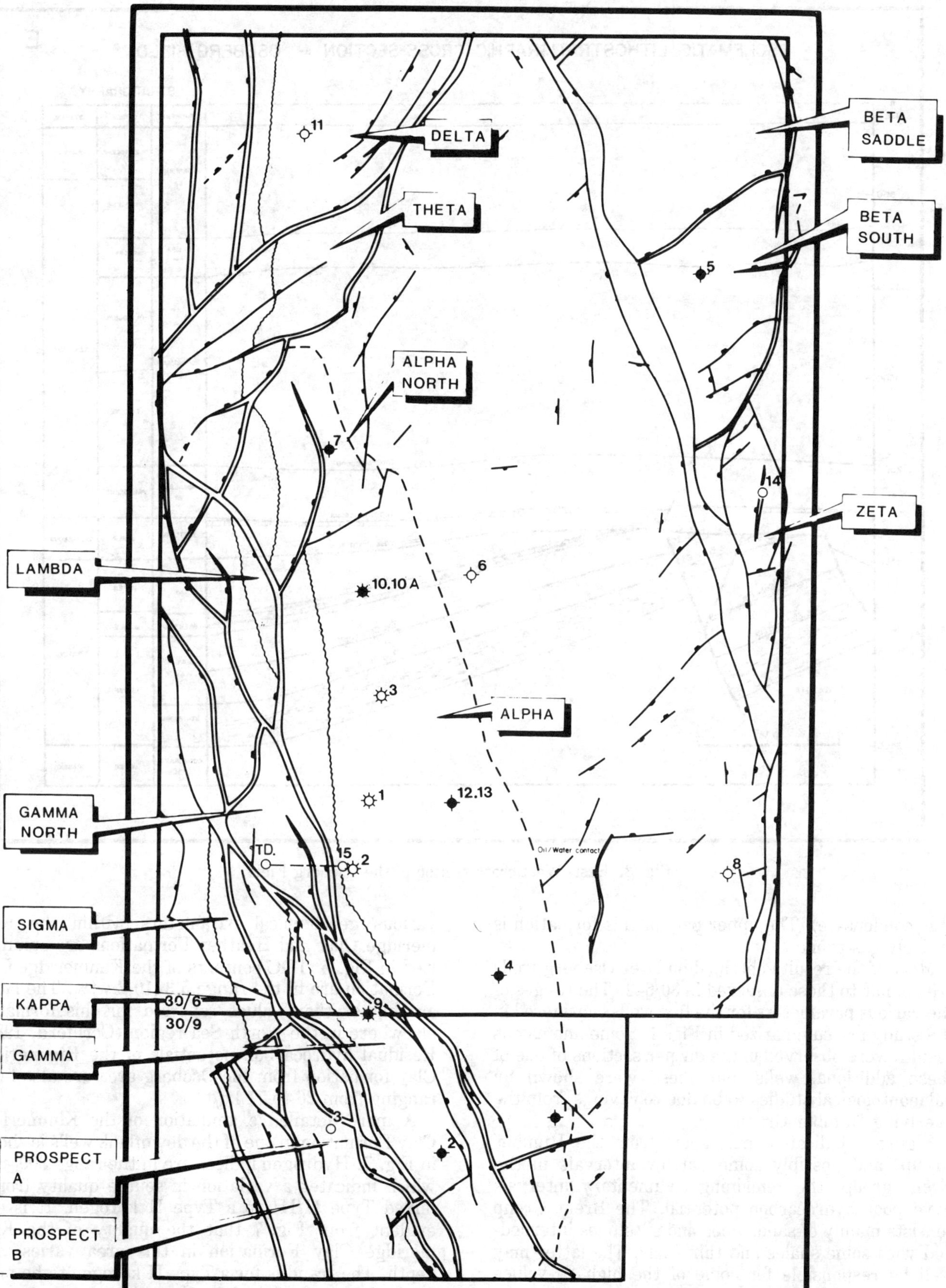

Fig. 2. Oseberg Field: structural map.

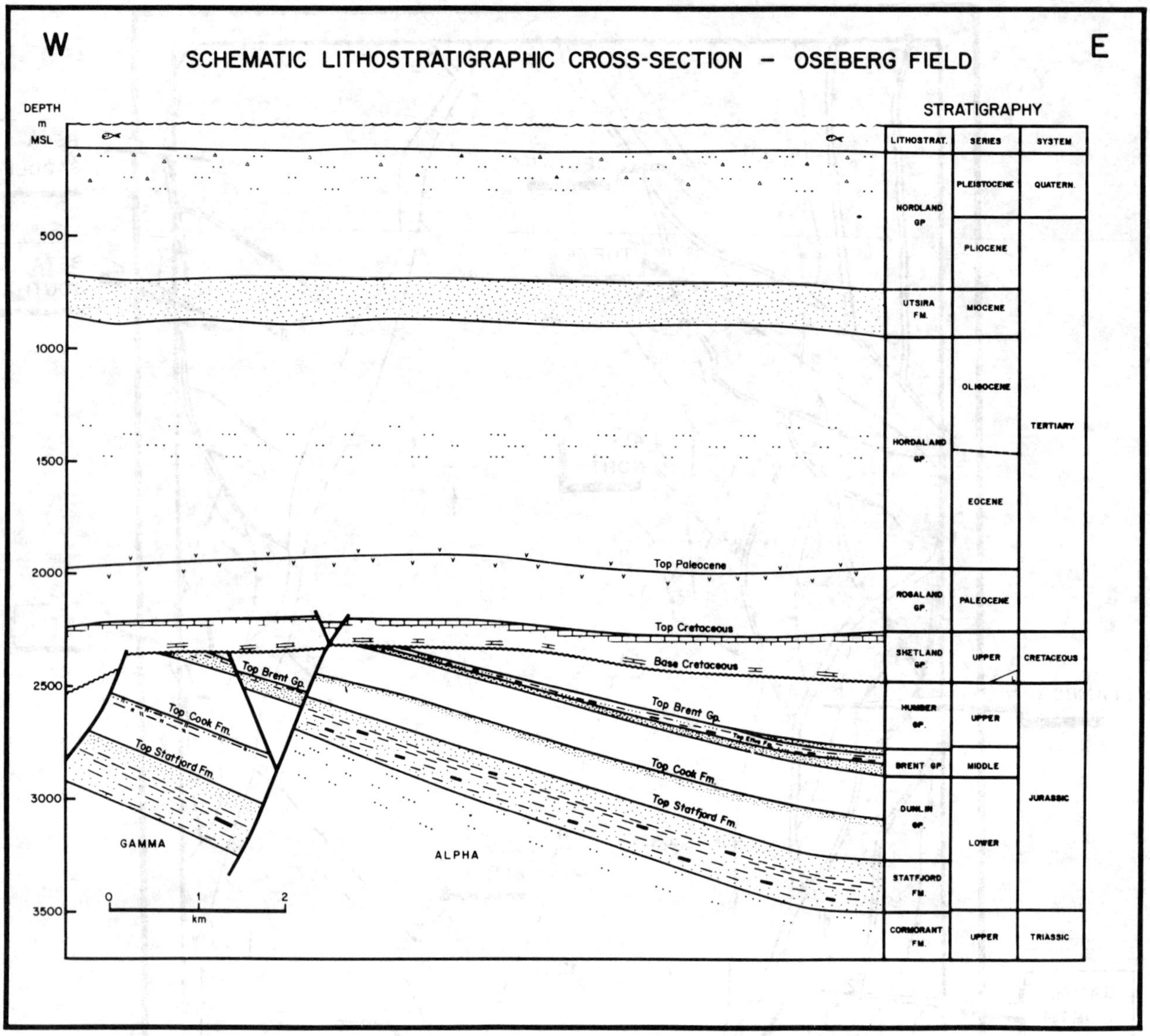

Fig. 3. East–West cross-section of the Oseberg Field.

(gas/condensate). The upper part of this formation is entirely gas-prone.

Most of the results obtained on later Oseberg wells are similar to those observed in 30/6–1. The ranges of the various parameters for the five wells considered in the study are summarized in Fig. 4. Some anomalous results were observed in the deeper sections of one of these additional wells, but these were shown by palaeontological studies to be due to cavings from the overlying Humber Group.

Figure 4 indicates that, apart from the Humber Group, and possibly some narrow intervals in the Brent group, the remaining sedimentary intervals have poor hydrocarbon potential. The Brent Group consists mainly of sandstones and siltstones interbedded with some shales and thin coals. The latter may well be responsible for some of the high S_2 values observed during screening studies.

The Statfjord Formation is even more shaly than the Brent Group, and also contains some thin coals. A poor, gas-prone potential is shown by the generally low S_2 values (1–2 kg/t) and hydrogen indices.

It has already been noted that the Kimmeridge Clay Formation is absent in well 30/6–1, but eastwards and downflank of the Oseberg structure the Upper Jurassic Humber Group thickens and up to 400 m of these sediments are encountered. The range of the various geochemical parameters within the Kimmeridge Clay and Heather Formations are summarized in Fig. 5. TOC contents of the Kimmeridge Clay Formation are in the range 5.3–10.2 wt%. The latter are higher than values reported for this formation elsewhere in the North Sea region (Cornford, 1984). Residual hydrocarbon potentials of the Kimmeridge Clay formation from the Oseberg area are also high, ranging from 26 to 71 kg/t.

A more detailed examination of the Kimmeridge Clay Formation in one of the downflank wells is shown in Fig. 7. Hydrogen indices are in the range 300–600, which indicates a variation in source quality from a mixed Type II/III to a Type II kerogen. It is also evident from Fig. 7 that the quality of the Kimmeridge Clay Formation in this area varies with depth. The trend is for a Type II kerogen to be present in the upper part of the formation, decreasing to a mixed Type II/III in the lower part of the section.

It should be noted that the results shown in Fig. 7 were obtained almost entirely on cuttings samples, and that the effect of cavings has probably resulted in an 'averaging' of quality indications throughout the formation. This process would tend to lower the apparent quality in the upper part of the formation, and increase it in the lower section. Unfortunately, only two sidewall cores were available for control pur-

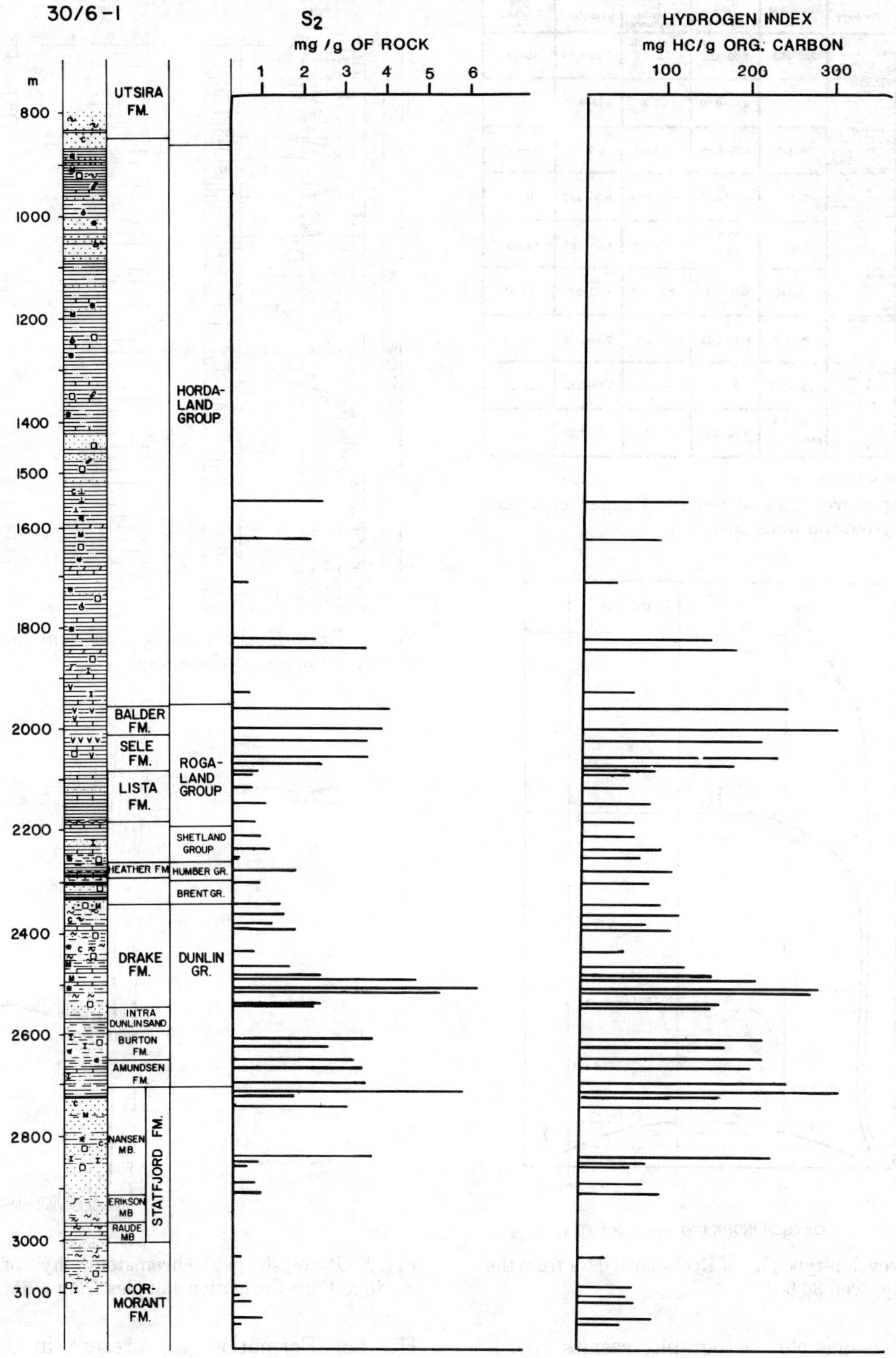

Fig. 4. Rock-Eval screening data from well 30/6–1.

poses, and these results are also included on Fig. 7. The sidewall core from the upper part of the formation gave an S_2 value higher than the associated cuttings, whereas in the lower part of the section the SWC was taken in a thin limestone band and, apart from giving a negligible S_2 value, was unrepresentative of the lower part of the formation. A range of sidewall cores from the Kimmeridge Clay Formation in another downflank well has also been examined, and this showed a similar trend to that observed in the previous well, but with a much more abrupt quality decrease with depth. These observations are in line with Kimmeridge Clay Formation quality variations reported by Barnard *et al.* (1981).

Quality decrease of the Kimmeridge Clay Formation with depth was also observed in visual kerogen studies of the Oseberg wells. The upper part of the formation had a high content of amorphous material, characteristic of a Type II kerogen, whereas the lower section displayed a significant increase in the amount of woody and herbaceous material.

Kerogen typing within the Kimmeridge Clay Formation was based initially upon Rock–Eval hydrogen indices, as noted above. These results are supported

Period Group/FM	Lithology	Thickness (m)	TOC (%WT)	HC Potential s_2 (kg/tonne)	Ro (%)	20S Steranes (%)
Paleocene Rogaland Group		225–288	1.00–1.50 *	3.0–3.7 *	0.30–0.61	10–20
Upper Cretaceous Shetland Group		26–909	0.60–0.80	< 1.0	0.32–0.77	—
Lower Cretaceous Cromer Knoll Group		0–143	0.25–0.50	<1.0	0.53	—
Upper Jurassic Kimmeridge Clay FM		0–143	5.3–10.20	26.0–71.0	0.36–0.47	16–24
Upper Jurassic Heather FM		0–541	4.00–5.00 occ < 1.0	1.4–15.0	0.39–1.02	12–19
✸ Middle Jurassic Brent Group		29–162	0.56–1.80	0.3–11.4	0.43–077	15–41
Lower Jurassic Drake FM		90–200	1.00–2.00	1.0–6.8	0.36–0.92	19–32
Lower Jurassic /Upper Triassic Statfjord FM		237–343	—	—	0.44–0.76	26–31
Upper/Middle Triassic Cormorant FM		Not fully penetrated	1.38–0.89	0.1–0.7	0.33–1.03	—

*Results from Balder and Sele Fms only.

Fig. 5. Osberg source rock summary. Ranges of values were compiled from the wells analysed.

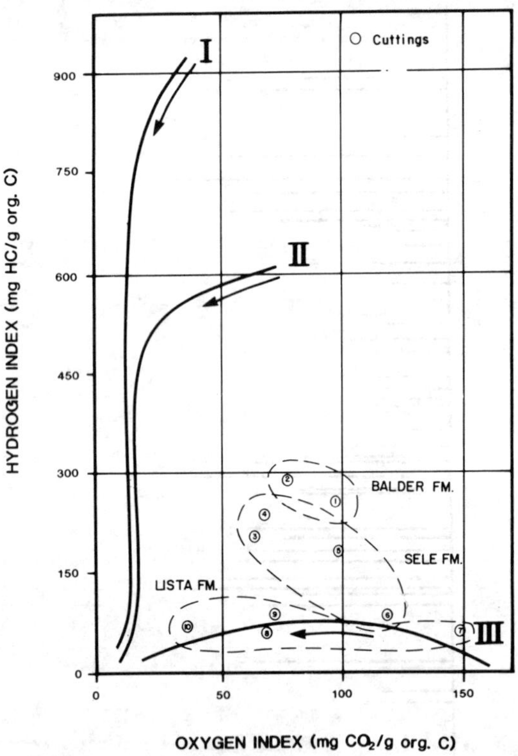

Fig. 6. Van Krevelen-type plot of Rock–Eval data from the Rogaland Group, well 30/6–1.

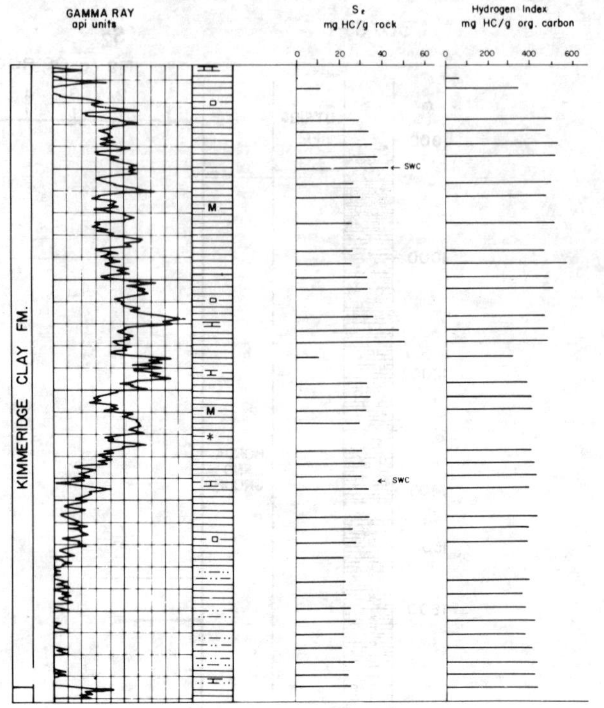

Fig. 7. Rock–Eval screening data from Kimmeridge Clay Formation in a downflank well.

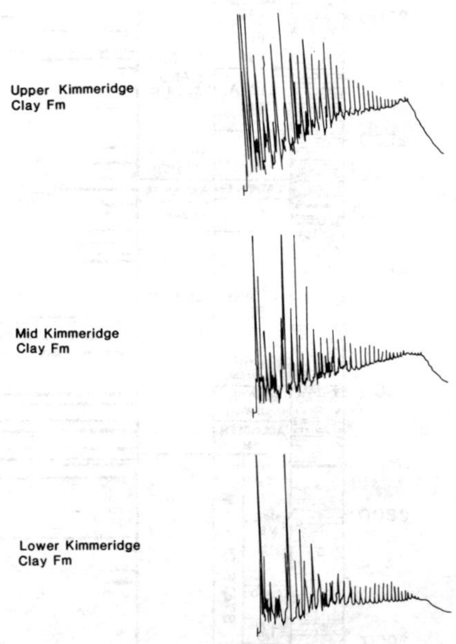

Fig. 8. Pyrolysis–gas chromatography of three Kimmeridge Clay Formation samples from a downflank well.

by the pyrolysis–gas chromatography results ('pyrograms') shown in Fig. 8. The three pyrograms were taken from the upper, middle and lower sections, and indicate a much more condensate-prone kerogen in the lower part of the formation.

The general observations on the Kimmeridge Clay Formation from these studies are not in agreement with the kerogen facies map of Barnard and Cooper (1984) in which they indicate Type III to be dominant in the Oseberg area.

The Middle–Upper Jurassic Heather Formation was penetrated in four out of the five wells used in this study. On the crest of the Oseberg structure this formation is very thin and it has not been sampled extensively for geochemical studies (e.g. see Fig. 5). As noted previously, the Humber Group thickens towards the east, and much greater thicknesses of the Heather Formation are present in the downflank wells. Geochemical data indicate that the source potential of the Heather Formation is highly variable, with the upper part being gas-prone and with low hydrocarbon yields. Quality improves in the middle section of the formation, which contains a more liquid-prone kerogen giving S_2 values up to 15 kg/t. The quality and hydrocarbon yield rapidly decreases towards the bottom of the section, which is gas-prone. These indications are supported by visual kerogen studies and pyrolysis–GC results.

Maturity

In well 30/6–1, vitrinite reflectance determinations were made on a number of sidewall cores from approximately 1900 m to total depth. A plot of the results

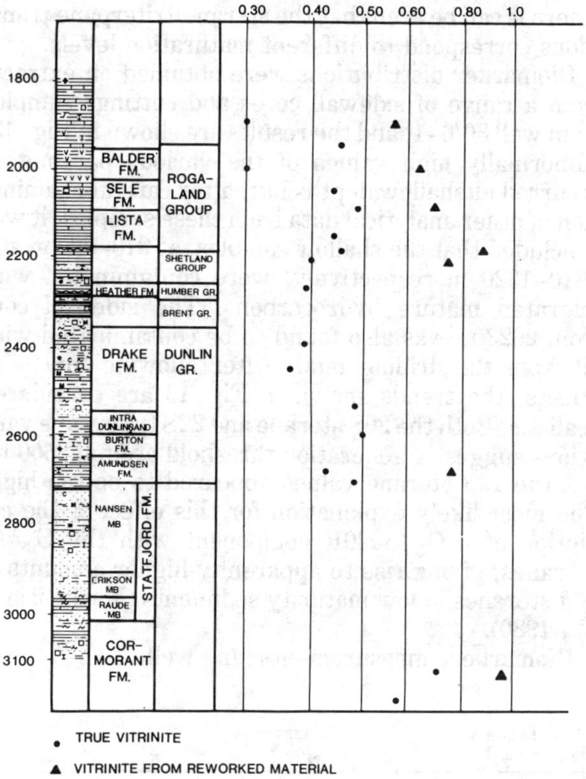

Fig. 9. Vitrinite reflectance versus depth in well 30/6–1 (sidewall samples).

against depth is shown in Fig. 9. Even after allowing for allochthonous (reworked) components, the spread of data points was very erratic and showed only a general increase with depth. A least mean squares fit of the vitrinite data was totally unsatisfactory for any reliable indications of maturity, so this parameter was not used in well 30/6–1. As sidewall cores were used for these determinations no problems exist with cavings, so the wide scatter of the vitrinite data was probably due to problems in differentiating autochthonous and allochthonous components, possibly combined with effects of bitumen 'staining'.

A more regular increase in maturation levels with depth was displayed by the TAI values of spores/pollen shown in Fig. 10. These results suggest that the hydrocarbon generation threshold is around 2000 m in this well. The two rather high TAI values observed at 2392 and 2849 m appear to be related to lithology, as both occur in limestone intervals.

Although only providing a rather approximate indication of maturity, Rock–Eval pyrolysis data (T_{max} values) suggest a generation threshold around 2400 m in well 30/6–1. However, as these results were obtained on cuttings, the indications may have been influenced by cavings.

Regional maturity studies (Barnard and Cooper, 1984) suggest that vitrinite reflectance values in the range 0.5–0.6% R_0 would be expected for the Base Cretaceous in the Oseberg area. In well 30/6–1, this would suggest a generation threshold around 2200–2300 m.

In view of the problems in assessing maturity variations in 30/6–1 by conventional parameters, a number of biomarker molecular variations, based upon regular steranes and hopane-type triterpanes were investigated (Mackenzie *et al.*, 1980; Mackenzie and Maxwell, 1981). Sterane transitions reflecting increasing maturity are shown in Fig. 11. The transition $b \rightarrow c$ corresponds to

$$C_{27}\alpha\alpha\alpha 20R \rightarrow C_{27}\alpha\alpha\alpha 20S,$$

and the parameter is expressed as

$$\text{percentage } 20S = \frac{100 \cdot \alpha\alpha\alpha 20S}{\alpha\alpha\alpha 20S + \alpha\alpha\alpha 20R}.$$

The transitions $(b) + (c) \rightarrow (d) + (c)$ are represented as:

$$C_{27}\alpha\alpha\alpha 20R + C_{27}\alpha\alpha\alpha 20S \rightarrow C_{27}\alpha\beta\beta 20R + C_{27}\alpha\beta\beta 20S,$$

And the second parameter is expressed as

$$\text{percentage } \alpha\beta\beta = \frac{\alpha\beta\beta(20R + 20S) \cdot 100}{\alpha\beta\beta(20R + 20S) + \alpha\alpha\alpha(20R + 20S)}.$$

As a result of resolution problems in determining these ratios, it is usual to use the C_{29} transitions rather than those of the C_{27} steranes. A third maturity para-

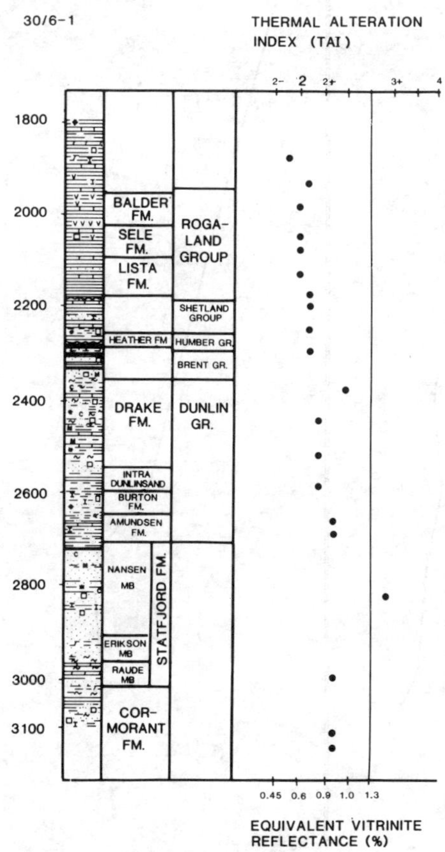

Fig. 10. TAI versus depth in well 30/6–1.

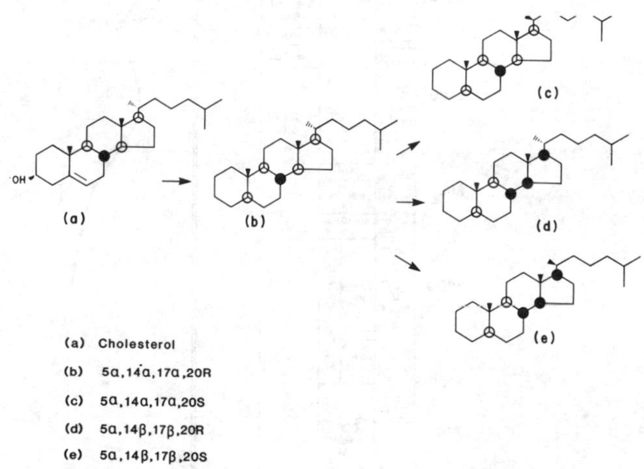

Fig. 11. Major transitions occurring in regular sterane molecules through increasing thermal stress.

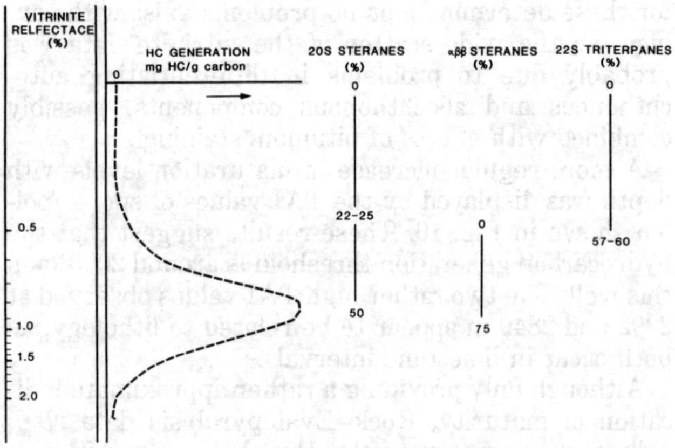

Fig. 12. Sterane and triterpane variations related to oil generation and vitrinite reflectance.

meter is based upon the relative amounts of the $C_{32}\alpha\beta$ 22R and $C_{32}\alpha\beta$ 22S triterpanes, and is calculated in the same way as the percentage 20S for the steranes. Variations of the three parameters with maturity are shown in Fig. 12 (Mackenzie and Maxwell, 1981) where it can be seen that the sterane/triterpane transitions correspond to different maturation levels.

Biomarker distributions were obtained on extracts from a range of sidewall cores and cuttings samples from well 30/6–1, and the results are shown in Fig. 13. Abnormally high values of the various parameters occurred at shallow depths but, after careful examination of other analytical data from these samples, it was concluded that the shallow samples, at 910–920 m and 1610–1620 m respectively, were contaminated with migrated mature hydrocarbons. The sidewall core from 2023 m was also found to be contaminated with oil from the drilling mud. After allowing for these effects, the trends shown in Fig. 13 are considered realistic. Both the 20S sterane and 22S triterpane variations suggest a generation threshold around 2500 m, but the $\alpha\beta\beta$ sterane values appeared to be too high. The most likely explanation for this effect is the coelution of a $C_{29}\beta\alpha\alpha$20R component with the $C_{29}\alpha\beta\beta$ steranes, giving rise to apparently higher amounts of $\alpha\beta\beta$ steranes in low maturity sediments (Mackenzie *et al.*, 1980).

Biomarker measurements in well 30/6–1 give

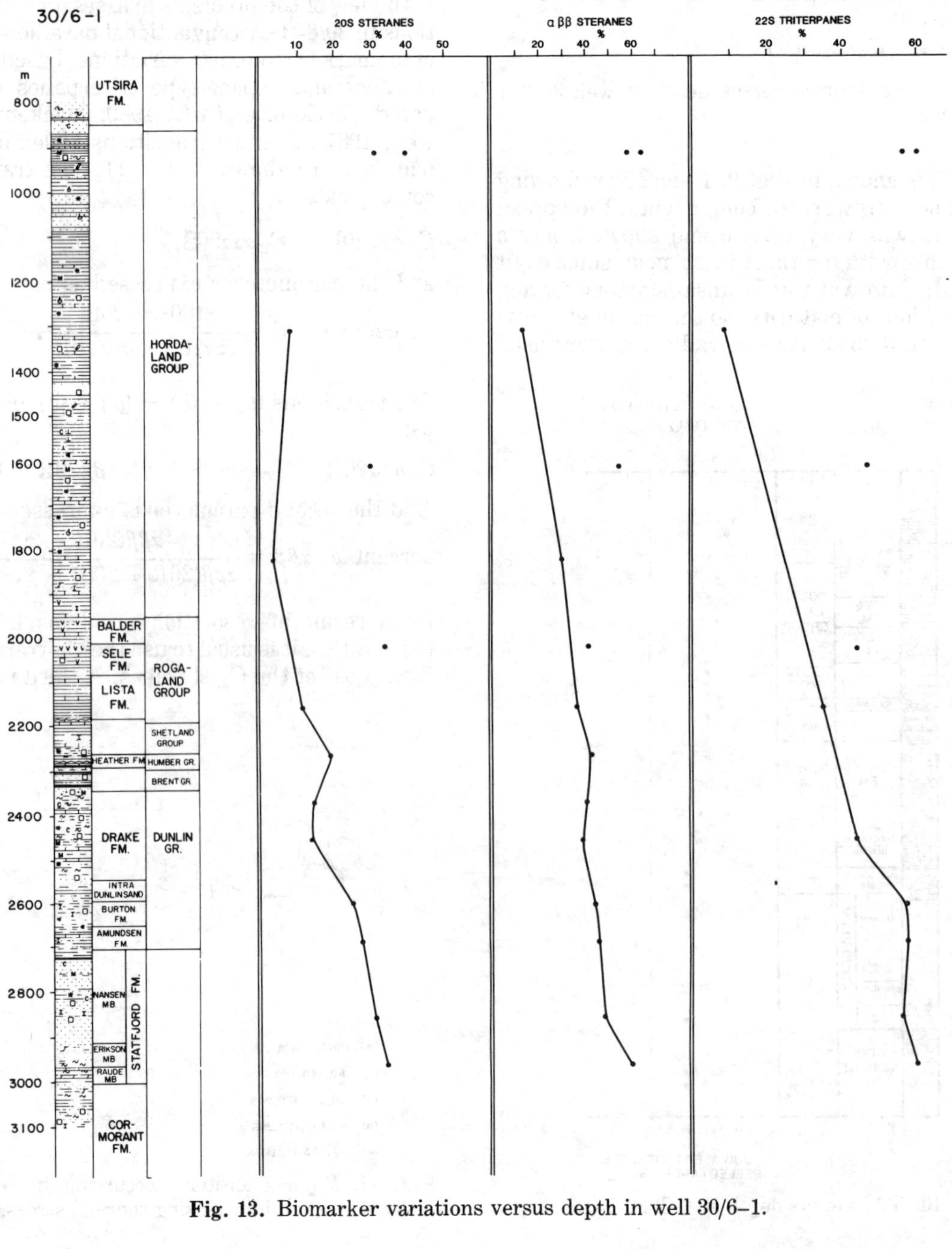

Fig. 13. Biomarker variations versus depth in well 30/6–1.

maturity indications which were later supported by much better quality vitrinite reflectance and other geochemical data from adjacent wells in the Oseberg area. Biomarker variations, when combined with other geochemical data, also give more definite indications of the presence of migrated hydrocarbons and contamination in well 30/6–1.

OSEBERG CRUDE OILS

Some properties of a sample of crude oil produced from the Alpha structure are given in Table 1. This oil is representative of the bulk of the crude present in the Oseberg Field and is typical of the highly paraffinic, low wax crudes, occurring in the Viking Graben/East Shetlands Basin area, which are considered to be sourced from Upper Jurassic shales (Barnard and Cooper, 1984). Asphaltene, sulphur and trace metal contents are all low, resulting in a high-quality crude oil.

The stable carbon isotope ratio of Oseberg oil ($\delta^{13}C = -28.5‰$) is slightly heavier than the range of −29.5 to −28.5‰ reported for oils in the East Shetlands basin area, but falls within the range of −27.5 to −31.0 quoted for Viking Graben oils (Barnard and Cooper, 1984). A pristane/phytane ratio of 1.6 is also consistent with values reported for North Sea crudes (Cornford *et al.*, 1983).

In the southern part of the Oseberg Field, a heavy, asphaltic oil has been encountered near the oil–water contact (OWC). Some properties of the residue >200°C from this oil are given in Table 2, together with the properties of a similar residue from the main producing interval in the same well. The most significant difference between these oil residues is the much higher content of asphaltenes (7 compared with 1.4 wt%) in the sample from near the OWC. However,

TABLE 1. Properties of an Oseberg oil

Specific gravity	0.8506
Density (°API)	34.9
Sulphur content (wt %)	0.29
Asphaltene content (wt %)	0.12
Wax content (wt %)	1.9
Vanadium (ppm)	1
Nickel (ppm)	1
$\delta^{13}C$ (‰)	−28.5*
Pristane/phytane	1.6
n-Alkane CPI	1.0

*NBS 22 = −29.8‰.

TABLE 2. Compositional variations in two crude oil samples from one of the Oseberg wells where a tar mat is encountered

Production crudes (residues >200 °C)	
Asphaltenes (wt %)	1.4
HPLC chromatographic fractions (wt %)	
SAC	50.1 (50.8)[a]
AROM	20.8 (21.1)[a]
NSO	27.7 (28.1)[a]
SAC/AROM ratio	2.4
Heavy oil (residue >200 °C)	*7.1*
Asphaltenes (wt %)	
HPLC chromatographic fractions (wt %)	
SAC	45.6 (49.1)[a]
AROM	17.3 (18.6)[a]
NSO	30.0 (32.3)[a]
SAC/AROM ratio	2.6

[a]Corrected for asphaltene content.

when the oils are compared on an asphaltene-free basis they are quite similar in composition, although the heavy oil has a somewhat larger NSO content and, possibly, a slightly decreased aromatic content.

Further study of the occurrence of the heavy oil (or 'tar mat') was undertaken by extraction of residual oil from reservoir core samples taken in wells which penetrated the OWC. Asphaltene contents of these residual oils were determined, and the results from one of these wells are shown in Fig. 14. In the upper part of the oil column, residual oil asphaltene contents are approximately 5 wt%, increasing very rapidly near 2730 m to approximately 50 wt%. Although not shown in Fig. 14, the bottom of the heavy oil zone in this well is shown by a very rapid decrease in the amount of residual oil present at approximately 2743 m. These observations indicate that the tar mat is approximately 13 m thick at this location, and that traces of heavy oil are present below the OWC.

Tar mats are usually formed as a result of biodegradation associated with extensive water washing, or by precipitation of an asphaltic 'sludge' which concentrates near the OWC as a result of gravity segregation. Destabilization of the reservoir crude can be brought about by the generation of light hydrocarbons as a result of oil maturation or, more probably of, a later stage addition of gas to the reservoir.

The Oseberg oils examined in this study show no evidence of biodegradation, as *n*-alkane and biomarker distributions are very similar in all of the residual oil samples. As noted previously in Table 2, the asphaltene-free oils from the main reservoir crude and the tar mat are quite similar, although the higher NSO content of the latter is probably due to incorporation of these components in the precipitated asphaltic sludge. The slightly lower aromatic content of the tar mat may also be the result of water washing processes. It seems likely, therefore, that the heavy oil, or tar mat, observed in the Oseberg Field is the result of *in situ* de-asphaltening of the crude, possibly associated with slight water washing, rather than biodegradation.

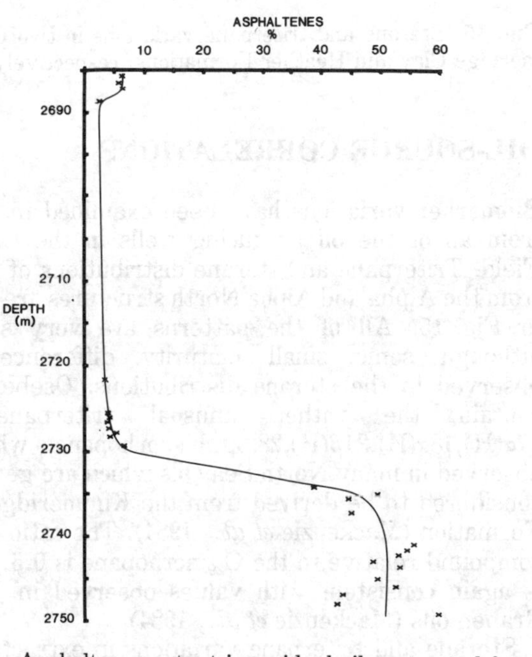

Fig. 14. Asphaltene content in residual oils extracted from cores in one of the Oseberg wells where a tar mat has formed.

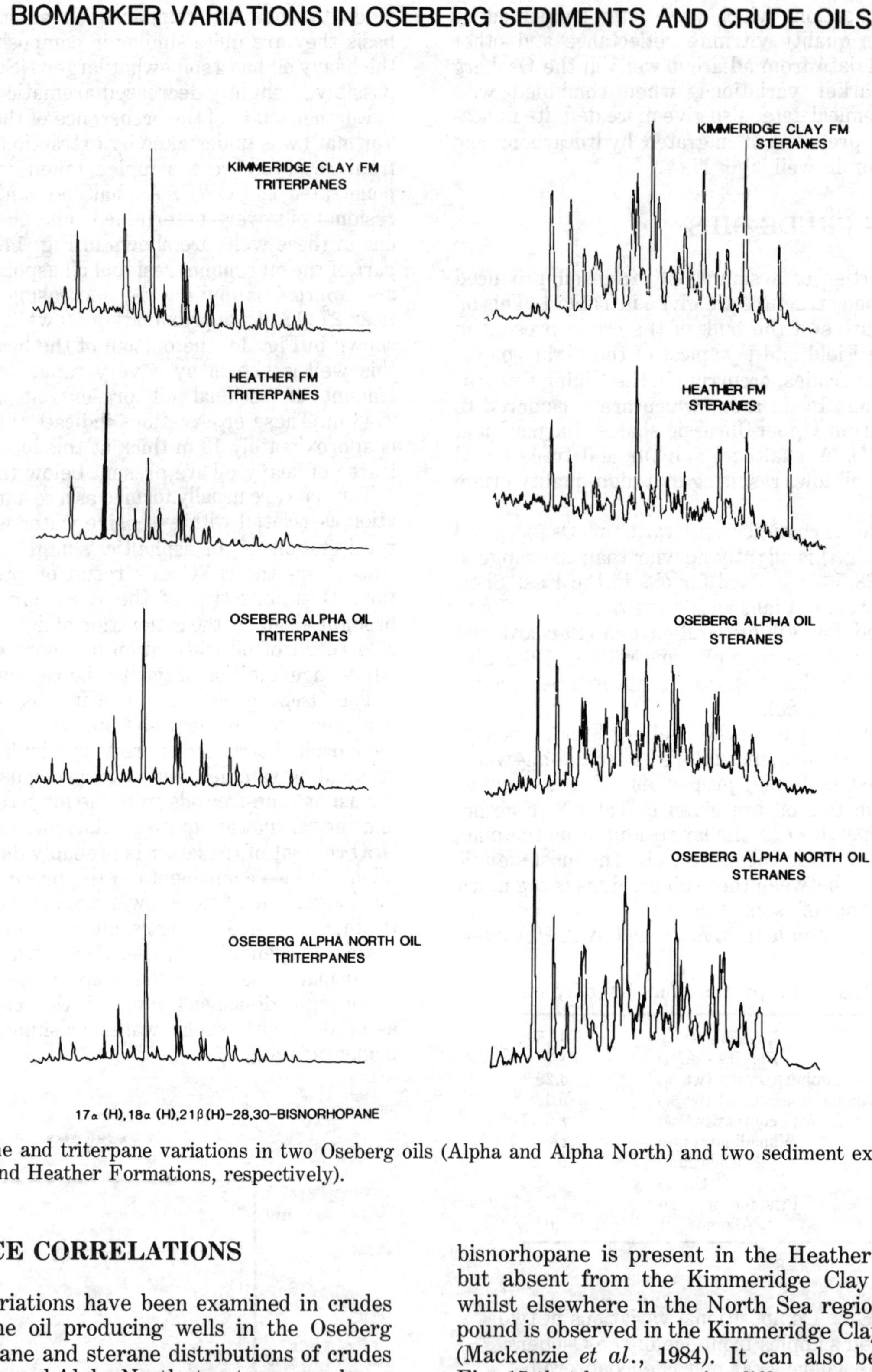

Fig. 15. Sterane and triterpane variations in two Oseberg oils (Alpha and Alpha North) and two sediment extracts (Kimmeridge Clay and Heather Formations, respectively).

OIL-SOURCE CORRELATIONS

Biomarker variations have been examined in crudes from all of the oil producing wells in the Oseberg Field. Triterpane and sterane distributions of crudes from the Alpha and Alpha North structures are shown in Fig. 15. All of the patterns are very similar, although some small maturity differences are observed in the sterane distributions. Oseberg oils contain the rather unusual triterpane C_{28} $17\alpha(H),18\alpha(H),21\beta(H)$,28.30 bisnorhopane, which is observed in many North Sea Oils which are generally considered to be derived from the Kimmeridge Clay Formation (Mackenzie *et al.*, 1984). The ratio of this compound relative to the C_{29} norhopane is 0.5, which is again consistent with values observed in Viking Graben oils (Mackenzie *et al.*, 1984).

Sterane and triterpane variations in extracts from the Kimmeridge Clay and Heather Formations in the Oseberg area are also included in Fig. 15. The C_{28}-bisnorhopane is present in the Heather Formation, but absent from the Kimmeridge Clay Formation, whilst elsewhere in the North Sea region, this compound is observed in the Kimmeridge Clay Formation (Mackenzie *et al.*, 1984). It can also be seen from Fig. 15 that large maturity differences exist between the on-structure sediment extracts and Oseberg crude oils, for example:

Osberg Alpha oil:
20S = 54%; $\alpha\beta\beta$ = 59%; 22S = 60%;
Kimmeridge Clay Formation:
20S = 24%; $\alpha\beta\beta$ = 28%; 22S = 46%.

As the Kimmeridge Clay and Heather Formations are not much more deeply buried in the eastern sub-basin, this region cannot be considered a major source area for Oseberg oil, which probably originates from the much more mature Humber Group shales located to the west and north-west of the Oseberg area.

CONCLUSIONS

(1) Source rock studies suggest that shales from the Humber Group are the most likely source of the oil found in the Oseberg structure. The richest oil-prone source rock within the Humber Group is the upper section of the Kimmeridge Clay Formation.

(2) Conventional parameters, e.g. vitrinite reflectance and TAI, were unsatisfactory for accurately assessing maturity in well 30/6–1. Molecular parameters based on percentage 20S steranes and 22S triterpanes are in good agreement and suggest a maturity level in this well which was confirmed by later results in adjacent wells.

(3) Biomarker studies also give more definitive indications of migrated hydrocarbons and contamination in sediment extracts from well 30/6–1.

(4) Oseberg crude oil is typical of the highly paraffinic, high quality North Sea crudes with a gravity of 35° API and low sulphur (0.29 wt%), asphaltene (0.12 wt%) and trace metal contents (V = 1 ppm, Ni = 1 ppm).

(5) A tar mat has been established in part of the Oseberg Field. This heavy oil is considered to result from *in situ* de-asphaltening of the crude followed by gravity segregation of the precipitated sludge and concentration near the oil–water contact. Some minor water washing of the heavy oil may also have occurred.

(6) Rocks of the Kimmeridge Clay and Heather Formations are not considered mature enough in the Oseberg area to have generated the reservoir crude oil. The major source area is probably the more mature Humber Group shales located to the west and north-west of the Oseberg area.

REFERENCES

Barnard, P. C. and Cooper, B. S. 1984. Source rocks and oils of the Central and Northern North Sea. In: Demaison, G. and Murris, R. J. (eds.), *Petroleum Geochemistry and Basin Evolution. AAPG Memoir*, **35**, 303–314.

Barnard, P. C., Collins, A. G. and Cooper, B. S. 1981. Identification and distribution of kerogen facies in a source rock horizon—examples from the North Sea Basin. In: Brooks, J. (ed.), *Organic Maturation Studies and Fossil Fuel Exploration.* Academic Press, pp. 271–282.

Cornford, C. 1984. Source rocks and hydrocarbons of the North Sea. In: Glennie, K. W. (ed.), *Introduction to the Petroleum Geology of the North Sea.* Blackwell Scientific, Oxford, pp. 171–204.

Cornford, C., Morrow, J. A., Turrington, A., Miles, J. A. and Brooks, J. 1983. Some geological controls on oil composition in the U.K. North Sea. In: Brooks, J. (ed.), *Petroleum Geochemistry and Exploration of Europe.* Blackwell Scientific, Oxford, pp. 175–194.

Deegan, C. E. and Scull, B. J. (Compilers) 1977. A standard lithostratigraphic nomenclature for the Central and Northern North Sea. *Rep. Inst. Geol. Sci. No. 77/25; Bull. Norwegian Petrol. Directorate, No. 1.* HMSO.

Katz, B. J. 1983. Limitations of 'Rock Eval' pyrolysis for typing organic matter. *Organic Geochem.*, **4**, 195–199.

Kirk, R. H.1980. Statfjord Field, a North Sea giant. In: Halbouty M. T. (ed.), *Giant Oil and Gas Fields of the Decade 1968–1978. AAPG Memoir*, **30**, 95–116.

Mackenzie, A. S. and Maxwell, J. R. 1981. Assessment of thermal maturation in sedimentary rocks by molecular measurements. In: Brooks, J. (ed.), *Organic Maturation Studies and Fossil Fuel Exploration.* Academic Press, pp. 239–254.

Mackenzie, A. S., Patience, R. L., Maxwell, J. R., Vandenbrouche, M. and Durand, B. 1980. Molecular parameters of maturation in the Toarcian shales, Paris Basin, France 1. Changes in the configuration of the acyclic isoprenoid alkanes, steranes and triterpanes. *Geochim. Cosmochim. Acta*, **44**, 1709–1721.

Mackenzie, A. S., Maxwell, J. R., Coleman, M. L. and Deegan, C. E. 1984. Biological marker and isotope studies of North Sea crude oils and sediments. *Proc. 11th World Petroleum Congress 1983*, Vol. 2. Heyden, pp. 45–56.

Van den Bark, E. and Thomas, O. D. 1980. Ekofisk: first of the giant oil fields in Western Europe. In: Halbouty, M. T. (ed.), *Giant Oil and Gas Fields of the Decade 1968–1978. AAPG Memoir*, **30**, 195–224.

16 Block 33/6 geochemical evaluation

B. Maragna, G. Zaro* and P. Pessina†*
*Agip S.p.A., 20097 S. Donato Milanese, Italy
†Norsk Agip a.s., Sandnes, Norway

A geological evaluation of Block 33/6 was performed using the direct knowledge of subcrops and subsurface conditions provided by the 33/6–1 wildcat well and by modelling the evolution of maturity in the nearby areas. Laboratory analyses (including optical typing of kerogen, TAI and vitrinite reflectance, Rock–Eval pyrolysis and stable isotopes) permit a definition of source rock characteristics and measurement of the present degree of thermal maturity.

The evolution and areal extent of maturity for Humber Group rocks, generally considered as the main source sequence in the North Sea, were studied applying the Lopatin method and the Tissot–Espitalié method.

INTRODUCTION

A production licence for petroleum exploration in Block 33/6 covering an area of 485.5 km^2 was granted by Royal decree on 23 December 1977 to a group consisting of Statoil (50%), Norsk Agip (30%, and operator) and Deminex (20%). The licence expired on 1 January 1984 when the block was released.

An exploratory well, 33/6–1, was drilled at 01°51′53″E, 61°32′14.58″N. The selected location was on the crestal area of a tilted fault block lying in a water depth of 306 m. The effective closure was considered dependent upon the sealing properties of a NE–SW fault immediately southward of the prospect. Target of the well was to test well documented Middle–Lower Jurassic reservoirs (Brent Group–Statfjord Formation) and to reach Upper Triassic sediments (Cormorant Formation) at total depth. The well was entirely water-bearing and was consequently abandoned. No other wells were drilled in the block.

The aim of the present study was to evaluate the petroleum potential of the block and, if possible, to discover whether geochemical and/or geological factors can explain the observation that no hydrocarbons were trapped.

REGIONAL GEOLOGICAL FRAMEWORK

Block 33/6 lies in the northern part of the North Sea, close to the UK–Norwegian border, which cuts the southwestern corner of the block (see Fig. 1). This area appears to be most attractive for petroleum exploration because it includes many important discoveries; the Statfjord field lies just inside the Norwegian border, Gullfaks and several smaller fields lie in the Norwegian sector and many others are found in UK waters (Brent, Hutton, Thistle, Magnus, Murchison, Dunlin) (see Fig. 2). All of these fields lie in the East Shetland Basin, which is covered in Norwegian waters by quadrants 33 and part of 34.

The tectonic evolution of the basin began in the Permo-Triassic when a system of Caledonian NE–SW-trending faults was reactivated. Tectonic activity continued throughout Jurassic times, characterized by mainly eastward block faulting and syndepositional westward rotation and subsidence of downthrown blocks. During 'Late Kimmerian' movements, the tectonic activity increased and caused the partial erosion of some previously uplifted Early to Middle Kimmerian fault blocks.

During the Cretaceous and Paleocene only minor tectonic activity is evident, with local reactivation of Kimmerian faults and the onset of a broad general subsidence.

The high well density in the North Sea has made possible exhaustive characterization of source rock occurrence in the area. The Kimmeridge Clay Formation seems to be the major source unit with a strongly oil-prone organic type, while the underlying Heather Formation has inferior characteristics, and is generally regarded as mainly gas-prone (Goff, 1983).

Several studies of the distribution of organic matter type have shown variations in composition which may be attributed to the relative distance of the paleoshore (Cooper and Barnard, 1984).

Petroleum Geochemistry in Exploration of the Norwegian Shelf © Norwegian Petroleum Society (Graham & Trotman, 1985) pp. 197–204.

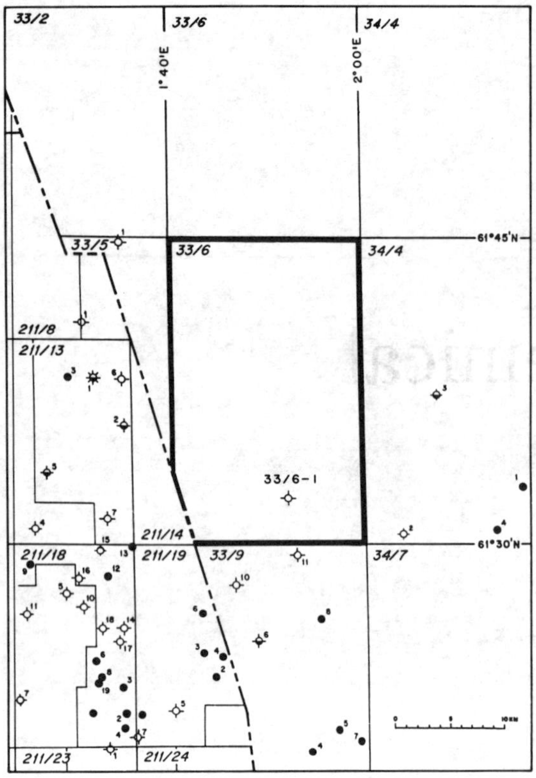

Fig. 1. Location map.

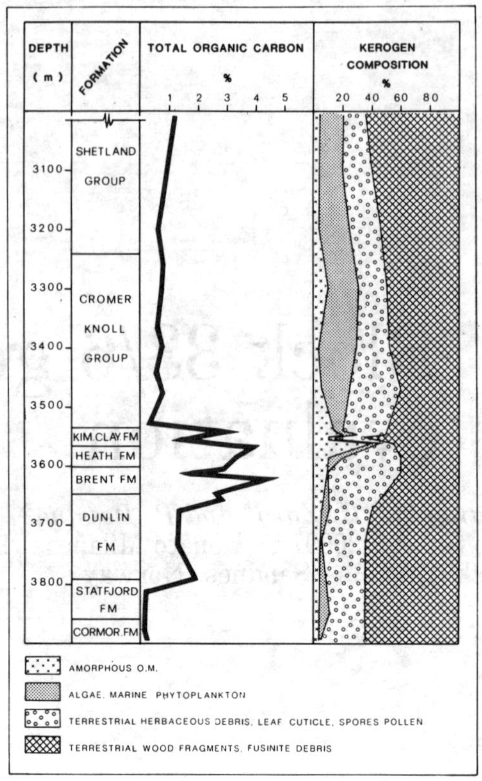

Fig. 3. Well 33/6–1: TOC and kerogen composition.

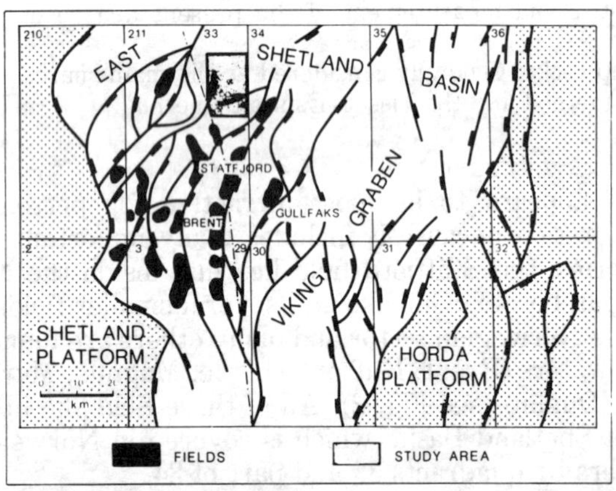

Fig. 2. Structural sketch of Northern North Sea (redrafted from Skarpnes *et al.*, 1982).

EXPERIMENTAL DATA FROM WELL 33/6-1

Samples (cuttings, sidewall samples and bottom cores) from well 33/6–1, from the Lower Miocene to the Upper Triassic, were analysed to measure the degree of maturity and to evaluate source potential. Total organic carbon contents of some interest were detected from 3536 m to 3647 m, where values ranging from 1% to 5% were recorded (see Fig. 3). This section corresponds to the Middle–Upper Jurassic Brent Group, Heather Formation and Kimmeridge Clay Formation. The Brent Group is characterized by a prevailing sandy lithology. Log response suggests that the Rannoch Formation is present, while a very thin, shaly Broom Formation can only be hypothesized. On the contrary, both the Heather and Kimmeridge Clay Formations consist of mainly claystone with local stringers of siltstone.

An optical study of kerogen quality, as summarized in Table 1 and Fig. 3, demonstrates that the Middle Jurassic Brent Group should be excluded as a significant source rock for oil. This is due to the poor quality of the organic matter, chiefly consisting of continental herbaceous fragments, woody inertinitic fragments and fusinite debris, which could, however, be a fair source for gas. Such an inference is in accordance with other data from the area reported by Goff (1983), Barnard and Cooper (1981) and Cooper and Barnard (1984).

A predominance of terrestrial organic matter persists in Humber Group rocks, although a small amount of marine phytoplankton and local higher concentrations of amorphous organic matter are present.

Rock–Eval pyrolysis results are summarized in Table 2. These are in agreement with the optical data, confirming a mixed quality of kerogen, largely tending towards type III. Humber Group formations display inferior source features (see Fig. 3) to those reported by Goff for the North Sea generally, and for the East Shetland Basin in particular.

The actual degree of maturity of organic matter was determined on the basis of both thermal alteration index (TAI) and mean vitrinite reflectance (see Fig. 4). These parameters suggest early generation in the well at a depth of 2300 m (R_0 % > 0.5) and full maturity onset (R_0% > 0.55) at 2900 m. Consequently, the Humber Group and Brent Group are considered to be mature in the study well.

Qualitative analyses on trace gas (head space and carbon isotopic composition of methane) agree with the evidence obtained by optical methods. These analyses were carried out on cutting samples collected at the well site in glass jars containing a small amount of bactericide to prevent any microbial activity. Samples were picked from the entire sedimentary section penetrated, and the relevant data are plotted in Fig. 5. Results indicate several different genetic origins for the analysed gas:

TABLE 1. Kerogen composition (all average values)

Unit	*Amorphous organic matter %)*	*Marine phytoplankton (%)*	*Continental herbaceous fragments (%)*	*Continental woody fragments and fusinite debris (%)*
Kimmeridge Clay Formation	19	6	22	53
Heather Formation	10	10	40	40
Brent Group	10	—	50	40

TABLE 2. Rock–Eval pyrolysis (average values)

Unit	*Samples*	*TOC (%)*	*S2 (‰)*	*HI*	*S1 (‰)*
Kimmeridge Clay Formation	Cuttings	1.89	2.83	150	0.48
Heather Formation	Cuttings	2.87	6.27	218	0.87
Brent Group	Bottom cores	1.47	1.47	100	0.18

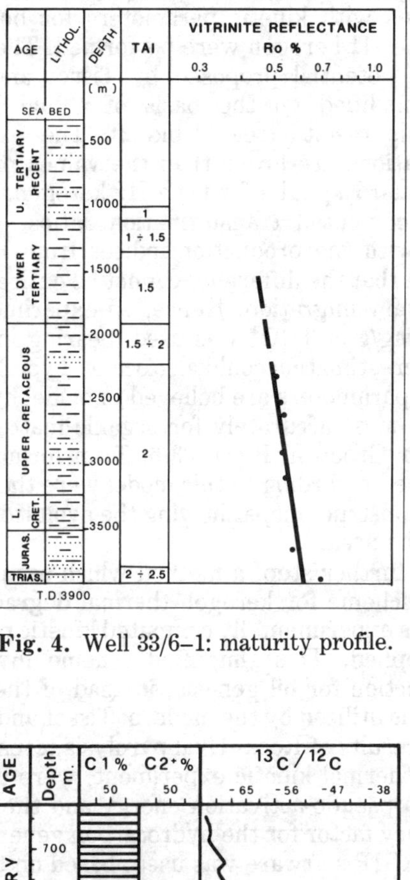

Fig. 4. Well 33/6–1: maturity profile.

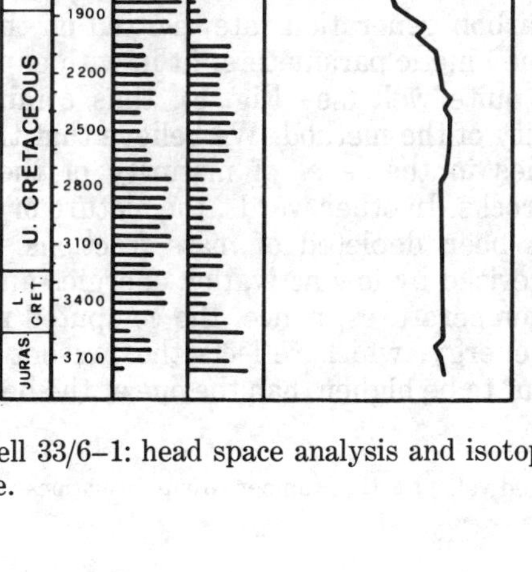

Fig. 5. Well 33/6–1: head space analysis and isotopic ratio of methane.

1. biogenic methane, distinguished by a very low ^{13}C content, down to approximately 1500 m;
2. thermogenic gas, with higher ^{13}C content and significant presence of C_2+, from approximately 2000 m to 3900 m T.D.;
3. mixed gas, with intermediate isotopic and chemical properties, from 1500 m to 2000 m approximately.

THERMAL MODELLING

The establishment of a model for hydrocarbon generation requires a definition of temperature distribution and of its evolution through time.

The geothermal gradient in well 33/6–1 was calculated from log-derived bottom hole temperatures, corrected according to the method described by Fertl and Wichmann (1977) to minimize the cooling influence of mud circulation. A value of 3.2 °C/100 m was thus determined.

To draw a mean temperature gradient map of regional significance, available data from 24 wells on the Tampen Spur, in the East Shetland Basin and on the East Shetland Platform, were taken and compared with other published data (Carstens and Finstad, 1981), see Fig. 6.

Gradient values range from 2.5 to 4°C/100 m; their distribution appears to be strongly affected by the tectonic setting of the basin. Lower gradients occur within the Viking Graben and in structurally depressed areas, while maximum values are predominantly associated with horsts and with the rims of structural highs. A similar relation exists in block 33/6, where maxima were generally recorded along the margin of structural highs in a transitional zone towards the northerly Møre–Trøndelag Basin.

A comparable regional pattern was previously reported by Carstens and Finstad (1981), who reviewed the geothermal gradients of the Northern North Sea. They suggested that subsurface fluid flow along major fault planes could be responsible for the observed situation. This explanation would also account for the predominance of high gradients in the Tampen Spur and Møre–Trøndelag areas.

Reconstruction of the palaeo-goethermal regime is obviously more difficult. Many authors have dealt with the problem of reconstructing the past thermal history of the Northern North Sea, either on the basis of experimental data, e.g. vitrinite reflectance (Yar-

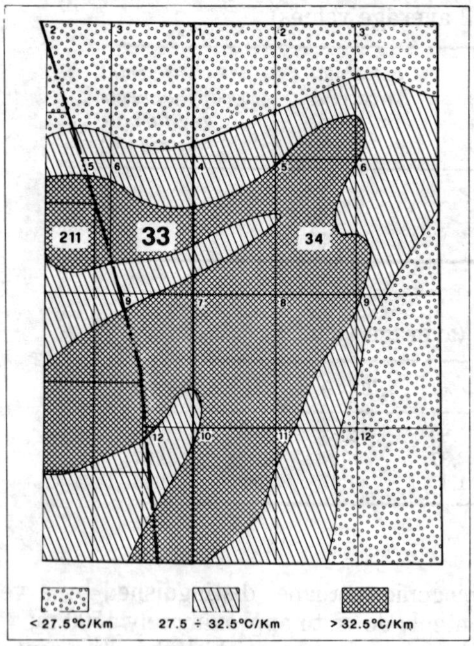

Fig. 6. Mean geothermal gradients in Northern North Sea.

zab *et al.*, 1983), or based on the sedimentary and structural evolution of the basin.

A heating event in the past must be hypothesized if the stretching model of McKenzie (1978) for the formation of the North Sea is accepted. In this case the age of the event can be determined by the occurrence of tensional movements, as expressed in periods of normal faulting. According to Goff (1983), some difficulties may arise in such analysis due to the presence of several phases of faulting between the Late Bathonian and Early Cretaceous. In effect, several authors have proposed different ages for this heating event, yet all accept that the heat flow has not changed in any systematic way during the last 100 million years. Before that time, it seems unlikely that the Upper Jurassic shales in the area (i.e. Heather Formation plus Kimmeridge Clay Formation), were buried sufficiently deeply to reach levels of temperature influencing hydrocarbon genesis. Therefore any assumption about the previous heat flow should not greatly affect the maturity level of the Upper Jurassic shales.

Moreover, accepting the theory that the thermal effects induced by stretching rapidly decline after the event, it seems reasonable to apply a constant heat flow when calculating the maturity of the Upper Jurassic shales in the block.

An attempt was made to reconstruct the temperature history of the sedimentary section, taking into account thermal conductivity variations due to compaction. However, a test performed applying thermal conductivity and porosity data from the literature did not give satisfactory results for the study well. Consequently, this aspect had to be ignored, resulting in a further simplification in the model.

KINETIC MODEL

To study the maturation history and the hydrocarbon generation in the study area two models were applied: the Lopatin method and the Tissot–Espitalié method.

The Lopatin method (Waples, 1980) was chosen as a first step in the study of the evolution of maturity. The relationship between the calculated maturity index

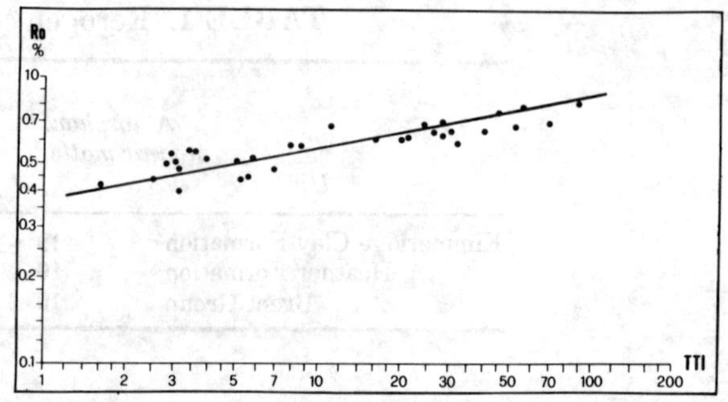

Fig. 7. Time–temperature index (TTI) versus vitrine reflectance (R_0%).

and the measured vitrinite reflectance of several wells from the region was examined to determine a general trend for the basin, which would then allow the interpretation of other computed maturity data. The calibration curve is shown in Fig. 7.

Using the Tissot and Espitalié (1975) method, computations with kinetic parameters for both type II and type III kerogen were performed. The values for genetic potential proposed by Tissot and Espitalié were modified on the basis of actual Rock–Eval pyrolysis results (see Table 2). The experimental observations are lower than the values proposed by Tissot and Espitalié for type III kerogen.

The computed transformation ratios, when compared with the production indices from Rock–Eval, suggest that the difference cannot be entirely ascribed to primary migration. Hence, an experimental value of 220 mg/g of TOC* was used, bearing in mind that an underestimation could also occur. Type III kerogen kinetic parameters are believed to model hydrocarbon genesis most accurately for organic matter from the Humber Group in Block 33/6. Transformation ratios calculated according to this model were therefore chosen to construct maps showing the evolution of maturity in the area.

As a further step, a method which assumes a simplified scheme for kerogen thermal degradation, but employs experimentally computed kinetic parameters, was applied. This simplified scheme involves only one reaction for oil genesis, instead of the set of six reactions utilized by the model of Tissot and Espitalié.

The results of Rock–Eval pyrolysis, considered as a non-isothermal kinetic experiment, were used to calculate a pseudo-activation energy and the respective frequency factor for the hydrocarbon generating reactions. AGIP software was used, based on the method developed by Jüntgen and co-workers (Jüntgen and Van Heek, 1968; Hanbaba *et al.*, 1968; Hanbaba and Jüntgen, 1968). Using the values obtained in this way, the hydrocarbon genesis model for the well gave underestimated results. Nevertheless, curves of the hydrocarbon generation rate, plotted on the basis of the same kinetic parameters, fit in with experimental curves quite well (see Fig. 8), thus confirming the reliability of the method. We believe that the explanation lies in the level of maturity of the Humber Group rocks. In other words, the mature organic matter has been depleted of mass fractions which are characterized by low activation energies and react at lower temperatures; hence, the computed mean activation energy, which reflects the present situation, turns out to be higher than the one at the beginning of

*Averaged value for the Humber Group formations.

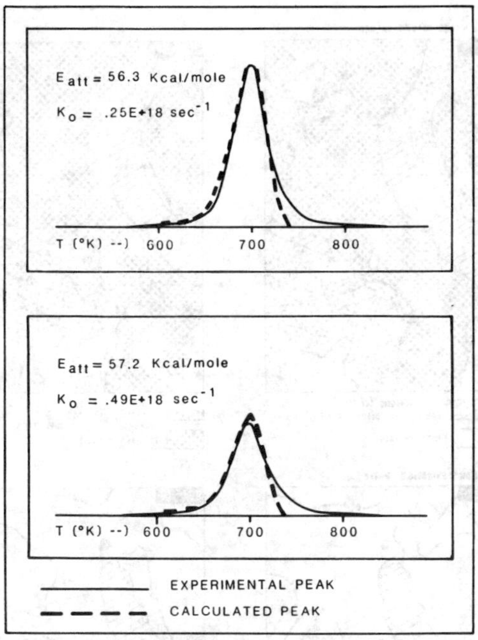

Fig. 8. Well 33/6–1: thermal decomposition of Heather Formation—comparison between experimental and calculated curve.

maturation, and cannot be used in modelling the complete genetic process.

MATURATION AND TIMING OF GENERATION

The reconstruction of the burial history of the source rocks in the study area using seismic data was a further step towards the modelling of maturation and hydrocarbon generation.

The quality of the seismic data forced two simplifications: firstly, the Kimmeridge Clay Formation and the Heather Formation had to be studied together, and are therefore referred to as 'Upper Jurassic shales' henceforth, as it is difficult to define the boundary between one formation and the other; secondly, no semi-quantitative evaluations could be carried out on the Brent Group since its base could not be recognized. In addition, the thickness and regional distribution of the source rocks were deduced from geophysical data.

On this basis, areas of non-deposition or total erosion of the Upper Jurassic shales and, in turn, sites of maximum accumulation were defined (see Fig. 9). Their distribution is characterized by a marked irregularity, which can be explained in terms of the structural evolution of the licence block. Extensional faulting, tilting and a Late Kimmerian erosional phase, in fact, promoted thickening of sediments deposited at the base of fault escarpments.

In order to describe the maturation of the Upper Jurassic shales present in Block 33/6, several maps which are relevant to the conditions at the top and the base of this unit at different times, were constructed by means of a grid of points, chosen to represent the various structural and stratigraphic features in the area.

Both R_0% equivalent values, calculated using the R_0%–TTI relation for the basin (see Fig. 7), and the transformation ratio computed applying the method of Tissot and Espitalié, were mapped.

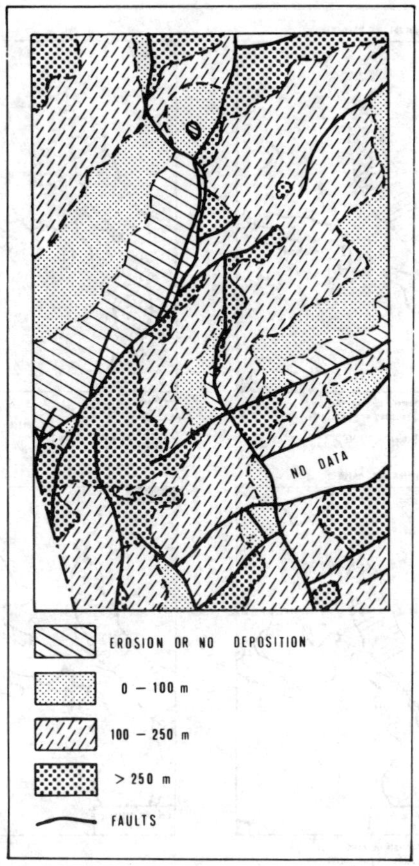

Fig. 9. Block 33/6: isopach map of Upper Jurassic shales.

The evolution of maturity and its distribution over the area, reconstructed on the basis of the two different methods, were studied separately and compared (see Figs. 10–13).

The transformation ratio values computed for reactions involving type III kerogen were mapped. The composition of this type of kerogen, in fact, parallels that of vitrinite; it therefore seems reasonable to consider the computed transformation ratio and the TTI values calibrated against vitrinite reflectance as properties of the same type of organic matter, and the two can easily be compared (Yükler and Kokesh, 1984).

The results achieved applying the Lopatin method show that the Upper Jurassic shales entered the mature facies (R_0% equivalent >0.55) during the Late Cretaceous–Early Paleocene. By the end of the Paleocene only the base of the unit had reached maturity over a significant area, in a belt crossing the block from SW to NE. This distribution was clearly controlled by the location of the main palaeo-highs (see Fig. 10).

The isolated mature zones existing at the top of the Upper Jurassic shales seem to be connected to areas of relatively higher subsidence; the irregular tectonics related to block faulting can account for their patchy distribution.

During the Paleocene–Oligocene interval maturity extended to the whole unit, the only exception being an immature area to the southeast, which still exists, though greatly reduced, and in central parts of the block the optimum maturity level (R_0% equivalent ~0.85) was reached (see Fig. 11). The whole central section is now at optimum maturity levels and even values close to the overmature facies have been recorded (see Fig. 12).

A reconstruction of the evolution of maturity based

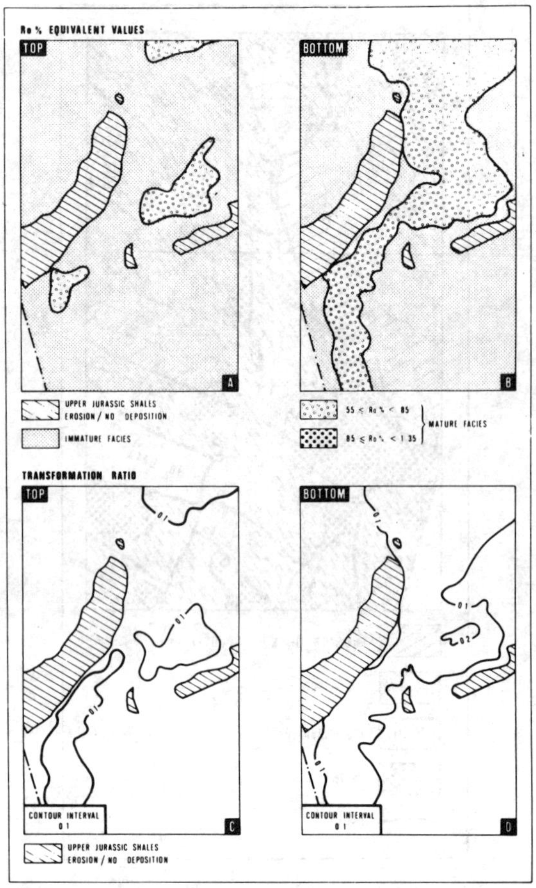

Fig. 10. Block 33/6: maturity maps of Upper Jurassic shales at the end of the Paleocene.

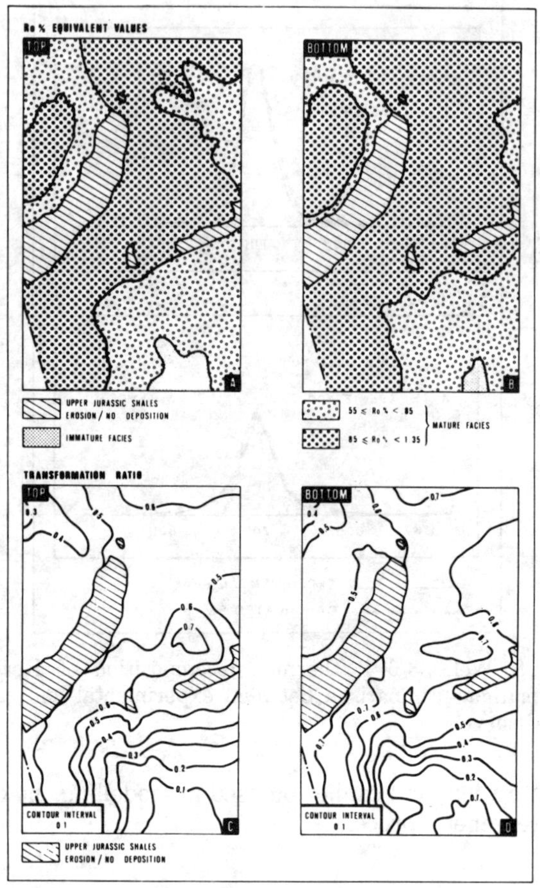

Fig. 12. Block 33/6: maturity maps of Upper Jurassic shales at the present time.

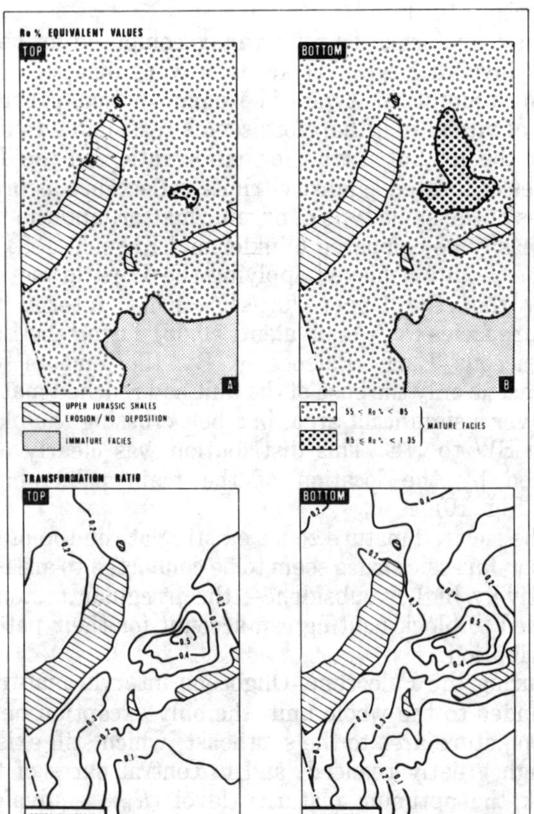

Fig. 11. Block 33/6: maturity maps of Upper Jurassic shales at the end of the Oligocene.

on the transformation ratio values (Tissot–Espitalié method) leads to a pattern substantially similar to the one shown by the vitrinite reflectance equivalent values (Lopatin method), see Fig. 10–13.

As a next step, the two sets of results were compared in order to find out just how they would correspond. It was found that a transformation ratio of 0.1 roughly corresponds to the onset of full maturity taken as 0.55 R_0% equivalent, while the optimum maturity level threshold (0.85 R_0% equivalent) falls between transformation values of 0.4 and 0.5.

A review of these figures compared with the relationship between vitrinite reflectance and the transformation ratio for type III kerogen proposed by Yükler and Kokesh (1984), on the basis of data from Tissot and Espitalié (1975), allows the following conclusions:

(1) at the early stages of generation (low transformation ratio values) the present findings correspond to those published by these authors; and

(2) at higher degrees of maturity (i.e. higher values of transformation ratio), a discrepancy is evident. In fact, much lower corresponding vitrinite reflectance equivalents were obtained than would have been obtained following the curve published by Yükler and Kokesh (1984).

In the opinion of the authors this behaviour can be explained by the fact that the vitrinite reflectance values used to deduce the R_0/TTI calibration are predominantly lower than 0.85; thus all the TTI values corresponding to a higher vitrinite reflectance have been interpreted according to an extrapolated

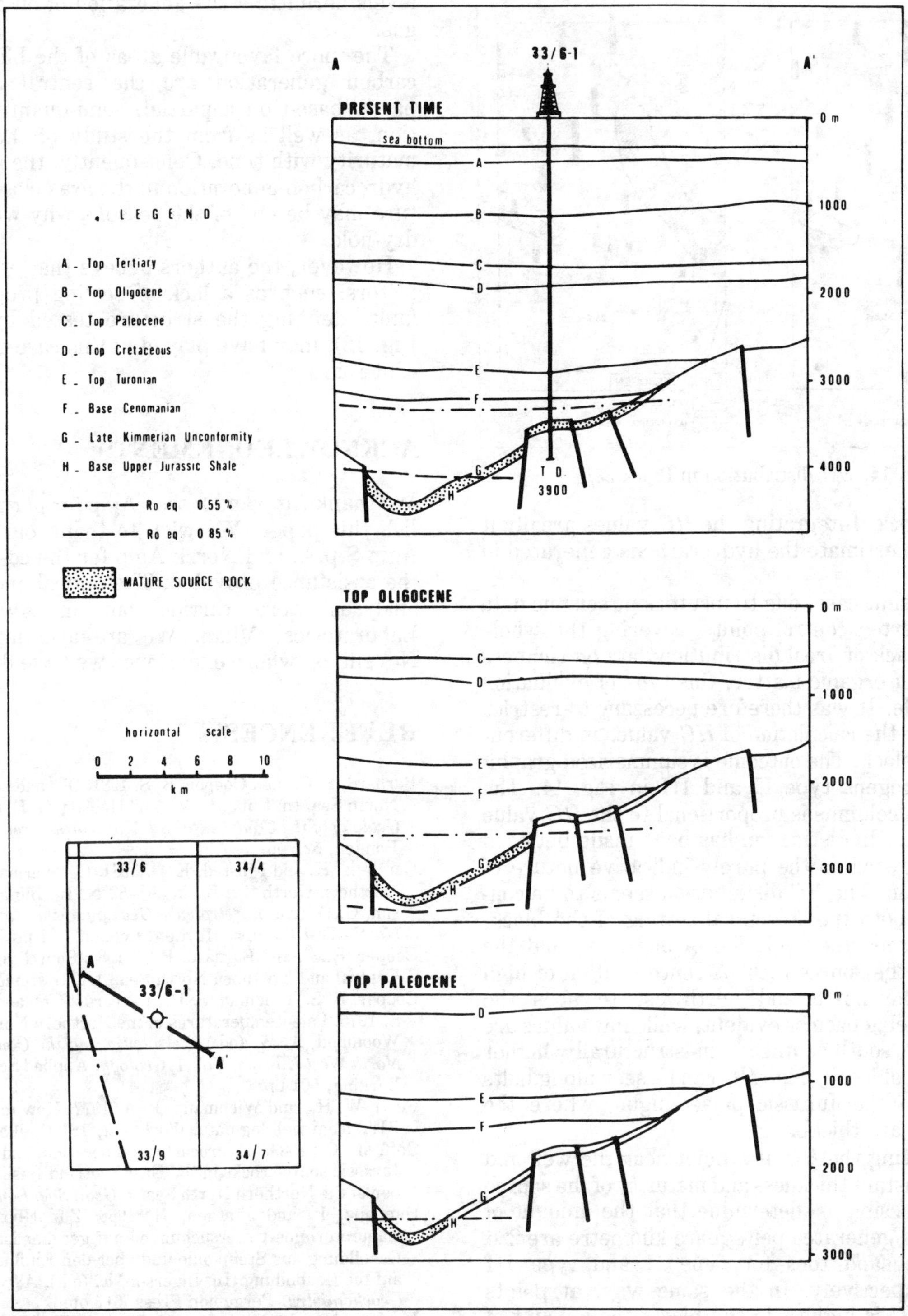

Fig. 13. Regional cross-section through well 33/6–1.

curve. Therefore, an underestimation of the R_0% equivalent is possible at high maturity levels.

Transformation ratios seem to be a more reliable means of investigating overmature situations. Zones of wet gas and condensate generation are to be expected where these values are higher than 0.6, as they are today in the central part of the block. The base of the unit has peak values of 0.7 within a basinal area to the north and, locally, in areas of thicker sedimentary accumulation at the base of fault escarpments. The values of transformation ratio suggest that significant gas generation did not occur in the past.

GENERATED HYDROCARBONS

It is possible to calculate the amount of hydrocarbons generated (HC) along a vertical profile at any given point of the basin by applying the method of Tissot and Espitalié. This is achieved by integrating the amount of hydrocarbons generated at each level (hc) over the thickness of the source rock using the following equation:

$$HC = \int_{z_1}^{z_2} hc(z)\, dz,$$

where z_1 and z_2 are, respectively, the top and base of

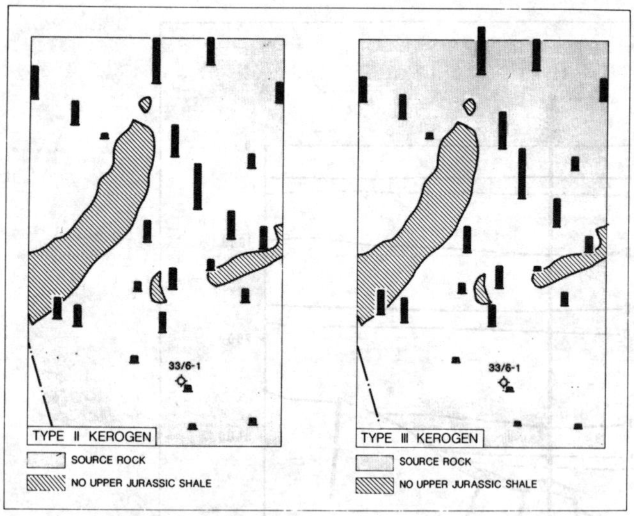

Fig. 14. *HC* distribution in Block 33/6.

the source rock. Integrating the *HC* values areally it is possible to estimate the hydrocarbons generated in the basin.

In the present case, due to insufficient seismic data (depth converted control points) covering the whole area, and a lack of areal distribution data on richness and quality of organic matter, this type of evaluation is not possible. It was therefore necessary to restrict evaluation to the calculation of *HC* values in different parts of the block. The outcome is summarized graphically for kerogens type II and III in Fig. 14. The height of the columns is proportional to the *HC* value in each point. No distinction has been made between oil and gas owing to the purely indicative nature of this evaluation. The *HC* distribution seems to be connected with both the structural setting of the block, which influenced the evolution of maturity, and the thickness of the source rock. A concentration of high values in the north and northwest towards the Møre–Trøndelag basin is evident, while low values are found in the south-southeast in structurally higher zones. Areas of locally high *HC* can be seen along belts bordering on the Jurassic palaeo-highs, where the source rocks are thicker.

By calculating the *HC* at a point near the well and assuming constant thickness and maturity of the source rock, it is possible to determine that the amount of hydrocarbons generated per square kilometre are 226 and 176 thousand tons for type II and type III kerogen, respectively. In the same way, at points where the *HC* is highest in the block, the amount of hydrocarbons generated per square kilometre are 1924 and 1597 thousand tons, respectively, assuming in these calculations a TOC content of 2.4% and a rock density of 2.3 g/cm^3.

CONCLUSIONS

The geochemical analyses carried out on samples taken from the well drilled in the study area have shown that the Kimmeridge Clay Formation and Heather Formation have, on the average, inferior source potential compared with other areas of the Northern North Sea, because of the quantity and, above all, the quality of the organic matter.

A study of the evolution of thermal maturity has demonstrated that vast areas of the Upper Jurassic shales reached maturity as far back as the Paleocene. At present, most of this unit has reached maturation levels permitting the generation of oil and, in places, gas.

The more favourable areas of the block for hydrocarbon generation are the central and northern parts, based on a partial, semi-quantitative evaluation, as well as from the study of the evolution of maturity with time. Consequently, the relatively lean hydrocarbon generation in the area close to the structure may be one of the reasons why well 33/6–1 is a dry-hole.

However, the authors believe that other geological factors, such as a lack of sealing properties in the faults defining the structure tested by 33/6–1 (see Fig. 13), may have played in this case a more significant role.

ACKNOWLEDGEMENTS

We thank Agip and Norsk Agip for permission to publish this paper. We wish to thank our colleagues in Agip S.p.A. and Norsk Agip for the co-operation and the assistance provided. The optical and geochemical analyses were carried out in Agip Geological Laboratories, Milan. We are also indebted to L. Novelli, on whose experience we have drawn heavily.

REFERENCES

Barnard, P. C., and Cooper, B. S. 1981. Oils and source rocks of the North Sea. In: Illing, L. V., and Hobson, G. D. (eds.), *Petroleum Geology of the Continental Shelf of North-West Europe*. Heyden, London, 521 pp.

Carstens, H. and Finstad, K. G. 1981. Geothermal gradients of the Northern North Sea Basin 59°–62°N. In: Illing, L. V. and Hobson, G. D. (eds.), *Petroleum Geology of the Continental Shelf of North-West Europe*. Heyden, London, 521 pp.

Cooper, B. S., and Barnard, P. C. 1984. Source rocks and oil of the Central and Northern North Sea. *AAPG Memoir*, **35**, 303–314.

Cooper, B. S., Coleman, S. H., Barnard, P. C. and Butterworth, J. S. 1975. Paleotemperatures in the Northern North Sea Basin. In: Woodland, A. W. (ed.), *Petroleum and the Continental Shelf of North West Europe*, Vol. 1, *Geology*. Applied Science Publishers, Barking, 501 pp.

Fertl, W. H., and Wichmann, P. A. 1977. How to determine static BHT from well log data. *World Oil*, **184**(1), 105–106.

Goff, J. C. 1983. Hydrocarbon generation and migration from Jurassic source rocks in the East Shetland Basin and Viking Graben of the Northern North Sea. *J. Geol. Soc. Lond.* **140**, 445–474.

Hanbaba, P. and Jüntgen, H. 1968. Zür Ubertragbarkeit von Laboratoriums-Untersuchungen auf geochemische Prozesse der Gasbildung aus Steinkohle und über den Einfluss von Sauerstoff auf die Gasbildung. In: Ingerson, E. (ed.), *Advances in Organic Geochemistry*. Pergamon Press, 617 pp.

Hanbaba, P., Jüntgen, H. and Peters, W. 1968. Nicht isotherme Reaktionskinetic der Kohlenpyrolyse. *Brennstoff-Chemie*, **49**(12), 368–376.

Jüntgen, H. and van Heek, K. H. 1968. Gas release from coal as a function of the rate of heating. *Fuel*, **48**, 103–117.

McKenzie, D. 1978. Some remarks on the development of sedimentary basins: *Earth Planet. Sci. Lett.*, **40**, 25–32.

Skarpnes, O., Briseid, E. and Milton, D. 1982. 34/10 Delta prospect of the Norwegian North Sea: Exploration study of an unconformity trap. In: Halbouty, M. T. (ed.), *The Deliberate Search for the Subtle Trap*. *AAPG Memoir*, **32**, 207–216.

Tissot, E., and Espitallé, J. 1975. L'évolution thermique de la matériel organique des sédiments: applications d'une simulation mathématique. *Inst. Français Petrole Rev.*, **30**, 743–777.

Waples, D. W. 1980. Time and temperature in petroleum formation: application of Lopatin's method to petroleum exploration. *AAPG Bull.*, **64**, 916–926.

Yarzab, R. F., Lerche, I. and Kendall, C. 1983. Relationship of vitrinite reflectance to heat flow history of North Sea. *AAPG Bull.*, **67**(3), 574.

Yükler, M. A. and Kokesh, F. 1984. A review of models used in petroleum resource estimation and organic geochemistry. In: Brooks, J., and Welte, D. H. (eds.), *Advances in Petroleum Geochemistry*, Vol. I. Academic Press, London, 344 pp.

17 Maturity studies of the Jurassic 'Coal Unit' in three wells from the Haltenbanken area

J. L. Pittion and J. Gouadain**
*Total–CFP 218/228, Avenue du Haut Lévèque, 33605 Pessac Cedex, France
†Total Marine Norsk, P.O. Box 4317, N.5013 Nygårdstangen, Norway

Source-rock analyses including Rock–Eval, elemental composition, maceral and extract analyses were carried out on coal samples from the Lower Jurassic 'Coal Unit' of the Haltenbanken area. Maturation level was determined using vitrinite reflectance, pyrolysis temperatures (T_{max}) and fluorescence of exinite. The source-rock potential of the coals appears to be fair (hydrogen indices between 100 and 350 mg/g) and similar to the oil and gas source-rock coals in the Mahakam Delta, Indonesia. The maturation stage of the 'Coal Unit' directly relates to its present depth, i.e. it depends upon the thickness of the post-Jurassic overburden. The oil window (0.6–1.3% R_0) lies between about 3500 m and 4800 m. Oil and gas are probably generated from both Upper Jurassic shales and the Coal Unit in this depth interval, which is most commonly found in the transition zone between the Trøndelag Platform and the Møre Basin. The same source rocks, where more deeply buried in the basin, have probably also generated oil at some time in the past, but this oil would have been cracked to gas unless it has migrated out of the cracking zone. From these maturation data, there is little or no chance of finding oil in reservoir sands below 5000 m.

INTRODUCTION

The Mesozoic sediments of the northern Norwegian continental shelf contain a coaly sequence about 400 m thick, Rhaetic to Hettangian in age, commonly called the 'Coal Unit'. This formation, where coal seams are interbedded with sandstones and shales, is most suitable for a maturity study since pure coals provide much more reliable data on vitrinite reflectance and Rock–Eval pyrolysis (Espitalié *et al.*, 1977) than does dispersed organic matter in rocks. Moreover, in coal analyses, there are no secondary influences on maturity indices due to changes in type of organic matter or mineral matrix effects.

GEOLOGICAL SETTING

In the Haltenbanken area (see Fig. 1), above marine Permian deposits, the Mesozoic starts with a thick continental and evaporitic Triassic section (Jørgensen and Navrestad, 1981). An overall transgressive sequence characterizes the Jurassic, in which coals—Rhaetic to Hettangian in age—were deposited which have a wide regional distribution. Above this coaly sequence, sands of Early to Middle Jurassic age are interbedded with shales. Organic-rich Upper Jurassic shales (equivalent to the Kimmeridge Clay Formation of the North Sea) overlie these sands. Above, Cretaceous and Tertiary formations are composed mainly of claystones which are occasionally calcareous (see Fig. 2).

Tectonic activity started at the end of the Triassic but the main 'Kimmerian' phases occurred at the end of the Middle Jurassic and at the beginning of the Cretaceous. These tectonic events were characterized by the formation of horsts and grabens and a general downfaulting of the Triassic–Jurassic formations between the Trøndelag platform and the Vøring and Møre Basins (see Fig. 1). During the Cretaceous and Tertiary, no more faulting activity occurred, and the differences in thickness of the post-Jurassic sediments were influenced by the relative rate of subsidence.

The area studied is characterized by two main structural elements: the Trøndelag Platform and the Halten Terrace (see Fig. 1). In the latter zone, the Triassic–Jurassic formations are downfaulted towards the Møre Basin. Well 'A' is at the western margin of the Trøndelag Platform where the coal Unit occurs between 2200 m and 2600 m. Well 'B' is in a faulted zone west of the Trøndelag Platform, where the Coal Unit lies between 2500 m and 2900 m. In well 'C', which is in the western part of the Halten Terrace, the Coal Unit is much deeper (4200–4600 m), see Fig. 3.

Petroleum Geochemistry in Exploration of the Norwegian Shelf © Norwegian Petroleum Society (Graham & Trotman, 1985) pp. 205–211.

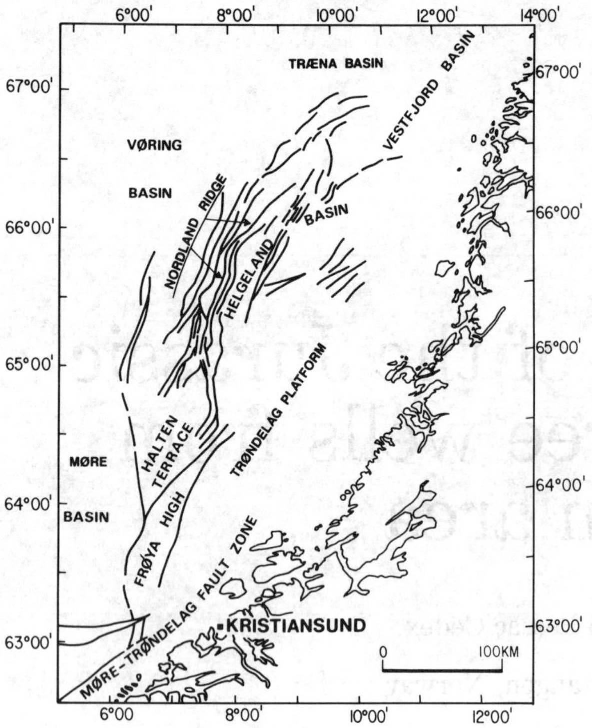

Fig. 1. Structural elements of the Haltenbanken area.

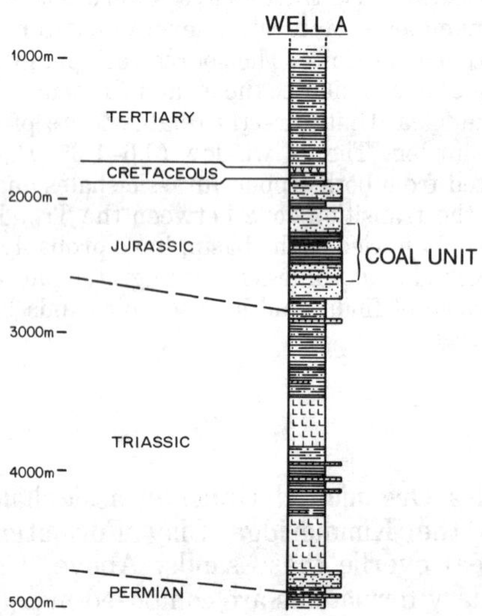

Fig. 2. Generalized stratigraphy, Haltenbanken area.

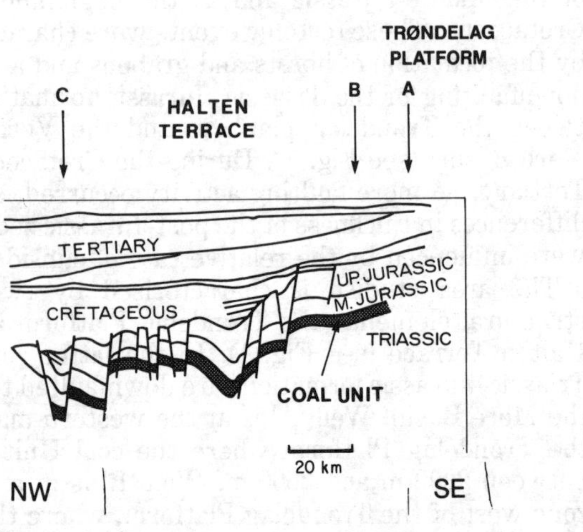

Fig. 3. Haltenbanken area cross-section, with approximate location of the studied wells (after Jørgensen and Navrestad, 1981).

MATURITY

Vitrinite reflectance

Coal beds are abundant in the Coal Unit, providing adequate and reliable material for vitrinite reflectance analyses. In the three wells studied, the petrographical composition of the coal is rather similar. The major component is vitrinite (75–90%) and inertinite represents 5–10% of the coal. Exinite is relatively abundant (5–20%) and is concentrated preferentially as alternations of pure vitrinite and exinite-rich beds. The exinite fraction consists mainly of sporinite. Liptodetrinite, resinite and cutinite are also present in small quantities.

In well A, the vitrinite reflectance of the Coal Unit ranges from 0.35% to 0.4%; in well B, it increases from about 0.4% to about 0.55% and in well C, from 0.75% to 1.10%.

In well C, below 4420 m, the exinite is not easily recognizable because its reflectance approaches that of the vitrinite, a classical feature of coals when reaching higher maturity (R_0 = 1.2–1.5%). Simultaneously, vitrinite becomes more fluorescent in a dark brown colour, making the recognition of exinite by fluorescence difficult also.

Reflectance data for the three wells were plotted against depth on the same graph (Fig. 4). Although no reflectance data are available between 3000 m and 4200 m, an extrapolated curve between the three sets of results could reasonably be drawn. It is obvious that reflectance is directly linked to present depth (see Fig. 4). From the general shape of the curve, the reflectance gradient is linear with depth in the less mature zone, and becomes exponential as maturity increases. This increase of the reflectance gradient with maturity is widely observed and is due to the exponential increase of the kinetic reaction rate with temperature.

From these data, 0.6% R_0 is reached at around 3500 m, and the value 1.3% near 4800 m within the Lower Jurassic Coal Unit. This is only an average relation for the three wells where individual trends may differ since in each well the burial history is not exactly the same. Thus, the 0.6% R_0 and 1.3% R_0 isoreflectance lines may be a little deeper in well C than in wells A and B because they occur in younger sediments.

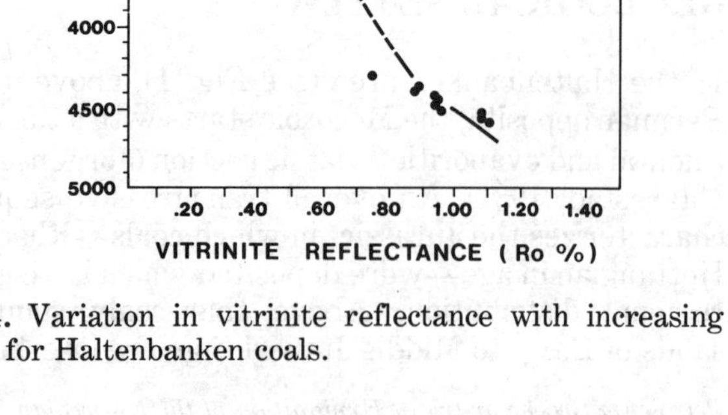

Fig. 4. Variation in vitrinite reflectance with increasing depth for Haltenbanken coals.

Fluorescence of exinite

Observations and measurements were made with a Zeiss u.v. illuminator Fl 400 and excitation filter UG 1. In wells A and B, the different exinite components show a strong fluorescence. Spores, which are the most abundant, show variety in colour and intensity of fluorescence. Moreover, there is often a change of colour beween the internal and the external part of the spores. Surprisingly, the external part shows brighter colour. In these wells, the wavelength corresponding to the maximum intensity in the spectrum (λ_{max}) varies between 560 nm and 630 nm, with a mode at 600 nm (Ottenjahn *et al.*, 1975).

In well C, the fluorescence colour of the spores is much darker: the intensity is too low for spectral analysis except in a sample at the top of the Coal Unit showing λ_{max} around 650 nm. When the maturation increases, spores are increasingly difficult to distinguish from vitrinite in their fluorescence because, as discussed previously, the vitrinite itself becomes more fluorescent. The dark brown colour of the fluorescent vitrinite is of too low an intensity for spectral analysis.

Rock–Eval pyrolysis temperature (T_{max})

In wells A and B, the T_{max} values obtained from the Rock–Eval pyrolysis of the coals do not vary much with depth, remaining between 420°C and 430°C. In well C, the T_{max} values are between 440°C and 450°C, and increase slightly with depth. As with vitrinite reflectance, a direct relationship between T_{max} and increasing depth is observed (see Fig. 5).

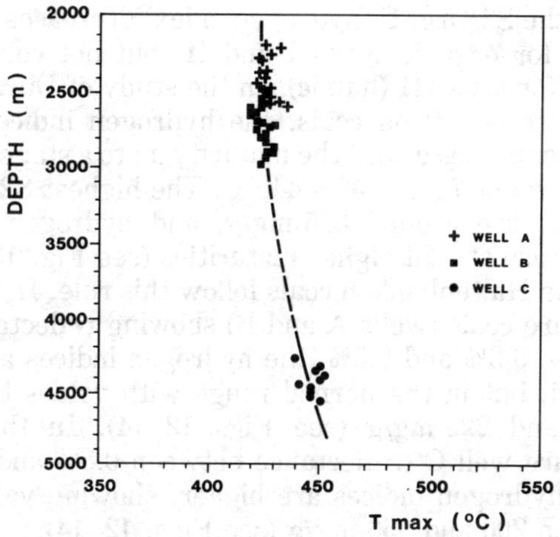

Fig. 5. Variation in Rock–Eval T_{max} with increasing depth for Haltenbanken coals.

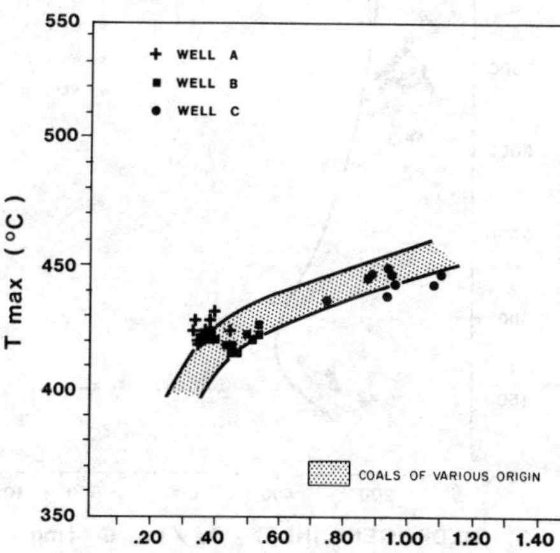

Fig. 6. Correlation between vitrinite reflectance and T_{max} for Haltenbanken coals.

Although the T_{max} gradient is lower than the reflectance gradient, these two parameters show a rather good correlation, as is seen in Fig. 6, where the data from Haltenbanken coals are plotted in comparison with other data from coals of various origins. All are in good agreement.

Other Rock–Eval pyrolysis data

In Rock–Eval pyrolysis, the S1/Org. C index is indicative of the amount of hydrocarbons already generated in the rock. This parameter is a good indicator of maturation, since the generation of hydrocarbons is directly dependent on the thermal evolution of the organic matter.

When only coals are analysed, three secondary factors usually influencing the S1/Org. C results can largely be ignored: (1) any change in organic type; (2) migration effects and (3) the mineral matrix effect. In the case of coals, the S1/Org. C is only dependent on the maturity, and can be considered as a direct parameter to detect the beginning of oil generation. Such an approach was used in the Miocene of the Mahakam Delta, Indonesia, where a drastic increase in the S1/Org. C ratio clearly indicates the beginning of the oil window (see Figs. 7 and 8).

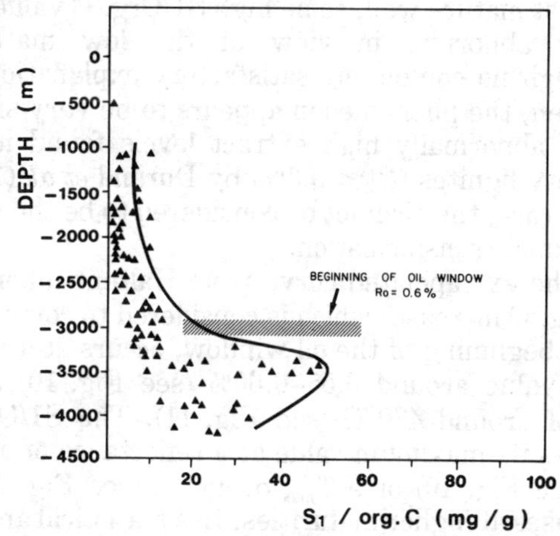

Fig. 7. Variation in the Rock–Eval S1/Org. C (mg hydrocarbons/g organic carbon) with increasing depth for Mahakam coals.

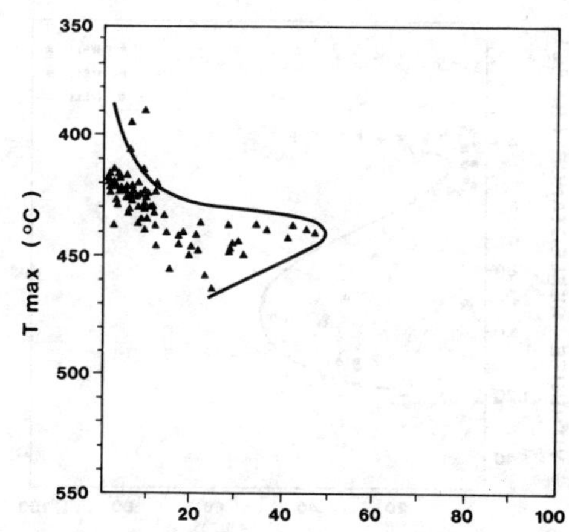

Fig. 8. Variation in the Rock–Eval S1/Org. C (mg hydrocarbons/g organic carbon) with increasing T_{max} for Mahakam coals.

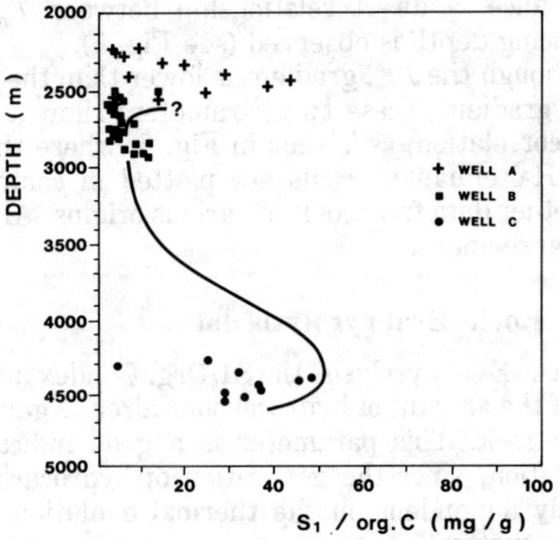

Fig. 9. Variation in the Rock–Eval S1/Org. C (mg hydrocarbons/g organic carbon) with increasing depth for Haltenbanken coals.

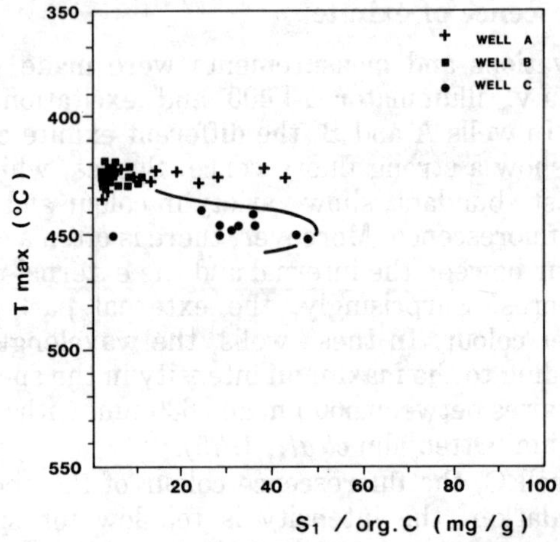

Fig. 11. Variation in the Rock–Eval S1/Org. C (mg hydrocarbons/g organic carbon) with increasing T_{max} for Haltenbanken coals.

In spite of a sampling gap between wells A, B and C, it seems to be acceptable to extrapolate a curve between the three wells for comparison with the Mahakam example. In the three Haltenbanken wells, the S1/Org. C ratio has values between zero and about 50 mg/g. The increase of S1/Org. C is clearly dependent on the depth (see Fig. 9) and thus on the maturation level, at least in wells B and C. In well A, which is the least mature well, some high S1/Org. C values are clearly abnormal in view of the low maturity. Although no completely satisfactory explanation can be given, the phenomenon appears to be very similar to the abnormally high extract levels found in low maturity lignites ($R_0 < 0.3\%$) by Durand *et al.*(1975). In any case, they cannot be considered to be the result of thermal transformation.

In the extrapolated curve from Haltenbanken, the S1/Org. C increase, which is considered to correspond to the beginning of the oil window, occurs at a reflectance value around 0.55–0.60% (see Fig. 10) and a T_{max} of around 430°C (see Fig. 11). The S1/Org. C reaches its maximum value at a reflectance of around 1% (see Fig. 10) or a T_{max} of 450°C (see Fig. 11). It decreases at higher maturities. In a statistical analysis of Rock–Eval pyrolysis data from coals of various origins, Durand and Paratte (1983) have shown highest S1/Org. C values of about 70 mg/g, which correspond to T_{max} = 450°C and R_0 = 1%. The decrease in S1/Org. C at higher maturities is explained by expulsion of hydrocarbons from the coals. Nevertheless, this decrease could also be due to part of the liquid hydrocarbons being transformed into gas, which would then escape more easily out of the coal. In this case, the decrease of the S1/Org. C as seen in Fig. 10 could be considered to be an indicator of the end of the oil window.

According to the concept of Rock–Eval pyrolysis, when the maturity increases, the S1/Org. C increases and the S2/Org. C (hydrogen index) decreases. This is true for organic types I and II, but not completely true for type III (humic). In the study of Durand and Paratte (1983) on coals, the hydrogen indices, as a whole, increase with the maturity up to a stage around 0.7% R_0 or T_{max} = 435–440°C. The highest S2/Org. C values are around 325 mg/g, and hydrogen indices decrease at still higher maturities (see Fig. 13).

The Haltenbanken coals follow this rule. In the less mature coals (wells A and B) showing reflectance between 0.3% and 0.5%, the hydrogen indices are scattered, but in the normal range with values between 100 and 225 mg/g (see Figs. 12–14). In the more mature well C (reflectance between 0.8% and 1.1%), the hydrogen indices are higher, showing values between 200 and 350 mg/g (see Figs. 12–14).

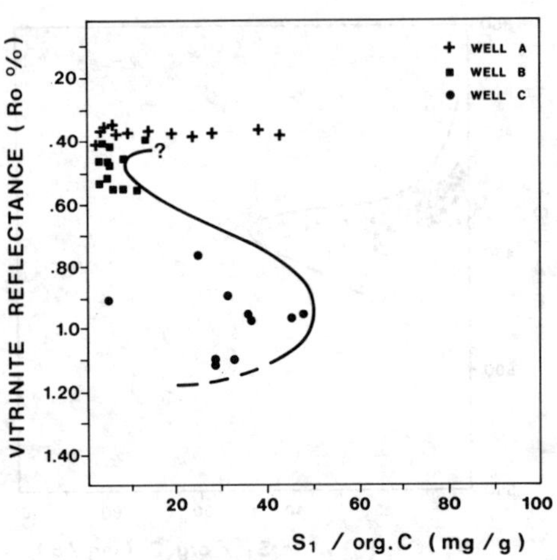

Fig. 10. Variation in the Rock–Eval S1/Org. C (mg hydrocarbons/organic carbon) with increasing vitrinite reflectance for Haltenbanken coals.

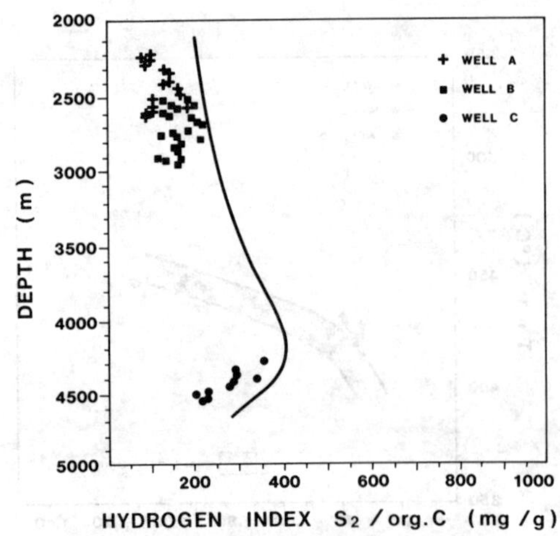

Fig. 12. Variation in the Rock–Eval S2/Org. C (mg hydrocarbons/g organic carbon) with increasing depth for Haltenbanken coals.

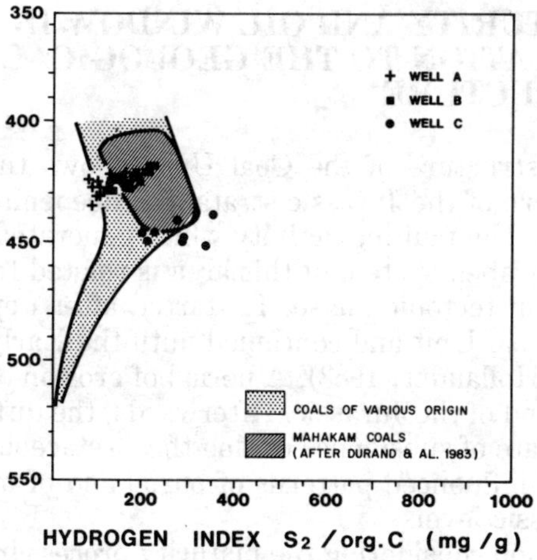

Fig. 13. Variation in the Rock–Eval S2/Org. C (mg hydrocarbon/g organic carbon) with increasing T_{max} for Haltenbanken coals (wells A, B and C) compared with coals of various origins.

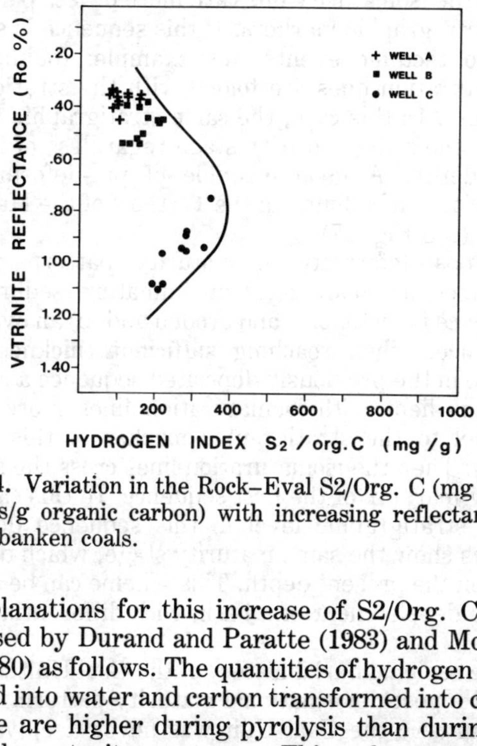

Fig. 14. Variation in the Rock–Eval S2/Org. C (mg hydrocarbons/g organic carbon) with increasing reflectance for Haltenbanken coals.

Explanations for this increase of S2/Org. C were proposed by Durand and Paratte (1983) and Monin *et al.* (1980) as follows. The quantities of hydrogen transformed into water and carbon transformed into carbon dioxide are higher during pyrolysis than during the natural maturity process. This phenomenon is emphasized in the immature oxygen-rich coals. In the more mature coals, which are poorer in oxygen, this distortion is smaller. Thus, the hydrogen indices in the mature coals are closer to the 'true' petroleum potential, whereas in the immature samples they are underestimated.

In the Haltenbanken coals, the highest S2/Org. C values are around 350 mg/g and correspond to a T_{max} of around 450°C (see Fig. 12) and a reflectance of around 0.9% (see Fig. 14). Due to the discontinuity in the sampling in Haltenbanken, it is not possible to confirm if the peak values correspond to a maturity stage of around 0.7% R_0, as indicated by Durand and Paratte (1983). Nevertheless, above 0.9% R_0, the hydrogen indices decrease clearly (see Fig. 14), indicating the proximity of the end of the oil window.

Elemental analyses (atomic H/C and O/C ratios)

In wells A and B, most of the H/C values for the Coal Unit are between 0.75 and 1, with O/C between 0.15

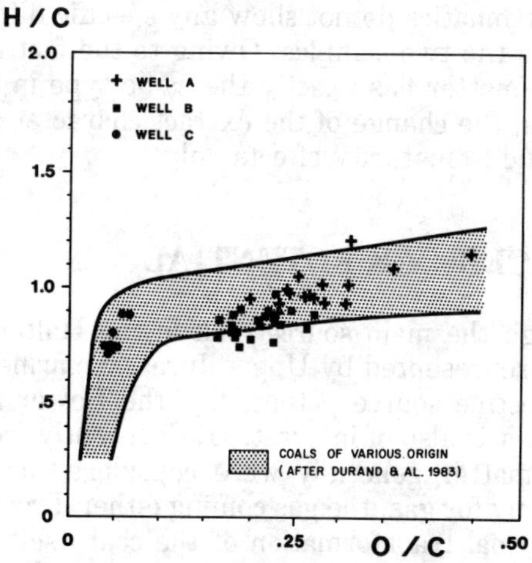

Fig. 15. Evolution of elemental analyses of Haltenbanken coals (wells A, B and C) compared with coals of various origins.

and 0.45. In well C, H/C is a little less, lying between 0.7 and 0.9, with O/C between 0.03 and 0.05. These results are in agreement with the normal effects of maturation on H/C and O/C values. As can be seen in Fig. 15, the Haltenbanken coals fall within the classical evolution path in a Van Krevelen diagram, although a few samples from well B have H/C values which are a little lower than average.

Extract analyses

Two extracts were performed on a sample of mixed coals and carbonaceous shales. One extract was taken in well B near 2600 m, i.e. in the immature zone (R_0 = 0.4%), and another in well C at 4457 m, i.e. in the mature zone near the end of the oil window (R_0 = 1.1%). The extracts from well B show typical features of immature humic matter: high CPI value (2.01), high pristane/C17 ratio (1.09) and abundance of saturates in the C25–C30 range (see Fig. 16).

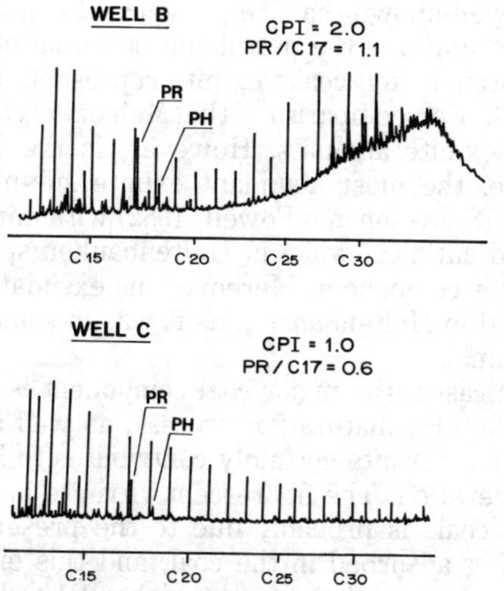

Fig. 16. Saturated hydrocarbon chromatograms from Haltenbanken coals; extracts from well B (R_0 = 0.4%) and well C (R_0 = 1.1%).

The extract from well C clearly shows more mature characteristics: CPI around 1 (1.03), lower pristane/C17 (0.62) and phytane/C18 (00.54). and predominance of light saturates (C15–C20) (see Fig. 16).

The aromatics do not show any specific differences between the two samples. Owing to the fact that the organic matter has exactly the same type in the two samples, the change of the extract character appears to be due to maturity effects only.

SOURCE-ROCK POTENTIAL

Although the main source-rock in the Haltenbanken area is represented by Upper Jurassic marine shales, the effective source potential of the Lower Jurassic Coal Unit is also of interest. Until recently, coals and humic matter generally were considered as source-rocks only for gas, the gas coming either directly from the thermal transformation of the coal itself or from the cracking of the oil remaining adsorbed in a coal. Geochemical studies and geological considerations have shown that, in some petroleum basins, the source-rocks for the oil found were coals or carbonaceous shales. Such examples include the Niger Delta (Nigeria), Mahakam Delta (Indonesia), and the Mackenzie Delta (Canada).

From the elemental analyses, the coals of the three Haltenbanken wells show the characteristics of classical humic coals at somewhat different maturity stages. Their hydrogen contents are not especially above the average. In the more mature well C, the H/C ratios are a little lower than in A and B. This decrease in hydrogen content is attributed to hydrocarbon generation. Nevertheless, it is to be recalled that the hydrogen indices from pyrolysis are higher at well C than in A or B. These two parameters, H/C and hydrogen index, which are dependent on the same process (i.e. hydrocarbon generation) vary here in an opposite way. As seen previously, the hydrogen indices are probably underestimated in the less mature coals for analytical reasons. The elemental analysis corresponds better to the natural evolution path. In any case, the hydrogen indices in well C are high values for typical coals (200–350 mg/g), even higher than those of the coals from the Mahakam Delta in the same range of maturity (see Fig. 12).

It is well-known that the presence of exinite contributes to improve the petroleum potential of coal. In the Haltenbanken coals, exinite represents 5–20% of the coal. For comparison, the Mahakam coals show similar exinite amounts. However, in the Mahakam example, the most abundant exinite sub-maceral is resinite (Snowdon and Powell, 1982) with minor sporinite and cutinite, while in Haltenbanken sporinite is the major component. Moreover, no exsudatinite was observed in Haltenbanken, as it was in some coals in Mahakam.

In all cases, the major coal component is vitrinite. In the natural maturation process, as well as during pyrolysis, vitrinite certainly contributes to hydrocarbon generation. The fluorescent vitrinite in the more mature coals is probably due to the presence of oil remaining adsorbed in the coal, and this oily micro-staining can be related to the higher S1/Org. C ratio seen in well C.

In conclusion, as noted previously, the best source-rocks of this area are probably the Upper Jurassic shales which contain an oil-prone organic matter in rather large quantities. Nevertheless, due to their relatively high hydrogen indices and exinite content, the contribution of the Coal Unit as a potential oil source-rock must also be taken into account.

MATURITY AND OIL WINDOW IN RELATION TO THE GEOLOGICAL STRUCTURE

The structure of the Coal Unit follows the regional pattern of the Jurassic strata, i.e. deepening towards west. The faulting activity which generated the horst and graben system of this area is related to the Kimmerian tectonic phase. It started after deposition of the Coal Unit and continued until the Early Cretaceous (Hollander, 1983). A period of erosion occurred at the end of the Jurassic. Afterwards, the differences in the rate of subsidence during the Cretaceous and Tertiary influenced patterns of burial and of maturity at Jurassic levels.

When considering the maturity process in a general way, two schemes are possible: (a) pre-deformational maturity patterns and (b) post-deformational maturity patterns. Pre-deformational maturity patterns can be summarized as follows. With a normal maturity gradient established from burial in a sedimentary sequence, the isomaturity lines are more or less parallel to the stratigraphic markers. If this sequence is submitted to tectonic events (for example, folding), the isomaturation lines are folded with the stratigraphic markers. In this case, the same stratigraphic marker shows the same maturity stage regardless of its present depth. A good example of pre-deformational maturity can be found in the Tertiary of the Mahakam Delta (see Fig. 17).

A post-deformational maturity pattern can be described as follows: (a) an immature sedimentary sequence is tectonized and eroded and (b) an overlying sequence, when reaching sufficient thickness, will induce in the previously deposited sequence a maturation gradient with isomaturation lines more-or-less parallel to the stratigraphic marker of this second series. Then the isomaturation lines cross the stratigraphic markers of the first sequence. In this case, the same stratigraphic level in this sequence does not always show the same maturity stage, which depends only on the present depth. This scheme can be applied to the development of a basin with differential subsidence.

The regional burial history of the Coal Unit has produced a post-deformational maturity pattern. During the Jurassic the Coal Unit was never deeply buried (probably not more than 1000 m). Downfaulting during the major Kimmerian tectonic events at the end of Jurassic and subsidence during Cretaceous and Tertiary induced differential burial of the Coal Unit (see Figs. 3 and 18) with consequent differences in maturity.

As seen previously from the Rock–Eval data, the

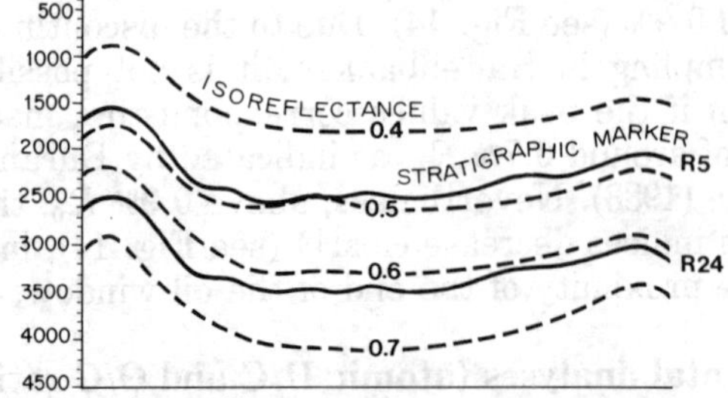

Fig. 17. Example of pre-deformational maturity in the Mahakam Delta, Indonesia.

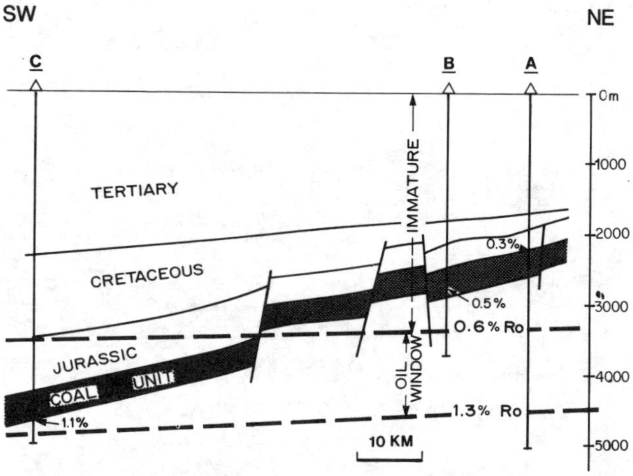

Fig. 18. Regional variation in the maturity and the oil window in relation to the geological structure, Haltenbanken area.

beginning of the oil window in the Coal Unit itself corresponds to 0.6% vitrinite reflectance. The reflectance values commonly accepted for the upper (shallow) limit of the oil window are between 0.5% and 0.7%. Although it is more difficult to detect it directly, the end of the oil window is probably around 1.3%.

The reflectance measurements show that in well A the Coal Unit is immature; in well B, it is at the transition stage between immature and the beginning of the oil window and, in well C, it is close to the end of the oil window. By extrapolation (see Fig. 4), the depth limits of the oil window in the Coal Unit are, respectively, 3500 m and 4800 m. These limits of maturity, established in the coals, can also be applied to the type II organic matter of the Upper Jurassic shales since the maturity needed to start hydrocarbon generation is about the same in these two types of organic matter.

Thus, in well C the lower part of the Coal Unit is close to the end of the oil window and has probably generated the major part of its hydrocarbons, while the Upper Jurassic shales have not yet reached the peak of oil generation. In this well (C), hydrocarbon generation in the Coal Unit started only during the Mid-Tertiary. Of course, in the Upper Jurassic source-rocks hydrocarbon generation began even later. In wells A and B, no major hydrocarbon generation has yet occurred, either in the Coal Unit or in the Upper Jurassic shales.

Deeper in the basin, i.e. west of well C, the source-rocks which are now over-mature have probably previously generated hydrocarbons. If these oils were trapped very close to their source-rocks, they were probably cracked into gas. However, if the oils generated have migrated into reservoirs which are still within the oil window, they will have been preserved from cracking.

For oil exploration, the most interesting area is probably the zone between the Trøndelag Platform and the Møre Basin, because source-rocks there have generated a great part of their hydrocarbons which could have migrated into existing reservoirs. On the other hand, if liquid hydrocarbons were generated at some past time in the now overmature Jurassic source-rocks of the Møre Basin, they could have reached this zone by long-distance lateral migration and may have been preserved from cracking.

In the deep Jurassic reservoirs of the Møre basin, below 5000 m, only gas would be expected. This gas would originate either from cracking of pre-existing oils or from thermal transformation of remaining kerogen.

ACKNOWLEDGEMENTS

We thank Saga and Statoil for permission to publish these results. We are grateful to M. Bellet for technical assistance and to Messrs Laporte, Oudin and Casanova for helpful discussions.

REFERENCES

Durand, B. and Paratte, M. 1983. Oil potential of coals: a geochemical approach. In: Brooks, J. (ed.), *Petroleum Geochemistry and Exploration of Europe*. Blackwell Scientific, Oxford, pp. 255–266.

Durand, B., Nicaise, G., Roucache, J., Vandenbroucke, M. and Hagemann, H. W. 1975. Etude géochimique d'une série de charbons. In: Campos, R. and Goni, J. (eds.), *Advances in Organic Geochemistry 1975*. Enadimsa, Madrid, pp. 601–632.

Espitalié, J., Laporte, J. L., Madec, M., Marquis, F., Leplat, P., Paulet, J. and Boutefeu, A. 1977. Méthode rapide de caractérisation des roches mères, de leur potentiel pétrolier et de leur degré d'évolution. *Rev. Inst. Français Petrol.* **XXXII**(1), 23–42.

Hollander, N. B. 1983. Geohistory and hydrocarbon evaluation of the Haltenbanken area. In: Spencer, A. M., *et al.*, (eds.), *Petroleum Geology of the North European Margin*. Graham and Trotman, London, pp. 383–388.

Jørgensen, F. and Navrestad, T. 1981. The geology of the Norwegian shelf between 62°N and the Lofoten Islands. In: Illing, L. V. and Hobson, G. D. (eds.), *Petroleum Geology of the Continental Shelf of North-West Europe*. Heyden, pp. 407–413.

Monin, J. C., Durand, B., Vandenbroucke, M. and Huc, A. Y., 1980. Experimental simulation of the natural transformation of kerogen. In: Douglas, A. G. and Maxwell, J. R. (eds.), *Advances in Organic Geochemistry 1979*. Pergamon Press, pp. 517–530.

Ottenjahn, K., Teichmuller, M. and Wolf, M. 1975. Spectral fluorescence measurements of sporinites in reflected light and their applicability for coalification studies. In: Albern, B. (ed.), *Petrographie de la matière organique des Sédiments, Relations avec la Paléotempérature et le Potentiel pétrolier*. CNRS, Paris.

Snowdon, L. R. and Powell, T. G. 1982. Immature oil and condensate, modification of hydrocarbon generation model for terrestrial organic matter. *Am. Ass. Petroleum Geologists Bull.*, **66**(6), 775–788.

18 Origin of the gas–condensate of the Midgard Field at Haltenbanken

Arvid Elvsborg, Terje Hagevang* and Torbjørn Throndsen†*
*Saga Petroleum A.S., Norway
†Institutt for Energiteknikk, Norway

The Haltenbanken area has proven to be a significant hydrocarbon province. The discoveries are in Jurassic sandstones, with gas–condensate in 6407/2-2 and 6507/11-1, oil in 6407/1-2, and oil with a gas cap in 6407/1-3. Significant gas shows have been experienced in 6407/2-1 and 6507/11-1 in Cretaceous siltstones. Recently (Summer 1984) Shell has discovered oil in block 6407/9.

Conventional geochemical techniques including carbon isotopes were used to determine the origin of the gas-condensate and the gas shows. A correlation is evident between the various gas–condensates, and between the gas-condensates and the gas shows. The bulk of the gasoline fraction and heavier hydrocarbons was generated during the main phase of hydrocarbon generation at a maturity of around R_0 = 0.95%. The major portion of the gas comes from a more advanced maturity level around the end of the oil window. The Lower Jurassic coal-bearing unit is preferred as the main source for the gas–condensate and the gas shows.

INTRODUCTION

The Haltenbanken licence area is located 150–200 km offshore mid-Norway in a tectonically complex zone that forms the transition between the stable Trøndelag Platform (landward) and the deep sedimentary Møre and Vøring Basins to the south and west (see Fig. 1).

The area is a proven hydrocarbon province, with so far a gas field (Midgard) and a gas/light oil field (Tyrihans) (see Fig. 2). Recently (Summer 1984) Shell discovered oil in block 6407/9.

The aim of this paper is to deduce the source rock and maturity of the gas–condensate in the Midgard Field, which straddles the border of blocks 6507/11 and 6407/2. The field has been tested by one well in each of the two blocks. The hydrocarbons are reservoired in Middle and Lower Jurassic sandstones in a structural trap configuration, as shown in Fig. 3. The reserves are estimated at 104 billion m^3 of gas and 21 million m^3 of condensate. Hence, the condensate yield is relatively low.

The structural and stratigraphic development of the area has been described by Rønnevik and Navrestad (1977), Bøen *et al.* (1984), Bukovics *et al.* (1984) and Hollander (1984). Details of the reservoir sequence can be found in Karlsson (1984) and Larese *et al.* (1984).

SOURCE ROCKS

Two potential source rock intervals are recognized in the Haltenbanken area, the Upper Jurassic Formation H2-2 marine source rock and the Upper Triassic to Lower Jurassic coal-bearing Formation H1-1 (see Fig. 4). The deep 6507/12-2 well, penetrating nearly 2500 m of Late and possibly Middle Triassic deposits, did not prove any deep Triassic source beds (Rønnevik *et al.*, 1983; Hollander, 1984; Wiik Jacobsen and van Veen, 1984). Consequently, it is likely that the gas–condensate is sourced from one or a combination of the two recognized source-rock intervals.

Formation H2-2

The Upper Jurassic Formation H2-2 is a black, organic-rich 'hot' shale equivalent to the North Sea Kimmeridge Clay (Draupne) Formation.

Average organic carbon content, pyrolysate yield (Rock–Eval S2-peak), hydrogen index, kerogen type and likely product are listed in Table 1. Pyrolysis data confirm a high potential for generation of hydrocarbons and correspondingly high hydrogen indices indicate oil source capacity. This is supported by the visual organic composition which is dominated by amorphous, liptinitic (oil-prone) material. This kerogen can be classified as type II.

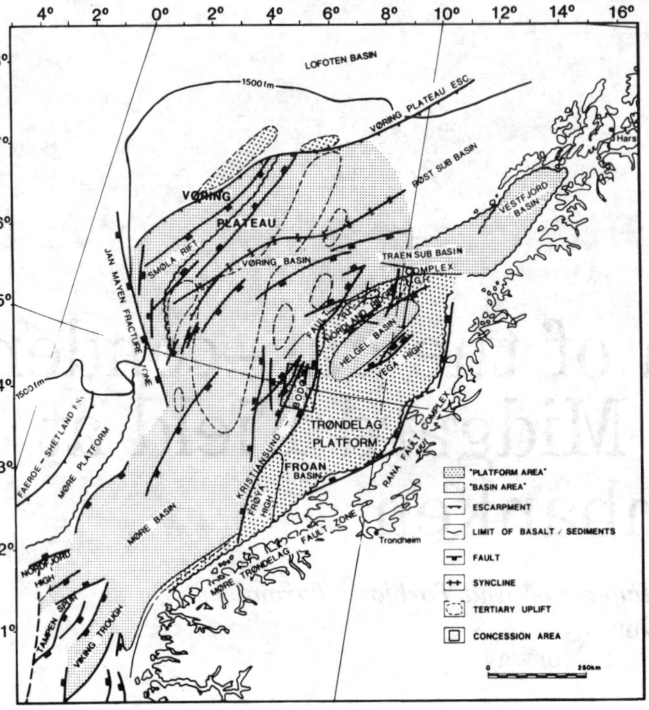

Fig. 1. Main tectonic elements, Møre–Lofoten area.

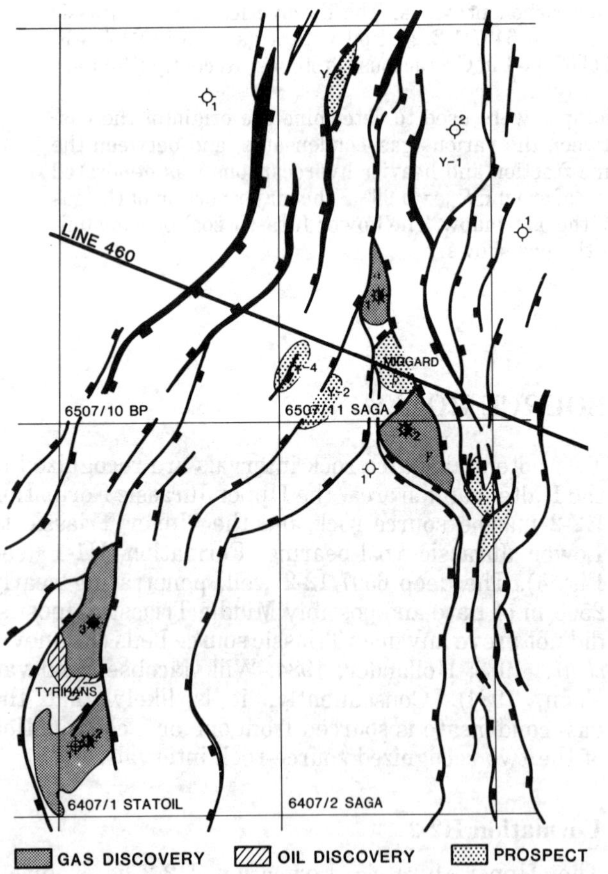

Fig. 2. Location of the Jurassic Midgard and Tyrihans Fields. Seismic line (Fig. 3) is indicated.

The unit is relatively thin in the drilled sections, with thicknesses up to around 60 m. However, sufficient volumes are expected in off-structure areas to make it a significant potential source rock.

A maturity map of Formation H2-2 is shown in Fig. 5. The maturity pattern shows regional variations from immature to peak mature, implying that an active oil source kitchen exists in the area. However, with respect to the drainage area for the Migard Field the area and probably also the volume of peak mature source rock appears to be too small to represent a major hydrocarbon supply (see Fig. 5).

Formation H1-1

The coal-bearing Formation H1-1 has been penetrated with thicknesses of around 400 m which appear to be relatively constant throughout the Haltenbanken area. The sequence consists of alternating sandstone, shale and coal lithologies deposited under paralic conditions.

The frequency and abundance of coal and carbonaceous shale are considerable. Carbon-enriched beds, marked by having densities less than 2.0 g/cm^3, make up a cumulative thickness of 30–60 m. In addition, the unit contains numerous shale intervals more or less enriched in organic carbon, grading from coal/carbonaceous shale to normal shales.

Average organic carbon content, pyrolysate yield (Rock–Eval S2-peak), hydrogen index, kerogen type and likely products are listed in Table 1. These are interpreted average values for the unit as a whole, and indicate a high hydrocarbon potential. The kerogen is dominated by humic organic matter and can confidently be classified as type III, with potential mainly for gas.

A maturity map of Formation H1-1 is shown in Fig. 6. The maturity pattern shows regional varia-

TABLE 1. Source rock data

	Source rock unit	
	FM H2-2	*FM H1-1*
Organic carbon (wt %)	10	8
Pyrolysate yield (mg/g rock)	35	12
Hydrogen index	350	150
Kerogen type	II	III
Predominant hydrocarbon product	Oil	Gas

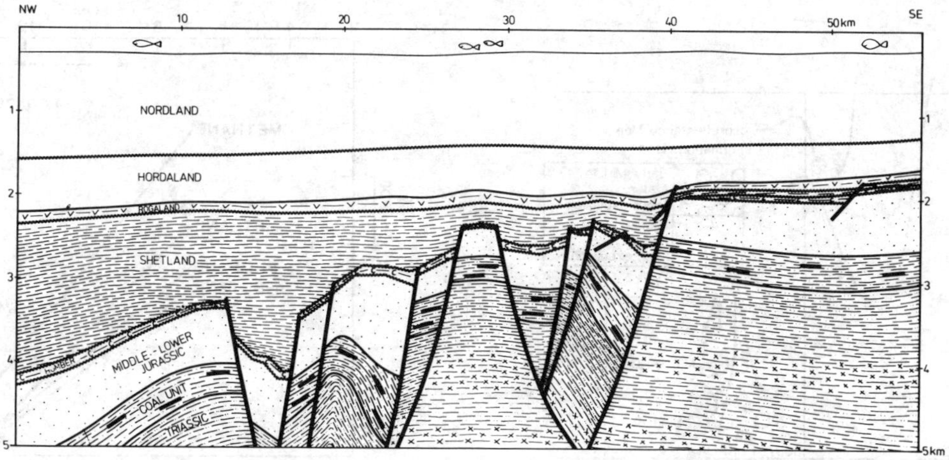

Fig. 3. Structural cross-section over the Midgard Field (from seismic line 460), see Fig. 2.

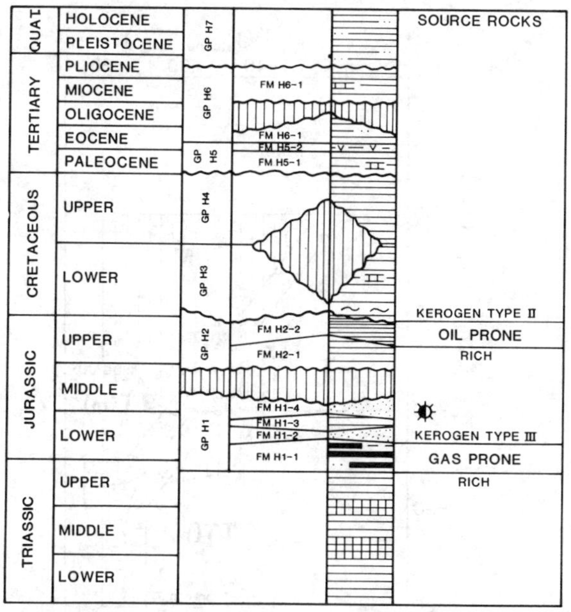

Fig. 4. General stratigraphy for the Haltenbanken area with source rock intervals indicated.

tions from immature to late mature, implying that an active gas generation kitchen exists in the area. Gas generation from Formation H1-1 is areally and volumetrically more important in the drainage area for the Midgard Field than oil generation from Formation H2-2. This is clearly evident from the hydrocarbon generation summary map shown in Fig. 7, which takes into account both oil generation from Formation H2-2 and gas generation from Formation H1-1.

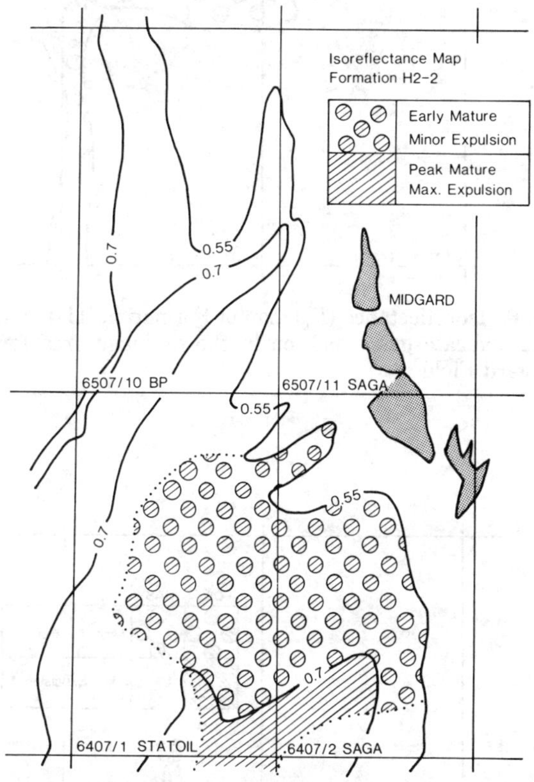

Fig. 5. Isoreflectance (R_0) map of Formation H2-2. Shaded areas indicate oil expulsion in the drainage area for the Midgard Field.

GEOCHEMISTRY OF THE MIDGARD GAS–CONDENSATE

Gaseous hydrocarbons

The molecular composition of the gas is dominated by methane accompanied by a significant proportion of ethane, propane, iso-butane and *n*-butane making up 16.2%. The iso-butane/*n*-butane ratio is 0.60, a value usually taken as indicating association with the main phase of oil generation from a source rock.

The isotopic composition of the methane, with a $\delta^{13}C$ value of −41.6‰, supports a thermogenic origin for the gas. The full suite of hydrocarbon gases has been analysed for carbon isotopic composition. The narrow separations recorded between the wet gas components fit the maturity diagram of James (1983) around a LOM (level of organic metamorphism) value of 13 (see Fig. 8), indicating that the major portion of the wet gas components comes from a fairly advanced maturity level in the source rock. If the gas is sourced from the coal-bearing Formation H1-1, it should be mentioned that James (1983) found that gases derived from coal frequently indicate a source maturity 1 to 2 LOM units too high. The methane value falls above the theoretical line. This discrepancy can, according to James, be interpreted in terms of a mixture between an immature possibly biogenic gas with a fully mature gas, or as a result of cracking of oil. An analogous explanation for the gas origin can be inferred using Schoell's (1983) diagrams shown in Figs. 9 and 10.

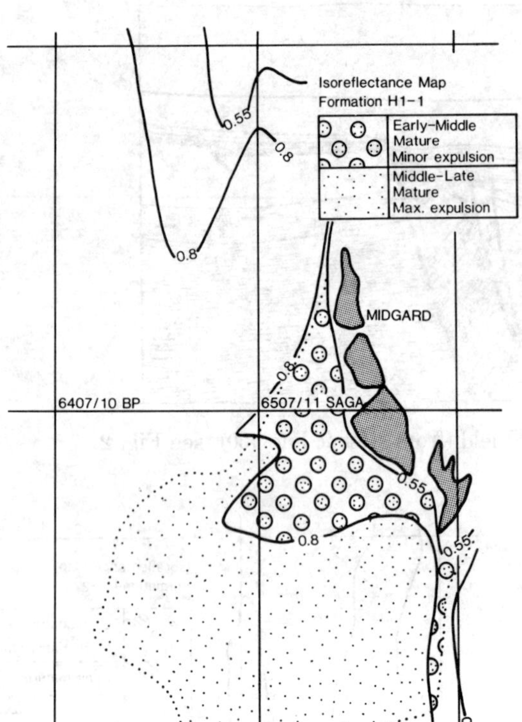

Fig. 6. Isoreflectance (R_0) map of Formation H1-1. Shaded areas indicate gas expulsion in the drainage area for the Midgard Field.

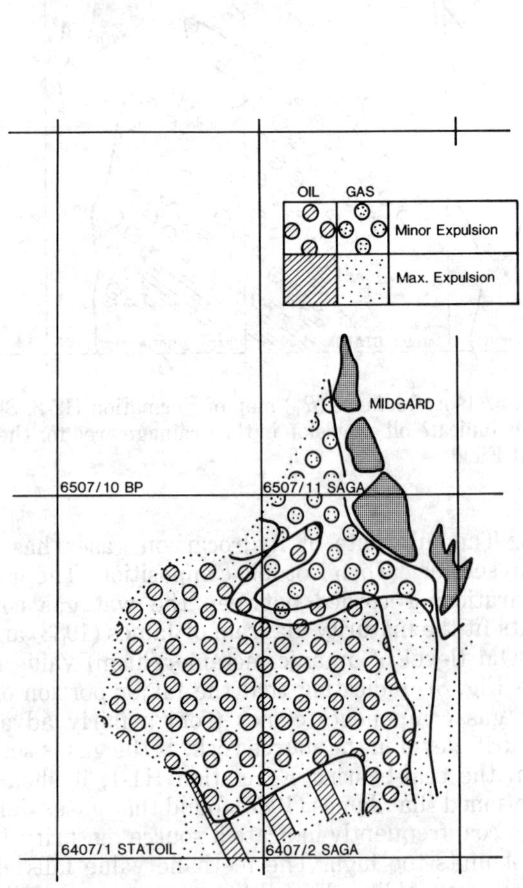

Fig. 7. Hydrocarbon generation summary map for the drainage area of the Midgard Field.

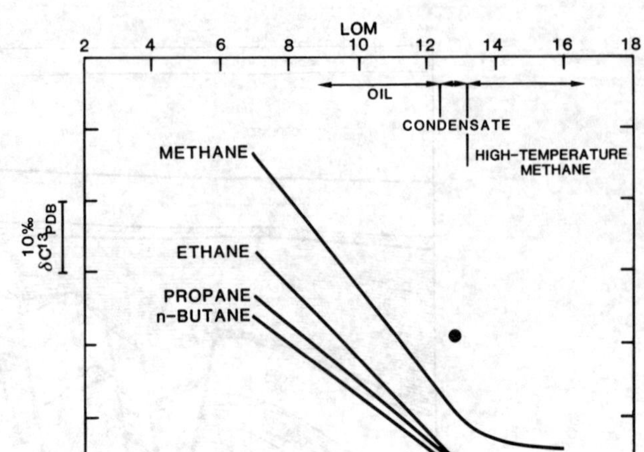

Fig. 8. Gaseous hydrocarbons from the Midgard Field plotted in a maturity diagram from James (1983).

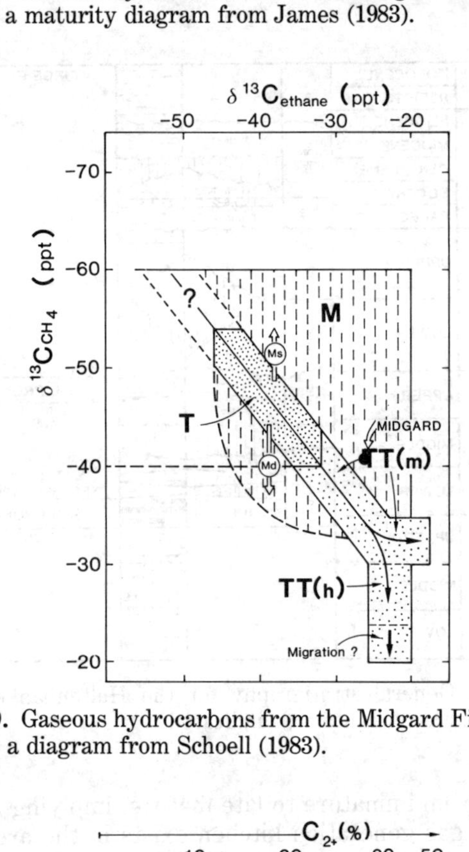

Fig. 9. Gaseous hydrocarbons from the Midgard Field plotted in a diagram from Schoell (1983).

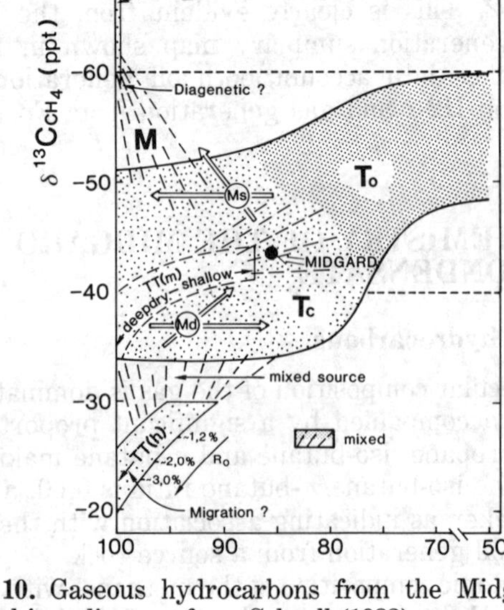

Fig. 10. Gaseous hydrocarbons from the Midgard Field plotted in a diagram from Schoell (1983).

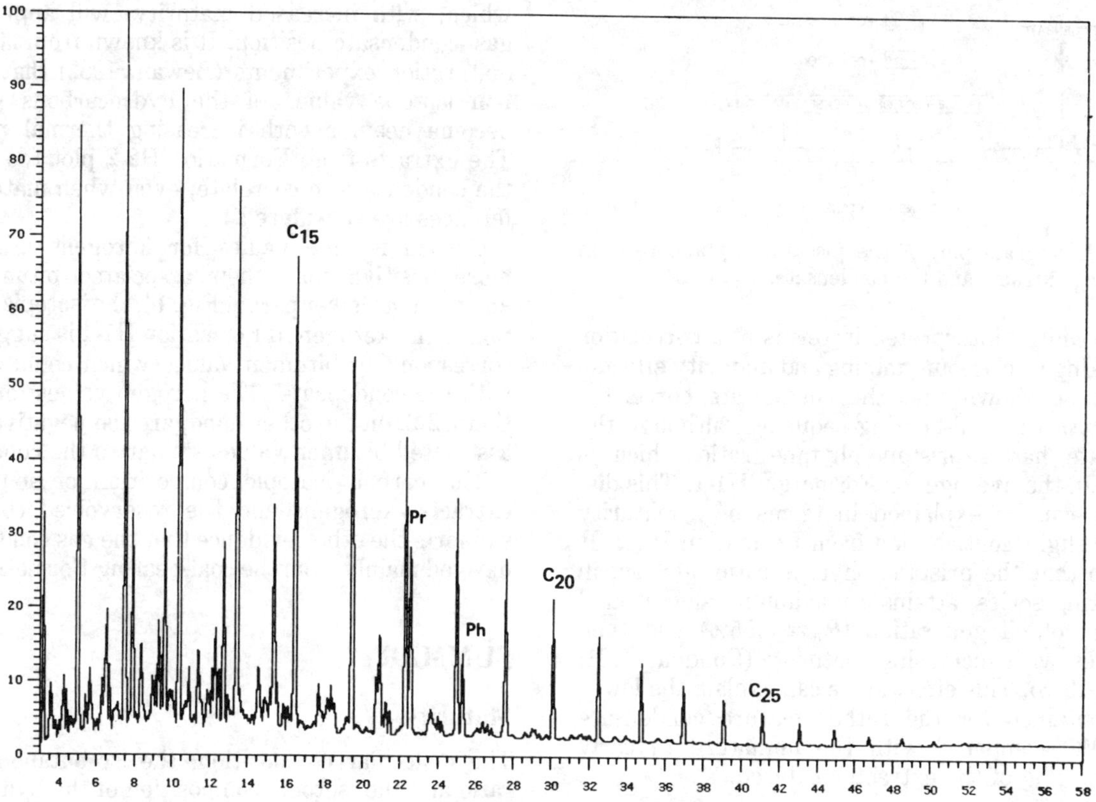

Fig. 11. Gas chromatogram of the saturated hydrocarbons in a stripped condensate sample from the Midgard Field.

C_{15+} hydrocarbons

Gross composition

The gross composition of the C_{15+} hydrocarbons in the condensate is highly paraffinic, consisting of 67.3% saturates, 25.2% aromatics and 7.8% non-hydrocarbons.

Gas chromatography

A gas chromatogram of the saturated hydrocarbons is shown in Fig. 11. The gas chromatogram shows a unimodal, smooth n-alkane distribution. The distribution achieves a maximum below n-C_{17}, followed by a continuous and steep decrease of n-alkanes versus increasing number of carbon atoms extending out to n-C_{30}. There is neither odd nor even predominance of n-alkanes (CPI = 1.0). Based on these features, the sample appears to be relatively mature. On the other hand, the pristane/n-C_{17} ratio is somewhat high (0.6) and could indicate a lower maturity. This can, however, also reflect the predominantly terrestrially-derived nature of the source-rock organic matter, a feature which is supported by the relatively high pristane/phytane ratio of 2.2.

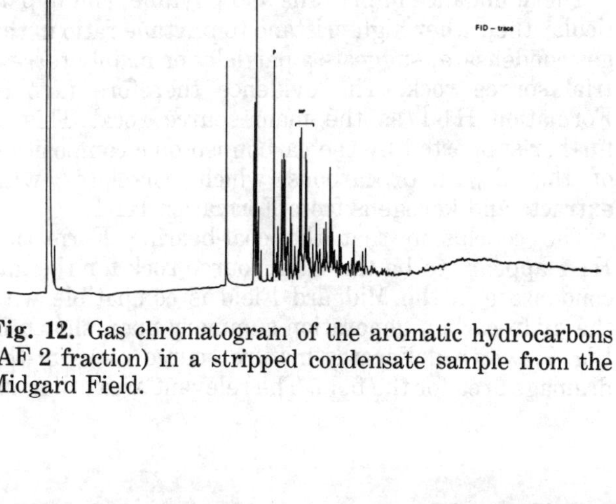

Fig. 12. Gas chromatogram of the aromatic hydrocarbons (AF 2 fraction) in a stripped condensate sample from the Midgard Field.

A gas chromatogram of the aromatic hydrocarbons (AF 2 fraction) is shown in Fig. 12. The methylphenanthrene index MPI 1 and the corresponding vitrinite reflectance value calculated according to the method of Radke *et al.* (1982a, b) and Radke and Welte (1983) indicate that the major portion of the condensate was released from the source rock at a maturity level of around R_0 = 0.9%.

CORRELATION

A correlation was attempted between the gas–condensate and the source rocks based on the C_{15+} hydrocarbons, using pristance/phytane ratios and carbon isotopes. The high carbon number biomarker compounds were not considered as the condensate yield is low and the inherently very low concentrations of high carbon number compounds are readily overprinted by trace amounts of less mature oil or biomarkers dissolved by the condensate on its migration route. They will therefore have characteristics of sediments other than the main source rock.

Pristane/phytane ratio

Pristane/phytane ratios of the condensate and source rock extracts are shown in Fig. 13. Extracts from Formation H2-2 give values ranging around 0.95, which is consistent with the marine anoxic depositional environment for this unit. The coal-bearing Formation H1-1, on the other hand, shows values averaging 4.5. Although neither of these average values is similar to the condensate, with a ratio of

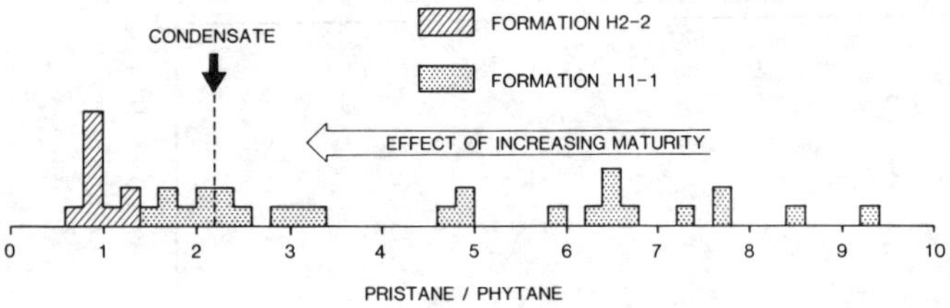

Fig. 13. Histogram plot of the pristane/phytane ratio in source-rock extracts and the condensate.

2.2, this can be interpreted in terms of a correlation when taking into account mixing and maturity effects.

It can be shown that the condensate correlates mainly with the coal-bearing sequence, although the condensate has a pristane/phytane ratio which is lower than the average for Formation H1-1. This discrepancy can be explained in terms of a maturity effect or slight contribution from Formation H2-2. It is known that the pristane/phytane ratio, at least in coal-bearing series, attains a maximum value around the onset of oil generation ($R_0 = 0.55\%$) and then diminishes with increasing maturity (Boudou, 1981; Durand, 1983). This effect alone can explain the lower values obtained for the rather mature condensate ($R_0 = 0.9\%$) compared with the immature to early mature setting of the extracts analysed.

Carbon isotopes

Figure 14 shows a cross-plot of the carbon isotopic composition of the C_{15+} saturate versus aromatic hydrocarbons. The extracts from Formation H1-1 plot in a different position from extracts from Formation H2-2. The Formation H1-1 extracts occur in a position

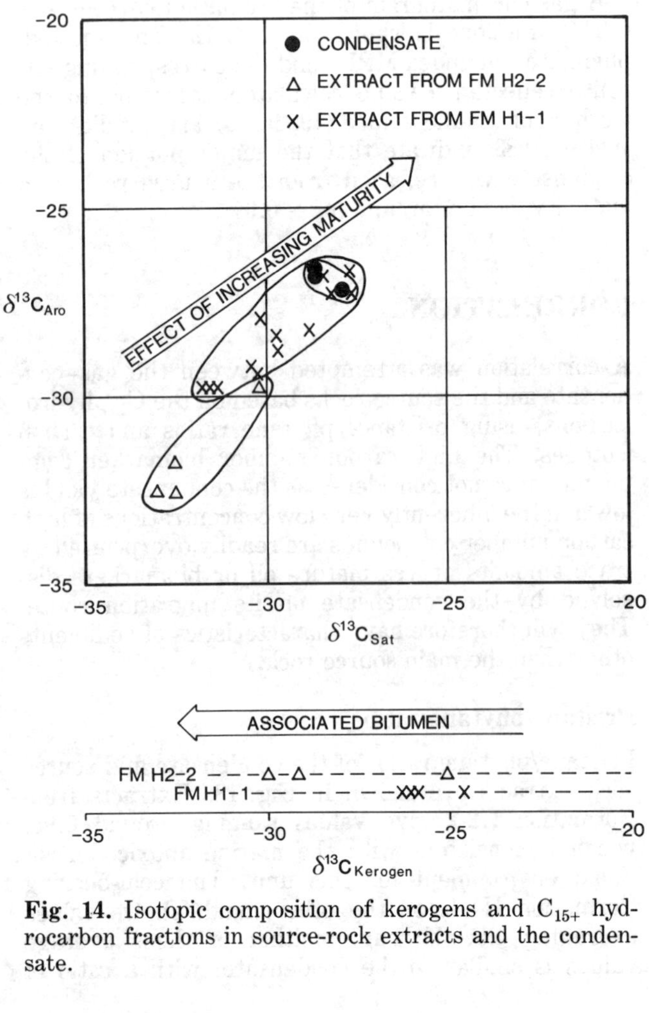

Fig. 14. Isotopic composition of kerogens and C_{15+} hydrocarbon fractions in source-rock extracts and the condensate.

which, with increased maturity, will approach the gas–condensate position. It is known from laboratory maturation experiments (Lewan, 1982) that the carbon isotope values of the hydrocarbons produced become heavier with increasing thermal maturity. The extracts from Formation H2-2 plot too far from the condensate to correlate, even when maturity differences are considered.

Carbon isotope values for kerogens are usually more positive than their associated bitumens. As shown in the lower part of Fig. 14, the isotopic composition of the kerogen in Formation H1-1 is suggestive of corresponding bitumen values which could correlate with the condensate. The kerogen values for Formation H2-2, on the other hand, are too negative to have associated bitumen values similar to the condensate.

The carbon isotopic composition of source rock extracts, kerogens and the reservoired condensate supports the other evidence that the gas condensate is derived mainly from the coal-bearing Formation H1-1.

SUMMARY

Maturity

The molecular composition, the iso-butane/n-butane ratio and the isotopic composition of the hydrocarbon gases clearly support a thermogenic origin for Midgard gas. The narrow isotopic separations between the wet gas components suggest that the gas was released from a source rock at a fairly advanced maturity level towards the end of the oil window.

The pristane/n-C_{17} ratio, on the other hand, points towards a more moderate maturity level. This is also supported by the C_{15+} aromatic hydrocarbons, which indicate a maturity level corresponding to a vitrinite reflectance of $R_0 = 0.9\%$.

A reasonable explanation for these observations is that the bulk of the gasoline range and heavier hydrocarbons were released from the source rock at moderate maturity levels during the main phase of hydrocarbon generation, whereas the major portion of the gas comes from a fairly high maturity close to the end of the oil window.

Source rocks

Two potential source rock intervals are positively identified in the Haltenbanken area. Formation H2-2, which is equivalent to the North Sea Kimmeridge Clay (Draupne) Formation, is a marine sapropelic source rock with type II kerogen and specifically high potential for oil generation. Formation H1-1 is a coal-bearing source rock containing type III kerogen and with high potential for gas and condensate generation.

The abundance of pristane and phytane, and in particular the rather high pristane to phytane ratio in the gas condensate, suggests a partially or mainly terrestrial source rock. The evidence therefore favours Formation H1-1 as the main source rock. This is further supported by the carbon isotopic composition of the C_{15+} hydrocarbons which correlates with extracts and kerogens from Formation H1-1.

The conclusion that the coal-bearing Formation H1-1 appears to be the main source rock for the gas condensate in the Midgard Field is compatible with the hydrocarbon generation summary map, shown in Fig. 7, which delineates mature source rocks in the drainage area for the field. The relevant area is dominated by gas-generative Formation H1-1, whereas the area of oil-mature Formation H2-2 is highly restricted.

ACKNOWLEDGEMENTS

The authors wish to thank Saga Petroleum, Statoil, Norsk Hydro, Shell, Arco, Agip and Deminex for their permission to present this study.

REFERENCES

Boudou, J. P. 1981. Diagenèse organique de sédiments deltaique, delta de la Mahakam, Indonésie. *Thesis*, University of Orleans.

Bukovics, C., Shaw, N. D., Cartier, E. G. and Ziegler, P. A. 1984. Structure and development of the Mid-Norway continental margin. In: Spencer, A. M. *et al.* (eds), *Petroleum Geology of the North European Margin*. Norwegian Petroleum Society/Graham and Trotman, London, pp. 407–424.

Bøen, F., Eggen, S. and Vollset, J. 1984. Structure and basins of the margin from 62 to 69 °N and their development. In: Spencer, A. M. *et al.* (eds), *Petroleum Geology of the North European Margin*. Norwegian Petroleum Society/Graham and Trotman, London, pp. 253–270.

Durand, B. 1983. Present trends in organic geochemistry in research on migration of hydrocarbons. In: Bjorøy, M. *et al.* (eds.), *Advances in Organic Geochemistry 1981*. Wiley, London, pp. 117–128.

Hollander, N. B. 1984. Geohistory and hydrocarbon evaluation of the Haltenbank area. In: Spencer, A. M. *et al.* (eds.), *Petroleum Geology of the North European Margin*. Norwegian Petroleum Society/Graham and Trotman, London, pp. 381–388.

James, A. T. 1983. Correlation of natural gas by use of carbon isotopic distribution between hydrocarbon components. *AAPG Bull.*, **67**, 1176–1191.

Karlsson, W. 1984. Sedimentology and diagenesis of Jurassic sediments offshore Mid-Norway. In: Spencer, A. M. *et al.* (eds.), *Petroleum Geology of the North European Margin*. Norwegian Petroleum Society/Graham and Trotman, London, pp. 389–396.

Larese, R. E., Haskell, N. L., Prezbindowski, D. R. and Beju, D. 1984. Porosity development of selected Jurassic sandstones from the Norwegian and North Seas, Norway—an overview. In: Spencer, A. M. *et al.* (eds.), *Petroleum Geology of the North European Margin*. Norwegian Petroleum Society/Graham and Trotman, London, pp. 81–96.

Lewan, M. D. 1982. Effects of thermal maturation on stable organic carbon isotopes as determined by hydrous pyrolysis of Woodford Shale. *Proceedings of Symposium 'Stable Isotopes in Hydrocarbon Exploration'*, Stavanger. Norwegian Petroleum Society, 15 pp.

Radke, M. and Welte, D. H. 1983. The methylphenanthrene index (MPI): a maturity parameter based on aromatic hydrocarbons. In: Bjorøy, M. *et al.* (eds.), *Advances in Organic Geochemistry 1981*. Wiley, London, pp. 504–512.

Radke, M., Welte, D. H. and Willsch, H. 1982a. Geochemical study of a well in the Western Canada Basin: relation of the aromatic distribution pattern to maturity of organic matter. *Geochim. Cosmochim. Acta*, **46**, 1–10.

Radke, M., Willsch, H., Leythaeuser, D. and Teichmüller, M. 1982b. Aromatic components of coal: relation of distribution pattern to rank. *Geochim. Cosmochim. Acta*, **46**, 1831–1848.

Rønnevik, H. and Navrestad, T. 1977. Geology of the Norwegian Shelf between 62 °N and 69 °N. *Geojournal*, **1**, 33–45.

Rønnevik, H., Eggen, S. and Vollset, J. 1983. Exploration of the Norwegian Shelf. In: Brooks, J. (ed.), *Petroleum Geochemistry and Exploration of Europe. Geol. Soc. Spec. Publ. No. 12*, 71–93. (Blackwell Scientific, Oxford.)

Schoell, M. 1983. Genetic characterization of natural gases. *AAPG Bull.*, **67**, 2225–2238.

Wiik Jacobsen, V. and van Veen, P. 1984. The Triassic offshore Norway north of 62 °N. In: Spencer, A. M. *et al.* (eds.), *Petroleum Geology of the North European Margin*. Norwegian Petroleum Society/Graham and Trotman, London, pp. 317–327.

Part IV
Surface Geochemistry

19 Surface geochemical exploration for hydrocarbons in offshore areas—principles, methods and results

E. J. Schiener, G. Stober* and E. Faber†*
*Deminex GmbH, Essen, FRG
†Federal Institute for Geosciences and Natural Resources (BGR), Hannover, FRG

Deminex has applied direct hydrocarbon detection methods in a number of surveys in the North Sea. Both regional open-grid surveys and detailed surveys with narrow sample spacing were conducted. The advantages and limitations of both bottom sampling and 'sniffer' are compared. The greater sensitivity of bottom sampling outweighs the operational advantage of the 'sniffer'. As most of the areas were surveyed before drilling activities, comparison of the early interpretations with drilling results permits a review of principles applied in interpretation. The necessity of integrating geological/geophysical subsurface information with the geochemical surface signature has emerged as the most important aspect. Interpretation compatible with realistic geological models reduces source- and migration-related risk in exploration by assessing:

presence and maturity of source rocks;
presence and type of hydrocarbons in seismically defined structures and
seal integrity.

As a case history, the results of a geochemical survey in the Trænabanken area are integrated with available subsurface information. The marginally mature Kimmeridge Clay Formation, as the principal oil source rock in the Helgeland Basin, is reflected in the near-surface hydrocarbon indications.

INTRODUCTION

The search for surface expressions of mineral deposits in the subsurface is as old as man's ability to utilze the resources. At the onset of the hydrocarbon age many of the early discoveries of oil were linked to surface seeps (Thompson, 1910). With increasing demand in the 1930s, attention was turning towards means of detecting less conspicuous hydrocarbon manifestations at the surface (Laubmeyer, 1933; Horvitz, 1939). From the early beginnings, the immediate goal was locating and delineating accumulations, mostly in or adjacent to areas where production had been established. Although the original aim has not been abandoned, recent emphasis has shifted towards regional evaluation of hydrocarbon potential and differentiation between oil or gas potential, specifically in offshore areas.

Compared with the multitude of approaches used in onshore investigations (Philp and Crisp, 1982), in offshore work the direct detection and analysis of hydrocarbons by exclusively organic geochemical methods has evolved (Faber *et al.*, 1983; Gevirtz *et al.*, 1983; Faber and Stahl, 1983).

All of these methods assume essentially vertical movement of hydrocarbons from depth, either from mature, generating source rocks or from accumulations. The basic assumptions have not changed during the half century of application. Despite significant advances in understanding and even quantifying some of the subsurface processes of migration, accumulation and retention (Roberts, 1980; Schowalter, 1979; Leythaeuser *et al.*, 1982) a large speculative element remains. This applies particularly to the effectiveness of various transport mechanisms, ranging from diffusion to bulk migration through microfracture systems.

Recent advances in analytical technology as well as in fundamental and applied research (Stahl and Carey, 1975; Bernard, 1978; Faber and Stahl, 1983) have contributed towards lowering detection limits. In particular, the measurement of stable carbon and hydrogen isotope ratios of gas quantities in the microlitre range has increased the value of the analyses. These advances have improved the resolving power of the method, but have also increased the cost of data acquisition.

Proprietary secrecy as well as scientifically underdocumented sales talk have not contributed to improving the reputation of geochemical prospecting for hydrocarbons.

Despite increased use of state-of-the-art science, these methods remain empirical. Only by a joint effort of integrating geological/geophysical information with geochemical data will optimal benefit be obtained.

Deminex has tested direct geochemical hydrocarbon detection methods, predominantly in the North Sea. The following text is a review, with reference to some case histories, of methods and of current ideas as developed through the co-operation between Deminex and the German Federal Institute of Geosciences and Natural Resources (BGR), partially supported by the German Federal Ministry of Research and Technology (BMFT).

DATA ACQUISITION

The standard approach in offshore work, as applied also by Deminex, consists of direct detection and measurement of hydrocarbon components in near surface sediments or in the watercolumn. Two different systems of investigation and data collection have developed:

—spot sampling of bottom/subbottom sediments by standard engineering or marine geological methods, with subsequent extraction of hydrocarbons adsorbed/dissolved in the mineral matrix and pore waters and
—continuous sampling of near-bottom waters by a pumping system suspended from a vessel with almost real-time shipboard extraction and analysis of dissolved hydrocarbons ('Sniffer').

Each of the systems has obvious advantages and shortcomings, which have to be considered together with the objectives of the planned survey.

Subbottom sediments

The least complicated sampling tools are gravity coring devices, possibly with added spring-operated piston devices for improved penetration and core retention. Although it is claimed that data quality is not influenced by depth below the water/sediment boundary (Gérard and Feugère, 1969), most investigators recommend sampling from as deep as technically feasible below the seafloor (Simoneit *et al.*, 1979; B. D. Carey, pers. comm.).

The optimum sampling depth varies, and depends on grain size, composition and related physical properties of the sediment. In clayey sediments, samples from below 1.0 m yield reproducible results, although deeper penetration is generally attempted (~2.0 m). Increasing content of silt/sand-sized components reduces penetration of the sampling tool (<1.0 m) and generates more 'noise' in the analytical results.

Recovered sample material is stored and transported under cryogenic conditions (in liquid nitrogen at between −196 °C and −80 °C) in order to prevent loss of volatile hydrocarbon components.

Bottom sampling is a stop–go operation with high demands on vessel manoeuvrability, station-keeping ability, and on-deck equipment in order to secure acceptable efficiency. There is no operational limit imposed by water-depth. In Deminex's experience the following limitations were found (given in order of decreasing severity):

—despite improved gear-handling equipment the procedure is strongly dependent on relatively calm sea state for progress (maximum force 4–5 Beaufort scale), unless a large vessel equipped with a moon pool is used;
—the sample recovery is strongly controlled by type of seafloor (poor or no recovery in areas with outcropping, consolidated sediments and sandy/gravelly cover) and
—as a minor (but nevertheless possibly irritating) limitation, complications and delays caused by shipping of sample material may occur, where analytical work is performed in onshore laboratories.

The advantages of bottom sampling compared with 'Sniffer' work are:

—sediment degassing and subsequent analysis of gases is highly sensitive, providing information at low gas concentrations;
—repeat analyses can be performed on sample splits;
—the standard analytical spectrum can be expanded if deemed necessary;
—type and character of recovered sediment yield additional information for data interpretation and assessment of data reliability;
—sample position is well-defined within the accuracy of the positioning system and
—sampling equipment can be shipped at lower cost, can be hired or be assembled locally in logistically difficult regions.

Near-bottom water column 'Sniffer'

Sampling is performed through a fish-mounted pump which supplies a continuous flow of water to the shipboard analytical instruments by means of the umbilical towline. Predetermined accumulated quantities of the water are stripped of dissolved gases by vacuum or inert gas assisted by mechanical agitation. At regular short intervals the gas so obtained is analysed for its hydrocarbon components by gas chromatography. In latest developments the accumulated methane fraction of a sufficient number of gas samples is oxidized to CO_2 for later mass spectrometric determination of the carbon isotope ratios, delta $^{13}C/^{12}C$ (Cornford and Dyson, 1984).

The tests performed by Deminex indicate the following limitations:

—maximum towing depth presently does not exceed 300–400 m;
—despite possible sonar-controlled automatic tow-depth adjustments, the height above bottom is kept at approximately 10 m to avoid snagging of tow body, with resulting loss of resolution (Schink *et al.*, 1971);
—positions of detected anomalies do not necessarily coincide with the actual seep, owing to possible displacement by marine/tidal currents;
—for low concentrations of dissolved hydrocarbons the technique is not sufficiently sensitive;
—range of analyses is limited to gas chromatography, possibly with mass spectrometry.

The obvious advantages listed below are not considered to outweigh those of bottom sampling in more-detailed investigations. For reconnaissance surveys covering large areas, the operational ease of the 'Sniffer' outweighs the lack of resolving power. It offers:

—continuous operation, allowing deployment parallel to (for example) seismic data;

—high sampling efficiency and density (at 5 knots, approximately 400 m horizontal distance between sample points) and
—sampling independent of type of seafloor sediments.

In test surveys where 'Sniffer' was run parallel to bottom sampling only pronounced concentration anomalies were detected by the 'Sniffer' method. When results were compared, the anomalies were either due to biogenic methane as shown by the $\delta^{13}C_1$ values in the subbottom samples ($<-60‰$ PDB) or were related to 'seepage' as proven by a dense pattern of subsurface faults (see Fig. 1).

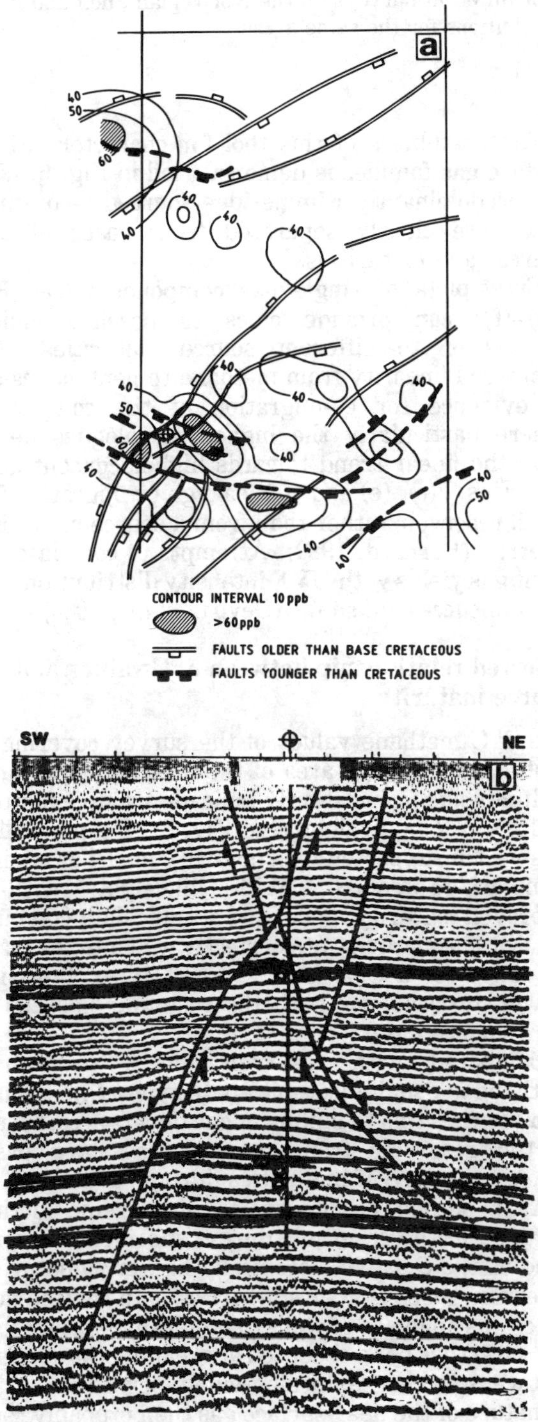

Fig. 1. (a) Surface anomaly (butane) as detected by 'Sniffer', with fault traces at base Cretaceous level. WNW–ESE faults penetrate into the Cretaceous sediments. (b) Corresponding seismic section with fault pattern and well location. ('O' indicates non-commercial oil.)

Analytical procedure

For the investigations presented in this paper the analytical instrumentation and procedure as described by Faber and Stahl (1983) were applied. The procedure consists of the following steps:

—exsolution and desorption of gas from the sediment;
—quantitative determination of hydrocarbon components C_1–C_6 by gas chromatography and
—determination of stable carbon isotope ratios of methane and, where quantities permit, of ethane.

Processing bulk samples in the standard gas stripping procedure frees the 'total gas', a mixture of free and adsorbed gas. It has been established empirically that data obtained by analysing only the adsorbed gas are more consistent in repeat analyses (Sokolov, 1955) and show to a lesser extent evidence of microbial modifications such as oxidation. Processing for adsorbed gas involves separation of the silt/clay fraction from the bulk sample by sieving. The fraction <63 μm is then treated according to the standard procedure.

For additional information on heavier hydrocarbons which may be present, u.v.-fluorescence (UVF) spectrophotometry (fluorimetry) of aromatic compounds in sediment extracts is performed. Two approaches are available:

—fixed-wavelength emission intensity measurement (excitation: 265 nm; emission: 320–365 nm) as pioneered in the USSR (Kartsev *et al.*, 1954) and described by Horvitz (1981) and Gevirtz *et al.* (1983). Aromatic compounds are recovered from the sediment by ultrasonic extraction instead of by the time-consuming exhaustive extraction method originally used and
—total-scanning fluorescence intensity measurement as described by Brooks *et al.* (1983).

RESULTS AND INTERPRETATION

The general principles applied in data interpretation are as follows (Jones and Drozd, 1983; Faber and Stahl, 1984):

—the chemical signature of the near-surface sediment gas, as determined mostly by the stable carbon isotope ratio of its methane (ethane) fraction and to some extent by its molecular composition, is indicative of its genetic origin (biogenic versus thermal) and
—elevated concentration values of thermal surface gas and elevated UVF-emission intensities are the surface expressions of migrated hydrocarbons and are indicative of either on-going generation (presence of mature source rocks) or of pooled hydrocarbons in the subsurface, particularly when related to faults or closed structures.

The geochemical data used in this paper for the interpretation of the near-surface manifestations of hydrocarbons consist of:

—distribution of the hydrocarbon gas components CH_4 through C_6H_{14} expressed in relative percentages and absolute values (ppb, w/w), (see Figs. 2(a) and 11, and Table 1);
—stable carbon isotope ratios of methane, and if possible of ethane, expressed as delta $^{13}C/^{12}C$ (‰ PDB standard) (see Fig. 2(b)) and
—intensities of UVF-emission spectra at wavelengths

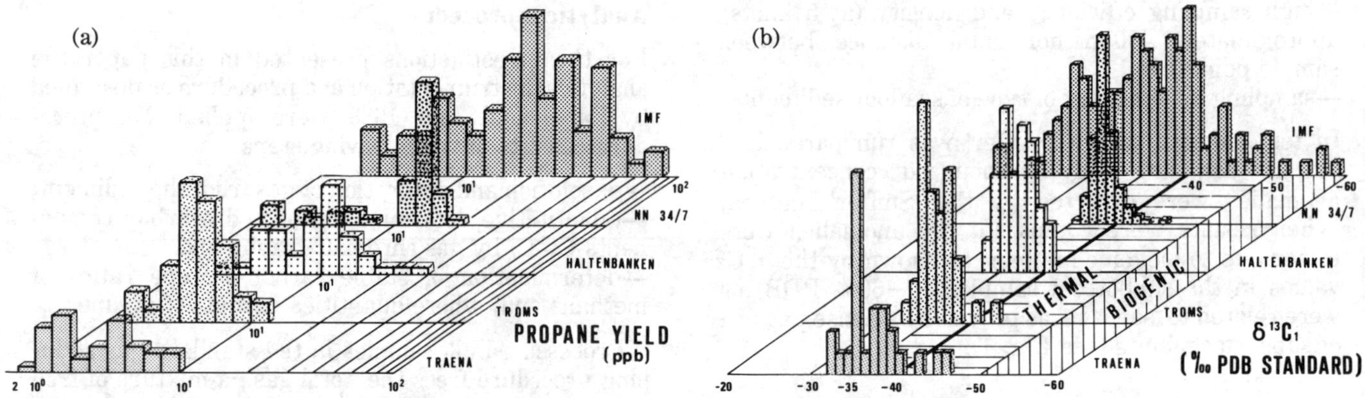

Fig. 2. Comparison of propane concentration distributions (in ppb) for different surveys on the Norwegian Shelf and in the UK Inner Moray Firth (IMF). (b) Comparison of $\delta^{13}C_1$ (‰ PDB) distributions for the same areas.

of 320 nm and 365 nm, with 270 nm as excitation wavelength; alternatively, total scanning fluorescence with numerical and 3-D graphic display (see Fig. 3).

Data statistics

Comparison of data sets is essential in a method that relies to a great extent on experience. However, as evident from Fig. 4 the statistical approach alone has only limited significance, despite claims to the contrary (Jones and Drozd, 1983). Experience on the Norwegian Shelf north of 62°N has shown that the occurrence of more than one major source rock and the burial of source-rock units to different depths, with resulting differences in maturity, generate data that cannot readily be accommodated in standard concepts which were developed for single source rocks.

Under favourable conditions (uniform surface lithologies, uniform age of sea-floor sediments, low sea-bottom temperatures, greater water depths) as found on the Norwegian Shelf, regional differences in magnitude and composition are already recognizable in statistical data sets. The differences, however, are rather subtle and are limited to isotope ratios for composition and gas yields for magnitude (see Figs. 2(a) and (b)). Reasons for the generally low yields on the Norwegian Shelf may be the low seismicity of the Scandinavian Shield which forms the base of the sedimentary section, combined with the widespread, thick and practically unfaulted Cretaceous and Tertiary cover sediments. In the Inner Moray Firth, where numerous faults cut the Cretaceous and where the Tertiary sediments are absent in the western parts, gas yields are higher by a factor of 2–3 (see Fig. 2(a)).

Cross-plots for genetic characterization of gases

Cross-plotting of data sets has been advocated for the genetic characterization of both near-surface gases and reservoir gases. Of particular value are those cross-plots that combine stable carbon isotope ratios with gas component ratios (Bernard, 1978; Faber *et al.*, 1983; Schoell, 1983). Figure 5 shows contrasting data distributions for areas with near-surface biogenic gas (Indonesia) and near-surface thermal gas (North Sea). Apart from a limited area in the Inner Moray Firth, the data of which do not significantly affect the plotted mean in Fig. 6, no evidence of biogenic methane has been seen in near-surface gases of North Sea surveys.

The usefulness of this tool for characterization of surface gas families is demonstrated in Fig. 6, where the predominantly Kimmeridge-sourced near-surface gases are clearly separated from gases of other sources and maturities.

Cross-plots of single gas components (see Figs. 7(a)–(f)) can provide clues to possible multiple sources or to different source maturities. Well-expressed linearity from methane to pentanes is seen as evidence for co-migration of the components. Where particularly the higher homologues deviate from the linear trend towards higher concentrations (see Figs. 7(c), (e) and (f)) leakage or migration from an oil reservoir rather than from a source rock can be inferred (Bernard, 1980). Attempts to correlate sediment gas yields with UVF intensity distributions have been unsuccessful so far (Gevirtz *et al.*, 1983).

Inferred relationship between $\delta^{13}C$ values and source maturity

The $\delta^{13}C$ methane values of the survey covering the postulated drainage area of the Midgard structure of Haltenbanken consistently indicate a source maturity of 1.8% VR equivalent (see Fig. 8). Before drilling, this was unsupported by a maturity model assuming a Kimmeridge source. With hindsight, with an improved maturity model and seismic data combined, the isotopic maturity fits perfectly well into the maturity range for a Coal Unit source (Upper Triassic to Lower Jurassic) for both the reservoir gas and the near-surface gas.

Compared with reservoir gases, near-surface methane is usually isotopically heavier. One explanation ascribes this to mixing of gas during accumulation in the reservoir from sources that have passed through a range of maturities during subsidence.

As a second possibility, fossil biogenic methane retained in the reservoir during subsidence is envisaged as the cause of isotopically lighter signatures of reservoir gases (Faber *et al.*, 1984). In routine headspace gas analyses of wells, isotopically light methane ($<-50‰$) has been observed at great depths in numerous wells in geographically widely spread areas. The signature of the near-surface gas then probably represents the most recent and most mature gas generation at depth without any biogenic input. The low pressure regime of biogenic gas formation precludes migration by pressure differential and favours retention in solution in the pore water (Barker, 1984).

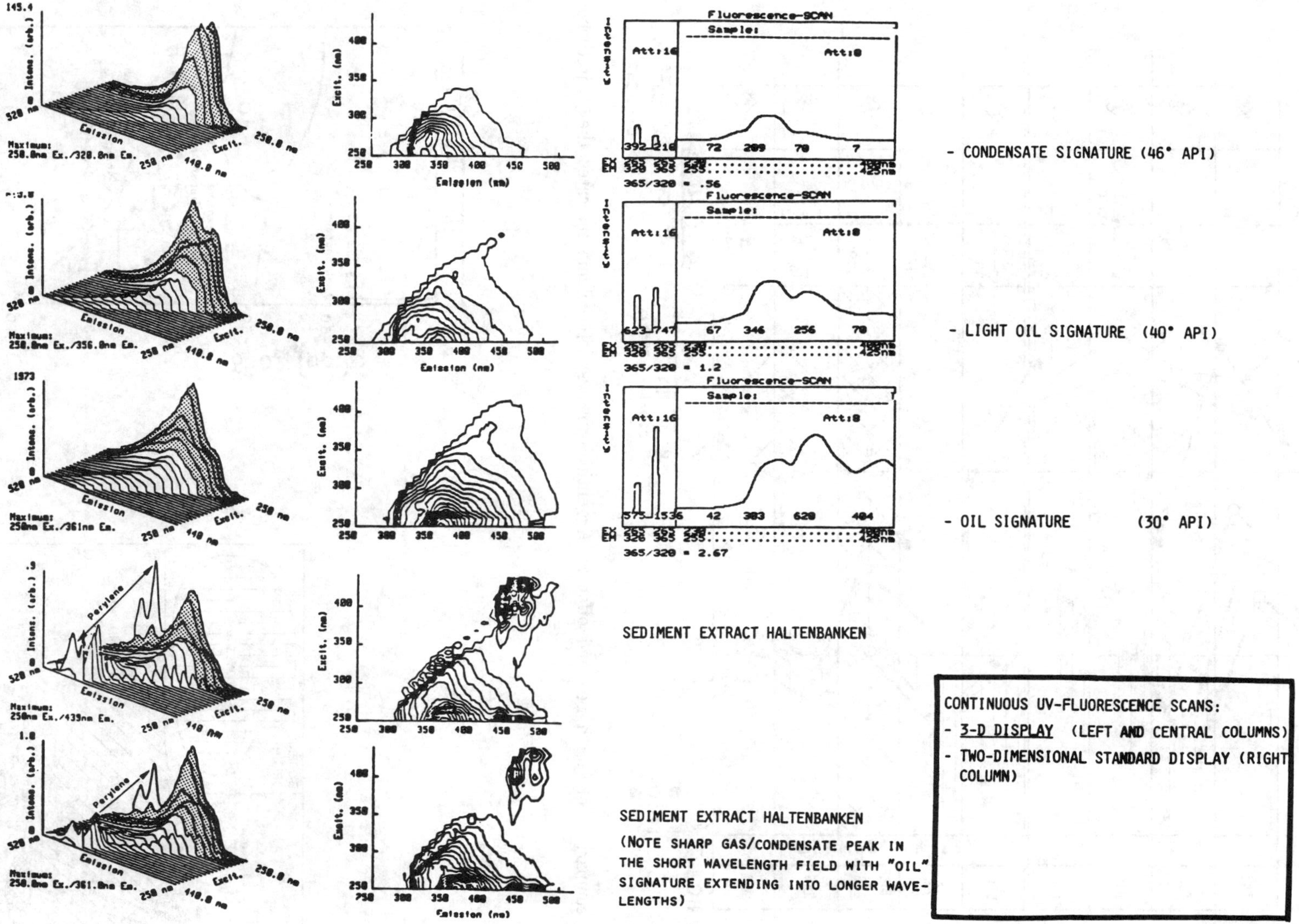

Fig. 3. Comparison of u.v.-fluorimetry measurements of condensate, oils and sediment extracts.

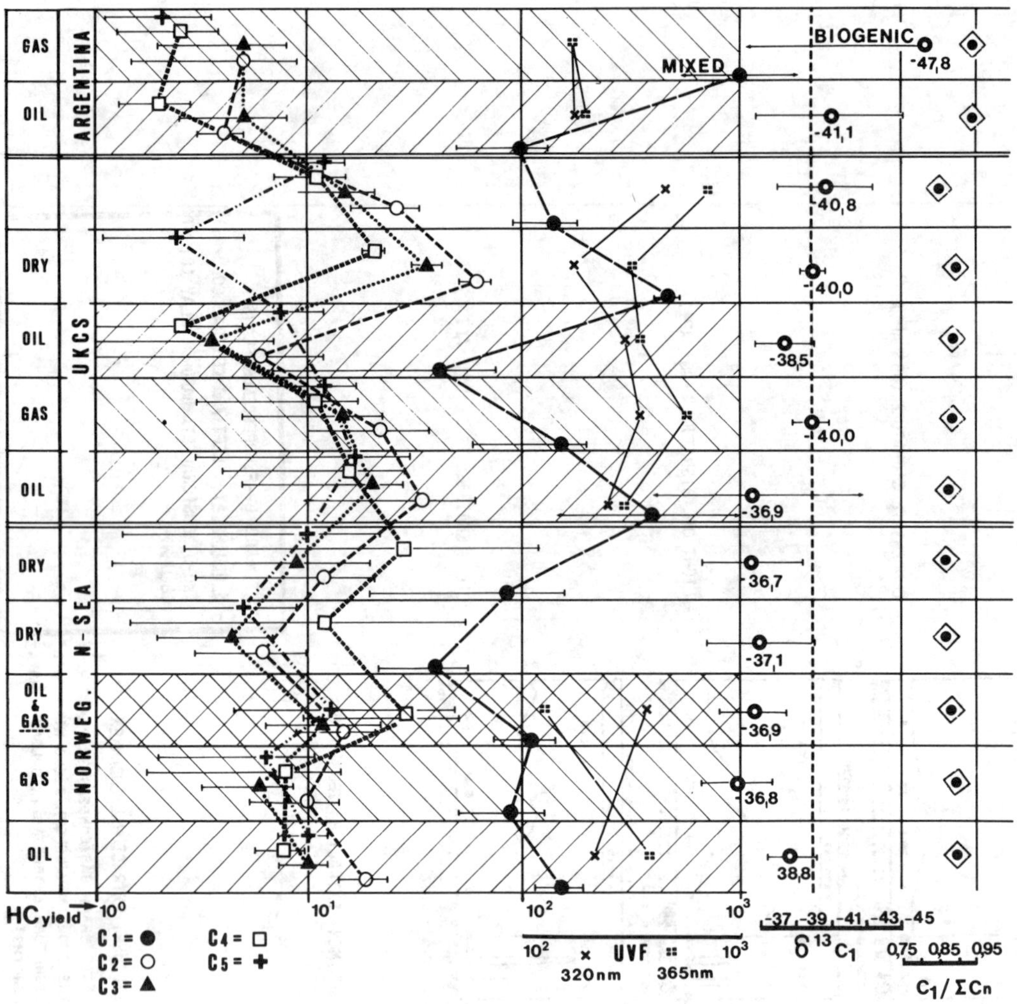

Fig. 4. Data comparison of North Sea and offshore Argentina surveys; left column indicates the hydrocarbon types discovered.

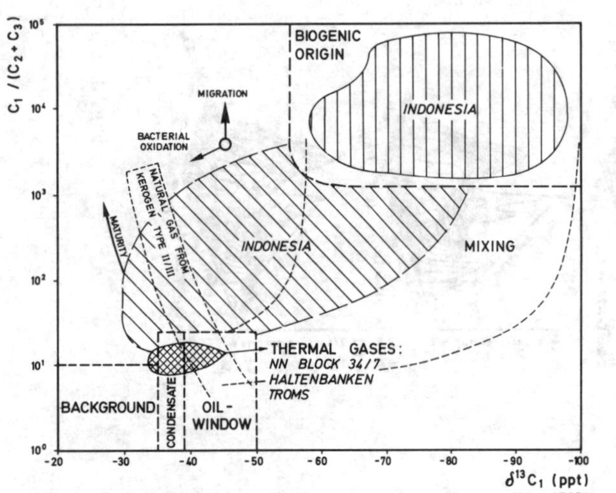

Fig. 5. Bernard diagram: $\delta^{13}C_1$ values versus ratio $C_1/C_2 + C_3$ of data from the Norwegian Shelf and Indonesia.

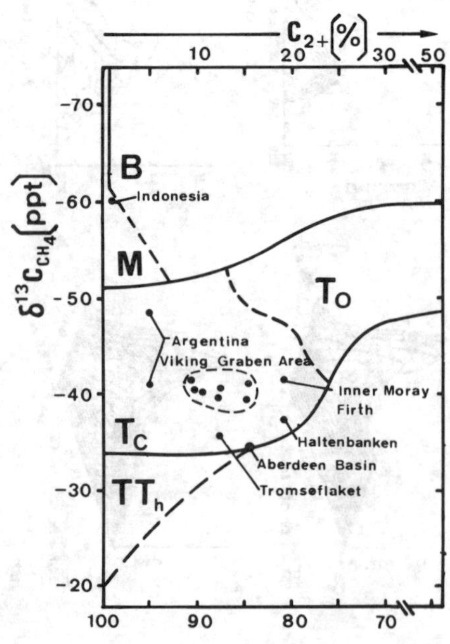

Fig. 6. Genetic characterization of sediment gases from different Deminex surveys, after Schoell (1983). B = biogenic origin; M = mixed biogenic and thermal origin; T_o = associated gases (oil); T_c = associated gases (condensate); TT_h = non-associated gases (coal/overmature gases).

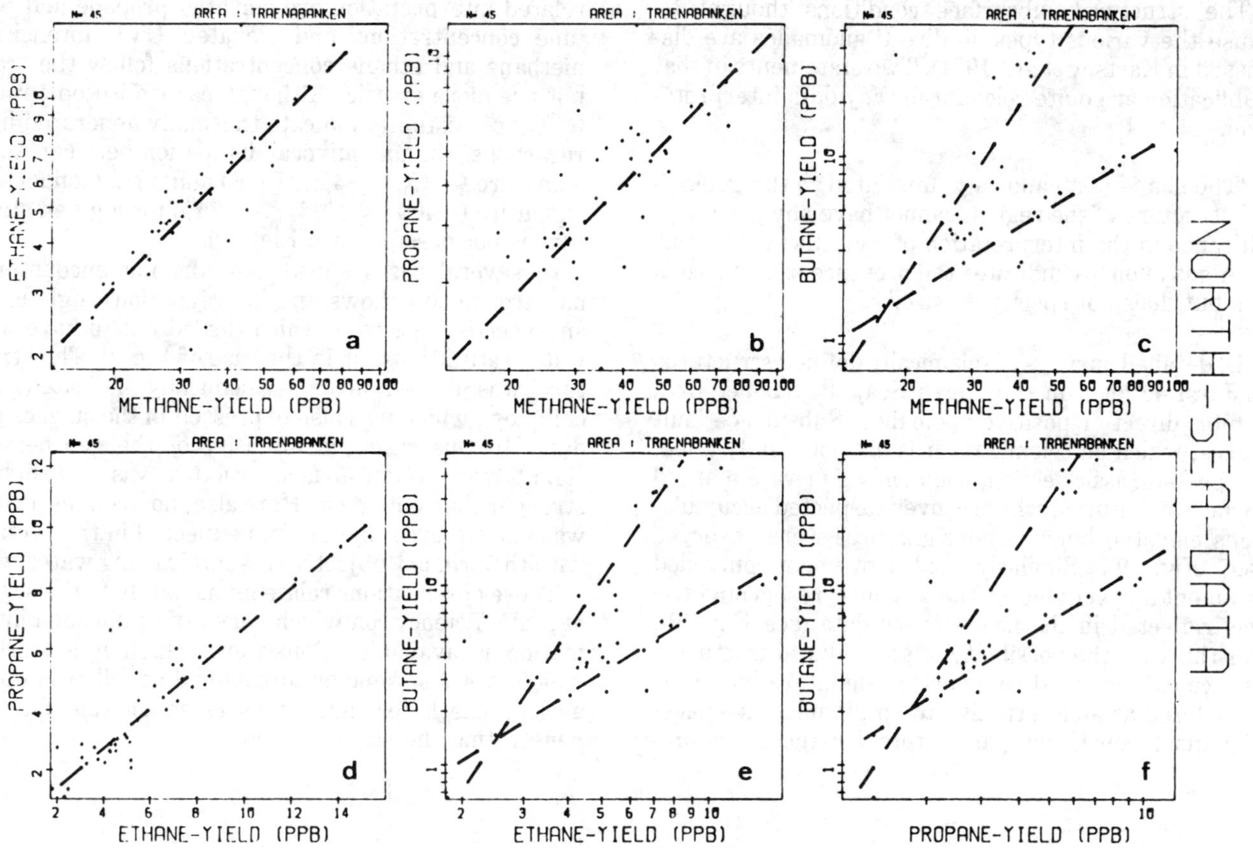

Fig. 7(a)–(f). Cross-plots of yield of sediment gas components, methane through butanes, with a linear relationship for methane through propane and a second concentration trend developing for the butanes.

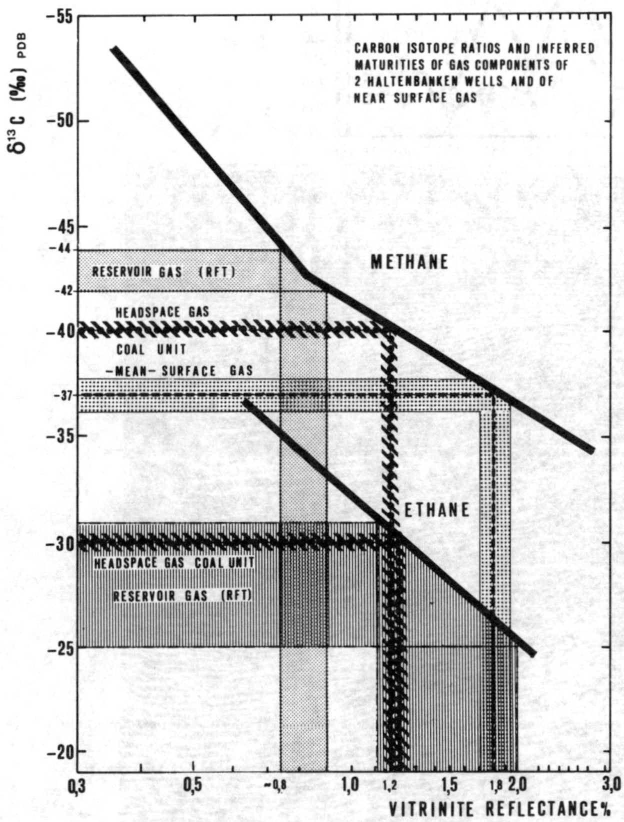

Fig. 8. Inferred relationship between $\delta^{13}C$ values (methane and ethane) and source rock maturity for reservoir, headspace and surface gases (Haltenbanken) after Faber *et al.* (1984).

Mapping

The traditional approach to evaluating near surface hydrocarbon data has been contouring of gridded data or their display as concentration profiles (Horvitz, 1939; Sokolov, 1955; Kartsev *et al.*, 1954; Jones and Drozd, 1983. Much of the early literature, but also more recent accounts, emphasized the 'halo' concept (Horvitz, 1980). According to this concept, hydrocarbon accumulations are supposed to be perfectly sealed by, amongst other effects, pronounced secondary cementation over the area of subsurface closure. Migrating hydrocarbons would 'spill' at the margins of the accumulation and create a halo-shaped rim of high concentrations of near-surface hydrocarbons enclosing an area of comparatively low concentrations. Published evidence generally consists of surface maps with producing wells encircled by the 'halo'. Subsurface structural relationships are rarely shown. As already stated by Jones and Drozd (1983), diapir-related structures with circular fracture systems might well also generate halo-shaped surface anomalies. The closest approximation to such a relation between subsurface structure and surface configuration was observed in a Deminex survey in the Senja Ridge area north of 62 °N (Norway), where shale diapirism is suspected.

An alternative approach, in contrast to the halo-concept, relies on imperfect seals as a working hypothesis. In this case subsurface accumulations are assumed to be expressed by higher than background concentrations of near-surface hydrocarbons over the accumulation at depth.

The structural subsurface conditions thought to cause the various types of 'direct' anomalies are discussed in Kartsev *et al*. (1954). Two statements in that publication are quite relevant for any data interpretation.

'The shape of an anomaly, unrelated to the geological structure of the region, cannot have any great significance in the interpretation of gas-survey results.'

'A gas anomaly indicates the presence of a source of gas but does not specify its size.'

In detailed surveys of seismically defined structures at Jurassic level in the North Sea, the surface data exhibit direct or positive anomalies. Subsurface fault traces, which on seismic records did not visibly affect the post-Jurassic section, commonly generate marked surface signatures, whereas over suspected accumulations elevated but weaker signatures were observed (see Fig. 9). Similarly, salt-movement-controlled sedimentary wedging at the assumed reservoir level was reflected in the near-surface data (see Fig. 10). For this case the possibility of salt-induced fracturing as a cause for the strong surface signature has to be considered as an alternative to up-dip micro-seepage. The most consistent parameters for the structure-related interpretation are elevated propane and pentane concentrations and elevated UVF intensities; methane and ethane concentrations follow the trend but are more erratic. Although carbon isotopic ratios ($\delta^{13}C_1$) consistently indicate thermally generated hydrocarbons, the unequivocal distinction between an oil signature ($\delta^{13}C_1 < -40‰$ PDB) and a gas/condensate signature ($-40‰ < \delta^{13}C_1 < -37‰$) for a given structure is not possible (see Fig. 10).

In several cases subsequent drilling encountered numerous good shows and impregnations high above an expected reservoir which turned out to have high water saturations, as in the case of Fig. 9. The structure was drilled near a seismically indicated 'gas chimney', which was also expressed in the surface gas data. In this case the intensity of the geochemical signals over the subsurface structure was only slightly stronger than outside it. Here also, non-commercial oil was encountered higher than expected in the section, with the primary objective reservoir being water-wet.

These observations relate immediately to the aspect of seal efficiency, on which very little published information is available. Consistently high hydrocarbon yields over a seismic feature should be followed up by a close check for indications of small-scale faulting penetrating the assumed seal.

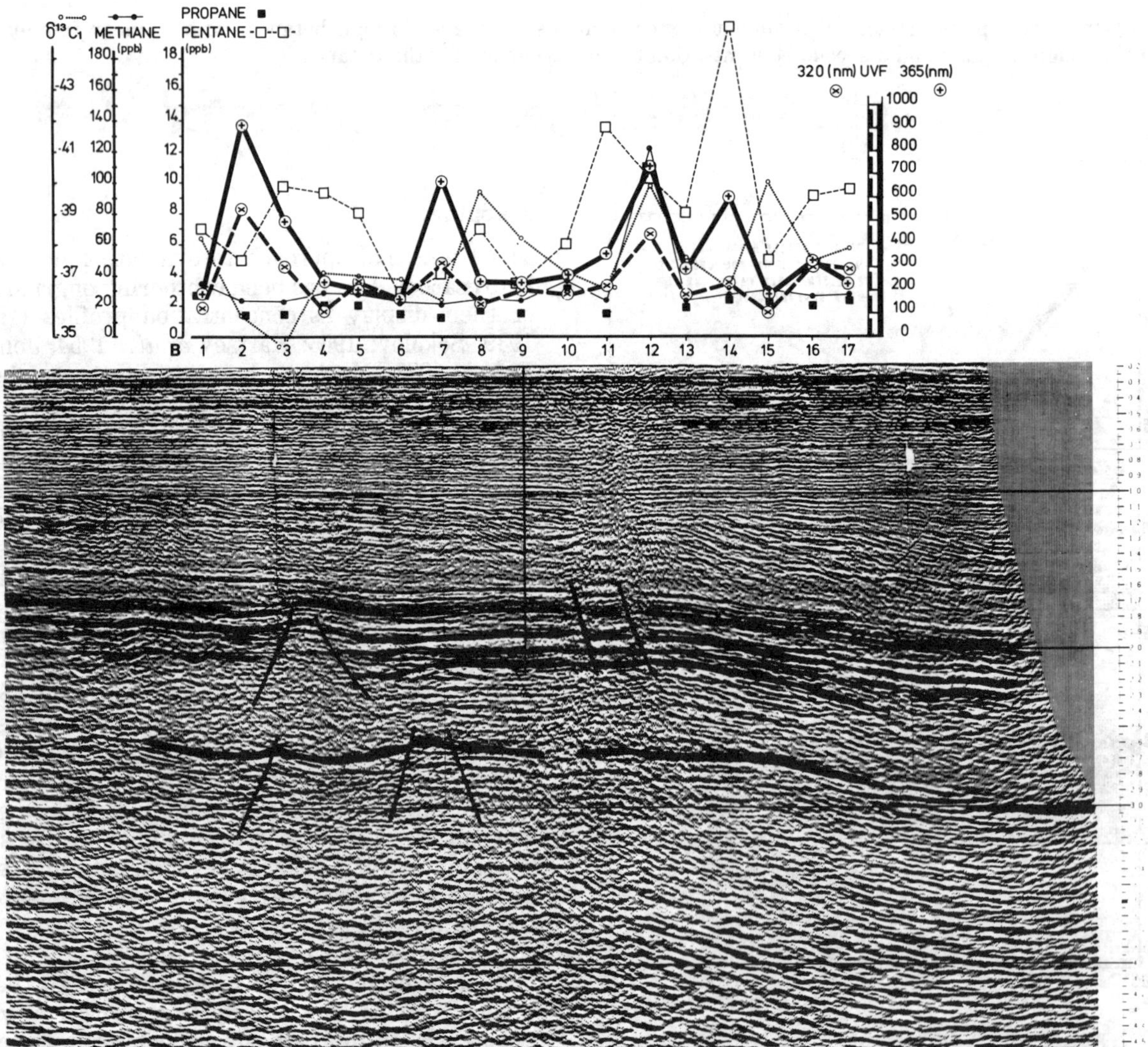

Fig. 9. Geochemical traverse for detailed structure related surface investigations (distance between sample stations 250 m).

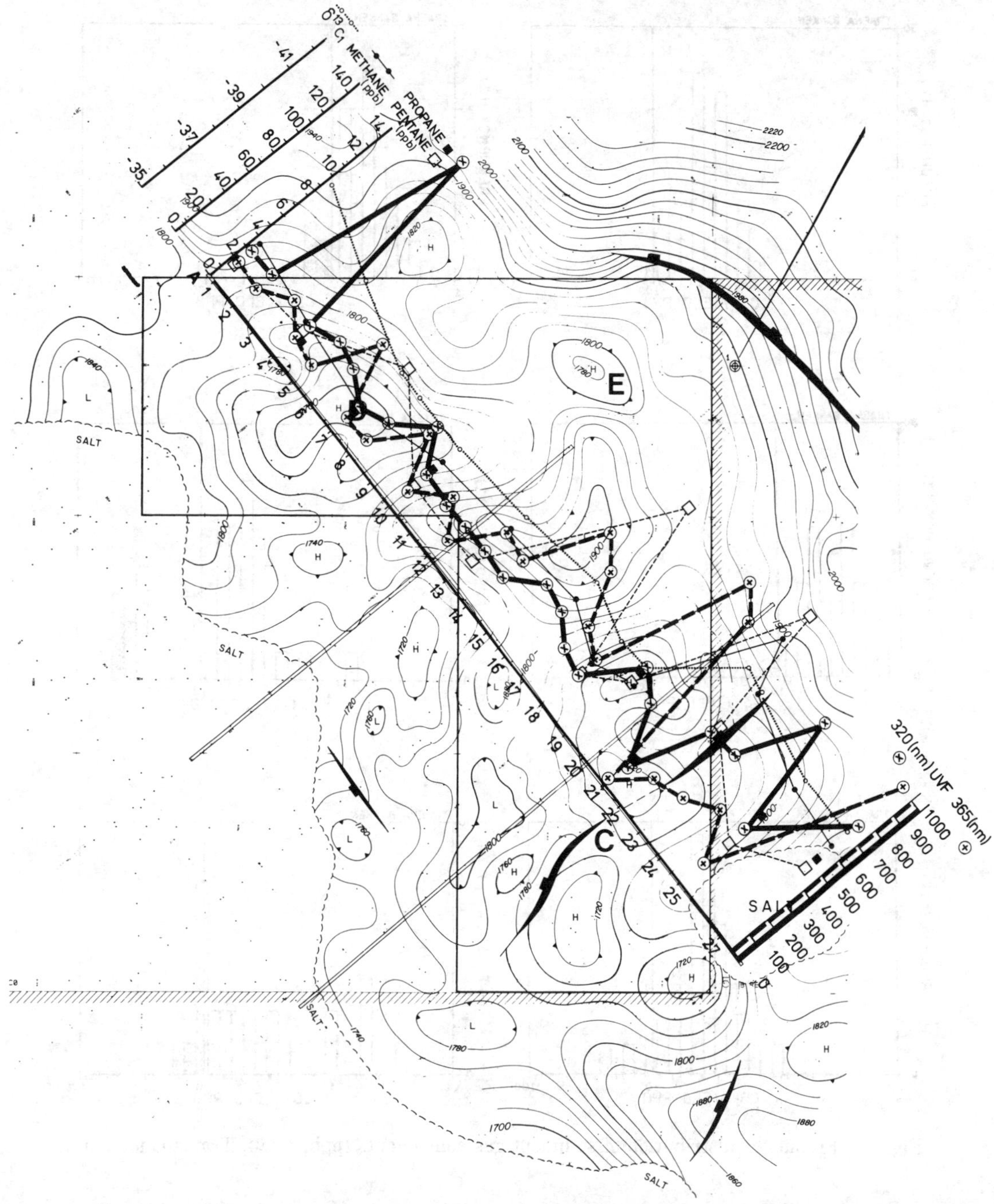

Fig. 10. Survey pattern for detailed investigations (distance between sample stations 250 m).

CASE HISTORY—INTEGRATION OF GEOCHEMICAL SURFACE DATA WITH GEOLOGICAL MODEL (TRÆNABANKEN AREA)

As an example of a regional interpretation integrated with available subsurface information, results of a survey over the Trænabanken area (north of 62 °N) are presented below.

Data statistics

The Trænabanken area is characterized by low hydrocarbon yields compared with other areas (see Fig. 2). In the frequency distribution a bimodality becomes increasingly better expressed from ethane through butanes (see Fig. 11). The bimodality is seen in the cross-plots of butanes versus the other components (see Figs. 7(c), (e) and (f)) whilst a good linearity between the plotted components methane through propane is maintained.

Delta $^{13}C_1$ values span a wide range, where the bulk is centred on values $>-40‰$ (PDB) (see Fig. 2(b)). Only a few data points are $<-40‰$ (PDB); they plot in the 'oil window' section of the Bernard plot (see Fig. 12). A synoptic interpretation of the statistical data indicates the following:

—overall low yields combined with the 'heavy' isotopic signature ($>-37\%$) for the majority of samples seem to suggest generally unfavourable source conditions (immature type III kerogen or alternatively overmature type II kerogen);

—the linear relationship between gas components up to propane suggests co-genesis and co-migration of the C_1 to C_3 components and

—the deviation from linearity of the butanes and the disproportionally high concentration of butanes in the

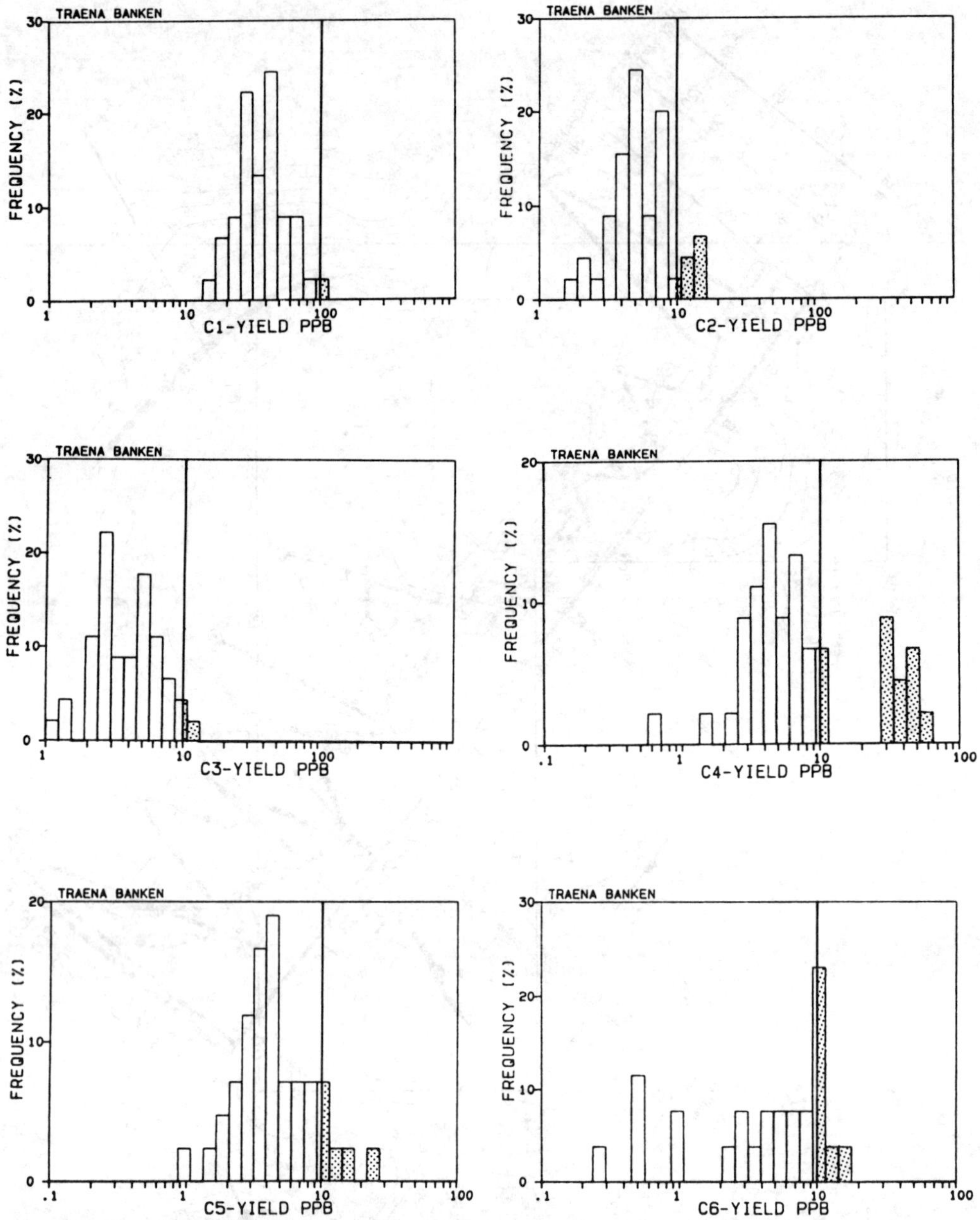

Fig. 11. Frequency distribution of sediment gas components (ppb, w/w). Trænabanken area.

few samples with 'light' isotopic ratios ($< -40\%$) suggests either a more favourable source rock facies (type I/II kerogen) and/or 'oil window' maturity of the source rock.

To resolve the ambiguity, the data set has to be integrated with a geological model.

Mapping of geochemical data

Analytical data or simple ratios are contoured (see Table 1). Methane, the most mobile gas component, exceeds the regional background of 30 ppb as an E–W aligned anomaly in the south of the survey area. A second trend of elevated values runs SW–NE across the central and northern parts (see Fig. 13).

Elevated butane yields essentially mirror the methane trends, but have their maximum in the south (see Fig. 14). Consequently, the distribution of the wetness parameter $C_1/\Sigma C_n$ demonstrates a combination of both the methane and the butane anomalies, where the 'wettest' signature coincides with the high yield area in the south. The regional background value of 0.8, still wet compared, for example, with Haltenbanken values ($\bar{x} = 0.86$), is assumed to be related to the low yields. The greater mobility of methane and ethane would tend to leave low concentration 'residues' of the higher homologues. In productive areas with strong hydrocarbon signatures it has been observed that the surface signatures generally appear 'drier' than the reservoir gases. This is thought to be due to the replenishment of methane from either mature source rocks or from hydrocarbon accumulations. Alternatively, the signatures may be caused by increased solubility of the C_2–C_4 components in the reservoired hydrocarbons.

The same systematic areal changes as in gas concentrations emerge from the distribution of $\delta^{13}C_1$ values (see Fig. 15). The areas with high gas yields and elevated wetness coincide with light isotope ratios. The background methane concentrations (<30 ppb) towards the north are too low for a reliable determination of the stable carbon isotope ratio. However, in samples with elevated methane yields the methane is also isotopically heavy ($>-37\%$) in the northern and

TABLE 1. Analytical data—Trænabanken

(a) Relative gas composition, Trænabankan, Norwegian Sea

Sample No.	C_1 (%)	C_2 (%)	C_3 (%)	n-C_4 (%)	iso-C_4 (%)	n-C_5 (%)	$\delta^{13}C_1$ (ppt)
t1	66.1	5.89	2.81	12.40	2.85	2.21	−38.7
t2	66.1	5.62	2.86	14.41	2.90	2.24	−45.2
t3	65.7	6.54	3.15	12.60	3.17	2.60	−44.1
t4	86.2	2.92	2.67	2.51	1.10	1.14	−38.8
t5	54.6	4.81	3.02	24.73	4.67	2.16	−35.9
t6	74.1	4.53	2.20	12.17	2.31	1.49	−37.2
t7	85.9	3.07	0.97	3.30	1.21	9.94	−38.6
t8	62.5	5.70	3.29	10.64	3.07	7.98	0.0
t9	56.6	11.19	5.99	7.82	2.42	4.14	0.0
t10	86.7	2.98	3.14	2.14	1.01	1.12	−37.6
t11	75.3	6.07	3.97	3.29	1.80	1.80	−39.4
t12	82.3	7.73	2.34	1.87	1.00	1.20	−35.9
t13	76.4	6.83	2.89	3.60	0.47	2.17	−35.3
t14	78.9	6.08	3.16	2.60	1.06	1.33	−36.8
t15	78.8	6.91	3.11	2.80	0.53	2.29	−35.8
t16	76.7	8.29	4.39	2.26	1.43	1.54	−38.0
t17	83.4	6.35	2.95	1.85	0.95	0.94	−41.0
t18	76.6	6.68	2.73	1.41	0.95	1.67	−35.7
t19	73.5	6.44	2.76	1.64	0.97	1.65	−35.5
t20	73.2	7.82	3.09	1.96	1.38	1.66	−35.9
t21	75.7	7.50	3.42	1.72	1.10	1.84	−37.4
t22	80.4	5.37	2.46	1.22	0.62	1.34	0.0
t23	85.5	5.02	2.21	1.53	0.91	1.15	−34.2
t24	72.8	6.63	2.61	1.86	0.94	1.97	−34.7
t25	67.6	4.74	2.42	0.88	0.00	1.77	−31.5
t26	71.1	5.90	3.12	1.64	1.26	1.98	−35.2
t27	75.3	6.39	2.66	1.68	0.94	1.54	−36.3
t28	69.4	6.08	2.73	1.79	1.06	1.88	0.0
t29	85.3	6.48	2.40	1.37	0.61	0.00	−35.2
t30	83.6	4.52	1.69	0.94	0.47	0.00	−34.7
t31	79.8	6.35	2.74	1.46	0.91	1.46	−37.8
t32	65.4	6.15	3.30	11.61	3.25	3.76	−40.3
t33	74.6	6.21	6.09	3.46	1.22	2.06	−35.0
t34/1	77.6	7.03	2.88	3.51	1.58	1.52	0.0
t34/2	84.2	7.38	3.01	1.01	0.48	0.39	−39.2
t35/1	70.1	6.24	3.08	10.90	2.45	1.76	−38.6
t35/2	76.6	7.05	1.90	3.21	4.18	0.00	−32.0
t36	64.5	6.26	3.30	11.11	3.25	5.26	−42.8
t37	80.6	7.78	3.00	0.51	1.49	2.05	−39.2
t38	78.3	6.75	3.06	3.17	1.45	2.17	−38.7
t39	70.4	6.24	2.59	1.58	0.89	1.72	−32.6
t40	76.4	6.09	2.55	1.48	1.10	1.50	−35.8
t41	77.9	6.89	3.20	1.60	1.04	2.11	−35.1
t42	77.9	6.93	2.83	1.86	0.80	1.37	−33.8
t43	76.5	5.41	2.46	1.61	1.02	3.49	−35.7

(b) Gas yield (ppb), Trænabanken, Norwegian Sea

Sample No.	C_1	C_2	C_3	C_4	C_5	C_6
t1	59.8	10.00	7.00	50.063	8.993	0.923
t2	46.0	7.34	5.48	43.889	7.008	0.523
t3	43.5	8.11	5.73	37.821	7.729	0.462
t4	32.1	2.04	2.73	4.871	1.907	0.000
t5	18.7	3.09	2.85	36.516	3.695	0.552
t6	74.7	8.56	6.10	52.926	6.751	0.000
t7	35.0	2.35	1.09	6.672	1.962	0.000
t8	41.6	7.10	6.01	33.037	24.284	5.209
t9	42.9	15.91	12.49	28.141	16.658	0.000
t10	66.4	4.28	6.62	8.753	4.685	0.000
t11	36.0	5.44	5.22	8.821	4.661	0.000
t12	29.6	5.21	2.31	3.740	2.374	0.000
t13	22.5	3.77	2.34	4.340	3.635	0.000
t14	24.9	3.60	2.75	4.192	2.356	2.492
t15	28.0	4.60	3.04	4.288	4.103	0.000
t16	27.3	5.54	4.30	4.765	2.465	0.000
t17	51.6	7.37	5.02	6.280	3.141	0.000
t18	43.8	7.16	4.29	4.892	4.780	8.717
t19	27.5	4.52	2.84	3.538	2.772	8.530
t20	41.7	8.35	4.84	6.892	4.246	10.875
t21	75.4	14.16	9.37	10.188	8.240	11.928
t22	52.4	6.55	4.40	4.342	3.920	9.608
t23	44.8	4.93	3.18	4.633	2.707	0.000
t24	26.6	4.54	2.62	3.704	3.230	7.481
t25	13.8	1.81	1.36	0.651	1.624	10.663
t26	19.9	3.10	2.40	2.942	4.929	5.092
t27	31.6	5.03	3.07	3.990	3.096	6.200
t28	26.3	4.32	2.84	3.911	3.198	10.911
t29	17.6	2.50	1.36	1.477	0.000	1.049
t30	40.1	4.07	2.23	2.454	0.000	9.690
t31	102.8	15.34	9.71	11.067	9.319	2.696
t32	39.5	6.97	5.48	32.541	10.206	2.983
t33	30.3	4.73	6.80	6.892	4.345	0.000
t34/1	25.3	4.29	2.58	6.013	2.958	0.000
t34/2	40.7	6.69	4.00	2.610	1.064	0.000
t35/1	64.8	10.82	7.84	44.774	7.317	0.000
t35/2	30.2	5.21	2.06	10.563	0.000	0.000
t36	34.1	6.20	4.79	27.501	12.487	3.885
t37	28.5	5.15	2.91	2.561	3.650	0.000
t38	34.0	5.49	3.65	7.261	4.831	0.000
t39	47.0	7.81	4.76	5.978	5.160	14.942
t40	28.6	4.27	2.62	3.497	3.831	0.281
t41	73.6	12.21	8.32	9.045	10.065	4.058
t42	51.2	8.54	5.11	6.337	4.045	10.122
t43	24.8	3.28	2.19	3.086	5.746	4.153

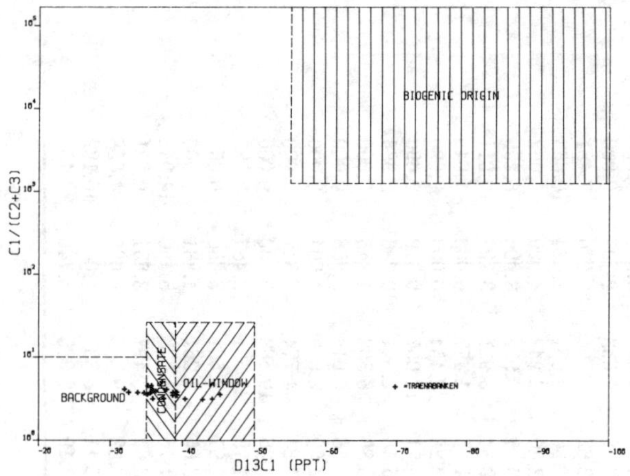

Fig. 12. Bernard Diagram, Trænabanken area.

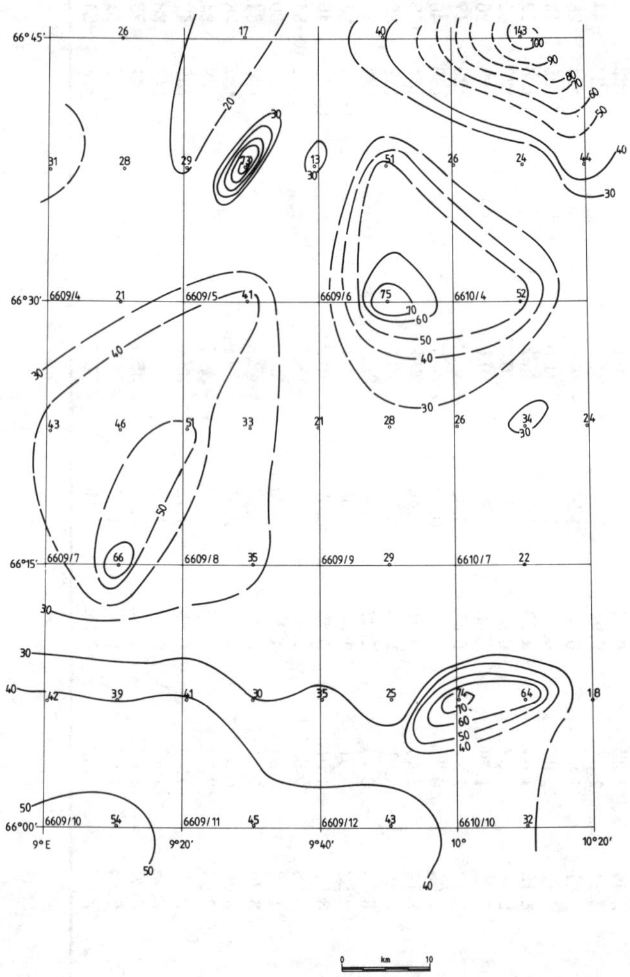

Fig. 13. Isoconcentration map: methane, Trænabanken.

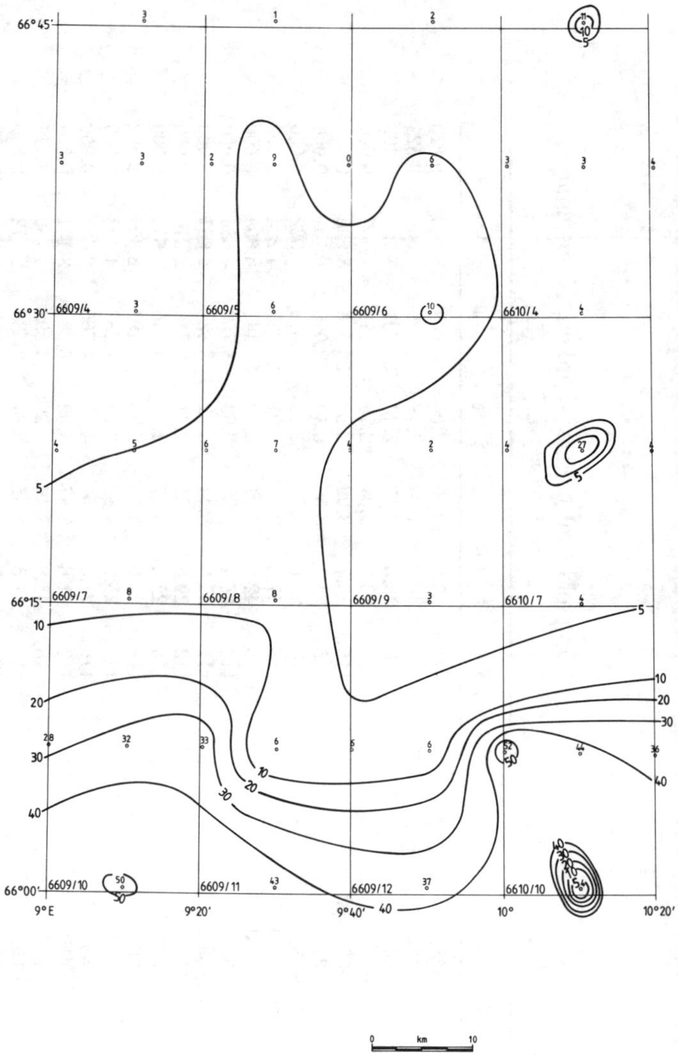

Fig. 14. Isoconcentration map: butane, Trænabanken.

northwestern parts of the area, suggestive of a deterioration of source quality.

A compilation of subsurface information is presented in the cross-section of Fig. 16.

On the structural map of base Jurassic (see Fig. 17) the main features are:

—a general SW–NE structural trend;
—the Helgeland Basin as a potential drainage area in the south and south-east;
—a structural and depositional terrace abutting the basement high of the Nordland Ridge and
—a major fault zone, steeply downthrowing the Mesozoic section towards the Træn Basin in the north-west.

For an assessment of source maturity, model calculations were performed, based on a modified Lopatin–Waples approach using geothermal gradients extrapolated from the Haltenbanken area. According to this calculation the Kimmeridge Clay as the potential oil source rock should barely reach a maturity of 0.7% vitrinite reflectance in the deepest parts of the Helgeland Basin (see Figs. 16 and 17).

None of the five wells drilled so far in the Trænabanken area seems to have had any significant indications of hydrocarbons. Explanations put forward (Curran, 1983; Karlsen, 1984) ascribe this to the regional immaturity of the Kimmeridge source rock.

The results of the open grid (15/7–5 × 15 km) geochemical survey of the Trænabanken area seem to corroborate the geological reasoning by outlining the marginally mature, but apparently rich, source rocks in the Helgeland Basin. The lack of significant hydrocarbon migration towards the Mesozoic structural terrace is borne out by the drastic drop in concentration of the C_{2+} fraction in the sediment gas. The methane anomalies, comparatively isotopically heavy, are tentatively related to Coal Unit-sourced gas. Their alignment approximately parallel to the

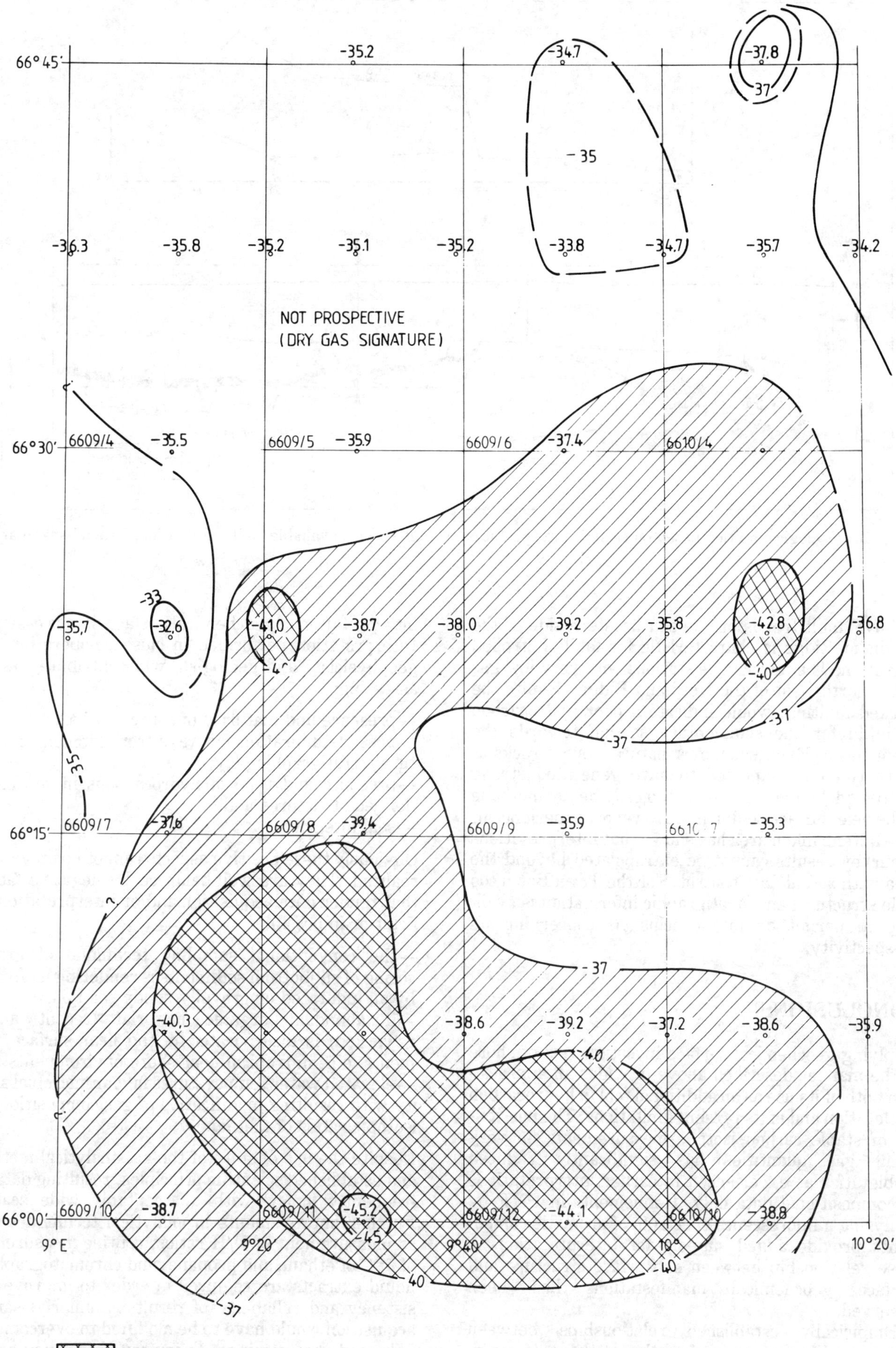

Fig. 15. Iso-ratios $\delta^{13}C_1$ (‰ PDB) Trænabanken.

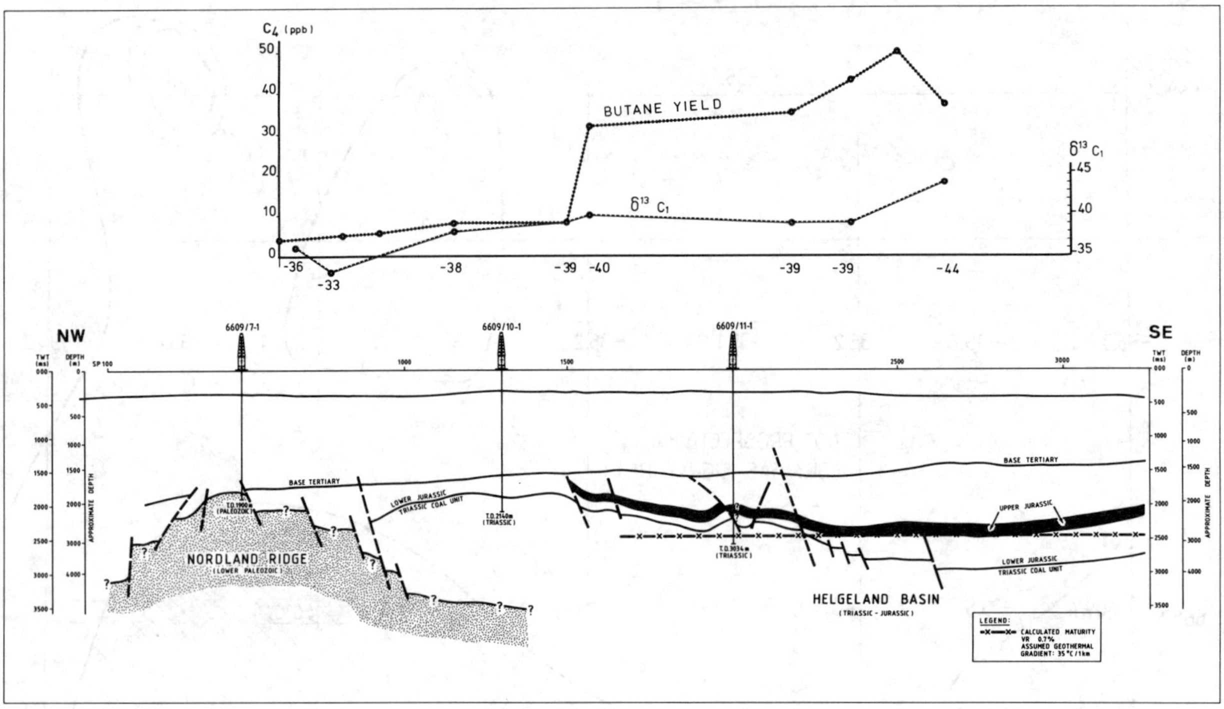

Fig. 16. Geological section based on seismic interpretation combined with available well information, Trænabanken area.

SW–NE fault trends on the southeastern side of the Nordland Ridge is characteristic for fault-controlled migration. The few sample stations with very low gas yields north-west of the Nordland Ridge, despite the marked faulting, would indicate a poor hydrocarbon potential for the sedimentary section towards the Træn Basin. Even when overmature, source rocks at depth could be expected to have generated strong positive yield anomalies of isotopically heavy methane in the near surface sediments. However, experience in the Haltenbanken area has shown that interpretations of survey results cannot be extrapolated beyond the area with actual data control. For the Træn Basin too little structural and stratigraphic information is available to permit definite conclusions concerning its prospectivity.

CONCLUSIONS

Applying geochemical surface investigations offshore in the manner described above provides an appreciation both of its use as an additional tool in hydrocarbon exploration and of its present limitations. The possibility of stable carbon isotope measurements on very small gas quantities has significantly reduced ambiguities as experienced in earlier interpretations of compositional data only. Refinements in UVF by analysing a fluorescence spectrum of aromatic components provide added support for interpretation. A close relationship between subsurface structure and surface geochemical manifestations has been observed.

Empirically established relationships between measured isotope ratios of methane/ethane of reservoir gases and maturity and type of respective source rocks are, with certain limitations, valid for near-surface gas.

The need for a thorough integration of all available subsurface information in order to achieve a realistic interpretation of geochemical surface data is essential. Interpretation performed in close co-operation with geophysicists and geologists will contribute towards assessing:

—presence and maturity of source rocks;
—possible migration fairways from kitchen to structural/stratigraphic play;
—presence and type of hydrocarbons in seismically defined structures and
—seal integrity.

Obviously these positive aspects cannot be universally realized. They are well-balanced by negative factors that render data acquisition and/or interpretation difficult or impossible:

—the conventional sampling technique of gravity coring of subbottom sediments is rendering ineffective by an indurated or gravelly/sandy sea-floor;
—if abundant organic material (recent and/or reworked) is incorporated in the near surface sediments, e.g. in tropical near shore environments and
—elevated water temperatures favour microbial activity with resulting modification of gas composition and isotopic ratios of methane.

Despite the use of state-of-the-art analytical methods and established scientific principles, resulting data are frequently 'noisy' with unacceptably wide scatter. Further improvements in analytical techniques such as measurement of D/H ratios, routine measurement of $\delta^{13}C$ of ethane and propane, and chromatography of liquid extracts are required in order to improve consistency and reliability of results. Similarly, sample acquisition would have to be modified to overcome the indicated shortcomings, to improve efficiency and to widen the application of the method.

Properly conducted and interpreted geochemical surveys have the potential for reducing the source- and migration-related risk elements in both regional and play-related exploration. Even at costs which

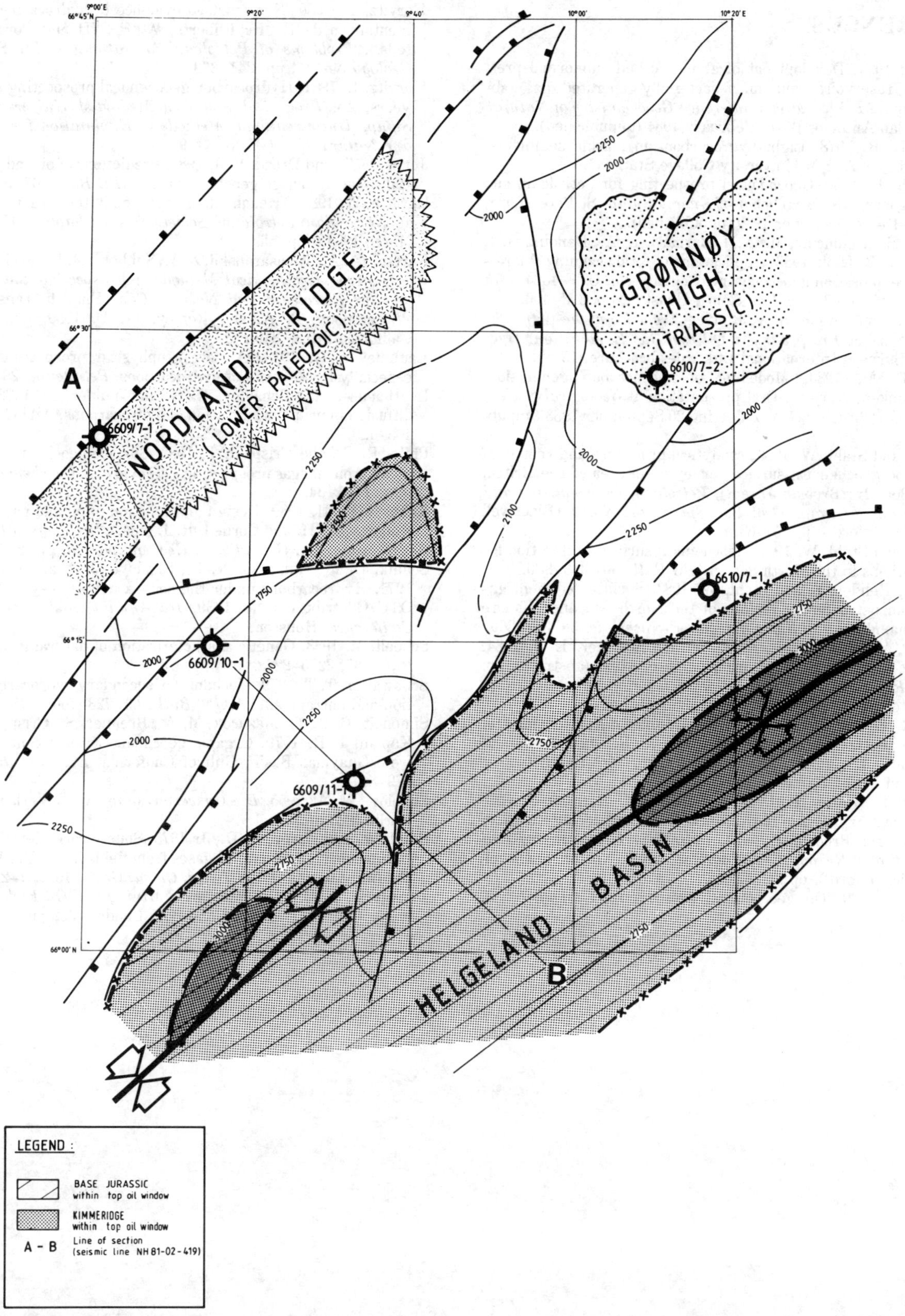

Fig. 17. Structural interpretation of base Jurassic with main structural features and potential hydrocarbon kitchen, Trænabanken area.

approach those of seismic data acquisition, their application as an additional exploration tool is considered an advantage.

ACKNOWLEDGEMENTS

The permission by Deminex management to publish this paper is gratefully acknowledged. We also appreciate the permission of the group consisting of Saga–Statoil–Norsk Hydro–Norske Shell–Arco Norge–Norsk Agip–Deminex Norge to publish stable carbon isotope data of gases from the Midgard discovery.

Gas analyses were performed by GCA-Laboratory, Manfred Schmitt, Lehrte, FRG and UVF spectrophotometry by Dr I. Dumke in the Federal Institute for Geosciences and Resources (BGR), Hannover, FRG.

REFERENCES

Barker, C. 1984. Development of abnormal and subnormal pressures in reservoirs containing bacterially-generated gas (abstract) In: *AAPG Research Conference 'Geochemistry of Natural Gases'*, San Antonio, 12–15 February 1984 (unpublished).

Bernard, B. B. 1978. Light hydrocarbons in marine sediments. *Thesis*, Texas A & M University College Station.

Bernard, B. B. 1980. Geochemical prospecting for petroleum and natural gases. *US Department of Energy Symp.* St Petersburg, Florida, 15–17 December 1980 (unpublished).

Brooks, J. M., Kennicutt, M. C.,II, Barnard, L. A., Denoux, G. J. and Carey, B. D. Jr 1983. Applications of total scanning fluorescence to exploration geochemistry. Contrib. No. OTC 4624, *15th Annual Offshore Technology Conf.*, Houston, pp. 393–400.

Cornford, C. and Dyson, A. V. R. 1984. *Seeps Point the Way. The Role of Surface Prospecting in Oil and Gas Exploration.* Hydrochem Surveys International Ltd, sales brochure, 15 pp.

Curran, P. M. 1983. Modelling of hydrocarbon generation, Trænabanken, Norway. Oral presentation at *Geological Society of London Petroleum Group* Meeting 20 September 1983 (unpublished).

Faber, E. and Stahl, W. 1983. Analytical procedure and results of an isotope geochemical surface survey in an area of the British North Sea. In: Brooks, J. (ed.), *Petroleum Geochemistry and Exploration of Europe. Geol. Soc. Spec. Publ. No. 12* (Blackwell Scientific, Oxford.), pp. 51–63.

Faber, E. and Stahl, W. 1984. Geochemical surface exploration for hydrocarbons in the North Sea. *AAPG Bull.*, **68**, 363–386.

Faber, E., Rehder, S. and Stahl, W. 1983. Surface sediment gas data from petroliferous regions of the North Sea and offshore California. *Erdöl und Kohle–Erdgas–Petrochemie*, **36**, 357–360.

Faber, E., Whiticar, M. J., Stahl, W. and Schiener, E. J. 1984. Biogenic methane in gases from oil reservoirs? (abstract). In: *AAPG Research Conference 'Geochemistry of Natural Gases'*, San Antonio, 12–15 February 1984 (unpublished).

Gérard, R. E. and Feugère, G. 1969. Results of an experimental offshore geochemical prospecting study. In: Schenck, P. A. and Havenaar, I. (eds.), *Advances in Organic Geochemistry* Pergamon Press, Oxford, pp. 355–371.

Gevirtz, J. L., Carey, B. D. Jr and Blanco, S. R. 1983. Surface geochemical exploration in the North Sea. Methodology with an example. In: Brooks, J. (ed.), *Petroleum Geochemistry and Exploration of Europe. Geol. Soc. Spec. Publ. No. 12* (Blackwell Scientific, Oxford.), pp. 35–50.

Horvitz, L. 1939. On geochemical prospecting. *Geophysics*, **3**, 108–109.

Horvitz, L. 1980. Near-surface evidence of hydrocarbon environment from depth. In: Roberts, W. H., III and Cordell, R. J. (eds.), *Problems of Petroleum Migration. AAPG Studies in Geology No. 10*, pp. 241–269.

Horvitz, L. 1981. Hydrocarbon geochemical prospecting after forty years. In: *Proc. 2nd Southern Methodist University Dallas Symp. 'Unconventional Methods in Exploration for Petroleum and Natural Gas II'*, pp. 83–95.

Jones, V. T. and Drozd, R. J. 1983. Predictions of oil and gas potential by near-surface geochemistry. *AAPG Bull.*, **67**, 932–952.

Karlson, W. 1984. Trænabanken's prospectivitet. Oral presentation at *Norwegian Petroleum Society (Oslo Chapter) Meeting*, 15 March 1984 (unpublished).

Kartsev, A. A., Tabasaranskii, Z. A., Subbota, M. I. and Mogilevskii, G. A. 1954. *Geochemical Methods of Prospecting and Exploration for Petroleum and Natural Gas.* English translation by Witherspoon, P. A. and Romey, W. D. 1959, University of California Press, 349 pp.

Laubmeyer, G. 1933. A new geophysical prospecting method, especially for deposits of hydrocarbons. *Petroleum*, **29**, 1–4.

Leythaeuser, D., Schaefer, R. G. and Yükler, A. 1982. Role of diffusion in primary migration of hydrocarbons. *AAPG Bull.*, **66**, 408–429.

Philp, R. P., and Crisp, P. T. 1982. Surface geochemical methods used for oil and gas prospecting—a review. *J. Geochem. Exploration*, **17**, 1–34.

Roberts, W. H. 1980. Design and function of oil and gas traps. In: Roberts, W. H. and Cordell, R. J. (eds.), *Problems of Petroleum Migration. AAPG Studies in Geology, No. 10*, pp. 217–240.

Schink, D. R., Guinasso, N. L., Sigalove, J. J. and Cima, B. E. 1971. Hydrocarbons under the sea—a new survey techn. signe. OTC Contribution No. 1339, *3rd Annual Offshore Technology Conference*, Houston.

Schoell, M. 1983. Genetic characterization of natural gases. *AAPG Bull.*, **67**, 2225–2238.

Schowalter, T. T. 1979. Mechanics of secondary hydrocarbon migration and entrapment. *AAPG Bull.*, **63**, 723–760.

Simoneit, B. R. T., Mazurek, M. A., Brenner, S., Crisp, P. T. and Kaplan, I. R. 1979. Organic geochemistry of recent sediments from Guaymas Basin, Gulf of California. *Deep Sea Res.*, **26A**, 879–891.

Sokolov, W. A. 1955. *Die Gasvermessung.* VEB Verlag Technik, Berlin, 228 pp.

Stahl, W. and Carey, B. D., Jr 1975. Source-rock identification by isotope analysis of natural gases from fields in the Val Verde and Delaware basins. West Texas. *Chem. Geol.*, **16**, 257–267.

Thompson, A. B. 1910. *Petroleum Mining and Oil Field Development.* Crosby, Lockwood and Son, London, 362 pp.

20 Geochemical exploration on the Norwegian Continental Shelf by analysis of shallow cores

R. H. Emmel, M. Bjorøy† and G. van Grass†*
*Texas Eastern Corporation, Houston, Texas, USA
†Continental Shelf Institute (IKU), Trondheim, Norway

A surface geochemical exploration survey has been carried out on 250 (near)-surface sediments which were collected in Quadrants 34 and 35 on the Norwegian Continental Shelf. This paper reports the results of the occluded and adsorbed gas analyses and the relationship between the gas data and several sediment parameters.

Variations in occluded gas concentrations and composition correspond to variations in sediment characteristics. This, coupled with the presence of significant amounts of unsaturated compounds, indicates that the occluded gas mainly has an *in situ* origin. Adsorbed gas data are not strongly related to sediment characteristics and neither does this gas fraction contain significant amounts of unsaturated compounds. This indicates the absence of a major contribution of *in situ* produced hydrocarbons to the adsorbed gas.

A preliminary interpretation of areal distributions of adsorbed gas concentrations shows some areas with high concentrations which can be related to different hydrocarbon sources in the deeper sediments. $\delta^{13}C$ values of methane to a certain extent correspond with the adsorbed gas concentrations. Finally, the combination of a seismic line with geochemical variations at the surface shows the relationship between changes in the deeper sediments and hydrocarbon data at the surface.

INTRODUCTION

Surface geochemical exploration for oil and gas is based on the concept that no reservoir cap rock is impermeable, and that hydrocarbons can escape from the reservoir and migrate towards the surface by a combination of diffusion and effusion mechanisms. Hydrocarbons reaching the surface can be measured directly—both in the sediment itself and in the overlying air or water—or through geochemical changes they induce (Gevirtz *et al.*, 1983; Jones and Drozd, 1983; Philp and Crisp, 1982).

Most reported studies on organic geochemical exploration concern the analysis of gaseous hydrocarbons in recent sediments because these compounds migrate easier than the larger, liquid hydrocarbons (Sokolov *et al.*, 1970). In marine sediments two gas fractions are generally distinguished: (1) interstitial gas which is dissolved in the pore water and can be released by mechanical disintegration of the sediment (Kvenvolden *et al.*, 1981) and (2) adsorbed gas which is bound on mineral surfaces and can be released by acid-treatment (Horvitz, 1939, 1980). If the interstitial gas is not removed first then the gas liberated upon acidification of the sediment is often referred to as total gas (Faber and Stahl, 1983, 1984).

This paper reports the results of a surface geochemical exploration survey carried out in Quadrants 34 and 35 on the Norwegian Continental Shelf. A total of 250 sediment samples was collected along a regular grid. These samples have been subjected to extensive analysis including adsorbed and occluded (interstitial) gas content, carbon isotope ratio of methane, sediment parameters such as grain-size class distribution, total organic carbon and carbonate content. Although a wider range of analytical techniques has been employed (see Experimental section) the discussion in this paper will focus on the occluded and adsorbed gas data and their correlation with the occurrence of different types of sediments in order to elucidate the extent to which variations in hydrocarbon gas data are influenced by variations in sediment characteristics. Additionally, some areal distributions of hydrocarbon concentrations which are thought to originate from migrated compounds are briefly discussed and the correlation between seismic data and geochemical variations observed at the surface is shown.

EXPERIMENTAL

Sampling

The samples were collected along a square grid with a spacing of 4.5 km (see Fig. 1). Sample stations are identified by a four-digit code with the first two digits

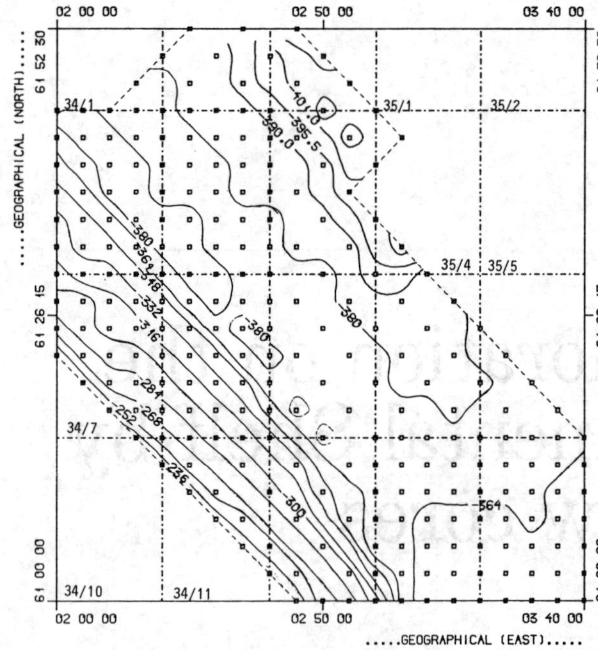

Fig. 1. Sample grid and bathymetry of survey area. Sample stations are indicated by small squares. Concession blocks are outlined by broken lines and numbered.

increasing from west to east (01 to 21) and the last two digits increasing from south to north (01 to 22). The sampling device used was a gravity corer equipped with a 3 m barrel. Core liners were made of chemically inert polypropylene.

On deck, the bottom part of the core liner was cut off and taken into an onboard transportable laboratory. After removal of the liner, the outside of the core, which had been in contact with the liner, was scraped off and four samples were taken for analysis. One sample for organic geochemical analyses (700–1000 g) was put in a can. Distilled water was added to leave a headspace volume of 100–150 ml. The can was sealed and stored in a freezer at −80 °C until analysis. Two samples (300–400 g each) were taken for isotopic analysis. Samples were put in plastic bags, which were then placed inside cloth bags. Two cloth bags containing duplicate samples were tied together, attached to a string and stored in liquid nitrogen. One sample (200 g) was taken for grain-size analysis and stored in a plastic bag.

Gas analyses

The cans were thawed out overnight. A septum was attached to the can, a sample of the headspace gas was taken and analysed for C_1, C_2, C_3, i-C_4, n-C_4 and C_5+ (see below). If any C_5+ was detected, a second sample was taken and analysed for C_4–C_{10} compounds (see below).

The can was opened, headspace volume, water volume and sample weight were measured, and subsamples for occluded and adsorbed gas analysis, water and carbonate determination, and fluorescence spectroscopy were taken. The remaining sample was stored in a freezer at −80 °C. For occluded gas analysis about 30 g of sample was milled in a ball mill together with water. The evolved gas was analysed as described for headspace gas. All samples were analysed for their adsorbed gas content using the method described by Faber and Stahl (1983). Isotopic analyses were carried out on adsorbed gas methane using a similar procedure to that of the GCA laboratory in Lehrte (West Germany).

GC conditions

C_1–C_5+ analysis

This analysis was performed on Carlo Erba Fractovap 2150 and 2350 gas chromatographs equipped with 2 m × $\frac{1}{8}$ inch stainless steel columns filled with Porapack Q on Chromosorb using nitrogen as carrier gas. The oven temperature was 150 °C. After elution of n-butane the column was back-flushed and C_5+ was recorded. A standard gas containing methane, ethane, propane, n-butane, n-pentane and n-hexane was used for quantification.

C_4–C_{10} analysis

C_4–C_{10} analysis was carried out on a HP 5880A gas chromatograph equipped with a 50 m × 0.2 mm (i.d.) fused silica column coated with OV-101. Helium was used as carrier gas at 1 ml/min. The inlet split ratio was 1 : 60. The temperature program was 35 °C (5 min)–8 °C/min–200 °C (2 min), which gave good resolution from C_3 upwards. Quantification was carried out using the same standard gas as for C_1–C_5+ analysis.

RESULTS AND DISCUSSION

The sampling grid (see Fig. 1) is located in Quadrants 34 and 35 on the Norwegian Continental Shelf. The water depth varies between 350 m and 412 m for the major part of the area, but in the south-west the slope of the Norwegian Trench is reached and water depths decrease to less than 220 m. The decreasing water depth coincides with the occurrence of a thick carpet

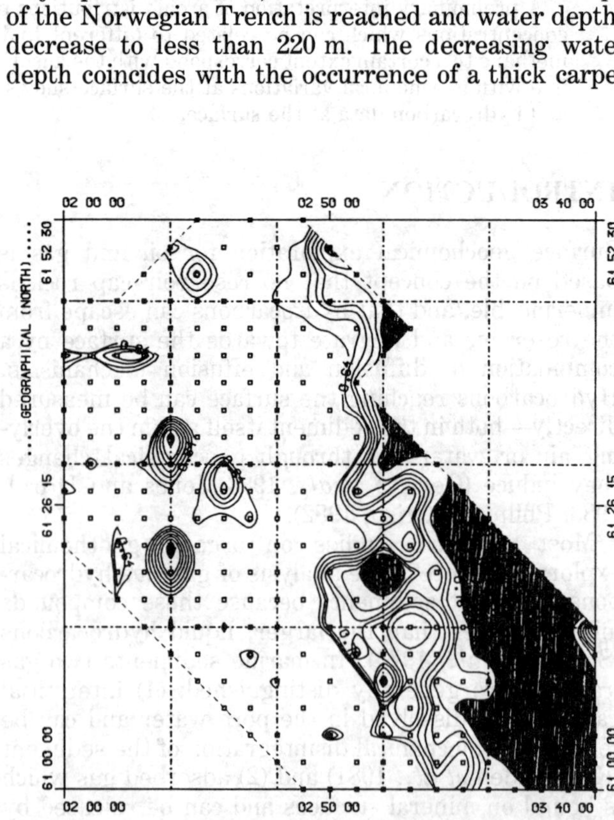

Fig. 2. Distribution of total organic carbon contents (wt%) in survey area. Values higher than the mean value ($\bar{x}$ = 0.6%) are indicated by contour lines. Values higher than the mean value plus one standard deviation ($\bar{x} + \sigma$ = 0.8%) are indicated by a shaded area.

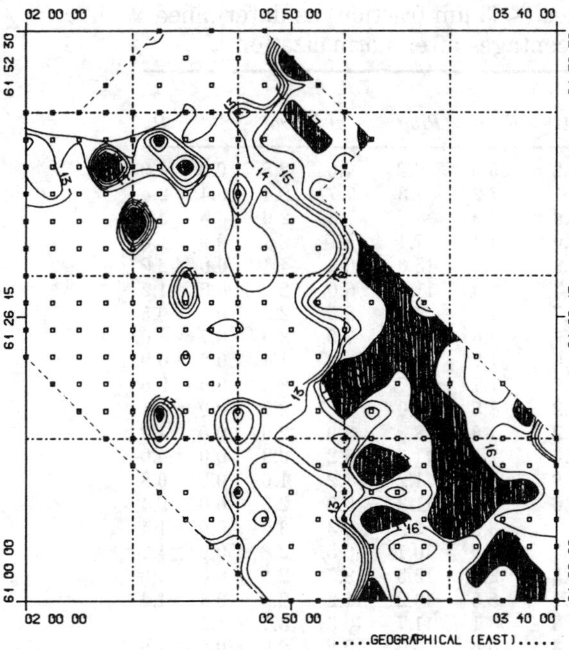

Fig. 3. Distribution of carbonate contents (wt%) in survey area. For explanation of contouring see Fig. 2 ($\bar{x}$ = 13%, $\bar{x} + \sigma$ = 17%).

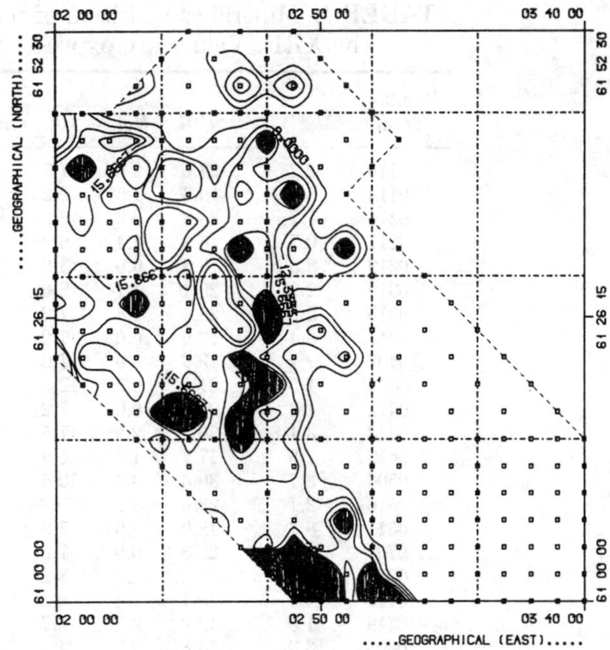

Fig. 5. Distribution of sand contents (wt%) in survey area. For explanation of contouring see Fig. 2 ($\bar{x}$ = 9%, $\bar{x} + \sigma$ = 19%).

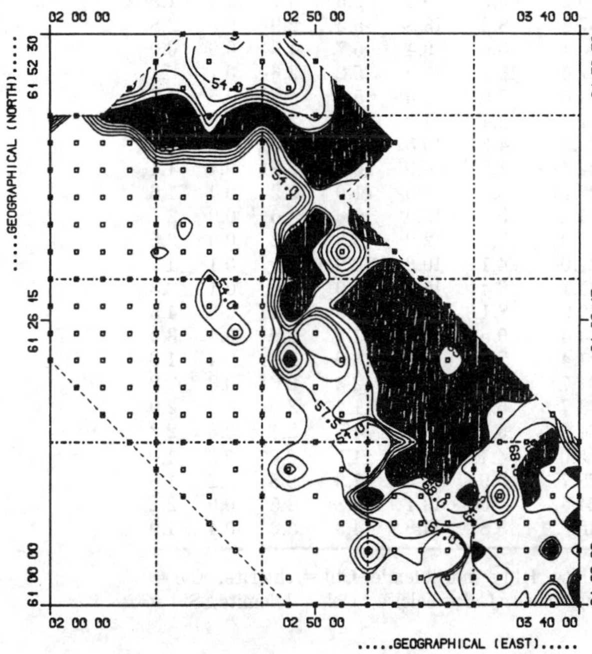

Fig. 4. Distribution of clay contents (wt%) in survey area. For explanation of contouring see Fig. 2 ($\bar{x}$ = 54%, $\bar{x} + \mu\sigma$ = 69%).

of cover sands which limits the extension of the sampling grid towards the southwest.

Figures 2–5 show the areal distributions of some sediment characteristics: total organic carbon (TOC), carbonate, clay and sand content (all expressed as wt%). In these and later figures the areas with values between the mean and the mean plus one standard deviation are contoured, while areas with higher values are shaded.

The data in Figs. 2–5 reveal the existence of two sediment types in the sampling area. The southwestern part consists of glacio-marine clays, which are relatively coarse (sand > 10%), low in carbonate (< 15%) and low in total organic material (TOC < 0.6%). The northern and eastern parts of the sample area contain post-glacial marine clays which are finer (sand <10%), carbonate-richer (>15%) and have higher TOC contents (TOC > 0.6%) than the glaciomarine clays.

The clay fractions from 50 stations have been subjected to XRD analysis (see Table 1). All of the samples examined have a rather similar, although complex, mineralogical composition. The frequency of each mineral phase may, however, differ to some extent. These differences are mainly a result of variation in grain sizes and early diagenesis. Any significant trend in mineralogical variations as a result of geographical positioning is not registered, except with respect to calcite, which varies in the way described above for carbonate.

The results from the gas analyses will now be discussed and correlations between gas data and the occurrence of the two sediment types will be checked for. For reasons of clarity a few words are necessary on the terms occluded and adsorbed gas that are employed here. Occluded gas is released by mechanical disintegration of the sample (see Experimental section) and should be very similar to what other authors indicate as interstitial gas. Adsorbed gas is released by acid-treatment of the total sediment (termed total gas by Faber and Stahl (1983)).

Areal distributions of occluded gas results give different results for methane compared with the C_2–C_4 hydrocarbons. Methane concentrations (see Fig. 6) are highest in the western part of the sample grid whereas the C_2–C_4 hydrocarbons (see Fig. 7; the diagrams for propane and the butanes are very similar to that of ethane, which is shown) have the highest concentrations in the eastern and northern part of the grid. A more detailed inspection of the gas chromatograms reveals the presence of significant amounts of unsaturated compounds. Due to insufficient GC-resolution, the alkenes are not detected as separate peaks but in many samples peak shapes indicate their

TABLE 1. Mineral composition of the samples (<63 μm fraction) as determined by XRD. Values are expressed as percentages after normalization

Sample ident.	*Mx.l*	*Mica/ill*	*Hbl*	*Chl*	*Kao*	*Qtz*	*K-fld*	*Plag*	*Cal*	*Dol*	*Sid*	*Py*
0110	2.7	27.6	2.0	7.7	6.4	30.2	5.5	6.2	7.0	2.8	0.0	1.9
0113	5.0	19.6	1.5	9.2	3.7	27.1	7.2	14.3	7.1	2.4	0.4	2.4
0214	1.7	23.3	0.0	9.3	4.7	36.4	2.9	8.7	7.3	2.6	0.5	2.6
0217	0.7	16.0	2.4	6.3	2.1	48.0	7.4	7.1	5.7	2.4	0.9	1.0
0219	3.3	9.3	0.0	3.0	6.0	41.9	7.9	15.0	6.2	4.4	0.0	2.9
0310	2.9	26.1	1.1	8.6	2.9	30.3	6.9	11.8	6.0	2.2	0.0	1.2
0315	2.9	28.4	1.8	7.2	7.2	29.7	4.4	7.0	8.5	2.3	0.0	0.5
0317	3.4	13.6	0.0	7.2	1.8	36.1	8.6	13.1	7.9	3.5	0.7	4.0
0407	2.1	19.9	0.0	7.6	3.8	38.0	4.7	11.3	10.2	1.4	0.2	0.9
0408	1.9	21.1	0.0	12.9	3.7	29.9	7.7	12.7	7.4	2.2	0.0	0.5
0412	1.8	24.6	0.0	7.2	4.8	31.2	8.8	10.0	7.8	2.7	0.0	1.1
0413	3.1	30.1	2.9	11.5	5.7	25.2	1.8	8.3	8.0	2.0	0.0	1.3
0418	2.9	17.7	1.3	6.4	3.2	41.0	7.5	11.5	5.2	3.3	0.0	0.0
0506	3.7	30.3	3.4	10.4	3.6	25.9	1.6	11.2	7.2	1.6	0.0	0.8
0516	2.1	23.6	4.8	6.8	6.8	25.6	8.2	11.3	5.9	2.5	0.0	2.4
0611	6.1	18.0	0.0	7.8	5.2	37.1	3.9	9.2	8.3	3.0	0.0	1.4
0707	2.9	22.8	0.0	7.6	5.0	37.3	3.9	10.1	6.5	2.8	0.0	1.2
0715	2.8	22.2	0.0	8.2	8.2	31.5	2.3	12.5	8.7	2.4	0.0	1.0
0717	4.1	19.1	1.4	4.2	2.8	34.1	5.5	13.2	11.2	2.8	0.4	1.4
0719	1.3	18.2	2.3	8.7	1.7	31.3	8.1	11.7	10.6	3.7	1.1	1.4
0804	3.3	13.8	2.3	4.7	1.9	34.3	20.6	9.6	5.2	2.6	0.0	1.6
0809	2.3	15.1	1.6	6.4	3.2	45.4	8.7	8.8	4.4	3.0	0.2	0.9
0814	3.5	19.2	0.0	6.8	4.5	37.6	5.5	11.3	7.4	3.1	0.0	1.1
0903	2.3	14.4	2.6	4.8	2.4	44.9	11.1	8.0	5.5	2.8	0.3	0.9
0911	3.6	19.5	5.3	8.6	2.1	31.3	8.5	11.1	5.0	3.1	0.3	1.6
0913	2.8	18.8	0.0	4.9	4.9	40.3	3.7	13.1	6.0	2.7	0.2	2.3
0918	3.0	22.6	0.0	7.0	3.5	33.9	7.5	9.8	7.7	2.8	0.2	2.0
0921	3.2	10.8	1.6	6.7	1.5	45.2	5.9	9.5	9.5	3.7	0.4	1.9
1001	3.8	16.8	0.0	5.2	2.6	37.2	8.9	13.8	5.5	2.9	0.5	2.8
1004	2.2	12.9	4.8	6.1	2.0	46.1	6.6	9.4	5.7	4.3	0.0	0.0
1008	2.3	16.2	2.0	5.9	1.5	37.6	13.5	11.0	5.2	2.6	0.1	2.1
1016	5.2	21.9	1.1	6.4	2.6	34.6	7.0	8.5	5.9	4.7	0.5	1.6
1022	3.6	22.7	0.0	9.5	1.9	27.3	8.1	11.9	9.0	2.5	0.3	3.1
1105	5.8	8.7	3.1	3.7	3.1	43.9	4.6	15.5	6.9	2.5	0.2	1.9
1119	7.2	18.6	1.9	7.3	3.7	34.4	1.8	10.9	10.6	2.2	0.4	1.1
1212	9.2	21.8	0.0	7.0	2.3	27.2	5.4	11.0	11.9	2.2	0.4	1.6
1304	2.6	10.7	2.5	4.6	0.9	44.4	10.0	12.0	6.6	3.6	0.0	2.0
1317	2.5	13.7	1.0	5.8	1.4	40.0	7.7	12.9	10.3	2.2	0.0	2.5
1407	10.6	20.4	1.5	6.7	2.7	27.0	4.1	10.9	12.0	2.3	0.1	1.7
1414	5.9	11.3	2.0	6.5	2.2	31.1	6.4	15.1	14.0	3.0	0.1	2.3
1501	4.0	13.5	0.0	6.0	1.5	36.9	8.1	13.2	8.9	2.9	0.6	4.5
1504	2.6	13.8	1.4	5.4	1.4	44.4	0.5	12.6	11.7	2.9	0.4	3.0
1610	6.9	14.9	2.2	5.3	2.1	31.4	7.5	11.1	13.9	2.5	0.4	1.9
1701	6.4	14.5	0.6	7.8	1.6	31.7	7.6	12.5	12.4	2.6	0.3	2.2
1706	6.5	13.9	1.1	6.0	2.0	33.7	6.2	14.4	11.4	2.6	0.2	2.0
1808	7.4	22.5	0.0	7.9	3.2	24.2	5.5	9.9	14.5	2.4	0.3	2.3
1905	5.1	18.3	2.0	6.7	2.7	35.0	4.1	10.2	11.9	2.8	0.2	1.0
2002	7.8	15.1	2.0	5.8	1.9	28.7	6.9	10.3	17.3	2.5	0.2	1.6
2103	8.8	19.4	1.2	7.7	3.8	24.5	4.0	13.1	12.8	2.5	0.0	2.2
2107	3.0	8.3	2.5	4.2	2.1	45.5	9.8	14.9	4.3	3.0	0.4	1.9

Mx.l = mixed layer clay minerals, Mica/ill = mica and illite, Hbl = hornblende, Chl = chlorite, Kao = kaolinite, Qtz = quartz, K-fld = K-feldspar, Plag = plagioclase, Cal = calcite, Dol = dolomite, Sid = siderite, Py = pyrite.

concentration to be equal to or larger than the coeluting alkanes. The area with high C_2–C_4 concentrations (see Fig. 7) is thus also rich in unsaturated compounds.

The presence of C_2–C_5 alkenes in surface sediments has been reported by several authors (Horvitz, 1972; Hunt and Whelan, 1979; Whelan and Hunt, 1983). These compounds are thought to be the result of *in situ* bacterial activity. In fact, ratios of alkanes to alkenes are used to distinguish biogenic from petrogenic gas (Cline and Holmes, 1977; Kvenvolden *et al.*, 1981; Sandstrom *et al.*, 1983). According to these criteria our occluded gas has a significant contribution from microbial sources. A comparison of the sediment parameters (see Figs. 2–5) and the occluded gas data (see Figs. 6 and 7) shows that the two areas with different occluded gas composition roughly coincide with the occurrence of the two sediment types.

Thus, the interstitial gas in sediments with low clay and TOC contents contains more methane and less C_2–C_4 hydrocarbons (western area), than is found in sediments with higher clay and TOC content (eastern to northern band). Furthermore, the C_2–C_4 hydrocarbons contain significant amounts of unsaturated hydrocarbons. The higher amounts of these are associated with the larger TOC and clay values. This is interpreted as indicating a higher biological activity. The difference in methane concentrations between the two gases cannot be explained by the effects of selective degassing as described by Faber and Stahl (1984).

In summary, the variations of the occluded gas data in this survey primarily reflect variations in the type of surface sediment. Additionally, they contain a significant contribution from microbial sources. For these reasons these data have not been used in further investigations.

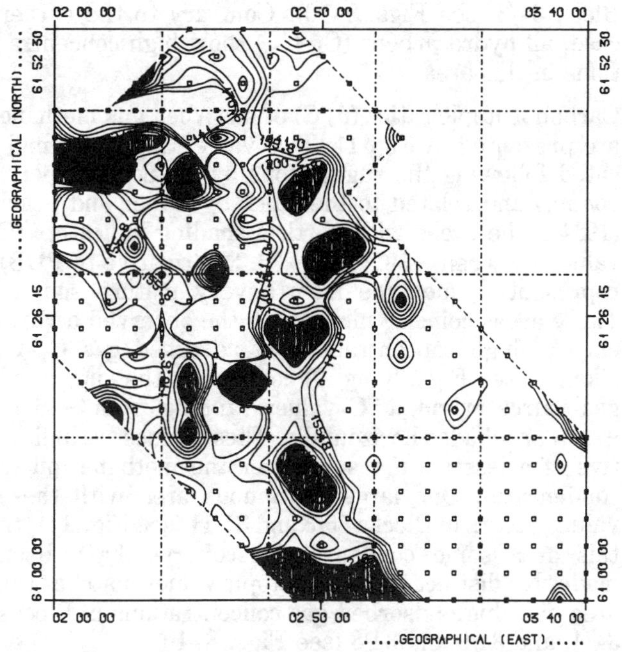

Fig. 6. Distribution of occluded gas methane concentrations (µl gas/kg dry sediment). For explanation of contouring see Fig. 2 ($\bar{x}$ = 141 µl/kg, $\bar{x} + \sigma$ = 215 µl/kg).

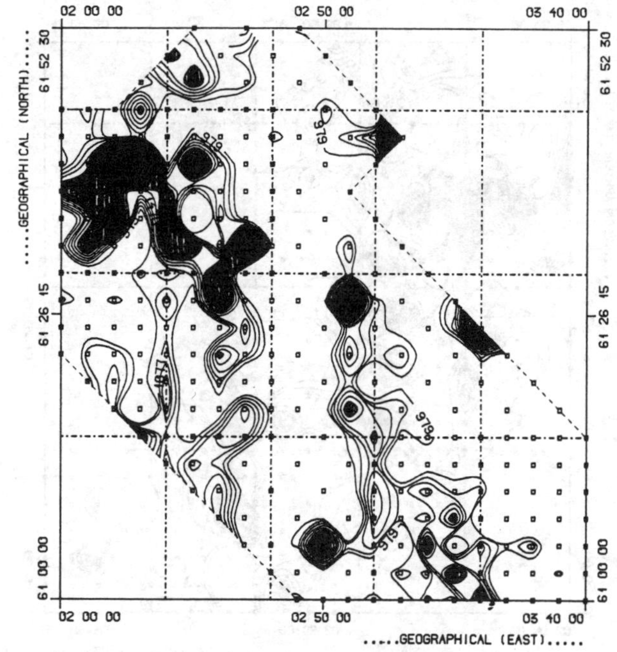

Fig. 8. Distribution of adsorbed gas methane concentrations (µl gas/kg dry sediment). For explanation of contouring see Fig. 2 ($\bar{x}$ = 979 µl/kg, $\bar{x} + \sigma$ = 1469 µl/kg).

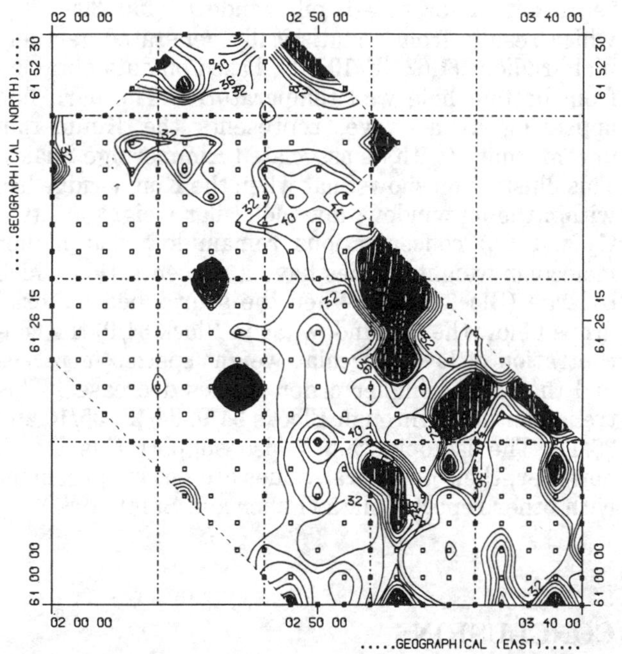

Fig. 7. Distribution of occluded gas ethane concentrations (µl gas/kg dry sediment). For explanation of contouring see Fig. 2 ($\bar{x}$ = 32 µl/kg, $\bar{x} + \sigma$ = 54 µl/kg).

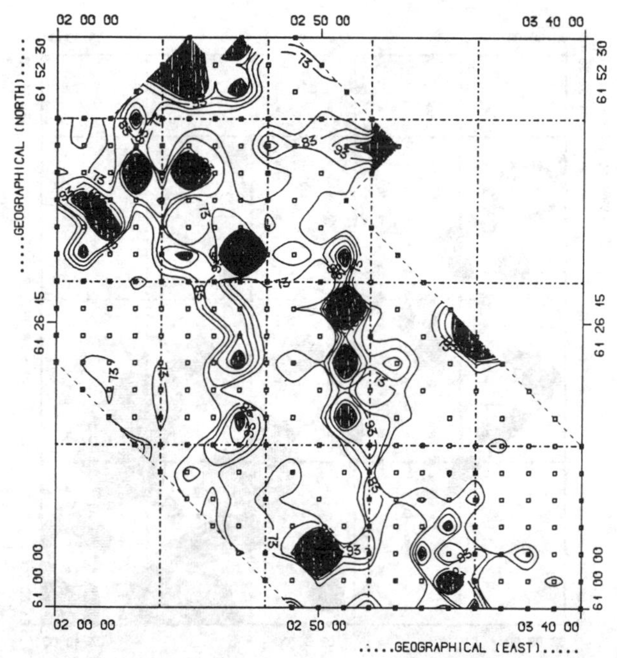

Fig. 9. Distribution of adsorbed gas ethane concentrations (µl gas/kg dry sediment). For explanation of contouring see Fig. 2 ($\bar{x}$ = 73 µl/kg, $\bar{x} + \sigma$ = 108 µl/kg).

The adsorbed gas compositions, yields and $\delta^{13}C$ of adsorbed gas methane have been compared with the data published by Faber and Stahl (1984). No evidence of sediment degassing fractionation is detected. Unsaturated compounds are rarely found, and those present were at a concentration approaching the sensitivity of the instrument. The concentrations of the saturated species are much larger (5×) than was measured for the occluded gases. Additionally, a number of higher alkanes and aromatics are detected. Figures 8–10 display the areal distributions of the concentrations of adsorbed gas methane, ethane, and pentanes and higher compounds (C_5+). Statistical testing of the adsorbed gas alkanes (C_1–C_5) shows them to be correlatable and to represent a homogeneous series. There is no strong correlation with the sediment parameters, such as exists for the occluded gas data. On the contrary, there is a moderate inverse correlation between adsorbed gas concentrations and TOC, clay and carbonate contents. This indicates that the adsorbed gas data are not primarily reflecting variations in sediment characteristics but that they are influenced by other processes, among which is the migration of hydrocarbons from deeper strata.

Although it is not the intent of this paper to present an in-depth interpretation of surface geochemical

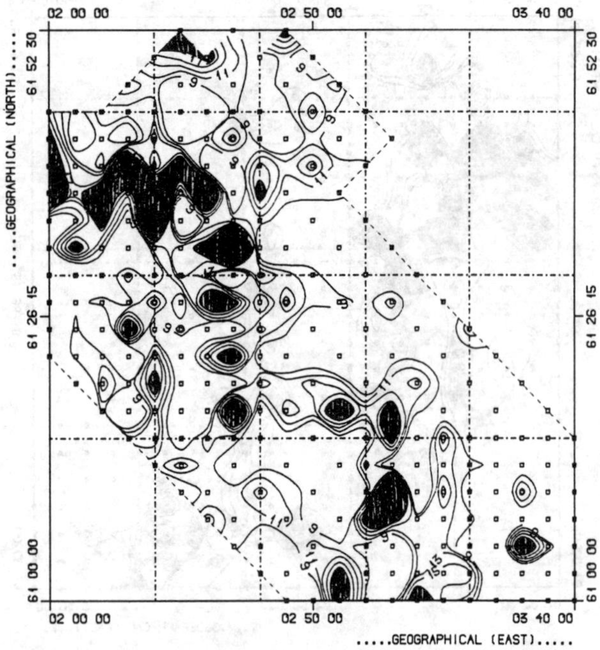

Fig. 10. Distribution of adsorbed gas C_5 and higher hydrocarbons (µl gas/kg dry sediment). For explanation of contouring see Fig. 2 ($\bar{x}$ = 9 µl/kg, $\bar{x} + \sigma$ = 17 µl/kg).

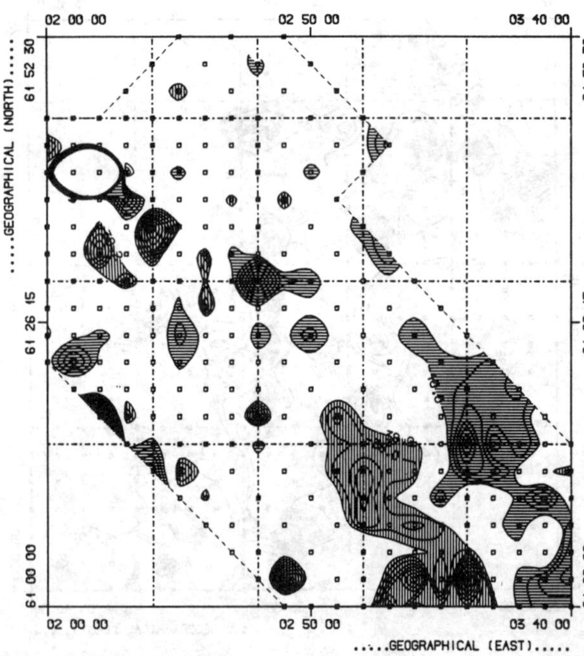

Fig. 11. Distribution of σ^{13}C values (‰ relative to PDB) of adsorbed gas methane. Values between −38.2 and −36.2‰ are shaded vertically, values between −46 and −40‰ are shaded horizontally.

data in an exploration context, several trends in the adsorbed gas data are interesting enough to be discussed in some detail.

An area with higher concentrations of C_1–C_3 alkanes stretches from Block 35/10 to the eastern part of Block 34/9 (see Figs. 8 and 9). This area is less distinct in the diagrams of the higher hydrocarbons (see Fig. 10) which can be indicative of a gas source. This area is located directly north of the Troll Field.

Another area with high adsorbed gas concentrations is located in Blocks 34/4 and 34/5, and continues into Block 34/8 (see Figs. 8–10). Contrary to the earlier case, all hydrocarbons (C_1–C_{10}) show high concentrations in this area.

Carbon isotopic ratios (δ^{13}C) of adsorbed gas methane are presented in Fig. 11. Two value ranges are indicated following the suggestions by M. Schoell (pers. comm.) and related to the work by Faber and Stahl (1984). The areas with vertical shading indicate δ^{13}C values between −38.2 and −36.2‰ (relative to PDB) representing methane from a very mature source. These areas coincide nicely with the observed occurrence of high concentrations of adsorbed gas C_1–C_3 alkanes (see Figs. 8 and 9) confirming the suggested gas-source origin. δ^{13}C values ranging from −46 to −40‰ are shaded horizontally. These values are indicative of a less mature source for the methane (oil to condensate). One large continuous area with these values occurs in Blocks 35/8 and 35/11 (see Fig. 11). In this area some condensate discoveries have been made. No distinct trend in isotopic values matches the area with high adsorbed gas concentrations in Blocks 34/4, and 34/5 and 34/8 (see Figs. 8–10).

In order to illustrate that the surface geochemical data reflect processes occurring in the deeper sediments, Fig. 12 combines a seismic lines with the variations of some geochemical parameters at the surface. The two parallel lines added to the seismic diagram represent a proposed oil window (150–300 °F), which results from a statistically generated geothermal gradient (1.62 °F/100 ft), based on data obtained from bottom hole well temperatures. The dark line appearing as a 'wave' represents the Kimmerian unconformity (with its associated Kimmeridge Shale). This illustration shows that when the Kimmeridge lies within the oil window, low molecular weight gas (C_1, C_2 and C_3) concentrations remain low and higher molecular weight species have a concentration which is large (Block 34/7). When the source bed material drops below the oil window (as in Block 34/9) the concentration of low molecular weight species increases and that of the higher order species decreases. This trend continues through Blocks 34/9, 34/12, 35/10 and 35/11. The isotopic values also support this trend; however, their numerical values are not in agreement with other reported data (Faber and Stahl, 1984).

CONCLUSIONS

One of the main objects of this study is the investigation of the relationship between sediment characteristics and the amount and composition of the occluded and the adsorbed gas phases. The results of sediment gas analyses can be affected by various sediment parameters (grain-size distribution, TOC, etc.), and this process interferes with the interpretation of surface geochemical data for exploration purposes.

The area under investigation contains two sediment types: glaciomarine clays—which are relatively coarse, low in carbonate and organic material—and marine clays—which are more fine-grained and have higher carbonate and TOC contents. The variations in the occluded gas data (amount and composition) correspond to a large extent with the occurrence of the two types of sediments. Furthermore, the occluded gas contains unsaturated hydrocarbons suggesting a significant microbial contribution to this gas fraction.

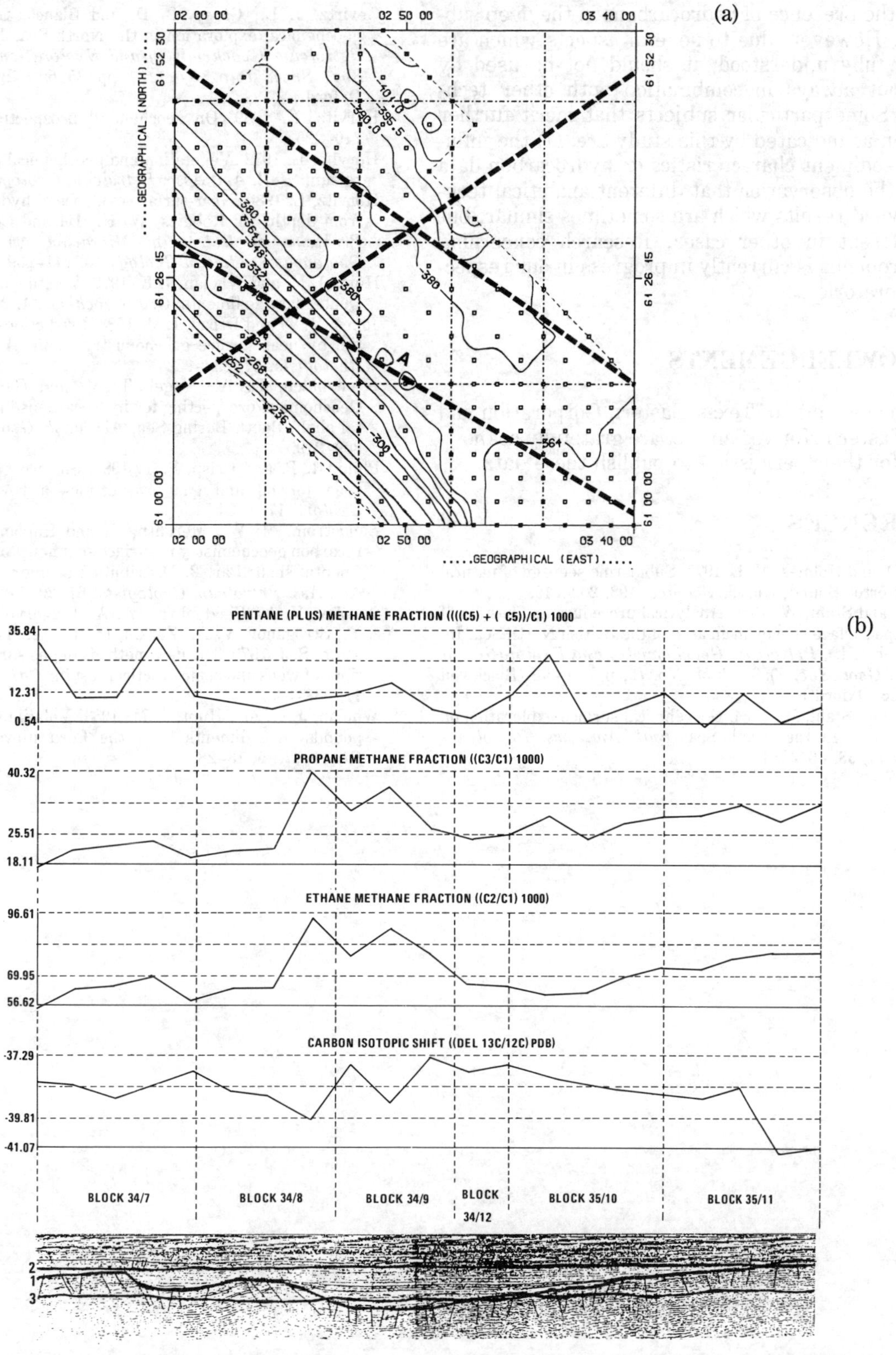

Fig. 12. (a) Location of three seismic lines that have been compared with geochemical data. (b) Combination of seismic line A with geochemical surface data. Line 1 in bottom diagram indicates the Kimmerian unconformity (base Cretaceous), lines 2 and 3 indicate, respectively, the top and bottom of the oil window. For further explanation see text.

As the occluded gas is to a large extent derived *in situ*, it has not been used for further interpretations.

The adsorbed gas fraction does not show a strong relationship with sediment characteristics, neither does it contain significant amounts of unsaturated compounds. Areal distributions of gas concentrations reveal the presence of several areas with high amounts of adsorbed gas. Depending on the hydrocarbon range which shows high concentrations, such areas can be related to oil or gas sources. Areal distribution of $\delta^{13}C$ of adsorbed gas methane corresponds in some areas to the gas concentration data, but gives different results in other areas.

A combination of geochemical surface data with a seismic line shows that changes at the surface can be related to variations in the deeper sediments, thus confirming that at least part of the hydrocarbons at the surface have migrated from below.

In conclusion it can be stated that surface geochemical exploration is a useful tool that can yield information on the presence of hydrocarbons in the deep subsurface. However, due to several aspects which are not yet fully understood, it should not be used by itself, but always in combination with other techniques. Some particular subjects that merit further attention as indicated by this study are: (1) the influence of sediment characteristics on hydrocarbon data and (2) the observation that different analytical techniques yield results which are sometimes similar, but are different in other cases. Research concerning these problems is currently in progress in our respective laboratories.

ACKNOWLEDGEMENTS

The managements of Texas Eastern Corporation and Texas Eastern Norwegian Inc. are gratefully acknowledged for their permission to publish these data.

REFERENCES

Cline, J. D. and Holmes, M. L. 1977. Submarine seepage of natural gas in Norton Sound, Alaska. *Science*, **198**, 209–219.

Faber, E. and Stahl, W. 1983. Analytical procedure and results of an isotope surface survey in an area of the British North Sea. In: Brooks, J. (ed.), *Petroleum Geochemistry and Exploration of Europe. Geol. Soc. Spec. Publ. No. 12*, pp. 51–63. (Blackwell Scientific, Oxford.)

Faber, E. and Stahl, W. 1984. Geocehmical surface exploration on hydrocarbons in the North Sea. *Bull. Am. Ass. Petroleum Geologists*, **68**, 363–384.

Gevirtz, J. L., Carey, B. D. and Blanco, S. R. 1983. Surface geochemical exploration in the North Sea. In: Brooks, J. (ed.), *Petroleum Geochemistry and Exploration of Europe. Geol. Soc. Spec. Pub. No. 12*, pp. 35–50. (Blackwell Scientific, Oxford.)

Horvitz, L. 1939. On geochemical prospecting. *Geophysics*, **4**, 210–225.

Horvitz, L. 1972. Vegetation and geochemical prospecting for petroleum. *Bull. Am. Assc. Petroleum Geologists*, **56**, 925–940.

Horvitz, L. 1980. Near-surface evidence of hydrocarbon movement from depth. In: Roberts, W. H., III and Cordell, R. J. (eds.), *Problems of Petroleum Migration Am. Ass. Petroleum Geologists Studies in Geology* **10**, 241–269.

Hunt, J. M. and Whelan, J. K. 1979. Volatile organic compounds in Quaternary sediments. *Org. Geochem.*, **1**, 217–224.

Jones, V. T. and Drozd, R. J. 1983. Predictions of oil or gas potential by near-surface geochemistry. *Bull. Am. Ass. Petroleum Geologists*, **67**, 932–952.

Kvenvolden, D. A., Vogel, T. M. and Gardner, J. V. 1981. Geochemical prospecting for hydrocarbons in the outer continental shelf, South Bering Sea, Alaska. *J. Geochem. Explor.*, **14**, 209–219.

Philp, R. P. and Crisp, P. T. 1982. Surface geochemical methods used for oil and gas prospecting—a review. *J. Geochem. Explor.*, **17**, 1–34.

Sandstrom, M. W., Meridith, D. and Kaplan, I. R. 1983. Hydrocarbon geochemistry in surface sediments of Alaskan outer continental shelf: Part 2. Distribution of hydrocarbon gases. *Bull. Am. Ass. Petroleum Geologists*, **67**, 2047–2052.

Sokolov, V. A., Geodenkyan, A. A., Grigoryev, G. G., Krems, A. Y., Stroganov, V. A., Zorkin, L. M., Aeiddson, M. I. and Vainsaum, S. J. 1970. The new methods of gas surveys, gas investigations of wells and some practical results. *Can. Inst. Min. Metall.*, **11**, 538–543.

Whelan, J. K. and Hunt, J. M. 1983. Volatile C_1–C_8 organic compounds in sediments from the Peru upwelling region. *Org. Geochem.*, **5**, 13–28.

21 Regional geochemical analysis of the southern portion of the Norwegian Sector of the North Sea

Joel L. Gevirtz, B. D. Carey, Jr* and Stephen R. Blanco†*
*Tenneco Oil E&P, Houston, Texas, USA
†Tenneco Oil Co. Norsk A/S, Sandnes, Norway

To test the usefulness of surface geochemical methods as regional evaluation tools, a study was undertaken by Tenneco Oil Company, comprising, in part, the southern quadrants of the Norwegian North Sea.

A suite of 345 samples collected from the bottoms of cores taken at block corners and centres over most quadrants south of 60 °N was analysed for adsorbed light hydrocarbons (C_1, C_2+ and C_5), indicators of aromatic compounds (fluorescence intensities at 320 nm and 365 nm wavelengths), and methane carbon isotopes. These data, when mapped, indicate patterns coinciding with major Jurassic structural trends. Application of a variety of bivariate and multivariate statistical techniques to these data to attempt to enhance the patterns further demonstrates relationships between geochemical signatures seen at the surface and underlying subsurface geology.

Geochemical parameters mapped at the surface represent a weighted average of generation and migration occurring at depth. The signature seen at the surface most closely represents the dominant process occurring at depth. At the sampling density used here, the data provides an idea of the type of hydrocarbons most likely to be found in a given basin. Because the sampling grid is widely spaced, the results of this study cannot be interpreted in terms of hydrocarbon accumulation in specific structures, but instead should be viewed as a surface representation of source rock and subsurface fluid phenomena occurring throughout the stratigraphic section. Maps developed here, then, represent a summation of hydrocarbon generation throughout the stratigraphic section, but weighted by dominant hydrocarbon-producing strata.

A regional study, such as the one described here, can provide important exploration data concerning the regional hydrocarbon potential in a basin. A study of this kind should ideally precede detailed seismic shooting in a region so that seismic acquisition can be concentrated in areas with the highest geochemical potential.

INTRODUCTION

Trace amounts of hydrocarbons collected from near surface marine sediments and soils are believed to be indicators of hydrocarbons reaching the surface from depth, and have been used by the oil industry as evidence of reservoired petroleum. However, until recently their origin has remained controversial. Recently, methane carbon isotopic ratios measured in sea-floor sediments show values that strongly suggest a catagenic origin for these compounds. The processes by which these compounds get to the surface and are retained there in fine-grained sediments are poorly understood; nevertheless, observations presented by earlier workers (Horvitz, 1972, 1980; Rosaire, 1940; Scull and Myhre, 1976) suggest that anomalously high concentrations of hydrocarbons recovered from sediments or soils below a level where pollution or other man-made phenomena can be ignored, represent microseeps. Often these microseeps are indicative of petroleum accumulation, but they can also indicate high leakage points along geological discontinuities. Therefore, proper interpretation of these data require their integration with geological and geophysical knowledge. In the past, surveys were designed to examine potential reservoirs and have been limited in scope, often covering only features of immediate interest with a closely spaced sample grid. Because virtually all samples contain some hydrocarbons and, based on existing evidence, at least some of these compounds originate by catagenic processes at depth, these data can also be used within a regional context for basin evaluation. This study, then, intends to demonstrate how near surface geochemical data can be used for evaluation of regional hydrocarbon potential.

Earlier workers (Horvitz, 1939, 1980; Rosaire, 1940) differentiate between 'anomalous' and 'background' yield values and, of course, are more interested in the high yield or anomalous samples. In this work, all values, high or low, are believed to carry information related to subsurface conditions. Subtle differences in source-rock facies and maturity are

Petroleum Geochemistry in Exploration of the Norwegian Shelf © Norwegian Petroleum Society (Graham & Trotman, 1985) pp. 247–261.

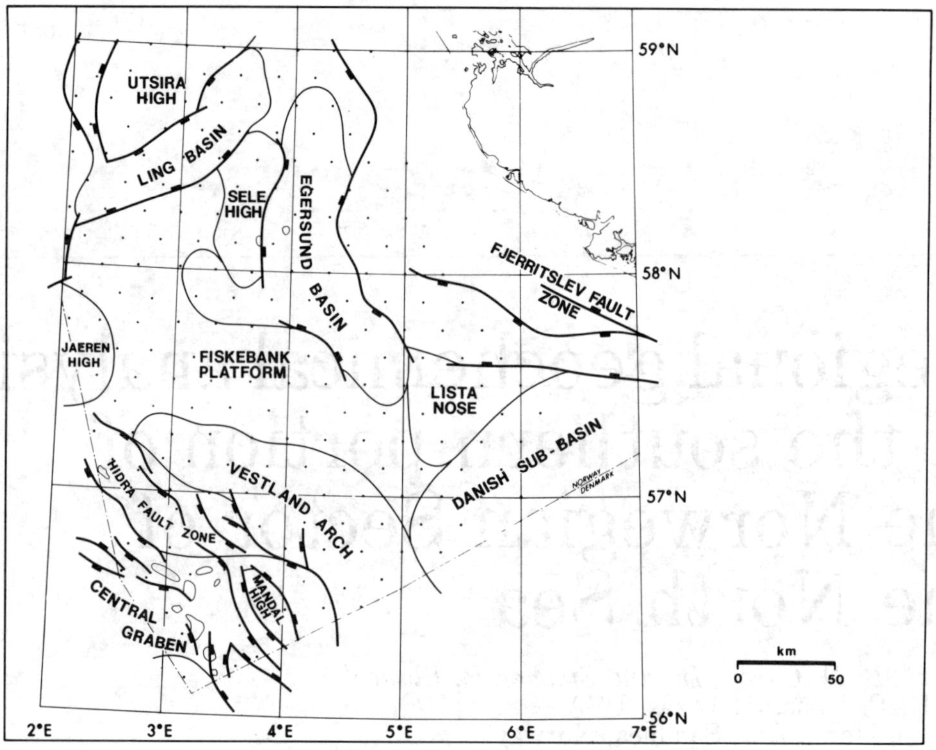

Fig. 1. Base map of Norwegian North Sea south of 59 °N, showing Jurassic structural elements and location of surface geochemical samples.

reflected by different gas compositions and fluorescence intensities. In a regional context, no differentiation can be made between anomalous and background concentrations; but instead, only the magnitudes and their relationship to each other have meaning.

There are several different gas populations in soils and sediments. These different populations often have different origins, and therefore convey different interpretive meanings. Collection procedures designed for each population have their own inaccuracies, and often different populations cannot be sampled in the same operation. Of importance to this study are two populations occurring in marine sediments. One population of gases occurs within sediment interstices, loosely bound, probably by Van der Waals' forces, to grain surfaces. This population, termed 'loosely bound' hydrocarbons by Horvitz (1969) represents both gases that are now reaching the surface and those being produced there. Because the mechanical process by which samples are collected from the sea-floor disturb the sediments and their interstitial water, gas is lost and the population is never reliably censused. Also, because marine samples are collected from depths below the sediment–water interface at which substantial anerobic and/or methanogenic activity occurs, these gases are often enriched in biogenic methane, and such mixtures are more difficult to relate to processes occurring at depth.

The second population, termed adsorbed gases by Horvitz (1972), are those gases sorbed within clay mineral interlayers (Barrer and MacLeod, 1954, 1955; Barrer and Perry, 1961; Eltantawny and Arnold, 1972). This population is unaffected by mechanical collection methods and storage, as evidenced by the fact that considerable chemical processing, designed to digest the clay lattice, is necessary to recover them. Experience has shown that for marine sediments, reproducible results are obtained for the adsorbed population. Loosely bound gases produce less reliable results. For this study, then, adsorbed gases will be used exclusively.

The study area covers southern Norwegian waters south of approximately 59 °N (see Fig. 1) and is part of a larger study carried out in the Norwegian and UK sectors of the North Sea by Tenneco Oil Company in 1979. Within the South Norway portion of the study area, 204 locations were sampled, generally at block corners and centres. Held acreage constraints prohibited regular sampling of Quadrants 1 and 2. Because these areas have an abundance of installations such as platforms and pipelines that have disturbed the sea floor, it is doubtful that they would yield representative samples in any case.

REGIONAL SUMMARY

Because it is tectonically and stratigraphically diverse, the Norwegian continental shelf is considered to be a good test area for surface geochemistry. Varied tectonic elements—including deep grabens, shallow basins, structural highs and stable platforms—yield a variety of source rock–reservoir relationships. Released well control is fairly evenly spread throughout the area, allowing calibration of surface geochemical data by correlation with available downhole geochemical information. Within the study area, three major tectonic/stratigraphic features can be delineated: the Central Graben, the Vestland Arch/Fiskebank Platform, and the Egersund Basin (see Figs. 1 and 2). This provides an essentially tripartite division of the southern Norwegian North Sea into the barren Vestland Arch/Fiskebank Platform separating two oil generating basins. The Central Graben is characterized by major oil and gas accumulations, thick sequences of mature and over-

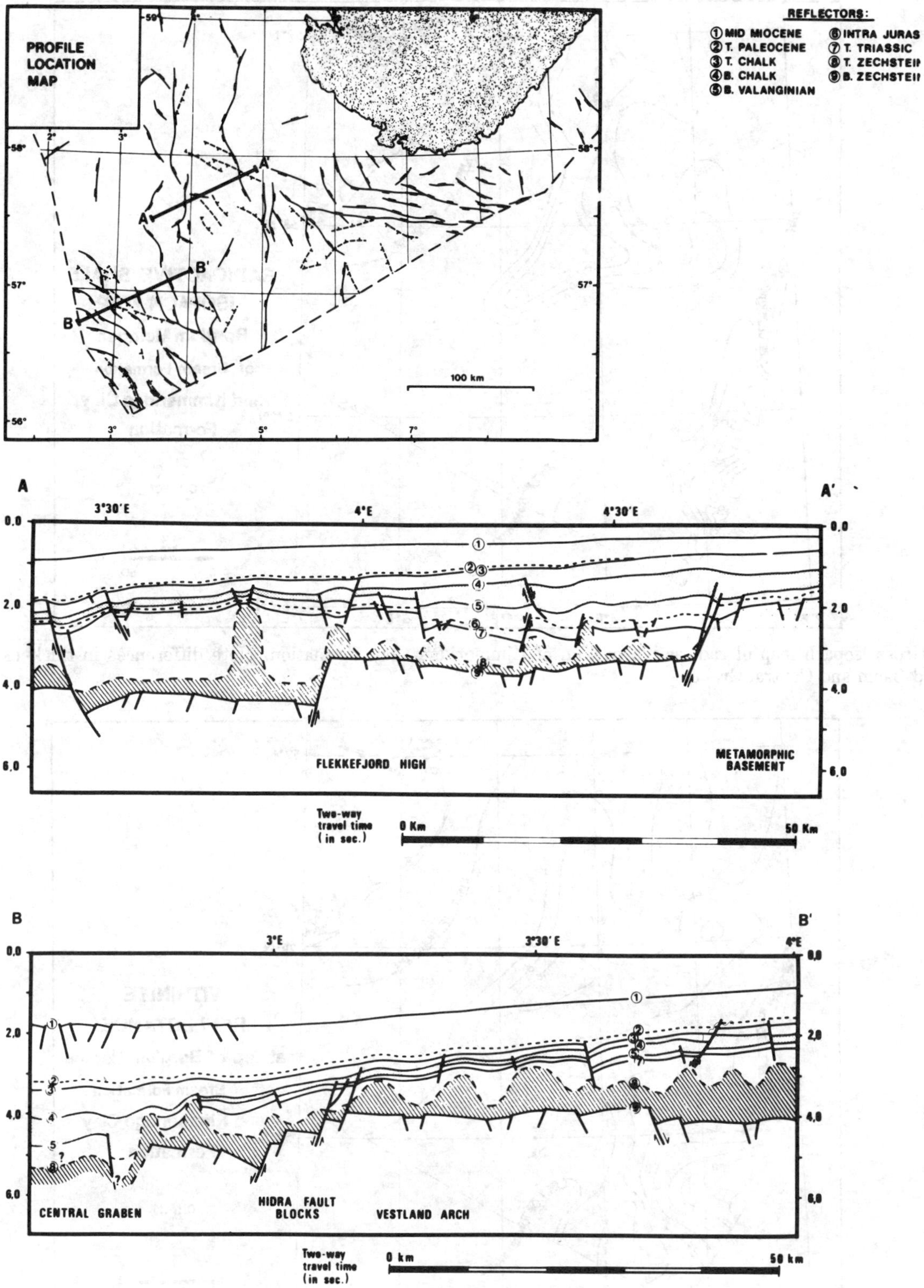

After Skjerven, Rijs & Kalheim, 1983.

Fig. 2. Regional cross-sections across Jurassic structural features of the Southern Norwegian North Sea, (after Skjerven *et al.*, 1983). Striped interval depicts salt.

mature oil-prone source rocks of the Kimmeridge Clay Formation, and several reservoir horizons occur, varying from Triassic through Paleocene. In contrast, the Kimmeridge Clay is thin or absent on the Vestland Arch; Jurassic reservoirs are generally absent. As the Vestland Arch grades north-eastward to the Fiskebank Platform, both Kimmeridge Clay and Middle Jurassic Haldager Sands thicken, but the Kimmeridge Clay is still generally immature. Further north-east, in the Egersund Basin, the Kimmeridge Clay and the Børglum Member of the Bream Formation become mature enough to have generated oil. To date, discovered accumulations have been small (Bream, Brisling, 18/10), and most structures do not appear to be filled to structural spill-point.

The gross isopach of the radioactive portion of the Kimmeridge Clay Formation and Borglum Member of Bream Formation shows the two basins separated by the intervening high (see Fig. 3). This is closely paralleled by the vitrinite reflectance trend indicating (see

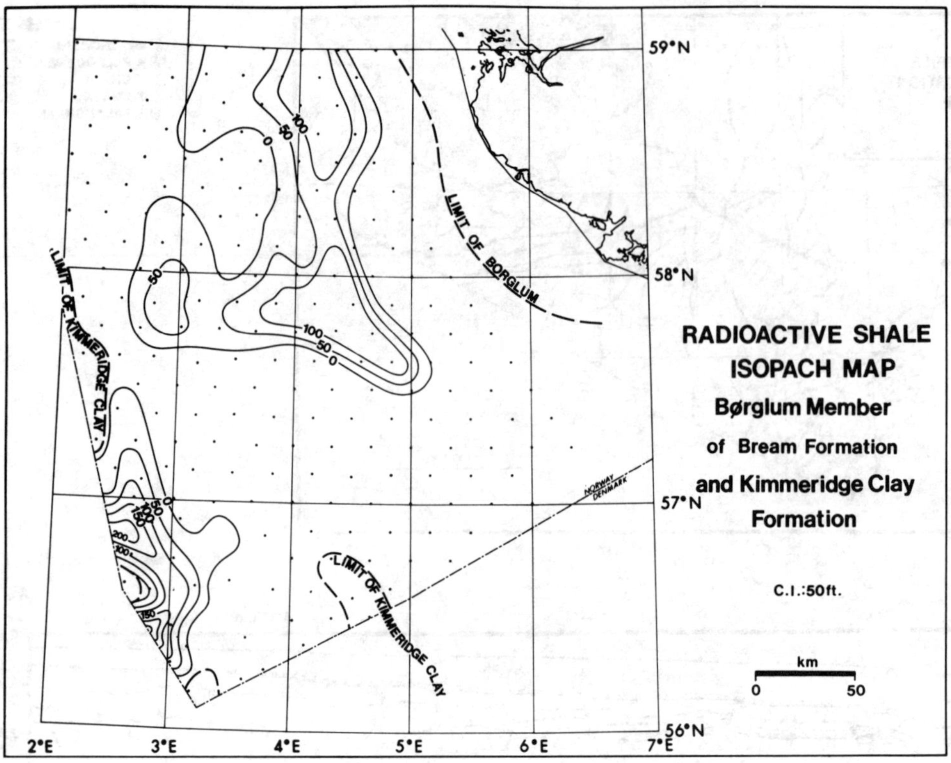

Fig. 3. Gross isopach map of radioactive portion of Kimmeridge Clay Formation. Note differences in thickness between Egersund Basin and Central Graben.

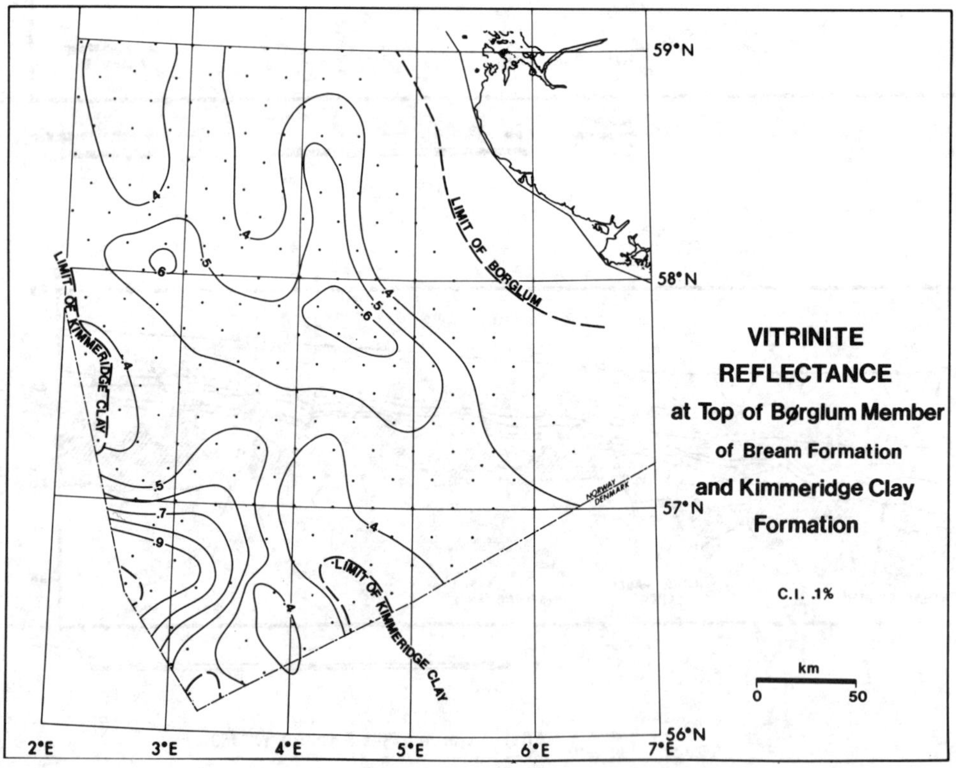

Fig. 4. Vitrinite reflectance at top Borglum Member (of Bream Formation) and Kimmeridge Clay Formation. Note that the Egersund Basin is less mature than the Central Graben.

Fig. 4) that, although both Central Graben and the Egersund Basin are at an oil generating stage of maturity, Jurassic source rocks of the Central Graben are more mature than those in the Egersund Basin.

METHODS

Samples were collected from the lower portion of shallow, 3 inch diameter cores obtained from the sea bottom using a modified gravity coring device that recovers up to 2 m of section. Core lengths varied from 0.5 m in hard sand bottoms to a full core barrel in soft clay. Two samples were collected from each core; one was packed in an inert plastic bag and shipped to Horvitz Research Laboratories in Houston, Texas, for analysis of adsorbed gases and fluorescence intensities. A second portion of each core, also packed in inert plastic bags, was frozen in liquid nitrogen (at − 196 °C) for later analysis by another laboratory for whole gas molecular composition and methane isotopic ratios. Of the 204 regional cores, 172 provided sufficient sample material for liquid nitrogen preservation. The details of this work have been reported elsewhere (Faber and Stahl, 1984).

In the laboratory, sediment samples for adsorbed gas analysis, stored at room temperature, were agitated in a blender to release any remaining interstitial gas and screened with a 62 μm sieve to remove coarse material. Resultant fines were dried by evaporation, were weighed and then digested in hot phosphoric acid under partial vacuum to desorb light hydrocarbons from the sediment (Horvitz, 1969, 1972). Evolved gas was purified and analysed by hydrogen flame chromatography to ascertain its relative composition (Horvitz, 1969, 1972). Results were expressed in parts per billion dry weight of sediment for methane (C_1), ethane and heavier hydrocarbons ($C_2 + C_3 +$ iso $- C_4 + n - C_4$), referred to hereinafter as C_2+, and pentanes (n+iso+neo+cyclo). To estimate the occurrence of heavier hydrocarbons, a portion of fine material was extracted using n-hexane, and was subjected to fluorimetric analysis using a 265 nm excitation wavelength and recording emission intensities at 320 nm and 365 nm.

Dudenbostel and Priestly (1954), Lloyd (1971*a*,*b*,*c*), Wakeham (1977) and Law (1981) indicate that u.v. emissions at 320 nm and 365 nm, respectively, represent two-ringed and three-ringed aromatic compounds including the triaromatic steroid hydrocarbons. However, these authors point out that recognition of individual compounds by fluorimetric methods can only be accomplished in simple mixtures. Because mixtures of aromatic compounds derived from petroleum are complex, fluorescence measurements obtained from surface sediments are reported as relative intensities. For this study, the 320 nm peak is taken as an index of lighter, two-ringed aromatic compounds and their derivatives occurring in gasoline fractions, and the 365 nm peak is taken as an index of three-ringed aromatic compounds and their derivatives found in higher boiling fractions of natural oils.

Samples preserved in liquid nitrogen were analysed for whole gas composition using gas chromatographic methods (Faber and Stahl, 1984). Methane thus recovered was also subjected to isotopic analysis using a mass-spectrograph. Gas yield data were expressed as a percentage of the gas fraction. Isotopic analysis yielded data expressed as the ratio δ^3C/δ^2C in parts per thousand PDB*. However, fractionation of gases recovered in some high sand samples resulted in a loss of methane, and therefore measurements for some samples were unreliable.

RESULTS

Data obtained from the various analytical techniques are presented as histograms and as a selection of bivariate cross-plots. Gas yield histograms are bimodal, representing a mixture of two distinct populations (see Fig. 5). These contrast strongly with those presented by Horvitz (1980) for the Gulf of Mexico, and those examined by Tenneco in other areas of the world where bottom lithology is uniform, which

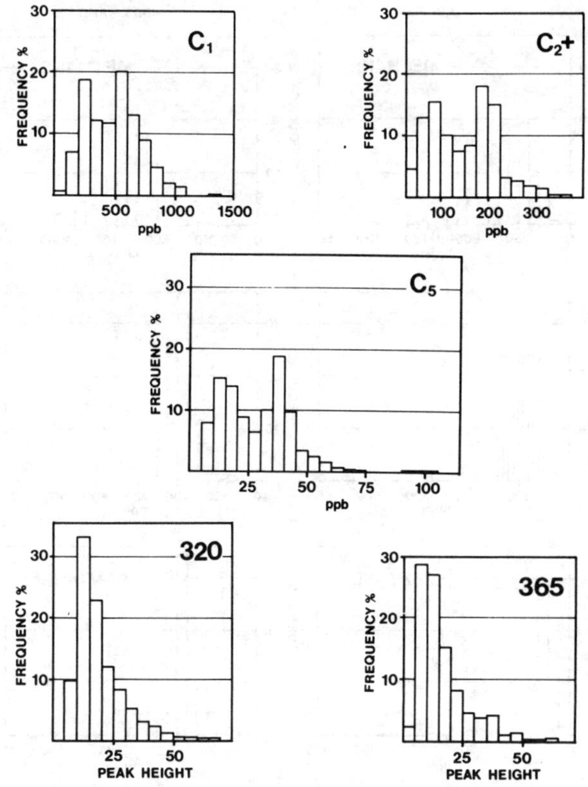

Fig. 5. Histograms showing distribution of yield values for adsorbed gases and intensities of two measured fluorescence wavelengths. Note that gas yield distributions are bimodally distributed.

indicate a single gas yield population. Histograms of fluorescence peak heights are unimodal and approximate a log-normal distribution (see Fig. 5).

Based upon analysis of the histograms, adsorbed gas yields can be separated into a population characterized by low gas yields from samples that contain high sand percentages and those containing low sand percentages and showing high gas yields (see Fig. 6). To determine the 'best' sand percentage at which to divide the samples into two distinct groups, a multivariate analysis of variance (MANOVA), using the three gas yields as discriminating variables, was performed. Wilks' Lambda is a statistic computed as a ratio of determinants of the total variance and the within-groups variance matrices, and is obtained as a result of the MANOVA (Cooley and Lohnes, 1971). As this ratio becomes smaller, the hypothesis that the adsorbed gas yields are sufficient discriminators of the two groups is rejected with increased confidence (Cooley and Lohnes, 1971). As the value of this statistic increases, the discrimination hypothesis is rejected with less and less confidence, indicating that the variables are not sufficient discriminators. Therefore, if the MANOVA is performed using groups separated at several different sand percentages, the 'best' division is obtained where Wilks' Lambda has the lowest value. The variation of Wilks' Lamdba with sand percentage is shown in Fig. 7, and reveals that the 'best' division ranges between 80% and 85% sand. Goodness-of-fit tests applied to each sediment-dependent population are significant at the 95% confidence level, strongly indicating that they are normally distributed. Because the populations overlap, it is difficult to assign to their proper population those samples containing adsorbed gas yields of 0–500, 125–150

$$* \; \delta^{13}C = \frac{(^{13}C/^{12}C)\text{ sample} - (^{13}C/^{12}C)\text{ standard}}{(^{13}C/^{12}C)\text{ standard}} \times 1000‰.$$

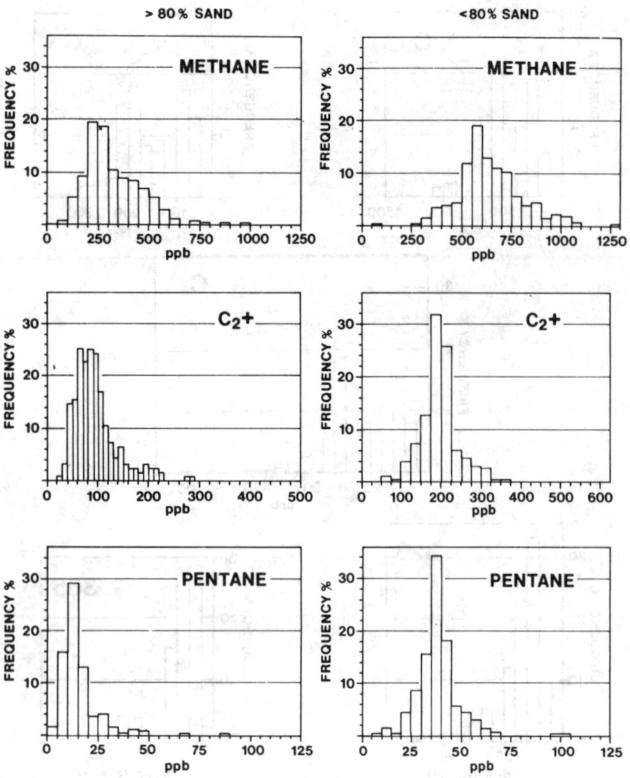

Fig. 6. Results of dividing sample set into two sediment-dependent groups at 80% sand. Note that these histograms approximate a normal distribution.

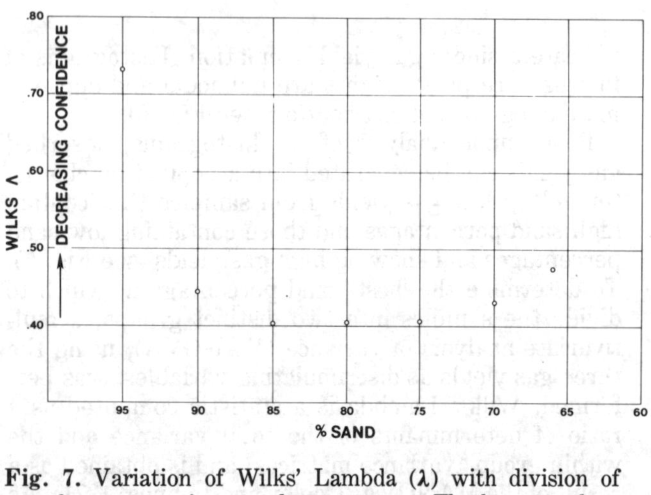

Fig. 7. Variation of Wilks' Lambda (λ) with division of sample set at various sand percentages. The lowest values occur at 80–85% sand, indicating that the 'best' division occurs there.

and 25–30 ppb for methane, C_2+, and pentane, respectively, and sand percentages of 75–85%.

In areas of contrasting bottom lithology, raw adsorbed gas yield maps partially resemble the sand percentage map. These maps partially reflect the relationship between surface sediments and gas yields, thus obscuring any relationship to subsurface phenomena. In order to examine any relationship between adsorbed gas yields and subsurface phenomena, values must first be corrected for these sediment effects. Several different statistical and graphical techniques have been tried. The technique described below is empirical, and is the most robust. It is essentially a transformation relying solely on the parameters (mean and standard deviation) describing each of the distributions. To place all yield values on a single basis, one population (usually the low sand) is selected as a standard, and values comprising the other are transformed so that these yields fall into the distribution of the standard group. The following equation has been derived and used to correct high sand values to approximate those values in the low sand population:

$$Y_{\text{corr}} = \frac{Y_s S_s \bar{Y}_m}{\bar{Y}_s S_m},$$

where $\bar{Y}_m$ = mean yield in mud, S_m = standard deviation in mud, $\bar{Y}_s$ = mean yield in sand, S_s = standard deviation in sand and Y_s = yield for given sand sample.

Application of this method to an area of mixed sediment type is fairly robust. However, there are several uncertainties involved. First, because distributions overlap, yields falling at the high tail of the high sand distribution and at the low tail of the low sand distribution are, at best, uncertain. Second, if the two component distributions are of widely differing statistical character, the conversion will be inaccurate for the entire range. Because light alkanes are most probably adsorbed differentially by different clay mineral species, and because the adsorbtivity of clays—especially those of the smectite group—depends on the condition of the lattice at the time of adsorption (Eltantawny and Arnold, 1972), a more realistic method of obtaining a basis on which to map adsorbed gas yields should be based on quantitative aspects of sample mineralogy, rather than on the crude grain size analyses as done here. However, until the adsorption mechanism becomes better understood, the normalization method derived and used in this paper must suffice, although some caution should be exercised when interpreting resultant maps.

To compare adsorbed gas data obtained from South Norway with those measured elsewhere, a cross-plot was prepared of methane versus C_2+ uncorrected yields for several areas of contrasting source-rock types, maturity levels and production potential. These include an area off southern California, six Haltenbanken (Trøndelag II) blocks in the Norwegian Sea, ECMA Trinidad, and several areas in the Gulf of Mexico (see Fig. 8). Southern California is underlain by

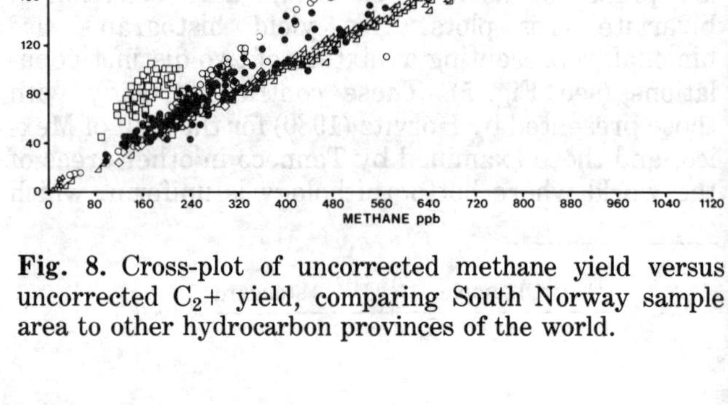

Fig. 8. Cross-plot of uncorrected methane yield versus uncorrected C_2+ yield, comparing South Norway sample area to other hydrocarbon provinces of the world.

TABLE 1. Area comparison, C_1/C_2+ ratios.

Area		*Source rock facies*
South Norway	2.60	Sapropelic–humic mixture
Santa Ynez	2.07	Sapropelic
Haltenbanken	3.28	Humic–sapropelic
ECMA Trinidad[a]	6.66	?
South Timbalier	3.63	Humic?
Green Canyon	2.77	Sapropelic–humic

[a]Ratio approximate because gas yields are extremely low.

Oligocene and Miocene source rocks composed entirely of type II (marine sapropelic) kerogen. Maturity varies slightly with structural depth and position. The six Haltenbanken blocks sampled are underlain by mature humic-sapropelic source rocks. No mature source rocks were penetrated to drilled depth in ECMA Trinidad, but mature and overcooked sources may exist at deeper horizons. Gas produced here is of biogenic origin ($\delta^{13}C$ of CH_4 = −60‰). The source sections underlying the sampled areas of the Gulf of Mexico have not yet been recovered by drilling, but regional geological considerations suggest that kerogens should be of mixed humic–sapropelic composition.

The methane versus C_2+ cross-plot comparing these diverse terrains shows several distinct trends expressing differences in both kerogen type and maturity level. Each trend can be approximated by a straight line through the origin, whose slope reflects the gross ratio between the gases. Average ratios for each area were given in Table 1 and range between 2.1:1 and 6.7:1.

Available evidence concerning kerogen facies and thermal maturation for these areas strongly suggests that average C_1/C_2+ ratios reflect the dominant kerogen facies. The plot demonstrates that for more sapropelic source rocks, the C_1/C_2+ ratio decreases.

For the South Norway area, the methane versus C_2+ cross-plot (Fig. 9) shows a trend indicating a C_1/C_2+ ratio of 2.6. This value suggests that most of the adsorbed gases recovered here originate from admixed sapropelic and humic kerogen with the sapropelic material predominating. Comparison of the surface data from which this diagram was constructed with source rock data obtained from well samples indicates that this interpretation is consistent and that the statistically derived value of 2.6 seen in the plot represents the dominant kerogen type.

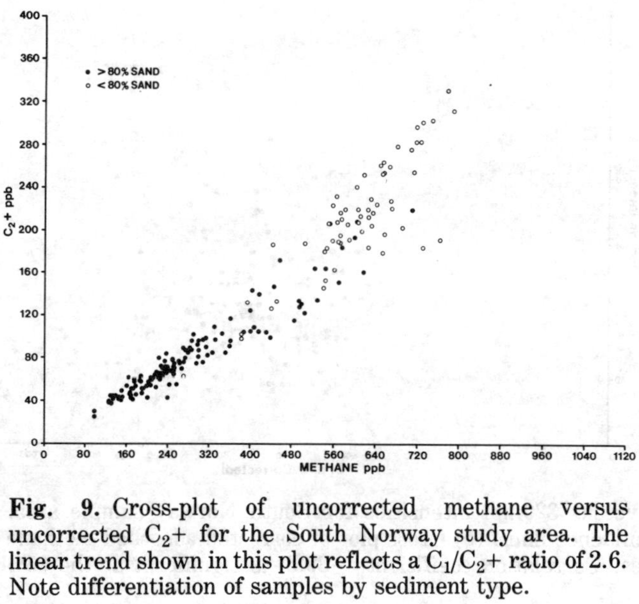

Fig. 9. Cross-plot of uncorrected methane versus uncorrected C_2+ for the South Norway study area. The linear trend shown in this plot reflects a C_1/C_2+ ratio of 2.6. Note differentiation of samples by sediment type.

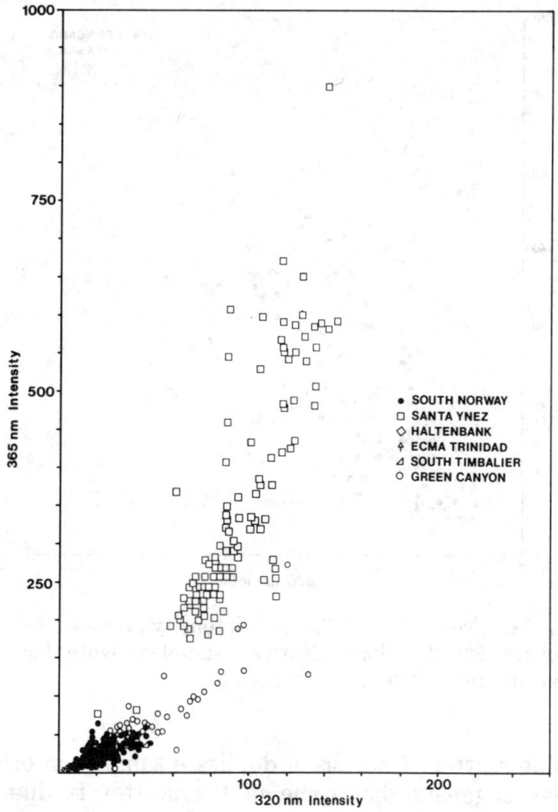

Fig. 10. Cross-plot of 320 nm intensity versus 365 nm intensity, comparing the South Norway study area to other hydrocarbon provinces of the world.

The range of gas yields exhibited in the methane versus C_2+ plot results in part from differences in surface sediment type. To demonstrate this, the methane versus C_2+ plot was constructed using different symbols for each of the two gas populations (see Fig. 6). The high sand samples are restricted to the low portions of the trend while the low sand samples are located in the high portions of the trend. However, differences in gas yields are not directly related to sand percentage, but rather to mineralogical composition of the fine fraction itself, as suggested earlier.

In contrast to gas yields, fluorescence intensities show no relationship to bottom sediments. Fluorescence intensity histograms are strongly unimodal and are lognormally distributed. To compare the South Norway region with others worldwide, a plot was prepared showing the variation of intensities for the two wavelengths for the same areas as those used for gas yields (Fig. 10). This plot shows that a relationship exists between the intensities, and also indicates that intensities differ by an order of magnitude with source facies. Southern California samples show fluorescence intensities up to 10 times greater than those found in other areas. Evidently, on a regional basis, fluorescence intensity is also related to source composition and thermal maturity. The plot suggests that sapropelic source rocks are represented by substantially higher intensities of both 320 nm and 365 nm wavelengths.

To compare fluorescence intensities measured in South Norway, a cross-plot was prepared for the two wavelengths (Fig. 11). This plot, in contrast with the gas yield plot presented above (Fig. 9), shows consid-

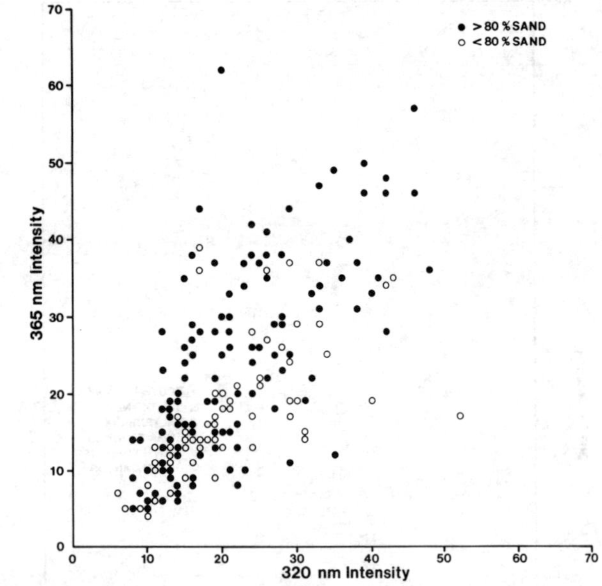

Fig. 11. Cross-plot of 320 nm intensity versus 365 nm intensity for the South Norway samples. Note lack of differentiation by sediment type.

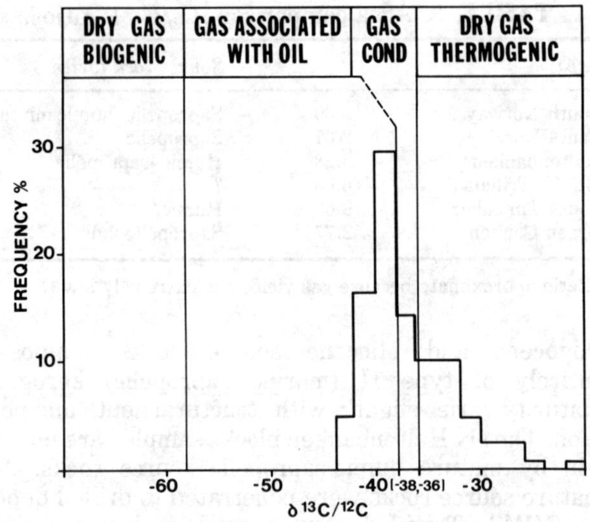

Fig. 13. Histogram showing the distribution of methane carbon isotopic ratios obtained from frozen portion of bottom samples from South Norway. Note that the bulk of the values fall within the liquid range.

erable scatter. Analysis of duplicate samples in other areas suggests that some of the scatter is due to laboratory and sampling errors. Because, as will be shown below, fluorescence measurements are valuable interpretive data, better laboratory technology is required. The remainder of the variation is related to maturity and source composition. To demonstrate the independence of fluorescence intensities and bottom lithology, these data were plotted using different symbols for each of the two sand populations (Fig. 11). High and low sand samples are intermixed in this plot and, in contrast to the gas plots, are not separable on the basis of fluorescence intensity.

Cross-plots were prepared to show the relationship between gas yields corrected for sand content and fluorescence intensities. Two of these plots are presented here (Figs. 12(a) and (b)). Although they appear similar, correlation analysis indicates that while methane shows no relationship to 320 nm intensity, a weak inverse relationship with 365 nm intensity is suggested. Correlations for corrected methane versus 320 nm and corrected methane versus 365 nm are −13.26% and −39.50%, respectively. The methane versus 320 nm correlation is not significant while the methane versus 365 nm is significant at the 95% confidence level. Plots of C_2+ and pentane against the two wavelengths show similar relationships, and have been therefore omitted for brevity.

Carbon isotopes measured for methane obtained from frozen samples collected on the regional grid are displayed in histogram form in Fig. 13. These values range between −42‰ and −20‰, thus presenting convincing evidence that the gases collected from these sediments are of catagenic origin. The bulk of the isotopic ratios fall in the oil region as defined by Faber and Stahl (1984).

A cross-plot prepared for sand percentage versus carbon isotopic ratio (Fig. 14) shows no obvious relationship. However, carbon isotopic ratios that are unreasonably heavy (−26‰ to −10‰) are those from high sand samples, suggesting that oxidation and fractionation are more likely in sandy samples.

To relate whole gas measurements obtained from frozen samples to adsorbed gases obtained from

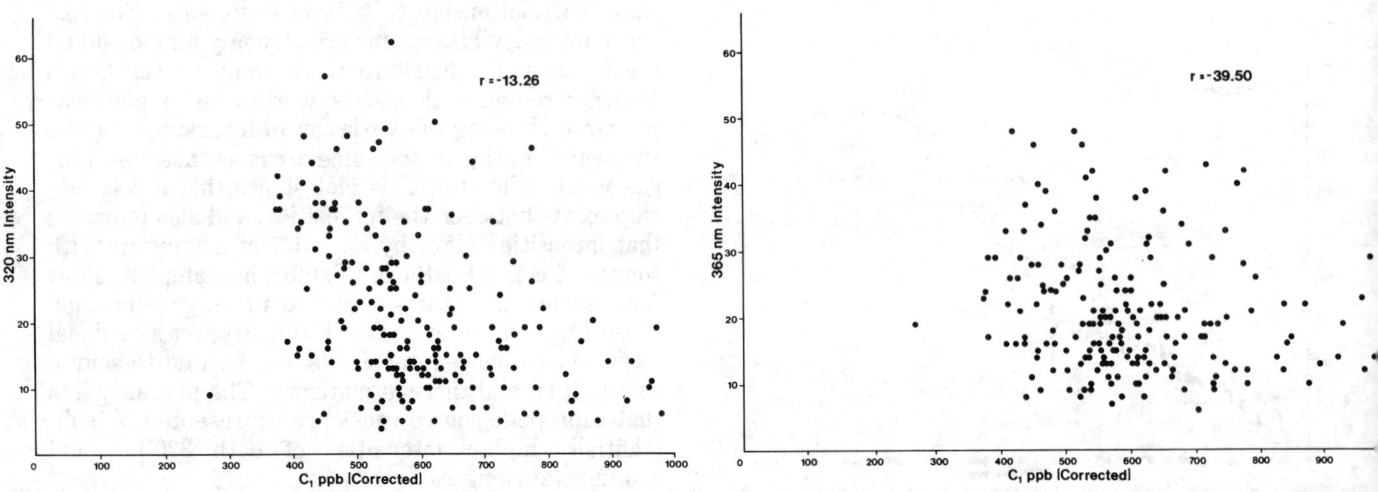

Fig. 12.(a) Cross-plot of corrected adsorbed methane yields versus 320 nm intensities for South Norway sample area. Correlation coefficient (−13.26) is not significant at the 95% confidence limit. (b) Cross-plot of corrected adsorbed methane yields versus 365 nm intensities for South Norway sample area. Correlation coefficient (−39.5) is significant at the 95% confidence interval, demonstrating an inverse relationship.

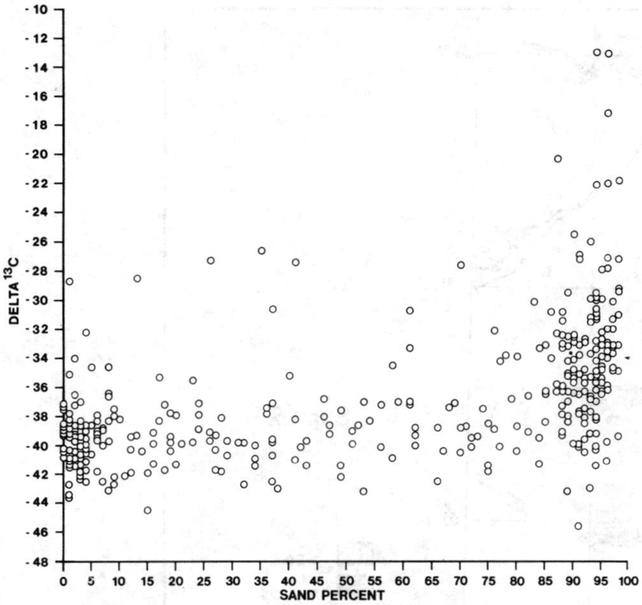

Fig. 14. Cross-plot of percentage sand versus methane carbon isotopic ratios obtained from frozen portions of South Norway samples. Samples showing abnormally heavy isotopic ratios in high sand percentage samples are probably fractionated.

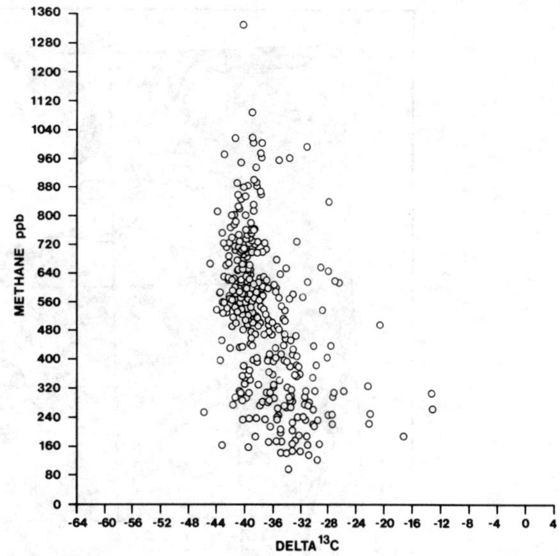

Fig. 15. Cross-plot of methane carbon isotopic ratio obtained from frozen portion of South Norway samples versus corrected adsorbed methane yield. Abnormally heavy isotopic ratios are associated with low adsorbed gas yields. The remaining samples form an almost vertical trend with larger methane yields having lighter isotopic composition.

unfrozen samples, a plot of $\delta^{13}C$ versus adsorbed methane yield (Fig. 15) was prepared. This plot shows that isotopic ratios become heavier as methane yields decrease, suggesting that low methane yields represent more-mature or coaly source rocks, or that many of these heavier values are fractionated. This diagram demonstrates that the variance in isotopic ratios is greatest at lower methane yields, lending support to the notion that low yield samples are fractionated. Cross-plots of $\delta^{13}C$ versus C_2+ and pentane (omitted for brevity) show similar relationships to those shown by the $\delta^{13}C$ versus methane plot, and can be interpreted similarly. Although it is not possible at this time to establish definitely the nature of the relationship between adsorbed gas yields and isotopic ratios, these ratios have proved valuable as independent methods of assessment in the North Sea and other areas.

Cross-plots prepared for methane carbon isotopes and the two fluorescence intensities show no obvious relationship (see Figs. 16(a) and (b)). However, correlation analysis reveals a weak negative correlation,

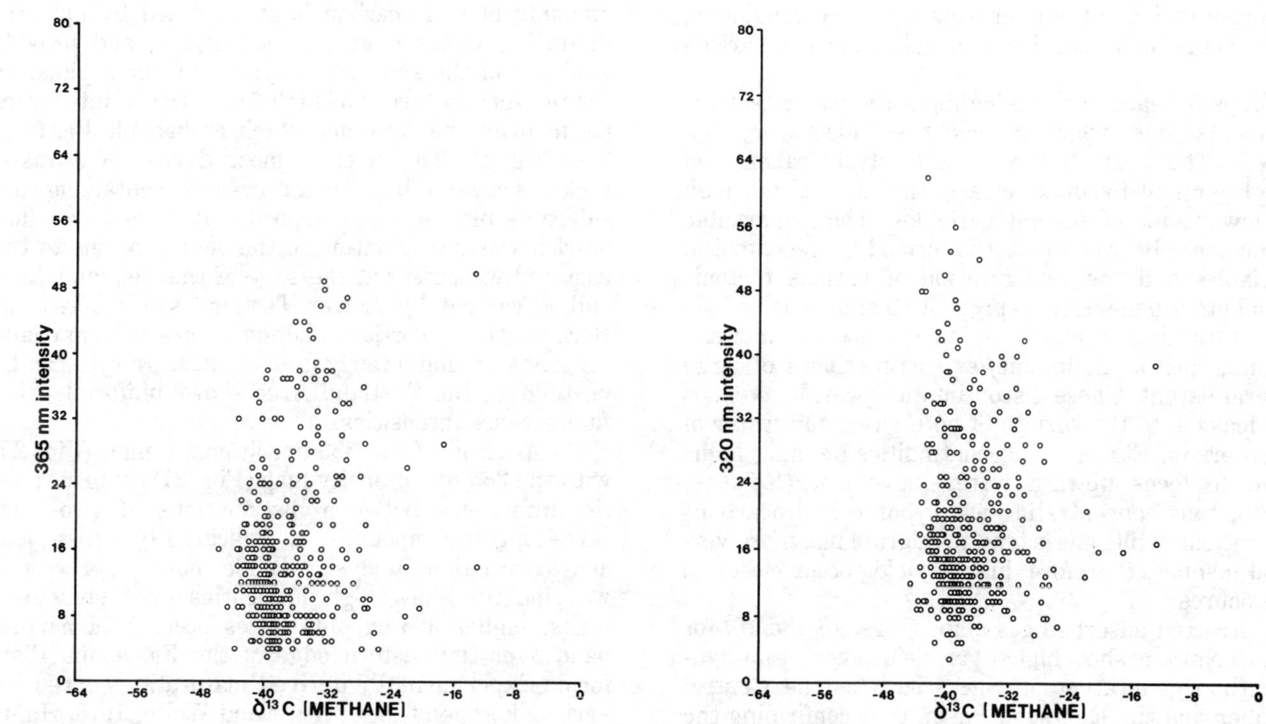

Fig. 16.(a) Cross-plot of methane carbon isotopic ratios obtained from frozen portions of South Norway samples versus 320 nm intensities. (b) Cross-plot of methane carbon isotopic ratios obtained from frozen portions of South Norway samples versus 365 nm intensities.

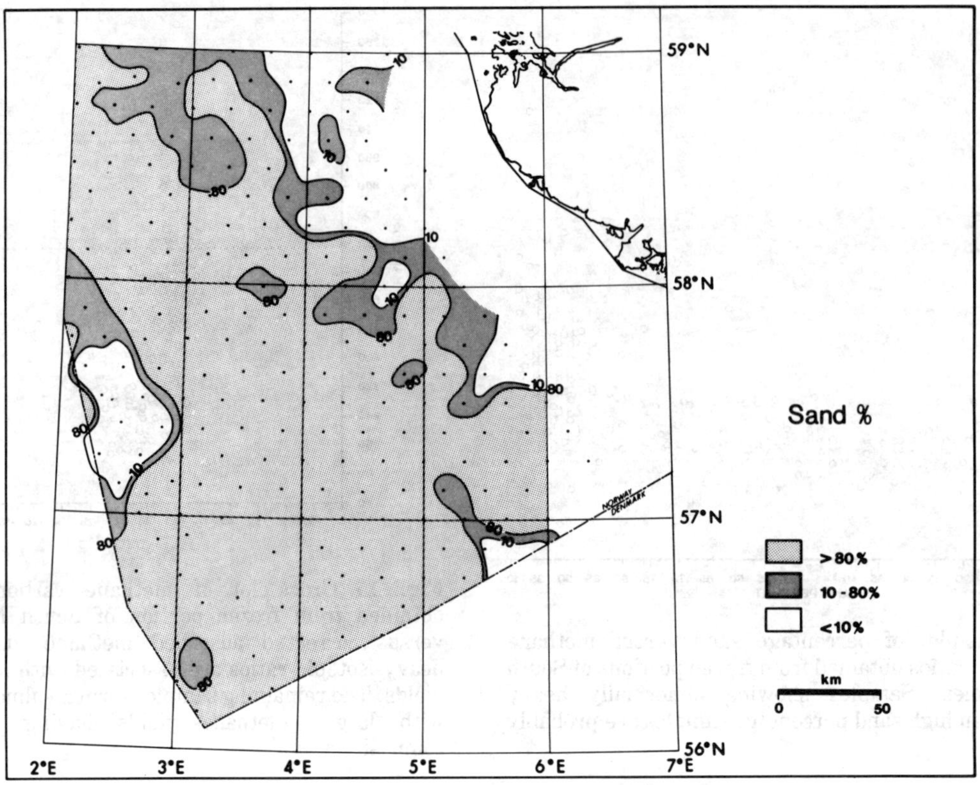

Fig. 17. Map of South Norway sample area showing distribution of sand.

indicating that light isotopic ratios occur with higher fluorescence intensities.

Corrected adsorbed gas yields, fluorescence intensities, and methane isotopic ratios obtained from regional samples were mapped over the southern Norwegian portion of the North Sea. These maps were compared with a regional tectonic and structural map (Fig. 1) and with several published cross-sections (Fig. 2) to relate surface geochemistry to geology. Source rock evaluations, where available, were also incorporated into the interpretation. A regional sand percentage map has been included for comparison (Fig. 17).

Maps for each variable highlight the major tectonic elements over which sufficient samples were collected. The Vestland Arch, the Central Graben, and the Egersund Basin are clearly shown by either high or low values of several variables. The remarkable coincidence between patterns formed by geochemical variables and the configuration of various tectonic elements is believed to represent an expression of the structural discontinuities defining these tectonic elements, such as faults and/or fracture zones of large lateral extent. These discontinuities provide avenues for leakage to the surface of both gases and water of deep origin. Structural discontinuities defining highs tend to focus upward water movement (Roberts, 1980), thus concentrating water-borne hydrocarbons over them, while gases tend to migrate in a more vertical manner; therefore, higher yields occur closer to the source.

Corrected adsorbed gas maps (Figs. 18 and 19) for South Norway show higher gas yields over major hydrocarbon generating provinces such as the Central Graben and the Egersund Basin, thus confirming the idea that these light hydrocarbons migrate in a more vertical direction. Areas that have been shown to be immature on the basis of well cuttings analysis, or in which major source section has been removed by erosion, show significantly lower adsorbed gas yields. These areas include the Vestland Arch/Fiskebank Platform.

In contrast, fluorescence intensity maps (Figs. 20 and 21) show high intensities over the platforms and relatively low intensities over areas highlighted as generating basins by adsorbed gases. The Beatrice Oil Field study (Gevirtz *et al.*, 1983) has demonstrated that fluorescence intensities indicate the types and amounts of hydrocarbon liquids carried by upward-migrating water reaching the surface, and provide evidence of the subsurface hydrodynamic regime. In South Norway, areas of high fluorescence intensities occur over the Vestland Arch/Fiskebank Platform (see Fig. 1). The Arch, almost devoid of Jurassic rocks, is covered by a Tertiary section containing considerable blanket sand deposits. It is believed that oil-rich water originating in the source basins to the east and west rises at the edge of the Vestland Arch and is carried by Lower Tertiary sands across it. Because these Tertiary sediments are only partially consolidated and a fairly uniform leakage rate can be postulated, the Vestland Arch shows uniformly high fluorescence intensities.

Comparison of the 320 nm intensity map (Fig. 20) with the 365 nm intensity map (Fig. 21) confirms subtle differences between distributions of two- and three-ringed compounds, as indicated by cross-plots and correlation analysis. While both fluorescence wavelengths show high intensities over structural highs, higher 365 nm intensities occur in a narrow band over the eastern edge of the Fiskebank Platform. Experimentally derived maturation curves for various kerogen types (Tissot and Welte, 1978; Hunt, 1979) suggest that for most oil-generating kerogens, three-ringed aromatic compounds predominate in less-mature source rocks while two-ringed aromatic

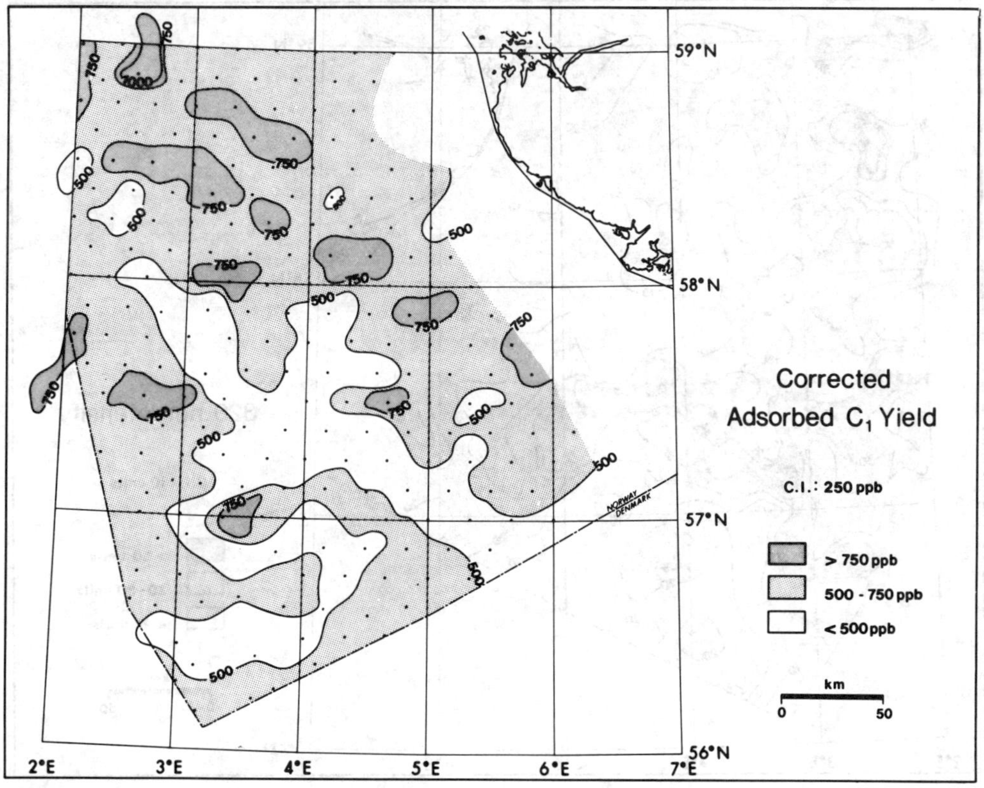

Fig. 18. Map showing methane yield distribution over sample area. High sand samples have all been corrected by method discussed in the text.

compounds are dominant at higher thermal maturity levels.

Very mature kerogens, those in the gas-generating range, produce little or no two- or three-ringed aromatic compounds. Therefore, assuming that kerogen types are grossly similar where they occur throughout the study area, these fluorescence intensity distributions should reflect relative maturity of the source rocks in which these fluorescing compounds originate. Therefore, because the Central Graben area is richer in 320 nm than the Egersund Basin is, the Central Graben is probably generating wet gas. This gas reaches the surface more or less vertically, and carries with it two-ringed aromatic compounds that show up in surface samples as higher 320 nm intensities. The Egersund Basin, on the other hand, does

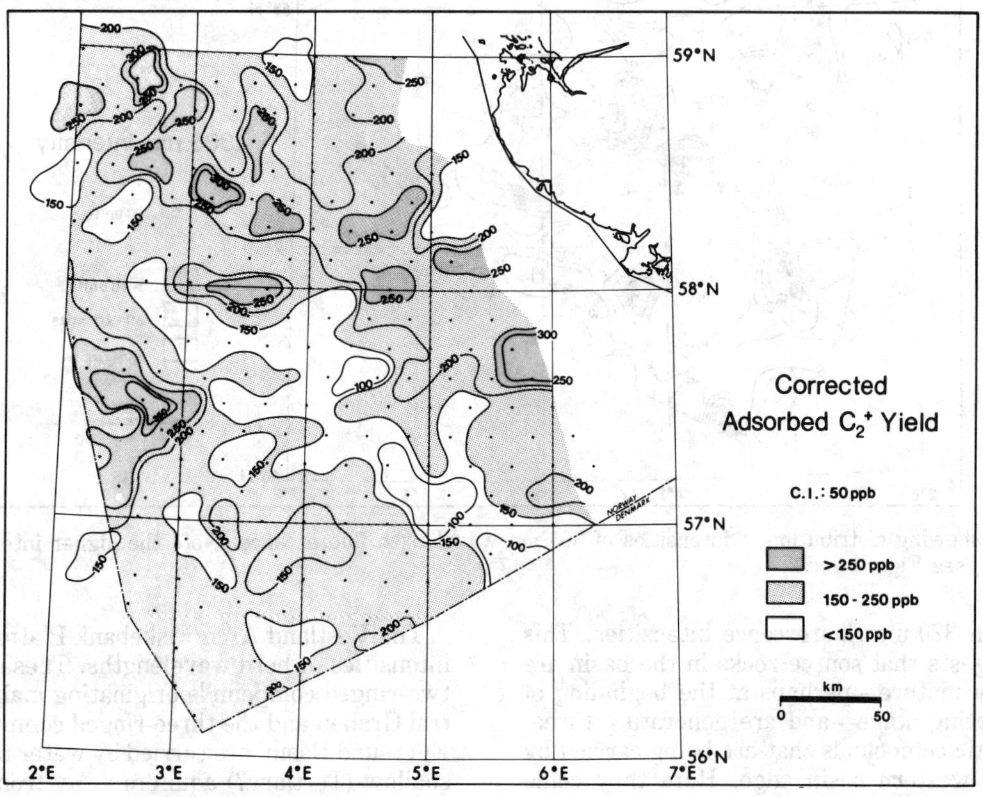

Fig. 19. Map showing C_2+ yield distribution over sample area. High sand samples have all been corrected by method discussed in text.

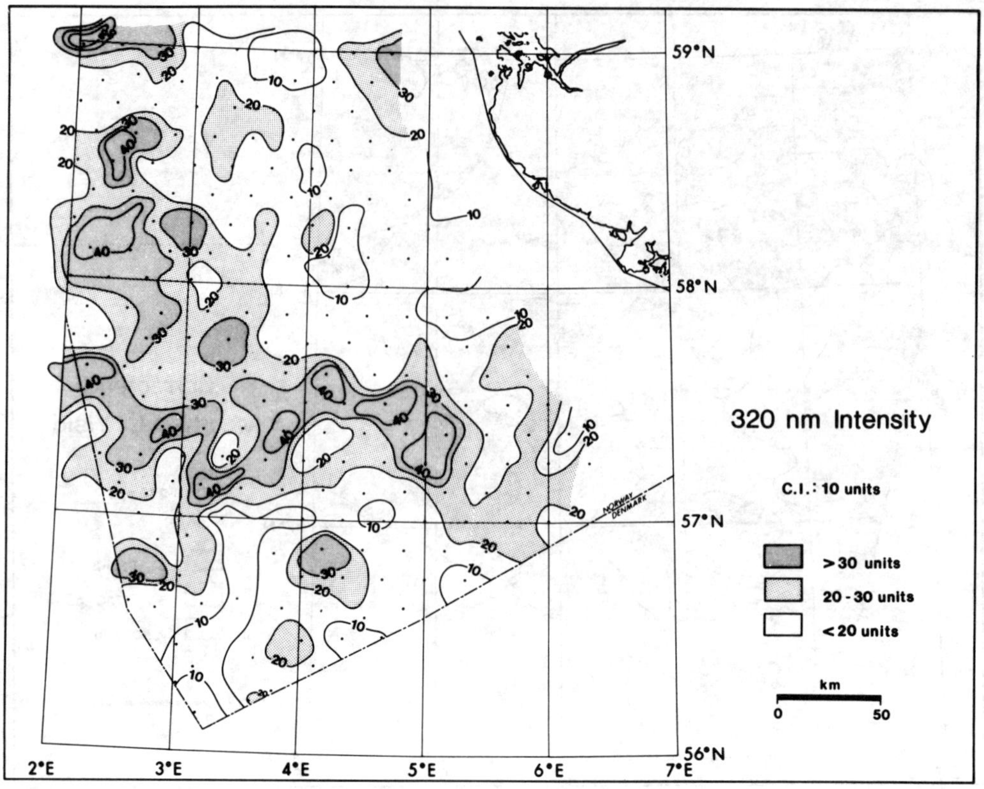

Fig. 20. Map showing distribution of intensities of 320 nm wavelength fluorescence. Note the higher intensities over the Vestland Arch and a portion of the Central Graben, see Figs. 1 and 2.

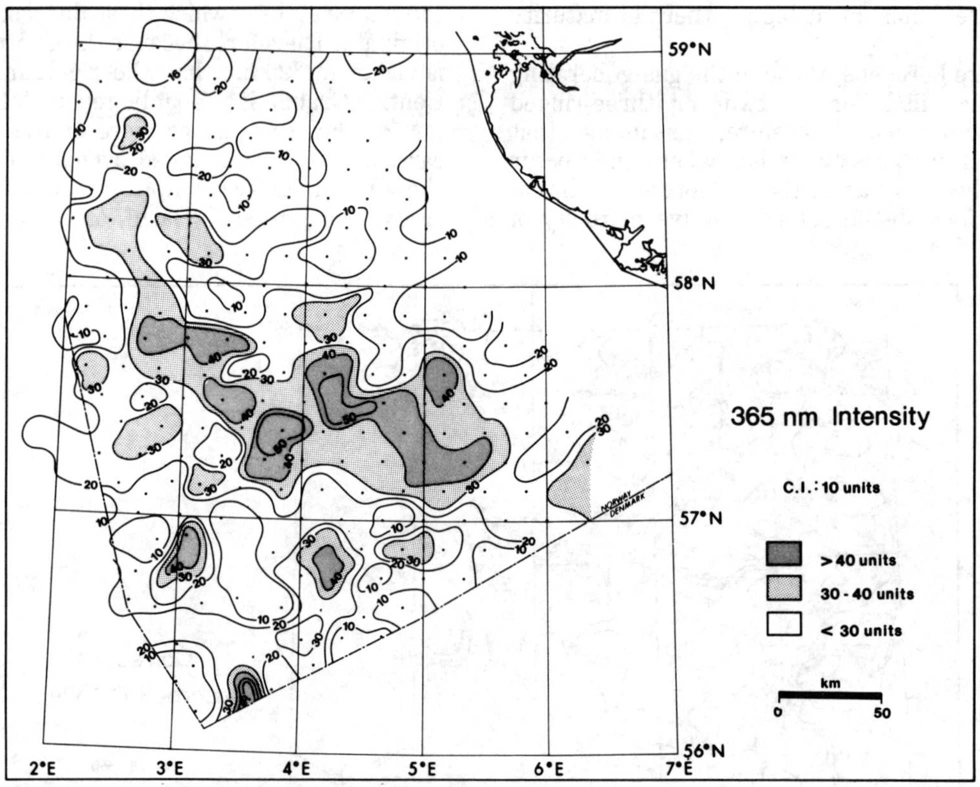

Fig. 21. Map showing distribution of intensities of 365 nm wavelength fluorescence. Note the higher intensities over the Vestland Arch, see Figs. 1 and 2.

not show high 320 nm fluorescence intensities. This evidence suggests that source rocks in the basin are less thermally mature—perhaps at the beginning of the oil-generating stage—and are generating three-ringed aromatic compounds that are being carried by water to the western basin edge. Here they come to the surface and are measured as high 365 nm intensities.

The Vestland Arch/Fiskebank Platform shows high intensities of both wavelengths. These materials, the two-ringed compounds originating mainly in the Central Graben and the three-ringed compounds from the Egersund Basin, are carried by water moving through shallow (Tertiary?) aquifers away from their respective generating sites, and reach the surface over this shallow basin–platform complex. It is important to

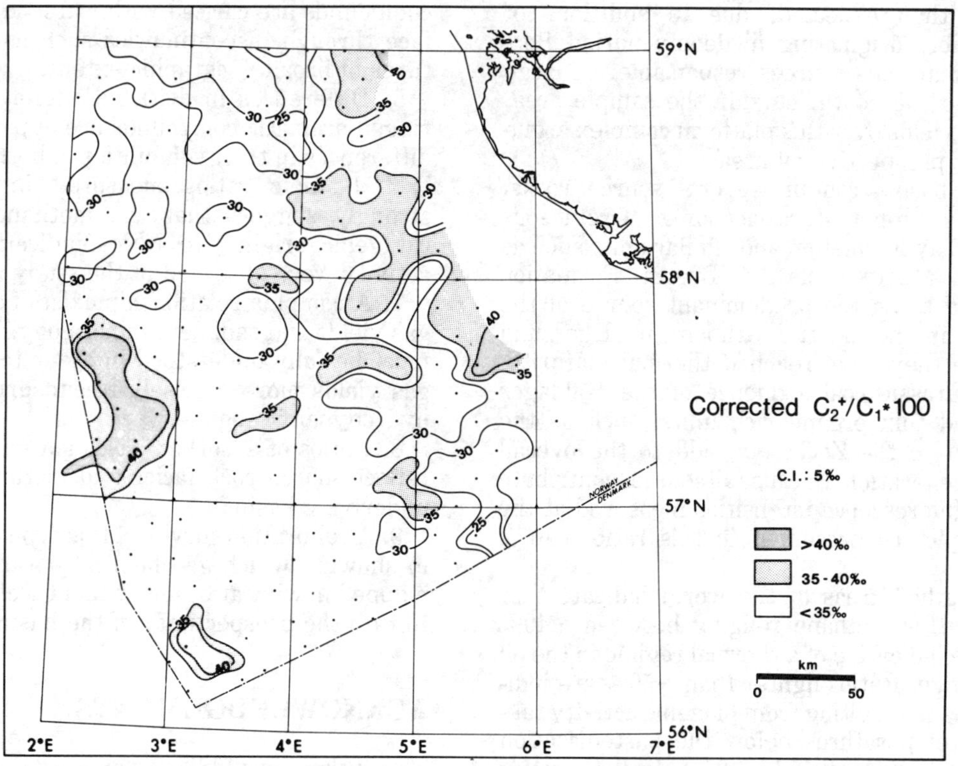

Fig. 22. Map showing the $C_2+/C_1 \times 100$ ratio. All adsorbed gases in high sand samples were corrected using method described in text. Note location of wetter (shaded) areas.

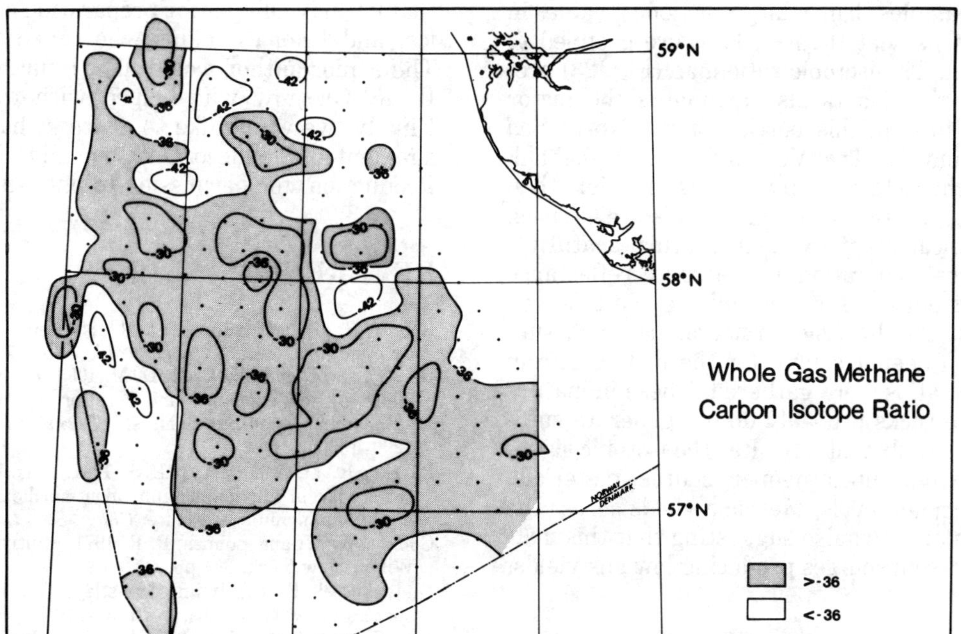

Fig. 23. Map showing distribution of methane carbon isotopic ratios obtained from frozen portions of South Norway samples. Note general conformation of heavy ratios with Fiskebank Platform/Vestland Arch.

recognize that high fluorescence intensities—particularly for the higher wavelengths—do not indicate a generating basin, but instead mark the basin edge.

The Beatrice Oil Field study (Gevirtz *et al.*, 1983) suggests that the expanded ratio $C_2+/C_1 \times 100$ (Fig. 22) expresses subtle variations from the relationship indicated in the cross-plot (Fig. 9). To examine the variation of this ratio in a regional context, a map was prepared for South Norway and compared with regional tectonic and isopach maps. The map shows that the South Norway basins sampled are generally in the oil to gas-condensate generating range. They are more mature over the deep graben, with maturity decreasing as structural highs are approached. The Vestland Arch/Fiskebank Platform, where Jurassic rocks are thin or absent, shows greater maturity than expected, due to additions of methane possibly originating in deeply buried Permian or Carboniferous sources responsible for large gas deposits to the south, outside the sample area. However, low yields over this platform complex argue that it is not a prospective oil area.

North Sea basins contain several source rocks. However, the Upper Jurassic/Lower Cretaceous Kimmeridge Clay Formation and similar units such as the Borglum Member of the Bream Formation are considered to be the predominant source of the hydrocarbons in the Central Graben and Egersund Basin. Where they have reached thermal maturity, the Middle Jurassic coal sequence of the Haldager Formation and thin organic-rich units, such as the Kupferschiefer in the Zechstein, add to the overall hydrocarbon generation. Because all sources contribute to gases and fluorescence intensities measured at the surface, the picture presented in this ratio map is unclear.

Studies in other areas of the world indicate that isotopic ratios for methane ranging between −40‰ and −50‰ are indicative of a thermal regime in the oil generation range. Ratios lighter than −50‰ are indicative of methane resulting from biogenic activity taking place at temperatures below the pasteurization point (87 °C) (Faber and Stahl, 1984). Isotopic ratios heavier than −35‰ are indicative of extreme thermal maturation in coaly material, and are usually found where catagenic gas is produced.

Because isotopic fractionation is more likely in small and/or sandy samples than in large clay ones, values in sandy samples heavier than −36‰ were grouped to avoid confusion. The isotopic ratio map (Fig. 23) prepared from regional data also highlights the major structural features in this portion of the North Sea (see Figs. 1 and 2). The Vestland Arch/Fiskebank Platform is characterized by values heavier than −36‰. Most of the scatter occurs in this area. These values are indicative of extreme thermal maturity. However, spore colouration and vitrinite reflectance data obtained from well samples indicate that this area is immature at the Jurassic stratigraphic level, suggesting a much deeper source for the methane from which isotopic ratios were gathered. These immature Jurassic source rocks probably do not generate sufficient methane to substantially alter the isotopic signature arising from much deeper source materials, perhaps at Permian levels. Methane yields across the Vestland Arch are low, also suggesting that this material is coming from sources producing low gas yields.

SUMMARY AND CONCLUSIONS

1. Surface geochemistry is an extremely valuable tool for regional evaluation when integrated with more conventional geological, geophysical and borehole geochemical data.

2. Several geochemical methods are far superior to one, and contribute to the understanding of hydrocarbon generation and migration throughout a series of basins in a given province.

3. A preliminary broad-brush approach such as the one described here can aid the explorationist in defining areas that merit further interest and those that show little hydrocarbon potential.

4. This study shows that gases migrate to the surface more or less vertically, while heavier aromatic compounds are carried with water and reach the surface through discontinuities such as faults and fractures of broad geographic extent.

5. Differences in distributions of two- and three-ringed aromatic compounds are apparently related to differences in their solubilities in both water and gas.

6. Isotopic ratios measured in South Norway strongly suggest that the methane sampled is of catagenic origin. No ratios indicative of biological activity were observed in the study area.

7. A complex relationship exists between adsorbed gas yields and sediment mineralogy. This relationship must be fully understood in order to relate adsorbed gas yields more precisely to subsurface geology and hydrodynamic regime.

8. Ratios of adsorbed gases gathered over a region reflect source rock facies and maturity levels in the underlying basins.

9. A calibration survey of this type can be invaluable in showing which geochemical parameters should be mapped in wildcat basins without well control in order to rate the prospectivity of the basin.

ACKNOWLEDGEMENTS

The authors would like to express thanks to the many people at Tenneco Oil Company, Exploration and Production, who helped us prepare data and maps, and for the numerous discussions on the geology and geophysics of this part of the North Sea. We would like to thank Cindy Olinger for preparation of the final graphics, and Rhona Jenkins who typed the manuscript. The senior author would like to further acknowledge P. R. Glasgow who helped with the final editing. Finally, we would like to express thanks to the management of Tenneco Oil Company, Exploration and Production, for permission to publish this work.

REFERENCES

Barrer, R. M. and MacLeod, D. M. 1954. Intercalation and sorption by montmorillonite. *Trans. Faraday Soc.* **50**, 980–989.

Barrer, R. M. and MacLeod, D. M. 1955. Activation of montmorillonite by ion exchange and sorption complexes of tetra-alkyl ammonium montmorillonites. *Trans. Faraday Soc.* **51**, 1290–1300.

Barrer, R. M. and Perry, G. S. 1961. Sorption of mixtures and selectivity in alkylammonium montmorillonites. Part I, Monomethyl ammonium bentonite. *Chem. Soc. Lond. J.*, 842–849.

Cooley, W. W. and Lohnes, P. R. 1971. Multivariate data analysis. Wiley, New York, 364 pp.

Dudenbostel, B. F., Jr and Priestley, W., Jr. 1954. Ultraviolet spectra of hydrocarbons, in Brooks, B. T., Boord, C. E., Kurtz, S. S. and Schmerling, L. (eds.), *The Chemistry of Petroleum Hydrocarbons*. Reinhold, New York, pp. 335–350.

Eltantawny, I. M. and Arnold, P. W. 1972. Adsorption of *n*-alkanes by Wyoming montmorillonite. *Nature*, **237**, 123–125.

Faber, E. and Stahl, W. 1984. Geochemical surface exploration for hydrocarbons in North Sea. *Bull. Am. Ass. Petroleum Geologists*, **68**(3), 363–386.

Gevirtz, J. L., Carey, B. D., Jr and Blanco, S. R. 1983. Surface geochemical exploration in the North Sea. In: Brooks, J. (ed.), *Petroleum Geochemistry and Exploration of Europe. Geol. Soc. Lond. Spec. Publ. 12*, pp.35–49. 11(Blackwell Scientific, Oxford.)

Horvitz, L. 1939. On geochemical prospecting for petroleum. *Geophysics*, **5**(10), 487–493.

Horvitz, L. 1969. Hydrocarbon geochemical prospecting after thirty years. In: *Unconventional Methods in Exploration for Petroleum and Natural Gas*, Southern Methodist University, Dallas, Texas, pp. 205–218.

Horvitz, L. 1972. Vegetation and geochemical prospecting for petroleum. *Bull. Am. Ass. Petroleum Geologists*, **56**(5), 925–940.

Horvitz, L. 1980. Near surface evidence of hydrocarbon movement from depth. In: Roberts, W. H. III and Cordell, R. J. (eds.), *Problems of Petroleum Migration. AAPG Studies in Geology, No. 10*, pp. 241–269.

Hunt, J. M. 1979. *Petroleum Geochemistry and Geology*. W. H. Freeman, San Francisco, 617 pp.

Law, R. J. 1981. Hydrocarbon concentrations in water and sediments from U.K. Marine waters, determined by fluorescence spectroscopy. *Marine Pollut. Bull.* **12**(5), 153–157.

Lloyd, J. B. F. 1971a. The nature and evidential value of the luminescence of automobile engine oil and related materials. I. Synchronous excitation of fluorescence emission. *J. Forensic Sci. Soc.* **VII**, 83–94.

Lloyd, J. B. F. 1971*b*. The nature and evidential value of the luminescence of automobile engine oil and related materials. II. Aggregate luminescence. *J. Forensic Sci. Soc.* **11**, 153–170.

Lloyd, J. B. F. 1971*c*. The nature and evidential value of the luminescence of automobile engine oil and related materials. III. Separated luminescence. *J. Forensic Sci. Soc.* **11**, 235–253.

Roberts, W. H., III 1980. Design and function of oil and gas traps. In: Roberts, W. H., III and Cordell, R. J. (eds.), *Problems of Petroleum Migration. AAPG Studies in Geology, No. 10*, pp. 217–240.

Rosaire, E. E. 1940. Geochemical prospecting for petroleum. *Bull. Am. Ass. Petroleum Geologists*, **5**, 1418–1426.

Skjerven, J., Rijs, F. and Kalheim, J. E. 1983. Late Paleozoic to early Cenozoic structural development of the south-southeastern Norwegian North Sea. *Geol. Mijnbouw*, **62**, 35–45.

Scull, B. J. and Myhre, L. A. 1976. A combination of geochemical and geophysical prospecting for petroleum, offshore Norway. In: *Empresu Nacional Adaro de Investigaciones Mineras*, pp. 561–569.

Tissot, B. P. and Welte, D. H. 1978. *Petroleum Formation and Occurrence: A New Approach to Oil and Gas Exploration*. Springer-Verlag, New York, 538 pp.

Wakeham, S. G. 1977. Synchronous fluorescence spectroscopy and its application to indigenous and petroleum-derived hydrocarbons in lacstrine sediments. *Envir. Sci. Technol.* **II**(3), 272–276.

22 Recently formed methane-derived carbonates from the North Sea floor

Martin Hovland, Mike Talbot,† Snorre Olaussen* and Lars Aasberg**
*Statoil, P.O. Box 300, N-4001 Stavanger, Norway
†Geologisk Institutt avd. A, University of Bergen, N-5000 Bergen, Norway

Numerous slabs and indurated crusts of limestones and carbonate cemented sandstone were found on the floor of pockmarks in the North Sea. Sea-floor pockmarks are local depressions found in unconsolidated fine-grained sediments. Their origin has been ascribed to percolation of gas or liquids through the sea floor. A group of five pockmarks believed to be associated with shallow gas, located in Norwegian Block 25/7, were studied by the use of a remotely operated vehicle (ROV). Petrographic studies of the cements in one of these slabs (the '25/7-slab') show a variety of aragonite and Mg-calcite crystals. Aragonite crystals are often perfectly preserved and are comparable with aragonite cements found in modern reef structures.

The $\delta^{13}C$-value of the aragonite cement in the 25/7-slab is −58.9‰ (PDB) which indicates a methanogenic origin. It is concluded that the 25/7-slab is a response to an active diagenetic zone on the floor of the pockmark, whereby a glauconitic and fossiliferous sand is transformed to a carbonate cemented sandstone and sandy limestone. Scanning electron microscopy has revealed globular and crystal structures in the slab's contact zone with the sea floor, which suggest that the diagenetic process was active when the slab was retrieved from the sea floor.

Sea-bed sediment samples were also obtained from the vicinity of the pockmarks. The yields of methane through hexane contained in the upper soft sediments together with the $\delta^{13}C_1$-value (−39‰ to −45‰ PDB) suggest a thermogenic origin for the shallow gas.

It is concluded that both the sea-floor pockmarks and the carbonate crusts are probably caused by the seepage of hydrocarbon gases through the sea-floor.

INTRODUCTION

During reconnaissance pipeline route surveys for the Statpipe project in 1981 an area with five isolated pockmarks was found south of the Heimdal Field in Block 25/7 (see Fig. 1). The area is of interest due to the associated occurrence of shallow gas in soft, layered sediments (Hovland, 1984*b*). The five pockmarks vary from 20 m to 40 m in diameter and are 2–4 m deep. They are situated above a possible gas 'cloud' seen on shallow seismic records between 38 m and 0 m beneath the sea floor (see Fig. 2). At the pockmark location in Block 25/7 the water depth is 127 m, and the sea floor, consisting of fine sand, is generally flat. Another characteristic of these pockmarks is their highly reflective floors which appear as dark patches on side-scan sonar records.

In order to study the pockmarks in more detail, a visual geological and geochemical survey employing an ROV (Remotely Operated Vehicle) was carried out in 1983.

POCKMARK CARBONATES

During the visual survey of the area, the floors of the pockmarks were found to be littered with blocks and protruding slabs of carbonate cemented sandstone and sandy limestone (see Fig. 3) (Hovland, 1984*a*). In some areas the pockmark floors seemed to be encrusted with an irregular layer of the material. The crust is dusted with finely grained sediments and has numerous holes and cavities. A slab about 30 × 20 × 10 cm and weighing about 10 kg in air (the '25/7-slab') was recovered from the centre of one of the pockmarks (see Fig. 4). It was located over an area where the subsurface gas anomaly apparently reaches the sea-floor surface. The slab protruded out over a cavity, or a minor crater, which seems to have been formed by some erosional process.

MINERALOGICAL AND GEOCHEMICAL RESULTS

Laboratory analysis of the 25/7-slab has provided a number of interesting results. A section of the slab is

Petroleum Geochemistry in Exploration of the Norwegian Shelf © Norwegian Petroleum Society (Graham & Trotman, 1985) pp. 263–266.

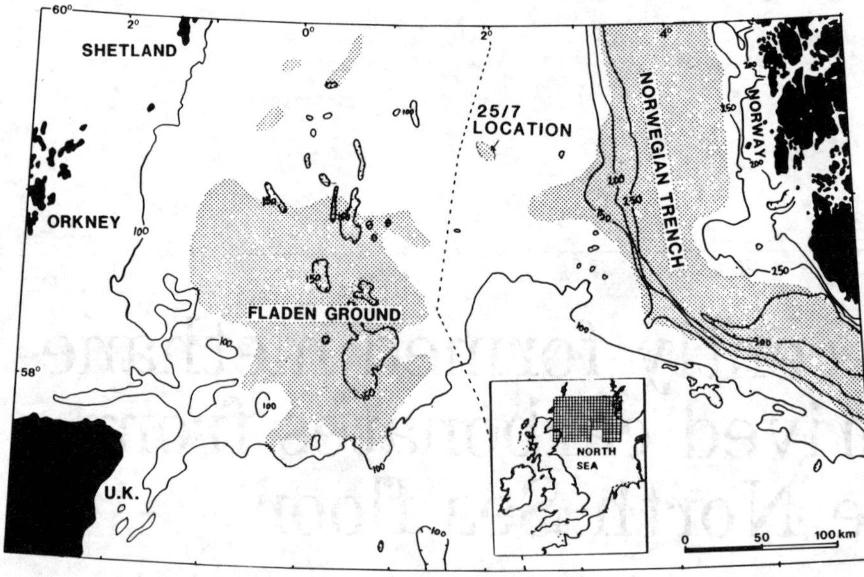

Fig. 1. Location map. Occurrence of pockmarks is indicated by stippled areas.

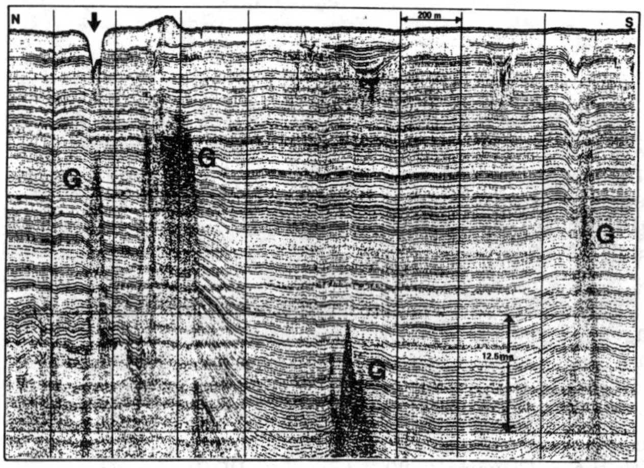

Fig. 2. Shallow seismic section across pockmarks and shallow gas clouds at the 25/7-location. G = shallow gas.

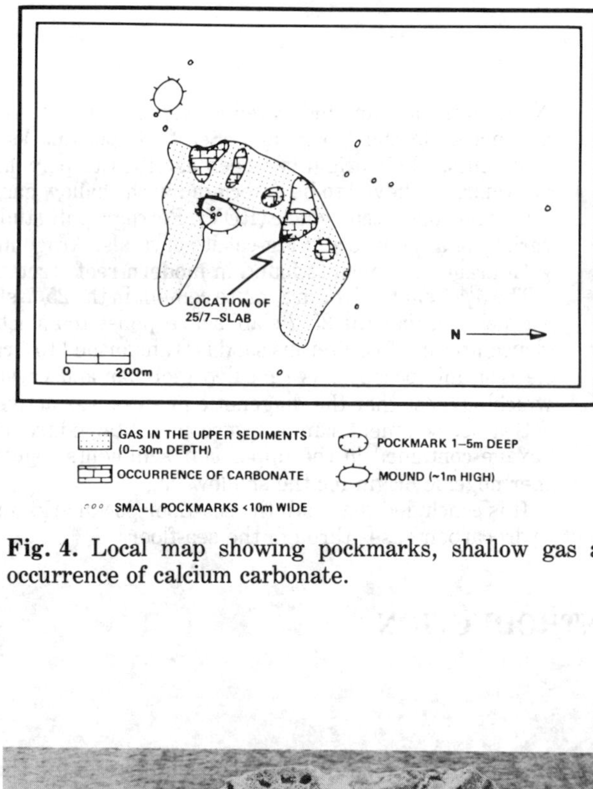

Fig. 4. Local map showing pockmarks, shallow gas and occurrence of calcium carbonate.

Fig. 3. Photograph of sea-floor inside pockmark where the 25/7-slab was retrieved. The slab is seen at the left, lower centre.

Fig. 5. Section of the 25/7 slab. The assumed active diagenetic zone is seen at the lower part of the slab.

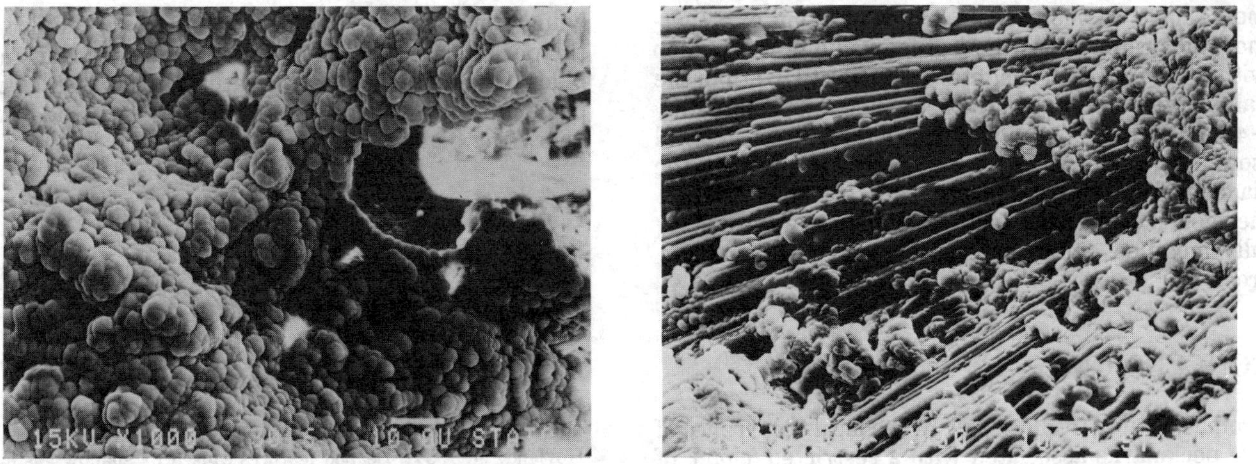

Fig. 6. SEM photograph showing (a) migration duct for solutions or gas in the active zone showing globular calcite or aragonite and (b) crystallization process in the assumed active zone (contact zone with sea-floor) of the 25/7-slab.

shown in Fig. 5. The main surface of the slab is very uneven and jagged, is full of cavities and has lobate protrusions (some of which suggest exposure to a reducing environment). A portion of the underside of the slab was, however, completely different. This zone, which was probably the contact zone embedded in the sea-floor sandy sediments, displays several uneven lithified sandstone ribs. Scanning electron microscope (SEM) studies of this zone show that it was probably an active zone of accretion and aragonite cementation (see Fig. 6).

The slab generally consists of irregularly shaped clasts of grey, very fine sandstone in a very fine to fine shelly sandstone matrix. Quartz is the dominant clastic mineral. Both the grey clasts and matrix contain a sprinkling of rounded glauconite grains. Feldspar, mica and amphibole are also present. While the clasts only contain poorly preserved foraminifera, the basal sandstone (the presumed active zone) has a rather diverse assemblage of foraminifera, molluscan and echinoderm shell fragments.

Clasts and matrix also differ in style of cementation. The clasts are cemented by a granular Mg-calcite microspar with crystals typically 5–30 μm in diameter, whereas the surrounding sandstone has a cement of acicular aragonite. Of particular interest is the presence, in large pores under sandstone clasts (but not in the active zone) and within shell fragments, of well-developed botryoidal aragonite crusts and spherulites. These are apparently closely comparable with some of the aragonite cements found in modern reef structures (James and Ginsburg, 1979; James, 1983). Some botryoids contain up to three generations of aragonite needles, each generation being separated by a brown stained, inclusion-rich zone which is sometimes also marked by possible micro-borings.

Aragonite from the 25/7-slab was analysed for carbon and oxygen stable isotope ratios. The $\delta^{13}C$-values were −58.9‰ and −59.0‰ PDB and the $\delta^{18}O$-values 2.52‰ and 2.65‰ PDB (Hovland, 1984*a*). The carbon atoms thus found are obviously not taken from a normal marine environment. The $\delta^{13}C$-values are, however, of the same order of magnitude as those found in carbonates of assumed methane-derived origin from other locations in the North Sea (Jørgensen, 1976) and on the continental margin off the northeastern USA (Hathaway and Degens, 1970; Allen *et al.*, 1969).

Additional geochemical information from the surrounding sea-floor sediments was gathered by dropcore sampling. Six dropcores were retrieved and 200 g sediment subsamples preserved from the lowest portion of the core (1–1.5 m sub sea-floor). The subsamples were packed in plastic bags and stored in liquid nitrogen (−196 °C) until geochemical analyses were carried out onshore. Samples were analysed for adsorbed and occluded concentrations of light hydrocarbon gases. Geochemical results are shown in Table 1.

The highest concentration of light hydrocarbons is found in core S16/2 which was retrieved from the soft sediments within the pockmark where the 25/7-slab was recovered. This is also within the location where the shallow gas apparently comes very close to the sea floor. The light hydrocarbon concentrations and $\delta^{13}C$ isotope values from the other five samples are of the same magnitude as those described by Faber and Stahl (1984) as being background values within this part of the North Sea. The gases found in the sedi-

TABLE 1. Geochemical results from the pockmarked area in Block 25/7

	(ppb)									
	Methane	*Ethane*	*Propane*	*Butane*	*Pentane*	*Hexane*	*Ethene*	*Propene*		
Sample	C_1	C_2	C_3	C_4	C_5	C_6	$\%C_2$	$\%C_3$	$\delta^{13}C_1$	$C_1/\Sigma Cn$
S12/2	43	6	3	3	9	0	0.25	0.00	−40.0	0.858
S16/2	128	21	10	8	7	0	0.09	0.09	−43.8	0.874
S16/1	50	7	4	3	10	1	0.34	0.09	−40.4	0.855
S17	58	8	5	4	11	0	0.45	0.19	−38.2	0.858
S18	42	6	3	3	6	0	0.19	0.00	−39.8	0.868
S19	49	7	4	3	10	0	0.23	0.14	−39.3	0.860
S21/2	92	15	9	7	9	0	0.12	0.00	−40.2	0.859
S22	41	6	3	2	5	0	0.18	0.13	−40.6	0.870
Stone									−58.9	

ments have all the characteristics of being of a thermogenic origin. Since the isotope value of the carbon within the 25/7-slab is lighter (−58.9‰ PDB) than the value of the methane in the sediments (approximately −40‰ PDB) it is assumed that either a geochemical isotope fractionation has occurred during the diagenesis of the carbonate. or that the original thermogenic methane in the sediments was biologically modified before participation in the diagenetic process.

CONCLUSIONS

From the study in Block 25/7 of the local occurrence of five pockmarks associated with a restricted 'cloud' of shallow gas the following conclusions may be drawn:

The five pockmarks in Block 25/7 have a causal relationship with the presence of the sub sea bed shallow gas. The pockmarks are thus probably formed by a gas-induced erosion process (Hovland, 1984*b*).

The occurrence of the lithified pockmark floor crusts and carbonate slabs has a causal relationship with the pockmark forming process. This means that thermogenic gas or a low density liquid solution containing light thermogenic hydrocarbons is emanating from or being pumped (by cyclic loading?) through the floor of the pockmarks. The cementation process is thus most likely a result of biogenic methane oxidation taking place in the unconsolidated sediments immediately below the benthic boundary.

REFERENCES

Allen, R. C., Gavish, E., Friedman, G. M. and Sanders, J. E. 1969. Aragonite-cemented sandstone from outer continental shelf off Delaware Bay: Submarine lithification mechanism yields product resembling beachrock. *J. Sediment. Petrology*, **39**, 136–149.

Faber, E. and Stahl, W. 1984. Geochemical surface exploration for hydrocarbons in North Sea. *AAPG Bull.*, **68**, 363–386.

Hathaway, J. C. and Degens, E. T. 1969. Methane derived marine carbonates of Pleistocene age. *Science*, **165**, 690–692.

Hovland, M. 1984*a*. Seabed features caused by shallow gas in the North Sea. *AAPG Research Conference*, San Antonio, Abstr. No. 19.

Hovland, M. 1984*b*. Gas-induced erosion features in the North Sea. *Earth Surface Processes and Landforms*, **9**, 209–229.

James, N. P. 1983 Reef environments. In: Scholle, P. A., Bebout, D. G. and Moore, C. H. (eds.), *Carbonate Depositional Environments. AAPG Memoir*, **33**, 346–462.

James, N. P. and Ginsburg, R. N. 1979. The Seaward margin of Belize barrier and atoll reefs. *Int. Ass. Sedimentologists, Spec. Publ.*, **3**, 191 pp.

Jørgensen, N. O. 1976. Recent high magnesian calcite/aragonite cementation of beach and submarine sediments from Denmark. *J. Sediment. Petrology*, **46**, 940–951.

Part V
Methods

23 Integrated kerogen typing in the recognition and quantitative assessment of petroleum source rocks

S. R. Larter
Union Oil Science and Technology Division, 376 South Valencia Avenue, Brea, California 92621, USA

Many of the kerogen typing procedures used in petroleum source-rock evaluation today are derived from basic principles which were developed before this decade. Quantitative high-resolution pyrolysis–gas chromatography is a recent technique which allows more detailed evaluation of kerogen in source rocks. Such detailed molecular kerogen typing data may now be integrated with the older established chemical and petrographic methods. An example is given of the application of integrated petrographic, bulk chemical and high-resolution chemical methods to the quantitative evaluation of 'actual source beds' (source beds that have expelled oil and gas) using data from the Viking Graben province of the North Sea. The limitations on the various kerogen typing methods are discussed and restrictions on the value of conventional source rock analysis are considered in the light of our limited understanding of the primary migration process.

INTRODUCTION

The characterization of kerogen, as the precursor for petroleum, is now a major part of many contemporary exploration programs. Whilst the most common routine chemical methods of kerogen analysis used in petroleum geochemistry date from the middle seventies (Durand, 1980; Tissot and Welte, 1978, and references therein), many of the basic concepts can be traced back to the coal science literature of previous decades. In particular, the petrographic maceral concept (Stopes, 1935) and the investigation of chemical variations between individual macerals and kerogens by CHO atomic ratio plots were well-established by the 1960s (Huntjen and Van Krevelen, 1957; Van Krevelen, 1961). Many current kerogen-typing procedures are the result of merging older concepts with the application of newer analytical technology.

Kerogen-typing studies may have a variety of objectives of which those related to petroleum geochemistry are listed in Fig. 1 and summarized below. Evaluation of the depositional environment of a sediment using reflected and transmitted light microscopy (Gutjahr, 1983) may have important implications either for source-rock potential or in terms of reservoir development and quality.

Kerogen analysis and typing related to organic diagenetic studies of the formation of chemical species in modern sediments (see Fig. 1), whilst of some theoretical importance, has not been a major area of interest for exploration geochemists. In contrast, analyses aimed at determining the gross origin and organic facies of older sediments, and their catagenetic hydrocarbon potential, have been an area of extensive study by geochemists in the previous decade (Tissot *et al.*, 1974; Durand, 1980; Tissot, 1984; Tissot and Welte, 1978; Demaison and Murris, 1984).

The concept of organic facies, which integrates the environment of deposition and type of organic matter in a sediment and its chemical composition, is now coupled with the 'generative basin' approach, to provide the basis for play and prospect evaluation in many exploration companies (Demaison, 1984). Whilst this approach is tried and tested (Demaison and Murris, 1984, and references therein) there seems to be a general movement towards application of more quantitative and absolute approaches to the evaluation of petroleum source beds. These more quantitative methods are aimed at a better understanding of the complexities of the petroleum migration process (Durand, 1983; Welte and Yükler, 1981; Ungerer *et al.*, 1984). In order to discuss the role of integrated kerogen typing in the evaluation of source rocks, objectives and definitions need to be clarified. In this context the dual objectives of kerogen typing are primarily considered to be discrimination of the source organic matter for a given facies, and its significance in terms of the actual source potential of the sediment (see Fig. 1).

The two major factors controlling the petroleum potential of any sediment at any rank level are (1) the quantitative fraction of the sediment that contains

OBJECTIVES

1. DEPOSITIONAL FACIES DETERMINATION — "PALYNOFACIES"

   APPLICATION — DETERMINATION OF SOURCE/RESERVOIR SYSTEM GEOMETRY AND PALAEOGEOGRAPHY

2. DIAGENETIC INTERPRETATIONS

   APPLICATION — UNDERSTANDING ORIGIN AND EARLY FATE OF SEDIMENTARY ORGANIC MATTER INCLUDING KEROGEN, UNDER A VARIETY OF DEPOSITIONAL ENVIRONMENTS

3. ORGANIC FACIES DETERMINATION

   APPLICATION — DETERMINATION OF THE NATURE OF THE ORGANIC MATTER IN A SEDIMENT. COUPLED WITH MATURITY AND RICHNESS DATA, USE OF THIS INFORMATION IN THE GENERATIVE BASIN APPROACH TO REGIONAL AND PROSPECT SCALE SOURCE POTENTIAL EVALUATION

SOME WORKING DEFINITIONS

SOURCE ROCK POTENTIAL: —

THE FRACTION OF A SEDIMENT, AT ANY MATURITY LEVEL, THAT CONTAINS RELEASABLE (BY CATAGENESIS OR OTHER MECHANISMS), EXPELLABLE PART STRUCTURES RELATED TO THOSE CHEMICAL SPECIES FOUND IN PETROLEUM

SOURCE ROCK POTENTIAL = f (ORGANIC RICHNESS, BITUMEN AND KEROGEN COMPOSITION)

MOLECULAR KEROGEN TYPE: —

A CHEMICAL DESCRIPTION OF A KEROGEN BASED ON THE QUANTITATIVE RELATIONSHIP BETWEEN THE DIFFERENT STRUCTURAL ELEMENTS COMPOSING THE MATERIAL

ACTUAL SOURCE ROCK

ANY SEDIMENT THAT HAS PRODUCED AND EXPELLED HYDROCARBONS AND RELATED SPECIES

Fig. 1. Kerogen typing—objectives and definitions.

catagenetically releasable structures related to those chemical species found in petroleum and (2) the expulsion efficiency of the individual bitumen structural types from the host sediment. The second factor, which is a complex function of organic content, organic type, host mineralogy, sediment rank level and hydrodynamic potential, is the ultimate parameter governing effective source potential.

Crude oils (and gases) are often described in precise compositional terms. In contrast, the inability of many of the historically established kerogen-typing techniques (such as elemental analysis) to define the detailed molecular nature of kerogens accurately has held back current typing schemes to general descriptions of bulk kerogen composition. Whilst these schemes, based mainly on the work of Tissot and co-workers (Tissot *et al.*, 1974; Durand, 1980) have provided a solid foundation for new approaches to typing kerogens, their usage alone as the major base for chemical typing is no longer consistent with the analytical resolution now obtainable. Ideally, the characterization of kerogens should attain the same level of molecular detail as is possible with crude oils or source-rock bitumens (Tissot and Welte, 1978).

It is the application of quantitative kerogen-typing studies to the evaluation of the source rock potential and recognition of actual source rocks (see Fig. 1) that is the purpose of this paper.

INTEGRATED KEROGEN TYPING

A complete, integrated description of a kerogen should include information relating to its visual composition (petrographic type), its bulk chemical composition (classical chemical type), the detailed molecular composition, its rank level (maturity) and the depositional environment of the host sediment (see Fig. 2).

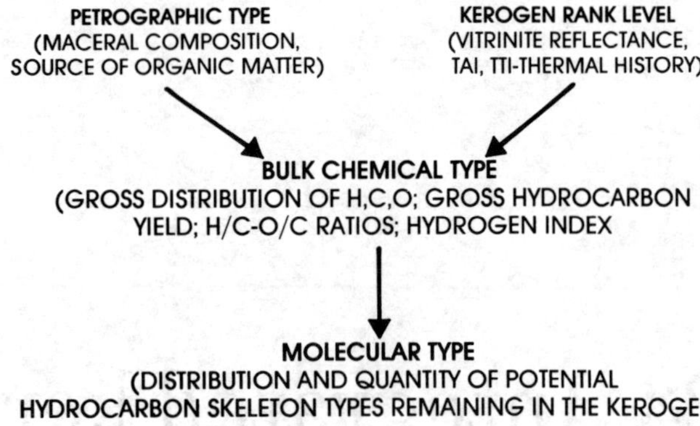

AFTER LARTER AND SENFTLE (1984)

Fig. 2. Elements of source-rock kerogen type.

Applications of kerogen typing methods to exploration geochemistry fall into two categories: (1) screening analyses aimed at identifying the gross organic type in a large number of sediment samples and (2) characterization in detail of relatively few samples of prime interest in the principal source intervals. The need to screen samples before time-consuming detailed analysis is made obvious when one considers that 5000 samples are available when 10 wells are analysed at 10 m sampling intervals assuming 5 km average total depth. In this context, petrographic and bulk chemical typing procedures (chiefly Rock–Eval) are the most suitable for screening kerogen studies. Elemental analysis, NMR and pyrolysis–GC methods are suitable for more detailed evaluations.

PETROGRAPHIC TYPE

The value of optical kerogen typing in source rock analysis has been well documented (Staplin, 1969; Robert, 1971; Powell *et al.*, 1982; Gutjahr, 1983). Whilst there are broad general relationships between petrographic kerogen typing and chemical kerogen composition parameters (Powell *et al.*, 1982), the wide range of chemical properties displayed within the same maceral group (Bailey, 1981; Horsfield, 1984; Larter, 1984) renders classification by solely petrographic means very hazardous, and often actually misleading. However, petrography (maceral description) currently provides the best way of identifying precursor organic matter, and provides the only practical way of attempting the typing of very high rank material. At high maturities the chemistry of the kerogen assemblages changes significantly and the values obtained from both bulk (Tissot *et al.*, 1974) and detailed chemical typing (Horsfield, 1984; Larter, 1984) may become misleading. With high rank sections in explorationally immature basins (where drilling of source rocks over the entire thermal evolutionary track has not occurred), often the only way to attempt reconstruction of the maturation pathway through time of a possible source rock is via petrographic typing, although this must be used with caution.

Whilst petrographic data cannot *a priori* evaluate organic matter chemically, once calibrated for a particular basin, petrographic data provides a rapid and inexpensive approach to determining the source of the

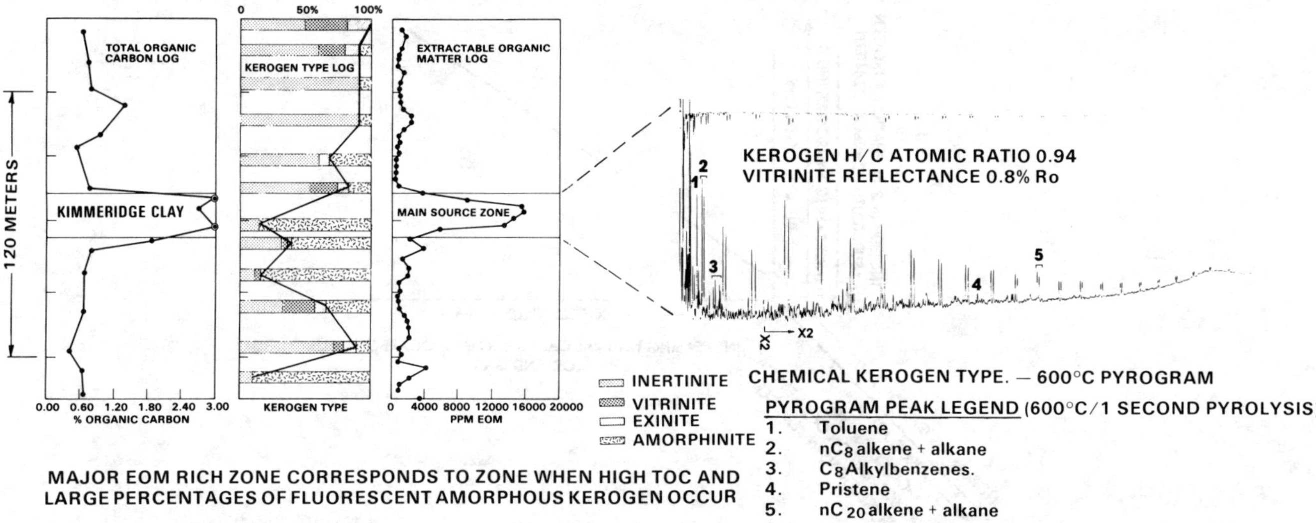

Fig. 3. Optical and chemical typing data for section near peak oil generation, Kimmeridge Clay Formation, Central Graben.

organic matter in a sediment, and helps in making preliminary predictions of the possible source potential. Frequently the chemical properties of a source bed are closely related to the amount and distribution of exinite and fluorescent (under u.v. illumination) amorphous kerogen types in a sediment. In this paper, the sum of the fluorescent amorphous kerogen species (presumed exinite/liptinite-derived) and the morphologically defined exinite macerals are summed into a parameter termed '%AMEX' (Senftle, 1983). This parameter is frequently found to be the most critical in relating petrographic and chemical source rock data. Figure 3 shows data from a well in the Central Graben of the Norwegian North Sea, illustrating an example of the correspondence between the chemical richness and petrographic parameters. This type of approach is of greatest value in the initial stages of basin exploration (Cooper and Barnard, 1984). It must be emphasized that the effective use of petrographic data is strongly dependent on a cross-calibration of petrographic and chemical data (see Fig. 3), and on the combined use of reflected light on whole rock samples and reflected and transmitted light on kerogen concentrates, both in white and u.v. illumination. It is the use of this multi-mode illumination (Senftle, 1983) that is enhancing the value of petrography in an exploration context. Simple white transmitted-light petrography does not have the resolution consistent with the chemical and petrographic variations evident in source-rock kerogen components.

BULK CHEMICAL TYPE

As commonly defined (Tissot *et al.*, 1974), the three main kerogen types are broadly analogous to the principal coal macerals alginite, exinite and vitrinite, the atomic ratio evolution paths for both kerogens and macerals being similar (Huntjen and Van Krevelen, 1957; Van Krevelen, 1961). A fourth kerogen, type 4 defined by Harwood (1977), is analogous to the inertinite coal maceral group with very low values of atomic H/C ratio at any rank level.

Most kerogens are poly-maceralic assemblages. Thus coal kerogens are frequently mixtures of components ranging from types 1 to 4, but for many Palaeozoic–Mesozoic coals bulk analyses would indicate a type 3 composition, this being consistent with their dominantly vitrinite-rich nature (Yarzab *et al.*, 1980). It is important to consider that many different petrographic compositions may result in a similar bulk chemical character.

The Brent Group coals in the Viking Graben of the Norwegian North Sea fall into type 3 although, as discussed below, the chemical variance in coals from different provinces is very great, and the Brent coals themselves have relatively high hydrogen indices for 'coals' and may not fall on the conventional Tissot *et al.* (1974) type 3 evolution curve. Brent coals may, indeed, have some liquid hydrocarbon generating potential (M. Bjorøy, pers. comm.; Ungerer *et al.*, 1984; Durand and Paratte, 1983).

MOLECULAR TYPE

The detailed molecular configuration of a kerogen cannot be obtained from examination of its bulk chemical properties, due to the isomeric complexity possible with high molecular weight macromolecules. The mechanically complex (poly-maceralic) nature of most kerogen assemblages further renders any 'average-structure' determination for a kerogen a somewhat academic exercise. However, the total kerogen reservoir of precursors of, for example, a petroleum species such as normal alkanes, is a primary factor in determining the molecular kerogen type. This is so even if the alkane precursors (such as kerogen-bound alkyl groups) are anisotropically distributed between different macerals. Thus, in a molecular chemical type it is the total quantity of a kerogen structural component in the kerogen mixture as a whole that is significant, rather than whether the parent alkyl group is bound to a sporinite, amorphinite or a vitrinite.

DETERMINATION OF MOLECULAR KEROGEN TYPE

In defining a basis for molecular kerogen typing any potential method should be based on elements (species) in a kerogen that will ultimately end up, after catagenesis, as a significant component of crude oils and gases (see Fig. 1). Secondly, the method must be based on kerogen molecular elements that can be satisfactorily and reproducibly analysed. The general

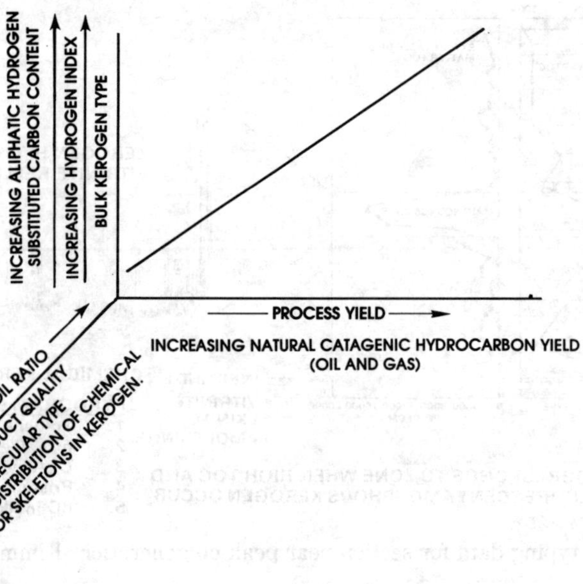

Fig. 4. Kerogen type and process yields.

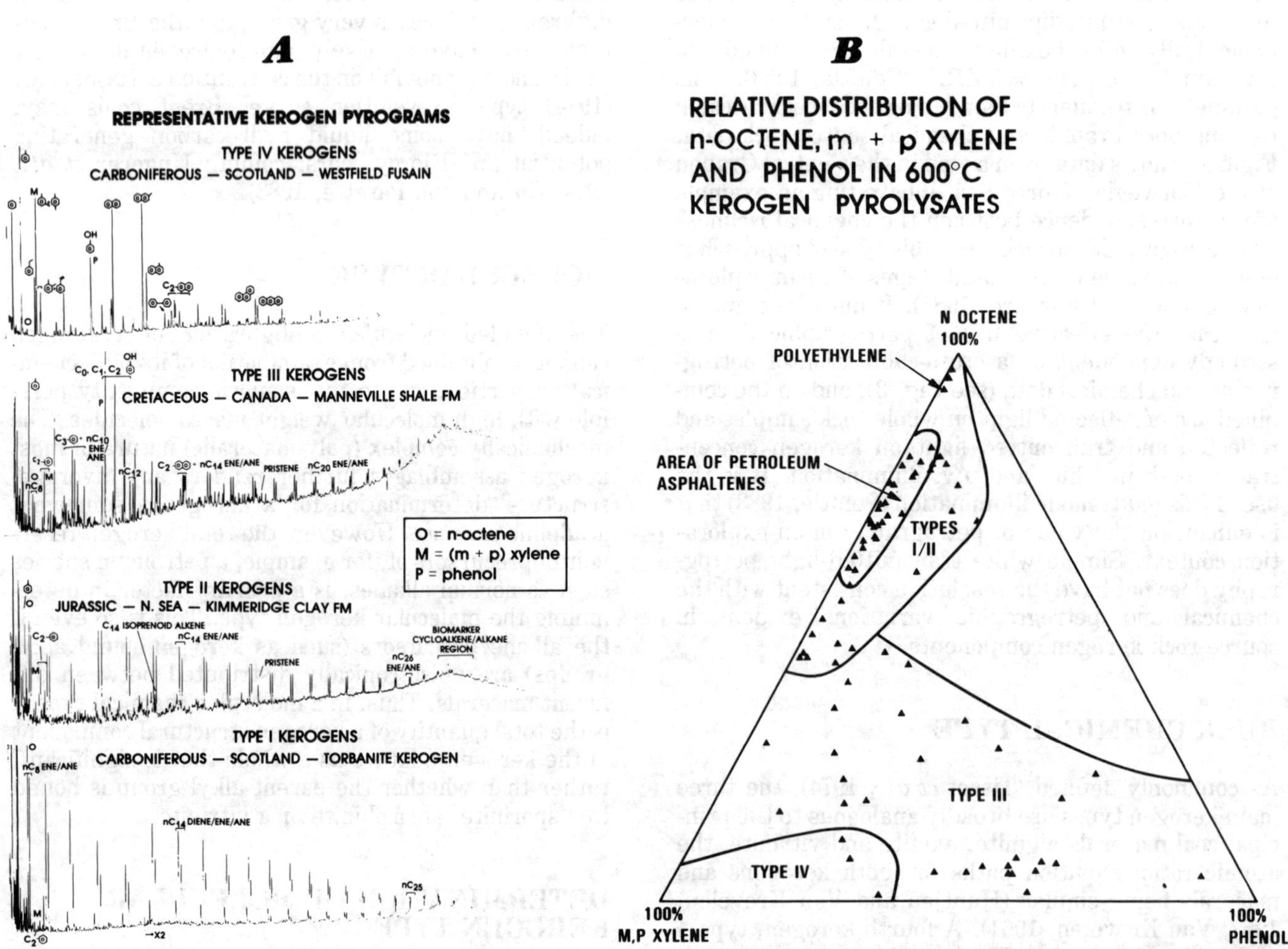

Fig. 5. Discrimination of different organic matter types using pyrolysis gas chromatography.

abundance and ease of resolving normal and other aliphatic hydrocarbons in flash kerogen pyrolysates (Larter, 1984; Horsfield, 1984) suggests that these components are a good candidate for such a method.

Miknis *et al.* (1982) have recently discussed the importance of the abundance of the aliphatic content of a kerogen (carbon-bound hydrogen content or the more easily determined and somewhat equivalent aliphatic carbon content) to high temperature thermal conversion to gases and liquids. In addition, extensive source-rock kerogen data, studies of various petroleum basins and the similarities between high temperature-induced laboratory experiments and natural catagenic processes (Tissot and Welte, 1978, and references therein) all suggest that the aliphatic carbon content of a kerogen is the critical factor in determining catagenic hydrocarbon yields in the Earth's crust. While total hydrocarbon yield (i.e. process yield in Fig. 4) is proportional to the aliphatic content of the kerogen, the product distribution (in gross terms the gas–oil ratio) is dictated by the distribution of the various structural elements in the kerogen, i.e. the molecular type (see Fig. 4).

The aliphatic fraction of kerogen has been determined by solid-state carbon NMR methods (Zilm *et al.*, 1981; Miknis *et al.*, 1982) and these methods are becoming more routinely used in the evaluation of source kerogens for exploration purposes. However, the most commonly used method for estimating the proportions of discrete structural elements in kerogens would seem, at present, to be pyrolysis methods, in particular pyrolysis–gas chromatography (Larter and Douglas, 1982; Horsfield, 1984; Larter, 1984).

Analytical kerogen typing methods have greater chemical resolution at the molecular level than routine methods such as elemental analysis, bulk flow pyrolysis (e.g. Rock–Eval) or optical spectroscopic methods. Further, since high-resolution methods directly monitor specific chemical species in a kerogen pyrolysate, such data are more pertinent to the determination of kerogen aliphatic fractions, and thus kerogen behaviour under natural processes. Traditional bulk chemical methods cannot determine the distribution of carbon types in a system, except in a most general way. Rock–Eval, in its most basic form, can determine the kerogen aliphatic content, but not its distribution.

The typing of kerogens by determination of the relative quantities of specific fragments in kerogen pyrolysates is an old idea, this concept having first been applied in coal studies (Romovacek and Kubat, 1968; Larter and Douglas, 1982). The more hydrogen-rich exinite coal maceral kerogens generate fewer aromatic pyrolysates than the more hydrogen-deficient vitrinite macerals. More recently this work has been extended to kerogens from a broader range of sedimentary facies (Van de Meent *et al.*, 1980; Larter and Douglas, 1980; McHugh *et al.*, 1978; Giraud, 1970; Horsfield, 1984; Larter, 1984).

Analytical pyrolysis techniques measure specific carbon : hydrogen : oxygen assemblages. They provide a reliable kerogen typing method in that the aliphatic carbon-bound hydrogen content of a kerogen, the factor primarily responsible for high hydrocarbon yield during catagenesis, is more directly assessed. Using pyrogram fingerprint data, it has been possible to distinguish the major variations between kerogens readily, even those of the same bulk chemical type. Figure 5 shows how the basic kerogen types can be distinguished in terms of both pyrogram fingerprint and the relative distribution of aliphatic, aromatic and phenolic (*n*-octene, xylene, phenol) species in the pyrograms. A complete treatment of this procedure is given in Larter (1984).

The kerogen fields shown are based on the classic scheme of Tissot *et al.* (1974). Some overlap of type 3 and 4 kerogens is evident, but a fairly clear discrimination between the hydrogen-rich types 1/2 and more hydrogen-deficient types 3/4 is evident. Intermediate kerogen compositions plot accordingly. Samples cover the vitrinite reflectance range 0.3–1.3% R_0 with most samples having values of <1.0% R_0. A major strength of pyrolysis methods is that while quantitative yields of kerogens are maturity-related (Espitalie *et al.*, 1977) the qualitative pyrogram fingerprints obtained are relatively rank independent over much of the oil window (Van Graas *et al.*, 1980).

The main distinguishing features of type 3/4 versus type 1/2 kerogens are the relative abundances in their pyrograms of aromatic and phenolic species versus aliphatic species, respectively (see Fig. 5). Most importantly, enormous variation in detailed composition can occur, even for kerogens of very similar elemental or Rock–Eval analysis. This is the major reason for calibrating bulk methods with high-resolution methods.

While discrimination within or between the various kerogen types is possible using either the pyrogram fingerprint or relative distributions of pyrolysate components, it is not easy to relate, directly and quantitatively, this information to source-rock potential. This is because, as with other contemporary chemical typing methods, no absolute measurement of the quantitative ability of a kerogen to generate specific petroleum hydrocarbons is readily available.

The major success of bulk flow methods such as Rock–Eval has been that they allow some estimate of the amount of hydrocarbons generated by a kerogen type. Orr (1983) suggested that hydrocarbon potential may be indirectly related to the difference in bulk flow pyrolysis yield for kerogens of different rank but constant chemical type. However, while combined chromatographic/bulk flow pyrolysis systems (Martin, 1977; Gormely and Mukhopadahay, 1983; Horsfield, 1984; Dembicki *et al.*, 1983) have enabled improved characterization, the ultimate aim of a completely quantified on-line analysis of kerogen pyrolysates on a routine basis has not yet been achieved.

Ultimately, as discussed earlier, classification of kerogens should be performed on the same molecular basis as that used for crude oils and rock extracts. Ideally kerogen analyses should enable the generation of data analogous to that used by Tissot and Welte (1978) in their proposed oil classification, namely the distribution of hydrocarbon and non-hydrocarbon types within a kerogen pyrolysate. Such data allows a direct assessment of the quantity and type of carbon skeletons potentially available in a kerogen, and allows the petroleum geochemist and explorationist to attempt more-sophisticated mass balance approaches to quantifying oil-expulsion phenomena.

Some idea of the need for such a quantitative basis for detailed molecular kerogen typing is illustrated in Fig. 6, which shows pyrograms of four kerogens containing predominantly non-marine organic matter. These are (A) a kerogen derived from the alga *Botryococcus*, (B) a kerogen derived from higher plant cutic-

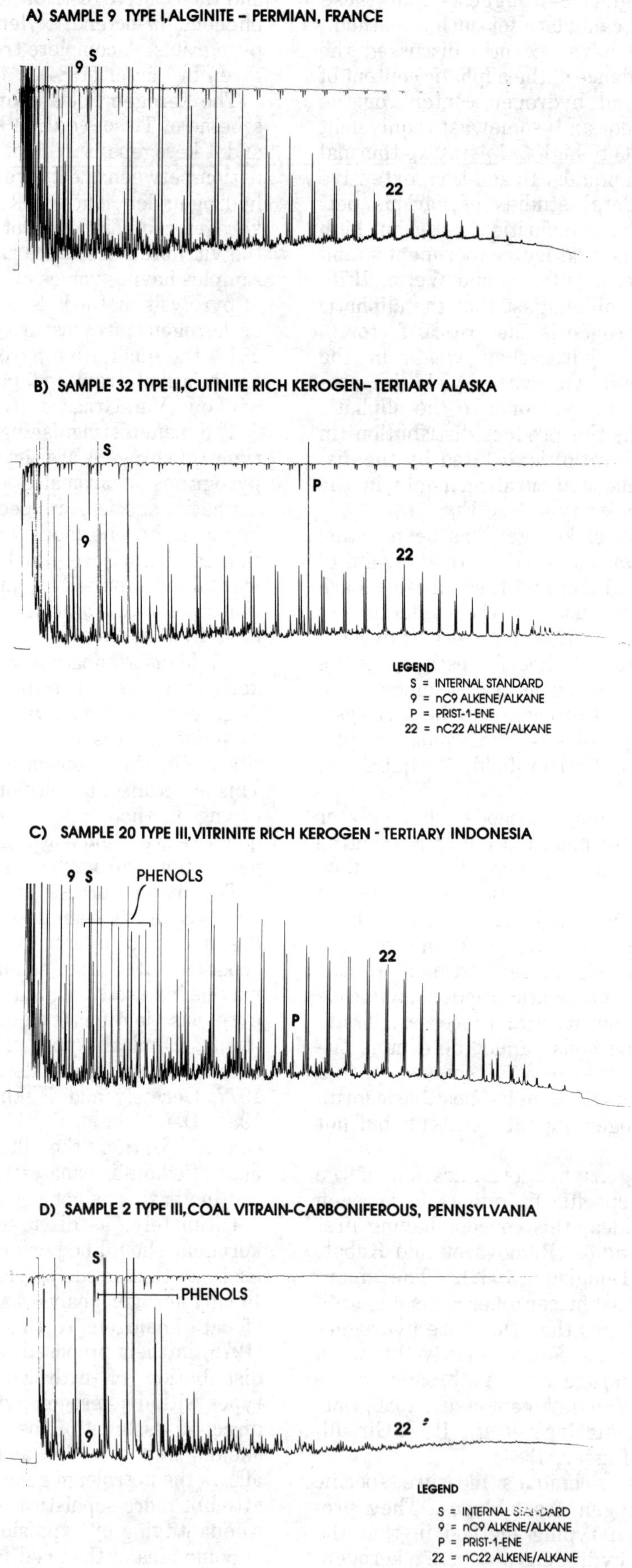

Fig. 6. Pyrograms of kerogens containing primarily non-marine organic matter.

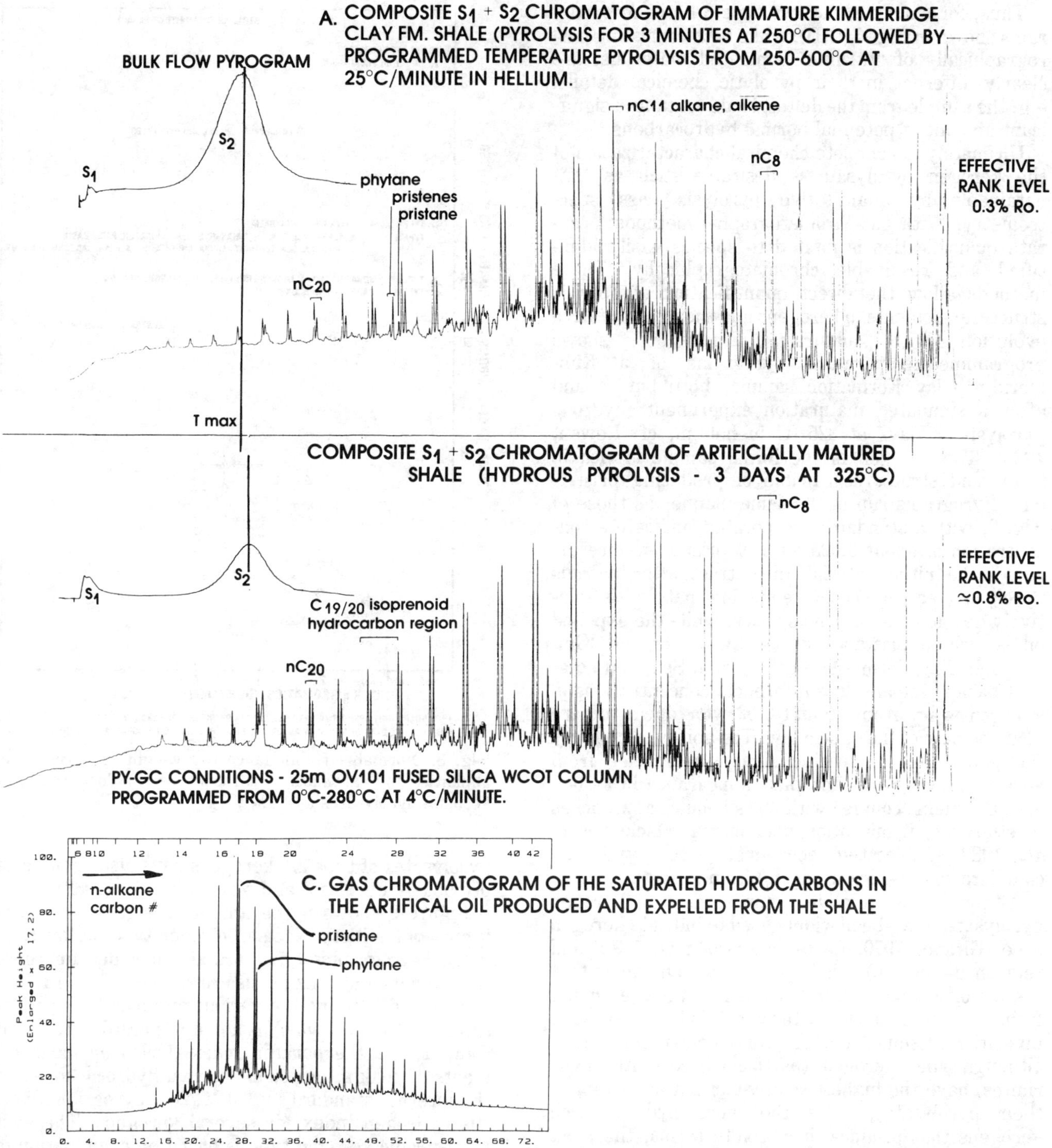

Fig. 7. Selective generation of isoprenoid species in artificial maturation experiments.

ular material and deposited in a marine setting, (C) a kerogen typed petrographically as dominantly vitrinite, deposited in a marine setting and (D) a kerogen consisting entirely of vitrinite picked from a coal seam. All the kerogens are at rank levels below peak oil generation (i.e. $<0.8\%R_0$). Samples C and D have very similar bulk chemical analyses.

The large peak in each pyrogram represents an internal standard (methylstyrene) generated in the pyrolysis process from a polymer (polymethylstyrene) quantitatively mixed with the kerogen before pyrolysis (Larter and Senftle, 1984). The absolute quantity (relative to starting kerogen weight) of the internal standard in each pyrogram is constant, approximately 0.25% by weight of the kerogen (see Appendix 1 for details). Using such a technique it is now possible to quantify, on a kerogen or organic carbon basis, the amount of specific components in a pyrolysate (Larter and Senftle, 1984). In the following discussions emphasis is placed on the quantification of normal hydrocarbons in kerogen pyrolysates, since these components are a major fraction of many crude oils, particularly those of higher plant origin (Tissot and Welte, 1978).

By directly determining the total reservoir of specific pyrolysate elements, direct assessment of actual petroleum potential for any kerogen becomes possible (see Fig. 1) if this type of information can be integrated into quantitative migration studies of the type discussed by Mackenzie *et al.* (1983) and Leythaeuser *et al.* (1983). Further, this method allows the systematic discrimination of kerogens previously grouped together, on the basis of elemental analysis or hydrogen/oxygen index, as of similar type.

Thus, for example, the two type 3 kerogen pyrograms shown in Fig. 6 (C,D), while both consisting petrographically of 'vitrinite'-dominated materials, are clearly different in their pyrolytic chemical details: e.g. the sample from the deltaic shale contains a significant amount of potential normal hydrocarbons.

Ultimately, a complete chemical characterization of the kerogen pyrolysate is desirable such as that obtainable by quantitative pyrolysis–mass spectrometry. With gas-chromatographic methods, accurate quantification is limited to species easily identified and resolvable chromatographically. These methods allow the direct quantification of specific structural elements in kerogens and their quantitative evolution with maturity. Figure 7 shows programmed-temperature pyrograms of a Kimmeridge Clay Formation sample, both before and after a simulated maturation experiment (hydrous pyrolysis, 3 days at 325 °C in helium, cf. Lewan, 1983). The kerogen in the initial sample is rich in isoprenoid structures capable of producing prist-1-ene. Pyrograms run in the same manner as those of Fig. 6, with a standard incorporated, indicate a concentration of about 0.025 wt% of prist-1-ene precursors in the initial material. In contrast, after hydrous pyrolysis (see Fig. 7) the remaining material is relatively devoid of isoprenoid species, while the expelled oil is rich in pristane and phytane. For the Kimmeridge Clay Formation in the North Sea, and adjacent basins, the absolute isoprenoid concentrations in kerogens vary from around 0.02 wt% down to zero near the base of the oil window. The ability to quantify the generation of specific structural species from kerogens and to monitor their appearance in associated bitumens, coupled with mass balance approaches to studying oil migration phenomena (Mackenzie *et al.*, 1983) enables the geochemist to now study the complete source-rock kerogen/bitumen system.

Although the relative compositions of kerogen pyrolysates have been widely used to indicate kerogen type (Giraud, 1970; Larter and Douglas, 1982, and references therein) it is important to remember that the major change in pyrolysis behaviour when going from, say, kerogen type 1 to type 3, is the increase in involatile residue during this transition. Thus, although type 3 kerogens (particularly coal vitrain vitrinites) have the highest relative aromatic content in their pyrolysates, it is the more hydrogen-rich kerogens that produce, in absolute terms, the most hydrocarbons of any type. In other words, the more hydrogen-rich a kerogen is, the more hydrocarbons of any type, including gases and aromatics, are produced under pyrolytic (or indeed catagenic) processes. However, the ratio of aromatic to aliphatic hydrocarbons in the thermal product always increases in the progression from type 1 to type 3. Figure 8 shows the relationship between the absolute yield of aromatic (expressed as total benzene, toluene and C_8 aromatics) and paraffinic (total C_7–C_{25} n-alkanes + n − alkenes) species for a variety of kerogens. Interestingly, the absolute paraffin content of a kerogen (on a pyrolysis–GC basis) seems relatively constant for many immature type 2/3 kerogens. It is the larger branched chain + naphthenic contributions of marine systems that provide the higher hydrogen indices of type 2 amorphous kerogen, such as are found in the Kimmeridge Clay Formation of the Central and Viking Grabens. Similar observations on the relative constancy of the normal hydrocarbon contents of kerogen

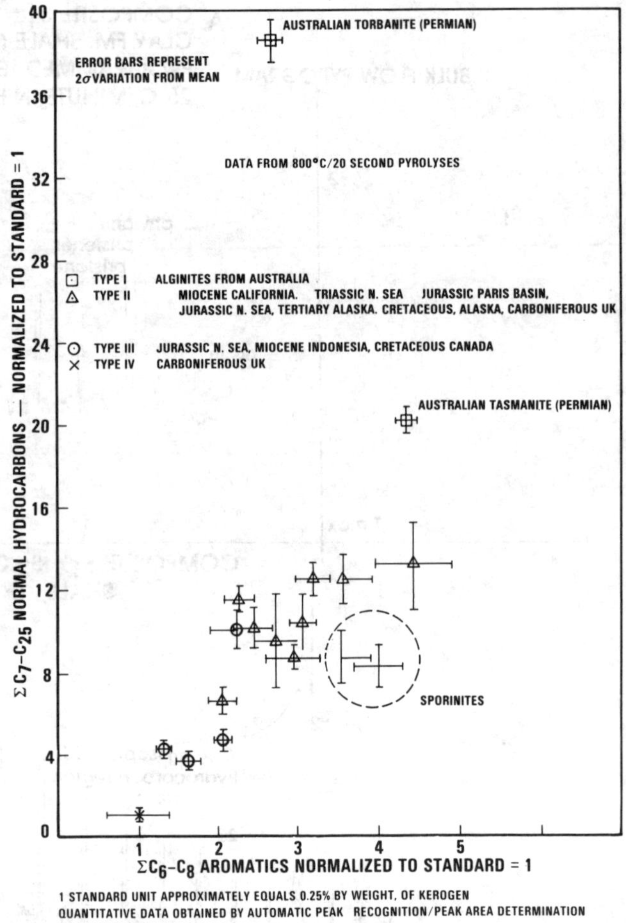

Fig. 8. Normalized abundance (by weight of kerogen) of aliphatic and aromatic hydrocarbons in low maturity kerogen pyrolysates (<0.8% VR).

pyrolysates of type 2/3 kerogens have also been made previously by Tissot and Vandenbrouke (1982).

Figure 9 shows pyrograms of kerogens which are typical of distinct molecular kerogen types at low rank level. Petrographic types are also included for these components and rank levels are shown in Table 1. The basis for identifying the various end-member types is shown in Fig. 10, which is a plot of paraffin precursor content, i.e. the paraffin index (based on measurements of peak intensity for normal hydrocarbons and the internal standard in 800 °C/20 s pyrograms) versus hydrogen index for several kerogens. The split into specific kerogen type is based on examination of many pyrograms (Larter and Senftle, 1984).

The recognized kerogen type end-members are as follows.

Type 1p (paraffinic) kerogens are characterized by having high hydrogen indices and highly paraffinic pyrolysates. They are typified by the *Botryococcus* -derived boghead coals (Larter and Douglas, 1980).

Type 1pn (paraffinic/naphthenic) kerogens, while hydrogen-rich, have the hydrogen distributed between paraffinic and naphthenic structures. Typical examples include kerogens derived from Tasmanitid algae and many mixed type 1 kerogens (Philp *et al.*, 1982). The term paraffinic/naphthenic is used as mass spectral data indicate that kerogens of this type have abundant naphthenic contributions to their pyrolysates (Philp *et al.*, 1982; Larter, unpubl. data).

Type 1n (wholly naphthenic) kerogens are unusual. The example shown here is a somewhat artificial illustration as the sample is a coal resinite.

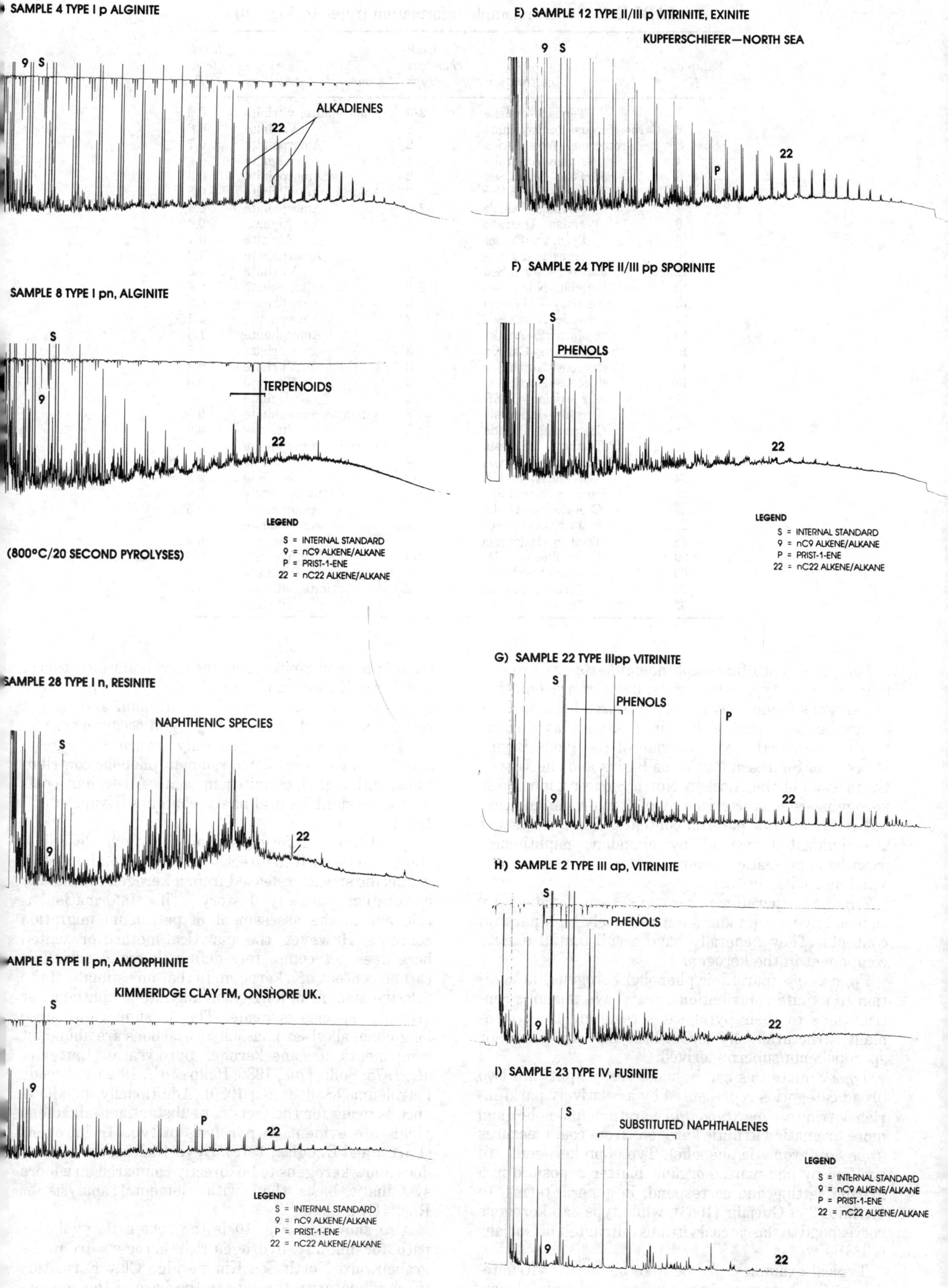

Fig. 9. Principal kerogen types as defined by pyrolysis–G.C. (a) Type I, (b) types I and II, (c) type II/III and (d) types III and IV.

TABLE 1. General sample information (refer to Fig. 10)

Sample No.	*Age/location*	*Bulk chemical type*	*Principal (>30%) maceral contributors*	*Rank level (VR)*
1	Tertiary–Alaska	2/3	Amorphinite, vitrinite	0.6
2	Carboniferous–Pennsylvania	3	Vitrinite	0.6
3	Cretaceous–West Africa	2	Amorphinite	0.7
4	Permian–Australia	1	Alginite	—
5	Jurassic–North Sea	2	Amorphinite	0.3
6	Jurassic–North Sea	2	Amorphinite, vitrinite	0.8
7	Tertiary–Canada	2/3	Amorphinite	0.4
8	Permian–Australia	1	Alginite	0.4
9	Permian–France	1	Alginite	0.4
10	Jurassic–France	2	Amorphinite	0.3
11	Jurassic–North Sea	3	Vitrinite	0.8
12	Permian–North Sea	2/3	Vitrinite, exinite	0.6
13	Tertiary–California	2	Amorphinite	0.3
14	Jurassic–North Sea	2	Amorphinite	1.1
15	Jurassic–Alaska	2	Amorphinite	1.3
16	Cretaceous–Canada	3	Vitrinite	0.5
17	Cretaceous–North Sea	3	Vitrinite	0.7
18	Cretaceous–North Sea	3	Vitrinite	0.6
19	Ordovician–USSR	2	Alginite	—
20	Tertiary–Indonesia	3	Vitrinite, amorphinite	0.6
21	Cretaceous–Canada	3	Vitrinite	0.6
22	Tertiary–Indonesia	3	Vitrinite, amorphinite	0.3
23	Carboniferous–UK	4	Fusinite	—
24	Carboniferous–UK	2/3	Sporinite	0.9
25	Permian–North Sea	2/3	Vitrinite, exinite	0.6
26	Cretaceous–Alaska	2	Amorphinite	0.5
27	Jurassic–France	2	Amorphinite	—
28	Tertiary–Indonesia	1	Resinite	0.4
29	Carboniferous–UK	2/3	Sporinite	0.6
30	Jurassic–North Sea	3	Vitrinite	0.5
31	Tertiary–Alaska	2	Cutinite, vitrinite	0.5
32	Tertiary–Alaska	2	Cutinite, vitrinite	0.5

Type 2pn (paraffinic/naphthenic) kerogens are the most common type 2 materials, being representative of kerogens found in rich marine source beds such as the grabenward part of the Kimmeridge Clay Formation in the North Sea, Toarcian of the Paris Basin, Miocene of Southern California basins and the Lower Cretaceous of the Alaskan North Slope region. Such kerogens are characterized by high hydrogen indices, but relatively low paraffin contents. The pyrolysates are dominated instead by abundant naphthenic/naphtheno-aromatic components (Tissot and Vandenbrouke, 1982).

Type 2p (paraffinic kerogens) are typified by reduced hydrogen indices but relatively high paraffin contents. They generally have a substantial exinite component in the kerogen.

Type 2/3pp (paraffinic/phenolic) kerogens, in addition to exinite contributions, may have phenolic contributions to their pyrolysates resulting from land-plant structural polymers which may be lignin or sporopollenin/suberin-derived.

Type 3 materials can be most easily split into two broad categories represented by a relatively paraffin-rich vitrinite type (*type 3pp* paraffinic/phenolic) and more aromatic vitrinitic kerogens from coal measures (*type 3ap*-aromatic/phenolic). Type 3pp kerogens are typified by non-marine organic matter deposited in a marine setting and correspond, in general terms, to vitrinite 2 of Gutjahr (1983), while type 3ap kerogens correspond to classic coal vitrains (vitrinite 1 of Gutjahr (1983)).

Typical examples of type 3 kerogens are vitrinite-dominated kerogens from marine or deltaic shales, e.g. Miocene of Indonesia, and many of the carbonaceous Middle Jurassic shales of the North Sea Basin (type 3pp) or vitrinites from the Carboniferous coals of Northern Europe and the USA (Type 3ap). These types are end members of a continuum and do not represent the entire field of kerogen occurrence.

Type 4 kerogens are generally aromatic in pyrolysate terms with occasional remnant phenolic contributions, although inertinites in some Australian coals may have significant aliphatic contents (Evans *et al.*, 1984).

Quantitative information of the type described above can be used to directly assess the total quantity of specific species released from a kerogen during any maturation pathway history. This information is relevant to the assessment of petroleum migration concepts. However, the analytical method presented here does not completely define the total paraffinic carbon content of a kerogen, in that no estimate of the relative amount of long-chain alkylated naphthenic or aromatic species is made. This is significant, since long-chain alkyl–aromatic hydrocarbons are abundant components of some kerogen pyrolysates (Larter *et al.*, 1978; Solli *et al.*, 1980; Philp *et al.*, 1982) and crude petroleums (Solli *et al.*, 1980). Additionally, no allowance is made for the fact that substantial *n*-alkadiene yields are evident on pyrolysis of type 1p kerogens (Larter and Douglas, 1978). Nevertheless, the method does allow kerogens to be directly compared on a more systematic basis than with elemental analysis or Rock–Eval alone.

Also shown in Fig. 10 is the schematic evolution path for initially hydrogen-rich kerogens from the grabenward North Sea Kimmeridge Clay Formation (from immaturity through to the base of the conventional liquid window, 0.5–1.3%R_0). Movement along the evolution track on the paraffin index–hydrogen

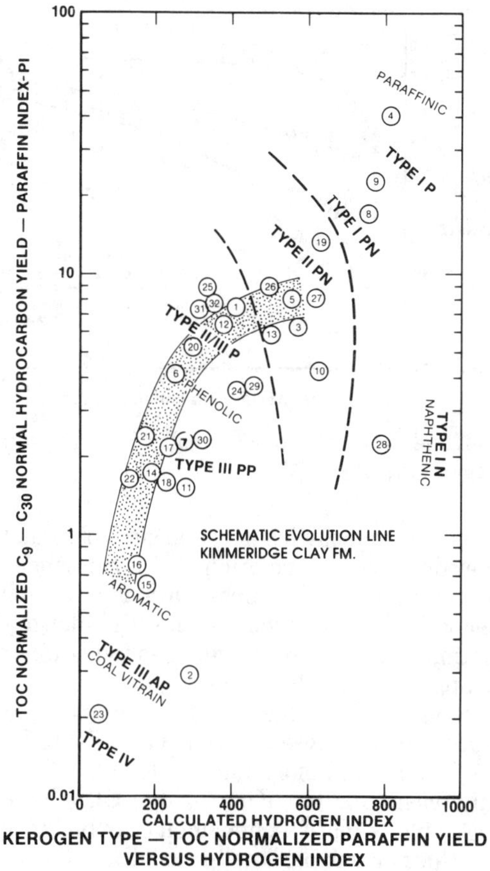

PARAFFIN INDEX—FROM PYROLYSIS-GC. 1 UNIT = 0.25% BY WT.
HYDROGEN INDEX—FROM ROCKEVAL.UNITS = mgs HC/gm TOC.

REPRESENTATIVE KEROGEN TYPES

TYPE Ip	— TORBANITE (BOTRYOCOCCUS) KEROGENS RICH IN ALGINITE.
TYPE Ipn	— TASMANITE (ALGINITE) RICH KEROGENS.
TYPE In	— RESINITE DERIVED MATERIALS.
TYPE II pn	— AMORPHINITE RICH KEROGENS FROM SOURCE ROCKS CONTAINING MARINE ORGANIC MATTER.
TYPE II/III p	— EXINITE RICH KEROGENS
TYPE III pp	— VITRINITE RICH KEROGENS FROM MARINE/DELTAIC SHALES.
TYPE III ap	— VITRINITE RICH KEROGENS FROM COAL SWAMPS.
TYPE IV	— INERTINITE RICH KEROGENS.

Fig. 10. Paraffin index–hydrogen index cross-plot.

index cross-plot allows estimates of the total yield of hydrocarbons (Δ hydrogen index) and the change in paraffin content with maturity for the generated materials (Δ paraffin index).

In defining source-rock kerogen type in terms of both bulk chemical composition (hydrogen/carbon ratio, hydrogen index), detailed chemical composition (paraffin index, i.e. summed normal hydrocarbons in kerogen pyrolysates) and the source of the organic matter (petrographic type) the requisite information is available to define the organic facies of the system. Coupling this information with maturity data enables an estimation of the petroleum potential.

The quantitative approach to defining the composition of a kerogen described here is believed to be directly relevant to defining better the petroleum potential of source kerogens. The model involves an assessment of the actual quantity of paraffins (and ultimately, using similar approaches, other kerogen pyrolysate products) potentially available in a source system. Clearly the method needs to be extended to the other hydrocarbon species in kerogen pyrolysates, but the abundance of normal alkanes in many crude petroleums suggests that even in the form described here the method is applicable in routine source-rock analysis.

Modern exploration geochemistry has advanced to the point where quantitative assessment of the local oil potential in a generative basin is now a feasible objective (Goff, 1983; Demaison and Murris, 1984). Thus the recognition of 'actual source rocks' (see Fig. 1) is becoming a reality.

RECOGNITION OF ACTUAL SOURCE ROCKS

Conventional source rock analytical methods have been very successful in recognizing areas of high petroleum potential. However, they are mainly aimed at determining the current richness of sediments, while ultimately it is the realized potential of the sediment that is of prime interest to the explorationist. In this context an *actual source rock* is defined as 'a sediment that has *produced* and *released* mobile hydrocarbons and related materials' (see Fig. 1).

Many previous studies (Brooks and Thusu, 1981; Barnard *et al.*, 1981; Goff, 1983; Cornford *et al.*, 1983; Cooper and Barnard, 1984) have demonstrated that the Kimmeridge Clay Formation is the principal source unit in the Central and Viking Grabens of the Norwegian North Sea. Using data from these provinces the conceptual problem of assessing the extent of 'actual source rocks' is addressed schematically within the context of available kerogen typing procedures.

In the simplest case the amount of organic material expelled from a source sediment (the expelled organic matter—'XOM') is simply represented by the difference in genetic potential of the sediment at the present time from its initial genetic potential. The genetic potential of a sediment is defined as the sum of free and potential hydrocarbons (and associated non-hydrocarbons) in the sediment at any condition of burial. Thus:

XOM = prematurity organic content − present organic content,

or, using Rock–Eval nomenclature,

$$\text{XOM} = (S_1i + S_2i) \text{ depth } i - (S_1x + S_2x) \text{ depth } x,$$

where S_1i is the corrected initial free 'hydrocarbon content', S_2i is the initial S_2 yield of the sediment, S_1x is the corrected free 'hydrocarbon content' at depth x and S_2x is the current S_2 yield of the sediment.

Thus, the basic concept is extremely simple; the expelled organic fraction is the difference between the present and original organic fractions. In the normal oil window the full genetic potential is never realized, and thus the genetic potential of a sediment is not a reliable indication of the actual yield. In addition two practical problems arise. Firstly, the original organic content is unknown and, secondly, several analytical limitations affect chemical data obtained from conven-

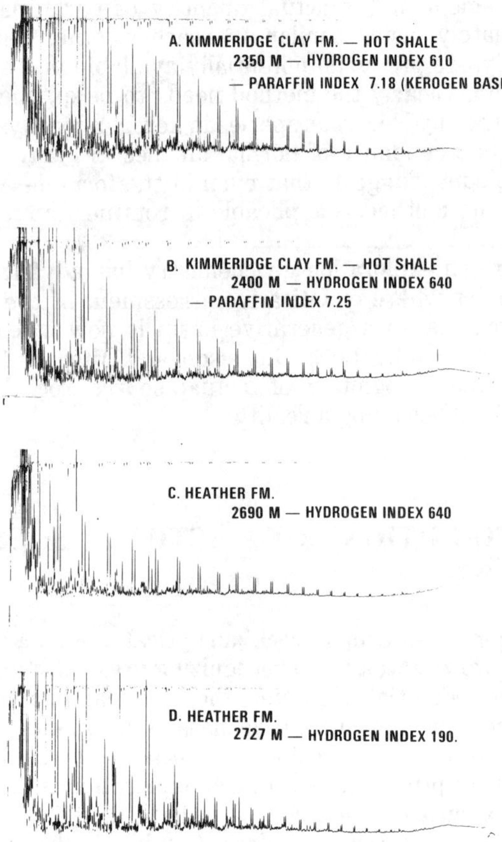

Fig. 11. Pyrograms of Upper Jurassic kerogens from the Viking Graben, North Sea.

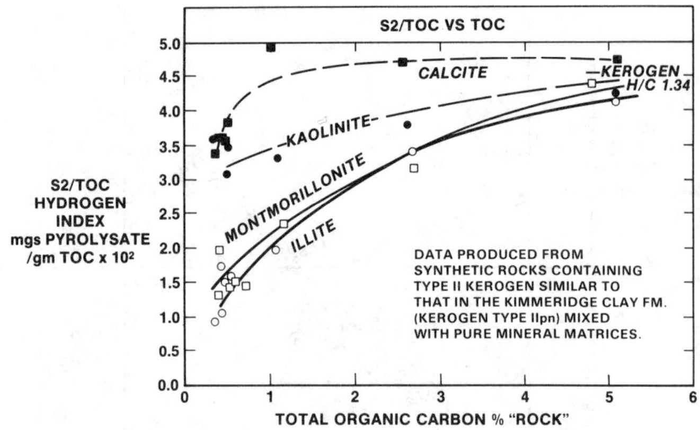

Fig. 12. Matrix effects—normalized pyrolysate yield (hydrogen index).

tional rotary cuttings. These problems, and possible solutions to them, are discussed below.

Two types of data relevant to this type of study are available in most oil company files. The first includes detailed kerogen analyses and chromatographic data from the solvent-soluble organic fractions of sediments. The second group includes less detailed, but more numerous, quantitative assessments of the volatile organic and kerogen contents (e.g. Rock–Eval data), and is generally more abundantly available.

Routine, repeated, cross-calibration of low-resolution kerogen typing methods using higher-resolution techniques (e.g. pyrolysis–GC) is mandatory as kerogen facies variations within lithostratigraphic units are the rule rather than the exception. Figure 11 shows pyrograms of kerogens from Humber Group sediments from a well in the Viking Graben, North Sea. While the Kimmeridge Clay Formation is of relatively uniform composition, Heather formation sediments are clearly quite variable. This type of information is critical to the accurate interpretation of 'average data' for any stratigraphic unit in a basinal context. Mapping of whole formation geochemical data in cases where extreme organic facies variations occur is potentially misleading. Once a calibration has been made, the quantitative use of screening data becomes useful in certain cases.

Several problems relate to the use of Rock–Eval data in a quantitative manner, the principal one being the 'matrix effect' as discussed by Espitalie *et al.* (1980); Horsfield and Douglas (1980) and Larter (1984).

Figure 12 shows the effect of mineral matrix and organic content on the measured yield of hydrocarbons during pyrolysis of synthetic rocks mixed with a type 2pn amorphous-rich kerogen similar to that in the Kimmeridge Clay Formation. Of importance to the present study is the observation that for very organic-rich shales, such as the Kimmeridge Clay Formation, the measured hydrogen index for the sediment approaches within 10% of the value for an isolated kerogen. Other matrix effects influencing yields in source-rock pyrolysates are reviewed and summarized in Fig. 13, and more fully by Larter (1984). Thus, in many cases for the Kimmeridge Clay Formation, hydrogen index matrix phenomena can be ignored and the S_2 Rock–Eval parameter used as a measure of potential hydrocarbons in the kerogen. For example, in the Viking Graben the Kimmeridge Clay Formation has a mean TOC of 4.6%, with local values as high as 10.8%.

Rock–Eval hydrogen index, in the absence of matrix effects, is a powerful typing tool, as many samples

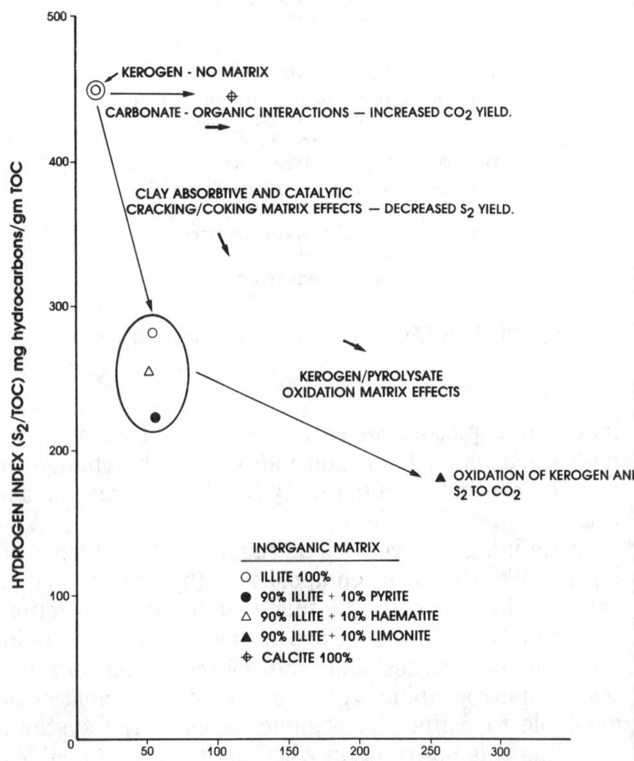

Fig. 13. Representative matrix effects, organically lean sediments. Rock–Eval data for a synthetic rock series containing 1.2–1.3% organic carbon of a type IIpm kerogen.

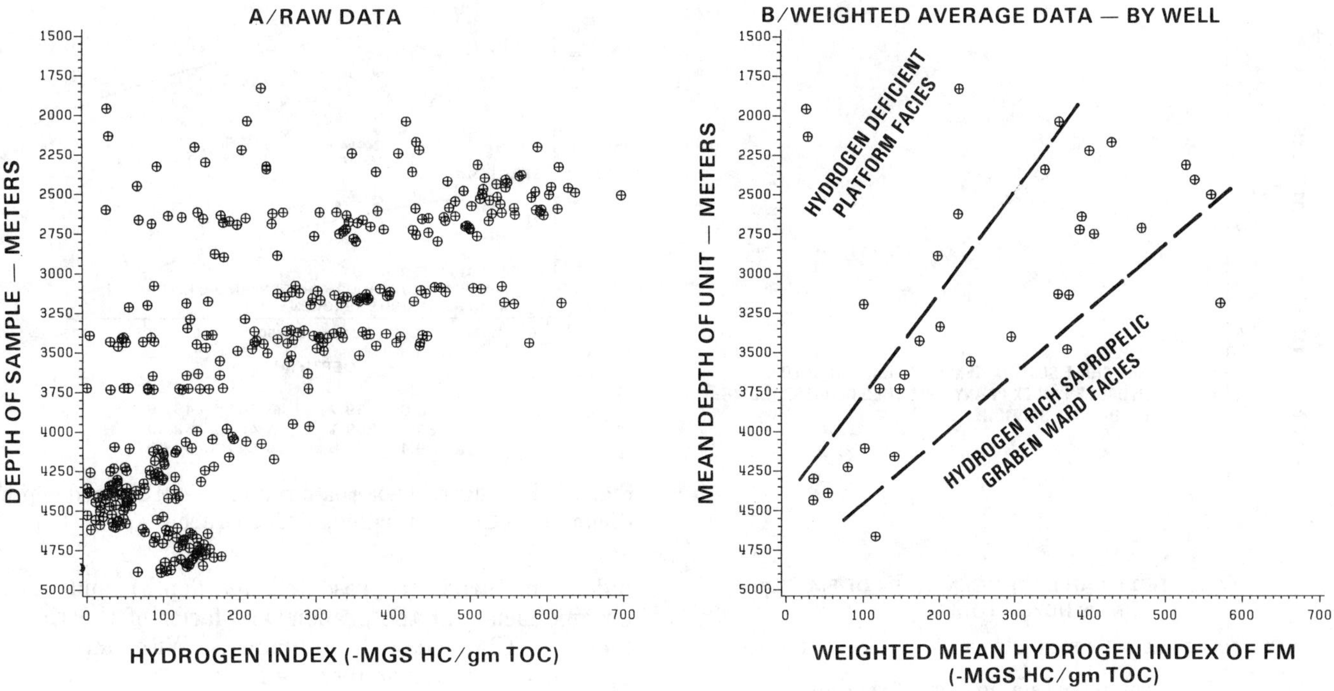

Fig. 14. Hydrogen index data, Kimmeridge Clay Formation, Viking Graben.

can be analysed on a regional basis and a significant data set obtained cheaply. Figure 14(A) shows hydrogen index data from the Kimmeridge Clay Formation in several wells in the Viking Graben which cover a range of maturities down to the base of the conventional oil window. A range of kerogen bulk compositions is evident, although the picture is substantially clarified if the data are reduced to weighted mean values. This procedure places greater weight on the relatively more uniform richer samples in the unit, as follows:

$$\text{weighted mean hydrogen index} = \frac{\left(\sum_{i=1}^{n} S_2 i\right)/n}{\left(\sum_{i=1}^{n} \mathrm{TOC}i\right)/n}$$

The weighted mean hydrogen index for a formation, in cases where the cuttings are well-picked and sampled at closely spaced intervals (10 m or less), best represents the hypothetical 'typical average formation sample'. Figure 14(B) shows data from several Viking Graben wells and illustrates the clarification obtained. Such plots indicate that the organic facies for the Kimmeridge Clay Formation varies, with a humic platform facies locally in the Horda Platform/Utsira High areas and with an initially hydrogen-rich facies in the Viking Graben and on its flanks. Further subdivision of the Kimmeridge Clay Formation is, of course, possible; Figs. 15 and 16 also show vertical facies variations in the kerogen quality in the Kimmeridge Clay Formation. These trends often, but not always, correlate with radioactivity (γ-ray) logs. Studies of the basin-wide variations in the properties of the 'hot shale' and 'cold shale' units allow more accurate assessment of the real source potential of the unit in any area (M. Bjorøy, pers. comm.). Figure 16 shows depth-related trends in the 'hot shale' and lower 'cold shale' units of the Kimmeridge Clay Formation.

During high temperature pyrolysis of kerogens, the operative processes are quite different from those

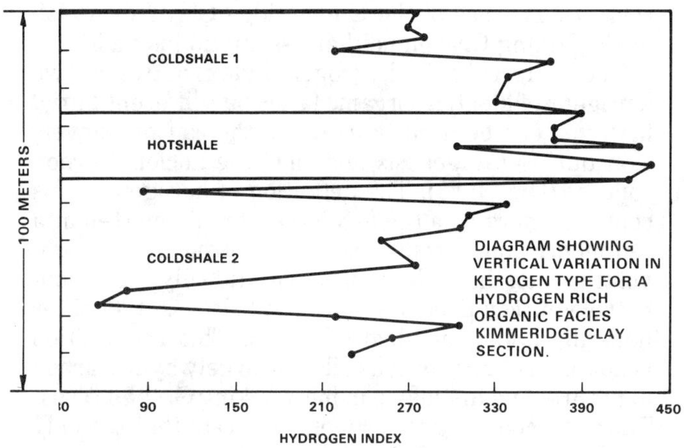

Fig. 15. Hydrogen index data for a well near peak hydrocarbon generation (0.8% R_0, 3500 m depth), Kimmeridge Clay Formation, Viking Graben.

occurring during natural catagenesis (Tissot and Vandenbrouke, 1982). However, differences in kerogen behavior under Rock–Eval pyrolysis between samples buried to different depths but of constant organic facies, still reflect the net catagenetic effect experienced by the more-mature kerogen. Thus, for kerogens of constant organic facies, the difference in the hydrogen index is proportional to the net amount of soluble species produced over the maturation pathway. This is so even though the analytical method itself is in no way simulating natural processes. This line of argument, of course, assumes that catagenesis involves extensively cleavage reactions rather than condensation processes. While both occur, the evidence is clear that normal catagenesis is dominated by fission rather than condensation processes (Tissot and Welte, 1978).

Many approaches for estimating the oil volume expelled from a source bed assume a constant kerogen type, and also generally an initial (pre-burial) value for the organic carbon content for the prospective source

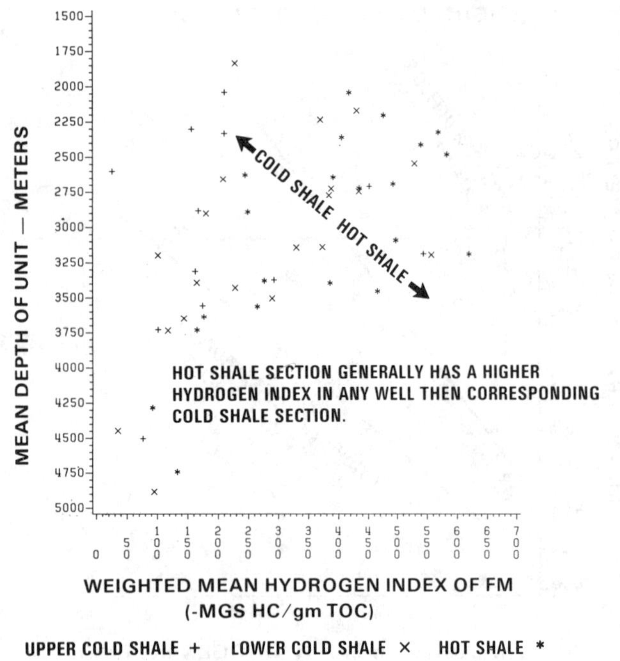

Fig. 16. Hydrogen index as a function of stratigraphy, Kimmeridge Clay Formation, Viking Graben.

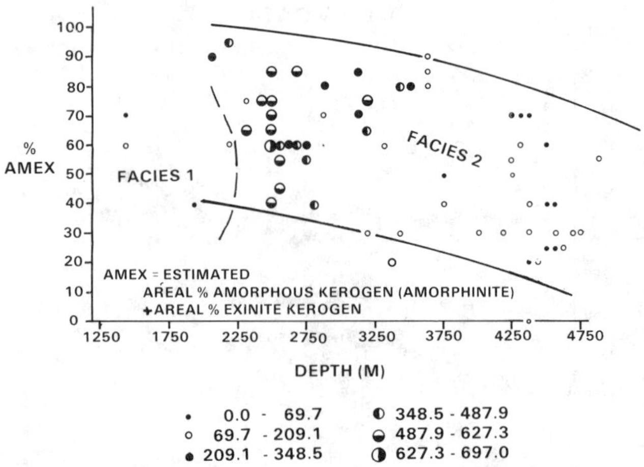

Fig. 17. Hydrogen index as a function of 'AMEX' and depth, Kimmeridge Clay Formation, Viking Graben.

rock. In the case of the Kimmeridge Clay Formation in the Viking Graben neither assumption is valid.

The difference in hydrogen index between two kerogens of identical organic facies but different burial histories can be used to estimate the carbon conversion during catagenesis and thus the ancient carbon content (Orr, 1983). The definition of kerogen in this context refers to all non-volatile organic matter in a sediment, and thus includes kerogen + non-volatile–non-distillable material (ideally all non-petroleum components) some asphaltenes and other non-migratable, non-hydrocarbons. This redefinition is not an arbitrary one, as oils are largely hydrocarbon in nature, and distillable in most Viking Graben cases. Thermal modelling experiments indicate that the only weakly migratable NSO and asphaltene fractions are considered to be an equivalently potent source of hydrocarbons as the conventionally defined kerogen in the Kimmeridge Clay Formation.

The variation in kerogen chemistry within a given formation is a function of both maturity and organic facies. A plot of data from many wells indicates a consistent relationship between vitrinite reflectance and depth of burial for the Kimmeridge Clay Formation of the Viking Graben (see also Goff, 1983). Similar conclusions were reached by Cornford *et al.* (1983) who indicated that locally, depth of burial (or more correctly, formation temperature) is a practical maturity indicator for the Kimmeridge Clay Formation in the UK sector.

The variation in source organic facies in a somewhat variable sediment such as the Kimmeridge Clay and over a range of maturity levels is best indicated by careful petrographic kerogen typing. This is because petrographic typing in transmitted light is less affected at low maturity levels (<0.8% R_0) by mass changes due to conversion of kerogen to liquid and gaseous hydrocarbons. The data in Fig. 17, however, indicate a trend of reducing 'AMEX' (areal % amorphous kerogen + exinite kerogen) with depth at higher maturity levels. Using a computerized regression procedure, the relationship between hydrogen index, maturity (represented as depth) and the organic facies for the grabenward facies of the Kimmeridge Clay may be obtained. With such an approach, a relationship of the form

hydrogen index = *f*(petrographic type)(depth)

can be obtained. Figure 18 shows the relationship between hydrogen indices calculated from petrographic data and those measured from Kimmeridge Clay Formation samples in six wells.

Unfortunately, this method is only found to be reliable for small, areally restricted data sets. On a regional scale an empirical approach was used to derive a maturation curve for any depth in the range 2500–4500 m. This range was chosen since in the North Sea it includes the most data from the Upper Jurassic, and thus the evolution curves are most reliable over this interval. The 2500 m depth is also the point at which significant hydrocarbon generation commences, corresponding to a true (vitrinite 1) maturation level of around 0.5%R_0. At shallower levels, major changes in kerogen composition involve heteroatom loss rather than hydrocarbon production, thus 2500 m is taken as the zero datum for calculation of generated hydrocarbon quantities. An initial hydrogen index can be defined for the kerogen in each sample at 2500 m.

HII represents the hydrogen index for each sample computed at a datum of 2500 m using predictive relationships such as those discussed above.

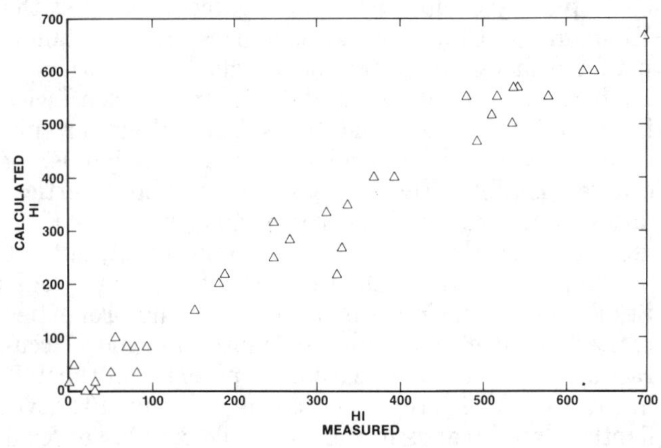

Fig. 18. Predicted versus measured hydrogen indices, based on petrographic data from six wells (Kimmeridge Clay Formation, Viking Graben).

For the kerogen-only system, the fractional loss of carbon through production of hydrocarbons from burial beyond 2500 m may be conservatively estimated by the method of Orr (1983), as follows:

conservative carbon loss (f) = (hydrogen index1 − hydrogen index2) × 0.85/1000,

where the hydrogen indices are for kerogens of the same facies at different maturities and the Rock–Eval evolved hydrocarbons consist of approximately 85% carbon. The 1000 represents a conversion factor from hydrogen index to absolute fractional units.

Analytical data provide current kerogen-based TOC contents and corrected hydrogen indices. Using modelled or observed kerogen composition data, an estimate of HII (hydrogen index at 2500 m) can be obtained, and thus the prematuration kerogen-based carbon content of the sediment can be approximately estimated.

Returning to the original mass balance equation, the expelled fraction of the sediment is estimated as follows:

$$\text{XOM} = (\text{S}_1 + \text{S}_2 \text{ at } 2500\text{ m}) - (\text{S}_1 + \text{S}_2 \text{ at current depth}).$$

Substituting,

$$\text{XOM} = (\text{S}_1 \text{ at } 2500\text{ m} + \text{ANCAR} \times \text{HII}/100) - (\text{S}_2 + \text{S}_1 \text{ at current depth}).$$

The hydrocarbon content at 2500 m can be estimated from plots of S_1/TOC for the immature Kimmeridge Clay Formation. This is a valid approach, as few low molecular weight species have been generated at 2500 m and thus the S_1 value is a true estimate of the distillable soluble fraction of the sediment at this low rank level.

In contrast, once hydrocarbons are being generated from kerogen and soluble non-hydrocarbon fractions, substantial amounts of gasoline range and gaseous molecules are produced. However, during return of cuttings to the surface and subsequent sample washing, handling and storage, substantial amounts of these components are lost (Espitalie, 1984). This results in a very low estimate of the total retained hydrocarbon content for the sediment, especially at depth, and thus in an overestimate of the expelled fraction. This is a critical observation. Additionally, the mode of operation of the Rock–Eval instrument and the presence in a sample of chromatographically active inorganic species such as clays, results in very high-boiling components being selectively retained by the sediment in the pyrolysis oven (Tarafa *et al*., 1984; Horsfield, 1984). Solli *et al*. (1984) and personal observations suggest that a typical S_1 fraction from a stored cuttings sample represents material in the boiling range of about C_{10}–C_{25} normal-alkanes, corresponding approximately to a kerosene through gasoil distillate cut for an oil.

The hydrocarbon-rich nature of Viking Graben oils, and also of the S_1 fraction of sediments, suggests that the 'petrogen' (kerogen + other non-migratable material) concept is applicable to quantitatively evaluating the retained organic matter (S_2 precursor) in a sediment.

The accurate estimation of the concentration of free gaseous and gasoline range components in a source rock is critical in assessing the relative contribution of the analyzable S_1 (or EOM) fraction to the total organic material retained in the source bed. This is critical since, if this relationship is evaluated wrongly, the value of the final migration efficiency will be in error. Using quantitative oil models, flash pyrolysis and hydrous pyrolysis data, the relationship between observed S_1 and actual hydrocarbon contents can be estimated as a function of maturity.

Once the relationship between measured hydrocarbon content (S_1 or C_{15}+ fraction) and the probable in-sediment value is obtained, the expelled organic matter in the sediment system approximates to

$$\text{XOM} = \left(\text{S}_1 i + \frac{(\text{ANCAR} \times \text{HII})}{100}\right) - (\text{S}_2 + \text{S}_1\text{F} \times \text{S}_1),$$

where S_1F is a factor relating observed S_1 to actual sediment C_1–C_n hydrocarbon content. S_1F is a function of maturity, and reaches high values at the base of the oil window. In a condensate-producing section, gas (C_1–C_4) may represent over 70% of the total migrating material, and S_1F would be correspondingly high.

General procedures of the type outlined above have substantial errors associated with their assumptions but they nevertheless provide the explorationist with useful estimates of mobile oil and gas in a basin. In general, the approach is more reliable when dealing with a source section of maturity up to around 0.8% R_0 ('black oil' generation and expulsion in the Viking Graben). At greater maturities, reliable estimates of the retained gas content of the sediments become more difficult to make.

Such simple models coupled with regionally extensive geochemical data bases, source-rock structure and isopach maps, can be used to assess readily charge availability to a given trap. Comparison of observed reservoir fills and calculated potentially migrated oil volumes permits the assessment of prospects in a systematic manner.

Such models further allow the geologist, for example, to evaluate whether the rotated fault block traps in the Viking Graben were filled from source beds largely within the block, the oil migrating up the dip slopes of the blocks and then laterally into the Middle Jurassic reservoirs, as proposed by Goff (1983) for the East Shetland Basin. Alternatively, sourcing may occur from Jurassic source sections in deeper downfaulted blocks, with migration along bounding fault planes. Such questions are relevant to the prospectiveness of untested fault block traps adjacent to the deeper Viking Graben/Møre Basin areas of the Norwegian North Sea.

Oil migration effects—restrictions on the use of geochemical data

Using a mass balance model conceptually similar to that described above, it is estimated that up to 25% of the organic carbon may have been expelled as hydrocarbons by peak oil generation* (0.8% R_0) in the Kimmeridge Clay Formation of the Viking Graben. High calculated expulsion efficiencies, high resistivities on E-logs, overpressuring, and the heterogeneous distribution of organic matter in Kimmeridge shales all strongly suggest that continuous oil phase migration is taking place throughout much of the generative zone beyond peak oil generation (Goff, 1983).

*Perhaps more aptly termed peak oil retention!

At shallow depths, the organic-rich sediment, containing up to 15% or more by volume of organic matter, supports part of the sedimentary load on its laminated organic layers. As burial continues, part of the kerogen is converted to fluids, chiefly oil below 0.8% R_0, resulting in overpressuring and lateral migration along kerogen sediment inferfaces (cf. Meissner, 1978; Du Rouchet, 1978; Durand, 1983).

With deeper burial (>3500 m) significant gas generation occurs along with liquid hydrocarbons, the increases in volume during this process (Ungerer *et al.*, 1983) possibly assisting in maintaining fluid overpressure and migration processes. Lateral oil and gas migration will occur until either the pressure field is exhausted or, more probably, until relatively permeable conduits such as faults are encountered. Once in the fracture system oil migration will follow local and regional fluid potential fields (Ungerer *et al.*, 1984).

As the sediments become more mature, the kerogens will inevitably become more hydrogen-deficient and less plastic (Van Krevelen, 1961). Further, as the organic contents decrease there will be reduced opportunity to sequence oil directly from the sediments, as grain–grain overburden load support becomes more important.

These processes, and others such as phase change and aqueous solubility behaviour relative to temperature and pressure (Goff, 1983; Ungerer *et al.*, 1984) may lead to changes in the effective migration mechanism and variations in the composition of the expelled fluids at depth. The nature of the expelled hydrocarbons cannot be predicted on the basis of maturity or kerogen type alone. Gas only may be expelled from more deeply buried 'oil-source' sediments even within the 'liquid windows'. Further, at the onset of oil expulsion, any compositional fractionation between organic matter expelled (XOM) and retained in the source bed (EOM) will be at a maximum (Kontorovich, 1984). Early generated oils will be the hardest to correlate to source beds using current approaches.

Durand (1983) concisely describes the limits which migration mechanisms may place on the composition of fluids expelled from source beds. Due to the principle of relative permeability, the effects of water and oil saturations in a shale pore fluid on oil expulsion efficiency are dramatic. Durand suggests, for example, that in the case of hydrocarbon generation in rapidly buried organically–lean, water–wet sediments, the expelled fluid may be dominantly gas, despite ample higher hydrocarbon generation in the source bed. The low relative permeabilities of the source section to oil (due to high water saturations in the source sediment) favour preferential gas migration in aqueous solution.

Durand (1983) clearly demonstrates that, despite the availability of accurate and sophisticated methods of characterizing the type and maturity of sediments, migration phenomena, understood so poorly at present, may be the prime factor in deciding the final composition of expelled and reservoired fluids. Thus, despite the ability to effectively type kerogens, and with some understanding of hydrocarbon generative mechanisms under some conditions, the composition of the fluid expelled from a source bed often cannot be predicted accurately using these data alone.

CONCLUSIONS

Petrographic, bulk chemical and detailed molecular kerogen typing can provide the explorationist with much relevant information on the nature and type of the organic matter in source rocks. Some preliminary quantitative estimate of actual oil yields from rich source beds is obtainable using these data. Constraints on the final prospectiveness of a source unit cannot be established using kerogen type and maturity data alone, as poorly understood migration phenomena add significant uncertainty to the effective interpretation of source rock data.

ACKNOWLEDGEMENTS

I am grateful to Union Norge A/S and Union Oil Science & Technology Division management for encouragement and permission to publish these ideas. Many scientists have been involved in discussions of many of the concepts described here, and I acknowledge the contributions made by Drs R. E. Sweeney, J. T. Senftle, B. W. Bromley, G. H. Smith, J. A. Curiale, J. C. Ellice-Flint and N. A. Briedis. Malvin Bjorøy (IKU), on a sabbatical research program at Brea, contributed greatly to ideas concerning Viking Graben geochemistry. Manuscript review was aided by comments from John Fox and John Ellice-Flint.

APPENDIX 1: PYROLYSIS METHODS

Kerogens were isolated and purified using standard methods (Durand, 1980). Pyrolysis analysis was carried out using a CDS Pyroprobe system modified for use with fused silica high-resolution capillary–GC columns (Larter, 1984). Unless indicated otherwise on the figures, pyrolysis conditions were as follows: pyrolysis of up to 500 μg of extracted kerogens at 800 °C for 20 s in flowing helium, followed by gas chromatography on an SP2100-coated, 50 m, Durabond fused silica capillary column programmed from 40 °C to 300 °C at 4 °C/min. The quantitative pyrolysis–GC experiments were carried out on samples pre-mixed with weighed amounts of poly(alpha)methyl styrene. The polymer was added quantitatively in methylene chloride solution to the weighed extracted dried kerogen and thoroughly mixed with the material. The solvent was then evaporated before pyrolysis. Programmed temperature pyrolysis was performed on a CDS 820GS geological sample analysis system.

REFERENCES

Bailey, J. J. L. 1981. Hydrocarbon potential of organic matter. In: Brooks, J. (ed.), *Organic Maturation Studies and Fossil Fuel Exploration*. Academic Press, New York, p. 283.

Barnard, P. C. and Cooper, B. S. 1981. Oils and source rocks of the North Sea area. In: Illing L. V. and Hobson, G. D. (eds.), *Petroleum Geology of the Continental Shelf of North West Europe*. Institute of Petroleum, London, pp. 169–175.

Barnard, P. C., Collins, A. G. and Cooper, B. S. 1981. Identification and distribution of kerogen facies in a source rock horizon—examples from the North Sea Basin. In: Brooks, J. (ed.), *Organic Maturation Studies and Fossil Fuel Exploration*. Academic Press, London, pp. 271–282.

Brooks, J. and Thusu, B. L. 1981. Oil-source identification and characterization of the Jurassic sediments in the Northern North Sea. *Chem. Geol.*, **20**, 283–294.

Cooper, B. S. and Barnard, P. C. 1984. Source rocks and oils of the Central and Northern North Sea. In: Demaison, G. and Murris, R. J. (eds.), *Petroleum Geochemistry and Basin Evaluation. AAPG Memoir*, **35**, 303–314.

Cornford, C., Morrow, J. A., Turrington, A., Miles, J. A. and Brooks, J. 1983. Some geological controls on oil composition in the UK North Sea. In: Brooks, J. (ed.), *Petroleum Geochemistry and Exploration of Europe. Geol. Soc. Spec. Pap. No. 12.*, pp. 255–265. (Blackwell, London.)

Demaison, G. 1984. The generative basin concept. In: Demaison, G. and Murris, R. J. (eds.), Petroleum Geochemistry and Basin Evaluation. *AAPG Memoir* **35**, 1–15.

Demaison, G. and Murris, R. J. (eds.) 1984. Petroleum geochemistry and basin evaluation. *AAPG Memoir* **35**, 426 pp.

Dembicki, H., Horsfield, B. and Ho, T. T. Y. 1983. Source rock evaluation by pyrolysis–gas chromatography. *AAPG Bull.*, **67**(7) 1094–1103.

Durand, B. (ed.) 1980. *Kerogen-insoluble Organic Matter From Sedimentary Rocks.* Editions Technip, Paris, 519 pp.

Durand, B. 1983. Present trends in organic geochemistry in research on migration of hydrocarbons. In: Bjorøy. M. *et al.* (eds.), *Advances in Organic Geochemistry 1981.* Wiley, London, pp. 117–128.

Durand, B. and Paratte, M. 1983. Oil potential of coals, a geochemical approach. In: Brooks, J. (ed.), *Petroleum Geochemistry and Exploration of Europe. Geol. Soc. Spec. Pap. No. 12,* pp. 255–265. Blackwell, London.

Du Rouchet, J. 1978. Eléments d'une théorie géomecanique de la migration de l'huile en phase constituée. *Bull. Cent. Rech. Explor.-Prod. Elf-Aquitaine*, **2**, 337–373.

Espitalie, J. 1984. Geochemical logging. In: Voorhees, K. (ed.), *Analytical pryolysis—Methods and Applications.* Butterworth, London, p. 275.

Espitalie, J., Laporte, J. C., Madec, M., Marquis, F., Le Plat, P., Paulet, J. and Boutefeu, A. 1977. Méthode rapide de charactérisation des roches meres, de leur potential pétrolier et de leur degré d'évolution. *Rev. Inst. Français Pétrole*, **32**, 23–42.

Espitalie, J., Madec, M. and Tissot, B. 1980. Role of mineral matter in kerogen pyrolysis. *AAPG, Bull*, **64**, 59–66.

Evans, E. J., Batts, B. D. and Smith, J. W. 1984. Determination of the hydrocarbon prospectivity of sediments by hydrogenation. *APEA Journal*, **24**(1), 222–229.

Giraud, A. 1970. Application of pyrolysis and pyrolysis–gas chromatography to the geochemical characterization of kerogen in sedimentary rocks. *AAPG Bull.*, **54**, 439–455.

Goff, J. C. 1983. Hydrocarbon generation and migration from Jurassic source rocks in the East Shetland Basin and Viking Graben of the North Sea. *J. Geol. Soc. Lond.*, **140**, 445–474.

Gormely, J. R. and Mukhopadahay, P. K. 1983. Hydrocarbon potential of kerogen types by pyrolysis–gas chromatography. In: Bjorøy M. *et al.* (eds.), *Advances in Organic Geochemistry 1981.* Wiley, London, pp. 597–606.

Gutjahr, C. C. M. 1983. Introduction to incident light microscopy of oil and gas source rocks. *Geol. Mijnbouw*, **62**, 417–425.

Harwood, R. J. 1977. Oil and gas generation by laboratory pyrolysis of kerogen. *AAPG Bull*, **61**, 2082–2102.

Horsfield, B. 1984. Pyrolysis studies and petroleum exploration. In: *Advances in Petroleum Geochemistry*, Vol. I. Academic Press, pp. 247–292.

Horsfield, B. and Douglas, A. G. 1980. The influence of minerals on the pyrolysis of kerogens. *Geochim. Cosmochim. Acta*, **44**, 2087–2095.

Huntjen, H. J. and Van Krevelen, K. 1957. Chemical structure and properties of coal—composition of the individual macerals (vitrinites, fusinites, micrinites, exinites). *Fuel*, **33**, 79.

Kontorovich, A. E. 1984. Geochemical methods for the quantitative evaluation of the petroleum potential of sedimentary basins. In: Demaison, G. and Murris, R. J. (eds.), *Petroleum Geochemistry and Basin Evaluation, AAPG Memoir*, **35**, 79–109.

Larter, S. R. 1984. Application of analytical pyrolysis techniques to kerogen characterization and fossil fuel exploration/exploitation. In: Voorhees, K. (ed.), *Analytical Pyrolysis—Methods and Applications.* Butterworth, London, pp. 212–275.

Larter, S. R. and Douglas, A. G. 1978. Low molecular weight aromatic hydrocarbons in coal maceral pyrolysates as indicators of diagenesis and organic matter type. In: Krumbein, W. E. (ed.), *Environmental Biogeochemistry and Geomicrobiology*, Vol. 1. Ann Arbor, Michigan, pp. 373–386.

Larter, S. R. and Douglas, A. G. 1980. A pyrolysis–gas chromatographic method for kerogen typing. In: Douglas, A. G. and Maxwell, J. R. (eds.), *Advances in Organic Geochemistry 1979.* Pergamon Press, Oxford, 579–584.

Larter, S. R. and Douglas, A. G. 1982. Pyrolysis methods in organic geochemistry—an overview. *J. Analyt. Appl. Pyrolysis*, **4**, 1–19.

Larter, S. R. and Senftle, J. T. 1984. Practical high resolution kerogen typing for petroleum source rock analysis. *AAPG Bull.* (submitted).

Larter, S. R., Horsfield, B. and Douglas, A. G. 1977. Pyrolysis as a possible means of determining the petroleum generating-potential of sedimentary organic matter. In: Jones, C. E. R. and Cramers, C. A. (eds.), *Analytical Pyrolysis.* Elsevier, Amsterdam, pp. 189–202.

Larter, S. R., Solli, H. and Douglas, A. G. 1978. Analysis of kerogens by pyrolysis–gas chromatography–mass spectrometry using selective ion detection. *J. Chromatogr.*, **167**, 421–431.

Lewan, M. D. 1983. Effects of thermal maturation on stable organic carbon isotopes as determined by hydrous pyrolysis of Woodford shale. *Geochim. Cosmochim. Acta.* **47**, 1471–1483.

Leythaeuser, D., Mackenzie, A. S., Schaefer, R. G., Altebaumer, F. J. and Bjorøy, M. 1983. Recognition of migration and its effects within two core-holes in shale/sandstone sequences from Svalbard, Norway. In: Bjorøy, M. *et al.* (eds.), *Advances in Organic Geochemistry 1981.* Wiley, London, pp. 136–146.

Mackenzie, A. S., Leythaeuser, D., Schaefer, R. G. and Bjorøy, M. 1983. Expulsion of petroleum hydrocarbons from shale source rocks. *Nature*, **301**, 506–509.

Martin, S. J. 1977. Thermally evolved hydrocarbons from the bitumen and kerogen constituents of whole rock. In: Campos, R. and Goni, J. (eds.), *Advances in Organic Geochemistry 1975.* Enadimsa, Madrid, pp. 677–692.

Maters, W. L., Van de Meent, D., Schuyl, P. J. W., De Leeuw, J. W. and Schenk, P. A. 1977. Curie point pyrolysis in organic geochemistry. In: Jones, C. E. R. and Cramers, C. A. (eds.), *Analytical Pyrolysis.* Elsevier, Amsterdam, pp. 203–214.

McHugh, D. J., Saxby, J. D. and Tardiff, J. W. 1978. Pyrolysis–gas chromatography of carbonaceous material from Australian sediments. *Chem. Geol.*, **21**, 1.

McIver, R. D. 1967. Composition of kerogen. *Proc. Seventh. World Petroleum Congr.*, **2**, 25–36.

Meissner, F. F. 1978. Petroleum geology of the Bakken Formation Williston Basin, North Dakota and Montana. *Montana Geol. Soc. Proc. 1978 Williston Basin Symp.* pp. 207–227.

Miknis, F. P., Netzel, D. A., Ward-Smith, J., Alisament, M. and Maciel, G. E. 1982. C–13 NMR measurements of the genetic potential of oil shales. *Geochim. Cosmichim. Acta*, **46**, 977.

Orr, W. L. 1983. Comments on pyrolytic hydrocarbon yields in source rock evaluation. In: Bjorøy, M. *et al.* (eds.), *Advances in Organic Geochemistry 1981.* Wiley, London, pp. 775–788.

Philp, R. P. and Russel, N. J. 1980. Pyrolysis–gas chromatography–mass spectrometry of batch autoclave products derived from coal macerals. In: Douglas, A. G. and Maxwell, J. R. (eds.), *Advances in Organic Geochemistry 1979.* Pergamon Press, Oxford, p. 653.

Philp, R. P., Gilbert, T. D. and Russell, N. J. 1982. Characterisation by pyrolysis–gas chromatography–mass spectrometry of the insoluble organic residues derived from hydrogenation of *Tasmanites* sp. oil shale. *Fuel*, **61**, 221.

Powell, T. G., Creaney, S. and Snowdon, L. R. 1982. Limitations of the use of organic petrographic techniques for identification of petroleum source rocks. *AAPG Bull.* **66**(4), 430–435.

Robert, P. 1971. Etude pétrographique des matières organiques insolubles par la mesure de leur pouvoir reflecteur. Contribution à l'exploration petrolière et à la connaissare des basins sédimentaires. *Rev. Inst. Français du Pétrole*, **26**(2), 106.

Romovacek, J. and Kubat, J. 1968. Characterization of coal substances by pyrolysis-gas chromatography. *Analyt. Chem.*, **40**, 1119.

Senftle, J. T. 1983. Optical analysis of kerogen—its role in an integrated approach for kerogen typing. Paper and Abstract Presented at the *North American Coal Petrographical Meeting*, Merrilville, Indiana.

Sigleo, A. C., Hoering, T. C. and Helz, G. R. 1982. Composition of estuarine colloidal material: organic components. *Geochim. Cosmochim. Acta*, **46**, 1619–1626.

Solli, H., Larter, S. R. and Douglas, A. G. 1980. Analysis of kerogens by pyrolysis–gas chromatography–mass spectrometry—Part I, Alkybenzenes. In: Douglas, A. G. and Maxwell, J. R. (eds.), *Advances in Organic Geochemistry 1979.* Pergamon Press, Oxford, p. 591.

Solli, H., Bjorøy, M. Leplat, P. and Hall, K. 1984. Analysis of organic matter in small rock samples using combined thermal extraction and pyrolysis–gas chromatography. *J. Analyt. Appl. Pyrolysis* (in press).

Staplin, F. L. 1969. Sedimentary organic matter, organic

metamorphism, and oil and gas occurrence. *Can. Petroleum Geologists Bull.*, **17**, 47–66.

Stopes, M. C. 1935. On the petrology of banded bituminous coal. *Fuel*, **14**, 4–13.

Tarafa, M. E., Hunt, J. M. and Whelan, J. K. 1984. *AAPG Bull.*, (in press).

Tissot, B. P. 1984. Recent advances in petroleum geochemistry applied to hydrocarbon exploration. *AAPG Bull.*, **68**(5), 545–563.

Tissot, B. P. and Vandenbrouke, M. 1982. Geochemistry and pyrolysis of oil shales. *Preprints Am. Chem. Soc., Div., Petroleum Chem.*, **28**(1), 92–99.

Tissot, B. P. and Welte, D. H. 1978. *Petroleum Formation and Occurrence, A New Approach to Oil and Gas exploration.* Springer-Verlag, Berlin, 583 pp.

Tissot, B. P., Durand, B., Espitalie, J. and Combaz, A. 1974. Influence of the nature and diagenesis of organic matter in the formation of petroleum. *AAPG Bull.*, **58**(3), 499–506.

Ungerer, P., Behar, E. and Discamps, D. 1983. Tentative calculation of the volume expansion of organic matter during hydrocarbon genesis from geochemistry data—implications for primary migration. In: Bjorøy, M. *et al.* (eds.), *Advances in Organic Geochemistry 1981.* Wiley, London, pp. 129–135.

Ungerer, P., Bessis, F., Chenet, P. Y., Durand, B., Nogaret, E., Chiarelli, A., Oudin, J. L. and Perrin, J. F. 1984. Geological and geochemical models in oil exploration, principles and practical examples. In: Demaison, G. and Murris, R. J. (eds.), *Petroleum Geochemistry and Basin Evaluation. AAPG Memoir*, **35**, 53–77.

Van de Meent, D., Brown, S. C., Philp, R. P. and Simoneit, B. 1980. Pyrolysis high resolution gas chromatography and pyrolysis–gas chromatography–mass spectrometry of kerogens and kerogen processes. *Geochim. Cosmochim. Acta*, **44**, 999–1014.

Van Krevelen, D. W. 1961. *Coal.* Elsevier, Amsterdam.

Van Graas, A., de Leeuw, J. W. and Schenk, P. A. 1980. Analysis of coals of different ranks by Curie-point pyrolysis–gas chromatography–mass spectrometry and Curie-point pyrolysis–mass spectrometry. In: Douglas, A. G. and Maxwell, J. R., (eds.), *Advances in Organic Geochemistry, 1979.* Pergamon Press, Oxford, p. 485.

Welte, D. H. and Yükler, M. A. 1981. Petroleum origin and accumulation in basin evoluation–a quantitative model. *AAPG Bull.*, **65**, 1387–1396.

Yarzab, R. F., Given, P. H., Spackman, W. and Davis, A. 1980. Dependence of coal liquefaction behavior on coal characteristics 4. Cluster analysis for characteristics of 104 coals. *Fuel*, **59**, 81.

Zilm, K. W., Pugmire, R. J., Larter, S. R., Allan, J. and Grant, D. M. 1981. Carbon-13 cross polarised/magic angle spinning NMR spectroscopy of coal macerals. *Fuel*, **60**, 717.

24 Hydrous pyrolysis of asphaltenes

N. Telnæs, G. C. Speers*, A. Steen* and A. G. Douglas†*
*Norsk Hydro Research Center, Box 4313, 5013 Nygaardstangen, Bergen, Norway
†Organic Geochemistry Unit, Drummond Bldg, University of Newcastle upon Tyne, Newcastle upon Tyne, NE1 7RU, UK

Asphaltenes are defined as small, soluble parts of the kerogen, and may be used as kerogen models.

Hydrous pyrolysis in small stainless steel bombs at elevated temperatures (280–370 °C) releases hydrocarbons that are very similar to the oil/bitumen from which the asphaltenes were precipitated. As asphaltenes are not affected by biodegradation, this is a possible method for generating an oil unaffected by biodegradation.

Hydrous pyrolysis of asphaltenes can also facilitate the study of mineral matrix effects in hydrocarbon generation. Asphaltenes can be dissolved and slurried with mineral matrices in order to obtain an intimate organic/inorganic contact.

The effect of hydrous pyrolysis on the sterane distribution is studied in detail using computerized GC–MS. The sterane distribution from the pyrolysate of the asphaltenes derived from a biodegraded oil have an unusually high contribution of the biologically derived R-epimer of $\alpha\alpha\alpha$ ethylcholestane.

INTRODUCTION

Asphaltenes are defined as that part of petroleum precipitating upon addition of excess non-polar solvents such as n-pentane or n-heptane. Asphaltenes are thus a highly complex solubility class, without a well-defined molecular structure.

The composition of asphaltenes may be studied by various spectroscopic techniques or by chemical/thermal degradation of either the total material or fractions separated by liquid chromatography. Earlier work on the pyrolysis of asphaltenes has been mainly concerned with flash-pyrolysis, and the use of the resulting pyrograms in oil–oil and oil–source rock correlations (Tissot, 1984; Huc *et al.* 1984).

In these studies, asphaltenes are regarded as small soluble fragments of kerogen with a composiiton inherited from the parent kerogen (Tissot, 1984). Preliminary investigations were concerned with the hydrous pyrolysis of asphaltenes, with special emphasis on biomarker compositions and their possible application in correlation studies. Asphaltene samples were derived either from Gilsonite or a range of North Sea crude oils.

EXPERIMENTAL

Gilsonite is a solid, black, naturally occurring 'bitumen' found extensively in the Uinta Basin, USA, especially in the upper Green River Formation. The sample of Gilsonite used in this study was from the 50 cm wide Nigger Bay Vein (Rainbow Group), located in Uinta County.

Asphalts and bitumens have been classified by Abraham (1945) mainly on the basis of solubility in carbon disulphide and their fusibility. According to this scheme, Gilsonite belongs to the soluble, but difficult to fuse, 'Asphaltite' class.

Crude oil samples were all from the same area of the North Sea, and are considered to be sourced from the Upper Jurassic Kimmeridge Clay Formation.

The Gilsonite was dissolved in dichloromethane and asphaltenes were precipitated by addition of a 40-fold excess of cold n-pentane, and left at room temperature overnight. Crude oil asphaltenes were precipitated with a 40-fold excess of cold n-hexane, and left at room temperature overnight. The crude asphaltenes were extracted in a soxhlet for 24 h with petroleum ether (40–60 °C). Gilsonite asphaltenes were reprecipitated from a clean solvent, and further soxhlet extracted. The Gilsonite asphaltenes were subjected to hydrous pyrolysis in small stainless steel bombs, one-third filled with water and in a nitrogen atmosphere, for 72 h at temperatures varying from 280 °C to 370 °C.

A solution of Gilsonite asphaltenes was slurried with calcium carbonate or sodium montmorillonite in a 1 : 3 ratio and, after evaporation of the solvent, the impregnated solids were subjected to hydrous pyrolysis as above.

Petroleum Geochemistry in Exploration of the Norwegian Shelf © Norwegian Petroleum Society (Graham & Trotman, 1985) pp. 287–292.

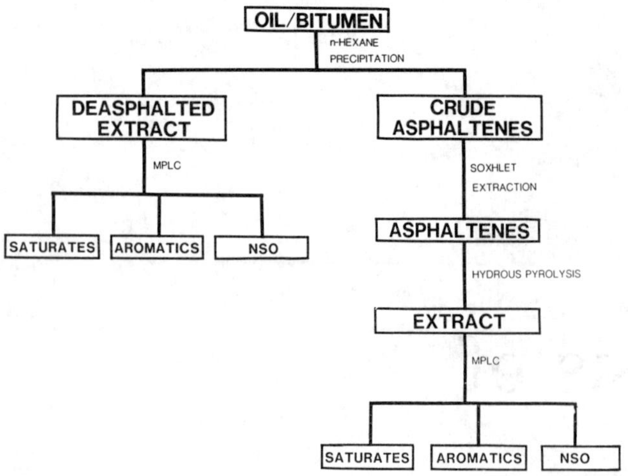

Fig. 1. Separation scheme for the precipitation of asphaltenes and fractionation of pyrolysates.

The crude oil asphaltenes were pyrolysed at 310 °C and 330 °C. After pyrolysis, the stainless steel bombs were extracted with dichloromethane, and the hexane soluble fractions were separated on a MPLC system (see Fig. 1). Gas chromatography was carried out on an OV-1 fused silica column, programmed from 50 °C to 320 °C with a ramp of 4 °C/min. GC–MS work was conducted on a VG7070E mass spectrometer in the multiple ion detection mode, with an electron impact energy of 70 eV.

RESULTS AND DISCUSSION

Gilsonite is a highly biodegraded and waterwashed natural asphalt. A gas chromatogram of the saturate fraction (see Fig. 2) shows almost complete depletion of the *n*-alkanes, and high concentrations of tetracyclic steranes and pentacyclic triterpanes. In addition to these compounds, β-carotane(s) are also abundant (marked with an asterisk in the chromatogram). One of the reasons the asphaltenes from this bitumen were chosen for these experiments is that co-precipitation of *n*-alkanes with the asphaltenes could be virtually neglected.

At a pyrolysis temperature of 280 °C, the saturate fraction is dominated by C-14 to C-20 isoprenoids, with pristane as the most abundant compound. Steranes and triterpanes are also present in abundance. A relatively uniform distribution of *n*-alkanes up to C-35, with a slight odd–even predominance, is also observed.

At 300 °C the relative amount of *n*-alkanes in the pyrolysate has increased, with a maximum around C-25. Pristane is still the most abundant component, and an odd–even predominance in the alkanes is still evident. The high abundance of triterpanes/steranes in the pyrolysates at 280 °C and 300 °C might indicate a major contribution from co-precipitated material in these pyrolysates. Heating the asphaltenes in the injection port of a gas chromatograph at 320 °C, however, gave no detectable release of any co-precipitated material (if present). When pyrolysing asphaltenes, coals and kerogens, it is generally very hard to distinguish between release of occluded material and true pyrolytic products.

At 330 °C the pyrolysate is dominated by an homologous series of *n*-alkanes with a maximum at C-30, but the pristane/*n*-C-17 and the phytane/*n*-C-18 ratios are still >1. Increasing the pyrolysis temperature to 350 °C gives rise to a bimodal *n*-alkane distribution with maxima at C-17 and C-25. The phytane/*n*-C-18 ratio approaches unity at this temperature. These results are indicative of a thermal cracking of the higher molecular weight *n*-alkanes observed in the pyrolysate at 330 °C. One of the interesting aspects of using asphaltenes as model compounds for kerogen is the possibility of studying the effects of mineral matrices on the generation of hydrocarbons. Asphaltenes in solution can be slurried with mineral matrices in order to achieve an intimate contact between the organic and inorganic phases followed by evaporation of the solvent. This will enable studies to be carried out on the catalytic effects of single minerals or representative mixtures of minerals on the conversion of asphaltenes to petroleum-like hydrocarbons under laboratory simulation experiments such as hydrous pyrolysis (Ivanov and Shcherban, 1983).

Hydrous pyrolysis in the presence of calcium carbonate or sodium montmorillonite as a mineral matrix gives pyrolysates with a lower abundance of *n*-alkanes relative to the pyrolysate at 330 °C without the mineral matrix. For both the carbonate and the clay mineral situations, there is a bimodal distribution of the *n*-alkanes in the saturate fractions of the pyrolysates, with maxima at C-17 and C-25. The odd–even predominance in the *n*-alkanes found in the pyrolysates at 280 °C and 300 °C is not apparent at 330 °C.

Steranes are known to be very sensitive to thermal stress (Seifert and Moldowan, 1978) and these compounds were used in order to study the maturity variations of the pyrolysates in more detail. Figure 3 gives the *m/z* 218 fragmentograms from the saturate fractions of some of the pyrolysis experiments, as well as the original bitumen. Table 1 is a summary of the GC–MS results for the C-29 sterane ethylcholestane. The steranes in the natural bitumen have a distribution which appears unusually mature for this low rank deposit (Grantham, 1974). With increasing pyrolysis temperatures the steranes start epimerizing from 5α(H),14α(H),17α(H) to 5α(H),14β(H),17β(H). The αββ/ααα + αββ ratio for the ethylcholestane changes from 0.23 at a pyrolysis temperature at 280 °C to 0.47 at 370 °C, reflecting the relative increase in the thermodynamically more stable αββ isomer. It is interesting that even at 370 °C this reaction has not reached its equilibrium value. With increasing maturation, the biologically inherited C-29 20R isomer (peak 8) is trans-

TABLE 1. A summary of the C-29 sterane GC–MS data for the Gilsonite bitumen, and the asphaltene pyrolysates

Sample name	*Ratio 1*	*Ratio 2*
G-BIT	0.21	0.25
G-280	0.23	0.33
G-300	0.28	0.39
G-310	0.29	0.42
G-330	0.31	0.36
G-340	0.42	0.51
G-350	0.46	0.56
G-370	0.47	0.51
G-330C	0.32	0.41
G-330M	0.39	0.45

Ratio 1: C-29 αββ/ααα + αββ ethyl cholestanes.
Ratio 2: C-29 20S/20R + 20S ααα ethyl cholestanes.

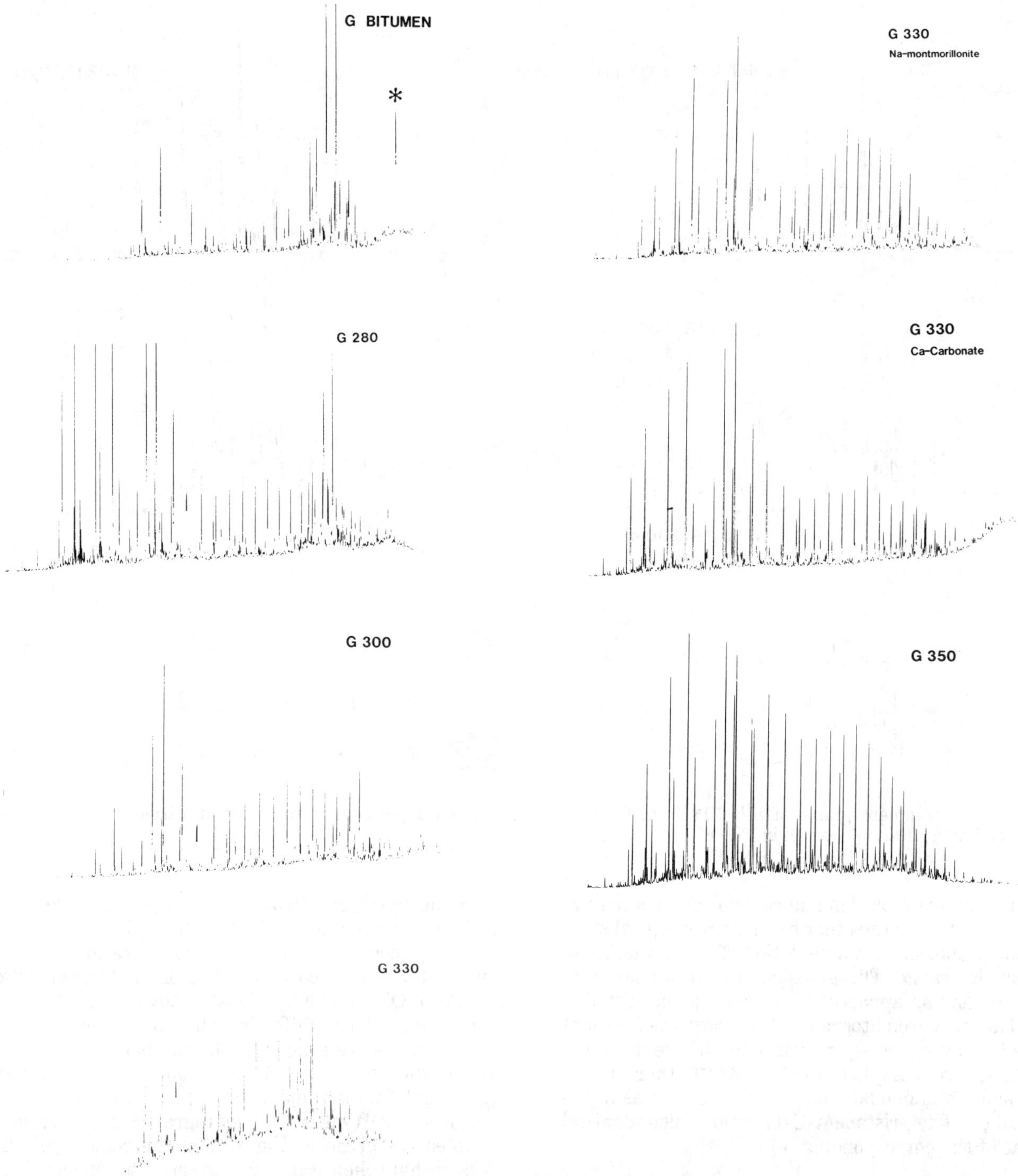

Fig. 2. Gas chromatograms of the saturate fractions from Gilsonite bitumen and asphaltene pyrolysates at various temperatures.

formed into the more stable C-29 20S isomer (peak 5). This leads to an increase in the 20S/20R + 20S ratio from 0.33 when pyrolysis is carried out at 280 °C, to 0.56 at 350 °C and 0.51 at 370 °C. The equilibrium for this reaction is reported to be 0.5 (MacKenzie and Maxwell, 1981) which is reached at maximum pyrolysis temperature. The high value at 350 °C may be due to some coelution of an unknown component. In Fig. 4, the $\alpha\beta\beta/\alpha\alpha\alpha + \alpha\beta\beta$ ratio is plotted against pyrolysis temperature. The effect of a mineral matrix is quite pronounced for the sodium montmorillonite, and shows that this mineral has a catalytic effect on the epimerization of steranes. The superacid effect of clay minerals on geochemical transformations is well documented (Sieskind *et al.*, 1979). The effect of calcium carbonate on the sterane epimerization seems to be negligible compared with the corresponding experiment without the presence of a mineral matrix.

Asphaltenes derived from crude oils have a more uniform composition than asphaltenes from source rocks. In the North Sea region, where most of the oils are sourced from the Upper Jurassic Kimmeridge Clay Formation, it was considered of interest to investigate whether hydrous pyrolysis of the corresponding asphaltenes would give significant differences in the composition of the pyrolysates.

The two crudes chosen for this part of the study have the same origin, but one is partially biodegraded (Oil B) and the other (Oil A) is not.

Gas chromatograms of the saturate fraction of these oils and the pyrolysates are given in Fig. 5. Oil A shows a smooth unimodal *n*-alkane distribution with-

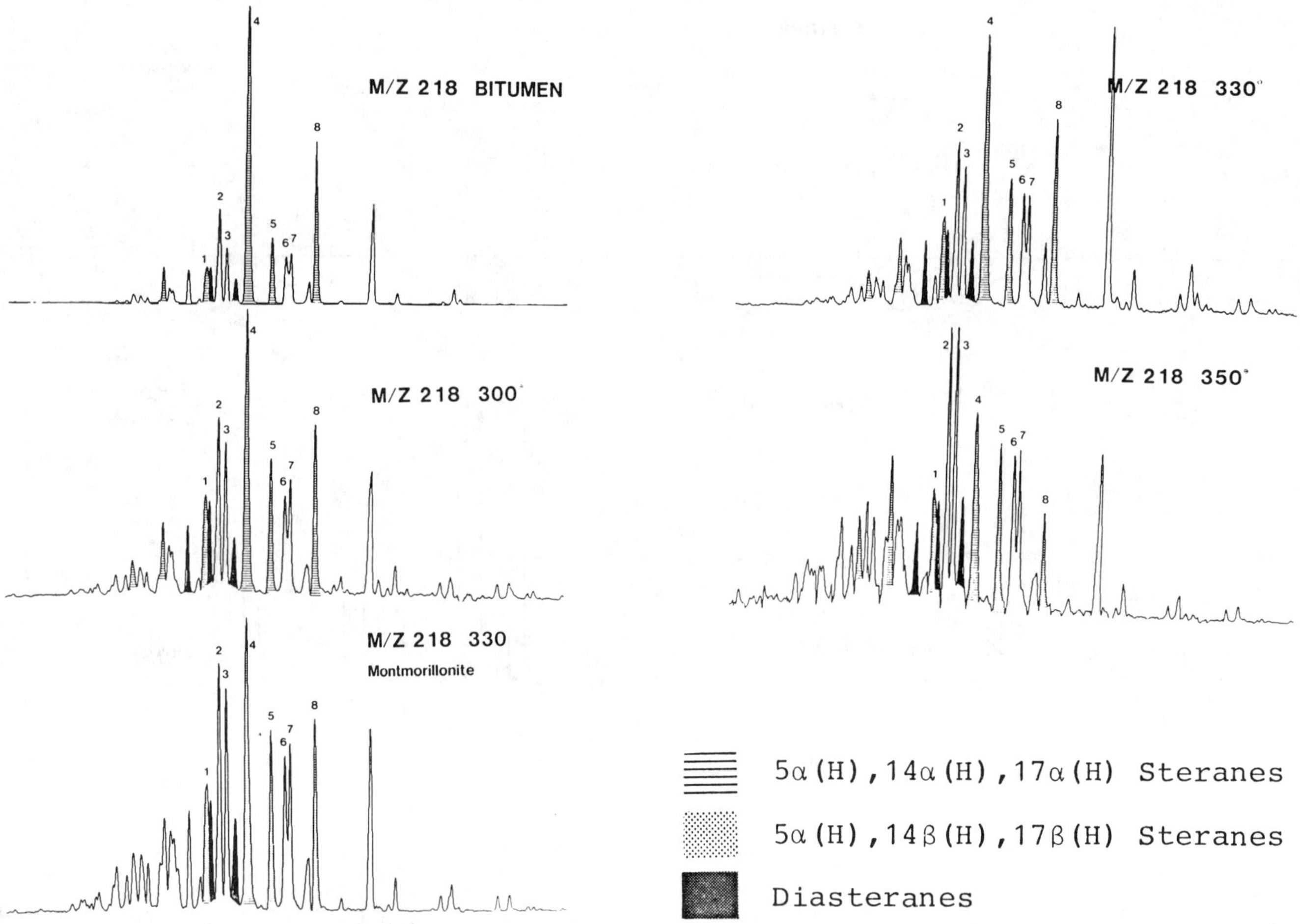

Fig. 3. Mass fragmentograms of m/z 218 from Gilsonite bitumen and asphaltene pyrolysates at various temperatures. Compound identification is given in Fig. 8.

out any odd–even predominance, typical for a mature crude oil. Oil B, on the other hand, shows typical signs of biodegradation with a depletion of the n-alkanes in the gasoline range. This gives rise to a pristane/n-C17 ratio >1, and an apparent bimodal n-alkane distribution. The gas chromatograms of the saturate fractions of the two pyrolysates are distinctly different. In the pyrolysate from Asphaltene A at 330°C there is the same smooth unimodal n-alkane distribution as in the original oil. The pristane/n-C-17 ratio is also identical to that of the corresponding oil.

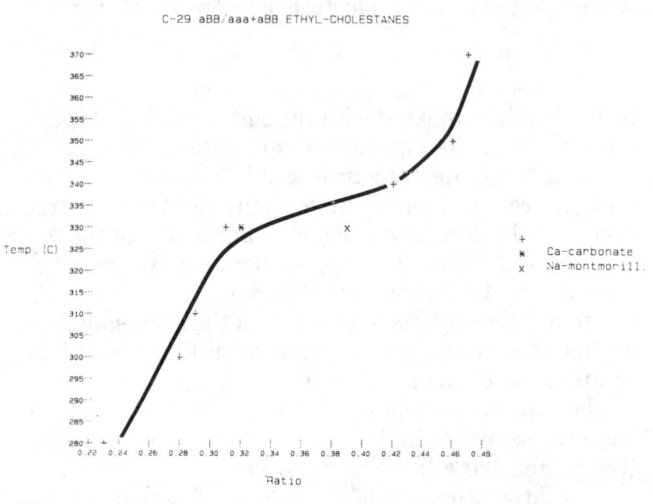

Fig. 4. The C-29 $\alpha\beta\beta/\alpha\alpha\alpha + \alpha\beta\beta$ ratio for Gilsonite asphaltene pyrolysates plotted against pyrolysis temperature.

In the pyrolysate from Asphaltene B, there is an n-alkane maximum at C-29 with a slight odd–even predominance. The pristane/n-C-17 ratio is much smaller than in the corresponding oil, and very similar to that of Oil A. The pyrolysate shows no evidence of the biodegradation affecting the oil from which the asphaltenes were precipitated, but there is a marked contribution to the pyrolysate from a higher molecular weight fraction, maximizing around C-29.

The m/z 218 mass fragmentograms of the various samples are given in Fig. 6, and the results for the C-29 ethyl cholestane ratios are summarized in Table 2.

The sterane distributions of the two original oils are very similar, lending support to the theory of a common origin for these crudes. The 20S/20R + 20S

TABLE 2. A summary of the C-29 sterane GC-MS data for the two crude oils, and the asphaltene pyrolysates

Sample name	*Ratio 1*	*Ratio 2*
Oil A	0.61	0.63
Asph. A 310	0.55	0.46
Asph. A 300	0.54	0.46
Oil B	0.64	0.59
Asph. B 330	0.40	0.22

Ratio 1: C-29 $\alpha\beta\beta/\alpha\alpha\alpha + \alpha\beta\beta$ ethyl cholestanes.
Ratio 2: C-29 20S/20S + 20R $\alpha\alpha\alpha$ ethyl cholestanes.

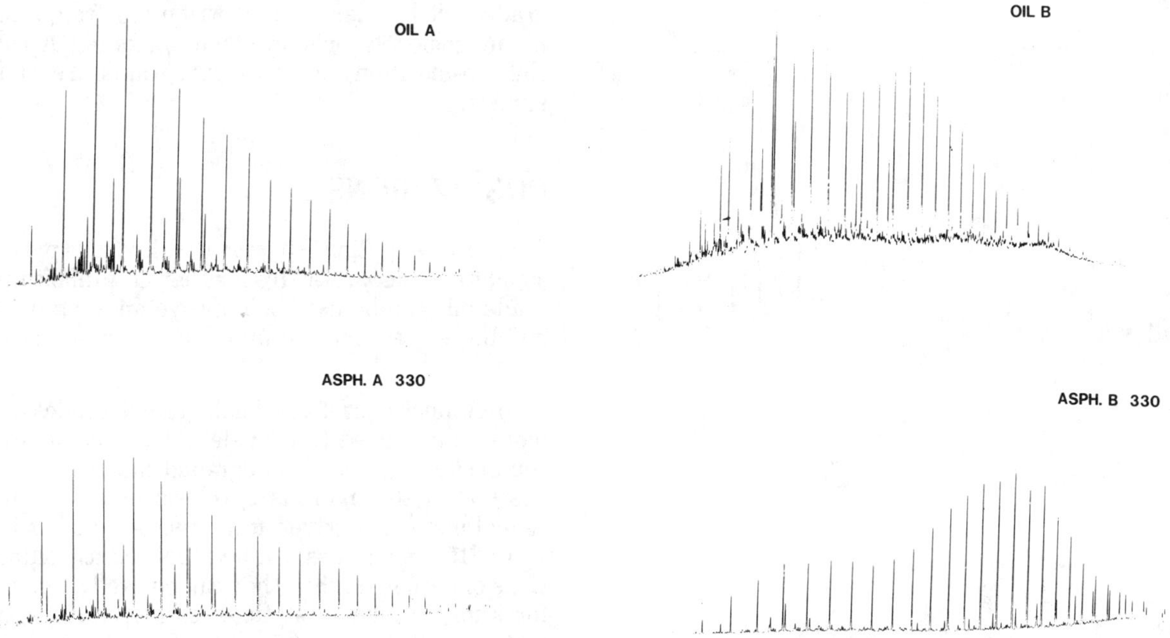

Fig. 5. Gas chromatograms of the saturate fractions of the crude oils and the asphaltene pyrolysates from the pyrolysis at 330 °C.

ratios (ratio 2) are 0.63 and 0.59, respectively, for Oils A and B, which is higher than the expected equilibrium value for this transformation. This may be due to co-elution of the 20S isomer with an unknown component.

The fragmentograms of the pyrolysates from Asphaltene A at 310 °C and 330 °C are very similar, and the maturity—as measured by sterane epimerization—is identical. There are some apparent differences in the sterane distributions in the oil and the asphaltene pyrolysates, particularly in the early part of the fragmentograms in the diasterane region. In both pyrolysates, ratio 2 has reached the expected equilibrium value of 0.5, but they are less mature than the original oil as measured by the $\alpha\beta\beta/\alpha\alpha\alpha + \alpha\beta\beta$ ratios. These results suggest that steranes released from asphaltenes during hydrous pyrolysis of asphaltenes have retained more of their biologically inherited stereochemistry.

The *m/z* 218 fragmentogram from the pyrolysis of

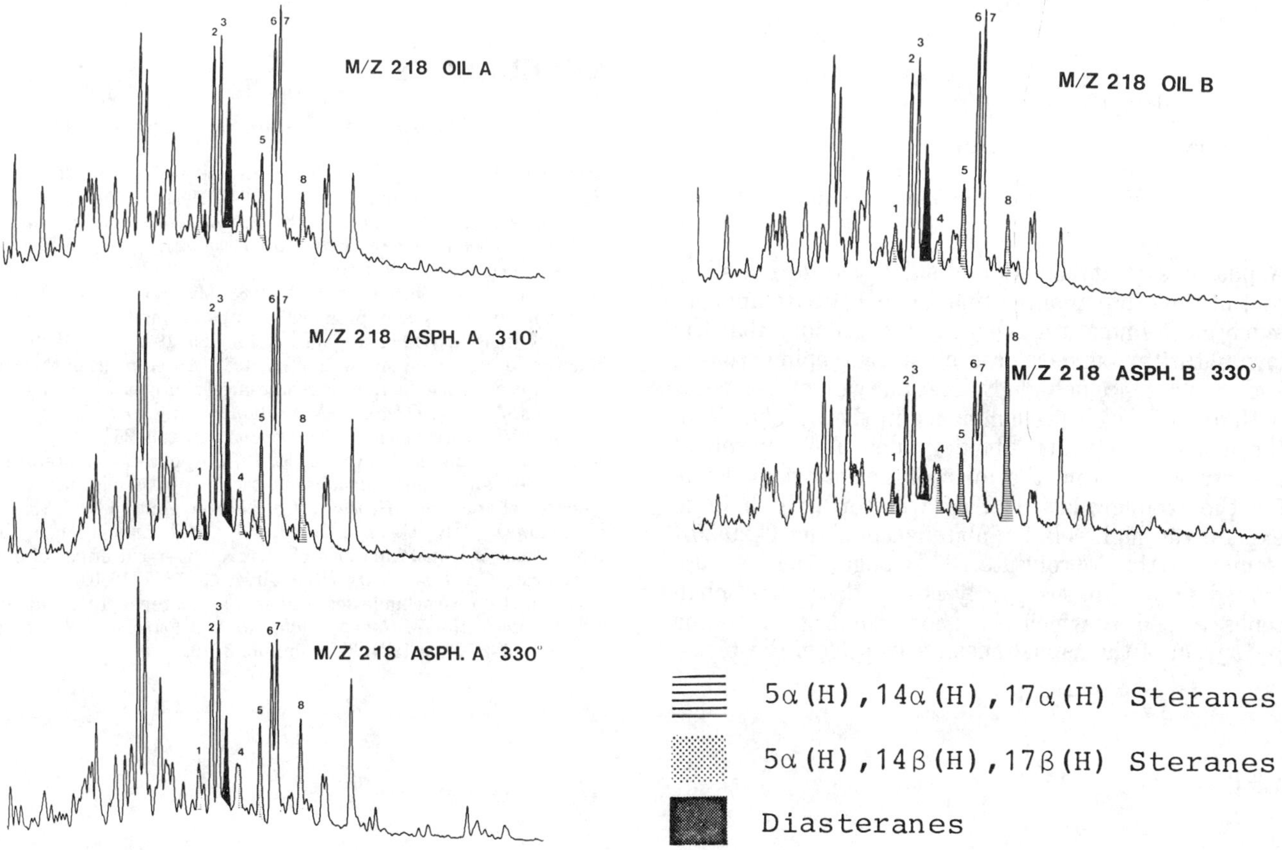

Fig. 6. Mass fragmentograms of *m/z* 218 from the crude oils and the asphaltene pyrolysates. Compound identification is given in Fig. 8.

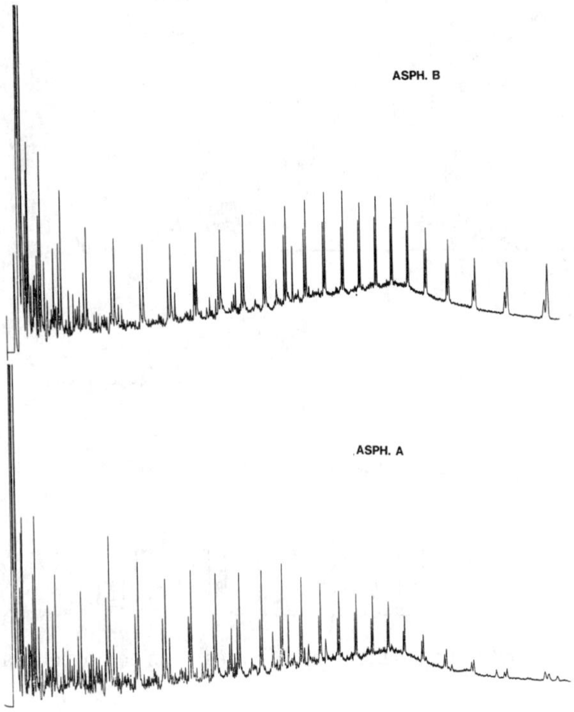

Fig. 7. Pyrograms from the anhydrous pyrolysis–gas chromatography of the two crude oil asphaltenes. Pyrolysis temperature 600 °C.

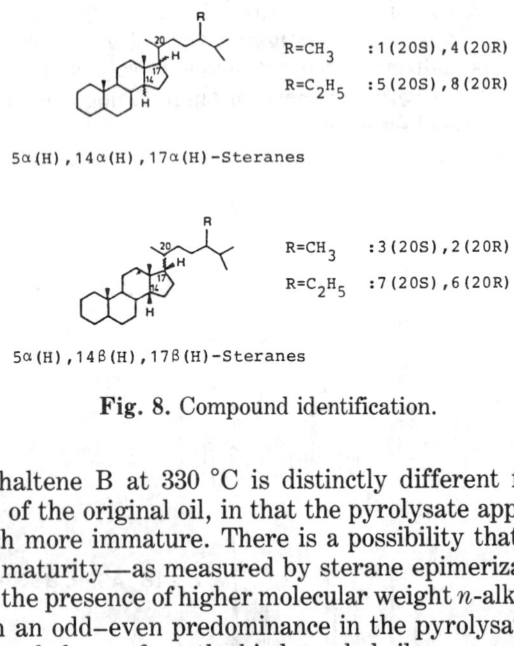

Fig. 8. Compound identification.

Asphaltene B at 330 °C is distinctly different from that of the original oil, in that the pyrolysate appears much more immature. There is a possibility that the low maturity—as measured by sterane epimerization and the presence of higher molecular weight *n*-alkanes with an odd–even predominance in the pyrolysate of the asphaltenes from the biodegraded oil—may reflect a contribution from the microorganisms responsible for the biodegradation. This will certainly help to explain the high relative abundance of the C-29 20R isomer in this pyrolysate. This suggestion is supported by anhydrous pyrolysis–GC data on Asphaltenes A and B which are shown in Fig. 7. In the pyrogram of the asphaltenes derived from the biodegraded oil B, there is a contribution from a suite of higher molecular weight alkenes/alkanes. Apart from this observation, the two pyrograms are strikingly similar.

CONCLUSIONS

These investigations suggest that hydrous pyrolysis of asphaltenes separated from either natural bitumens or crude oils can be used to improve understanding of a number of geochemical processes, such as the following.

(i) Asphaltenes from biodegraded crudes or bitumens can be used to provide indications of the likely composition of the non-degraded material. The composition of the hydrous pyrolysis products may also reveal material derived from micro-organisms.

(ii) If asphaltenes can be regarded as soluble fractions of kerogens, then they can be used as a substrate for studying the catalytic effects of various mineral matrices. The present studies showed very large differences in the effect of clays and carbonates on hydrous pyrolysis products.

(iii) The relative ease of separating asphaltenes in comparison with kerogens provides a simpler means of simulating the process of hydrocarbon generation from kerogen-like material.

(iv) Biomarker compositions of hydrous pyrolysis products may provide another means of assessing crude oil maturity, as well as improving oil correlation studies.

A significant problem in the use of asphaltenes in hydrous pyrolysis studies is the difficulty of distinguishing occluded material from 'true' pyrolytic products, and further work is necessary on this problem.

REFERENCES

Abraham, H. 1945. *Asphalts and Allied Substances*. Van Nostrand, New York.

Grantham, P. 1974. *Ph.D. Thesis*, University of Newcastle.

Huc, A. Y., Behar, F. and Roussle, J. -C. 1984. Geochemical variety of asphaltenes from crude oils. In: *Symp. on Characterization of Heavy Crude Oils and Petroleum Residues*, Lyon, 25–27 June 1984. Technip, pp. 99–108.

Ivanov, V. V. and Shcherban, O. V. 1983. Mineral matrix influence on the dynamics and products of organic matter catagenetic transformation. *Org. Geochem.*, 4(3/4), 185–194.

MacKenzie, A. S. and Maxwell, J. R. 1981. Assessment of thermal maturation in sedimentary rocks by molecular measurements. In: Brooks, J. (ed.), *Organic Maturation Studies and Fossil Fuel Exploration*. Academic Press, London, pp. 239–254.

Seifert, W. K. and Moldowan, J. M. 1978. Application of steranes, terpanes and monoaromatics to the maturity migration and source of crude oils. *Geochim. Cosmochim. Acta*, 43, 77–95.

Sieskind, O., Joly, G. and Albrecht, P. 1979. Simulation of the geochemical transformations of sterols: superacid effect of clay minerals. *Geochim. Cosmochim. Acta*, 43, 1675–1680.

Tissot, B. 1984. Geochimie des resines et asphaltenes. In: *Symp. on Characterization of Heavy Crude Oils and Petroleum Residues*, Lyon, 25–27 June 1984. Technip, pp. 3–18.

25 Aromatic hydrocarbon variations in North Sea Wells

P. B. Hall, L. Schou and M. Bjorøy
Continental Shelf Institute (IKU), Trondheim, Norway

An investigation was made into the use of aromatic hydrocarbons in routine geochemical analysis of North Sea wells.

Generally, in routine geochemical analysis only saturated hydrocarbons are analysed by gas chromatorgraphy. However, aromatic hydrocarbons can also supply useful information on maturity, and also on environment of deposition and organic input. Aromatic maturity ratios such as methyl phenanthrene index have already been employed in other areas. Certain aromatic hydrocarbons in some oils have been related to specific organic inputs; for example, cadalene from land plant sources. Observations are made on some of the variations in aromatic hydrocarbon distributions and the possible causes of those variations.

INTRODUCTION

Aromatic hydrocarbons have been identified in all recent and ancient sediments (Maxwell *et al.*, 1971; Anders *et al.*, 1973; Giger and Blumer, 1974; Blumer and Youngblood, 1975; White and Lee, 1980; Radke and Welte, 1981; Radke *et al.*, 1982*a*,*b*). However, unlike saturated hydrocarbons, aromatic hydrocarbon distribtuions were until recently only treated in general terms, and related to gross changes rather than individual compound variations. Early investigations of aromatic hydrocarbons in source rocks have concentrated mainly on the types of aromatic hydrocarbons in coals (e.g. Hofmann and Damm, 1922; Sakabe and Sassa, 1952; Danyuskerskaya, 1959; Streibl and Herout, 1969; Birkofer and Pauly, 1969). More recently, general trends in aromatic hydrocarbon distributions with maturity for both oil- and gas-prone source rocks have been assessed (Tissot *et al.*, 1971, 1974; Albrecht *et al.*, 1976). Specific aromatic ratios, particularly of mono-aromatized steranes, in both oils and oil-prone source rocks have also been shown to be maturity-sensitive (Seifert and Moldowan, 1978; Mackenzie *et al.*, 1981). Artificial maturation experiments on recent marine kerogens by Ishiwatari and Fukushima (1979) showed that the relative distribution of alkylphenanthrenes changes with increased maturity. Radke *et al.* (1981, 1982*a*,*b*) have shown that several aromatic hydrocarbon ratios can be used as maturity parameters.

Biosynthesis of polycyclic aromatic hydrocarbons has been shown to be unlikely (Hase and Hites, 1976). The aromatic hydrocarbons found in recent and ancient sediments are, in part, derived from non-aromatic biogenic precursors (Laflamme and Hites, 1978, Wakeham *et al.*, 1980). Some of the processes involved in the formation of the polycyclic aromatic hydrocarbons in recent sediments during early diagenesis have been elucidated (Blumer and Youngblood, 1975; Youngblood and Blumer, 1975; Greiner *et al.*, 1976; Laflamme and Hites, 1978, 1979; Wakeham *et al.*, 1979, 1980).

The suggested precursor compounds for aromatic hydrocarbons include unsaturated fatty acids (Breger, 1960; Erdman, 1961), carotenoids (Day and Erdman, 1963; Ikan *et al.*, 1975; Gallegos, 1981), polyhydroxyquinones (Blumer, 1965) and terpenoids (Mair, 1964; Albrecht *et al.*, 1976; Greiner *et al.*, 1976).

Polycyclic terpenoids and steroids produced biogenetically do not have methyl groups in positions which would correspond to the C-4 or C-5 positions of phenanthrene (Mair, 1964). Until recently the 4- and 5-methyl phenanthrenes had not been identified in petroleum. Recent studies using high resolution spectrofluorimetry (Ewald *et al.*, 1983) suggest that 4-methyl phenanthrene is present in minor amounts in some petroleums. However, the proposition that the phenanthrenes in petroleum are derived to some extent from polycyclic terpenoids and steroids is able to account for the general low concentrations of 4- and 5-alkyl derivatives. The relatively small content of compounds with an anthracene structure is an additional support for the claim that the terpenoids and related steroids (containing the phenanthrene structure) are major precursors (Mair, 1964; Greiner *et al.*,

Petroleum Geochemistry in Exploration of the Norwegian Shelf © Norwegian Petroleum Society (Graham & Trotman, 1985) pp. 293–301.

1976). One specific source for alkylated phenanthrenes may come from the terpenoids derived from constituents of higher plants, such as the resins of conifer species. Diterpenoid acids are major constituents of higher plant resins (Thomas, 1969), and several recent studies have indicated that aromatic hydrocarbons including retene and pimanthrene have been derived by diagenetic aromatisation of such acids (Skrigan, 1964; Stonecipher and Turner, 1970; Laflamme and Hites, 1978; Wakeham *et al.*, 1980). Radke *et al.* (1982*a*) have suggested that a substantial source for alkyl phenanthrenes is through direct methylation of phenanthrene during catagenesis.

Other tricyclic aromatics, such as alkylated fluorenes, have been identified in petroleum (Mair, 1964). Sources for such compounds would come from alkaloids or steroids (Mair, 1964) or at very high temperatures from phenanthrenes (Yu and Hites, 1981).

Another possible source of di- and tricyclic aromatic compounds may come through cyclization and dehydrogenation of unsaturated compounds such as certain fatty acids and carotenoids (Erdman, 1961; Ikan *et al.*, 1975; Cane, 1976; Gallegos, 1981). Alkylated naphthalenes are common constituents of coals (Birkofer and Pauly, 1969; Radke *et al.*, 1982b), petroleum (Carruthers and Douglas, 1955; Yew and Mair, 1966) and shales (Murphy, 1969; Anders *et al.*, 1973; Radke *et al.*, 1982a). It has been postulated that naphthalenes, e.g. eudalene, are dehydrogenation products of sesquiterpenoids (Murphy, 1969; Bendoraitis, 1974). Alternatively, alkylated naphthalenes could be the products of diagenetic aromatization of carotenoids, as mentioned above. Mair and Douglas (1965) have shown that the reaction of elemental sulphur with terpenoids and steroids produces alkylated naphthalenes, sulphur aromatics plus other aromatic hydrocarbons.

Polycyclic aromatic hydrocarbons with more than three rings which have been identified in recent sediments and coals include fluoranthene, pyrene, benzanthracene and perylene (Aizenshtat, 1973; Orr and Grady, 1967; Wakeham *et al.*, 1980). Non-alkylated polycyclic aromatic hydrocarbons such as fluoranthene, pyrene and chrysene may require longer exposure to higher temperatures for formation, and involve aromatization and dealkylation of partially hydrogenated aromatics (White and Lee, 1980).

Variations in aromatics in sediments have been mostly related to variations in source and maturity rather than in depositional environment. In type III kerogens (as defined using Rock–Eval hydrogen indices) early aromatization during deposition or organic matter in relatively acid, mildly reducing of oxidizing environments leads to the early formation and dominance of pure aromatic compounds. Thus Vandenbroucke *et al.* (1976) have observed that in the Douala shales with type III kerogens the aromatics consist dominantly of pure aromatic molecules. These are mainly naphthalene and phenanthrene types with short alkyl chains, which show little evolution with depth regarding different cyclic structures or carbon numbers. However, in type II kerogens, as found in the Toarcian shales of the Paris Basin, there is considerable evolution in pure aromatic compounds with depth due to rearrangement and breaking of alkyl chains in polycyclic naphthenic aromatics which are abundant in the immature samples.

The abundance of sulphur aromatics certainly appears to be related to environmental factors (White and Lee, 1980). Sulphur aromatics, such as benzothiophene and dibenzothiophenes found in coals, can be produced by the reaction of elemental sulphur or pyrite with hydrocarbons (Deroo and Hodgson, 1978). The abundance of sulphur aromatic compounds in marine type II kerogens, compared with type I and III kerogens, is probably related to environmental conditions during deposition. Type II kerogens generally form in anoxic marine waters with abundant available sulphur in the form of sulphates, and in which sulphur bacteria play an important role.

The main objective of this investigation has been to assess the usefulness of aromatic hydrocarbon ratios as maturity parameters in North Sea wells. These ratios are known from other studies to give useful information about maturity (Radke and Welte, 1981; Radke *et al.*, 1982*a*,*b*).

This paper deals firstly with the general changes in aromatic hydrocarbon profiles and secondly with variations in individual aromatic hydrocarbon ratios (MNR, MPI 1, etc.).

Nine wells were chosen for the study and basic organic geochemical data were obtained for all the wells to evaluate the source rock potential, maturity and possible migration effects.

RESULTS AND DISCUSSION

The samples used in this study for evaluation of aromatic hydrocarbons were mainly of Jurassic age from nine different wells in the North Sea, together with some outcrop samples to cover the whole maturity range and spread in kerogen types.

When dealing with cutting samples from wells it is extremely important to evaluate problems arising from caved material, turbodrilling, mud additives and migrated hydrocarbons. This will not be discussed in this paper, but reference will be made to these problems where they arise.

Variations in aromatics with kerogen type/depositional environment

There are three main groups of peaks seen in the aromatic hydrocarbon gas chromatograms of most samples (aromatic fraction obtained by extract fractionation by MPLC). These are the following.

Group 1. Naphthalene and alkyl naphthalenes.
Group 2. Phenanthrene and alkyl phenanthrenes.
Group 3. Aromatised steranes and triterpanes plus other polycyclic compounds.

Figure 1 is a typical example showing the three main groups, and tentative identifications for some of the main peaks in groups 1 and 2. Another common group of compounds are the organic sulphur compounds such as dibenzothiophene and methyldibenzothiophenes.

If only the groups consisting of fully aromatized components (i.e. 1 and 2) are considered, there are several useful maturity parameters which can be used. However, before discussion of variations within the groups there are many variations in the relative amounts of the three different groups which can be roughly related to a combination of maturity and type of kerogen.

(a) Group 1 compounds (naphthalene and alkylated naphthalenes) are absent or in low concentration in type IV kerogens and in highly mature (>2% R_0) kerogens of all types (see Fig. 2). The level of matur-

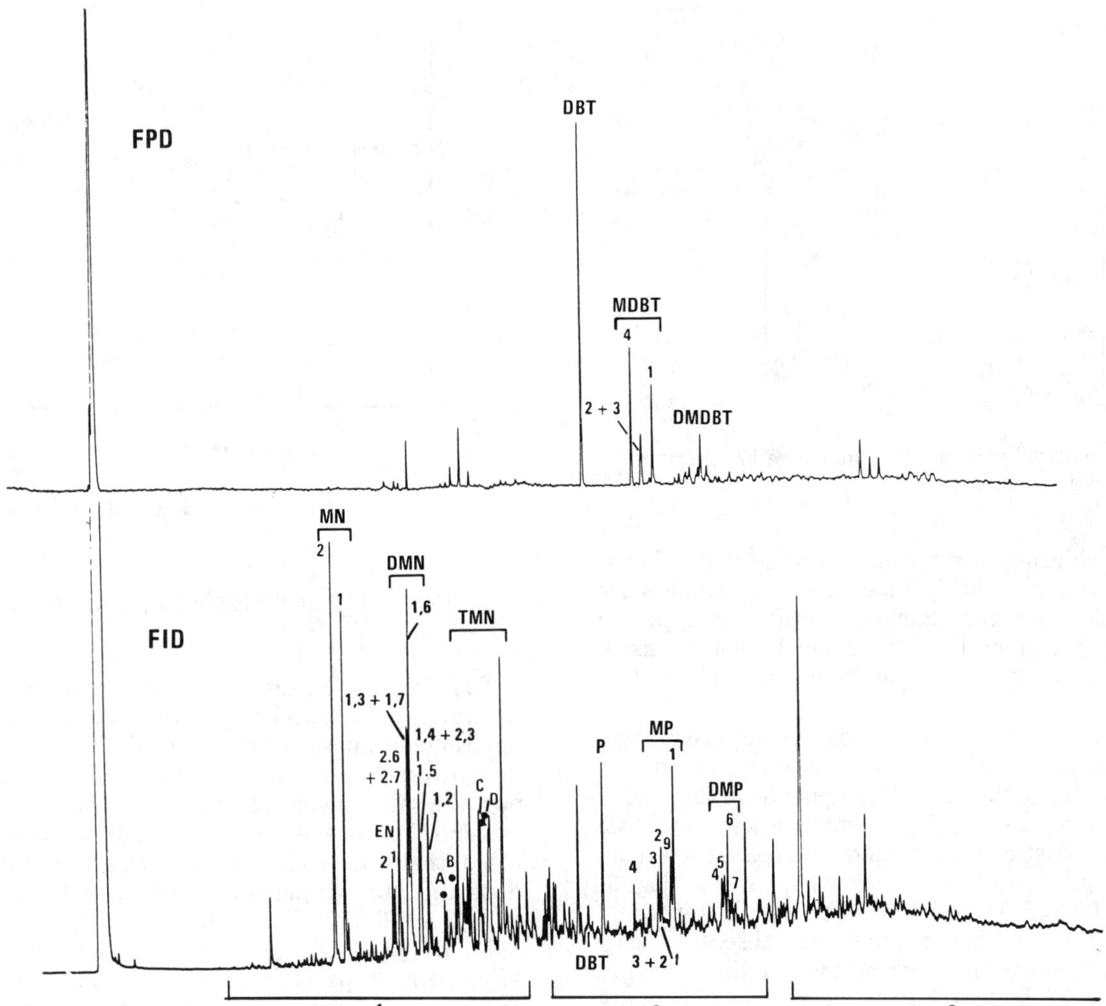

Fig. 1. identification of main aromatic components. Vitrinite reflectance $R_0 = 0.6\%$.

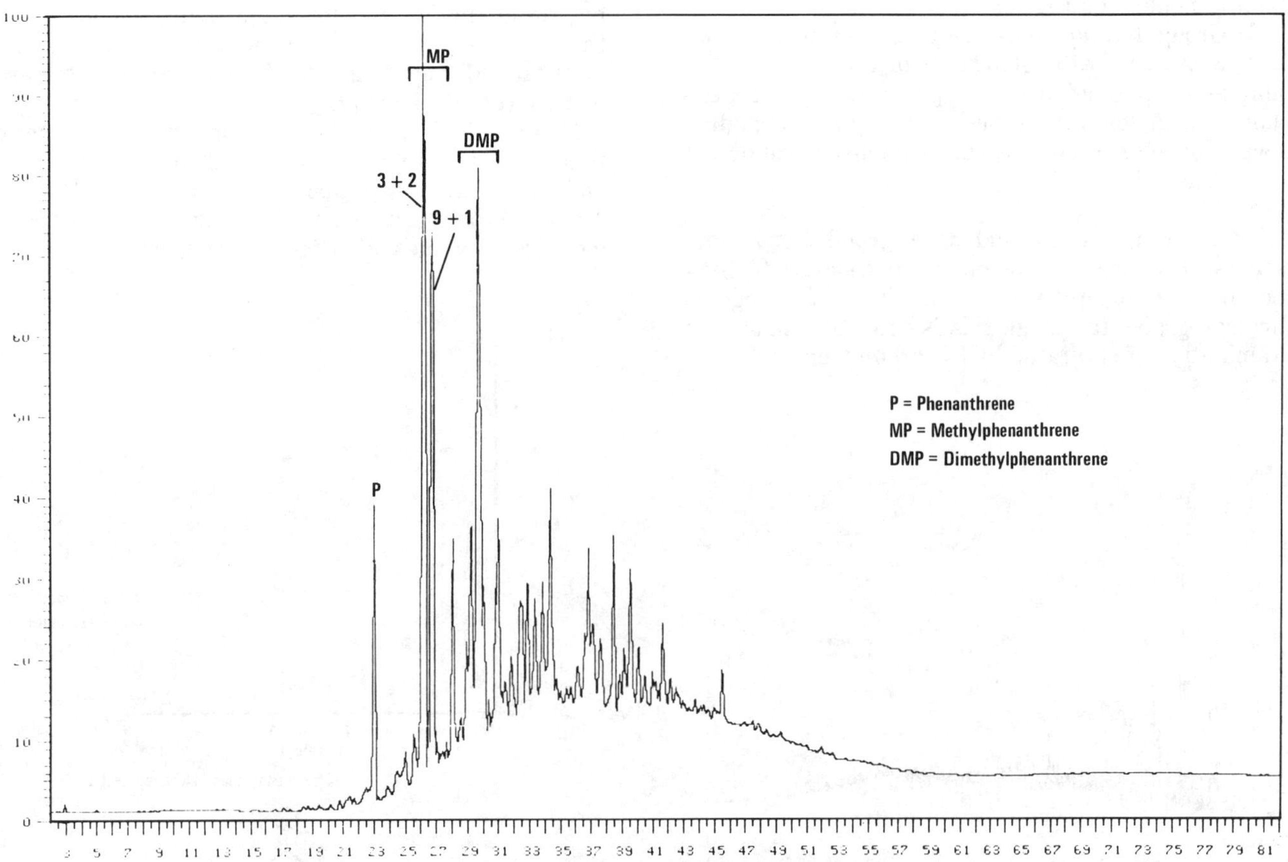

Fig. 2. Overmature type III kerogen.

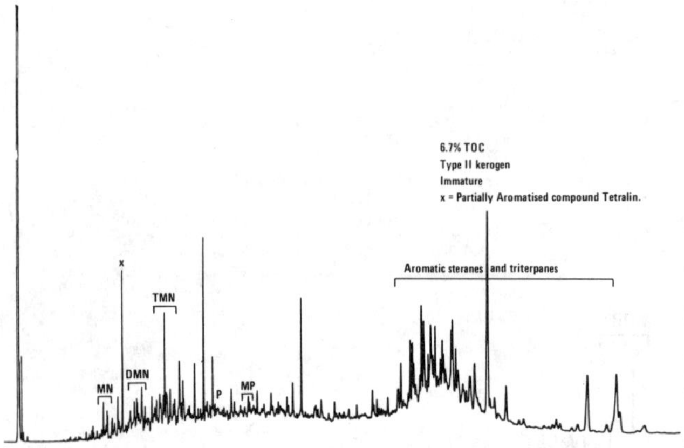

Fig. 3. Aromatic hydrocarbons, immature Kimmeridge Clay Formation shale.

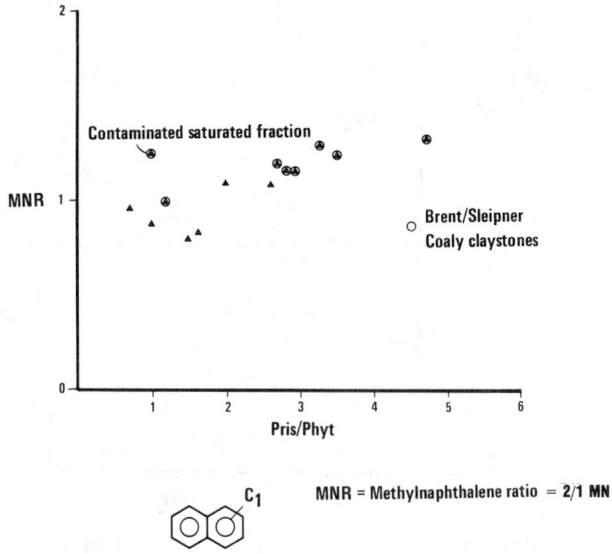

Fig. 5. Plot of MNR versus pristane/phytane.

ity at which group 1 compounds become absent in gas chromatograms probably depends on the richness and type of the kerogen. Samples with poor type III kerogens probably lose all group 1 compounds at a lower maturity level than type I and II kerogen samples.

(b) Group 3 compounds, which include aromatized steranes and triterpanes are absent or in low concentration relative to the group 2 compounds in the case of type IV kerogens and highly mature kerogens (see Fig. 2). Aromatic steranes and triterpanes are particularly prominent in immature samples (e.g. Fig. 3, less than 0.5% R_0) which have type I or II kerogen.

(c) Group 1 and 2 compounds are absent or very minor components in very immature sediments, i.e. less than 0.4% R_0 (see Fig. 3). Common components in such sediments are hydro-aromatic (partially aromatized) compounds. Groups 1 and 2 are mainly formed during thermal degradation of the kerogen in the sediments (see Fig. 4).

(d) Group 1 compounds are relatively more abundant (compared with compounds of groups 2 or 3) in samples containing mature type I and II kerogens than in samples with type III kerogens, including coals. However, there are some samples which do not follow this rule.

(e) Oils, and particularly condensates, are much richer in group 1 compounds than group 2 or 3 compounds. The presence of migrated condensate hydrocarbons in immature shales can give misleading gas chromatograms (although this is probably clearer on examination of saturated hydrocarbon traces).

To indicate the degree to which aromatic hydrocarbon ratios vary with kerogen type and/or depositional environment, plots have been made for certain aromatic hydrocarbon maturity ratios versus the pristane/phytane ratio calculated from the saturated hydrocarbon fractions of the same sample. It is already recognized that pristane/phytane ratios are affected by depositional environment as well as maturity (Brooks *et al.*, 1969; Powell and McKirdy, 1973; Lijmbach, 1975; Didyk *et al.*, 1978). If aromatic ratios are not affected by type differences, then any variations in pristane/phytane ratio will not reflect in the aromatic ratios and a straight-line plot should result. These plots are shown for one well, chosen because the Jurassic interval analysed appears to be least affected by contamination and migrated hydrocarbons. Also, there are only minor changes in maturity over the interval under consideration from moderately to early mature (i.e. 0.5–0.6% R_0).

Figures 5–7 show the variation in methyl naphthalene ratio (MNR), ethyl naphthalene ratio (ENR) and dimethyl naphthalene ratio (DMNR)—see Table 1 for actual ratios used. These show a noticeable increase in values with increasing pristane/phytane,

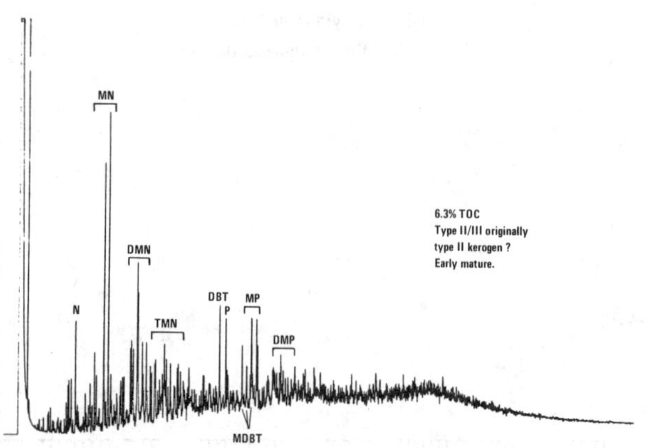

Fig. 4. Aromatic hydrocarbons, early mature Kimmeridge Clay Formation shale.

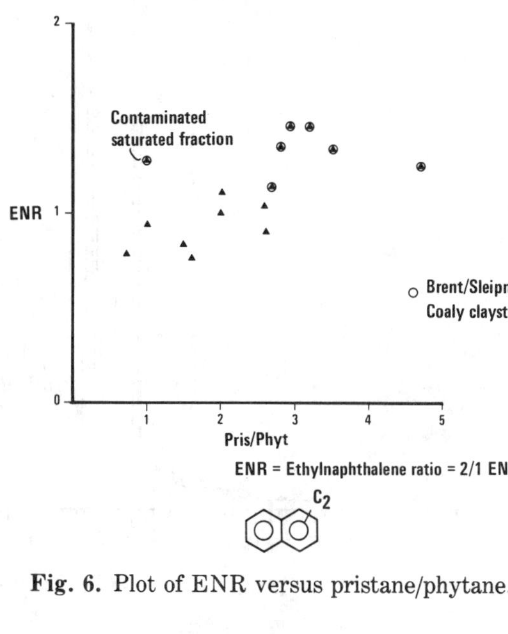

Fig. 6. Plot of ENR versus pristane/phytane.

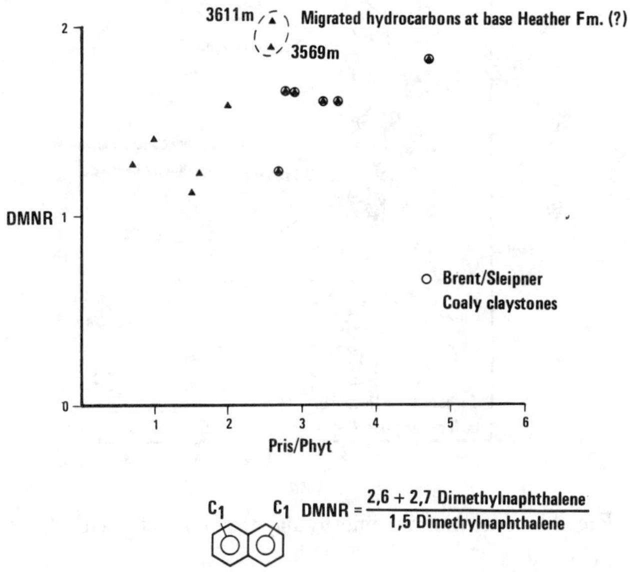

Fig. 7. Plot of DMNR versus pristane/phytane.

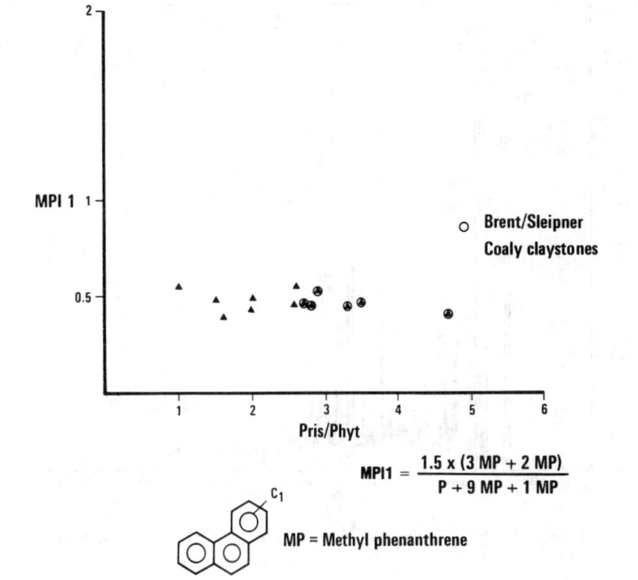

Fig. 8. Plot of MPI 1 versus pristane/phytane.

particularly when a comparison is made between the non-marine coaly shale and marine shales. This suggests that there are slight differences in these ratios between the mainly non-marine Brent coaly claystones and the marine Kimmeridge claystones used in this study.

Figures 8 and 9 show the variation in MPI 1 and 2/1 MP versus pristane/phytane. In both wells there is very little change in these parameters, regardless of organic matter type/depositional environment (particularly in the MPI 1 values). The only problem with MPI 1 is that at low maturities there appears to be interference from abundant sulphur aromatics present in extracts from marine shales with type II kerogens.

A plot of the variation in the sulphur aromatic ratio 4/1 MDBT versus pristane/phytane was also made (Fig. 10). This indicates a fairly good correlation be-

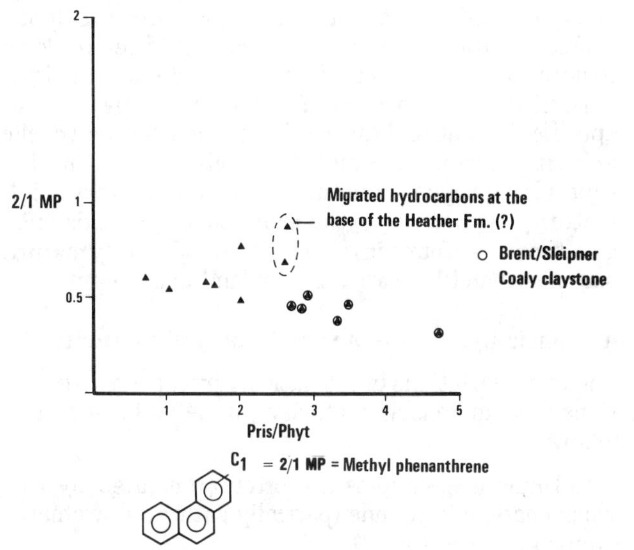

Fig. 9. Plot of 2/1 MP versus pristane/phytane.

TABLE 1. List of maturity ratios, indicating those used in this paper

Ratio		Definition
*MNR	=	$\frac{\text{2-methylnaphthalene}}{\text{1-methylnaphthalene}}$
*BPHR	=	$\frac{\text{biphenyl}}{\text{1,6 dimethyl naphthalene}}$
*ENR	=	$\frac{\text{2-ethyl naphthalene}}{\text{1-ethyl naphthalene}}$
*DMNR	=	$\frac{\text{2,6 + 2,7-dimethyl naphthalene}}{\text{1,5-dimethyl naphthalene}}$
$\frac{\text{4MDBT}}{\text{DBT}}$	=	$\frac{\text{4-methyl dibenzothiophene}}{\text{dibenzothiophene}}$
$\frac{\text{*4MDBT}}{\text{1MDBT}}$	=	$\frac{\text{4-Methyl dibenzothiophene}}{\text{1-methyl dibenzothiophene}}$
$\frac{\text{3/2MDBT}}{\text{1MDBT}}$	=	$\frac{\text{3 + 2-methyl dibenzothiophene}}{\text{1-methyl dibenzothiophene}}$
*2/1 MP	=	$\frac{\text{2-methyl phenanthrene}}{\text{1-methyl phenanthrene}}$
*MPI 1	=	$\frac{1.5 \times (3MP + 2MP)}{P + 9MP + 1MP}$ where P = phenanthrene and MP = methylphenanthrene

*Ratios used.

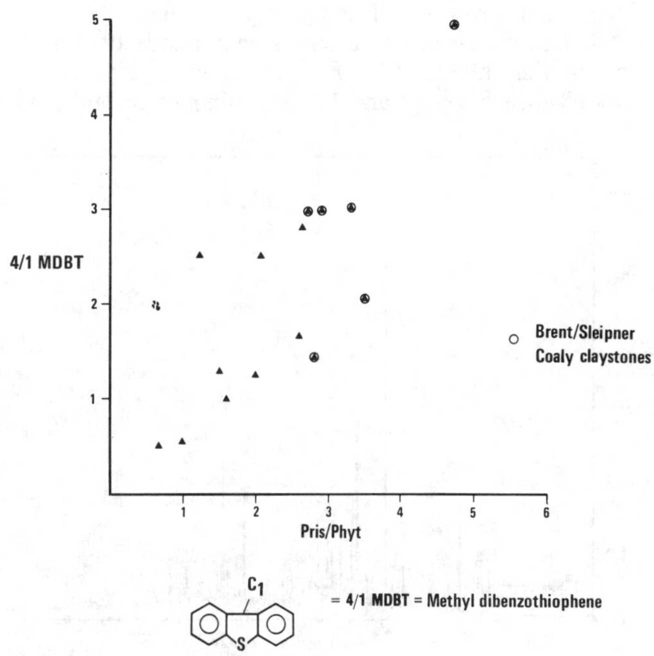

Fig. 10. Plot of 4/1 MDBT versus pristane/phytane.

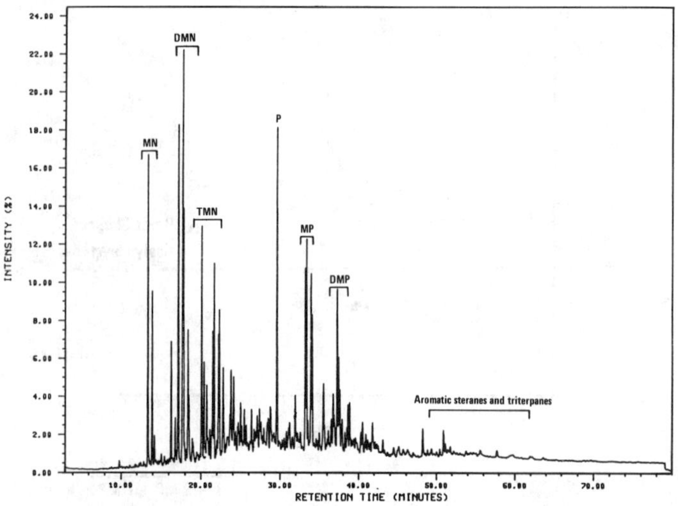

Fig. 11. Post-mature Kimmeridge Clay Formation.

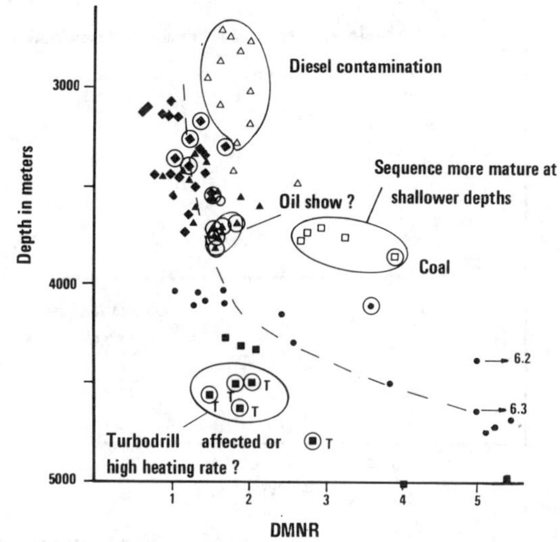

Fig. 13. Variation in dimethylnaphthalene ratio with depth.

tween the two ratios, and suggests that the 4/1 MDBT ratio is affected by variations in organic matter and/or depositional environment. However, for Kimmeridge Clay Formation samples this and other sulphur aromatic ratios show a marked change after peak oil generation, which is a useful rough guide to maturity.

Individual compounds are known to be source-specific. Pyrene and fluoranthene and other polycyclic aromatics are most common in coals and non-marine type III kerogens and in sediments of very high maturity. Other polycyclic aromatic compounds such as retene were found in immature to moderately mature sediments and indicate a major land plant input.

Aromatic hydrocarbon variations and maturity

The main variations in aromatic hydrocarbon distributions with increasing maturity can be summarized as follows.

1. Immature samples are often dominated by hydroaromatic compounds (partially saturated aromatic compounds, see Fig. 3).
2. The characteristic aromatic hydrocarbon gas chromatogram is seen in moderately mature to oil window mature samples (e.g. see Fig. 4). Particularly prominent are group 1 and group 2 compounds.
3. Disappearance of group 1 compounds begins at maturities above 1.3% R_0 and particularly above 2% R_0 (see Figs. 11 and 12). Prominence of biphenyl (BPH) and methylbiphenyl (MBPH) is also seen at this high maturity level.

A list of the various maturity ratios that were applied is given in Table 1. The various parameters will be discussed separately, and plots of some of the parameters versus depth have been made for the data in all the wells in the study.

Methyl naphthalenes. Two ratios were calculated from the distribution of the methyl naphthalenes, the methyl naphthalene ratio (MNR) and dimethylnaphthalene ratio (DMNR) (see Table 1 for actual compounds used in determination of ratios). They are both fairly simple to calculate, but are susceptible to inaccuracies due to work-up procedures, and are affected by hydrocarbon contaminants and migrated hydrocarbons. They are therefore only rough maturity parameters. The MNR changes from less than 1 in immature to early mature samples ($<0.7\%\ R_0$) to greater than 1, increasing to 2 from 0.7 to 1.3% R_0. The DMNR changes from values between 1 and 1.5 in immature to early mature samples to 2+ in mature samples. A depth plot of the DMNR shown in Fig. 13 shows a large difference between coal and claystones over a narrow depth interval within one well. This probably indicates that the type of kerogen affects these maturity ratios.

Biphenyl. The ratio of biphenyl (BPH) and one of the dimethyl naphthalenes (1,6 DMN) was seen to

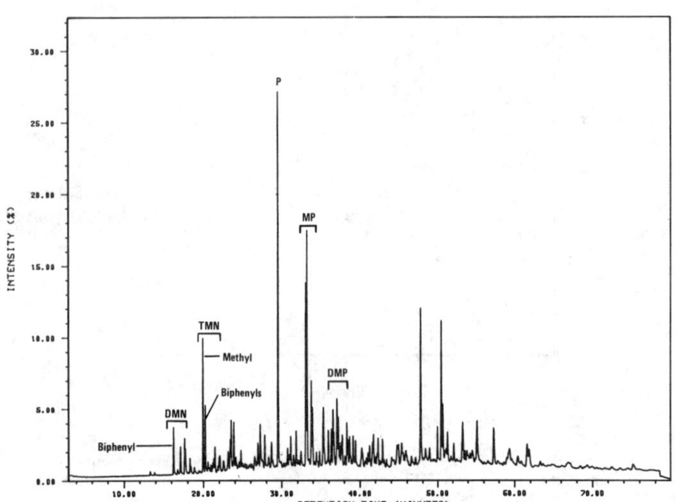

Fig. 12. Dry gas zone, Kimmeridge Clay Formation.

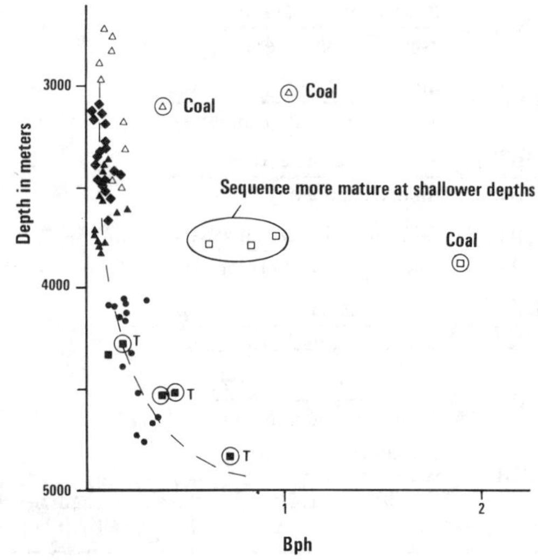

Fig. 14. Variation in biphenyl ratio with depth.

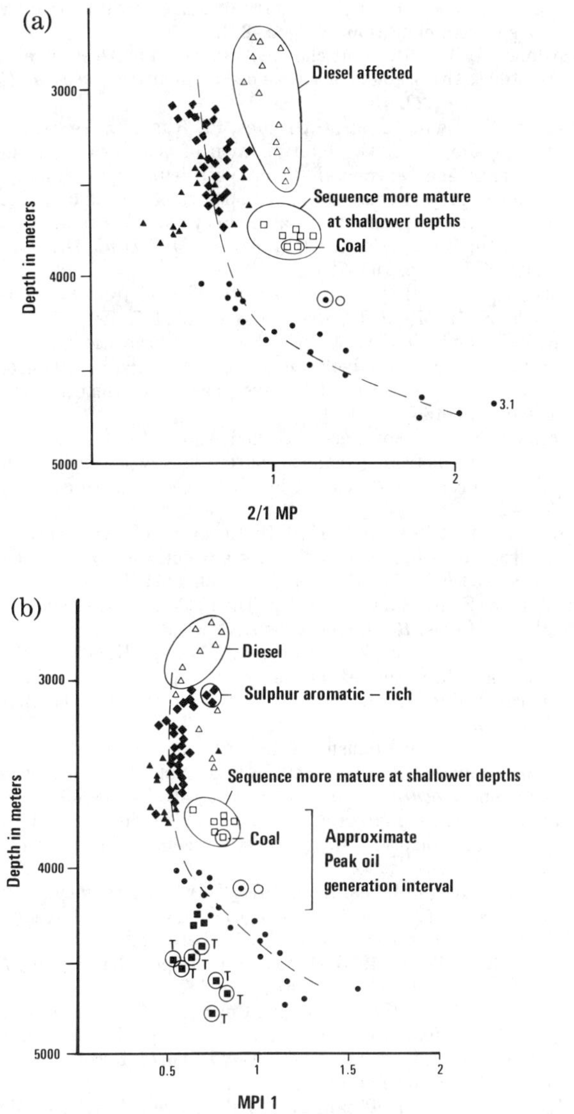

Fig. 15. Variation in methylphenanthrene ratio with depth. (a) 2/1 Methylphenanthrene; (b) MPI 1.

change with maturity. However, the difference in the ratio for coal and claystone samples from the same depth interval suggests that type of organic matter plays an important role in this context. Values in claystones increase from <0.1 in immature to moderately mature samples to 0.3 at about 1% R_0, and rapidly increase to values greater than 1 above 1% R_0 (see Fig. 14).

Methyl phenanthrenes. Two ratios were determined from the distribution of phenanthrenes, the 2/1 methylphenanthrene, and the MPI 1. The general trend observed in most samples is for a decrease in phenanthrene and the 9- and 1-methyl phenanthrenes with increasing maturity relative to the 3- and 2-methyl isomers (at least to about 1.3% R_0) (see Figs. 15(a) and (b)).

If the important process in the formation of methyl phenanthrenes is by methylation of phenanthrene, then in this instance the 1-methyl phenanthrene is preferred. Radke *et al.* (1982*b*) indicate that intense generation of phenanthrene occurs before methyl phenanthrene generation and during peak oil generation. Therefore migration of phenanthrene before methyl phenanthrene generation could account for phenanthrene depletion relative to the methyl phenanthrenes.

The methyl phenanthrene ratios are easy to measure often without requiring further group separation, although sulphur aromatics make measurement difficult in moderately to early mature samples. These ratios suffer less from evaporation losses than do the naphthalene ratios, but are affected by contaminants such as diesel as are the other ratios.

Sulphur aromatics. The sulphur aromatic ratios are easy to measure (if the chromatograph is equipped with a sulphur-sensitive detector, e.g. flame photometer), but they are affected by type of kerogen and environment. They show a general change in marine type II kerogens from values less than 2 in moderately-early mature samples to values greater than 10 past peak oil generation, i.e. >0.8% R_0. The ratios which we used are the 4/1 methyldibenzothiophene and 2 + 3/1 methyldibenzothiophene.

It is apparent that many of the ratios discussed change radically in value only after peak oil generation. This is, perhaps, a feature similar to that seen on pyrolysis of kerogens and analysis of biomarker compounds produced (Seifert and Moldowan, 1978). This has shown that whilst 'free' (i.e. not kerogen-bound) hydrocarbons are able to isomerise with increasing maturity, the kerogen-bound compounds are not. Similarly, for aromatic compounds isomerization may occur in the soluble bitumen and when released from the kerogen. When most of the hydrocarbons have been generated, then isomerization processes dominate.

There appear to be variations in the trend for ratios such as the MPI 1 with vitrinite reflectance due to differences in heating rate. Figure 16 shows the trends for a Western Canada well (Radke *et al.*, 1982*a*), from the Bramsche Massif (Radke and Willsch, 1981), from the North Sea wells used in this study (excluding those samples considered to be contaminated and/or turbodrill affected) and artificially matured Kimmeridge type II samples also from this

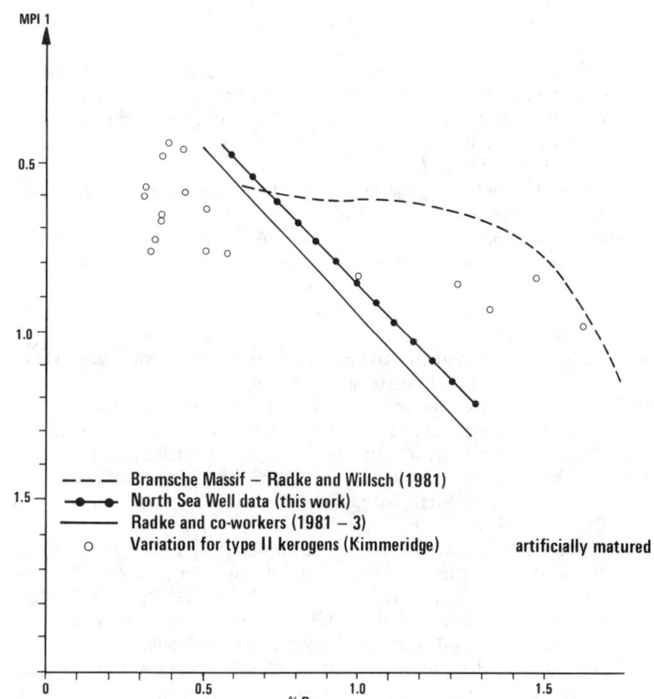

Fig. 16. Variation in MPI 1 index with vitrinite reflectance.

study. This suggests that in areas of high heating rates the vitrinite reflectance increases before MPI 1 values change, unlike in more normal thermal regimes. It is therefore possible that rapid heating due to turbo-drilling may also have the same effect as natural high heating rates, e.g. due to intrusives.

CONCLUSIONS

Both source input and heating rate affect aromatic hydrocarbon distributions. The naphthalene ratios MNR and ENR are useful, although changes from the true values can occur due to work-up procedures, presence of migrated light hydrocarbons and contamination. The most useful ratios employed appear to be DMNR and the methyl phenanthrene ratios, particularly the MPI 1 index. At higher maturities the biphenyl ratio (BPHR) appears to be useful, although there are not enough data to confirm the trend.

ACKNOWLEDGEMENTS

We thank the following oil companies for supplying the samples and for their permission to publish this paper: Mobil Exploration Norway Inc., Norsk Hydro a.s., Statoil, Unionoil Norge A/S, Elf Aquitaine Norge A/S, Phillips Petroleum Co. Norway, FINA Exploration Norway U/A and Texaco Exploration Norway A/S.

REFERENCES

Aizenshtat, Z. 1973. Perylene and its geochemical significance. *Geochim. Cosmochim. Acta*, **37**, 559–567.

Albrecht, P., Vandenbroucke, M. and Mandengue, M. 1976. Geochemical studies on the organic matter from the Douala Basin (Cameroon)—I. Evolution of extractable organic matter and the formation of petroleum. *Geochim. Cosmochim. Acta*, **40**, 791–799.

Anders, D. E., Doolittle, F. G. and Robison, E. W. 1973. Analysis of some aromatic hydrocarbons in a benzene-soluble bitumen from the Green River Shale. *Geochim. Cosmochim. Acta*, **37**, 1213–1228.

Bendoraitis, J. G. 1974. Hydrocarbon of biogenic origin in petroleum-aromatic triterpenes and bicyclic sesquiterpenes. In: Tissot, B. and Bienner, F. (eds.), *Advances in Organic Geochemistry, 1973*. Editions Technip, Paris, pp. 209–224.

Birkofer, L. and Pauly, W. 1969. Gas chromatographische und massenspektroskopische Untersuchung von Kohleextrakten. *Brennstoff-Chem.*, **50**, 376–382.

Blumer, M. 1965. Organic pigments: their long-term fate. *Science*, **149**, 722–726.

Blumer, M. and Youngblood, W. W. 1975. Polycyclic aromatic hydrocarbons in soils and recent sediments. *Science*, **188**, 53–55.

Breger, I. A. 1960. Diagenesis of metabolites and a discussion of the origin of petroleum hydrocarbons. *Geochim. Cosmochim. Acta*, **19**, 297–308.

Brooks, J. D., Gould, K. and Smith, J. W. 1969. Isoprenoid hydrocarbons in coal and petroleum. *Nature*, 222, 257.

Cane, R. F. 1976. The origin and formation of oil shale. In: Yen, T. F. and Chilingarian, G. V. (eds.), *Oil Shale. Developments in Petroleum Science*, **5**, Chapter 3, pp. 27–60. (Elsevier.)

Carruthers, W. and Douglas, A. G. 1955. The constituents of high boiling petroleum distillates. Part II. Trimethylnaphthalenes in a Trinidad oil. *J. Chem. Soc.*, 1847–1850.

Danyusherskaya, A. J. 1959. The chromatography of hydrocarbons found in the bitumen of the Sangarsk coal. *Nauchn Issled. Inst. Geol. Arktiki, Mini. Geol., Okhrany, Nedr. SSSR*, **98**, 120–129.

Day, W. C. and Erdman, J. G. 1963. Ionene, a thermal decomposition product of β-Carotene. *Science*, **141**, 808–809.

Deroo, J. and Hodgson, G. W. 1978. Geochemical origin of organic sulfur compounds; thiophene derivatives from ethylbenzene and sulfur. *Chem. Geol.*, **22**, 71–78.

Didyk, B. M., Simoneit, B. R., Brassel, S. C. and Eglinton, G. 1978. Organic geochemical parameters of palaeoenvironmental conditions of sedimentation. *Nature*, **272**, 216.

Erdman, J. G. 1961. Some chemical aspects of petroleum genesis as related to the problem of source bed recognition. *Geochim. Cosmochim. Acta*, **22**, 16–36.

Ewald, M., Lamotte, M., Garrigues, P., Rima, J., Veyres, A. and Laponyade, R. 1983. Determination of isomers of monomethyl phenanthrene extracted from petroleum by capillary gas chromatography coupled to mass spectrometry and by high resolution spectrofluorimetry in alkane crystals at a temperature of 15 K. In: Bjorøy, M. (ed.), *Advances in Organic Geochemistry 1981*. Wiley, pp. 705–709.

Gallegos, E. J. 1981. Alkyl benzenes derived from Carotenes in coals by GC/MS. *J. Chromatog. Sci.*, **19**, 177–182.

Giger, W. and Blumer, M. 1974. Polycyclic aromatic hydrocarbons in the environment. Isolation and characterization by chromatography, visible, ultra-violet and mass spectrometry. *Analyt. Chem.*, **46**, 1663–1671.

Greiner, A. Ch., Spyckerell, C. and Albrecht, P. 1976. Aromatic hydrocarbons from geological sources 1. New naturally occurring phenanthrene and chrysene derivatives. *Tetrahedron*, **32**, 257–260.

Hase, A. and Hites, R. A. 1976. On the origin of polycyclic aromatic hydrocarbons in recent sediments: biosynthesis by anaerobic bacteria. *Geochim. Cosmochim. Acta*, **40**, 1141–1143.

Hofmann, F. and Damm, P. 1922. The pyridine extraction of Upper Silesian Coals. *Brennstoff-Chem.*, **3**, 81–91.

Ikan, R., Aizenshtat, Z., Baedecker, M. J. and Kaplan, I. R. 1975. Thermal alteration experiments on organic matter in recent marine sediment—I. Pigments. *Geochim. Cosmochim. Acta*, **39**, 173–185.

Ishiwatari, R. and Fukushima, K. 1979. Generation of unsaturated and aromatic hydrocarbons by thermal alteration of young kerogen. *Geochim. Cosmochim. Acta*, **43**, 1343–1349.

Laflamme, R. E. and Hites, R. A. 1978. The global distribution of polycyclic aromatic hydrocarbons in recent sediments. *Geochim. Cosmochim. Acta*, **42**, 289–303.

Laflamme, R. and Hites, R. A. 1979. Tetra- and pentacyclic, naturally occurring, aromatic hydrocarbon in recent sediments. *Geochim. Cosmochim. Acta*, **43**, 1687–1691.

Lijmbach, G. W. M. 1975. On the origin of petroleum. *Proc. Ninth World Petroleum Cong., Spec.* **5**, p. 357.

Mackenzie, A. S., Maxwell, J. R., Coleman, M. L. and Deegan, C. E., 1983. Biological marker and isotope studies of North Sea crude oils and sediments. *Eleventh World Petroleum Congr.* Panel Discussion 1. London, 12 pp.

Mackenzie, A. S., Hoffman, C. F. and Maxwell, J. R. 1981. Molecular parameters of maturation in the Toarcian shales, Paris Basin, France—III. Changes in the aromatic steroid hydrocarbons. *Geochim. Cosmochim. Acta*, **45**, 1345–1355.

Mair, B. J., 1964. Terpenoids, fatty acids and alcohols as source materials for petroleum hydrocarbons. *Geochim. Cosmochim. Acta*, **28**, 1303–1321.

Mair, B. J. and Douglas, A. G. 1965. Sulfur: role in genesis of petroleum. *Science*, **147**, 499–501.

Maxwell, J. R., Pillinger, G. T. and Eglinton, G. 1971. Organic geochemistry. *Chem. Soc. Lond. Q. Rev.*, **25**(4), 571–628.

Murphy, M. T. J. 1969. Analytical methods in organic geochemistry. In: Eglinton, G. and Murphy, M. T. J. (eds.), *Organic Geochemistry*. Springer-Verlag.

Orr, W. L. and Grady, J. R. 1967. Perylene in basin sediments of southern California. *Geochim. Cosmochim. Acta*, **31**, 1201–1209.

Powell, T. G. and McKirdy, M. D. 1973. A relationship between ratio of pristane to phytane, crude oil composition and geological environment in Australia. *Nature Phys. Sci.*, **243**, 37.

Radke, M. and Welte, D. H. 1981. The methylphenanthrene index (MPI): a maturity parameter based on aromatic hydrocarbons. In: Bjorøy, M. *et al.* (eds.), *Advances in Organic Geochemistry 1983*. Wiley, Chichester, pp. 504–512.

Radke, M. and Willsch, H. 1981. Maturity parameters based on aromatic hydrocarbons from the Bramsche Massif area, NW-Germany. *KFA Report No. 502681*. (unpublished).

Radke, M., Welte, D. H. and Willsch, H. 1982*a*. Geochemical study on a well in the Western Canada Basin: relation of the aromatic distribution pattern to maturity of organic matter. *Geochim. Cosmochim. Acta*, **46**, 1–10.

Radke, M., Willsch, H., Leythaeuser, D. and Teichmüller, M. 1982*b*. Aromatic components of coal: relation of distribution pattern to rank. *Geochim. Cosmochim. Acta*, **46**, 1831–1848.

Sakabe, T. and Sassa, R. 1952. An aromatic hydrocarbon isolated from coal bitumen. *Bull. Chem. Soc. Jpn*, **25**, 353–355.

Seifert, W. K. and Moldowan, J. M. 1978. Application of Steranes, Terpanes and Monoaromatics to the maturation and source of crude oil. *Geochim. Cosmochim. Acta*, **42**, 77–95.

Skrigan, A. J. 1964. Preparation and utilization of fichtelite and retene. *Trudy. Vses. Nauchn. Techn. Soveshch. Gorki* **1963**, 108–115; *Chem. Abstr.*, **62**, 10664.

Stonecipher, W. D. and Turner, R. W. 1970. Rosin and rosin derivates. *Encycl. Polymer Sci. Technol.*, **12**, 139–161.

Streibl, M. and Herout, V. 1969. Terpenoids-especially oxygenated mono-, sesqui-, di-, and triterpenes. In: Eglinton, G. and Murphy, M. T. J. (eds.), *Organic Geochemistry*. Springer-Verlag, pp. 401–424.

Thomas, B. R. 1969. Kauri resins—modern and fossil. In: Eglinton, G. and Murphy, M. T. J. (eds.), Springer-Verlag, pp. 599–618.

Tissot, B., Califet-Debyser, Y., Deroo, G. and Oudin, J. L. 1971. Origin and evolution of hydrocarbons in early Toarcian shales, Paris Basin, France. *Bull. Am. Ass. Petroleum Geologists*, **55**, 2177–2193.

Tissot, B., Durand, B., Espitalie, J. and Combaz, A. 1974. Influence of the nature and diagenesis of organic matter in the formation of petroleum. *Bull. Am. Ass. Petroleum Geologists*, **58**, 499–506.

Vandenbroucke, M., Albrecht, P. and Durand, B. 1976. Geochemical studies on the organic matter from the Douala Basin (Cameroon) comparison with the Early Toarcian shales, Paris, Basin, France. *Geochim. Cosmochim. Acta*, **40**, 1241–1249.

Wakeham, S. G., Schaffner, C. and Giger, W. 1979. Perylene in sediments from the Nambian Shelf. *Geochim. Cosmochim. Acta*, **43**, 1141–1144.

Wakeham, S. G., Schaffner, C. and Giger, W. 1980. Polycyclic aromatic hydrocarbons in recent lake sediments. II. Compounds derived from biogenic precursors during early diagenesis. *Geochim. Cosmochim. Acta*, **44**, 415–430.

White, C. M. and Lee, M. L. 1980. Identification and geochemical significance of some aromatic components of coal. *Geochim. Cosmochim. Acta*, **44**, 1825–1832.

Yew, F. H. and Mair, B. J. 1966. Isolation and identification of C_{13} to C_{17} alkylnaphthalenes, alkylbiphenyls, and alkyldibenzofurans from the 275 ° to 305 °C dinuclear aromatic fraction of petroleum. *Analyt. Chem.*, **38**, 231–237.

Youngblood, W. W. and Blumer, M. 1975. Polycyclic aromatic hydrocarbons in the environment: homologous series in soils and recent marine sediments. *Geochim. Cosmochim. Acta*, **39**, 1303–1314.

Yu, Ming-Li and Hites, R. A. 1981. Identification of organic compounds on Diesel engine soot. *Analyt. Chem.* **53**, 95–954.

26 Quantification of steranes and triterpanes by gas chromatographic–mass spectrometric analysis

B. Dahl, G. C. Speers,* A. Steen*, N. Telnæs* and J. E. Johansen†*
*Norsk Hydro Research Centre, Bergen, Norway
†Chiron Laboratories, Trondheim, Norway

Quantitative determination of C_{27} $\alpha\alpha\alpha$ 20R cholestane has been investigated using a deuterated internal standard and combined gas chromatography–mass spectrometry. Low resolution Voltage Selected Ion Recording (VSIR) of a major fragment ion and parent ion have been compared. The new technique of Selected Metastable Ion Monitoring (SMIM) has also been investigated and compared with VSIR results.

The deuterated cholestane $3D_2$-C_{27} $\alpha\alpha\alpha$ 20R was also used as an internal standard in comparisons with a synthesized triterpane, 30,29,22 Trisnor 17α(H)-Hopane (T_m). Improvements should be made in quantitative GC–MS before applying statistical techniques for interpretation.

INTRODUCTION

The use of steranes and triterpanes in petroleum geochemistry is now well established (e.g. Mackenzie, 1984; Philp, 1982). By determining the composition, isomerization, rearrangement, aromatization and degradation of these compounds, it is possible to obtain information on their environment of deposition, maturation and migration history, biodegradation and stage of thermal decomposition.

Early methods for the separation and analysis of these compounds were complex and time-consuming (Speers and Whitehead, 1969) and tended to rely upon either gas chromatography (Pym and Whitehead, 1975) or mass spectrometry for determination of the various components. Improvements in equipment and techniques resulted in simplification of the analytical method and the widespread acceptance of the use of combined gas chromatography–mass spectrometry for the characterisation of these compounds (Kimble *et al*, 1974; Wardroper *et al*., 1977).

As part of a programme studying the systematic variations of biomarkers in a number of sedimentary sequences, it was necessary to examine existing methods for the quantitative determination of steranes and triterpanes. Relative concentrations are usually determined by integrating peak areas of mass fragmentograms (SIR traces), but absolute concentrations are only infrequently determined. Some indication of concentrations can be obtained using a combination of Mass Spectrometric Group Type Analysis (MSGTA) and mass fragmentograms (Seifert and Moldowan, 1978), but MSGTA is not considered to be a particularly accurate method. A much better procedure for estimating absolute concentrations of various biomarkers has recently been outlined by Rullkötter *et al*. (1984) using a commercially available alkylperhydroanthracene as an internal standard.

Use of an internal standard for quantification will compensate for variations in sample preparation techniques, e.g. losses during separation procedures, as well as for instrument response. Results are clearly dependent upon the similarity in behaviour between the internal standard and the compound, or compounds, being determined. By analogy with the technique of isotope dilution analysis, the best internal standard would be a stable isotope-labelled version of the compound being determined, followed by the use of SIR for analysis.

A small number of deuterium-labelled biomarkers have recently been produced by Chiron Laboratories in Trondheim, including cholestane, $C_{27}5\alpha(H),14\alpha(H),17\alpha(H)$ 20R ($C_{27}\alpha\alpha\alpha$ 20R) with two deuterium atoms in the 3-position (see Fig. 1). The mass spectra of C_{27} $\alpha\alpha\alpha$ 20R and $3D_2$-C_{27} $\alpha\alpha\alpha$ 20R are shown in Fig. 2.

As an initial step in establishing an analytical method, it was necessary to construct a calibration curve for the determination of cholestane, and to examine any factors influencing the results. An alkane fraction from the natural bitumen Wurtzilite was then used as an 'unknown' sample of cholestane as it contained a relatively high concentration of this compound.

Quantification was achieved by adding a fixed amount of the internal standard to the sample and

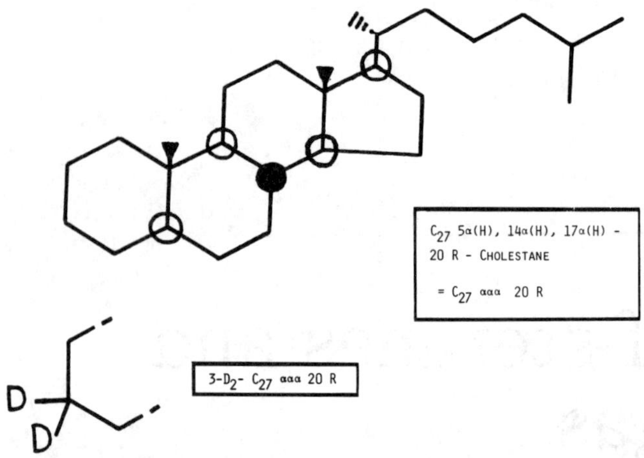

Fig. 1. Molecular structure of C_{27} ααα 20R and the deuterated internal standard, 3-D_2-C_{27} ααα 20R.

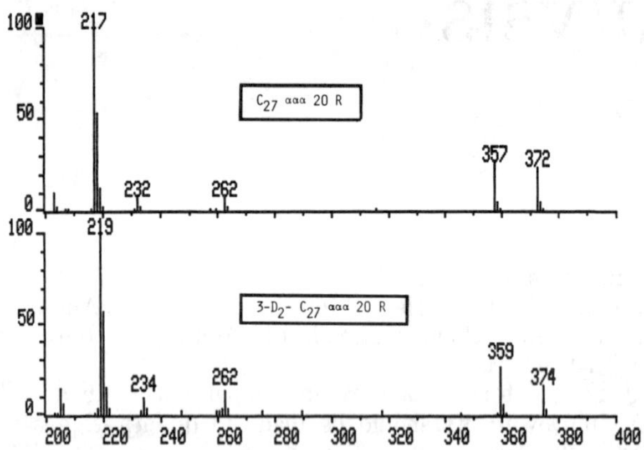

Fig. 2. Mass spectra of C_{27} ααα 20R and 3-D_2-C_{27} ααα 20R.

measuring the ratios of the corrected ion intensities which are proportional to the amounts of the components present. Both Voltage Selected Ion Recording (VSIR) and the new technique of Selected Metastable Ion Monitoring (SMIM), see Warburton and Zumberge (1983), were used in these studies.

EXPERIMENTAL

High purity samples of C_{27} ααα 20R and 3D_2-C_{27} ααα 20R were provided by Chiron Laboratories, Trondheim.

The sample used for quantification studies was an alkane concentrate obtained from a sample of Wurtzilite from the Uinta Basin. Following exhaustive extraction with dichloromethane and pentane deasphaltening, the sample was separated into broad compound types using a conventional HPLC Group Type separation.

The saturated alkane (SAC) fraction was analysed using a combination gas chromatography (HP 5790) double focussing magnetic mass spectrometer (VG 70E) coupled to a data handling system (VG 11/250). The solvent used was iso-octane, and gas chromatographic separation was carried out on a 25 m OV-101 bonded stationary phase capillary column. Inlet to the mass spectrometer was via a 70 eV 6kV Electron Impact Source.

Quantitative estimations were made using both low-resolution Voltage Selected Ion Recording (VSIR) and Selected Metastable Ion Monitoring (SMIM). The latter method is based upon metastable transitions occurring in the first field-free region of the instrument. To be sure of comparable results, all analyses were at the same sampling rate.

RESULTS AND DISCUSSION

As noted previously, a calibration curve for C_{227}ααα 20R was first prepared using varying concentrations of cholestane and a fixed amount of the deuterated cholestane internal standard (107.4 pg) and measuring the ratios of the appropriate ion intensities. Detector linearity was checked over the range 10–550 pg. Quantification and repeatability checks were carried out using a sequence of six samples of the SAC fraction from the Wurtzilite. The contribution of the isotopic signal to the internal standard was established by analysing cholestane standards in the absence of the deuterated internal standard. The respective contributions were found to be 16.9% of the *m/z* 217 fragment ion and 0.9% of the *m/z* 372 parent ion.

Figure 3 shows that the calibration curves for VSIR determinations of the *m/z* 217 fragment ion and the *m/z* 372 parent ion displayed good linearity over the concentration range investigated. In Fig. 3, the lines are fitted to computer-determined algorithms and the derived functions are displayed in the figure. Using the *m/z* 217 fragment ion, the cholestane concentration was found to be 97.3 pg/μl, and six consecutive runs gave a standard deviation of 2.7%. The corresponding values for the *m/z* 372 parent ion were, respectively, 104.7 pg/μl and 2.0% standard deviation.

Deviation of the two lines in Fig. 3 could indicate that fragmentation depends upon concentration. This may be due to the concentration range affecting source vacuum and ion beam focussing. Higher concentrations might also result in ion–molecule interactions similar to chemical ionisation effects. However, in the low and limited concentration ranges, the deviation is more likely the result of experimental error

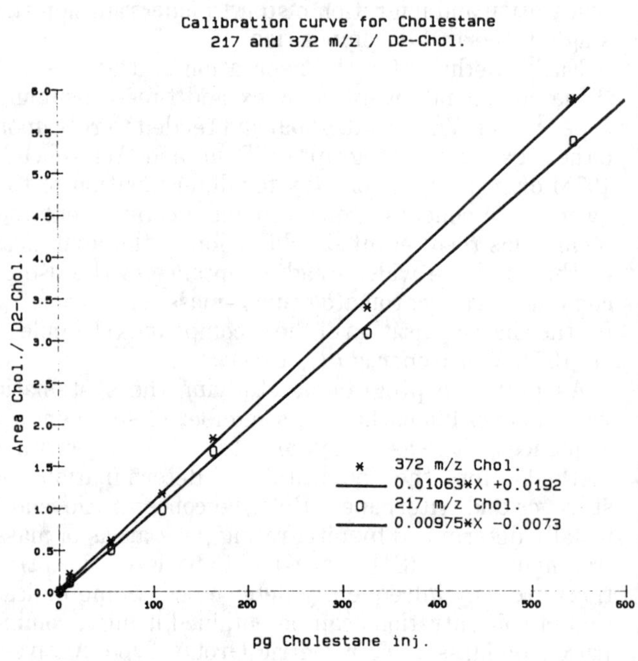

Fig. 3. Calibration curve for steranes. VSIR fragment ion, *m/z* 217, and parent ion, *m/z* 372.

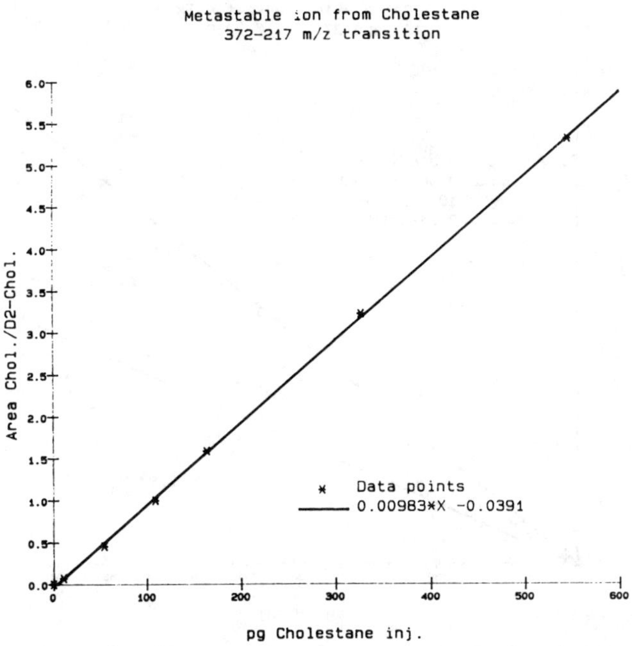

Fig. 4. Calibration curve for steranes. SMIM metastable transition ion m/z 372–217.

resulting from subtraction of the very large isotopic contribution to the m/z 219 internal standard peak.

Figure 4 shows a calibration curve derived from SMIM measurements of the transition m/z 372–217. The determined concentration was 103.3 pg/μl and the excellent linearity resulted in a standard deviation of only 1.0%. The metastable fragmentation should be very sensitive to instrumental variations and ion interactions, but the observed 1% standard deviation indicates exceptionally good repeatability. In SMIM measurements, it should also be noted that resolution is defined by the instrument conditions used for this type of work, including open slit mode, and not by deflection of a homogeneous ion beam through narrow slits. The method is, therefore, more open to false signals and, hence, the observed repeatability is quite outstanding. The sensitivity for SMIM determination of steranes is, however, only in the range of about 1% of the m/z 217 base peak from low-resolution VSIR.

Figure 5 shows a VSIR m/z 217 trace for the

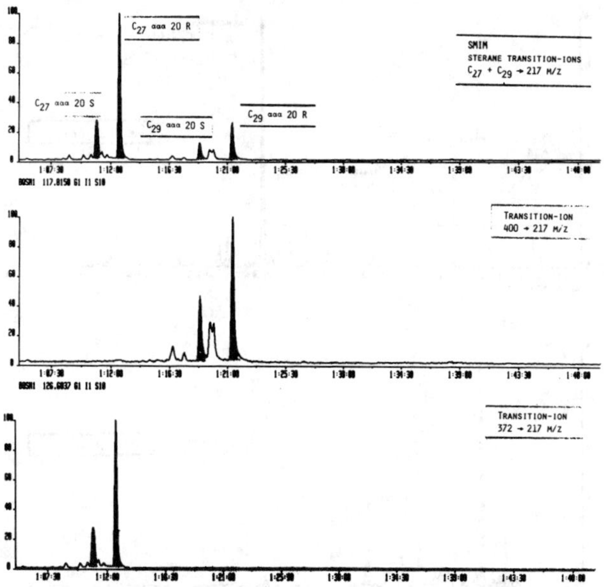

Fig. 6. SMIM transitionograms of C_{27}, C_{29} and C_{27} + C_{29}.

steranes present in the Wurtzilite SAC fraction. It would obviously be unwise to assume that the calibration curve for C_{27} ααα 20R could be used to obtain quantitative values for C_{27} ααα 20S, C_{29} ααα 20S and C_{29} ααα 20R, but determination of the response values for these various compounds showed them to have a better than 6% standard deviation.

Figure 6 shows the SMIM traces for C_{27} and C_{29} steranes, together with the sum of these 'transitionograms'. It is interesting to note that the responses relative to C_{27} ααα 20R are not the same for VSIR and SMIM. The probable explanation is that two different types of reactions are being observed. VSIR is detecting the ionization of molecules by electron impact, whereas SMIM is detecting the collapse of a metastable ion. In the results shown in Fig. 6, the C_{27} ααα 20S response has increased relative to VSIR, but for C_{29} ααα 20R and C_{29} ααα 20S the situation is reversed (cf. Fig. 5). The difference between VSIR and SMIM results is also illustrated in Fig. 7.

This quantification study also included examination of the response of the triterpane 30,29,22-Trisnor 17α(H)-Hopane (T_m), also synthesized by Chiron

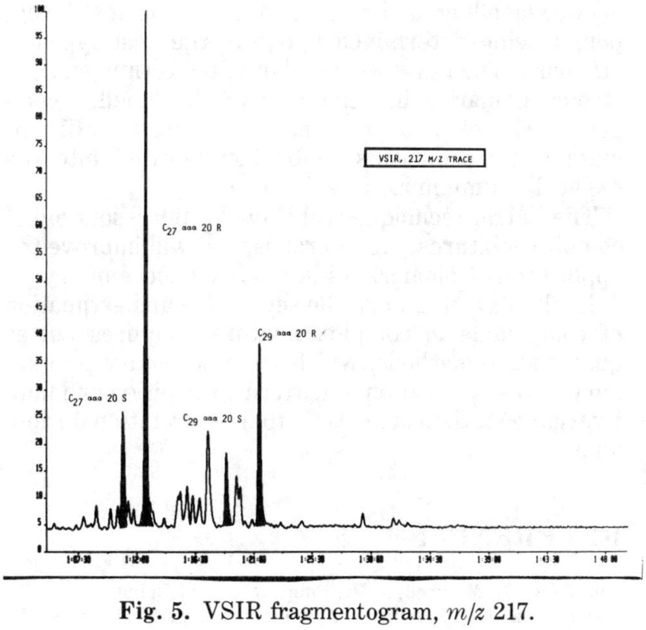

Fig. 5. VSIR fragmentogram, m/z 217.

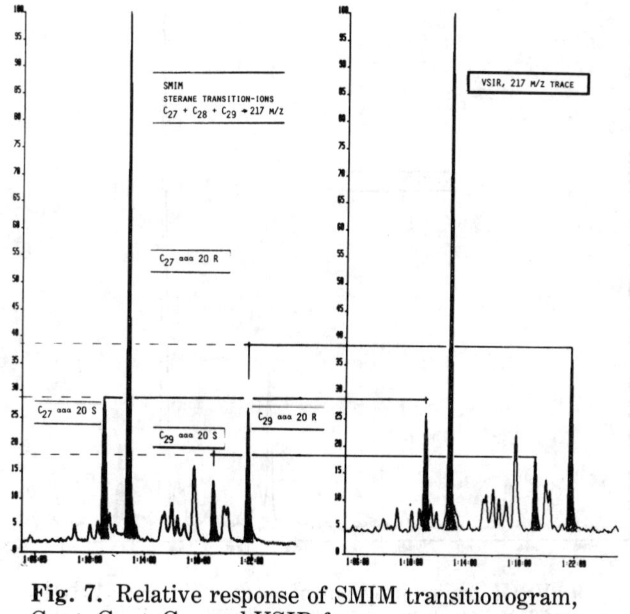

Fig. 7. Relative response of SMIM transitionogram, C_{27} + C_{28} + C_{29}, and VSIR fragmentogram, m/z 217.

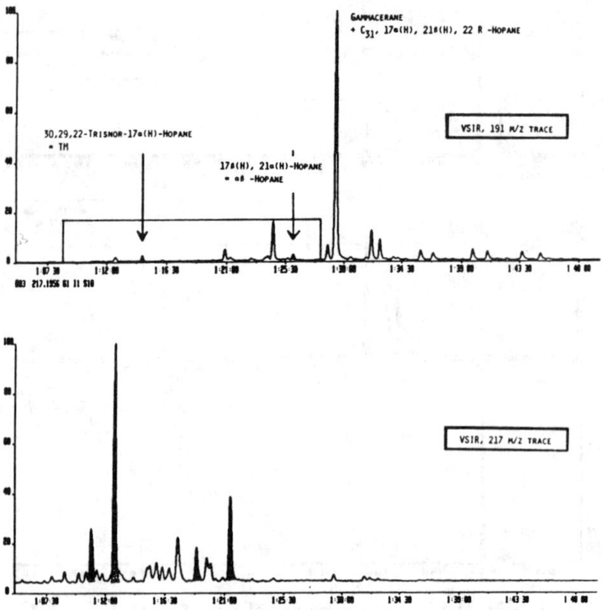

Fig. 8. VSIR fragmentograms of steranes, m/z 217, and triterpanes, m/z 191.

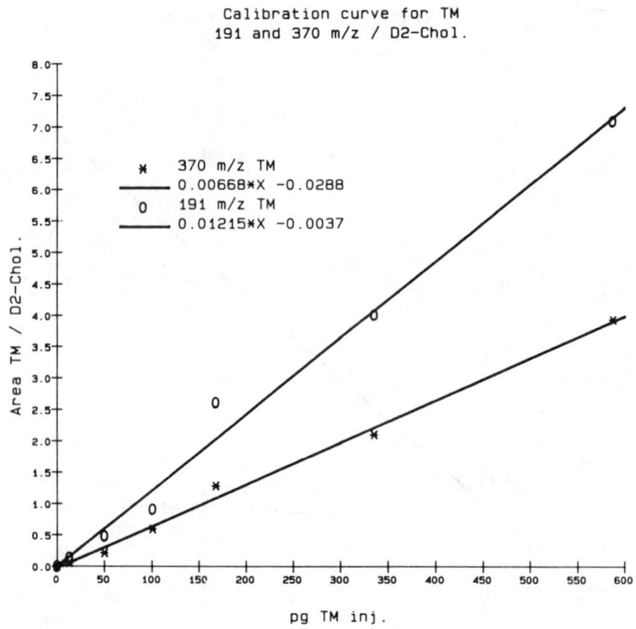

Fig. 10. Calibration curve for triterpane (T_m). VISR fragment ion, m/z 191, and parent ion, m/z 370.

Laboratories. Concentrations were varied over the range 13–590 pg injected. In the absence of a deuterated internal standard, $3D_2$-C_{27} $\alpha\alpha\alpha$ 20R was used. Figure 8 shows the m/z 217 and m/z 191 VSIR traces for the Wurtzilite saturate alkane fraction. The area outlined in Fig. 8 containing T_m is shown enlarged in Fig. 9.

The use of an unrelated internal standard is strictly incorrect. However, this situation resulted in the calibration curves illustrated in Fig. 10. The variation in line slopes for fragment and parent ions can readily be explained by the different fragmentation patterns for C_{27} $\alpha\alpha\alpha$ 20R and T_m. Divergence of the curves is more likely the result of an unsuitable internal standard, but a noteworthy observation from the results shown in Fig. 10 was a repeatability with less than 5% standard deviation.

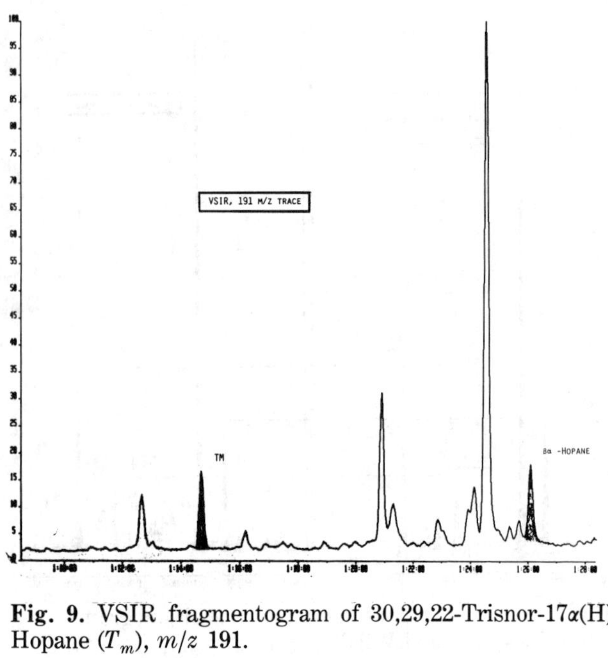

Fig. 9. VSIR fragmentogram of 30,29,22-Trisnor-17α(H)-Hopane (T_m), m/z 191.

CONCLUSIONS

The selectivity and sensitivity of GC–MS has resulted in its widespread acceptance as the main tool for the analysis of complex organic mixtures. In organic geochemistry it is the single, most useful technique for a wide range of analyses, especially in the biomarker field. However, use of VSIR of a major fragment ion, e.g. m/z 217 in sterane analyses, often results in a very complex distribution. Improved selectivity can be achieved by detecting parent ions, and the reduced intensity of the latter can be enhanced using chemical ionization techniques. SMIM has been shown to be a simple and repeatable method for resolution of complex sterane mixtures. Improved specificity in this type of analysis can be obtained using the more-sophisticated high resolution VSIR.

In any quantitative method it is important to use a suitable internal standard to compensate for instrumental and other variations, e.g. separation/recovery efficiencies. The MS-technique allows the use of stable isotope labelling, and a deuterated version of the compound being determined is one of the best types of standard. The low mass number of deuterium ensures strong similarity in behaviour of the labelled compound. However, as outlined in this paper, sufficient mass difference needs to be incorporated into the molecule to minimize interference.

The SMIM technique will allow better resolution of complex mixtures, e.g. steranes, and will improve the application of biomarkers in organic geochemistry.

In all MS techniques, the separation and estimation of compounds in complex mixtures requires better quantitative methods, which are a necessary preliminary to the application of pattern recognition and multivariable statistical analysis to interpretational problems.

REFERENCES

Kimble, B. J., Maxwell, J. R., Philp, R. P. and Eglinton, G. 1974. Identification of steranes and triterpanes in geolipid extracts by high-resolution gas chromatography and mass spectrometry. *Chem. Geol.*, **14**, 174–198.

Mackenzie, A. S. 1984. Applications of biological markers in petroleum geochemistry. In: Brooks, J. and Welte, D. H. (eds.) *Advances in Petroleum Geochemistry*, Vol. I. Academic Press, pp. 115–215.

Philp, R. P. 1982. Application of computerised gas chromatography/mass spectrometry to fossil fuel research. *Spectra*, 8(2/3), 6–31 (Finnigan MAT Publ.).

Pym, J. G. and Whitehead, E. V. 1975. Petroleum triterpane fingerprinting of crude oils. *Analy. Chem.*, **47**, 1617–1620.

Rullkötter, J., Mackenzie, A. S., Welte, D. H., Leythauser, D. and Radke, M. 1984. Quantitative gas chromatography–mass spectrometry analysis of geological samples. In: *Advances in Organic Geochemistry 1983* (in press).

Seifert, W. K. and Moldowan, J. M. 1978. Application of steranes, terpanes and monoaromatics to the maturation, migration and source of crude oils. *Geochim. Cosmochim. Acta*, **42**, 77–95.

Speers, G. C. and Whitehead, E. V. 1969. Crude petroleum. In: Eglinton, G. and Murphy, M. T. J. (eds.), *Organic Geochemistry*. Springer-Verlag, pp. 638–675.

Warburton, G. A. and Zumberge, J. E. 1983. Determination of petroleum sterane distributions by mass spectrometry with selective metastable ion monitoring. *Analyt. Chem.*, **55**, 123–126.

Wardroper, A. M. K., Brooks, P. W., Humberston, M. J. and Maxwell, J. R. 1977. Analyses of steranes and triterpanes in geolipid extracts by automatic classification of mass spectra. *Geochim. Cosmochim. Acta*, **41**, 499–510.

27 Characterization of sedimentary organic matter using nuclear magnetic resonance and pyrolysis techniques

Hauk Solli[a], *Liv Schou,* Jostein Krane, Tore Skjetne† and Paul Leplat‡*
*Continental Shelf Institute (IKU), P.O. Box 1883, 7001 Trondheim, Norway
†Department of Chemistry, AVH, University of Trondheim, Norway
‡Petrofina, Exploration Department, Rue de la Loi 33, B-1040 Brussels, Belgium

The aromatic hydrocarbon fractions from a series of sediment extracts have been isolated and subjected to proton nuclear magnetic resonance spectroscopy (^{1}H NMR). The NMR results were correlated with maturity data such as Rock–Eval T_{max} values, vitrinite reflectance and specific biomarker ratios. Evidence is presented to indicate that the percentage of aromatic protons in the aromatic fraction (PAP) is proportional to the level of maturity of the sediment. Some kerogens have been subjected to ^{13}C NMR (CP/MAS) and flash pyrolysis–gas chromatography (Py–GC) analysis to provide complementary structural information about the kerogens. The ^{13}C NMR and pyrolysis results are compared with maturity and other organic geochemical data.

INTRODUCTION

Optical techniques are well-established, and are used to characterize the type and evolutionary stage of organic matter in sediments. Vitrinite reflectance is one of the most widely used parameters for measuring source rock maturity (Dow, 1977; Alpern, 1980; Powell *et al.*, 1982). The maturation stage is also obtained from pyrolysis methods, for example the Rock–Eval T_{max} parameter, the temperature corresponding to the peak of kerogen pyrolysis (Espitalié *et al.*, 1977).

Recently chemical parameters based on alterations of biological markers (isomerization and aromatization of steranes/triterpanes) have been proposed for estimation of thermal maturation in the subsurface (Mackenzie *et al.*, 1980, 1981). Radke *et al.* (1982) have introduced the use of specific aromatic ratios as another chemical maturity parameter. Alexander *et al.* (1980) have used proton nuclear magnetic resonance (^{1}H NMR) to monitor the maturity changes of the extractable aromatic compounds and evaluated their potential as maturation indicators.

NMR has also been used in coal and kerogen studies by using Cross Polarization/Magic Angle Spinning (CP/MAS) ^{13}C NMR operating on the solid phase and providing structural information about the carbon distribution in the kerogen macromolecular matrix (Miknis, 1982; Solli *et al.*, 1984*a*). However, at present pyrolysis techniques are the most commonly used methods for characterization of kerogens and their applications in organic geochemistry and hydrocarbon exploration have recently been reviewed (Larter and Douglas, 1982; Larter, 1984; Horsfield, 1984).

The authors have recently reported on the analysis of kerogens in a homogeneous sedimentary column (Miocene, Angola) by using pyrolysis–gas chromatography (Py–GC), pyrolysis–mass spectrometry (Py–MS) and CP/MAS ^{13}C NMR. The type of organic material which has been deposited is expected to be uniform and the extractable aromatic hydrocarbons from this sedimentary column appear to be suitable for the study of the maturity-dependent changes in the NMR properties of aromatic compounds. A GC–MS study of Middle Triassic source rocks from Svalbard has been reported recently (Schou *et al.*, 1984) and the ^{1}H NMR analysis of the aromatic fractions of these samples is also included in this study.

In connection with ongoing Py–GC studies of kerogens and asphaltenes we have used ^{13}C NMR as a complementary technique in order to obtain additional structural information about these macromolecules and a few examples are included in this paper.

The purpose of the present paper is to report some preliminary results obtained by using NMR techniques in combination with analytical pyrolysis and gas chromatography methods to study maturity changes

[a]Present address: Department of Geology, University of Geology, University of Oslo, P.O. Box 1047, Blindern, Oslo 3, Norway.

Petroleum Geochemistry in Exploration of the Norwegian Shelf © Norwegian Petroleum Society (Graham & Trotman, 1985) pp. 309–317.

and to characterize sedimentary organic matter. However, the authors wish to emphasize that the results presented here were obtained on a few sets of samples only, and that refinement and evaluation of the methods is currently being investigated.

EXPERIMENTAL

Sample information and details of the experimental conditions have been described elsewhere (Solli *et al.*, 1984*a*, *b*; Schou *et al.*, 1984) and are also given in the text and figure legends.

RESULTS AND DISCUSSION

Representative gas chromatograms of the aromatic fractions from a homogeneous sedimentary column (Miocene Angola; Fu 2-series) are shown in Fig. 1. The chromatogram of the shallow immature sample (801 m; $R_0 = 0.35\%$) shows a high abundance of hydroaromatic compounds (e.g. tetrahydronaphthalenes and aromatized steranes and triterpanes) together with alkylated naphthalenes. The abundance of phenanthrenes is low. This is in contrast to the GC trace of the deeper-lying early mature sample (2483 m; $R_0 = 0.60\%$) which is dominated by alkylated naphthalenes and phenanthrenes. This difference

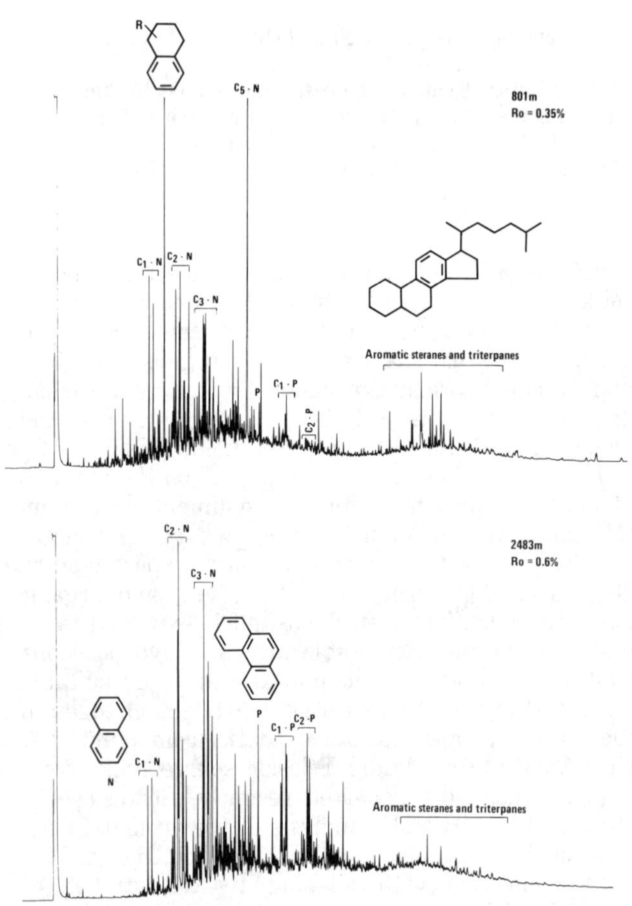

Fig. 1. Gas chromatograms of aromatic fractions (Miocene Angola; Fu 2 series). Sample depths and maturity are indicated. Conditions: Varian 3700; Carrier gas H_2 with inlet pressure 12 psi; 50 m × 0.25 mm I.D. fused silica capillary column coated with Se-54; split ratio 1 : 10. Peak identification: N = Naphthalene; P = phenanthrene; C_1, C_2, etc., are alkyl substituents of that carbon number.

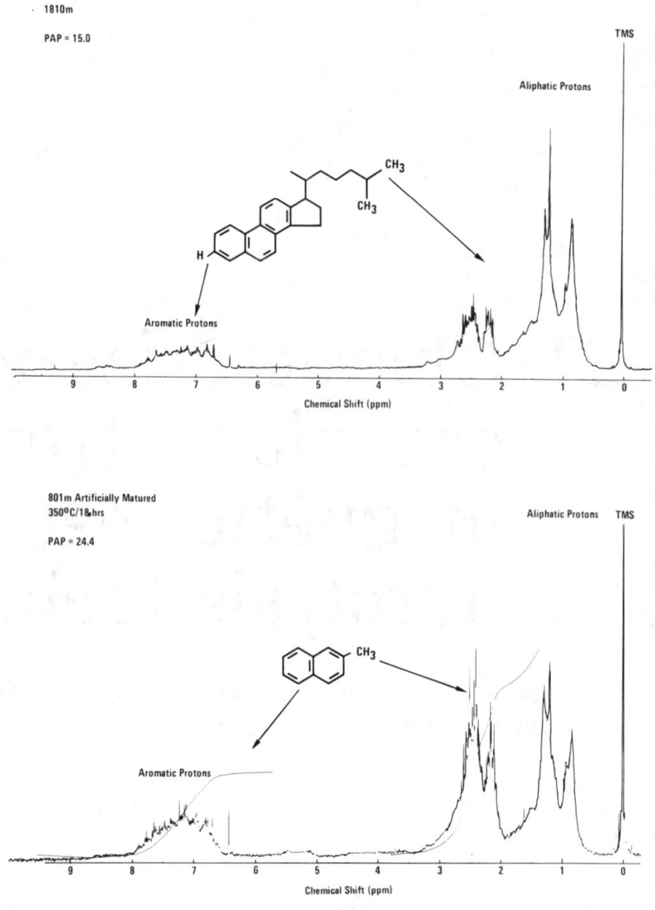

Fig. 2. 400 MHz ^{1}H-NMR spectra of typical aromatic fractions. Sample identification and percentage of aromatic protons, PAP, are indicated.

reflects the aromatisation of cycloalkyl aromatics (hydroaromatics) and the breaking of long alkyl chains which take place with increasing maturation. The chromatograms of samples from intermediate depths (not shown here) show a gradual change between the two extremes which are shown.

^{1}H NMR has been used to monitor the changing proton chemical environments caused by increased maturation (Alexander *et al.*, 1980). A proton NMR spectrum of a typical aromatic fraction shows well-separated signals from aromatic and aliphatic protons (see Fig. 2). The NMR spectrum of the aromatic fraction of an artificially matured sample (801 m heated in nitrogen for 18 h) is also shown in order to demonstrate significant differences. The integral due to aromatic protons (6.7–9.0 ppm) has increased relative to that of non-aromatic protons. Changes are also observed in the aliphatic region compared with the unheated sample; signals from methyl and methylene protons adjacent (α- and β-positions) to an aromatic nucleus (approximately 2–3 ppm) have increased relative to other aliphatic protons, reflecting a relative decrease in chain length of *n*-alkyl substituents with increasing maturation.

The integral due to aromatic protons is expressed as a percentage of the total integral from the spectrum, and the percentage of aromatic protons in the aromatic fraction (PAP) is shown in Tables 1 and 2 for the Fu 2 and Svalbard samples, respectively. The PAP-values for sediment extracts were plotted against T_{max} and vitrinite reflectance values (reflectance data were not obtained for the Svalbard samples) and their relationships are shown in Figs. 3 and 4. The regular

TABLE 1. PAP: percentage aromatic protons of total amount of protons, Fu 2 samples

Depth (m)	*R_0 (%)*	*T_{max} (°C)*	*PAP (%)*
801	0.35	418	11.2
1075	0.37	424	13.2
1401	0.37	426	14.1
1810	0.38	432	15.0
2483	0.60	436	20.6
Artificial maturation			
801 m/300 °C/40 h	—	—	19.7
801 m/350 °C/18 h	—	—	24.4

TABLE 2. Percentage aromatic protons, Svalbard samples

Location (number of samples)	*T_{max}, mean (S.D.)*	*PAP, mean (S.D.)*
I (5)	429 (1.7)	12.8 (1.5)
II (6)	436 (3.3)	13.2 (0.7)
III (9)	439 (2.7)	19.8 (2.6)
IV (12)	450 (3.0)	17.6 (5.4)
V (4)	>463 (—)	21.3 (4.4)

increase in PAP with increasing maturity is approximately linear and is in accordance with the results obtained by Alexander *et al.* (1980) and with the GC results described above. The difference in slope of the PAP versus T_{max} and vitrinite reflectance is probably caused by a combination of difference in organic source material and reaction kinetics (heating rates and catalytic effects) in the sediments. Standard PAP scales probably cannot be developed; however, a wider range of organic matter types (kerogens) in different geological situations needs to be investigated before a definite conclusion can be drawn regarding their affect on the slope of the curves. The same considerations are also relevant when biomarkers are used as a maturity parameter (Tissot, 1984).

The use of NMR to evaluate percentage of aromatic protons can be regarded as a bulk chemical aromatic maturity parameter. In order to compare the PAP-values with other chemical maturity parameters, some GC–MS biomarker ratios for the Fu 2 samples

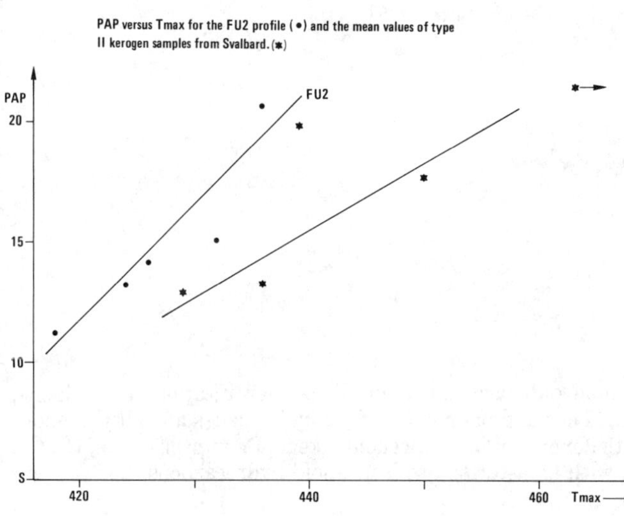

Fig. 3. Plot of PAP-values against T_{max}. Sample code is indicated.

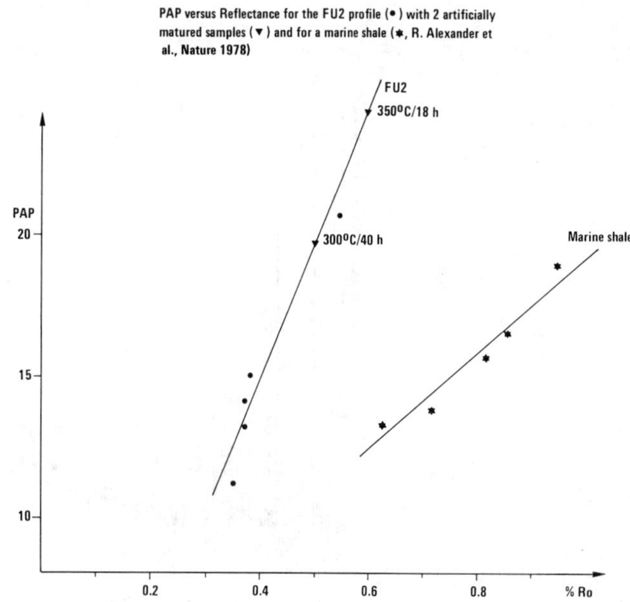

Fig. 4. Plot of PAP-values against vitrinite reflectance (R_0). Sample code is indicated. Marine shale data are taken from Alexander *et al.* (1978).

were plotted versus their PAP-values. These data are presented in Fig. 5, and show reasonably good linear relationships in the range below the individual equilibrium values of these parameters.

In the present work CP/MAS ^{13}C NMR and Py–GC have also been used, in order to study some oil shales and a kerogen from the North Sea. The present and other authors have shown that the Py–GC characteristics of kerogens change with increasing maturation, i.e. the abundance of aromatics and isoprenoids is reduced and the aliphatic *n*-alkene/*n*-alkane homology becomes more dominant with increasing sample depth (Solli *et al.*, 1984*a*; van Graas *et al.*, 1981). By using CP/MAS ^{13}C NMR it has also been observed that the relative carbon distribution in kerogens becomes more aromatic with increasing maturity (Solli *et al.*, 1984*b*).

Figure 6 shows the pyrograms of kerogen concentrates isolated from three samples of Kimmeridge Clay. Basic organic geochemical data and peak identification are noted on the figure and in the figure legends. Two of the kerogens (K7 and K5, Dorset, South England) are genetically similar (input of the same type of organic source material) with different maturity (R_0 = 0.35% and 0.45%, respectively).

The pyrogram of K7 shows an *n*-alkene/*n*-alkane homology ranging from C_7 to about C_{27}. The abun-

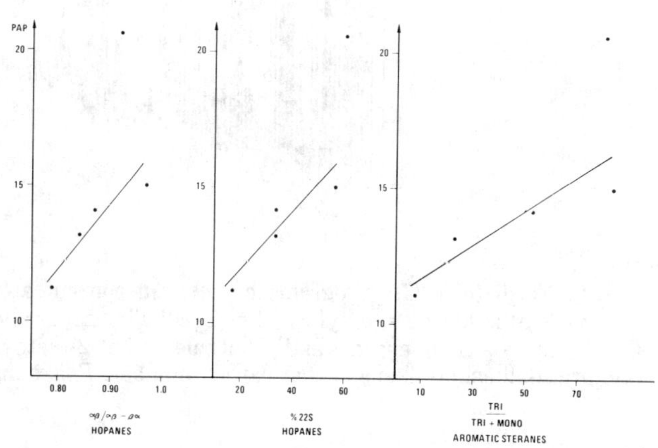

Fig. 5. Plot of PAP-values against various biomarker ratios.

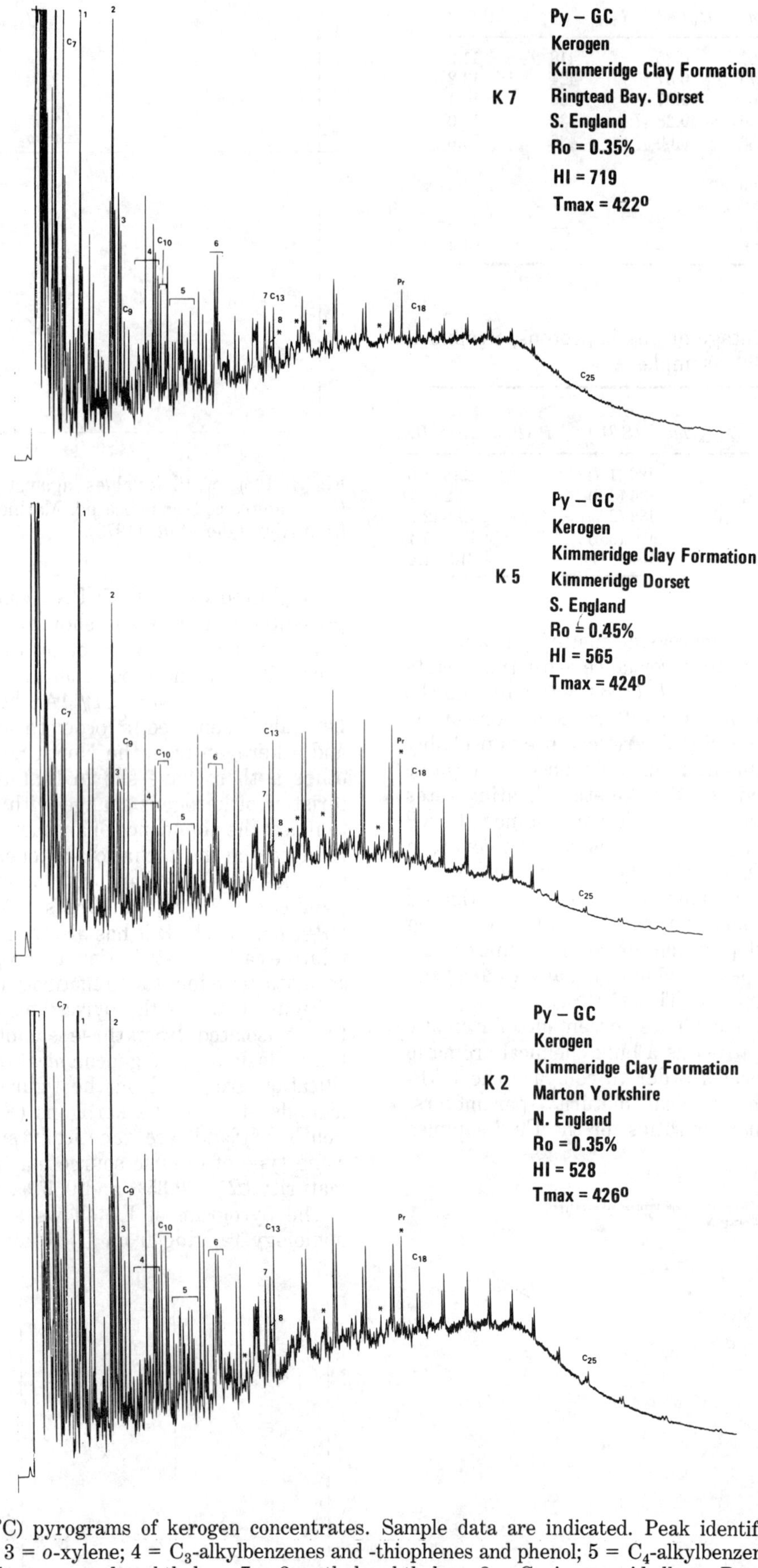

Fig. 6. Flash (600 °C) pyrograms of kerogen concentrates. Sample data are indicated. Peak identification: 1 = toluene; 2 = (*m* + *p*)-xylene; 3 = *o*-xylene; 4 = C_3-alkylbenzenes and -thiophenes and phenol; 5 = C_4-alkylbenzenes and alkylphenols; 6 = C_4- and C_5-alkylbenzenes and naphthalene; 7 = 2-methylnaphthalene; 8 = C_{14}-isoprenoid alkene; Pr = prist-1-ene; C_7, C_8, etc. are *n*-alkene/*n*-alkane of that carbon number. Peaks marked with an asterisk are isoprenoid hydrocarbons.

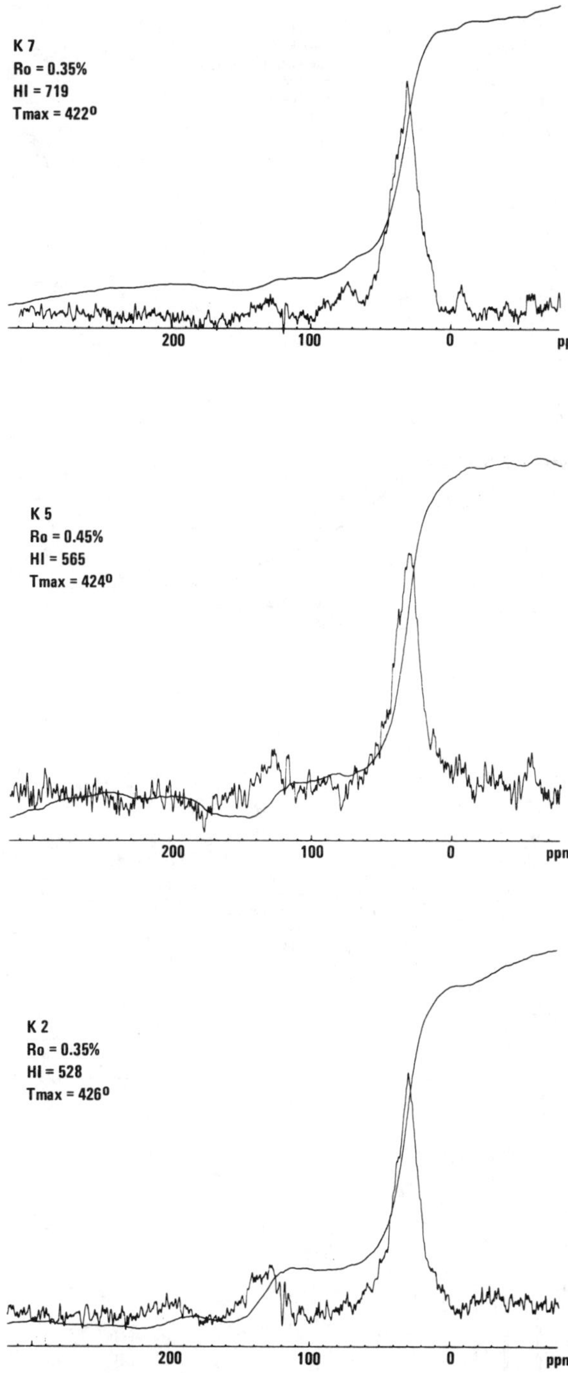

Fig. 7. 50 MHz CP/MAS ^{13}C NMR spectra of kerogen concentrates. Sample data are indicated.

dance of aromatic and isoprenoid compounds is relatively high, and the Py–GC trace shows an unresolved naphthenic envelope in the C_{13}–C_{27} range. These features are characteristic of the pyrograms of immature marine type II kerogens. The pyrogram of K5 is dominated by an aliphatic homology ranging from C_7 to C_{27}. The pyrograms of K7 and K5 demonstrate the maturity-dependent Py–GC characteristics which were observed for the kerogens in a homogenous sedimentary column, i.e. the relative abundance of aromatics and prist-1-ene is reduced and the n-alkene/n-alkane homology becomes more dominant with increasing maturity. It is also observed that the abundance of the naphthenic envelope is reduced in the more mature sample.

The Kimmeridge Clay kerogen K2 (Yorkshire North England) has the same maturity level as K7. However, the pyrogram of K2 shows a higher relative abundance of the n-alkene/n-alkane homology compared with the pyrogram of K7, reflecting a difference in the organic source material. This is also consistent with the difference in Rock–Eval hydrogen index (HI) and T_{max} parameters.

The three Kimmeridge Clay kerogen concentrates were subjected to solid state CP/MAS ^{13}C NMR analysis and their spectra are shown in Fig. 7. The spectra are dominated by a resonance band in the aliphatic region (0–50 ppm) with a maximum at 30 ppm and with only a minor band in the aromatic region (100–180 ppm). However, some differences are observed in the spectra; the spectrum of K7 shows a small shoulder in the 50–100 ppm region which includes aliphatic carbons bound to oxygen and nitrogen (not carbonyl and carboxylic acid derivatives). This shoulder is not observed in the spectrum of the more mature sample K5, and might be explained in terms of loss of functional groups with increasing maturity.

The most common type of information provided by ^{13}C CP/MAS studies of coals and kerogens is their apparent aromaticities (f_a'), the fraction of organic carbon atoms that are part of aromatic rings (Miknis *et al.*, 1981). The f_a'' values were calculated from the integrated spectra (see Table 3) and show a slight increase in aromaticity going from K7 to the more mature K5. This is in accordance with the generally accepted concept that kerogens become more aromatic with increasing maturation (Tissot and Welte, 1978), i.e. the relative carbon distribution in kerogens becomes more aromatic. It is also observed that the NMR spectrum of K2 (same maturity as K7) shows a higher aromaticity than that of K7. This is in agreement with the Py–GC and Rock–Eval data suggesting a difference in the organic source material for the two kerogens.

TABLE 3. Aromaticities (f_a') from CP/MAS ^{13}C NMR spectra

Sample	f_a'
K7	0.08
K5	0.13
K2	0.17
GRS	0.05
HS22	0.51
HS22 (350 °C)	0.61

By visual inspection of the spectra of K7 and K2 it appears that the spectrum of K7 possesses a broader triangulated aliphatic resonance band than that of K2 which shows a narrower aliphatic band with peak maximum at 30 ppm. It might be somewhat speculative to draw too broad conclusions based on these small spectral differences. However, the sharp resonance band at 30 ppm can be assigned to methylene groups $(CH_2)_n$ in long-chain normal alkanes and n-alkyl substituents on aromatic rings, while the broader aliphatic band at lower and higher ppm values (relative to 30 ppm) indi-

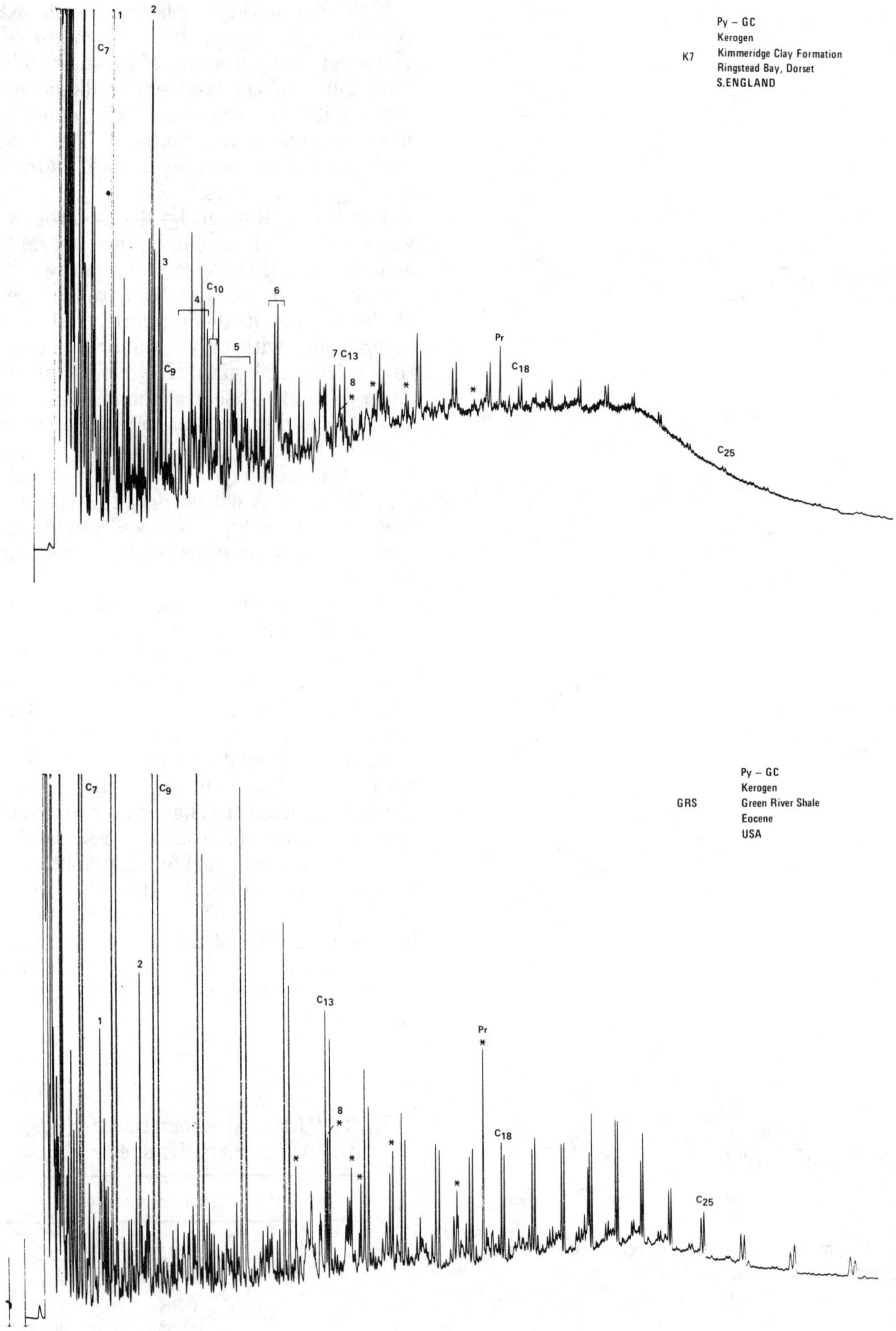

Fig. 8. Pyrograms of kerogen concentrates. Sample data are indicated. For peak identification see Fig. 6.

cates methyl branching and condensed alicylic (naphthenic) structural moieties in the kerogen matrix (Hagaman *et al.*, 1984). This observation is consistent with the Py–GC results; the pyrogram of K2 shows a higher relative abundance of the n-alkene/n-alkane homology compared with that of K7, suggesting a higher content of long-chain alkanes relative to naphthenic structures in the macromolecular matrix of K2 compared with K7. The same difference is observed in the spectra of K7 and K5; K5 shows a narrower aliphatic resonance band than K7. This is in agreement with the view that branched and cyclic hydrocarbons (isoprenoids) are preferentially generated during the early stages of diagenesis/catagenesis (Tissot and Welte, 1978).

Alginites are known to produce mainly normal hydrocarbons upon pyrolysis (Larter, 1984; Douglas *et al.*, 1983). In this study a Green River Shale kerogen (GRS) was used as a representative for algal type I kerogens. For the purpose of comparison, the pyrogram of GRS is shown together with that of K7 in Fig. 8. The Py–GC trace of GRS is totally dominated

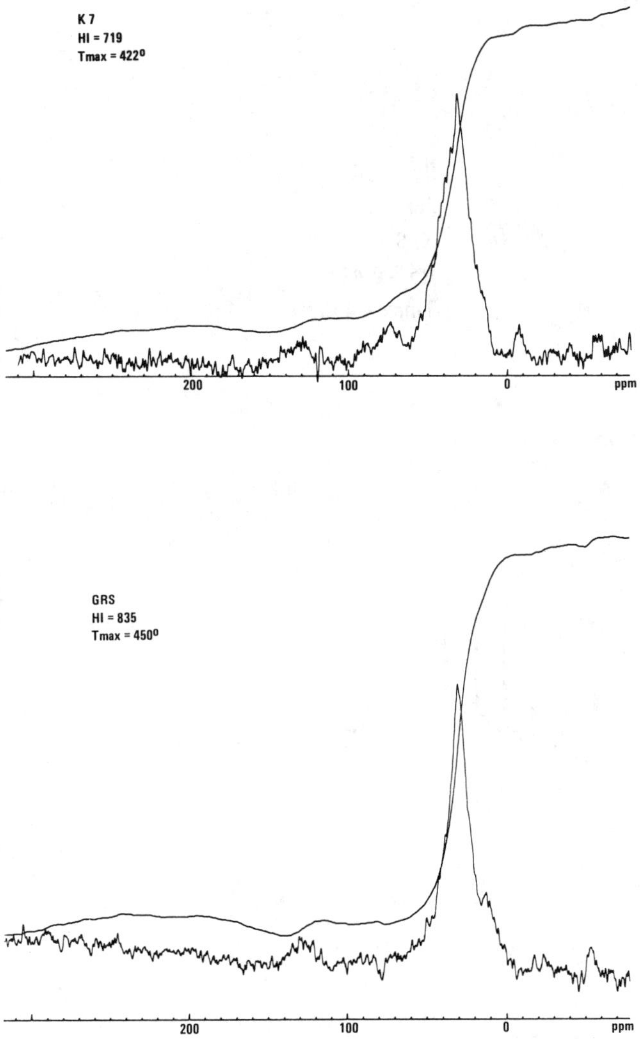

Fig. 9. 50 MHz CP/MAS ^{13}C spectra of kerogen concentrates. Sample data are indicated.

by the *n*-alkene/*n*-alkane homology ranging from C_7 to C_{28}, with only a minor unresolved naphthenic hump. This is in contrast to the pyrogram of K7, which is described above. Using the same arguments as above, the pyrolysis results are confirmed by the C_{13} NMR data (see Fig. 9). The spectrum of GRS shows a much narrower resonance band with maximum at 30 ppm compared with that of K7 (Fig. 9). The GRS spectrum is very similar to that of the alginite maceral (Zilm *et al.*, 1981). It is also relevant to note the significantly higher Rock–Eval T_{max} value of GRS compared with that of K7 (450 °C and 422 °C, respectively), which reflects the higher activation energy required to break the aliphatic C—C bonds in the GRS matrix than that required to break the labile mixed bond types in K7 (Tissot and Welte, 1978).

In connection with a model study concerning the influence of thermal stress on the kerogen structure, a large number of different kerogen types were heated to different temperatures for specified time intervals (Solli *et al.*, 1984*a*). As a final example of the complementary information which is provided by Py–GC and CP/MAS studies, a type III kerogen from the North Sea (HS 22) was selected, and its artificial maturation residue (HS 22 heated at 350 °C for 18 h). The pyrogram of HS 22 (Fig. 10) shows an aliphatic *n*-alkene/*n*-alkane homology ranging from C_7 to C_{29} with a high abundance of aromatic hydrocarbons and phenols (peak identification is noted in the figure legend). The pyrogram is characteristic for terrestrial type III kerogens deposited in a marine environment. The main distinguishing feature compared with pyrograms of type I and II kerogens is the high abundance of phenols in the pyrograms of type III kerogens (Larter, 1984). The Py–GC trace of HS 22 also shows a high abundance of *n*-alkenes/*n*-alkanes in the C_{18} to C_{25} region which indicates an input of lipid-rich plant wax material.

The pyrogram of the artificially matured sample (350 °C/18 h) shows the same trend as for naturally matured kerogens; a reduced relative abundance of aromatics and prist-1-ene compared with the unheated sample, and also a reduced abundance of phenols. The abundance of *n*-alkenes/n-alkanes in the high molecular end of the pyrogram is reduced drastically compared to the unheated sample.

The CP/MAS ^{13}C spectra of HS 22 and the heated residue (see Fig. 11) show a drastically higher aromaticity than the oil shale spectra which have been discussed previously. The spectrum of HS 22 shows a sharp resonance band at 30 ppm which is due to methylene groups (see above) and a significant shoulder in the 50–100 ppm range which might be assigned to C—O bonds in phenols and ether linkages. The spectrum of the heated sample shows an increased relative aromatic carbon distribution which is caused by the loss of the sharp component at 30 ppm, suggesting that alkane chains have been released during the simulated maturation process. This is also in agreement with the Py–GC results which suggests a reduced relative chain length of *n*-alkyl groups attached to the heated kerogen matrix.

Reports have shown that the aromatic carbon content of oil shales is relatively unchanged following retorting (Hagaman *et al.*, 1984) supporting the hypothesis that it is the aliphatic carbon which is largely responsible for, and the critical factor in determining, the quality and amount of oil which is produced during retorting and catagenesis in the sediments.

Using HS 22 as an example, the Py–GC and NMR data show that the increased aromaticity of the heated sample is caused by the loss of aliphatic structural moieties and demonstrates the high potential of this combined kerogen analysis approach.

CONCLUSIONS

In the present study it has been demonstrated that non-destructive NMR techniques provide information on sedimentary organic matter which is complementary to that obtained by the more commonly used gas chromatography and pyrolysis–gas chromatography methods.

Proton NMR of the aromatic fraction isolated from sediment extracts gives information about the relative amounts of aromatic and aliphatic protons, and it is shown that the percentage of aromatic protons in the aromatic fraction (PAP) is proportional to the level of maturity of the sediment.

Solid-state CP/MAS ^{13}C NMR represents a good complementary tool to pyrolysis–gas chromatography in kerogen studies, as it provides information on the

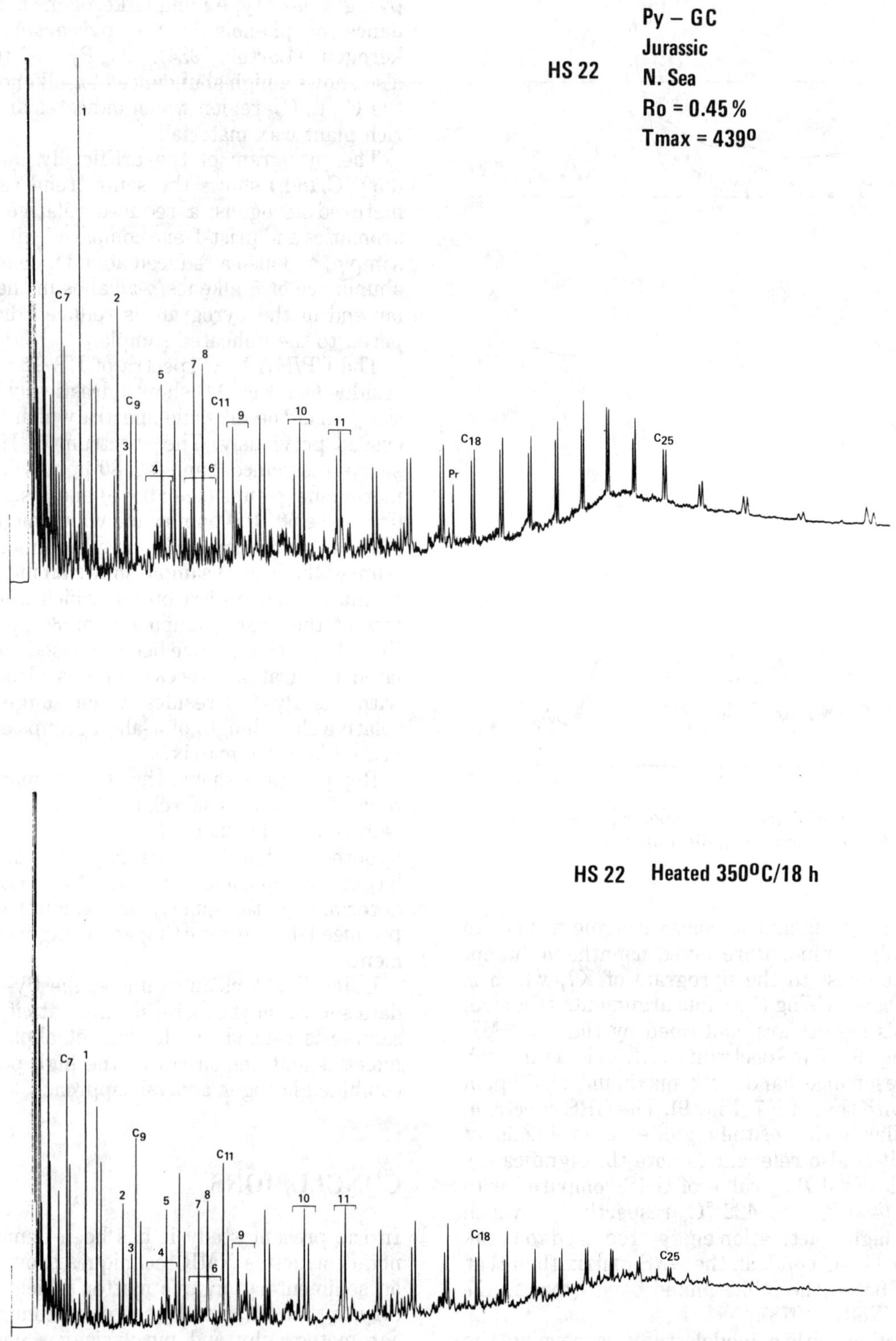

Fig. 10. Pyrograms of kerogen concentrates. Sample data are indicated. Peak identification: 1 = toluene; 2 = (*m* + *p*)-xylenes; 3 = *o*-xylene; 4 = C_3-alkylbenzenes; 5 = phenol; 6 = C_4-alkylbenzenes; 7 = *o*-cresol; 8 = (*m* + *p*)-cresol; 9 = dimethylphenols and naphthalene; 10 = C_{13} + methylnaphthalenes; 11 = C_{14} + dimethylnaphthalenes; Pr = prist-1-ene.

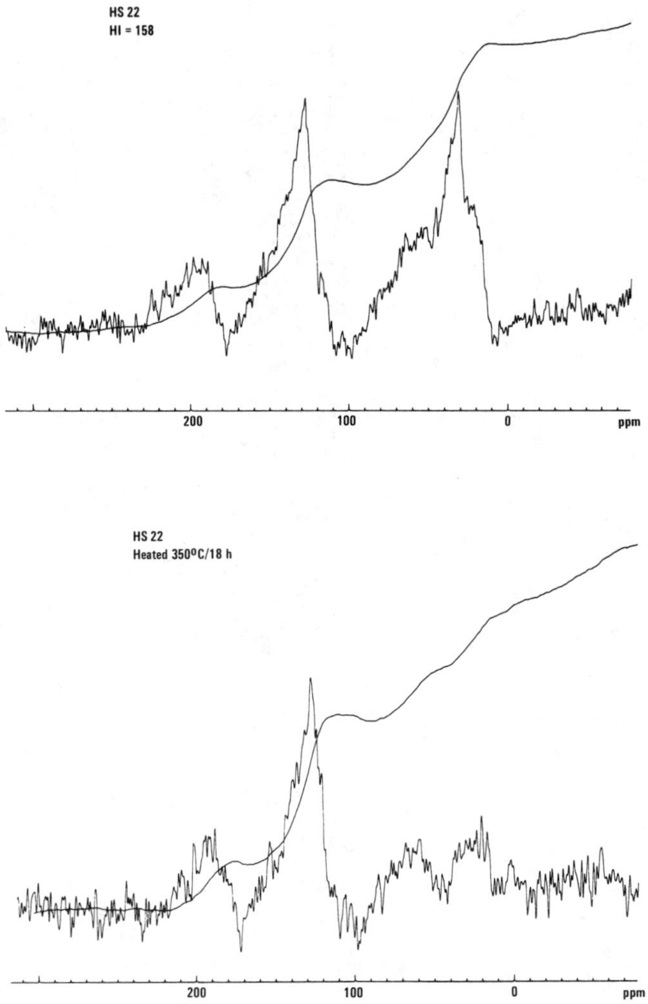

Fig. 11. 50 MHz CP/MAS [13]NMR spectra of kerogen concentrates. Sample data are indicated.

distribution of the total amount of aliphatic and aromatic carbon atoms in the kerogen whereas Py–GC only analyses the volatile pyrolysis products.

Both NMR techniques are still at the development stage and further refinement and evaluation of the methods with a large number of samples is still required.

ACKNOWLEDGEMENTS

The authors would like to thank the management of Petrofina for permission to publish this paper, and R. L. Skogstad for recording the CP/MAS ^{13}C NMR spectra at the National NMR Laboratory, University of Oslo. The spectra were recorded on NMR spectrometers obtained through a grant from the Norwegian Research Council for Science and Humanities (NAVF).

REFERENCES

Alexander, R., Gray, M. D., Kagi, R. I. and Woodhouse, G. W. 1980. Proton magnetic resonance spectroscopy as a technique for measuring the maturity of petroleum. *Chem. Geol.*, **30**, 1–14.

Alpern, B. 1980. Pétrographie du kérogéne. In: Durand, B. (ed.), *Kerogen*. Editions Technip, Paris, pp. 339–371.

Douglas, A. G., Hall, P. B. and Solli, H. 1983. Comparative organic geochemistry of some European oil shales. In: Miknis, F. P. and Mackay, J. F. (eds.), *ACS Symposium Series, No. 230. Geochemistry and Chemistry of Oil Shales*, pp. 59–84.

Dow, W. G. 1977. Kerogen studies and geological interpretations. *J. Geochem. Exploration*, **7**, 79–99.

Espitalie, J., Laporte, J. C., Madec, M., Marquis, F., Leplat, P., Paulet, J. and Boutefey, A. 1977. Méthode rapide de characterisation des roches meres de leur potential pétrolier et de leur degré d'évolution. *Rev. Inst. Français Pétrole*, **32**, 23–42.

Hagaman, E. W., Schell, F. M. and Cronauer, D. C. 1984. Oil shale analysis by CP/MAS ^{13}C NMR spectroscopy. *Fuel*, **63**, 915–919.

Horsfield, B. 1984. Pyrolysis studies and petroleum exploration. In: *Advances in Petroleum Geochemistry*, Vol. I. Academic Press, pp. 247–292.

Larter, S. R. 1984. Application of analytical pyrolysis techniques to kerogen characterization and fossil fuel exploration/exploitation. In: Voorhees, K. (ed.), *Analytical Pyrolysis—Methods and Applications*. Butterworth, London, pp. 212–275.

Larter, S. R. and Douglas, A. G. 1982. Pyrolysis methods in organic geochemistry. An overview. *J. Analyt. Appl. Pyrolysis*, **4**, 1–19.

Mackenzie, A. S., Patience, R. L., Maxwell, J. R., Vandenbroucke, M. and Durand, B. 1980. Molecular parameters of maturation in the Toarcian shales, Paris Basin. France, I. Changes in the configuration of acyclic isoprenoid alkanes, steranes, and triterpanes. *Geochim. Cosmochim. Acta*, **44**, 1709–1721.

Mackenzie, A. S., Hoffman, C. F. and Maxwell, J. R. 1981. Molecular parameters of maturation in the Toarcian shales, Paris Basin, France. III. Changes in the aromatic steroid hydrocarbons. *Geochim. Cosmochim. Acta*, **45**, 1345–1355.

Miknis, F. P. 1982. NMR studies of solid fossil fuels. *Magnetic Resonance Rev.*, **7**, 87–121.

Miknis, F. P., Sullivan, M., Bartuska, V. J. and Maciel, G. E. 1981. Cross-polarization magic-angle spinning ^{13}C NMR spectra of coals of varying rank. *Org. Geochem.*, **3**, 19–28.

Powell, T. G. S., Creany, and Snowdon, L. R. 1982. Limitations of use of organic petrographic techniques for identification of petroleum source rocks. *AAPG Bull.* **66**, 430–435.

Radke, M., Welte, D. H. and Willsch, H. 1982. Geochemical study on a well in the Western Canada basin: relation of the aromatic distribution pattern to maturity of organic matter. *Geochim. Cosmochim. Acta*, **46**, 1–10.

Schou, L., Mørk, A. and Bjorøy, M. 1984. Correlation of source rocks and migrated hydrocarbons by GC-MS in the Middle Triassic of Svalbard. In: *Advances in Organic Geochemistry 1983* (in press).

Solli, H., Graas, G. van, Leplat, P. and Krane, J. 1984*a*. Analysis of kerogens of Miocene shales in an homogenous sedimentary column. A study of maturation using flash pyrolysis techniques and carbon-13 CP/MAS NMR. In: *Advances in Organic Geochemistry 1983* (in press).

Solli, H., Bjorøy, M. Leplat, P. and Hall, K. 1984*b*. Analysis of organic matter in small rock samples using combined thermal extraction and pyrolysis–gas chromatography. *J. Analyt. Appl. Pyrolysis* (in press).

Tissot, B. P. 1984. Recent advances in petroleum geochemistry applied to hydrocarbon exploration. *Bull. Am. Ass. Petroleum Geologists*, **68**, 545–561.

Tissot, B. P. and Welte, D. H. 1978. *Petroleum Formation and Occurrence*. Springer-Verlag, Berlin.

van Graas, G., de Leeuw, J. W., Schenck, P. A. and Haverkamp, J. 1981. Kerogen of Toarcian shales of the Paris Basin. A study of its maturation flash pyrolysis techniques. *Geochim. Cosmochim. Acta*, **45**, 2465–2474.

Zilm, K. W., Pugmire, R. J., Larter, S. R., Allan, J. and Grant, D. M. 1981. Carbon-13 CP/MAS spectroscopy of coal macerals. *Fuel*, **60**, 717–722.

28 Study of the diagenesis, catagenesis and metagenesis of coals and dispersed organic matter with a high-temperature modified Rock–Eval

Paul Leplat and Jacques Paulet
Petrofina Exploration, Brussels, Belgium

A series of model coal samples ranging in rank from high volatile bituminous to anthracite was studied for the relationship between vitrinite reflectance and Rock–Eval pyrolysis. A small modification in the electronic part of the temperature programmer allows it to reach a temperature of 750 °C. The study of the hydrocarbons and carbon dioxide released, as well as the increase of T_{max} of the S_2 peak in borehole samples, shows a relationship with burial depth. This modified Rock–Eval enables a rapid estimation of diagenesis and catagenesis, as well as metagenesis, of dispersed organic matter in small quantities of whole rock samples. The oil and gas potential on whole rock samples has been evaluated by a combination of thermal extraction and pyrolysis–gas chromatography.

INTRODUCTION

Rock–Eval studies on various types of coals and mature sediment samples have provided the basis for a rapid method for characterization of the various types of organic matter and estimation of their degree of maturation. However, in the case of highly mature organic matter, it is necessary to heat the sample at higher temperatures for total and quantitative cracking of the kerogen by using a modified Rock–Eval principle.

EXPERIMENTAL

Sample description

With a modified Rock–Eval (Fig. 1), the pyrolysis of various coals with different maturities (R_0 = 0.5–4.2%) and maceral compositions was performed at high temperatures. The same procedure was also applied to 'dispersed organic matter' in rock samples taken from a petroleum exploration well.

Instrumentation

Modified high temperature Rock–Eval

(a) The pyrolysis device is the standard 'Rock–Eval' working in normalized conditions which has been modified as follows:

—to reach a temperature of 750 °C the electronics of temperature programmer (PTF-7) were modified and
—to avoid damage of a special joint for closing the oven and to obtain an adequate sealing of the mechanical system, an improvement in the water-cooling system was necessary.

(b) Preliminary tests were performed as follows:

—checking the temperature resistance of the oven and its heating element (pyrotenax);
—ensuring good response of the FID detector;
—checking optimal FID response (no baseline drift) and
—calibrating the oven temperature programming (25 °C/min, see Fig. 2).

(c) the CO_2 trap was replaced by an infrared analyser for CO_2 in series with a second infrared analyser for CO_2 (Leplat *et al.*, 1983).

(d) Operating procedure was as follows.

A known quantity of crushed sample (20 mg is used for coals, and 100 mg for rock samples) was weighed into the crucible.

Starting at 250 °C with a heating rate of 25 °C/min, the final temperature of 750 °C is reached after approximately 20 min. Including oven-cooling time, the complete analysis takes half an hour (see Fig. 3).

The different analytical steps are completely automatic and the principal ones are:

Petroleum Geochemistry in Exploration of the Norwegian Shelf © Norwegian Petroleum Society (Graham & Trotman, 1985) pp. 319–326.

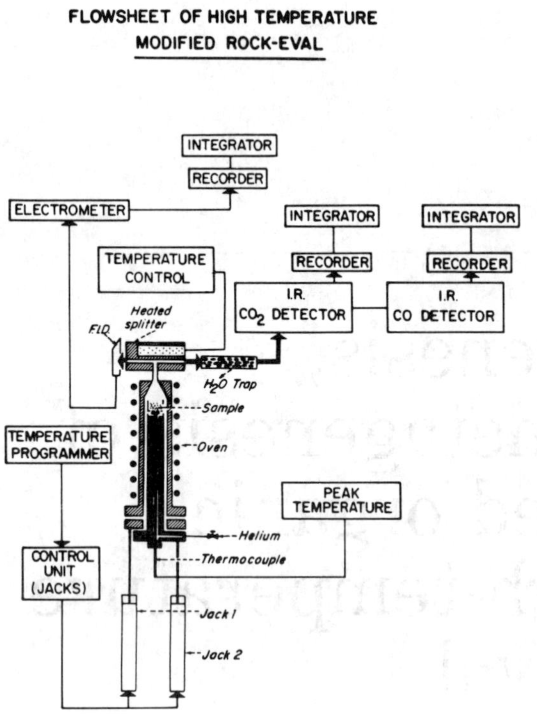

Fig. 1. Operational flowsheet of the apparatus.

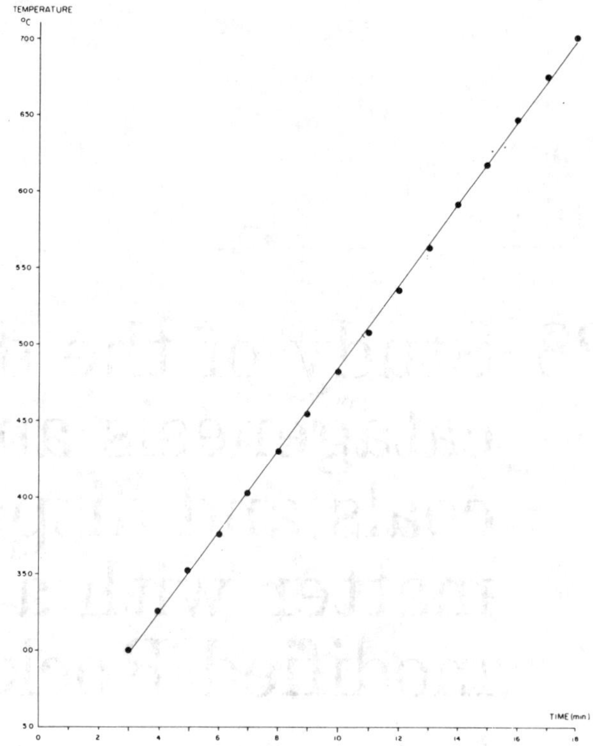

Fig. 3. Analytical cycle.

peak P1 = free hydrocarbons, which are vaporized at moderate temperatures (250 °C);
peak P2 = hydrocarbon resulting from the cracking of kerogen (or coal);
peak P3 = Oxygenated compounds evolved as CO_2 which is desorbed at the end of the cycle; in some cases, the CO_2 trap was replaced by an IR analyser not shown in Fig. 1;
T_{max} = maximum of peak P2.

Pyrolysis–gas chromatography

The instrumentation has been described in detail (Solli *et al.*, 1984). The system combines a rapid simultaneous analysis of C_8–C_{15} free hydrocarbons by thermal extraction and temperature-programmed furnace pyrolysis–gas chromatography using focused cryogenic trapping on capillary columns. The thermally extracted whole-rock samples or coals are pyrolysed.

Conditions: temperature-programmed furnace pyrolysis (320–550 at °C/min) in helium (30 ml/min) followed by focused cyrogenic trapping (liquid nitrogen) and gas chromatography on a fused-silica capillary column (20 m × 0.25 mm I.D.) coated with OV-1; sample split ratio 1 : 30; oven temperature programme from 40 °C to 280 °C at 4 °C/min; helium carrier gas inlet pressure 8 psi.

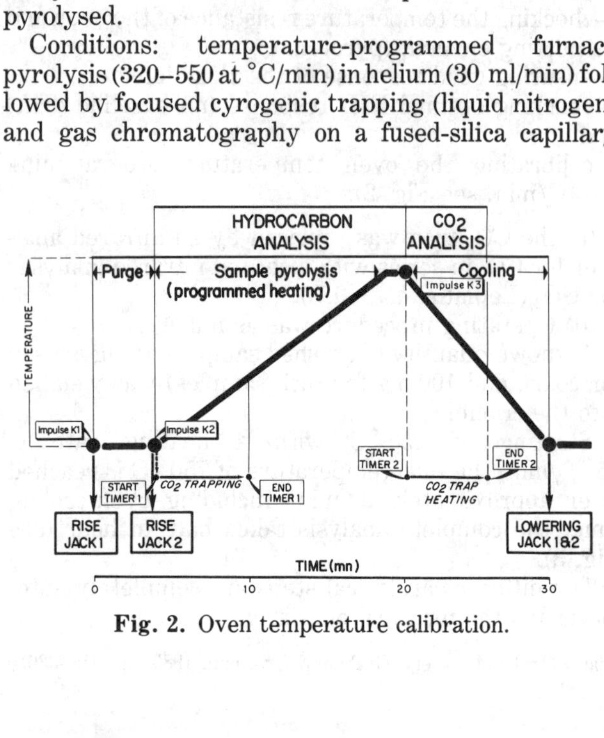

Fig. 2. Oven temperature calibration.

RESULTS AND DISCUSSION

A series of coals (Table 1) with known vitrinite reflectance (R_0) and volatile matter content (VM) have been analysed with the modified high-temperature Rock–Eval (see Fig. 4).

The data reported in Table 2 show an increasing T_{max} with respect to coal rank up to R_0 = 4.45%, and its significance in terms of hydrocarbon generation.

As described by Alpern (1970), Robert (1974), Teichmüller (1971) and Teichmüller and Durand (1983):

—the first coalification jump ($R_0 \cong 0.6°$) corresponds to the onset of catagenetic generation of hydrocarbons in source rocks.
—the second coalification jump ($R_0 \cong 1.2\%$) is the beginning of oil cracking and its transformation into condensate and gas.

By plotting T_{max} values (in °C) of coal as well as of rock samples against vitrinite reflectance a good correlation throughout the evolutionary path is shown in Fig. 5.

The geochemical method is simple and is applied directly on sediments without pretreatment as in the standard Rock–Eval method (Espitalié *et al.*, 1977). In a short time this analytical procedure and the use of Fig. 5 provides a preliminary estimate of the hydrocarbon potential of possible source rocks.

As an example, a sedimentary series of rock samples with depths ranging from 296 m to 2810 m were analysed with the high-temperature modified Rock–Eval. The samples, which contain total organic carbon contents of less than 1%, have already gener-

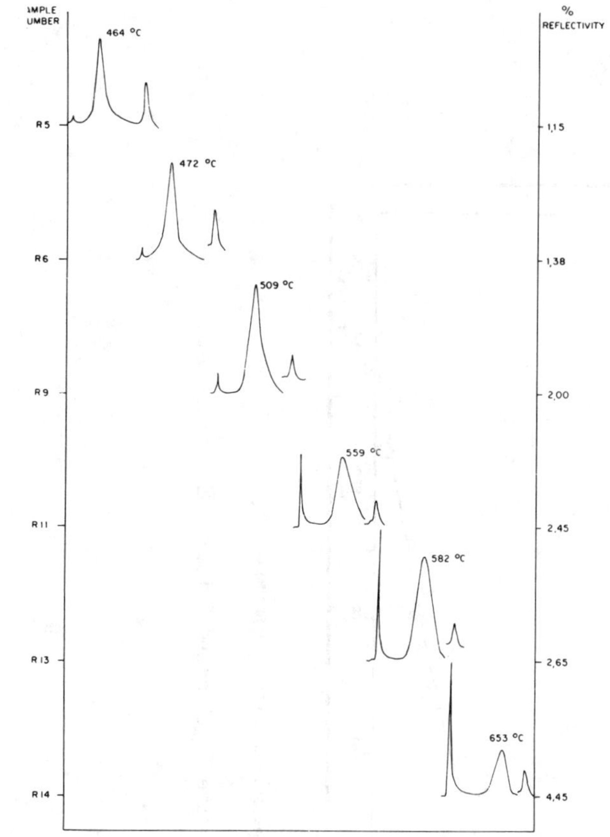

Fig. 4. Characterization of the degree of evolution in a series of coal samples.

ated some hydrocarbons, and the T_{max} values are in good agreement with the vitrinite reflectance data (see Table 3).

Figure 6 shows the Rock–Eval traces of the rock samples with increasing depth and the vitrinite reflectance. It seems important to have a closer look at the morphology of the peaks because, as shown for example in Fig. 7, the 'hump' is due to the fact that this sample is a mixture of 95% of a mature (R_0 = 0.60%) sedimentary rock (T_{max} = 438 °C), with 5% of a high volatile bituminous coal (R_0 = 1.15%, T_{max} = 464 °C).

The Rock–Eval data (see Table 3) for three rock samples (out of the series of seven) are complemented with temperature-programmed pyrolysis–gas chromatographic traces (Figs. 8, 9 and 10); as can be seen, the mature sample (R_0 = 0.87%) taken at 296 m depth shows a complete chromatographic picture with the *n*-alkene/*n*-alkane doublet homology dominating in the range from C_8 to C_{19}. The prominence of *n*-alkanes in the whole rock sample pyrogram can be explained in terms of its maturity.

By visual inspection of the pyrograms of mature to over-mature samples (see Figs. 8–10), it appears that dominance of the aliphatic doublet homology decreases with depth and maturation. The normalized data of the *n*-C_{11} and *n*-C_{12} *n*-alkane relative concentrations are given in Table 4. Table 5 shows the relative concentrations of aromatic compounds (benzene, toluene, xylenes). As can be seen in these tables, the *n*-C_{11} and *n*-C_{12} *n*-alkane concentrations in the thermal extracts (S1 detailed peak) increase as they decrease in the pyrolysates (S2 detailed peak). The sum of *n*-C_{11} and *n*-C_{12} *n*-alkane concentrations produced in the two peaks (S1 + S2) are similar; they increase a little with maturation (see Table 4, columns 2 and 6).

This trend is in accordance with results obtained on samples from Miocene Shales (Solli *et al.*, 1985), and suggests an apparent absence of migration of the *n*-alkanes. The concentration of the *n*-alkanes in the S2 peak decreases as it increases in the S1 peak; thus the sum of *n*-C_{11} + *n*-C_{12} in the S1 + S2 peaks remains almost the same with increasing depth.

The relative concentrations of the aromatics decrease from mature to over-mature samples (see

TABLE 1. Rank data for a set of reference coals

Coal sample	*Organic carbon (%)*	*Vitrinite reflectance, R_0 (%)*	*Volatile matter (% d.a.f.)*
R1	57.8	0.31	54.7
R2	81.9	0.77	41.0
R3	67.1	0.75	39.0
R4	79.2	0.92	35.1
R5	76.6	1.15	30.2
R6	75.6	1.04	29.6
R7	58.0	1.38	25.3
R8	88.4	1.70	20.3
R9	89.1	2.00	15.1
R10	86.7	1.83	12.9
R11	90.5	2.45	10.0
R12	84.7	2.25	9.3
R13	91.8	2.65	7.9
R14	69.8	4.45	3.6

TABLE 2. Stages of evolution and hydrocarbon generation for reference coals

Vitrinite reflectance (R_0 (%)	*Volatile matter (% d.a.f.)*	*T_{max} (°C)*	*Evolution stages*	*Hydrocarbon generation*
0.31	54.7	425	Early diagenesis	Biogenic gas
(0.55) - - -	- - -	(435) - - -	- - -	- - -
0.77	41.0	443	↑	
0.75	39.0	441		
0.92	35.1	444		Oil
1.04	29.6	447		
(1.10) - - -	- - -	(455) - - -	Catagenesis	- - -
1.15	30.2	464		
1.38	25.3	472		Wet gas
1.70	20.3	492		
1.83	12.9	529		
2.00	15.1	509	↓	
(2.00) - - -	- - -	(530) - - -	- - -	- - -
2.25	9.3	588	↑	
2.45	10.0	559		
2.65	7.9	582		Dry
(3.00) - - -	- - -	- - -	Metagenesis	- - - gas
4.45	3.6	653	↓	

TABLE 3. Stages of evolution and hydrocarbon generation for a series of sediment samples

Depth (m)	*T_{max} (°C)*	*Vitrinite reflectance (%)*	*Evolution stages*	*Hydrocarbon generation*
296	450	0.83	↑	Oil (+gas)
- - -	- - -	- - -		- - -
570	464	1.11	Catagenesis	
786	469	1.29		Wet gas
960	479	1.37		
1190	504	1.73	↓	
- - -	- - -	- - -	- - -	- - -
1658	589	2.48	↑	
- - -	- - -	- - -	Metagenesis	- - - Dry gas
2810	630	3.32	↓	

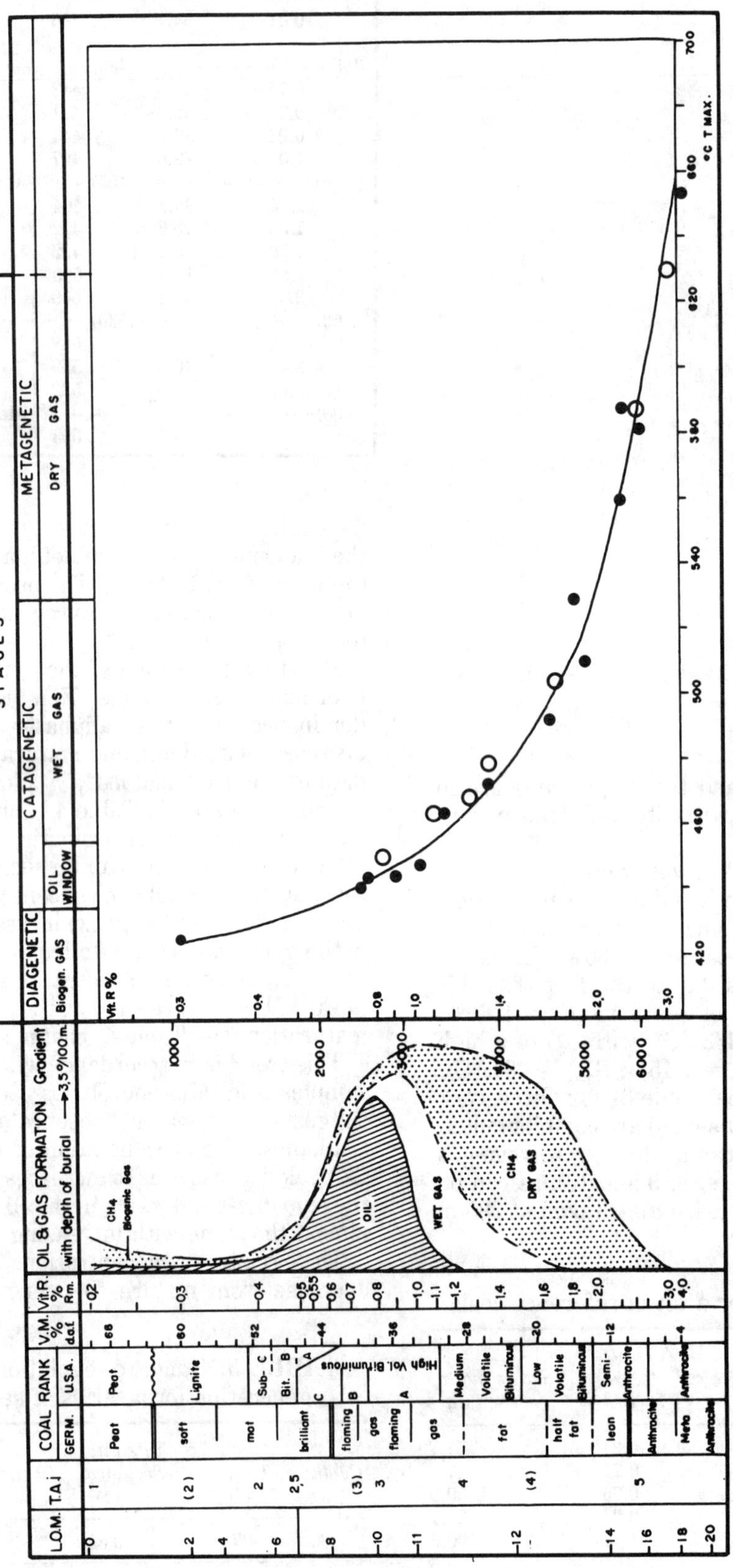

Fig. 5. Relation of physical and optical parameters to maximum temperature (T_{max}) in modified Rock–Eval procedure.

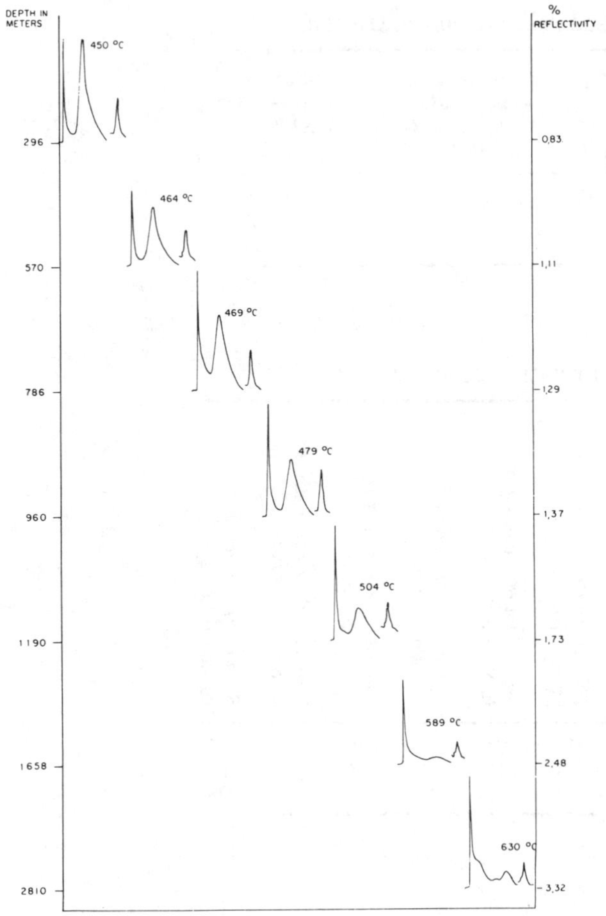

Fig. 6. Characterization of the degree of evolution in a series of sediment samples.

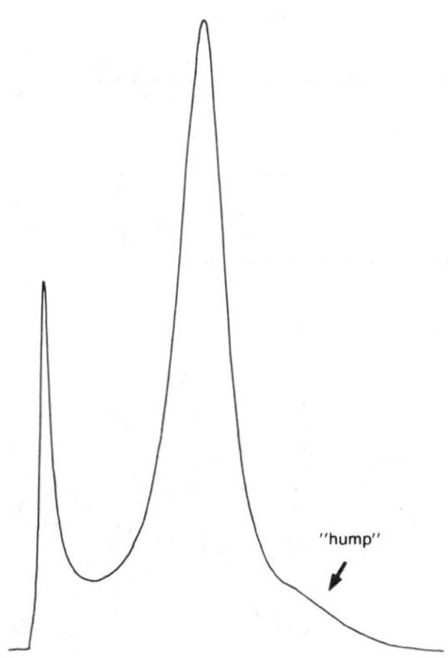

Fig. 7. Mixture of rock sample with coal.

Table 5, column 13). The drastic change from 71 to 38 is the result of a modification of structure in the kerogen during metagenesis over a vitrinite reflectance range from $R_0 = 0.87$ to $R_0 = 3.37$ (Table 5, column 2). The sample taken at 2810 m depth is probably polluted with gas-oil; indeed, the n-alkane range from n-C_{14} to n-C_{20} is typical of gas-oil.

In order to compare the effects of the evolution from 'mature' to 'over-mature', coal samples of similar maturities to the rock samples were analysed (Table 6).

The characteristic 'oil' pyrogram fingerprint of coal B (see Fig. 11) shows a continuous n-alkene/n-alkane

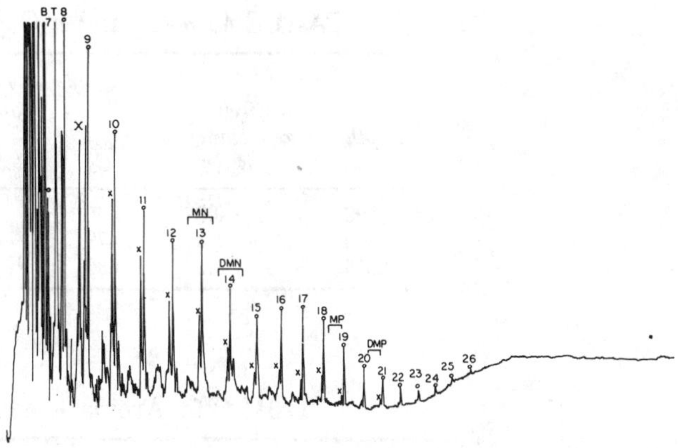

Fig. 8. Pyrogram (temperature-programmed) of thermally extracted whole rock sample taken at 296 m depth. *Conditions*: temperature-programmed furnace pyrolysis (320–550 °C at 30 °C/min) in helium (30 ml/min) followed by focused cryogenic trapping (liquid nitrogen) and gas chromatography on fused silica capillary column (20 m × 0.25 mm I.D.) coated with OV-1; sample split ratio 1 : 30; oven temperature programme from 40 °C to 280 °C at 4 °C/min; helium carrier gas inlet pressure, 8 psi. *Peak identification*: B = benzene, T = toluene, X = xylenes, °7, °8, etc. are n-alkanes, X, X, etc. are n-alkenes, MN = methyl naphthalene, DMN = dimethyl naphthalene, P = phenanthrene, MP = methyl phenanthrene, DMP- = dimethyl phenanthrene, DBT = dibenzothiophene.

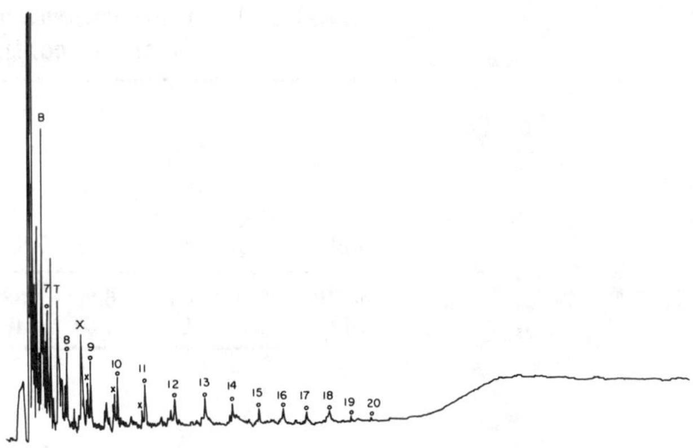

Fig. 9. Pyrogram (temperature programmed) of thermally extracted whole rock sample taken at a depth of 960 m. Conditions and peak identification as in Fig. 8.

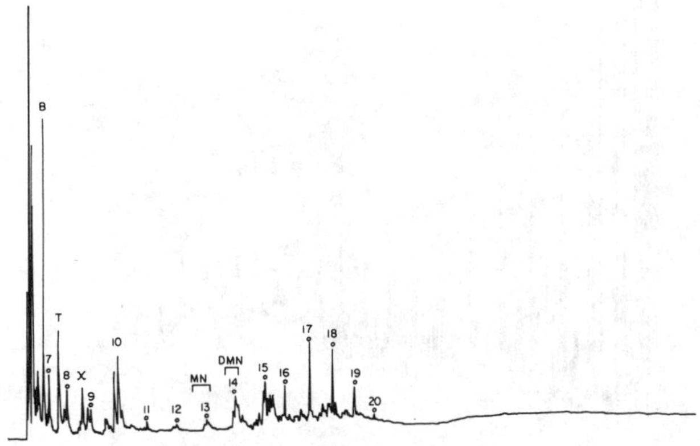

Fig. 10. Pyrogram (temperature-programmed) of thermally extracted whole rock sample taken at a depth of 2810 m. Conditions and peak identification as in Fig. 8.

TABLE 4. n-C_{11} and n-C_{12} n-alkane relative concentrations

Depth (m)	Vitrinite reflectance R_0 (%)	T_{max} (°C)	n-C_{11}–n-C_{12} concentration in S1 detailed peak	n-C_{11}–n-C_{12} concentration in S2 detailed peak	Sum of n-C_{11}–n-C_{12} in S1 + S2 peaks
296	0.87	450	11	9	20
960	1.37	479	16	6	22
2810	3.32	630	25	1	26

TABLE 5. Aromatic compound relative concentrations

Depth (m)	Vitrinite reflectance R_0 (%)	T_{max} (°C)	Benzene concentration in S1 detailed peak	Benzene concentration in S2 detailed peak	Sum of benzene concentration in S1 + S2 peaks	Toluene concentration in S1 detailed peak	Toluene concentration in S2 detailed peak	Sum of toluene concentration in S1 + S2 peaks	Xylene concentration in S1 detailed peak	Xylene concentration in S2 detailed peak	Sum of xylene concentration in S1 + S2 peaks	Sum of aromatics concentration
296	0.87	450	2	25	27	8	16	24	4	16	20	71
960	1.37	479	3	20	23	8	9	17	9	11	20	60
2810	3.32	630	7	4	11	12	2	14	11	2	13	38

TABLE 6. Vitrinite reflectance, high-temperature modified Rock–Eval and continuous IR analysis data

			Rock–Eval						Continuous Ir analysis CO_2–CO	
Sample	TOC (%)	R_0 (%)	S1	S2	S3	HI	OI	T_{max} (°C)	OI CO_2	OI CO
Coal B	73.50	0.87	6.20	138	2.65	189	4	443	7	2
Coal C	66.72	1.37	4.60	46	1.70	69	3	480	2	1

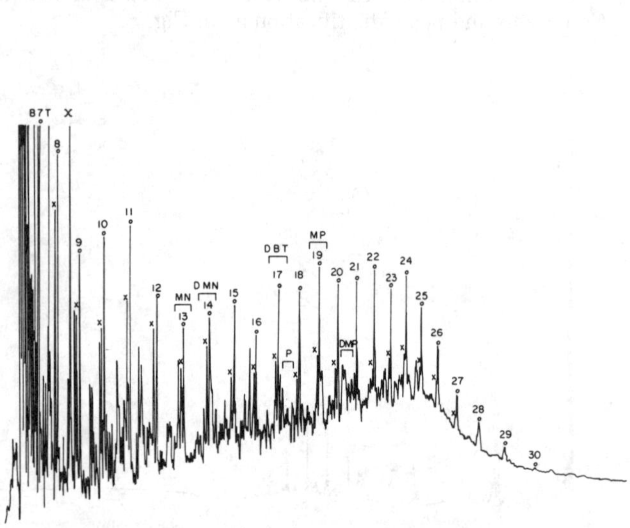

Fig. 11. Pyrogram (temperature-programmed) of thermally extracted coal sample B. Conditions and peak identification as in Fig. 8.

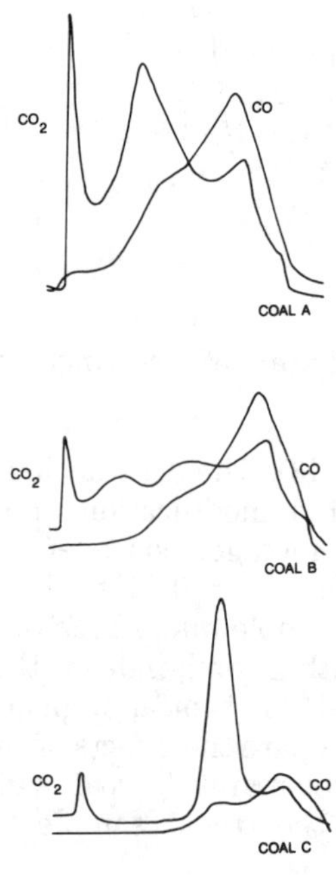

Fig. 12. CO_2 and CO pyrograms for coal samples A, B and C.

TABLE 7. Maceral compositions for three coal samples and corresponding high-temperature modified Rock–Eval and continuous IR analysis data

Sample	*Maceral composition*			*Vitrinite reflectance*	*Rock–Eval*		*Continuous IR analysis CO_2-CO*	
	Exinite	*Inertinite*	*Vitrinite*	R_0 *(%)*	T_{max} *(°C)*	*HI*	*OI* CO_2	*OI CO*
Coal A	17	36	47	0.54	429	150	30	7
Coal B	10	64	26	0.87	443	189	7	2
Coal C	—	87	13	1.37	480	69	2	1

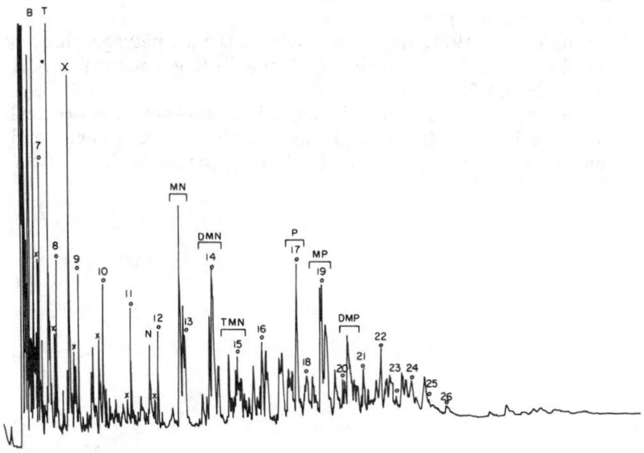

Fig. 13. Pyrogram (temperature-programmed) of thermally extracted coal sample C. Conditions and peak identification as in Fig. 8.

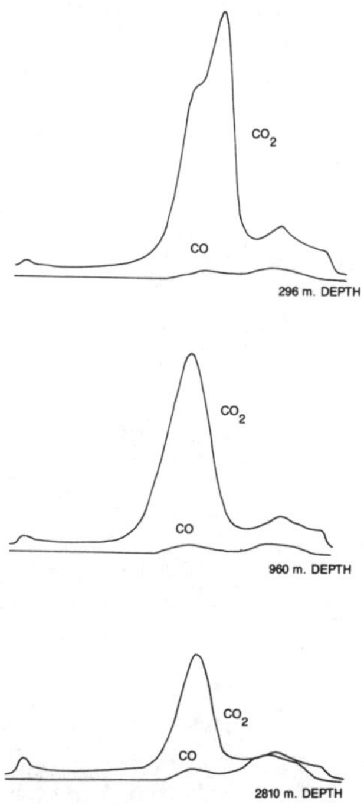

Fig. 14. CO_2 and CO pyrograms for three sediment samples.

distribution between n-C_7 and about n-C_{30}. This coal B indeed has some 'oil potential' for three reasons.

(1) The n-alkane distribution has a typical 'oil-like fingerprint'. The odd carbon number preference in the n-C_{23} range is considered to be due to a higher land plant input.

(2) The reflectance ($R_0 = 0.87$) is situated within the range of the 'oil window'.

(3) The similarity of the present CO pyrogram (Fig. 12(b)) to one of the generalised CO pyrograms in Daly and Peters (1982).

Significant also in the pyrogram fingerprint of the pyrolysate of coal B is the presence of aromatic compounds such as naphthalenes, phenanthrenes and, particularly, methyl and dimethyl naphthalenes. These aromatic compounds are not visible in the pyrogram of the rock sample taken at a depth of 296 m, where only the width at the base of the n-C_{13} and n-C_{14} peaks gives some indication of aromatic compounds (see Fig. 8).

It must be kept in mind that vitrinite reflectance (R_0) only gives data relating to the vitrinite maceral; in contrast the apparent scattering of T_{max} is thought to be due to variations in maceral composition (see Table 7). This is also indicated by the CO_2 and CO pyrogram fingerprints (see Fig. 12) and the pyrogram of coal C (see Fig. 13), with a vitrinite reflectance of $R_0 = 1.40$, which has few alkanes; most of the peaks are representative of aromatic compounds. This contrasts with the coal B fingerprint (Fig. 11) and the rock sample taken at a depth of 960 m, which are quite paraffinic. Fig. 14 shows the CO_2 and CO pyrograms of the sediment samples taken, respectively, at depths of 296, 960 and 2810 m (refer Table 3), showing the effects of increasing maturity with depth (most evident in the CO_2 trace).

CONCLUSIONS

(1) The high-temperature modified Rock–Eval device represents an important screening method which is complementary to optical studies on concentrated as well as disseminated organic matter.

(2) The high-temperature modified Rock–Eval device is one way of estimating the evolutionary stage of highly mature samples, compared with the time-consuming vitrinite reflectance analysis.

(3) The fingerprints of CO_2 and CO can be used to distinguish quickly the general maturity levels of organic matter.

(4) The combined thermal extraction and pyrolysis–chromatography data are available within a period of one hour of the high-temperature Rock–Eval data.

(5) Aromatics, if present in the pyrogram, may supply useful information about specific input of organic material and specific environment. This conclusion is necessarily tentative at this stage and studies in this direction are now in progress in the authors' laboratories.

ACKNOWLEDGEMENTS

We thank Mr G. Fortems, manager of Petrofina Exploration and Mr P. Rousseau for the permission to publish this paper.

REFERENCES

Alpern, B. 1970. Classification pétrographique des constituants organiques fossiles des roches sédimentaires. *Rev. Inst. Français Pétrole*, **25**(11), 1233–1266.

Daly, A. R. and Peters, K. E. 1982. Continuous detection of pyrolytic carbon monoxide: a rapid method for determining sedimentary organic facies. *AAPG Bull.*, **66**(12), 2672–2681.

Espitalie, J., Laporte, J. C., Madec, M., Marquis, F., Leplat, P., Paulet, J. and Boutefeu, A. 1977. Méthode rapide de caractérisation des roches mères, de leur potentiel pétrolier et de leur degré d'évolution. *Rev. Inst. Français Pétrole*, **32**(1), 23–42.

Leplat, P., Paulet, J. and Melotte, M. 1983. Continuous infrared analysis of temperature programmed pyrolysis products of laboratory-oxidized kerogens and borehole samples. In: Bjorøy, M. (ed.), *Advances in Organic Geochemistry 1981*. Wiley, pp. 613–619.

Robert, P. 1974. Analyses microscopiques des charbons et des bitumes dispersés dans les roches et mesure de leur pouvoir réflecteur. Application à l'étude de la paléogéothermie des bassins sédimentaires et de la genèse des hydrocarbures. In: Tissot, B. and Bienner, F. (eds.), *Advances in Organic Geochemistry 1973*. Technip, pp. 549–569.

Solli, H., Bjorøy, M., Leplat, P. and Hall, K. 1984. Analysis of organic matter in small rock samples using combined thermal extraction and pyrolysis-gas chromatography. *J. Analyt. Appl. Pyrolysis*. (submitted).

Solli, H., van Grass, G., Leplat, P. and Krane, J. 1985. Analysis of kerogens of Miocene Shales in an homogeneous Sedimentary Column. A study of maturation using flash pyrolysis techniques and carbon-13 CP/MAS NMR. In: *Advances in Organic Geochemistry 1983*. Pergamon Press (in press).

Teichmüller, M. 1971. Application des méthodes pétrographiques des charbons à la prospection de l'huile et du gaz naturel. *Erdöl Khole*, **24**, 69–76.

Teichmüller, M. and Durand, B. 1983. Fluorescence microscopical rank studies of liptinites and vitrinites in peat and coals, and comparison with results of the Rock–Eval pyrolysis. *Int. J. Coal Geol.*, **2**, 197–230.

29 Analysis of source rocks, reservoir rocks and cap rocks by combined thermal extraction and pyrolysis–gas chromatography

*Malvin Bjorøy**[a], *Hauk Solli**[b], *Keith Hall*† *and Paul Leplat*‡
*Continental Shelf Institute (IKU), P.O. Box 1883, 7001 Trondheim, Norway
†(GC)^2 Ltd, P.O. Box 30, Altrincham, Cheshire, UK
‡Petrofina, Exploration Department, Rue de la Loi 33, B-1040 Brussels, Belgium

A combined thermal extraction/pyrolysis–gas chromatography system has been developed for the multiple evaluation of potential source rocks, variation of hydrocarbons within a reservoir, residual oil and the effectiveness of a cap rock. The determination of maturity and organic richness is based on previously published methods. The evaluation of the type of hydrocarbons a source rock is capable of producing is based on the relative abundance and distribution of aliphatic and aromatic compounds in the pyrolysis–gas chromatography trace of the kerogen and the gas chromatogram of the thermally extracted free hydrocarbons in the sample.

Hydrocarbon variations within a reservoir are detected by the gas chromatographic analysis of the thermal extract. If care is taken during sampling, a complete gas chromatogram (C_6–C_{32}) can be obtained by this method.

The sealing effectiveness of a cap rock and residual oil in a reservoir is evaluated by a combination of thermal extraction and pyrolysis–gas chromatography. The distance that hydrocarbons from a reservoir penetrate into a cap rock can be detected by comparing the thermal extract from the top of the reservoir with those from different intervals within the cap rock, together with the use of pyrolysis–GC to evaluate whether the thermally extracted material was produced *in situ*.

INTRODUCTION

The petroleum generation process, i.e. production of bitumen from kerogen, can be studied by various methods. The type of kerogen is normally defined either by microscopic or by pyrolytic methods. Bitumen is usually studied either by gas chromatographic analysis of the pore gas or by solvent extraction followed by liquid and gas chromatographic techniques, or a combination of both. Due to evaporation losses, the extraction and fractionation technique restricts the analyses to compounds containing 15 carbon atoms or more, which are normally classified as the 'C_{15}+ fraction'. The analysis of pore gas normally gives information on C_1–C_7 compounds and there is, therefore, a lack of information of the C_8–C_{14} compounds. These techniques will normally be used on different subsamples, and a direct comparison of the pore gases, extract and kerogen is therefore difficult.

The C_1–C_{30} fraction comprises those hydrocarbons which are most mobile during the migration process, and it is therefore of prime interest to have techniques that will analyse the whole range of hydrocarbons present in a rock sample at the same time as the kerogen composition is determined. The gaseous compounds (C_1–C_5) can be analysed by headspace analysis (Hunt, 1975). Heating and grinding, either under reduced pressure (Gedenk, 1966) or with the aid of an auxiliary gas (Durand and Espitalié, 1972; Philippi, 1975; Le Tran, 1975) has given satisfactory results in studying the thermal mobilization of hydrocarbons from the rock matrix at elevated temperatures. All these techniques have large sample requirements, and they will only analyse hydrocarbons up to C_{15} due to use of lower temperatures. Several micro-scale techniques of combined thermal mobilization and gas chromatographic analysis have been developed (Jonathan *et al.*, 1975; Schaefer *et al.*, 1978; Saint-Paul *et al.*, 1980) as

[a]*Present address*: Geolab Nor., Hornebergveien 5, 7000 Trondheim, Norway.
[b]*Present address*: Department of Geology, University of Oslo, P.O. Box 1047, Blindern, Oslo 3, Norway.

Petroleum Geochemistry in Exploration of the Norwegian Shelf © Norwegian Petroleum Society (Graham & Trotman, 1985) pp. 327–337.

well as analysis of the C_1–C_{15} hydrocarbon range in the pore water by a combined thermal mobilization/ gas capillary GC system (Schaefer *et al.*, 1983).

The application of pyrolysis methods in organic geochemistry and petroleum exploration has recently been reviewed by Larter and Douglas (1980), Larter (1984) and Horsfield (1984). Pyrolysis methods for the investigation and characterization of organic matter in sediments can be broadly split into two categories: (a) bulk-flow methods in which only quantity or rate of evolution of the pyrolysates is monitored with little or no detailed separation, as used in the well-known IFP–Fina Rock–Eval process (Espitalié *et al.*, 1977) and (b) analytical pyrolysis techniques in which kerogen or whole rock pyrolysates are characterized in detail. Two methods that have been developed and employed in a number of laboratories are pyrolysis–gas chromatography (Py–GC), alone, or in combination with mass spectrometry (Py–GC–MS), and pyrolysis–mass spectrometry (Py–MS) (Larter and Douglas, 1980). The success of the commercial Rock–Eval instrument as a preliminary method for determining petroleum source rock potential attests to the widespread acceptance of pyrolysis methods by the petroleum industry (Orr, 1981). The Rock–Eval instrument uses the bulk-flow approach monitoring the amounts of thermally extracted free hydrocarbons (bitumen) and kerogen pyrolysates in rock samples without prior detailed separation. Using the terminology associated with the Rock–Eval method, the thermally extracted hydrocarbons at 250–300 °C are represented by a peak, S_1, and the high-temperature (300–550 °C) kerogen pyrolysate by the S_2 peak.

Martin (1977) used a thermal extraction procedure coupled with Py–GC to evaluate the source rock potential of sediments. Dembicki *et al.* (1983) and Whelan *et al.* (1980) used a similar technique to study source rocks and marine pollution.

Py–GC analysis of coal macerals and a large number of kerogens from different sedimentary facies have been shown to produce characteristic fingerprint pyrograms (Larter and Douglas, 1980; Douglas *et al.*, 1983).

Solli *et al.* (1984) have recently reported the development and application of a pyrolysis–GC system which provides simultaneous analyses of the released hydrocarbons during thermal extraction (to C_{35}) and the pyrolysis products of the kerogen in small size rock samples by using cryogenic trapping on two capillary columns (see 'Experimental' below). This paper describes the application of the system on selected samples to estimate (a) source rock quality, (b) generation of hydrocarbons from a source rock, (c) remaining hydrocarbons in a source rock, (d) hydrocarbon variations within a reservoir and (e) the effectiveness of a cap rock. All of the analyses for this paper have been undertaken on whole rock samples from the North Sea area and UK coast. The effect of minerals in whole rock pyrolysis has been discussed elsewhere (Larter, 1984; Solli *et al.*, 1984) where it is suggested that, due to possible mineral–organic matter interactions, the Py–GC character of kerogens in whole rocks is somewhat more difficult to determine than with kerogen concentrates. On the other hand, whole rock pyrolysis might, in some cases, give a better indication of the type and distribution of hydrocarbons likely to be produced under natural catagenic conditions, but more work needs to be done in this area by studying well-documented geological models.

EXPERIMENTAL

Samples

A number of claystone samples were used in these studies, spanning a range of immature to lower oil window maturity, as determined by vitrinite reflectance (Vassoevich *et al.*, 1974) and T_{max} on Rock–Eval pyrolysis (Espitalié *et al.*, 1977). Various types of kerogen were determined by visual kerogen examination (Staplin, 1969; Burgess, 1974) and Rock–Eval pyrolysis. The samples were also extracted by a rapid rotary system using dichloromethane as solvent, chromatographically separated into saturated, aromatic and NSO fractions by using medium pressure liquid chromatography (MPLC) (Radke *et al.*, 1977), and the saturated fraction was analysed by capillary gas chromatography. The results from these routine organic geochemical analyses are shown in Table 1.

Instrumentation

The combined thermal extraction and temperature-programmed furnace pyrolysis system, shown diagrammatically in Fig. 1, comprises a sample probe, furnace tube with controls and a micro-switching valve/splitting system coupled to two fused-silica capillary columns. The programmed pyrolysis assembly is mounted into a Varian 3700 gas chromatograph fitted with three flame ionization detectors.

The sample probe (see Fig. 1) was constructed from a 3.2 × 1.8 mm I.D. stainless steel tube with boat-shaped section formed at one end by cutting the tube over a 1 cm section to expose the bore of the tube. The bore at both ends of the section was brazed to form a sample cup. An alignment bar was brazed in a horizontal plane to the sample cup section to allow correct alignment of the sample cup in the furnace tube. Nitrogen is passed over the sample (30 ml/min) through a gas inlet in the furnace tube.

The furnace tube was made from 6.4 × 4 mm bore HPLC-grade stainless steel. The furnace assembly was fabricated from an aluminium block heated by a 150 W cartridge heater. The furnace tube assembly was mounted into a stainless steel enclosure to ensure a good draught-free environment. The temperature control of the furnace was achieved via a laboratory-constructed temperature programmer which enabled variable ramp rates between 5 and 150 °C/min to be selected. The temperature range normally used was 320 °C for thermal extraction and from 320 °C to 550 °C at 30 °C/min for temperature-programmed pyrolysis.

Dried, finely-ground whole rock (20–30 mg) was placed in the sample cup, inserted into the furnace tube and held at 320 °C for about 3 min during the thermal extraction step (S_1). The furnace temperature was then increased to 550 °C at 30 °C/min during the pyrolysis step (S_2). The effluent from the furnace was directed to the required column system via a six-port microvalve (Valco) which was operated at 300 °C using the injector temperature control of the gas chromatograph. The two split exit lines (sample/split ratio = 1 : 30) were heated by a heater tape during the thermal desorption and pyrolysis stages and were led to flame ionization detector C. The output from flame ionization detector C recorded the bulk, unresolved, thermally extracted material and kerogen pyrolysis products on a pen-recorder as two broad peaks, S_1 and S_2 (see Fig. 1). The thermally extracted and kerogen

TABLE 1. Basic organic geochemical data on analytical samples

Origin of sample	*Sample No.*	*Kerogen type*	*Vit. Refl.*	T_{max} *(°C)*	*S1* $\left(\frac{mg\ Hc}{g\ C\ org}\right)$	*S2* $\left(\frac{mg\ Hc}{g\ C\ org}\right)$	*HI*	*OI*	*EOM* $\left(\frac{mg}{g\ C\ org}\right)$	*EHC* $\left(\frac{mg}{g\ C\ org}\right)$	*TOC (%)*
Kimmeridge Clay Fm. (Dorset)	K-5	II	0.42	424	0.46	202.21	565	70	21.9	3.1	18.1
Kimmeridge Clay Fm. (North Sea)	M-386	II	0.51	429	1.56	17.42	396	12	81.1	44.3	4.4
Lower Jurassic (North Sea)	M-5141	III	0.37	426	0.19	33.74	284	11	51.3	11.3	11.8
Lower Jurassic (North Sea)	M-5247	III	0.50	434	0.17	2.51	95	12	49.8	2.8	2.6

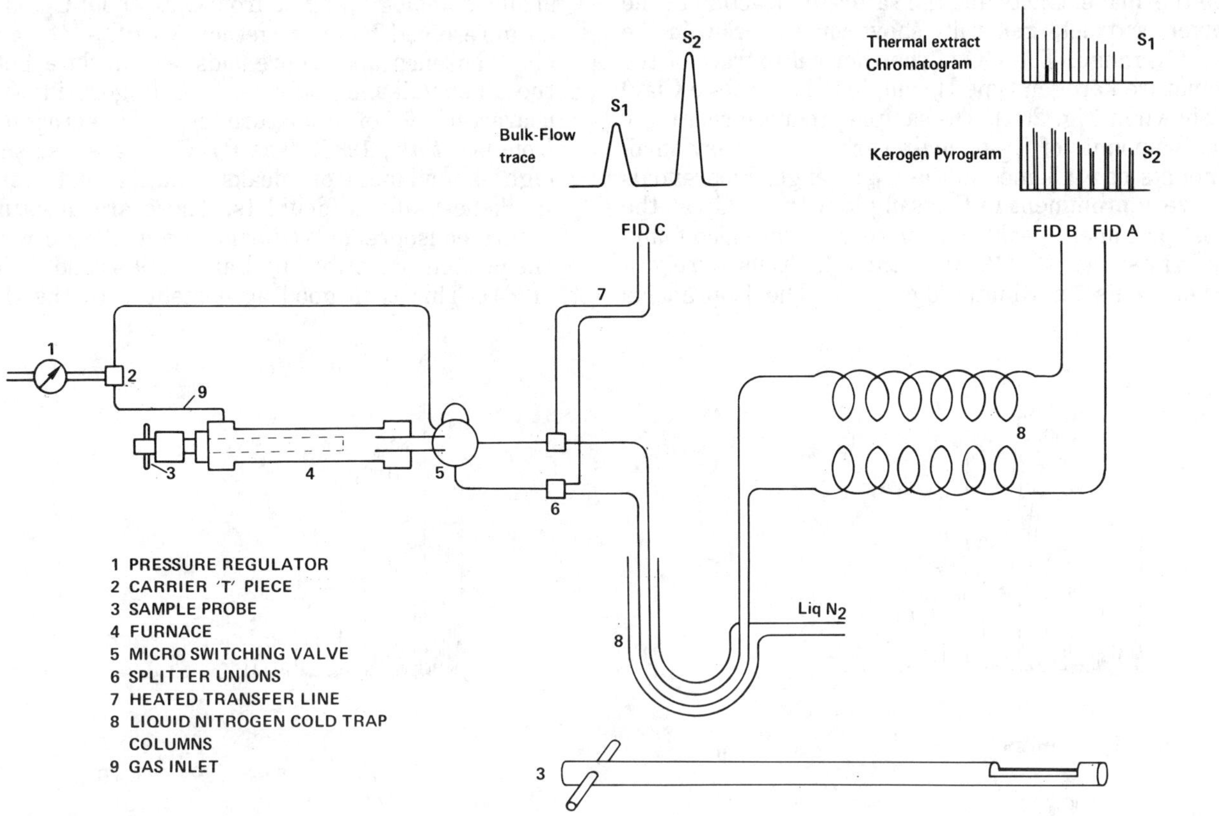

Fig. 1. Schematic diagram of the thermal extraction and pyrolysis–gas chromatography system.

pyrolysis products were cold-trapped in matched bonded OV-1 fused silica capillary columns using a thin-wall PTFE tubing ($\frac{1}{8}$ inch O.D.) which was flushed with liquid nitrogen during the desorption and pyrolysis stages. GC analysis of the trapped material was achieved by turning off the stream of liquid nitrogen and starting the GC oven programme. GC conditions are noted in the figure legends. The output from flame ionization detectors A and B and C was monitored on two recorders. The data from the three detectors was also acquired in raw data format into a V.G. Laboratory Systems Minichrom Chromatography data system for further manipulation.

Gas chromatographic conditions and peak identifications (see Figs. 2–9)

Thermal extraction

Conditions: 20 m × 0.25 mm I.D. fused-silica capillary column coated with OV-1; sample split ratio 1 : 30; oven temperature-programmed from 40 °C to 280 °C at 4 °C/min; helium carrier gas inlet pressure 8 psi. Peak identification : X = (m + p)-xylenes; Pr = pristane; Ph = phytane; numbers are n-alkanes and isoprenoids of that carbon number.

Saturated fraction

Conditions: 20 m × 0.25 mm. I.D. fused-silica capillary column coated with OV-1; sample split ratio 1 : 20; oven temperature-programmed from 80 °C to 280 °C at 6 °C/min; hydrogen carrier gas inlet pressure 5 psi. For peak identification see above.

Pyrolysis–gas chromatography

Conditions: temperature-programmed furnace pyrolysis (320–550 °C at 30 °C/min) in helium (30 ml/min) followed by focused cryogenic trapping (liquid nitrogen) and gas chromatography on a fused-silica capillary column (20 m × 0.25 mm I.D.) coated with OV-1; sample split ratio 1 : 30; oven temperature-programmed from 40 °C to 280 °C at 4 °C/min; helium carrier gas inlet pressure 8 psi. Peak identification: T = toluene; X = (m + p)-xylenes;

C_3-, C_4- and C_5B are alkylbenzenes of that carbon number; P = phenol; C_1P = alkylphenols; N = naphthalene; C_1N = methylnaphthalenes; numbers are *n*-alkenes/*n*-alkanes of that carbon number with the *n*-alkenes eluting before the corresponding *n*-alkanes; Pr = pristene.

RESULTS AND DISCUSSIONS

Source rocks

Capillary gas chromatograms of the thermal extract, C_{15}+ fraction and pyrolysate of immature samples containing kerogen of types II and III are shown in Figs. 2 and 3 respectively. The gas chromatograms of the thermal extracts and the saturated fraction of the solvent extracts generally show some overlap in the C_{15}–C_{25} range. The GC of the thermal extract of the immature kerogen type II sample (Kimmeridge Clay) is shown in Fig. 2(A). The sample produced relatively small amounts of C_1–C_8 hydrocarbons and very small amounts of *n*-alkanes in the C_{15}+ range. Isoprenoids are very prominent in this sample with iso-C_{16} as the most prominent peak. Similar results were also found by Winters *et al.* (1983) in their hydrous pyrolysis experiments on Kimmeridge Clay. The isoprenoids are far more prominent in the thermal extracts than in the solvent extracts for this sample (see Fig. 2(B)). Analysis of solvent extracts of Kimmeridge Clay samples have shown large variations in the relative abundances of *n*-alkanes and isoprenoids in samples collected within short intervals (Williams and Douglas, 1980) showing these differences to be clearly source dependent. Some of the variations found in the relative abundance of isoprenoids relative to *n*-alkanes for the two extraction techniques could therefore be due to inhomogeneities in the samples. However, this cannot explain fully the large differences observed, and it is therefore believed that solvent extraction will yield relatively less isoprenoids than thermal extraction.

The pyrogram (see Fig. 2(C)) shows an alkene/alkane homology ranging from C_8 to about C_{30} on top of an unresolved hump, together with C_3-, C_4- and C_5-alkylbenzenes and isoprenoids which elute between the alkene/alkane doublets. This fingerprint is quite characteristic of immature type II kerogen (see, Douglas *et al.*, 1983). The Py–GC trace also shows a high abundance of peaks eluting between the *n*-alkene/*n*-alkane doublets. These are unsaturated/saturated isoprenoids (Burnham *et al.*, 1982) similar to the pristene described by Larter (1984) and Solli *et al.* (1984). This is in good agreement with the thermal

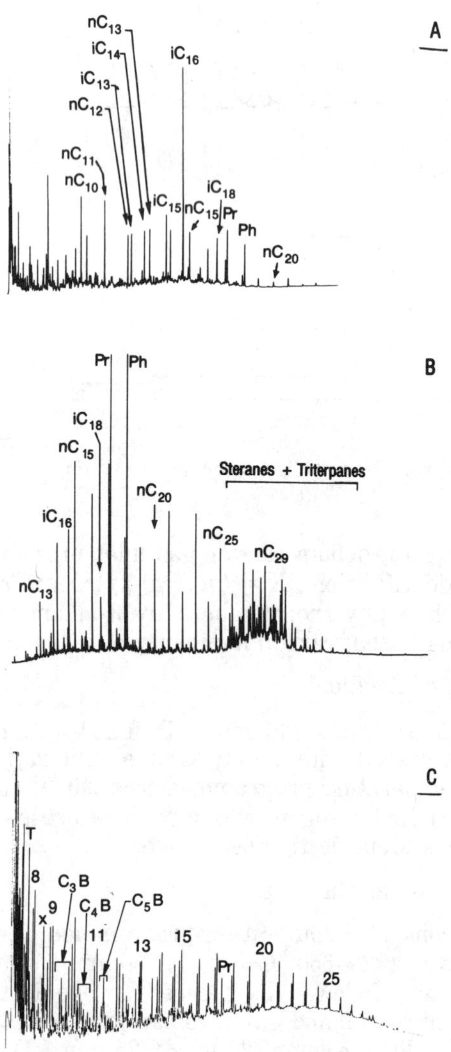

Fig. 2. Gas chromatograms of thermal extract (A), saturated hydrocarbons of solvent extract (B) and pyrolysate (C) of immature kerogen type II (Kimmeridge Clay Formation, Dorset).

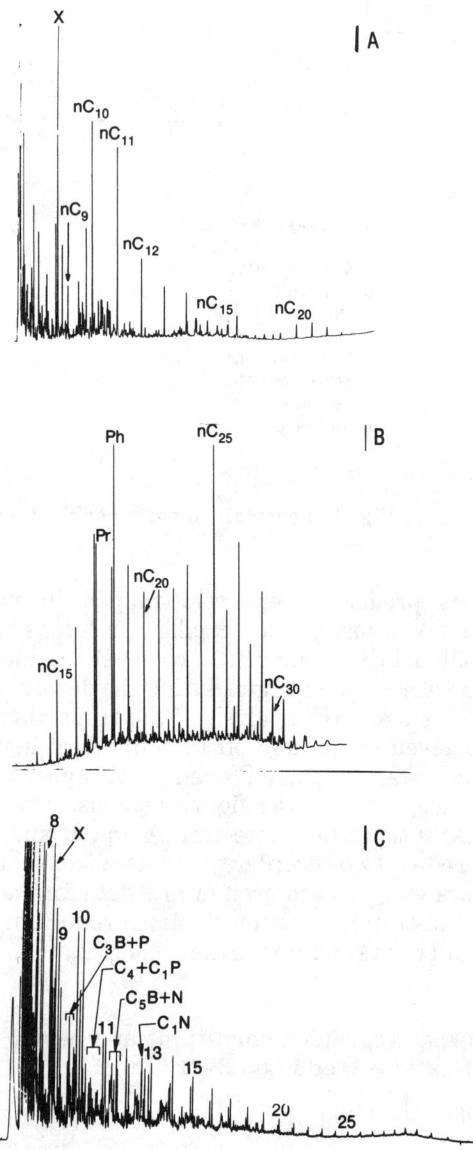

Fig. 3. As Fig. 2 of immature kerogen type III (Lower Jurassic, North Sea).

extract shown in Fig. 2(A) and hydrous pyrolysis of the same type of material (Winters *et al.*, 1983) which indicates an abundance of isoprenoids during the first stages of maturation. It is concluded that type II kerogen therefore generates unsaturated/saturated isoprenoids during pyrolysis. The solvent extract (see Fig. 2(B)) shows a large abundance of biomarkers in the sterane/triterpane range. These are not detected in the thermal extract gas chromatogram, due to possible mineral retention effects (Larter, 1984).

The immature sample of type III kerogen is of Early Jurassic age. The sample has a relatively high hydrogen index (see Table 1) as determined by Rock–Eval pyrolysis, but visual kerogen examination did not show any indication of amorphous material. The gas chromatogram of the thermal extract of this sample is shown in Fig. 3(A). This shows the sample to contain a large abundance of light C_1–C_4 hydrocarbons and *m*-/*p*-xylenes. The abundance of C_{10} and C_{11} *n*-alkanes is high with a sharp decline for C_{12} and higher. Due to the large abundance of *m*-/*p*-xylenes the higher hydrocarbons can hardly be detected in the thermal extract, but these can easily be seen in the saturated fraction of the solvent extract (see Fig. 2(B)) which contains abundant high molecular weight *n*-alkanes with a high CPI value which is typical of immature kerogen type III (Albrecht *et al.*, 1976).

The pyrogram (see Fig. 3(C)) shows an alkene/alkane homology ranging from C_8 to about C_{30}; however the intensity of the doublets is drastically reduced after C_{15}. This is in contrast to the type II kerogen pyrogram (see Fig. 2(C)). The high abundance of phenol, alkylbenzenes and naphthalenes is characteristic of type III kerogens derived from terrestrial material (Larter and Douglas, 1980; Douglas *et al.*, 1983). It must be emphasized that the pyrograms which are discussed above were obtained from whole rock samples. However, the results are also in general in good agreement with Py–GC analysis of kerogen concentrates (Larter and Douglas, 1980; Douglas *et al.*, 1983).

Now that the characteristics of immature samples of kerogen types II and III have been established, the effects of increasing maturity must also be considered.

A sample of more-mature kerogen type II from the Kimmeridge Clay Formation in the North Sea has been analysed and the results are shown in Fig. 4(A). The increase in the maturity is easily detected in the thermally extracted product, Fig. 4(A). The amount of product has increased markedly, i.e. the kerogen in this sample has produced a significant amount of hydrocarbons even at a relatively low maturity (R_0 = 0.5%) (see Table 1). The distribution of the hydrocarbons in the thermal extract is more oil-like than was observed at a maturity of R_0 = 0.4%, but distinct features of low maturity, e.g. high pristane/*n*-C_{17} ratio, are still observed.

The gas chromatogram of the saturated hydrocarbon fraction of the solvent extract (Fig. 4(B)) shows a distribution of the C_{15}–C_{30} range similar to the thermal extract. The main difference is the larger abundance of biomarkers (steranes/triterpanes) in the solvent extract, which are released during thermal extraction, and a higher ratio of *n*-alkanes to isoprenoids in the solvent extract. Typical examples are pristane/*n*-C_{17} and phytane/*n*-C_{18}. This phenomenon is registered for all the samples examined, and there is presently no adequate explanation for it. However, a possible cause might be that the thermal extraction releases entrapped isoprenoids which are not extracted by solvents (Solli *et al.*, 1984).

The pyrogram (see Fig. 4(C)) shows an alkene/alkane homology ranging from C_8 to C_{30} and the aliphatic homology is more pronounced compared with the less mature Kimmeridge type II kerogen. This difference is in agreement with previous observations of Py–GC characteristics with increasing maturity (van Graas *et al.*, 1981; Solli *et al.*, 1984), i.e. the abundance of isoprenoids and aromatics is reduced with increasing maturity resulting in a more pronounced aliphatic doublet homology. This is also consistent with the early generation of isoprenoids and aromatics during maturation (Burnham *et al.*, 1982).

A sample of approximately the same maturity but with kerogen type III is shown in Fig. 5. The gas chromatogram of the thermal extract shows *o*/*m*-xylenes to be the largest peak, with low concentrations of isoprenoids and *n*-alkanes from C_{12} upwards. This is similar to the less mature sample of type III kerogen (see Fig. 3) and suggests that there is hardly any production of hydrocarbons from this type of kerogen at such a low maturity (R_0 = 0.50%). This is also shown in the extractable hydrocarbons; the gas chromatogram of the saturated fraction of the extractable hydrocarbons is shown in Fig. 6(B). The GC trace shows a higher proportion of high molecular

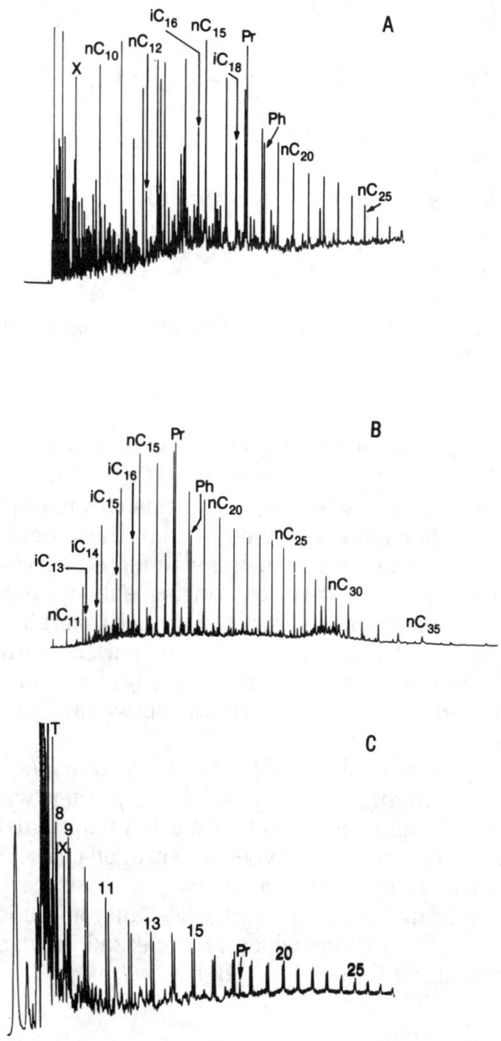

Fig. 4. As Fig. 2, of moderately mature kerogen type II (Kimmeridge Clay Formation, North Sea).

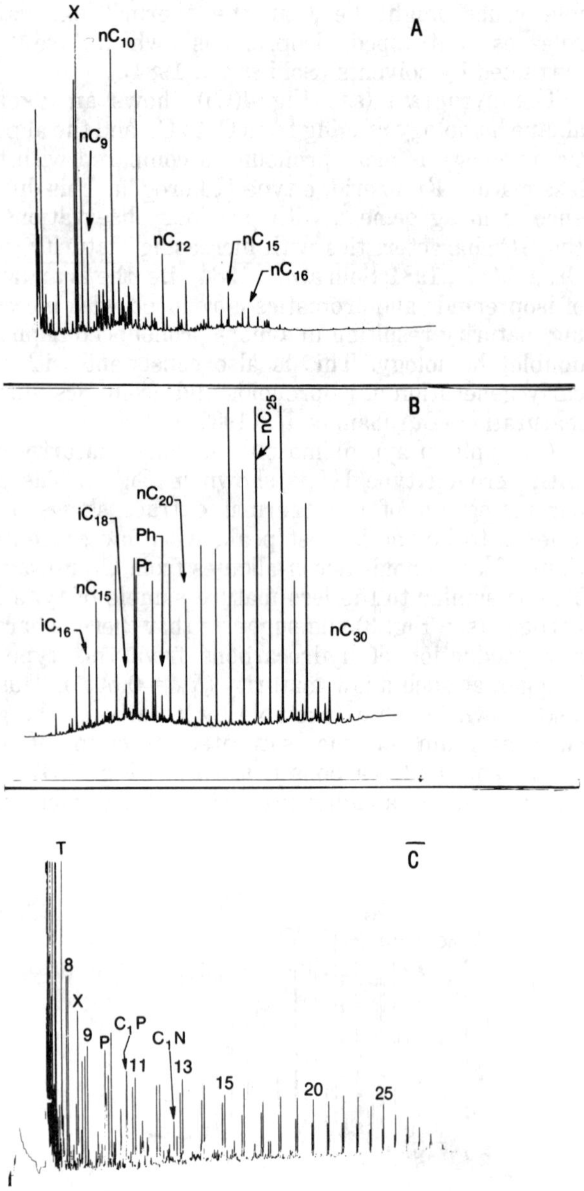

Fig. 5. As Fig. 2, of moderately mature kerogen type III (Lower Jurassic, North Sea).

weight n-alkanes with a high CPI value which is typical of kerogen type III (Albrecht *et al.*, 1976). The main fractions of hydrocarbons in this chromatogram are of a higher molecular weight than those detected in the thermal extract, which is dominated by C_1–C_{15} hydrocarbons which are lost during solvent extraction. This sample also shows a high pristane/phytane ratio, far higher than expected by comparison with the slightly less mature sample (see Fig. 3), and may indicate some type differences between the two samples.

The pyrogram (Fig. 5(C)) shows an alkene/alkane homology ranging from C_8 to C_{30}, together with a high abundance of phenols and methyl naphthalenes. This is characteristic of type III kerogens. The high abundance of high molecular weight hydrocarbons might indicate cuticular material- and/or mineral-retained hydrocarbons which are released during the pyrolysis step (Solli *et al.*, 1984).

Reservoir rocks

When fluid samples are collected from wide intervals within a reservoir, these most often represent an average content and will not give any indication of variations within the reservoir. Analyses of extracts from core samples will, however, give more-detailed information on the compositional changes throughout the reservoir. A solvent extraction of a core sample provides good chromatographic results for the C_{15}+ fraction, but the lighter components will be lost during evaporation of the solvent. It will be shown how a combination of thermal extraction, solvent extraction and pyrolysis with gas chromatographic analysis can complement each other in reservoir rock analysis. To illustrate the variations often found within reservoir fluids, samples from a number of different reservoir zones were selected.

The first suite of samples represents a Middle Jurassic reservoir from a well in the North Sea. This reservoir covers a depth range of 90 m, and consists of high porosity, high permeability sandstones with some thin layers of coal and claystone. The reservoir produced condensate during testing which, upon gas chromatographic analysis was found to contain n-alkanes up to C_{40}. The sample, which was almost colourless, went solid when kept in a refrigerator. A total of 20 samples of core chips from the reservoir zone were analysed by the three methods described above. All of the samples showed an almost complete absence of hydrocarbons lower than n-C_{14}, as seen in the gas chromatograms of the thermal extracts (Figs. 6(A), (B), (E) and (F)). The reservoir is in a high pressure zone and the cores were reported to 'bleed' hydrocarbons while being handled on the rig. Only small differences were found in the hydrocarbons detected by thermal extraction, mainly in the n-alkane distribution from the C_{15}–C_{25} region, as illustrated in Figs. 6(A), (B), (E) and (F). Some of the samples show a CPI (C_{17}–C_{25}) greater than 1.0, while others are less than 1.0. This feature was found repeatedly in analyses of samples from this reservoir which have good reproducibility. The samples show a decrease in the CPI with increasing depth which probably represents a fractionation effect within the reservoir. Similar but not so evident variations in n-alkane distributions are also noted in the gas chromatograms of the saturated hydrocarbon fraction of the solvent extracts (see Figs. 6(C), (D), (G) and (H)). The samples from this reservoir section were also pyrolysed, but produced no pyrolysate, thus showing that all of the organic matter in these samples is mobilized at the temperature used for thermal extraction.

A second suite of reservoir samples was chosen from another Middle Jurassic section in the North Sea. Core chips covering an interval of 40 m were analysed as described above. The gas chromatograms of the thermal extracts show large variations in the hydrocarbon distributions above C_{10} for the different samples analysed. Below C_{10} only traces of hydrocarbons were detected, probably due to losses during sampling and transport. Some of the samples, especially at the top of the reservoir, showed C_{10}–C_{16} to be the major components with a significant decline in C_{17} and higher n-alkanes. This would indicate a condensate-type field or a sample from near a gas/oil contact (see Fig. 7(A)). A number of the analysed samples had gas chromatographic distributions for the thermal extract as shown in Fig. 7(B), with large unresolved envelopes in the range n-C_{24}–n-C_{30} and with a large abundance of high molecular weight n-alkanes. The light end of the gas chromatogram is similar to that in Fig. 7(A), suggesting that the free

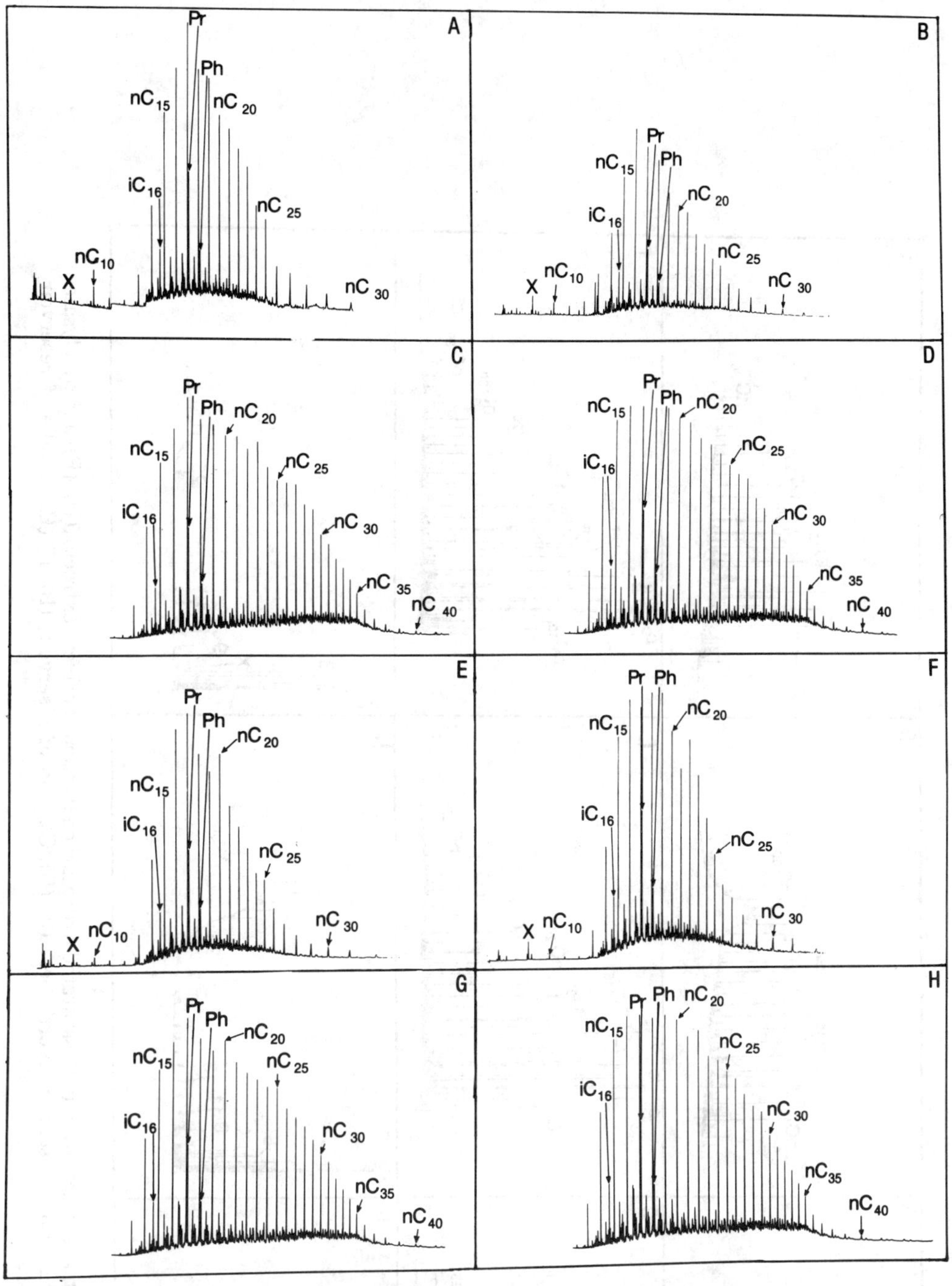

Fig. 6. Gas chromatograms of thermal extracts ((A), (B), (E) and (F)) and saturated hydrocarbon fraction of solvent extracts ((C), (D), (G) and (H)) of four different intervals in a Middle Jurassic sandstone reservoir, North Sea.

hydrocarbons detected from thermal extraction of sample B may be a mixture of those present in sample A and another fraction with components of much higher molecular weight. The third type of gas chromatogram for the thermal extracts is shown in Fig. 7(C). This shows the normal distribution for a mature oil with a slight even/odd n-alkane predominance.

The gas chromatograms of the saturated hydrocarbon fraction of the solvent extracts are almost identical for the various samples from the reservoir (see Figs. 7(D), (E) and (F)). The gas chromatograms of the pyrolysates show significant differences. The samples from the top of the reservoir show some very light compounds together with very small amounts of the alkane/alkene doublets and an unresolved envelope of high molecular weight n-alkanes. These are, however, small and can only be detected at the highest sensitivity. The pyrolysate of the sample from the middle part of the reservoir shows a similar large unresolved envelope as seen in the thermal extract of the sample (see Figs. 7(B) and (H)). The samples from the lower part of the reservoir have a pattern for the pyrolysate similar to those from the upper part of the core (see Figs. 7(G) and (I)). Solvent extraction with liquid chromatographic separation showed that the samples from the upper and lower part of the reservoir contained 9.1% of non-hydrocarbons, whilst the samples from the middle part of the reservoir contained 35–40% non-hydrocarbons. Thermal evaporation with subsequent pyrolysis will release this heavy material and generate new hydrocarbons. This residue probably represents asphalthenes or heavy oil which normally were not produced, but may subsequently clog the pores of the reservoir rock.

The above technique was also tried on samples close to an oil/water contact, which were taken at 0.5 m intervals from the same reservoir as those shown in

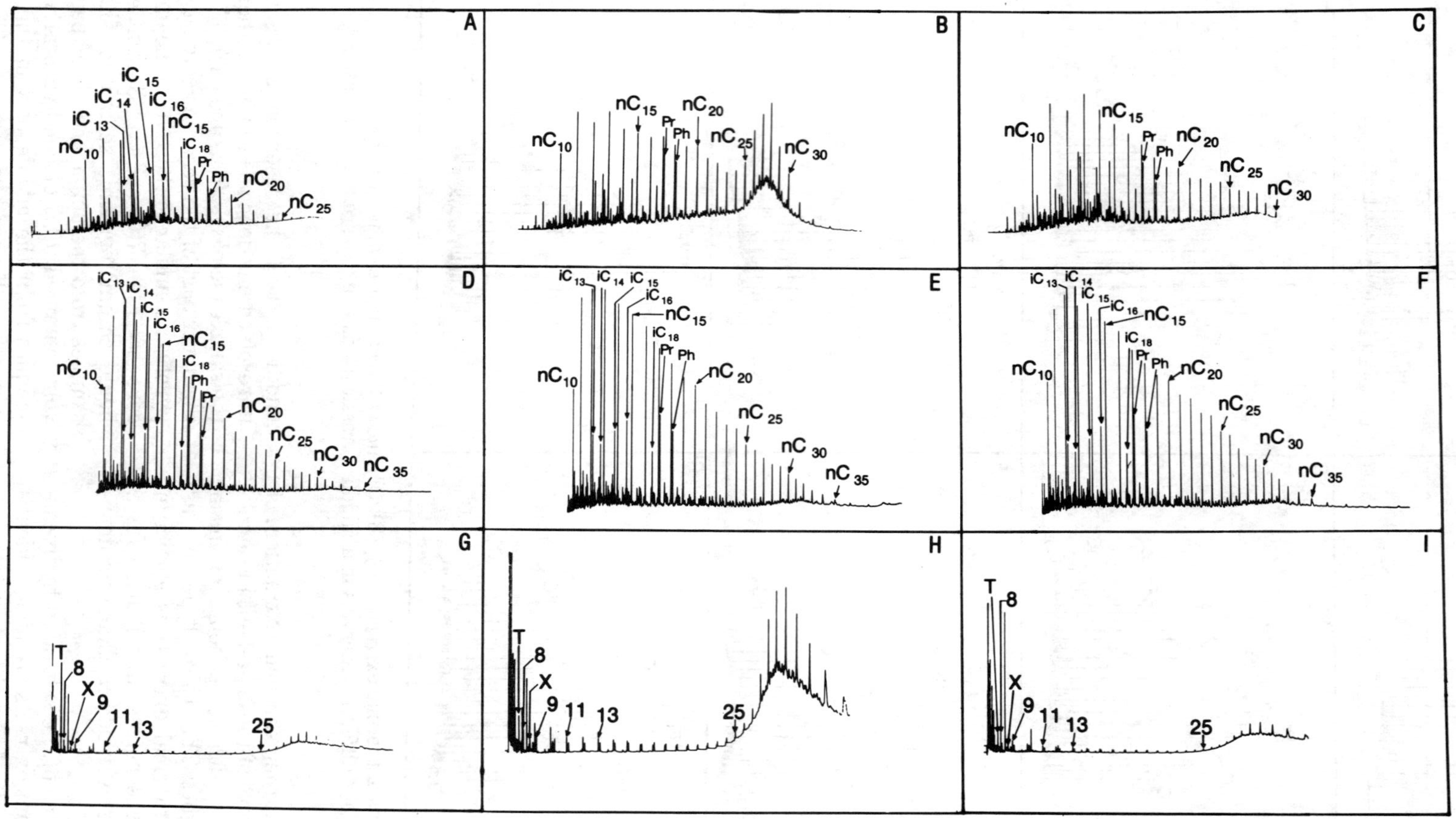

Fig. 7. Gas chromatograms of thermal extracts ((A)–(C)), saturated hydrocarbon fraction of solvent extracts ((D)–(F)) and pyrolysates ((G)–(I)) from a Middle Jurassic sandstone reservoir, North Sea. (A),(D),(G): Top of reservoir; (B),(E),(H): middle of reservoir, (C),(F),(I): lower part of reservoir.

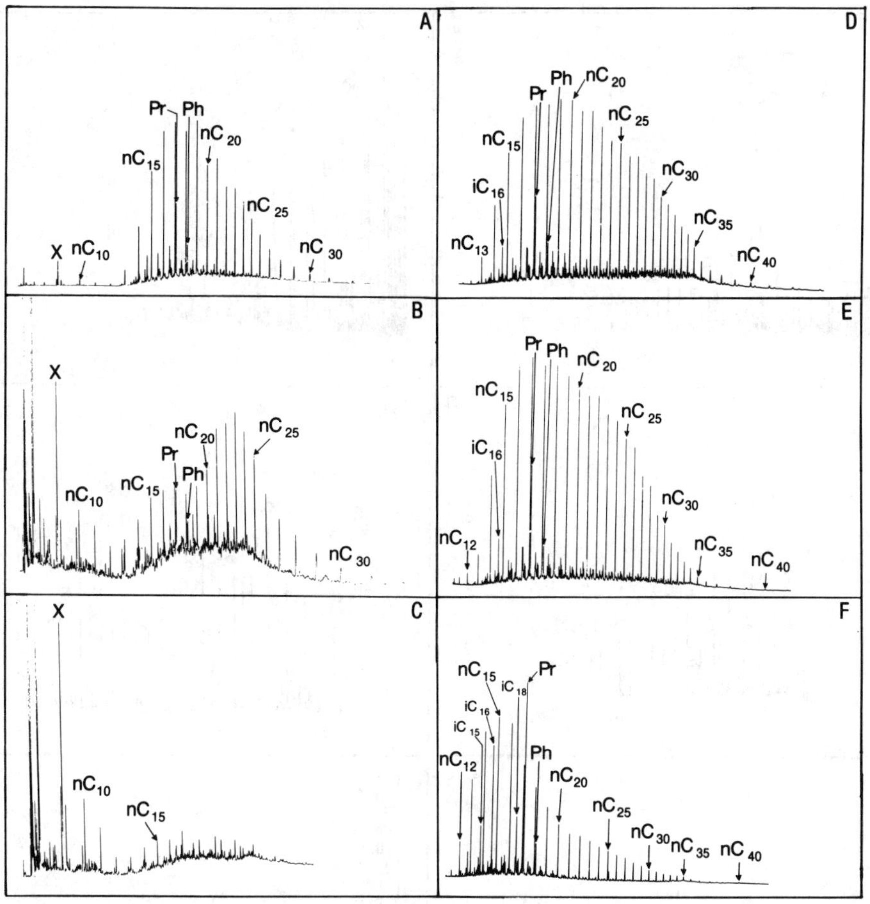

Fig. 8. Gas chromatograms of thermal extracts ((A)–(C)) and saturated hydrocarbon fraction of solvent extracts ((D)–(F)) from a Middle Jurassic sandstone reservoir, North Sea. (A),(D): 65.5 m below top of reservoir; oil zone; (B),(E): 66.0 m below top of reservoir; oil/water contact; (C),(F): 66.5 m below top of reservoir; water zone.

Fig. 6. The gas chromatograms of the thermal extract and the saturated hydrocarbon fraction of the solvent extract of a sample 65.5 m below the top of the reservoir are shown in Figs. 8(A) and (B). These have similar distributions to those found higher up in the reservoir (see Fig. 6). The gas chromatograms of the samples from 66.0 m are shown in Figs. 8(B) and (E), while those from 66.5 m are shown in Figs. 8(C) and (F).

The samples from 66.0 m (Figs. 8(B) and (E)) have a distribution of heavier hydrocarbons similar to those found higher in the reservoir. This is particularly seen in the saturated hydrocarbon fraction of the solvent extract (see Fig. 8(E)). The light hydrocarbons are significantly different, as seen in the gas chromatogram of the thermal extract, Fig. 8(B). There is a significant loss of C_{13}–C_{22} *n*-alkanes while there is some increase in lighter hydrocarbons compared with higher in the reservoir. The latter could be due to slightly different permeability in the reservoir rock, while the former effect could be due to the sample being just below the oil/water contact (assuming that the oil/water contact as reported from logs is slightly inaccurate). The sample from 65.5 m below the top of the reservoir has a completely different character from that found higher in the reservoir. Figures 8(E) and (F) clearly show that this sample is below the oil/water contact. The main reason for the large difference between the thermal extract (see Fig. 8(C)) and the saturated fraction of the solvent extract (see Fig. 8(F)), is that the thermal extract contains mainly aromatic hydrocarbons and NSO compounds which have been dissolved in the water phase. The saturated components will only be a minor fraction in the thermal extract.

Cap Rock

The effectiveness of the caprock for a reservoir is an important factor in exploration. This can be investigated in terms of the amount and type of hydrocarbons found within a caprock and how this varies with distance from the top of the underlying reservoir. An example is given in Figs. 9(A) and (D), where Fig. 9(A) represents the thermal extract from a caprock sample 5 m above the top of the reservoir, and Fig. 9(D) represents a sample 10 m above the top. The gas chromatogram of the thermal extract of a sample from the top of the reservoir is shown in Fig. 6(A). These examples clearly show how the hydrocarbons from the reservoirs have penetrated somewhat into the cap rock, but they cannot be detected 10 m above the reservoir. The kerogens in the two cap rock samples are almost identical. The thermal extract shown in Fig. 9(D) represents *in situ* hydrocarbons related to the kerogen, while those shown in Fig. 9(A) represent a mixture of *in situ* hydrocarbons and those migrated in from the reservoir rock.

Gas chromatograms of the saturated hydrocarbon fraction of the solvent extract of the two caprock samples are shown in Figs. 9(B), and (C). These show the same trend as seen in the thermal extracts.

The chromatograms of the pyrolysates indicate some differences, particularly in the high molecular

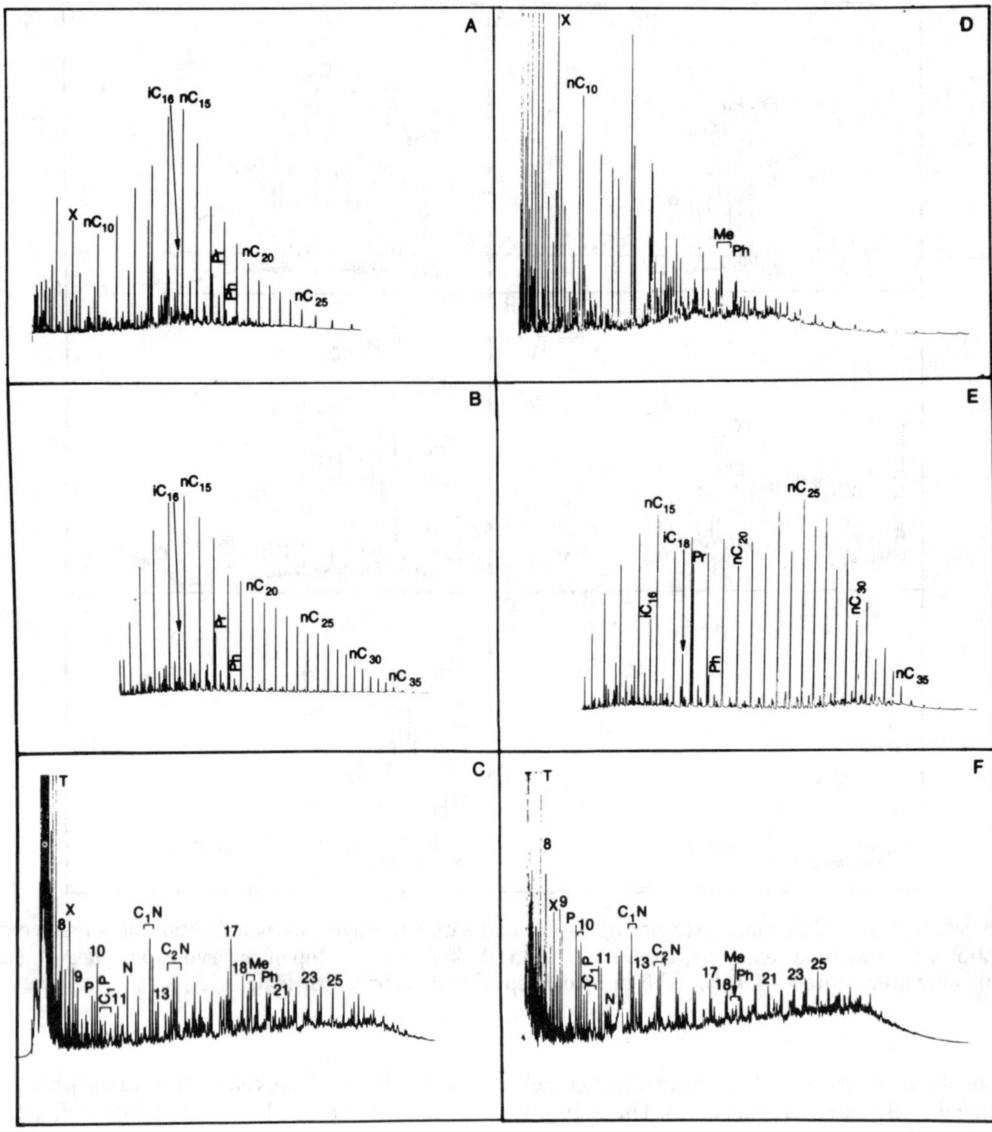

Fig. 9. Gas chromatograms of thermal extracts ((A) and (D)), saturated hydrocarbon fraction of solvent extracts ((B) and (E)) and pyrolysates ((C) and (F)) of a caprock. (A)–(C): 5 m above top of reservoir; (D)–(F): 10 m above top of reservoir.

weight region where the sample containing migrated material (see Fig. 9(C)) has a large abundance of high molecular weight alkanes. These have probably been retained in the sample during thermal extraction, and are now found in the pyrolysate. The results from the pyrolysates are in good agreement with what was found in the thermal extracts and solvent extracts, if the migrated hydrocarbons found in the sample closest to the reservoir are excluded.

CONCLUSIONS

The combination of thermal evaporation and pyrolysis gas chromatography is a rapid method for screening both source rocks and reservoir rocks. This paper has shown how, with as little as 20–30 mg of rock, the amount and type of hydrocarbons present in the rock, the type of hydrocarbons which can be generated from the rock and the maturity of the rock can be evaluated. The analyses will also show if the hydrocarbons present in the rock are produced *in situ* or if they have migrated into the rock. The system is also a good tool for detection of contamination.

The first part of the present study confirmed that branched and cyclic paraffins together with low molecular weight aromatic compounds, are produced at an earlier stage of maturity than the n-alkanes. The n-alkanes can only be detected by overloading the system. Kerogen type III has a large relative abundance of low molecular weight aromatics than found in other kerogen types. Variations in the thermal extract, i.e. the hydrocarbons already produced from the kerogen, and in the pyrolysate, i.e. the hydrocarbons which the kerogen is still capable of generating at increased maturity, are clearly illustrated. With increasing maturity the percentage of higher molecular weight compounds, especially n-alkanes, increases for all types of kerogen. At higher maturities, an increase in light and medium molecular weight relative to the high molecular weight hydrocarbons is observed. The pyrolysates exhibit similar changes. The amount of high molecular weight hydrocarbons in the pyrolysates decreases with increasing maturity, clearly showing that these are transferred from the kerogen.

In analyses of core samples from a reservoir, variations in the reservoir fluid composition are easily detected, as well as residual oils and the oil/water contact. With the amount of material needed for the analyses being so small, cuttings, sidewall cores and core chips are all suitable for analysis which takes only 1–1½ h per sample, although care must be taken in sampling, especially of coarse material, since light hydrocarbons are easily lost.

The thermal extraction/pyrolysis–gas chromatography system may be used on its own for the types of problems described in this paper, but it is advisable that such analyses are backed up by conventional solvent extraction and gas chromatographic analyses, at least for selected samples. This will give a complete coverage of all the hydrocarbons in the sample.

REFERENCES

Albrecht, P., Vandenbroucke, M. and Mandengue, M. 1976. Geochemical studies on the organic matter from the Douala basin (Cameroon); 1. Evolution of the extractable organic matter and the formation of petroleum. *Geochim. Cosmochim. Acta*, **40**, 791–799.

Burgess, J. D. 1974. Microscopic examination of kerogen (dispersed organic matter) in petroleum exploration. *Geol. Soc. Am. Spec. Paper 153*, pp. 19–30.

Burnham, A. K., Clarkson, J. E., Singleton, M. F., Wong, C. M. and Crawford, R. W. 1982. Biological markers from Green River kerogen decomposition. *Geochim. Cosmochim. Acta*, **46**, 1243–1251.

Dembicki, H., Jr, Horsfield, B. and Ho, T. T. Y. 1983. Source rock evaluation by pyrolysis–gas chromatography. *Bull. Am. Ass. Petroleum Geologists*, **67**, 1094–1103.

Douglas, A. G., Hall, P. B. and Solli, H. 1983. Comparative organic geochemistry of some European oil shales. In: Miknis, F. P. and Mackay, J. F. (eds.), *ACS Symposium Series, No. 230, Geochemistry and Chemistry of Oil Shales*, pp. 59–84.

Durand, B. and Espitalié, J. 1972. Formation et évaluation des hydrocarbones de C_1 à C_{15} et des gas permanents dans les angiles du toarcian du bassin de Paris. In: von Gaertner, H. and Weiner, H. (eds.), *Advances in Organic Geochemistry 1971*. Pergamon Press, Oxford, pp. 455–468.

Espitalié, J., Laporte, J. C., Madec, M., Marquis, F., Leplat, P., Paulet, J. and Boutefeu, A. 1977. Méthode rapide de charactérisation des roches mères, et de leur potential pétrolier et de leur degré d'évaluation. *Rev. inst. Français Pétrole*, **32**, 23–42.

Gedenk, R. 1966. Auf geochemischen Beurteilung und Genese von Kohlenwasseransammelungen. *Erdøl, Kohle, Erdgas Petrochem.*, **19**, 720–727.

Horsfield, B. 1984. Pyrolysis studies and petroleum exploration. In: *Advances in Petroleum Geochemistry*, Vol. I. Academic Press, pp. 247–292.

Hunt, J. M. 1975. Origin of gasoline range alkanes in the deep sea. *Nature*, **254**, 411–413.

Jonathan, D., l'Hote, G. and du Rouchet, J. 1975. Analyse géochimique des hydrocarbures légers par thermoévaporation. *Rev. Inst. Français Pétrole*, **30**, 65–88.

Larter, S. R. 1984. Application of analytical pyrolysis techniques to kerogen characterization and fossil fuel exploration/exploitation. In: Voorhees, K. (ed.), *Analytical Pyrolysis—Methods and Applications*. Butterworths, London, pp. 212–275.

Larter, S. R. and Douglas, A. G. 1980. Typing of kerogens by pyrolysis capillary-gc. In: Douglas, A. G. and Maxwell, J. R. (eds.), *Advances in Organic Geochemistry 1979*. Pergamon Press, Oxford, pp. 579–583.

Le Tran, K. 1975. Analyse et étude des hydrocarbures gayeux occlus dans les sédiments. Exemple d'application à l'éxploration pétrolière. *Bull. Centre Rech. Pau, SNPA* (1975), 223 pp.

Martin, S. J. 1977. Thermally evolved hydrocarbons from the bitumen and kerogen constituents of whole rock. In: Campos, R. and Goni, J. (eds.), *Advances in Organic Geochemistry 1975*. Enadimsa, Madrid, 677 pp.

Orr, W. L. 1983. Comments on pyrolytic hydrocarbon yields in source-rock evaluations. In: Bjorøy, M. (ed.), *Advances in Organic Geochemistry 1981*. Wiley, p. 775.

Phillipi, G. T. 1975. The deep subsurface temperature controlled origin of the gaseous and gasoline-range hydrocarbons of petroleum. *Geochim. Cosmochim. Acta*, **39**, 1353–1373.

Radke, M., Sittard, H. G. and Welte, D. H. 1978. Removal of soluble organic matter from rock samples with a flow through extraction cell. *Analyt. Chem.*, **50**, 663–665.

Saint-Paul, C., Monin, J. C. and Durand, B. 1980. Méthode de caracterisation rapide des hydrocarbures de C_1 à C_{35} contenus dans les roches sédimentaires et dans les huiles. *Rev. Inst. Français Pétrole* **35**(6), 1065–1078.

Schaefer, R. G. and Poock, H. 1983. Thermal mobilization of hydrocarbons from small-size rock samples. Application in petroleum geochemistry. *Chromatographica*, **16**, 257–260.

Schaefer, R. G., Weiner, B. and Leythauser, D. 1978. Determination of subnanograms per gram quantities of light hydrocarbons (C_2–C_9) in rock samples by hydrogen stripping in the flow system of a capillary gas chromatograph. *Analyt. Chem.*, **50**, 1848–1854.

Solli, H., Bjorøy, M., Leplat, P. and Hall, K. 1984. Analysis of organic matter in small rock samples using combined thermal extraction and pyrolysis–gas chromatography. *J. Analyt. Appl. Pyrolysis* (in press).

Staplin, F. L. 1969. Sedimentary organic matter, organic metamorphism and oil and gas occurrence. *Bull. Can. Petrol. Geol.*, **17**(1), 47–66.

Van Graas, G., de Leeuw, J. W. and Schenck, P. A. 1980. Analysis of coals of different rank by Curie-point pyrolysis–mass spectrometry. In: Douglas, A. G. and Maxwell, J. R. (eds.), *Advances in Organic Geochemistry 1979*. Pergamon Press, Oxford, pp. 485–494.

Vassoevich, N. B., Akramkhodaghaev, A. M. and Geodekyan, A. A. 1974. Principal zone of oil formation. In: Tissot, B. and Bienner, F. (eds.), *Advances in Organic Geochemistry 1973*. Technip, Paris, pp. 309–314.

Whelan, J. K., Hunt, J. M. and Huc, A. Y. 1980. Applications of thermal distillation–pyrolysis to petroleum source rock studies and marine pollution. *J. Analyt. Appl. Pyrolysis*, **2**, 79–96.

Williams, P. F. W. and Douglas, A. G. 1980. A preliminary organic geochemical investigation of the Kimmeridgian oil shales. In: Douglas, A. G. and Maxwell, J. R. (eds.), *Advances in Organic Geochemistry 1979*. Pergamon Press, Oxford, pp. 531–546.

Winters, J. C., Williams, J. A. and Lewan, M. D. 1983. A laboratory study of petroleum generation by hydrous pyrolysis. In: Bjorøy, M., *et al.* (eds.), *Advances in Organic Geochemistry 1981*. Wileu, Chichester, pp. 524–533.